Feature Papers in Materials Physics (2nd Edition)

Feature Papers in Materials Physics (2nd Edition)

Guest Editors

Vlassios Likodimos
Andres Sotelo

MDPI

Basel • Beijing • Wuhan • Barcelona • Belgrade • Novi Sad • Cluj • Manchester

Guest Editors
Vlassios Likodimos
Department of Physics
National and Kapodistrian
University of Athens
Athens
Greece

Andres Sotelo
Instituto de Ciencia de
Materials de Aragón
Universidad de Zaragoza
Zaragoza
Spain

Editorial Office
MDPI AG
Grosspeteranlage 5
4052 Basel, Switzerland

This is a reprint of the Special Issue, published open access by the journal *Materials* (ISSN 1996-1944), freely accessible at: https://www.mdpi.com/journal/materials/special_issues/GEU8U60KS3.

For citation purposes, cite each article independently as indicated on the article page online and as indicated below:

Lastname, A.A.; Lastname, B.B. Article Title. *Journal Name* **Year**, *Volume Number*, Page Range.

ISBN 978-3-7258-6632-8 (Hbk)
ISBN 978-3-7258-6633-5 (PDF)
https://doi.org/10.3390/books978-3-7258-6633-5

Contents

Preface

This Reprint presents a collection of feature papers originally published in the Special Issue "Feature Papers in Materials Physics (2nd Edition)". The included contributions present recent advances in materials physics, providing new insights into the physical properties of materials and their relevance to targeted applications. The volume comprises original research articles reporting both experimental and theoretical/computational studies. The papers address a wide range of physical properties, including structural, mechanical, optical, electronic, thermal, magnetic, and superconducting properties, as well as exciton dynamics. Advanced characterization techniques and modeling approaches are employed to elucidate structure–property relationships in functional materials. The materials systems investigated include metal oxides, transition metal mono- and dichalcogenides, ferroelectric and multiferroic materials, semiconductors, and metals and alloys. These studies cover both fundamental aspects and application-oriented performance considerations and devices. The contributions are relevant to a broad spectrum of current and emerging applications, including next-generation electronics and mechanics, optoelectronics and photonics, magnetocaloric and thermoelectric technologies, optical data storage, sensors, solar energy conversion, and environmental engineering. This Reprint provides a representative overview of current research trends in materials physics and highlights the field's continued importance in advancing functional materials and future technologies.

Vlassios Likodimos and Andres Sotelo
Guest Editors

materials

MDPI

Article

Theoretical Study of the Magnetic and Optical Properties of Ion-Doped LiMPO$_4$ (M = Fe , Ni, Co, Mn)

Iliana N. Apostolova [1], Angel T. Apostolov [2] and Julia Mihailowa Wesselinowa [3,*]

[1] University of Forestry, Kl. Ohridsky Blvd. 10, 1756 Sofia, Bulgaria; inaapos@abv.bg
[2] University of Architecture, Civil Engineering and Geodesy, Hr. Smirnenski Blvd. 1, 1046 Sofia, Bulgaria; angelapos@abv.bg
[3] Faculty of Physics, Sofia University "St. Kliment Ohridski", J. Bouchier Blvd. 5, 1164 Sofia, Bulgaria
* Correspondence: jwesselinowa@gmail.com

Abstract: Using a microscopic model and Green's function theory, we calculated the magnetization and band-gap energy in ion-doped LiMPO$_4$ (LMPO), where M = Fe, Ni, Co, Mn. Ion doping, such as with Nb, Ti, or Al ions at the Li site, induces weak ferromagnetism in LiFePO$_4$. Substituting Li with ions of a smaller radius, such as Nb, Ti, or Al, creates compressive strain, resulting in increased exchange interaction constants and a decreased band-gap energy, E_g, in the doped material. Notably, Nb ion doping at the Fe site leads to a more pronounced decrease in E_g compared to doping at the Li site, potentially enhancing conductivity. Similar trends in E_g reduction are observed across other LMPO$_4$ compounds. Conversely, substituting ions with a larger ionic radius than Fe, such as Zn and Cd, causes an increase in E_g.

Keywords: LiMPO$_4$; ion doping; magnetization; band gap; microscopic model

1. Introduction

Lithium-ion batteries have garnered significant interest for portable devices due to their superior energy density compared to other systems. With a growing demand for rechargeable batteries in electric vehicles, lithium transition metal phosphates, denoted as LiMPO$_4$ (LMPO), where M = Fe, Ni, Co, Mn, are being explored as replacements for the prevalent LiCoO$_2$ cathodes in lithium-ion cells. LMPO batteries find extensive applications in energy storage systems. LMPO has emerged as a crucial cathode material for next-generation power lithium-ion battery applications due to its abundant raw materials, eco-friendliness, excellent cycling performance, and safety attributes. However, the commercial utilization of LMPO cathode material has been impeded thus far by its low electronic conductivity. Strategies such as ion doping or synthesizing smaller particles have proven effective in enhancing electronic conductivity and reducing the band-gap energy [1–6].

The magnetic properties of many compounds of the LMPO family have been widely studied, mainly because of their frustration effects. In the present paper, using the s-d model, we show that the magnetic and electronic properties in LMPO are connected. This means that the magnetic properties can affect the electronic ones, and vice versa. Zaghib et al. [7] reported that in LiFePO$_4$ (LFPO), there is indirect evidence of the existence of small polarons, showing the interplay between the electronic and magnetic properties. Therefore, before delving into methods to enhance conductivity and reduce the band-gap energy, it is crucial to underscore that compounds within the LMPO family exhibit antiferromagnetic properties [8–12]. These compounds, comprising corner-sharing MO$_6$ octahedra with high-spin M^{2+} ions, exhibit an antiferromagnetic ground state typically below a transition temperature, T_N, ranging from 30 to 50 K [13]. Notably, weak ferromagnetism has been observed in LiMnPO$_4$ (LMPO), as reported by Arcon et al. [14]. The ground state of LFPO is characterized as a collinear antiferromagnet, whereas LMPO exhibits a weak ferromagnetic ground state with a transition temperature, T_N, of approximately 42 K. LFPO boasts the highest Neel temperature (T_N = 50 K), below which it demonstrates antiferromagnetic ordering [15]. These magnetic property disparities stem from differences in electronic structures, which could have implications for the electrochemistry of LFPO and LMPO. Spin-wave dispersions in LFPO's antiferromagnetic state have been elucidated through a linear spin-wave theory by Li et al. [16]. Additionally, antiferromagnetic behavior was observed in $LiCoPO_4$ (LCPO) and $LiNiPO_4$ (LNPO) by Vaknin et al. [10], showcasing characteristics akin to weakly coupled two-dimensional Ising antiferromagnets. LNPO exhibits a spontaneous first-order magnetic phase transition. Dai et al. [17] conducted an analysis of LMPO's spin-exchange interactions utilizing density functional theory (DFT) calculations, shedding further light on their magnetic properties.

Mercier et al. [18] elucidated the intriguing magnetoelectric properties of isostructural transition-metal lithium orthophosphates, denoted as LMPO systems. These materials exhibit a significant interplay between their magnetic and electric properties, rendering them promising candidates for advanced functional applications. In parallel investigations, Khrustalyov et al. [9] and Vaknin et al. [10] independently documented the magnetoelectric effect within the antiferromagnetic framework of LNPO. This phenomenon underscores the intricate coupling between magnetic and electric degrees of freedom in these compounds. Moreover, recent investigations by Fogh et al. [19] have provided valuable insights into the magnetic field-induced electric polarization in LFPO, offering a comprehensive examination of this phenomenon across a range of temperatures and applied magnetic field strengths. Notably, their findings unveiled a previously unreported diagonal magnetoelectric tensor element, thus enriching our understanding of the underlying mechanisms governing the magnetoelectric behavior in these systems. Building upon these foundational studies, future research endeavors are poised to delve deeper into the magnetoelectric properties of LMPO materials, promising further advancements in the field of multifunctional materials science.

Undoped LFPO typically exhibits semiconductor behavior with a considerable band gap ranging from 3.8 to 4.0 eV, as established by previous research [20]. However, modifications through doping, particularly on the lithium or iron sites, have been demonstrated to reduce the band gap, as observed with dopants such as V [21] or Nb [22]. The introduction of high-valent metal ions, such as Nb^{5+}, Al^{3+}, Ti^{4+}, K^{+}, and Na^{+}, among others, at the Li site has been shown to enhance the conductivity and hole mobility in LFPO, thereby positively influencing its electrochemical performance [1]. DFT calculations, specifically involving Nb substitution for Fe or Li in LFPO, have elucidated the electronic structure of the material, indicating a decrease in the band gap post-doping, leading to enhanced electronic conductivity [5]. Furthermore, investigations by Gao et al. [23] into Zr and Co co-doped LFPO, utilizing first-principle calculations, revealed a similar trend of a reduced band gap and improved electrochemical properties, underscoring the potential of co-doping strategies in enhancing LFPO performance. Other studies, such as those by Xu et al. [24], have explored the effects of doping on LFPO electronic properties, highlighting the band-gap reduction associated with dopants like Mn at the iron site or Na at the lithium site. Additionally, ion doping with elements such as La, Y, and Na has been reported to decrease the band gap and increase conductivity in LFPO [5,25]. Moreover, research by Zhang et al. [1,26] employed DFT to compute the band gap of LNPO doped with transition metal atoms. Furthermore, investigations conducted by various researchers [27–30] have explored the impact of substituting Mn in LMPO with Co, Cr, Cu, Fe, Ni, or V on a range of properties, including structural, magnetic, electronic, and conductivity characteristics.

The present study seeks to explore a novel approach by investigating the band-gap energy and conductivity in ion-doped LMPO, using, for the first time, a microscopic model and Green's function theory. This method allows for a more comprehensive understanding of the dynamic behavior of ion-doped LMPO, offering insights into their properties across a range of temperatures and excitation conditions. We look for doping ions at the Li or Fe sites that can increase or decrease the magnetization and band-gap energy. Due to the different ionic radii of the doping and host ions, which lead to different strains, the exchange interaction constants in the doped material are modified compared to the undoped one. We explain, on a microscopic level, the doping mechanism. The obtained decrease in the band-gap energy for some doping ions could be used in applications for enhancing the electronic conductivity in L*M*PO$_4$ compounds. Moreover, we observe that by doping, for example, with Nb, Ti, or Al ions at the Li site, the magnetization increases, i.e., weak ferromagnetism appears in LiFePO$_4$. Unfortunately, there are no experimental data for this statement.

It should be noted that the majority of studies on ion-doped L*M*PO utilize DFT [1,5,20–24,26–28]. DFT is a very powerful tool for investigating many-body problems. However, DFT is mostly concerned with ground-state properties at zero temperature. In our approach, we are able to cover the whole temperature regime. It is a finite temperature analysis and includes the entire excitation spectrum. In particular, the method allows us to study the total phase diagram, which is based on the different excitation energies realized in the system. The disadvantage of our approach lies in the consideration of collective properties from the beginning. Our basic quantities are not the naked electrons but the effective spins of the underlying quasi-particles. However, with DFT, all parameters of the system can, at least in principle, be calculated, so we are forced to use additional models to determine these parameters. We are convinced that both approaches, DFT and the Green's function method, are appropriate and, to a certain extent, can be alternatives for describing many-body systems.

2. Model and Method

The olivine-type compounds, Li*M*PO$_4$ (M = Mn, Fe, Co, Ni), consist of MO_4 layers made up of corner-sharing MO_6 octahedra of high-spin M^{2+} ions (see Figure 1). The magnetic properties, for example, of ion-doped LiFePO$_4$ with doping concentration, x, are described by the Heisenberg Hamiltonian:

$$H_{sp} = -\frac{1}{2}\sum_{i,j}(1-x)J_{ij}\mathbf{S}_i^{Fe}\cdot\mathbf{S}_j^{Fe} - \frac{1}{2}\sum_{i,j}xJ_{dij}\mathbf{S}_i^{DI}\cdot\mathbf{S}_j^{DI} - D_z\sum_i(S_i^z)^2, \quad (1)$$

where $\mathbf{S}_i$ and its z component S_i^z are the spin operators for the localized Fe^{2+} spins at site i. J_{ij} stands for the spin-exchange interactions, including J_1 and J_2, which are the in-plane nearest-neighbor (nn) and next-nearest-neighbor (nnn) coupling constants, respectively, as well as J_3, which is the inter-plane nn coupling constant (see Figure 1). J_{dij} is the exchange interaction constant between the doping ions (DI), x is the ion-doping concentration, and D_z is the single-site anisotropy parameter of the easy-axis type.

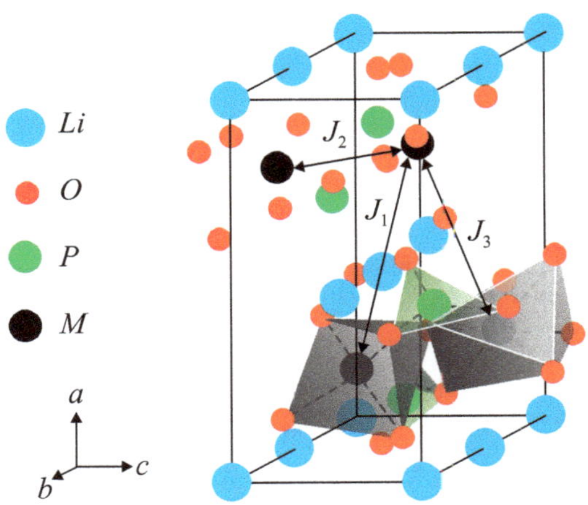

Figure 1. Structure of the unit cell of Li*M*PO$_4$.

For the investigation of the band-gap energy, we use the s-d model:

$$H_m = H_{sp} + H_{el} + H_{sp-el}. \tag{2}$$

H_{sp} is the Heisenberg model of the localized spins, which is described by Equation (1). H_{el} represents the usual Hamiltonian of the conduction band electrons

$$H_{el} = \sum_{ij\sigma} t_{ij} c_{i\sigma}^{+} c_{j\sigma}, \tag{3}$$

where t_{ij} is the hopping integral and $c_{i\sigma}^{+}$ and $c_{i\sigma}$ are the Fermi-creation and -annihilation operators.

The Hamiltonian, H_{sp-el}, couples the two subsystems—Equations (1) and (3)—by an intra-atomic exchange interaction, I_i:

$$H_{sp-el} = \sum_{i} I_i \mathbf{S}_i \mathbf{s}_i. \tag{4}$$

The spin operators, $\mathbf{s}_i$, of the conduction electrons at site i can be expressed as $s_i^{+} = c_{i+}^{+} c_{i-}$, $s_i^{z} = (c_{i+}^{+} c_{i+} - c_{i-}^{+} c_{i-})/2$.

The band-gap energy

$$E_g = \epsilon^{+}(\mathbf{k} = 0) - \epsilon^{-}(\mathbf{k} = \mathbf{k}_\sigma) \tag{5}$$

is defined as the energy difference between the valence and conduction bands:

$$\epsilon_{ij}^{\pm} = \epsilon_{ij} - \frac{\sigma}{2} I \langle S^z \rangle. \tag{6}$$

The electronic energies are obtained from the Green's functions $g_{ij\sigma} = \ll c_{i\sigma}; c_{j\sigma}^{+} \gg$, where $\sigma = \pm 1$. ϵ_{ij} is the conduction band energy in the paramagnetic state, $\langle n_{m\sigma} \rangle$ is the occupation number distribution, and $\langle S^z \rangle$ is the magnetization.

The magnetization, M, for an arbitrary spin, S, is given by

$$M = \langle S^z \rangle = \frac{1}{N} \sum_{i} \Big[(S + 0.5) \coth[(S + 0.5)\beta E_{mi}] - 0.5 \coth(0.5\beta E_{mi}) \Big], \tag{7}$$

where E_{mi} is the spin-wave energy, calculated from the poles of the Green's function $G_{ij}(t) = \ll S_i^{+}(t); S_j^{-} \gg$.

3. Numerical Results and Discussion

The numerical calculations are performed in the JAVA programming environment using simple iterative procedures and summation over nearest neighbors with the following model parameters: $J_1 = -0.662$ meV, $J_2 = -0.27$ meV, $J_3 = 0.021$ meV, $D = -0.37$ meV [16], $S = 2$, $t = 1$ eV, and $I = 0.5$ eV.

Let us emphasize that J depends on the distance between the spins and on the lattice parameters. Due to the different strains caused by different doping ions, J_d in the doped state can be changed compared to the undoped state, J; that is, it can differ from J.

3.1. Dependence of the Magnetization on the Doping with Nb, Ti, and Al at the Li site in LFPO

Pure lithium iron phosphate (LFPO) exhibits a substantial band-gap energy of 3.763 eV [21], resulting in low conductivity. To address this limitation and potentially reduce the band-gap energy, we explore the effects of ion doping on LFPO. The calculation of the band-gap energy, E_g, necessitates knowledge of the magnetization, M, as per Equation (6). LFPO, being an antiferromagnetic compound, undergoes initial substitution at the Li site with ions such as Nb^{5+}, Ti^{4+}, or Al^{3+}, possessing smaller ionic radii of 0.64 Å, 0.745 Å, and 0.675 Å, respectively, compared to Li^{+} (0.9 Å). This substitution induces compressive strain, thereby enhancing the exchange interaction constants, J_d, with J_d surpassing J. Notably, J is contingent upon the spin distance and lattice parameters inversely proportional to it. The magnetization's dependence on the doping concentration is illustrated in Figure 2, revealing a trend of increasing M with higher dopant concentrations. Additionally, the substitution of Li with ions featuring smaller ionic radii than Li results in enhanced M, indicating the emergence of weak ferromagnetism upon ion doping. It can be assumed that the electron carrier doping effects in the antiferromagnetic LFPO indicate that a transition from an antiferromagnetic to a weak ferromagnetic phase occurs upon ion doping.

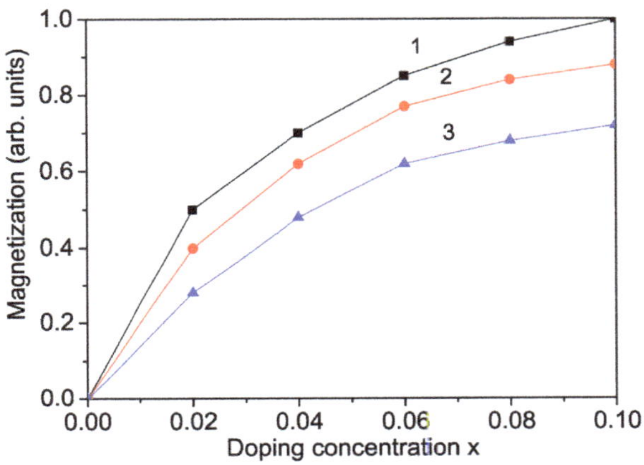

Figure 2. Doping concentration dependence of the magnetization, M, in LFPO for different dopants at the Li site: (1) Nb; (2) Ti; (3) Al.

It should be noted that the doping concentration in the experimental works, in general, is small, for example, for Al doping, x = 0.02; for Co, it is 0.03; for Nb, it is 0.01; for Ru, it is 0.05; for Mg, it is 0.1, etc. A high doping concentration would lead to negative effects. With the increase in the dopant content, impurity phases appear. Moreover, excessive ion doping could lead to a decrease in the Fermi energy.

3.2. Dependence of the Band-Gap Energy on the Doping with Nb, Ti, and Al at the Li Site in LFPO

Observing Equation (6), we note a reduction in the band-gap energy, E_g, upon substituting Nb, Ti, or Al ions at the Li site, as depicted in Figure 3, curves 1–3. These findings align closely with the experimental data from Chung et al. [31] and Zhang et al. [1], suggesting that the decrease in the band-gap energy, E_g, could potentially enhance LFPO's conductivity, rendering it a promising cathode material. The introduction of Nb, Ti, or Al ions induces Li vacancies to maintain the charge balance, consequently improving material conductivity. Our model predicts a similar decrease in the band-gap energy, E_g, upon Mg doping in LFPO at the Li site, consistent with previous research by Yao et al. [32].

Additionally, substitution at the Fe site [22], as exemplified by Ru (r = 0.705 Å) or Nb (r = 0.64 Å), with smaller ionic radii compared to Fe ions (0.75 Å), induces compressive strain and requires the condition $J_d > J$. The decrease in E_g, as illustrated in Figure 3, curve 4, for Nb-doped LFPO at the Fe site, is more pronounced than that observed for Li ion substitution. This observation is consistent with the findings of previous studies by Gao et al. [33], Zhang et al. [1,22], and Karimzadeh et al. [34], suggesting that Ru or Nb doping at the Fe site effectively enhances LFPO's electronic conductivity. In summary, doping LFPO with rare-earth ions increases the carrier concentration, shifting the Fermi level and reducing the band-gap energy, E_g, thereby enhancing conductivity. Co-doping of LFPO with La^{3+} and Y^{3+}, as demonstrated by Zhang et al. [5], also improves electronic conductivity, corroborating our model predictions.

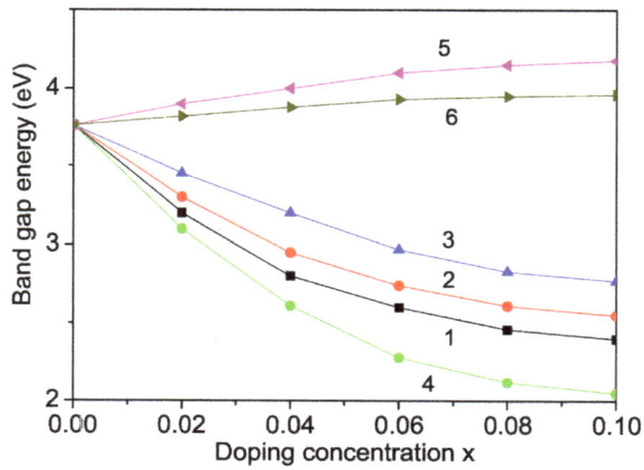

Figure 3. Doping concentration dependence of the band-gap energy in LFPO for different dopants: At the Li site—(1) Nb; (2) Ti; (3) Al—and at the Fe site—(4) Nb; (5) Zn; (6) Cd.

3.3. Dependence of the Band-Gap Energy on the Doping with Nb, Ti, and Al at the Li Site in LNPO

Let us broaden our investigation to encompass other members of the LiMPO_4 family and compare their characteristics with those of LFPO. Pure compounds like LMPO and LNPO exhibit relatively high band-gap values: LMPO has a band-gap value of 3.7 eV [35] and LNPO has a band-gap value of 2.77 eV [27]. We once again turn our attention to the band-gap energy, E_g, a crucial determinant of electronic conductivity, which can be computed from the difference between the conduction band minimums and valence band maximums using Equation (6). Doping with smaller ions such as Nb^{5+}, Nd^{3+}, Ti^{4+} or Al^{3+} at the Li site in both LNPO and LMPO mirrors the trend observed in LFPO, resulting in a reduction in the band-gap energy, E_g. This phenomenon is exemplified by the results of Nb, Ti, and Al doping in LNPO, as illustrated in Figure 4, curves 1–3. Correspondingly, Karthickprabhu et al. [36] documented an increase in electronic conductivity upon Nd^{3+} doping in LNPO, indicating a decrease in the band gap, consistent with our model predictions.

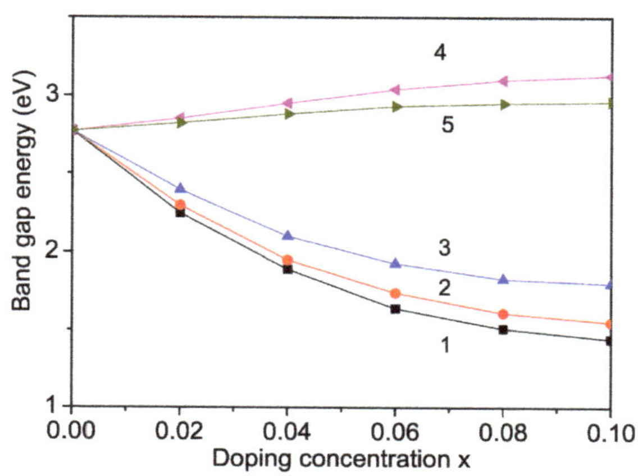

Figure 4. Doping concentration dependence of the band-gap energy in LNPO for different dopants: At the Li site—(1) Nb; (2) Ti; (3) Al—and at the Fe site—(4) Zn; (5) Cd.

Moving forward, we explore the effect of substituting Fe, Ni, or Mn ions with similar doping ions, as in the LFPO case, on the band-gap energy, E_g. The ionic radii of Ni^{2+} and

Mn^{2+} ions are 0.83 Å and 0.97 Å, respectively, whereas that of Fe^{2+} ion is 0.75 Å. Doping with Ru ions at the Ni and Mn sites also results in a reduced band-gap energy, E_g, in LNPO and LMPO, akin to LFPO. Similarly, substitution with V (r = 0.72 Å), Ti (r = 0.745 Å), or Cr (r = 0.74 Å) ions at the Fe, Ni, or Mn site induces a decrease in E_g, thereby enhancing the conductivity in all three compounds—LFPO, LNMO, and LMPO. This enhancement in electronic conductivity has been observed in V- and Cr-doped LFPO [21,26], as well as in Cr-doped LMPO [29], corroborating our model predictions.

3.4. Dependence of the Band-Gap Energy on the Doping with Zn and Cd on the Ni or Fe Site in LNPO and LFPO

It is crucial to note that not all doping ions induce a decrease in the band-gap energy, E_g. Taking LNPO as an example, when the Ni ion is substituted with ions like Zn and Cd, with larger ionic radii (0.88 Å and 1.09 Å, respectively) compared to Ni (0.83 Å), it results in a tensile strain, leading to an increase in the lattice parameters. In this scenario, the relationship between the exchange interaction constants in the doped and undoped states follows $J_d < J$. As depicted in Figure 4, curves 3 and 4, the band-gap energy, E_g, increases with higher doping concentrations, a trend corroborated by Zhang et al. [26], who found that transition metal ions (Zn, Cd, and Hg) substituted at the Ni site do not decrease the band-gap energy, E_g.

Similar behavior is observed in Zn- and Cd-doped LFPO at the Fe site, as shown in Figure 3, curves 3 and 4. This is because the ionic radius of Fe is smaller than that of the substituted ions, resulting in a tensile strain and an increase in the lattice parameters and cell volume. It is noteworthy that the behavior of the band gap strongly depends on the compound's structure.

Moreover, experimental data on Zn-doped LFPO exhibit discrepancies. Some studies by Shenouda et al. [37], Bilecka et al. [38], and Liu et al. [39] reported an increase in the cell volume, whereas others by Zhao et al. [40] and Yiming et al. [41] suggested a decrease. Unfortunately, experimental data on the band-gap energy or conductivity in ion-doped LMPO are limited.

Observations indicate that the doping effects of Nb, Ti, and Al ions on the Li site in LMPO and LCPO, as well as Zn and Cd ions on the Mn or Co sites, mirror those observed in LFPO and LNPO [29,42]. This consistency underscores the importance of considering the structural effects when assessing doping-induced changes in the band-gap energy across the LiMPO$_4$ family.

4. Conclusions

In summary, our investigation focused on studying the magnetization, M, and band-gap energy, E_g, in ion-doped LiMPO$_4$ (LMPO), where M represents various transition metals such as Fe, Ni, Co, and Mn, utilizing a combination of the s-d model and Green's function theory. We delved into understanding these macroscopic properties from a microscopic perspective. The primary objective of ion doping is to augment the intrinsic electronic conductivity of LMPO, although the precise mechanism remains subject to debate within the scientific community. Notably, doping with elements like Nb, Ti, or Al at the Li site in LFPO induces weak ferromagnetism, which intensifies with increasing dopant concentrations. The band gap serves as a pivotal determinant of the electronic conductivity of solid materials, particularly significant in the context of battery materials. Substituting Li with ions possessing smaller ionic radii induces compressive strain, thereby increasing the exchange interaction constants and decreasing the band-gap energy, E_g. We observed a more pronounced decrease in E_g with Nb ion doping at the Fe site compared to the Li site, a trend consistent across various LMPO compounds. This reduction in E_g signifies a potential enhancement in conductivity, suggesting the potential suitability of LMPO as a cathode material. Conversely, substitution with ions of larger ionic radii, such as Zn and Cd, compared to Fe or Ni ions, induces tensile strain, resulting in an increase in E_g. Overall, our findings align qualitatively with experimental data, further validating our theoretical framework.

Let us emphasize that multi-element co-doping may contribute to better electrochemical performance compared to single doping [1,43]. The substitution on different sites has different effects; for example, on the *M* site, it leads to enhancement of the electronic conductivity, on the Li site it leads to a decrease in the charge transfer resistance, and on the O site, it facilitates the migration of Li ions. Therefore, the combined modification of Li-, *M*-, and O-site co-doping is an effective method to improve the electrochemical properties of the L*M*PO compounds. This problem will be considered in a future paper.

Author Contributions: Conceptualization, J.M.W.; Methodology, A.T.A.; Software, I.N.A.; Formal analysis, A.T.A.; Investigation, I.N.A., A.T.A. and J.M.W.; Writing—original draft, J.M.W. All authors have read and agreed to the published version of the manuscript.

Funding: A.T.A. and I.N.A. acknowledge financial support from the Bulgarian National Science Fund (contract number KP-06 PN68/17/BG-175467353-2022-04-0232).

Institutional Review Board Statement: Not applicable.

Informed Consent Statement: Not applicable.

Data Availability Statement: Derived data supporting the findings of this study are available from the corresponding author upon reasonable request.

Conflicts of Interest: The authors declare no conflicts of interest.

References

1. Zhang, H.; Zou, Z.; Zhang, S.; Liu, J.; Zhong, S. A review of the Doping Modification of $LiFePO_4$ as a Cathode Material for Lithium Ion Batteries. *Int. J. Electrochem. Sci.* **2020**, *15*, 12041–12067. [CrossRef]
2. Wani, T.A.; Suresh, G. A comprehensive review of $LiMnPO_4$ based cathode materials for lithium-ion batteries: Current strategies to improve its performance. *J. Energy Storage* **2021**, *44*, 103307. [CrossRef]
3. Yi, T.-F.; Li, X.-Y.; Liu, H.; Shu, J.; Zhu, Y.-R.; Zhu, R.-S. Recent developments in the doping and surface modification of $LiFePO_4$ as cathode material for power lithium ion battery. *Ionics* **2012**, *18*, 529–539. [CrossRef]
4. Shang, S.; Wang, Y.; Mei, Z.; Hui, X.D.; Liu, Z.K. Lattice dynamics, thermodynamics, and bonding strength of lithium-ion battery materials $LiMPO_4$ (M = Mn, Fe, Co, and Ni): A comparative first-principles study. *J. Mater. Chem.* **2011**, *22*, 1142–1149. [CrossRef]
5. Zhang, Q.; Zhou, J.; Zeng, G.; Ren, S. Effect of lanthanum and yttrium doped $LiFePO_4$ cathodes on electrochemical performance of lithium-ion battery. *J. Mater. Sci.* **2023**, *58*, 8463–8477. [CrossRef]
6. Liu, H.; Yang, H.; Li, J. A novel method for preparing $LiFePO_4$ nanorods as a cathode material for lithium-ion power batteries. *Electrochim. Acta* **2010**, *55*, 1626–1629. [CrossRef]
7. Zaghib, K.; Mauger, A.; Goodenough, J.B.; Gendron, F.; Julien, C.M. Electronic, Optical, and Magnetic Properties of $LiFePO_4$: Small Magnetic Polaron Effects. *Chem. Mater.* **2007**, *19*, 3740. [CrossRef]
8. Arcon, D.; Zorko, A.; Dominko, R.; Jaglicic, Z. A comparative study of magnetic properties of $LiFePO_4$ and $LiMnPO_4$. *J. Phys. Condens. Matter* **2004**, *16*, 5531.
9. Khrustalyov, V.M.; Savytsky, V.M.; Kharchenko, M.F. Magnetoelectric effect in antiferromagnetic $LiNiPO_4$ in pulsed magnetic fields. *Low Temp. Phys.* **2016**, *42*, 1126–1129. [CrossRef]
10. Vaknin, D.; Zarestky, J.L.; Rivera, J.-P.; Schmid, H. Antiferromagnetism in $LiCoPO_4$ and $LiNiPO_4$. *ChemInform* **2006**, *37*. [CrossRef]
11. Masrour, R.; Jabar, A.; Benyoussef, A.; Hamedoun, M.; Hlil, E.K. Magnetic properties of $LiFePO_4$ compound: A Monte Carlo study. *Chem. Phys. Lett.* **2015**, *635*, 268–272. [CrossRef]
12. Jena, A.; Nanda, B.R.K. Unconventional Magnetism and Band Gap Formation in $LiFePO_4$: Consequence of Polyanion Induced Non-planarity. *Sci. Rep.* **2016**, *6*, 19573. [CrossRef]
13. Ofer, O.; Sugiyama, J.; Brewer, J.H.; Mansson, M.; Prsa, K.; Ansaldo, E.J.; Kobayashi, G.; Kanno, R. The magnetic phase of Lithium transition metal phosphates $LiMPO_4$ (M = Mn, Co, Ni) detected by μ+SR. *Phys. Procedia* **2012**, *30*, 160–163. [CrossRef]
14. Arcon, D.; Zorko, A.; Cevc, P.; Dominko, R.; Bele, M.; Jamnik, J.; Jaglicic, Z.; Golosovsky, I.V. Weak ferromagnetism of $LiMnPO_4$. *J. Phys. Chem. Solids* **2004**, *65*, 1773–1777. [CrossRef]
15. Werner, J.; Neef, C.; Koo, C.; Ponomaryov, A.; Zvyagin, S.; Klingeler, R. Exceptional field dependence of antiferromagnetic magnons in $LiFePO_4$. *Phys. Rev. B* **2021**, *103*, 174406. [CrossRef]
16. Li, J.; Garlea, V.O.; Zarestky, J.L.; Vaknin, D. Spin-waves in antiferromagnetic single-crystal $LiFePO_4$. *Phys. Rev. B* **2006**, *73*, 024410. [CrossRef]
17. Dai, D.; Whangbo, M.-H.; Koo, H.-J.; Rocquefelte, X.; Jobic, S.; Villesuzanne, A. Analysis of the spin exchange interactions and the ordered magnetic structures of lithium transition metal phosphates $LiMPO_4$ (M = Mn, Fe, Co, Ni) with the olivine structure. *Inorg. Chem.* **2005**, *44*, 2407. [CrossRef]
18. Mercier, M.; Gareyte, J. Un nouveau corps magneto-electrique: $LiMnPO_4$. *Solid State Commun.* **1967**, *5*, 139–142. [CrossRef]

19. Fogh, E.; Klemke, B.; Pages, A.; Li, J.; Vaknin, D.; Ronnow, H.M.; Christensen, N.B.; Toft-Petersen, R. The magnetoelectric effect in $LiFePO_4$—Revisited. *Phys. B Condens. Matter* **2023**, *648*, 414380. [CrossRef]
20. Zhou, F.; Kang, K.; Maxisch, T.; Ceder, G.; Morgan, D. The electronic structure and band gap of $LiFePO_4$ and $LiMnPO_4$. *Solid State Commun.* **2004**, *132*, 181–186. [CrossRef]
21. Shen, C.; Lin, W.; Hu, H.; Yang, P.; Wang, L. The electronic and geometric structure modifications of $LiFePO_4$ with vanadium doping to achieve ultrafast discharging capability: The experimental and theoretical investigations. *J. Alloys Compd.* **2023**, *936*, 168035. [CrossRef]
22. Zhang, P.; Zhang, D.; Huang, L.; Hui, W.; Wei, Q.; Song, S. First-Principles Study of Electronic Structure of Nb-Doped $LiFePO_4$. *Rare Met. Mater. Eng.* **2013**, *42*, 718–723.
23. Gao, L.; Xu, Z.; Zhang, S. The co-doping effects of Zr and Co on structure and electrochemical properties of $LiFePO_4$ cathode materials. *J. Alloys Compd.* **2018**, *739*, 529–535. [CrossRef]
24. Xu, J.; Chen, G. Effects of doping on the electronic properties of $LiFePO_4$: A first-principles investigation. *Phys. B Condens. Matter* **2010**, *405*, 803–807. [CrossRef]
25. Zhu, Y.R.; Zhang, R.; Deng, L.; Yi, T.-F.; Ye, M.-F.; Yao, J.-H.; Dai, C.-S. Lithium-Ion Insertion Kinetics of Na-Doped $LiFePO_4$ as Cathode Materials for Lithium-Ion Batteries. *Metall. Mater. Trans. E* **2015**, *2*, 33–38. [CrossRef]
26. Zhang, B.; He, Y.; Wang, A.; Xiao, X.; Liu, J.; Wu, Y.; Wang, L.; Xu, H.; Li, J.; Zhang, H.; et al. Regulation voltage of $LiNiPO_4$ by density functional theory (DFT) calculation to move towards practical application. *Interdiscip. Mater.* **2023**, *2*, 443–458. [CrossRef]
27. Oukahou, S.; Elomrani, A.; Maymoun, M.; Sbiaai, K.; Hasnaoui, A. Investigation of $LiMn_{1-x}M_xPO_4$ (M = Ni, Fe) as cathode materials for Li-ion batteries using density functional theory. *Comput. Mater. Sci.* **2022**, *202*, 111006. [CrossRef]
28. Sgroi, M.F.; Lazzaroni, R.; Beljonne, D.; Pullini, D. Doping $LiMnPO_4$ with Cobalt and Nickel: A Firstc Principle Study. *Batteries* **2017**, *3*, 11. [CrossRef]
29. Sukkabot, W. Effect of transition metals doping on the structural and electronic properties of $LiMnPO_4$: Spin density functional investigation. *Phys. Scr.* **2020**, *95*, 045811. [CrossRef]
30. Kellerman, D.G.; Chukalkin, Y.G.; Medvedeva, N.I.; Semenova, A. Effect of vanadium doping on the magnetic properties of $LiMnPO_4$. *Phys. Status Solidi B* **2016**, *253*, 965–975. [CrossRef]
31. Chung, S.Y.; Bloking, J.T.; Chiang, Y.M. Electronically conductive phospho-olivines as lithium storage electrodes. *Nat. Mater.* **2002**, *1*, 123. [CrossRef] [PubMed]
32. Yao, J.; Konstantinov, K.; Wang, G.; Liu, H.K. Electrochemical and magnetic characterization of $LiFePO_4$ and Li0.95Mg0.05FePO4 cathode materials. *J. Solid State Electrochem.* **2007**, *11*, 177–185. [CrossRef]
33. Gao, Y.; Xiong, K.; Zhang, H.; Zhu, B. Effect of Ru Doping on the Properties of $LiFePO_4$/C Cathode Materials for Lithium-Ion Batteries. *ACS Omega* **2021**, *6*, 14122–14129. [CrossRef] [PubMed]
34. Karimzadeh, S.; Safaei, B.; Huang, W.; Jen, T.-C. Theoretical investigation on niobium doped $LiFePO_4$ cathode material for high performance lithium-ion batteries. *J. Energy Storage* **2023**, *67*, 107572. [CrossRef]
35. Han, J.; Yang, J.; Xu, Z.; Li, H.; Wang, J. Dramatic improvement in high-rate capability of $LiMnPO_4$ nanosheets via crystallite size regulation. *J. Alloys Compd.* **2022**, *894*, 162510. [CrossRef]
36. Karthickprabhu, S.; Vikraman, D.; Kathalingam, A.; Prasanna, K.; Kim, H.S.; Karuppasamy, K. Electrochemical and cycling performance of neodymium (Nd^{3+}) doped $LiNiPO_4$ cathode materials for high voltage lithium-ion batteries. *Mater. Lett.* **2019**, *237*, 224–227. [CrossRef]
37. Shenouda, A.Y.; Liu, H.K. Studies on electrochemical behaviour of zinc-doped $LiFePO_4$ for lithium battery positive electrode. *J. Alloys Compd.* **2009**, *477*, 498–503. [CrossRef]
38. Bilecka, I.; Hintennach, A.; Rossell, M.D.; Xie, D.; Novak, P.; Niederberger, M. Microwave-assisted solution synthesis of doped $LiFePO_4$ with high specific charge and outstanding cycling performance. *J. Mater. Chem.* **2011**, *21*, 5881–5890. [CrossRef]
39. Liu, H.; Cao, Q.; Fu, L.J.; Li, C.; Wu, Y.P.; Wu, H.Q. Doping effects of zinc on $LiFePO_4$ cathode material for lithium ion batteries. *Electrochem. Commun.* **2006**, *8*, 1553–1557. [CrossRef]
40. Zhao, Y.; Chen, L.; Chen, X.; Kuang, Q.; Dong, Y. Crystal structure and electrochemical properties of $LiFe_{1-x}Zn_xPO_4$. *Powder Diffr.* **2011**, *26*, 238. [CrossRef]
41. Yiming, W.; Giuli, G.; Moretti, A.; Nobili, F.; Fehr, K.T.; Paris, E.; Marassi, R. Synthesis and characterization of Zn-doped $LiFePO_4$ cathode materials for Li-ion battery. *Mater. Chem. Phys.* **2015**, *155*, 191–204. [CrossRef]
42. Lin, Z.-P.; Zhao, Y.-J.; Zhao, Y.-M. Li-Site and Metal-Site Ion Doping in Phosphate-Olivine $LiCoPO_4$ by First-Principles Calculation. *Chin. Phys. Lett.* **2009**, *26*, 038202.
43. Xu, Z.; Gao, L.; Liu, Y.; Li, L. Review–Recent Developments in the Doped $LiFePO_4$ Cathode Materials for Power Lithium Ion Batteries. *J. Electrochem. Soc.* **2016**, *163*, A2600. [CrossRef]

materials

MDPI

Article

Chalcogen Doping in SnO_2: A DFT Investigation of Optical and Electronic Properties for Enhanced Photocatalytic Applications

Nikolaos Kelaidis [1,2,*], Yerassimos Panayiotatos [2] and Alexander Chroneos [3,4,*]

1 Theoretical and Physical Chemistry Institute, National Hellenic Research Foundation, Vass. Constantinou 48, 11635 Athens, Greece
2 Department of Mechanical Engineering, University of West Attica, 12241 Athens, Greece; gpana@uniwa.gr
3 Department of Electrical and Computer Engineering, University of Thessaly, 38221 Volos, Greece
4 Department of Materials, Imperial College, London SW7 2AZ, UK
* Correspondence: nkelaidis@eie.gr (N.K.); alexander.chroneos@imperial.ac.uk (A.C.)

Abstract: Tin dioxide (SnO_2) is an important transparent conductive oxide (TCO), highly desirable for its use in various technologies due to its earth abundance and non-toxicity. It is studied for applications such as photocatalysis, energy harvesting, energy storage, LEDs, and photovoltaics as an electron transport layer. Elemental doping has been an established method to tune its band gap, increase conductivity, passivate defects, etc. In this study, we apply density functional theory (DFT) calculations to examine the electronic and optical properties of SnO_2 when doped with members of the oxygen family, namely S, Se, and Te. By calculating defect formation energies, we find that S doping is energetically favourable in the oxygen substitutional position, whereas Se and Te prefer the Sn substitutional site. We show that S and Se substitutional doping leads to near gap states and can be an effective way to reduce the band gap, which results in an increased absorbance in the optical part of the spectrum, leading to improved photocatalytic activity, whereas Te doping results in several mid-gap states.

Keywords: SnO_2; electronic properties; optical properties; density functional theory

1. Introduction

SnO_2 is an n-type wide band gap (~3.6 eV) semiconductor that has attracted much interest in the scientific community due its abundance, non-toxicity, and interesting and tunable properties. Potential applications include gas sensing [1,2], photocatalysis [3], energy storage [4], and use as a transparent conducting oxide (TCO) in touchscreens, flat-panel displays, and photovoltaic cells. Various new and efficient preparation methods [5] have been developed for SnO_2 growth, such as magnetron sputtering [6] and electrochemical deposition [7], which are expected to facilitate scalability and lead to large-area, uniform SnO_2 films. Although undoped SnO_2 has a rather wide band gap and primarily absorbs shortwave ultraviolet light, this can be addressed by tuning its properties through established methods such as varying its morphology [8,9] and introducing doping [10,11]. The modification of band gaps and band edges has been a matter of research, particularly from a computational perspective [12,13].

Here, we investigate the impact of doping with the chalcogen elements S, Se, and Te. S, Se, and Te (though mainly S) have been examined for sensing applications, but it is also interesting to investigate the effect of their doping on optical properties when they are used in optoelectronics or as a transport layer in photovoltaics, given the limited theoretical studies on these dopants.

Experimental studies have shown that doping with S (as a surface dopant) significantly improves the sensitivity of resistive NO_2 gas sensors [14]. In the experimental study by Ma et al. [15], S-doped SnO_2 nanoparticles were successfully prepared and characterised. These nanoparticles exhibited increased photocatalytic activity due to a significantly improved visible light response, leading to the effective separation of photo-generated electron–hole pairs.

In a recent computational study, Yu et al. [16] calculated the effect of non-metal doping of SnO_2, including sulphur. They modelled S at an oxygen substitutional position and used the PBE functional to calculate the electronic properties and band gap. The calculated band gap in this study was 1.28 eV, and the effect of S was examined, mainly being the introduction of impurity levels into the band gap.

The effect of Se as a dopant has been also examined for sensing applications [17]. Se-doped SnO_2 was found to exhibit a 2 to 3 times higher response towards certain gases attributed to a considerable electron transfer by the dopant Se, which was found to be prone to act as a substitutional dopant on the Sn site. DFT calculations were performed on $Sn_{16}O_{32}$ slabs to examine the activation energy barrier of the CO oxidation, which was found to be significantly lower for Se-doped than pristine SnO_2. The density of states was computed with the PBE functional.

Regarding Te as a dopant, a recent study [18] prepared Sn-O-Te layers and examined them as humidity sensors. The preparation method involved thermal co-evaporation of Sn and TeO_2 in a vacuum, resulting in a nanosized SnO_2 matrix with a finely dispersed phase of Sn, Te, TeO_2, and SnTe. Elemental Te exhibits high mobility and begins to diffuse from the bulk and segregate on the layer surface at temperatures above 100 °C. At room temperature, as-deposited layers with Sn/Te ratios between 0.4 and 0.9 show excellent characteristics as humidity sensors. It was determined that both electron and ionic conduction simultaneously drive the overall conduction mechanism of the sensor response.

A large-scale computational work [19] examined the screening of a total of 63 SnO_2 dopants, all positioned in a Sn substitutional site, including S, Se, and Te. PBE calculations were applied as a screening method. The importance of a high level of theory, such as the PBE0 hybrid functional for the calculations of the electronic properties of defected structures, was pointed out. The authors concluded that p-type doping is not possible via elemental substitution of Sn. Nevertheless, that study was not extended to oxygen substitutional or interstitial dopants.

In the present study, we employ DFT calculations to examine the electronic, structural, and optical properties of SnO_2 doped with S, Se, and Te at both oxygen or sulphur substitutional and interstitial sites, which seems to be lacking in the literature. The electronic and optical properties of doped SnO_2 are systematically examined with a higher level of theory, employing hybrid functionals. Additionally, we have employed a 'harder' Sn projector wave potential (Sn_d) that provides higher accuracy and correctly predicts the SnO_2 band gap. We believe that this study can be of interest for potential application in solar cells and photocatalysis.

2. Materials and Methods

The crystal structure of SnO_2 is that of rutile. It crystallises in the tetragonal $P4_2/mnm$ space group with the lattice constants of a = 4.737 Å and c = 3.186 Å, as determined by X-ray diffraction (XRD) experiments [20]. For the defect calculations, we used a $2 \times 2 \times 2$ supercell approximation, which consists of 48 atoms in the perfect cell configuration (see Figure 1). We substituted either the Sn atom (for substitutional cation doping) or the O atom (for substitutional anion doping) with members of the chalcogen group (excluding polonium), namely S, Se, and Te. All Sn positions in the crystal structure are equivalent and occupy the 2a Wyckoff site, as do all oxygen positions, which occupy the 4f Wyckoff site. Additionally, we examined these dopants at interstitial positions. To account for size-related errors, we also performed calculations using a larger $3 \times 3 \times 3$ supercell [21]. However, when comparing defect formation energies with those from the $2 \times 2 \times 2$ supercell, the differences were minimal, on the order of about 0.1 eV.

The calculations were performed with the aid of the Cambridge Serial Total Energy Package (CASTEP) code using the PBE functional. Following the convergence tests, 800 eV was chosen as the cutoff energy and a Brillouin zone $3 \times 3 \times 3$ k-point sampling was chosen for the geometry optimisation. In order to account for the impact of localised electrons and the underestimation of the band gap that is typically observed in GGA and LDA, we applied the hybrid functional PBE0 [22] for the electronic and optical property calculations applied on the 2 × 2 × 2 supercell with a denser k-point mesh of 5 × 5 × 5. Regarding convergence criteria for our calculations, the self-consistent field (SCF) method tolerance was 1.0×10^{-5} eV/atom, the force tolerance was set at 0.05 eV/Å, and the max displacement tolerance was set at 0.001 eV/Å.

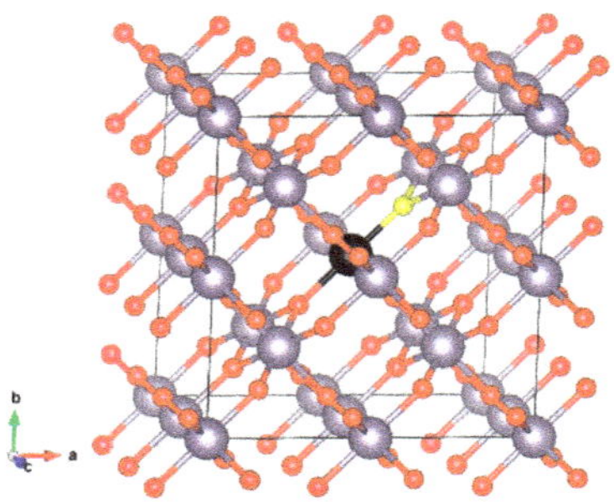

Figure 1. SnO_2 2 × 2 × 2 supercell (a) initial (perfect) structure. Sn occupies the 2a Wyckoff site and O occupies the 4f site. The S atom that is substituted in our model (black) and the oxygen atom (yellow) are shown in different colours.

3. Results and Discussion

3.1. Formation Energies and Structural Optimisation

For the calculation of the formation energies for the various doping cases, geometry optimisation was performed using the PBE functional. The following formula was applied for the calculation of the defect formation energy:

$$E_d^f = E_T^{defect} - E_T^{perfect} + \sum_i \mu_i n_i \quad (1)$$

where E_d^f is the defect formation energy, $E_T^{perfect}$ is the total energy of the supercell with no defects, and E_T^{defect} is the total energy of the supercell where n_i atoms of chemical potential μ_i have been added or removed.

Since we are examining doping with a single element, this formula can be reduced to a simpler form:

$$E_{formation} = E_{perfect} - E_{doped} - \left(\mu_{dopant} + \mu_{substituted}\right) \quad (2)$$

The chemical potential of each element was determined by calculating the total energy per atom in its most stable, standard state. The calculated defect formation energies are presented in Table 1, along with the DFT calculated lattice constants, cell volume, and relative volume expansion.

An important conclusion from the calculation of the defect formation energies is that doping with S appears energetically favourable on the oxygen site, although it is found to be so in the Sn position in some computational studies. Here, we find that the formation energy of S as an oxygen substitutional atom is 3.95 eV, which is significantly less than that at the Sn substitutional position, which is 5.43 eV. We also find that Se and especially Te are preferably Sn-substitutional dopants, with a formation energy of 3.90 eV for Se (less than the 5.04 eV at the oxygen substitution site) and just 2.88 eV for Te (7.66 eV at the oxygen substitutional position, which renders this case unfeasible).

The minimum energy configuration structures are shown in Figure 2 for the energetically preferable doping cases. In Figure 2a, the structure with S as an oxygen substitution is presented. The Sn-S atomic distances are calculated as 2.34 Å. In Figure 2b, the case of doping with Se as a Sn substitution is shown. The Se-O bond is at 2.064 Å, slightly smaller than the Sn-O bond, which is at 2.098 Å. The Se-Sn bond is at 3.233 Å, very similar to the Sn-Sn bond (calculated for this structure as 3.236 Å). In Figure 2c, the Te atom is at a Sn substitutional site. The Te-O distance is calculated as 2.16 Å and the Te-Sn distance as 3.253 Å, the same as the Sn-Sn bond length. Structures were drawn using VESTA visualisation software (version 3) [23].

Table 1. Calculated lattice constants and cell volume for the examined dopants at different positions.

SnO_2	a (Å)	c (Å)	Vol (Å^3)	ΔV (%)	$E_{formation}$
perfect	4.818	3.236	75.121	n/a	n/a
S_{Sn}	4.816	3.232	74.958	−0.218	5.43
S_{ox}	4.833	3.240	75.681	0.745	3.95
S_{int}	4.843	3.251	76.427	1.739	4.64
Se_{Sn}	4.820	3.234	75.128	0. 009	3.90
Se_{ox}	4.839	3.241	75.892	1.027	5.04
Se_{int}	4.843	3.252	76.533	1.879	4.73
Te_{Sn}	4.826	3.241	75.468	0. 462	2.88
Te_{ox}	4.848	3.246	76.272	1.532	7.66
Te_{int}	4.838	3.251	76.786	2.216	5.62

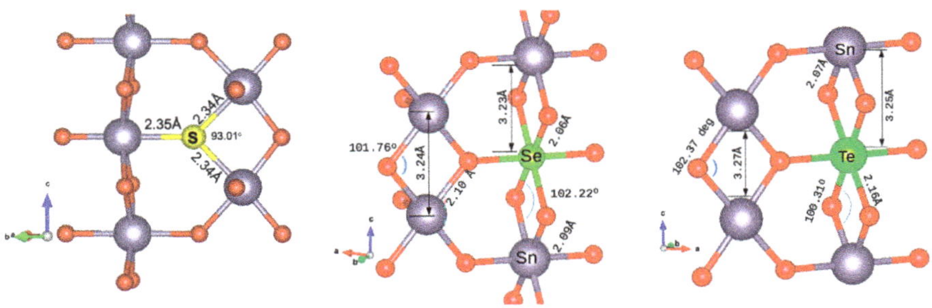

(**a**) (**b**) (**c**)

Figure 2. Dopant positions in the SnO_2 2 × 2 × 2 supercell of (**a**) S substituting an oxygen atom, (**b**) Se substituting a Sn atom, and (**c**) Te substituting a Sn atom.

In Table 1, the relative volume change is shown. The substitution of Sn with S is the only case of a decrease in total volume, attributed to the smaller atomic radius of S compared to Sn (100 pm vs. 140 pm). When substituting oxygen with S, Se, or Te, the total cell volume is increased. When dopants are located in an interstitial position, the cell volume will be increased in all cases, with Te showing the highest increase in volume as it has the highest atomic radius.

3.2. Electronic Properties

Herein we will examine the electronic properties of the doped SnO_2 by calculating the density of states (DOS) and partial DOS. The doping level is at approx. 2% per atom. The DOS and PDOS plots for the undoped cell are shown in Figure 2.

The band gap has been calculated as 3.6 eV using the PBE0 functional with a fairly 'harder' and quite accurate potential [24], included in the CASTEP code. The use of the Sn_d PAW pseudopotential, which includes 14 valence electrons and accounts for d-orbitals, yields a band gap for the undoped structure that is almost identical to the experimental value (3.62 eV) [25]. This approach appears to perform better than other computational works, which typically calculate the band gap as around 3.2 eV (with PBE0 or HSE06). Using the same code with standard pseudopotentials yields a band gap of 3.18 eV [9].

The DOS plots of the undoped structure are shown in Figure 3. The valence band of SnO_2 is dominated by oxygen p-states, while the conduction band mostly comprises tin s- and p-states. The electrical conductivity of SnO_2 is largely controlled by the movement of electrons in the tin d-orbitals.

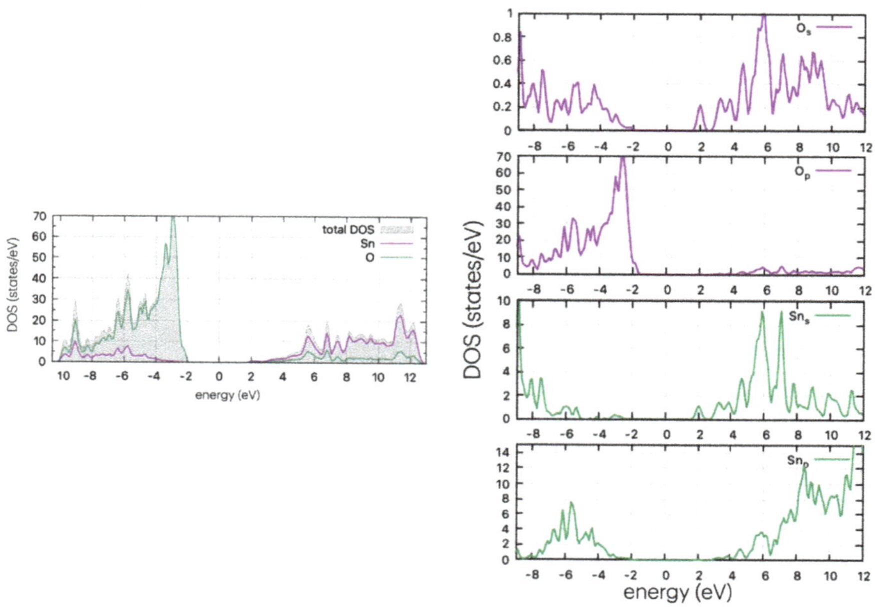

(a) (b)

Figure 3. (**a**) DOS and (**b**) PDOS plots for the undoped SnO_2.

Using the same method as for the pristine supercell, we calculated the DOS of the doped structures. These are presented in Figures 4–6 for the S, Se, and Te dopants, respectively.

As Figure 4 indicates, the insertion of S in the supercell leads to band gap states that are just above the valence band maximum (VBM). In the case of substitutional S in the oxygen position, the calculations show band tail-like behaviour and a reduction of the band gap at ~2.8 eV. The S peaks arise at 0.77 eV above the valence band maximum (VBM) for the oxygen substitutional case, and a second neighbouring peak is present for the interstitial case, at approx. 1.45 eV above the VBM. The band tail is expected to vary the optical properties of the material, producing a red shift in the optical absorption. This is examined in the next section. The formulated peak is due to the hybridisation of s-orbitals of S and O, as confirmed by partial DOS analysis (Figure 4b).

In Figure 5, the DOS of Se-doped SnO_2 with Se replacing a Sn atom is presented. In this case (Se_{Sn}), a Se peak is present as a band tail, reducing the band gap. The calculated results are very similar to the S case, which is to be expected as both dopants have four p-electrons in their outer shell. According to partial DOS analysis, these peaks are attributed to p-orbitals of Se and O. Se peaks are present at 0.77 eV above the VBM, and the band gap is at approximately ~2.7 eV, although this may vary slightly on the Gaussian smearing parameter (here set at 0.2). This decrease in value is consistent with the experimental work of Kumar et al. [26] who determined a decrease in the band gap (depending on the Se%), reaching a minimum of 2.9 eV and a red shift in the absorption spectrum.

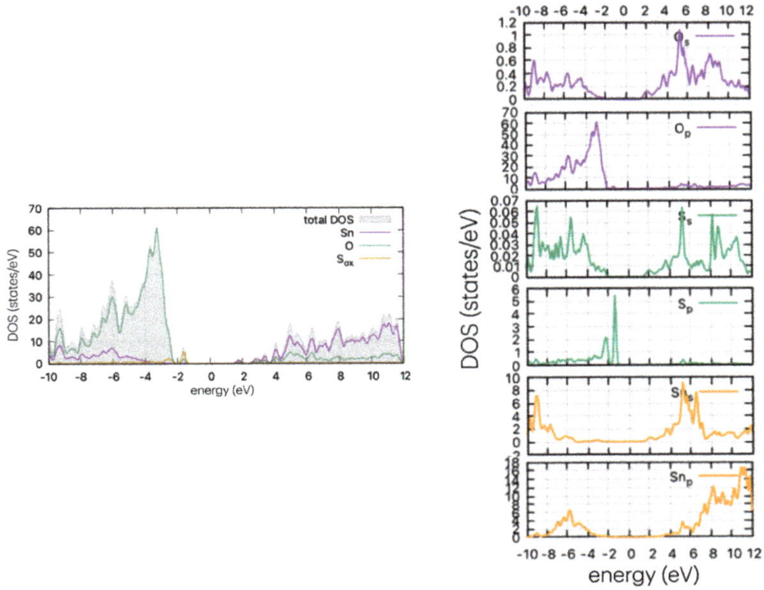

(**a**) (**b**)

Figure 4. (**a**) DOS and (**b**) PDOS of S-doped SnO_2 for oxygen substitutional position.

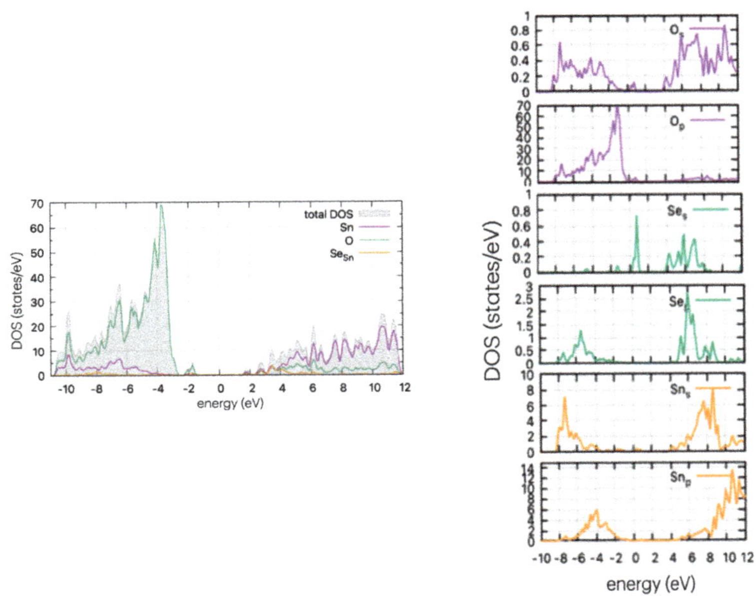

(**a**) (**b**)

Figure 5. (**a**) DOS and (**b**) PDOS of Se-doped SnO_2 for Sn substitutional position.

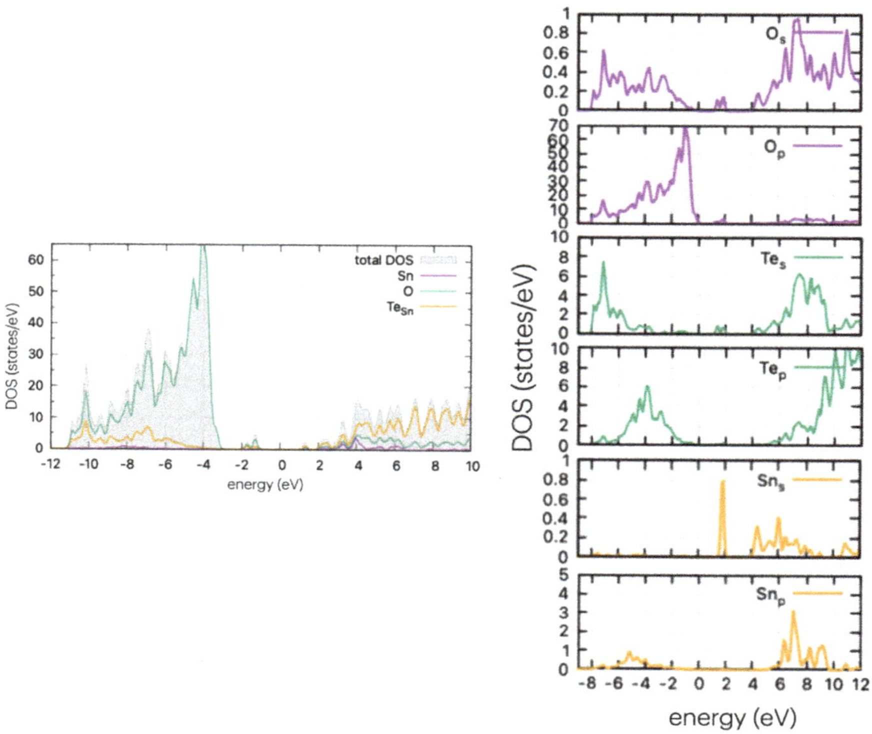

(**a**) (**b**)

Figure 6. (**a**) DOS and (**b**) PDOS of Te-doped SnO_2 for Sn substitutional position.

In Figure 6, the calculated DOS of Te-doped SnO_2 is shown for the Sn substitutional case. A strong and wide peak is evident below the middle of the band gap, at approx. 1.6 eV from the VBM, as well as a band tail nearing the conduction band edge. In contrast with the cases of S and Se doping, the calculations show that Te does not significantly reduce the band gap with the introduction of states near the band edges, but instead shows considerable mid-gap states. These can enhance photocatalytic processes by stabilising intermediate species or by providing additional reaction pathways. However, they can also reduce efficiency at high concentrations by increasing recombination rates. Therefore, understanding and controlling mid-gap states is important for optimising photocatalytic materials. For completeness, the DOS of all doping cases (interstitial and substitutional of Sn or O) are shown in Figure S1.

3.3. Optical Properties

Examining the optical properties can provide important information for comprehending the electronic structure of materials. These properties are derived from the complex dielectric function $\varepsilon(\omega) = \varepsilon_1(\omega) + i\varepsilon_2(\omega)$, a quantity that characterises the material reaction to an external electric field. Using the CASTEP DFT code, we obtain the optical properties resulting from electronic transitions of the examined structures. We performed DFT calculations of the optical properties using LDA, PBE, and non-local PBE0 and HSE06 functionals. Hybrid functionals PBE0 and HSE06 yield, at low frequencies, a refractive index of approx. 1.52 for undoped SnO_2, while non-local functionals LDA and PBE produce a refractive index of approx. 2 (Figure S2).

The refractive index of the undoped SnO_2 may vary depending on the growth conditions and crystal structure of the sample. It has been reported to be about 1.9 [27] or 2.0 [28] at the incident photon wavelength of 550 nm (photon energy of 2.48 eV). Here we will present the DFT results obtained using the hybrid functional PBE0 (very similar to HSE06).

To present an overview of the impact of doping on the optical properties of the SnO_2-doped system, the effect of doping on (a) the real part ε_1 and (b) the imaginary part ε_2 of the **dielectric function** is presented in Figure 7 (curves are smoothed for visibility).

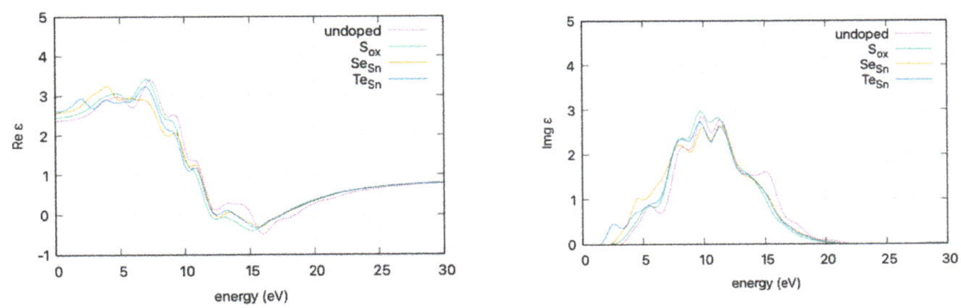

(**a**) (**b**)

Figure 7. DFT calculated (HSE06) **dielectric function** vs. incident photon energy for doped and undoped SnO_2: (**a**) real part ε_1 and (**b**) imaginary part ε_2. Graphs smoothed for visibility.

The calculated static dielectric function, $\varepsilon_1(0)$, increases with doping, from 2.381 for undoped SnO_2 to 2.466 for S doping, 2.576 for Se doping, and 2.604 for Te doping. A similar trend is observed for the $\varepsilon_2(\omega)$ parameter in the low energy region, where the parameter exhibits a transition to non-zero values at lower energy levels for Te, Se, and S. This is consistent with the band gap calculations, as a greater static dielectric constant indicates a lower band gap [29]. At higher energy values, the $\varepsilon(\infty)$ has a maximum at almost ~4, consistent with wide band gap oxides. In Figure 8, the calculated refractive index is shown. Its values at low frequencies range from 1.54 for undoped SnO_2 to 1.58 for S doping, 1.605 for Se doping, and 1.614 for Te doping. In Figure 9, the calculated absorption coefficients and loss energy function vs. energy are plotted. It can be observed that at higher energies (lower wavelengths), doping increases the absorbance of the film, making it less transparent.

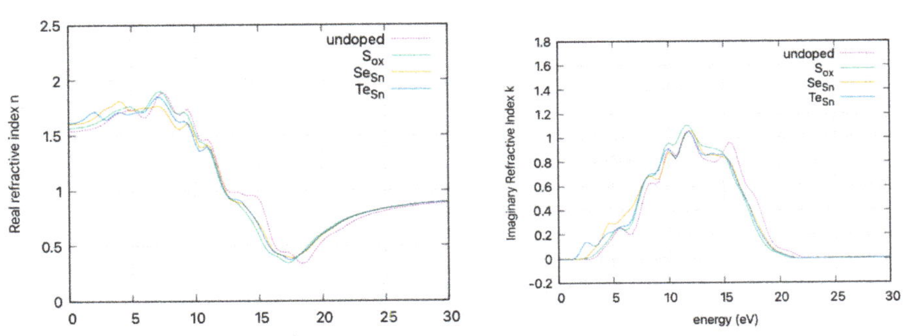

(**a**) (**b**)

Figure 8. DFT calculated (HSE06) **refractive index** (smoothed curves) vs. incident photon energy for doped and undoped SnO_2. (**a**) Real refractive index and (**b**) imaginary refractive index

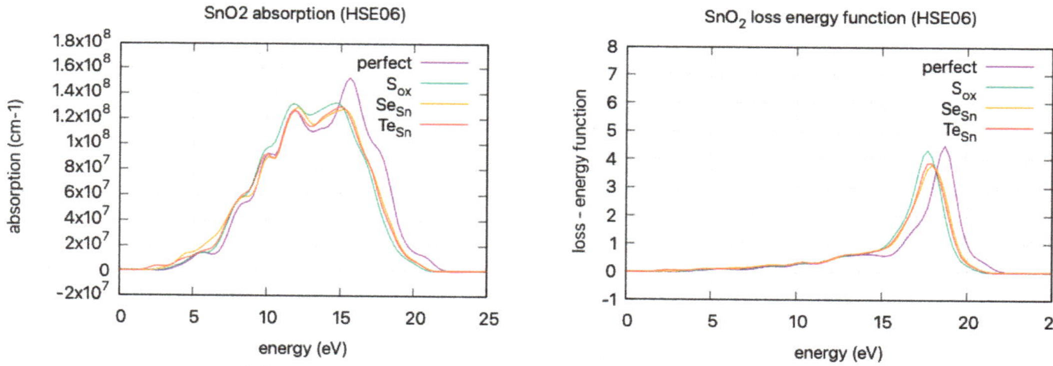

Figure 9. DFT calculated (HSE06) (**a**) absorption coefficient vs incident photon energy (**b**) loss energy function vs energy.

The refractive index, calculated at low frequencies, varies from 1.54 for undoped SnO_2 to 1.61 for Te doping. The static dielectric constant increases with doping. This increase is consistent with a lower band gap and is supported by band gap calculations. At higher energies, the $\varepsilon(\infty)$ approaches ~4, characteristic of wide band gap oxides. Additionally, the absorption coefficients and loss energy function indicate that doping results in increased absorbance, diminishing transparency at higher energies, leading to a red shift in the absorption spectrum.

4. Conclusions

The above findings can offer valuable guidance for tailoring SnO_2 properties for applications ranging from photocatalysis to energy harvesting. The reduction of its bandgap, as demonstrated through S and Se doping, can enhance its suitability for such applications by increasing its capability to absorb light, particularly in the longer wavelengths of the spectrum (red shift), thus enabling the material to utilise a broader spectrum of sunlight. This can also have applications as an electron transfer layer (ETL) in photovoltaic devices, where a smaller band gap can improve band alignment with adjacent layers, which can significantly influence the overall performance of a solar cell. In the case of Te doping, the presence of energy levels in the middle of the bandgap could be beneficial in the context of photocatalytic or sensing applications as it provides a site where chemical reactions can take advantage of the trapped electrons. However, these mid-gap states can be detrimental for photovoltaic applications when Te-doped SnO_2 is used as an ETL. Nevertheless, Te doping has theoretical potential for applications in photocatalysis, such as hydrogen production. In conclusion, this study provides additional insight into the effects of chalcogen doping of SnO_2. The suitability of these doping techniques depends on various parameters, including overall device architecture, material compatibility, and performance requirements.

Supplementary Materials: The following supporting information can be downloaded at: https://www.mdpi.com/article/10.3390/ma17163910/s1, Figure S1: The DOS of all doping cases (interstitial, substitutional of Sn or O). Figure S2: Comparison of refractive index calculated with LDA, GGA-PBE and hybrid functionals PBE0 and HSE06.

Author Contributions: The contributions of the authors to this research are as follows: Author N.K. conducted the DFT calculations, wrote the manuscript, and contributed to the conceptualisation and design of the study. Author Y.P. provided support and assistance in the interpretation of results. Author A.C. provided valuable insights, contributed to the interpretation of results, and reviewed and revised the manuscript. All authors have read and agreed to the published version of the manuscript.

Funding: This research received no external funding.

Institutional Review Board Statement: Not applicable

Informed Consent Statement: Not applicable.

Data Availability Statement: The datasets used and/or analysed during the current study are available from the corresponding author upon reasonable request.

Acknowledgments: We acknowledge support by computational time granted by the National Infrastructures for Research and Technology S.A. (GRNET S.A.) in the National HPC facility—ARIS—under project ID pa240203.

Conflicts of Interest: The authors declare no conflicts of interest.

References

1. Batzill, M.; Diebold, U. The surface and materials science of tin oxide. *Prog. Surf. Sci.* **2005**, *79*, 47–154. [CrossRef]
2. Li, B.; Zhou, Q.; Peng, S.; Liao, Y. Recent Advances of SnO_2-Based Sensors for Detecting Volatile Organic Compounds. *Front. Chem.* **2020**, *8*, 321. [CrossRef] [PubMed]
3. Khan, M.M.; Adil, S.F.; Al-Mayouf, A. Metal oxides as photocatalysts. *J. Saudi Chem. Soc.* **2015**, *19*, 462–464. [CrossRef]
4. Dixon, D.; Ávila, M.; Ehrenberg, H.; Bhaskar, A. Difference in Electrochemical Mechanism of SnO_2 Conversion in Lithi-um-Ion and Sodium-Ion Batteries: Combined in Operando and Ex Situ XAS Investigations. *ACS Appl. Energy Mater.* **2020**, *3*, 5552–5562.
5. Wu, P.; Wang, S.; Li, X.; Zhang, F. Advances in SnO_2-based perovskite solar cells: From preparation to photovoltaic applica-tions. *J. Mater. Chem. A* **2021**, *9*, 19554–19588. [CrossRef]
6. Ma, Y.; Li, L.; Qian, J.; Qu, W.; Luo, R.; Wu, F.; Chen, R. Materials and structure engineering by magnetron sputtering for advanced lithium batteries. *Energy Storage Mater.* **2021**, *39*, 203–224. [CrossRef]
7. Ko, Y.; Kim, Y.R.; Jang, H.; Chanyong, L.; Kang, M.G.; Jun, Y. Electrodeposition of SnO_2 on FTO and its Application in Planar Heterojunction Perovskite Solar Cells as an Electron Transport Layer. *Nanoscale Res. Lett.* **2017**, *12*, 498. [CrossRef] [PubMed]
8. Periyasamy, M.; Kar, A. Modulating the properties of SnO_2 nanocrystals: Morphological effects on structural, photolumi-nescence, photocatalytic, electrochemical and gas sensing properties. *J. Mater. Chem. C* **2020**, *8*, 4604–4635. [CrossRef]
9. Kelaidis, N.; Zervos, M.; Lathiotakis, N.; Chroneos, A. Vapor-Liquid-Solid Growth and Properties of One Dimensional PbO_x and PbO_x/SnO_2 (x = 1, 2) Nanowires for Energy Conversion and Storage. *Mater. Adv.* **2022**, *3*, 1695–1702. [CrossRef]
10. Sun, C.; Yang, J.; Xu, M.; Cui, Y.; Ren, W.; Zhang, J.; Zhao, H.; Liang, B. Recent intensification strategies of SnO_2-based photocatalysts: A review. *Chem. Eng. J.* **2022**, *427*, 131564. [CrossRef]
11. Filippatos, P.-P.; Kelaidis, N.; Vasilopoulou, M.; Davazoglou, D.; Chroneos, A. Impact of boron and indium doping on the structural, electronic and optical properties of SnO_2. *Sci. Rep.* **2021**, *11*, 13031. [CrossRef] [PubMed]
12. Filippatos, P.-P.; Sharma, R.; Soultati, A.; Kelaidis, N.; Petaroudis, C.; Alivisatou, A.-A.; Drivas, C.; Kennou, S.; Christopoulos, S.-R.G.; Davazoglou, D.; et al. Optimization of the hydrogen response characteristics of halogen-doped SnO_2. *Sci. Rep.* **2023**, *13*, 2524. [CrossRef] [PubMed]
13. Filippatos, P.-P.; Kelaidis, N.; Vasilopoulou, M.; Davazoglou, D.; Chroneos, A. Defect Processes in Halogen Doped SnO_2. *Appl. Sci.* **2021**, *11*, 551. [CrossRef]
14. Xu, K.; Tian, S.; Zhu, J.; Yang, Y.; Shi, J.; Yu, T.; Yuan, C. High selectivity of sulfur-doped SnO_2 in NO_2 detection at lower operating temperatures. *Nanoscale* **2018**, *10*, 20761. [CrossRef]
15. Ma, L.; Xu, L.; Xu, X.; Zhou, X.; Zhang, L. One-Pot Hydrothermal Synthesis of Sulfur-Doped SnO_2 Nanoparticles and their Enhanced Photocatalytic Properties. *Nano* **2016**, *11*, 1–14. [CrossRef]
16. Yu, J.; Wang, Y.; Huang, Y.; Wang, X.; Guo, J.; Yang, J.; Zhao, H. Structural electronic properties of SnO_2 doped with non-metal elements. *Beilstein J. Nan-otechnol.* **2020**, *11*, 1321–1328. Available online: https://www.beilstein-journals.org/bjnano/articles/11/116 (accessed on 1 January 2023). [CrossRef] [PubMed]
17. Zhao, L.; Gong, X.; Tao, W.; Wang, T.; Liu, X.; Liu, F.; Yan, X.; Wang, C.; Sun, P.; Lu, G. Cationic Se-Doped SnO_2 for Superior Acetone Sensing with a Moderate Baseline Resistance Rise. *SSRN Electron. J.* **2021**. [CrossRef]
18. Georgieva, B.; Podolesheva, I.; Spasov, G.; Pirov, J. Nanosized Thin SnO_2 Layers Doped with Te and TeO_2 as Room Temperature Humidity Sensors. *Sensors* **2014**, *14*, 8950–8960. [CrossRef] [PubMed]
19. Graužinyte, M.; Goedecker, S.; Flores-Livas, J.A. Computational screening of useful hole-electron dopants in SnO_2. *Chem. Mater* **2017**, *29*, 10095–10103. [CrossRef]
20. Lu, P.F.; Shen, Y.; Yu, Z.Y.; Zhao, L.; Li, Q.Y.; Ma, S.J.; Han, L.H.; Liu, Y.M. Electronic structure and optical properties of antimony-doped SnO_2 from first-principle study. *Commun. Math. Phys.* **2012**, *57*, 145.
21. Castleton, C.W.M.; Höglund, A.; Mirbt, S. Density functional theory calculations of defect energies using supercells. *Model. Simul. Mater. Sci. Eng.* **2009**, *17*, 084003. [CrossRef]
22. Adamo, C.; Barone, V. Toward reliable density functional methods without adjustable parameters: The PBE0 model. *J. Chem. Phys.* **1999**, *110*, 6158–6170. [CrossRef]
23. Momma, K.; Izumi, F. VESTA 3 for three-dimensional visualization of crystal, volumetric and morphology data. *J. Appl. Crystallogr.* **2011**, *44*, 1272–1276. [CrossRef]

24. Lee, M.-H. Advanced Pseudopotentials for Large Scale Electronic Structure Calculations With Application to A Study of Weakly Ordered Material—g-Al_2O_3. Ph.D. Thesis, Clare College, University of Cambridge, Cambridge, UK, 1995.
25. Nagasawa, M.; Shionoya, S. Exciton structure in optical absorption of SnO_2 crystals. *Phys. Lett.* **1966**, *22*, 409–410. [CrossRef]
26. Kumar, S.; Chauhan, P.; Kundu, V. Sol–gel synthesis, structural, morphological and optical properties of Se-doped SnO_2 nanoparticles. *J. Mater. Sci. Mater. Electron.* **2016**, *27*, 3103–3108. [CrossRef]
27. Afify, H.H.; Momtaz, R.S.; Badawy, W.A.; Nasser, S.A. Some physical properties of fluorine-doped SnO_2 films prepared by spray pyrolysis. *J. Mater. Sci. Mater. Electron.* **1991**, *2*, 40–45. [CrossRef]
28. Manifacier, J.; De Murcia, M.; Fillard, J.; Vicario, E. Optical and electrical properties of SnO_2 thin films in relation to their stoichiometric deviation and their crystalline structure. *Thin Solid Films* **1977**, *41*, 127–135. [CrossRef]
29. Ravichandran, R.; Wang, A.X.; Wager, J.F. Solid State Dielectric Screening Versus Band Gap Trends and Implications. *Opt. Mater.* **2016**, *60*, 181–187. [CrossRef]

materials

MDPI

Article

A Novel Approach for Evaluating the Influence of Texture Intensities on the First Magnetization Curve and Hysteresis Loss in Fe–Si Alloys

Daniele Carosi [1,*], Alessandro Morri [1], Lorella Ceschini [1] and Alessandro Ferraiuolo [2]

1 Department of Industrial Engineering—DIN, Metallurgy Group, Alma Mater Studiorum—Università di Bologna, 40126 Bologna, Italy; alessandro.morri4@unibo.it (A.M.); lorella.ceschini@unibo.it (L.C.)
2 Marcegaglia Ravenna s.p.a., Via Baiona, 141, 48123 Ravenna, Italy; alessandro.ferraiuolo@marcegaglia.com
* Correspondence: daniele.carosi2@unibo.it

Abstract: This paper examines the relationship between the magnetization behavior and crystal lattice orientations of Fe–Si alloys intended for magnetic applications. A novel approach is introduced to assess anisotropy of the magnetic losses and first magnetization curves. This method links the magnetocrystalline anisotropy energy of single crystal structures to the textures of polycrystalline materials through a vectorial space description of the crystal unit cell, incorporating vectors for external applied field and saturation magnetization. This study provides a preliminary understanding of how texture influences magnetic loss rates and the first magnetization curves. Experimental results from Electron Back-Scattered Diffraction (EBSD) and Single-Sheet Tests (SSTs), combined with energy considerations and mathematical modeling, reveal the following key findings: (i) a higher density of cubic texture components, whether aligned or rotated relative to the rolling direction, decreases magnetic anisotropy, suggesting that optimizing cubic texture can enhance material performance; (ii) at high magnetic fields, there is no straightforward correlation between energy losses and polarization; and (iii) magnetization rates significantly impact magnetization loss rates, highlighting the importance of considering these rates in optimizing Fe–Si sheet manufacturing processes. These findings offer valuable insights for improving the manufacturing and performance of Fe–Si sheets, emphasizing the need for further exploration of texture effects on magnetic behavior.

Keywords: hysteresis; mathematical modeling; recrystallization; texture; electrical steels

1. Introduction

Magnetic materials are extensively used in commercial applications, including transformers, electrical motors, solenoids, and generators. Among the soft magnetic materials, Non-Oriented Grain (NGO) Fe–Si alloys are the most widely used due to their cost-effectiveness and availability [1–7]. Recently, the demand for higher magnetic performance in automotive electric motors has driven research towards reducing the magnetic energy loss in these steels when subjected to an external magnetic field.

It Is well known that microstructural features such as mean grain size, defects, crystallographic texture, and precipitates influence the magnetic domain movement and, consequently, the magnetic behavior of the material [1–7]. Researchers are thus intensively studying these effects, particularly how to improve the microstructure of Fe–Si alloys during production.

Several studies have highlighted that in Fe–Si steels, crystallographic textures significantly impact the permeability of the first magnetization curve and the area of the hysteresis loop. Specifically, a higher number of crystals with <111> directions aligned with the external applied field results in lower magnetization at a given field intensity and reduced maximum permeability. Conversely, the <100> directions have the opposite effect [1,3,5,7,8].

Research on hysteresis loss has considered various aspects of material textures, examining losses at specific values of external fields or magnetic polarization. Key parameters include the following: a generic energy texture parameter, which correlates linearly with loss [9]; average magnetocrystalline energy [10]; texture parameter A_θ, dependent on the intensities of relevant magnetic texture components [11]; texture parameter A_α, related to the rolling direction and magnetization vector [12]; anisotropy parameter ε [13]; and weighted A_θ parameter [14].

This paper introduces a novel method for studying the influence of crystalline anisotropy on the first magnetization curve and total magnetic energy loss. Specifically, the total magnetic energy loss curve is derived from the areas of the hysteresis loops, with the peak polarization values at given external magnetic fields defining the first magnetization curve. It is important to note that magnetic behavior varies between the rolling and transverse directions.

The proposed method includes the following: (i) saturation magnetization; (ii) external applied field vectors; and (iii) energies associated with these vectors within the vectorial space of the crystal unit cell.

This method is implemented through the following steps: (i) Electron Back-Scattered Diffraction (EBSD) analyses and Single-Sheet Tests (SSTs); (ii) EBSD data processing using an appropriate toolbox; (iii) energy considerations related to the magnetization process; and (iv) mathematical modeling of energy loss.

Figure 1 presents a conceptual map of the work.

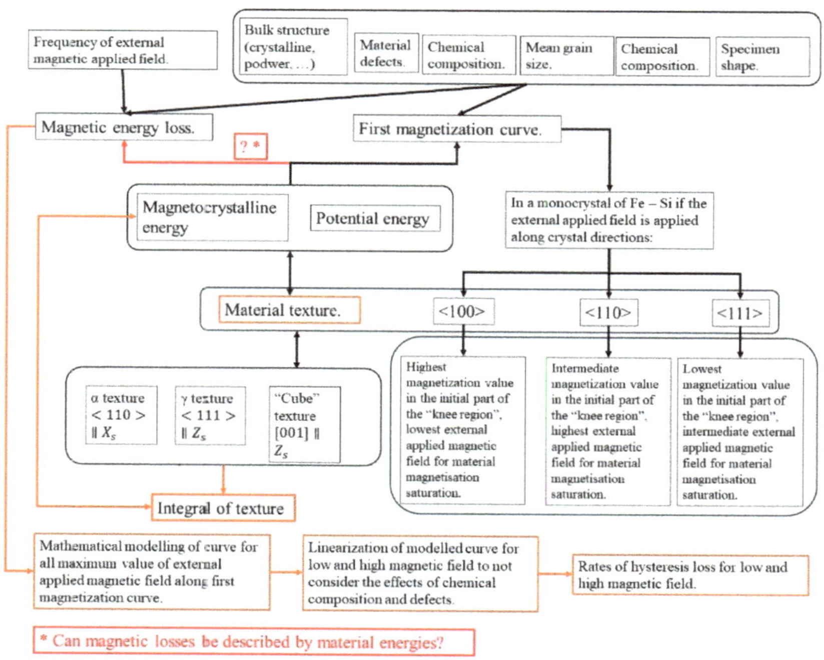

Figure 1. Conceptual map of the paper's work.

This approach could enable the development of more accurate models for predicting the magnetization curve and magnetic energy losses compared to existing methods. Furthermore, it will provide preliminary data to help establish guidelines for optimizing rolling processes and heat treatments in industrial plants.

2. Theoretical Bases

In the initial part of the first magnetization curve up to the "knee region", as shown in Figure 2, the magnetization process is primarily governed by the motion of domain walls. As the curve progresses through the "knee region", the saturation magnetization vectors of the newly formed domains begin to rotate, aligning with the direction of the externally applied magnetic field. Eventually, at high external magnetic field values, the domain configuration reaches a state where all saturation magnetization vectors are parallel to the applied field direction.

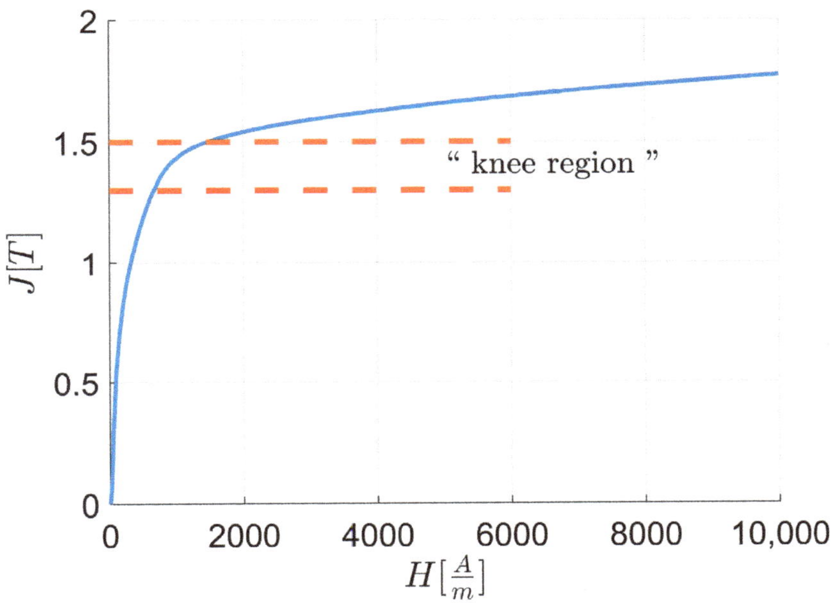

Figure 2. Magnetic polarization J vs. external applied magnetic field H and "knee region" highlighted.

For a single bcc crystal structure material subjected to an external magnetic field applied along one of the crystal directions <100>, <110>, and <111>, the magnetization value at the initial part of the "knee region" varies as follows: it is highest for <100>, lowest for <111>, and intermediate for <110>. However, the external magnetic field required to align all saturation magnetization vectors parallel to the applied field is lowest for <100>, highest for <110>, and intermediate for <111>.

The mean magnetization of a material is related to the variation in the Landau free energy. Specifically, under constant entropy and temperature conditions, it is given by the Helmholtz free energy, which reflects changes in the system's internal energy.

In an ideal bcc crystal, the potential energy of the saturation magnetization vector E_p, which drives the magnetization process up to the "knee region," is given by:

$$E_p = -\mu_0 \, \boldsymbol{M_s} \cdot \boldsymbol{H} \tag{1}$$

where $\boldsymbol{M_s}$ is the saturation magnetization vector; $\boldsymbol{H}$ is the external applied magnetic field vector; and μ_0 is the vacuum permeability; the symbol denotes the scalar product between vectors.

However, the magnetization process up to saturation is also influenced by the magnetocrystalline anisotropy energy E_a.

$$E_a = K_0 + K_1(\alpha_1\alpha_2 + \alpha_1\alpha_3 + \alpha_2\alpha_3) + K_2\alpha_1^2\alpha_2^2\alpha_3^2 \tag{2}$$

Here, α_i represents the direction cosine, which is the angle between the saturation magnetization vector and a specific crystal direction during its rotation toward the direction of the external applied field, where it reaches the minimum value of E_a. Specifically, the magnetocrystalline anisotropy energy E_a follows the order $E_{a,<100>} < E_{a,<110>} < E_{a,<111>}$ when the saturation magnetization vectors are fully aligned with the external applied field [1,2,5,7,8].

Moreover, in materials like iron or Fe–Si alloys, the term K_2 could be neglected since $K_1 \gg |K_2|$ and $K_0 \cong 0$ [1,8,15].

The total energy of the ideal material during the magnetization process up to the "knee region" ($E_{t,1}$), neglecting other energy contributions like the shape one, is as follows:

$$E_{t,1} = E_p \tag{3}$$

While over the "knee region", considering also the anisotropy magnetocrystalline energy, the total energy ($E_{t,2}$) is calculated as follows:

$$E_{t,2} = E_a + E_p \tag{4}$$

In real polycrystalline materials, the magnetization process is more complex due to factors such as the frequency of the external magnetic field and the specific crystal orientations of each grain. Additionally, materials often contain various defects—such as non-metallic inclusions, voids, and second phases—that act as pinning sites for domain wall motion. Despite these complexities, this study focuses on crystal orientations. Consequently, factors such as frequency, shape anisotropy, and inclusions are considered constant, as they are consistent across the specimens studied.

In polycrystals, lattice orientations can be described using Euler angles, which represent the rotation of the reference frame fixed to the unit crystal cell relative to the reference frame fixed to the specimen. The most comprehensive information about the distribution of crystal orientations in the material is provided by orientation distribution function (ODF) maps [15–19].

The ODFs for each orientation g are given by the orientation function distribution density $f(g)$.

$$\frac{dV}{V} = f(g)dg \tag{5}$$

where dV is the volume of the crystal with orientation g; V is the total volume of the crystal; and dg is the orientation variation from g [15].

It is also possible to plot the ODF along a particular set of directions, obtaining a so-called fiber plot.

To assess the relationships between energy contributions and texture analyses, two reference frames are considered: first, the sheet reference frame (RD, TD, and ND), as shown in Figure 7 (d2); second, the specimen reference frame (X_s, Y_s, Z_s), as shown in Figure 7 (d3). The reference frame axes are aligned such that $RD \parallel X_s$, $ND \parallel Y_s$, and $TD \parallel Z_s$.

The most important rolled fibers for bcc crystal materials are α, with $< 110 > \parallel X_s \parallel RD$, and γ, with $< 111 > \parallel Z_s \parallel TD$. Specifically, the orientations are defined by a set of Euler angles that describe the rotation from the initial orientation g_i to the final orientation g_f. For α and γ fibers the set of orientations are (Table 1):

Table 1. Set of Euler angles of fibers α and γ.

Fiber	$g_i - g_f$
α	$[0°, 0°, 45°] - [0°, 90°, 45°]$
γ	$[60°, 54.7°, 45°] - [90°, 54.7°, 45°]$

In recrystallized Fe–Si alloys, the commonly considered texture components include the Goss component {110}<001> and the cube component {100}<001> [16,17,19,20]. However, this study will focus on analyzing the set of cubic orientations, also known as the "fiber *Cube*" (Table 2).

Table 2. Set of Euler angles of fiber *Cube*.

Fiber	$g_i - g_f$
Cube	$[0°, 0°, 0°] - [90°, 0°, 0°]$

$Cube : [001] \parallel Z_s \parallel TD$, considering all the rotated cubes with respect to the $X_s \parallel RD$ and considering the magnetic behavior along the [100] and [010] axes.

This assumption enables the calculation of $E_{t,2}$ along each principal crystal axis while varying only one Euler angle. Table 3 provides the anisotropy energy E_a and potential energy E_p along the rolling direction (RD):

Table 3. Anisotropy (E_a) and potential energies (E_p) along RD.

	E_a	E_p
Cube	$K_0 + K_1 \sin^2(2\,\theta_1)$	$-M_s\, H\cos(\beta_1 - \theta_1)$
α	$K_0 + K_1 \sin^2(2\,\theta_2)$	$-M_s\, H\cos(45° - \theta_2)$
γ	K_0	$-M_s\, H\cos(\delta_1)$

In this context, β_1 is the angle between the external applied field $\boldsymbol{H}$ and the saturation magnetization $\boldsymbol{M}_s$ along one of the easy axes (<100>), with $0 \leq \beta_1 \leq 45°$, accounting for the symmetry of $\boldsymbol{M}_s$ along the crystal axes; θ_1 is the angle between $\boldsymbol{M}_s$ and the <100> directions, with $0 \leq \theta_1 \leq 45°$; θ_2 is the angle between $\boldsymbol{M}_s$ and the <100> directions, with $0 \leq \theta_2 \leq 45°$; and δ_1 is the misorientation angle between $\boldsymbol{M}_s$ and $\boldsymbol{H}$, given that $\boldsymbol{H}$ belongs to the {111} planes.

The definitions of the angles θ_1 and θ_2 imply the following relationships: $E_{a,Cube} < E_{a,\alpha}$; $E_{a,\alpha} = cost. = \max(E_a)$; and $E_{a,\gamma} = cost. = K_0$, $K_0 \cong 0$.

Thus, for the anisotropy energy: $E_{a,\gamma} < E_{a,Cube} < E_{a,\alpha}$.

Therefore, the α fiber is the one that generates the highest value of magnetocrystalline anisotropy energy opposing the alignment with the external magnetic field.

The potential energy is influenced by the β_1 angle. If $\beta_1 < 45°$, then $E_{p,Cube} > E_{p,\alpha}$.

Thus, it follows that: (i) the fewer the <100> directions that are misoriented with respect to the rolling direction (RD), the higher the potential energy of the magnetic moments and the lower the anisotropy field. As a result, the magnetization value at the beginning of the "knee region" is higher, and saturation magnetization can be achieved with a lower external magnetic field. (ii) The E_a contribution of γ fiber can be neglected because $K_0 \cong 0$ and only $E_p \neq 0$. (iii) $E_{a,\,\alpha}$ provides the highest energetic contribution. The higher its value, the lower the magnetization value at high external magnetic field intensities H.

Table 4 reports anisotropy E_a and potential E_p along TD.

Table 4. Anisotropy (E_a) and potential (E_p) energies along TD.

	E_a	E_p
Cube	K_0	$-M_s\, H$
α	$K_0 + K_1 \sin^2(2\,\theta_3)$	$-M_s\, H\cos(\beta_2 - \theta_3)$
γ	$K_0 + K_1(\alpha_1\alpha_2 + \alpha_1\alpha_3 + \alpha_2\alpha_3) + K_2\alpha_1^2\alpha_2^2\alpha_3^2$	$-M_s\, H\cos(\delta_2)$

Where β_2 is the angle between $\boldsymbol{H}$ and $\boldsymbol{M}_s$ along one of the easy axes, with $0 \leq \beta_2 \leq 45°$ due to the symmetry of $\boldsymbol{M}_s$ along the crystal axes; θ_3 is the angle between $\boldsymbol{M}_s$ and the <100> directions, with $0 \leq \theta_3 \leq \beta_2$; and δ_2 is the misorientation angle between $\boldsymbol{M}_s$ and $\boldsymbol{H}$, where $\boldsymbol{H}$ is parallel to <111>.

The definition of the angles θ_3 implies that $E_{a,Cube} < E_{a,\alpha} < E_{a,\gamma}$.

Thus, the γ fiber generates the highest magnetocrystalline anisotropy energy, opposing the alignment with the external magnetic field.

Regarding the potential energy: if $0 \le \beta_2 < 45°$, then $E_{p,Cube} \ge E_{p,\alpha}$.

Thus, the γ produces the highest magnetocrystalline anisotropy energy opposing the external applied field. This results in a reduced magnetization value at the beginning of the "knee region", and the saturation magnetization can only be achieved at higher external magnetic field values compared to those along the rolling direction (RD). Additionally, the E_a contribution of the "*Cube*" fiber can be neglected because $K_0 \cong 0$ and $E_{p,\ Cube} = \max(E_p)$.

Ultimately, because $E_{a,\ \gamma}$ provides the highest anisotropy contribution among the fibers, a higher value of $E_{a,\ \gamma}$ results in a lower magnetization value at high external magnetic field intensities H.

From these observations, anisotropy energy contributions can be assessed for an external magnetic field applied along specific crystal directions or within particular crystal planes. This leads to a simplification in studying the effects of anisotropy energies.

To qualitatively assess the influence of textures on energy contributions, we consider a simplified analysis of the function $f(g)$ along a fiber.

Given that $g \in \mathbb{R}^3,\ f: \mathbb{R}^3 \to \mathbb{R}$, and since two angles remain constant while one angle varies along the fiber, we can define a function $h(\theta)$ where θ is the varying angle along the fiber, with $\theta \in \mathbb{R}$ and $h: \mathbb{R} \to \mathbb{R}$.

The function $h(\theta)$ is not defined but is derived from $f(g)$ along a fiber.

To account for all fiber orientation contributions with one specific direction parallel to the specimen-fixed reference frame, a numerical integration of the function $h(\theta) - I$ is performed from the initial varying angle θ_i to the final angle θ_f:

$$I = \int_{\theta_i}^{\theta_f} h(\theta)\ d\theta \tag{6}$$

The study of the influence of crystallographic anisotropy on total energy loss can be conducted by describing the phenomenon using a mathematical function, following an identification process [21]. This function does not directly represent the physical processes of magnetism but provides a mathematical model useful for a preliminary analysis of how various factors affect the material's behavior.

Given the shape of the total energy loss, a double exponential function is selected as the fitting model:

$$y(x) = ae^{b\ x} + ce^{d\ x} \tag{7}$$

In this study, the rates of variation of total energy loss up to the "knee region" and in the rotating regions were chosen since the function could be linearized in two main regions.

The first region is around $H = 0\left[\frac{A}{m}\right]$, with a McLaurin series until the first degree:

$$y_{lin,1}(x) = (ab + cd)\ x + (a + c) \tag{8}$$

The coefficient of the linearized function, which represents the rate of variation in the first region, is:

$$m_{lin,1} = (ab + cd) \tag{9}$$

The second region corresponds to the high value of H where the function can be approximated with a line with starting point $P_1 = \left(H \equiv H_{ref};\ P_{s,fit}\left(H_{ref}\right)\right)$ and ending at point $P_2 = \left(H \equiv H_{end} = 10000\left[\frac{A}{m}\right];\ P_{s,fit}(H_{end})\right)$:

The linear approximating function is as follows:

$$y_{lin,2}(x) = p_1 x + p_2 \tag{10}$$

The coefficient of the linearized function, which represents the rate of variation in the second region, is as follows:

$$m_{lin,2} = p_1 \tag{11}$$

3. Materials and Methods

The materials studied were three rolled Fe–Si alloys of grade M350—50A, in accordance with EN 10106. Their chemical compositions were analyzed using an Optical Emission Spectrometer (OES ARL 346 (Thermo Fisher Scientific, 168 Third Avenue, Waltham, MA USA 02451)). The acronyms and compositions of the analyzed sheets are detailed in Table 5. The key alloying elements are C, Si, Mn, and Al, as they significantly impact the magnetization process, saturation polarization, energy loss, and magnetocrystalline anisotropy. Additionally, S and N influence the total energy loss [1,2,6,8,10].

Table 5. Chemical composition (weight %) of the studied Fe—Si alloys.

Sample Acronym	% C	% Si	% Mn	% Al	% S	% N	% Other	% Fe
L1	0.0036	1.879	0.245	0.403	0.004	0.0086	0.0383	Bal.
L2	0.0062	1.718	0.270	0.397	0.005	0.003	0.1031	Bal.
L3	0.0041	1.971	0.290	0.353	0.004	0.0042	0.0581	Bal.

Magnetic characterization was performed using a Brockhaus Messtechnik Single-Sheet Tester (SST) (BROCKHAUS MEASUREMENTS, Dr. Brockhaus Messtechnik GmbH & Co. KG, Lüdenscheid, Germany) model MPG100 D DC/AC. The device features an external magnetic field frequency range from 3 Hz to 10 kHz, with a maximum polarization of 2 T and measurement repeatability within ±2%.

The external magnetic field was applied along the rolling direction (RD) and transverse direction (TD) of the sheets at a frequency of 50 Hz.

Microstructural characterization was conducted with a scanning electron microscope (TESCAN MIRA3 FEG-SEM (TESCAN ORSAY HOLDING, Brno, Czech Republic)) equipped with energy-dispersive X-ray spectroscopy (EDS, Bruker Quantax 200/30 mm^2 (Bruker Nano GmbH, Berlin, Germany)) and an electron backscattered diffraction (EBSD) detector (e–Flash HD (Bruker Nano GmbH, Berlin, Germany)). The crystallographic analyses were performed on the longitudinal cross-sections of the sheets and processed using the MATLAB R2024a Toolbox MTEX v5.10.0 [22]. Specifically, the EBSD data were pre-processed with the Half Quadratic filter to denoise and complete the data [23].

Before investigation, samples were embedded in conductive resin and polished according to standard metallographic procedures, achieving a finish with 50 nm colloidal silica.

4. Experimental Results

4.1. Magnetic Characterization

Figure 3 illustrates the energy losses $P_s\left[\frac{W}{kg}\right]$ of all the materials in function of the intensity of the externally applied $H\left[\frac{A}{m}\right]$ field along RD and TD, as evaluated by the SST method. It highlights that at H_{end}: $P_s(L_1) > P_s(L_3) > P_s(L_2)$ in RD and TD.

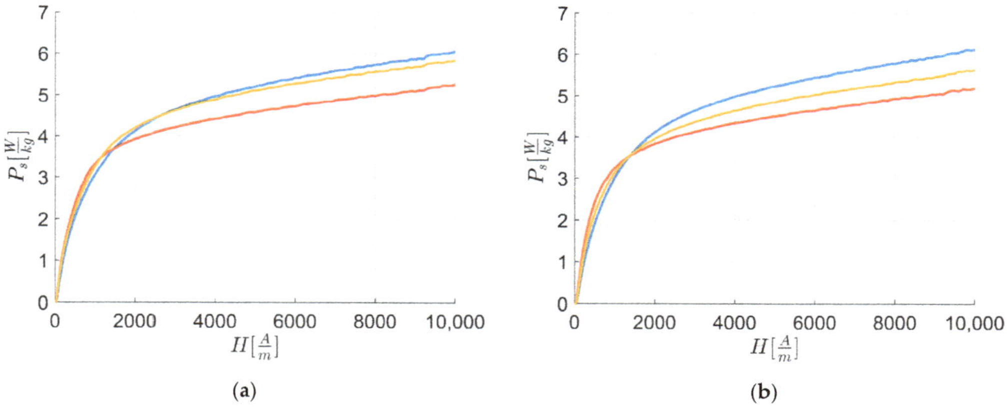

Figure 3. Magnetic characterization experimental energy loss P_s vs. external applied magnetic field H: energy losses along RD (**a**) and TD (**b**). The colors are related to the behavior of: L1—blue; L2—orange; and L3—yellow.

The energy loss curves are derived from the value areas of hysteresis loops, whose magnetic polarization tips and corresponding external magnetic field intensities generate the first magnetization curve (Figure 5).

Figure 4 presents the experimental energy losses for each material along the respective RD and TD. The curves show that P_s is higher at H_{end} only for L1 along TD than RD.

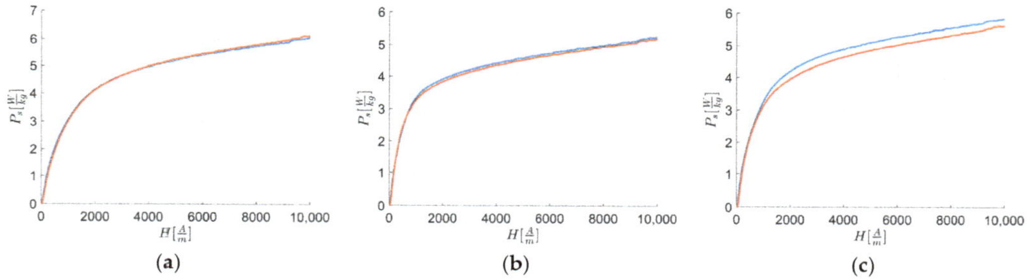

Figure 4. Magnetic characterization experimental energy loss P_s vs. external applied magnetic field H: energy losses along RD (blue) and TD (orange). The figures are related to: (**a**) L1; (**b**) L2; and (**c**) L3.

Table 6 summarizes the differences in magnetic energy loss along RD and TD at H_{end}. L1 exhibits a negative difference, indicating higher energy loss along TD compared to RD, while L3 has the highest difference between RD and TD.

Table 6. Total energy loss differences between RD and TD for all the sheets.

	$\Delta P_{s,fit}(H_{end})$
L1	−0.0728
L2	0.0639
L3	0.2057

Figure 5 displays the experimental first magnetization curves along RD and TD up to H_{end}.

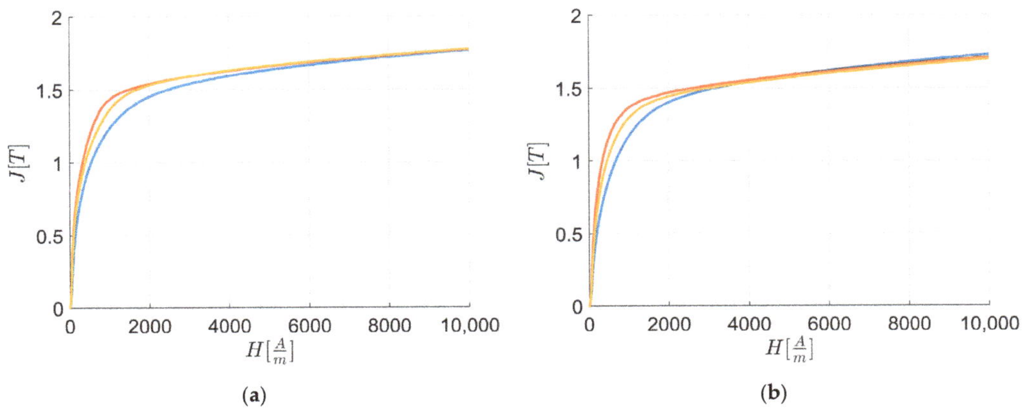

Figure 5. Magnetic characterization polarization J vs. external applied magnetic field H: first magnetization curves along RD (**a**) and TD (**b**). The colors are related to the behavior of: L1—blue; L2—orange; and L3—yellow.

Table 7 provides the polarization J[T] at H_{end}. The data show that $J(L_3) > J(L_2) > J(L_1)$ along RD while $J(L_1) > J(L_2) > J(L_3)$ along TD.

Table 7. Polarization values at the maximum value of the externally applied field along RD and TD.

	$J[T]$	
	RD	TD
L1	1.7684	1.7302
L2	1.7740	1.7092
L3	1.7771	1.6976

Figure 6 presents the experimental first magnetization curves for each material along RD and TD. The graphs show that for all the sheets $J[T]$ at H_{end} is higher along RD compared to TD.

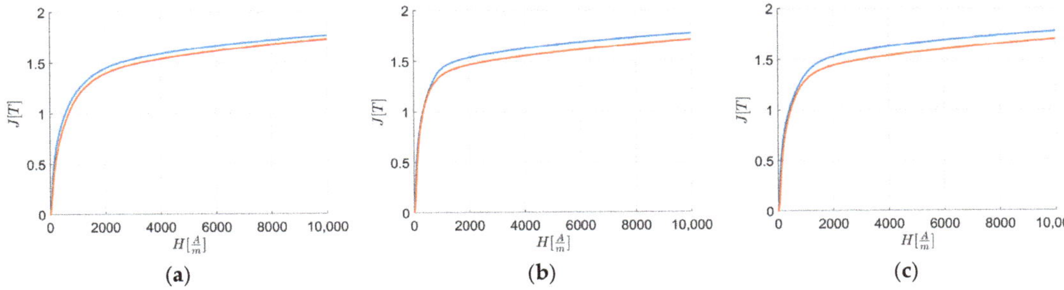

Figure 6. Magnetic characterization polarization J vs. external applied magnetic field H: first magnetization curves of L1 (**a**); L2 (**b**); and L3 (**c**) along RD (blue) and TD (orange).

Table 8 reports the differences in polarization ΔJ along RD and TD at $H = H_k = 700\left[\frac{A}{m}\right]$, considered the initial point of the "knee region". It shows that polarization along RD is consistently higher than along TD at H_k.

Table 8. Differences of polarizations between RD and TD for each sheet.

	$\Delta J(H_k)$
L1	0.0595
L2	0.0637
L3	0.0617

Table 9 presents the differences in polarization between all sheets along RD and TD at H_k. The results indicate that L2 shows the greatest differences in magnetization compared to the other sheets.

Table 9. Differences of polarization along RD and TD between all the sheets.

		$\Delta J(H_k)$
L1–L2	RD	−0.2192
	TD	−0.2578
L1–L3	RD	−0.1242
	TD	−0.1545
L2–L3	RD	0.0950
	TD	0.1033

4.2. Crystallographic Characterisations

Figure 7 summarizes the results of EBSD analyses, processed with the MATLAB Toolbox MTEX. It shows that the [111] direction (blue) aligned with the normal direction of the transverse section is predominant in all sheets. Additionally, some grains with [001] (red) and [011] (green) directions are present in L2 and L3.

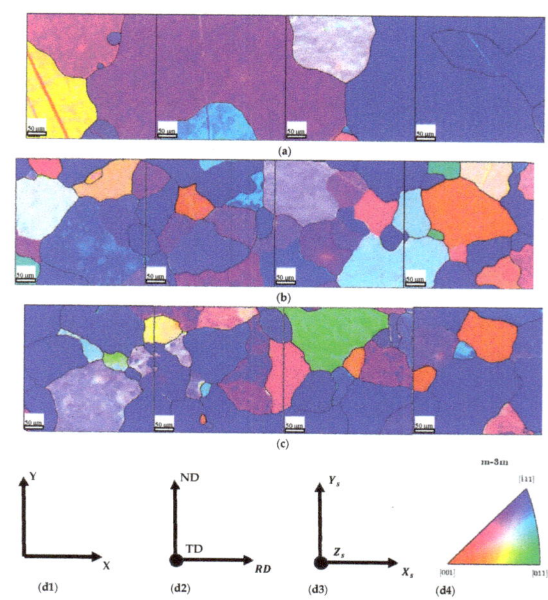

Figure 7. Orientations reconstruction of sheets, micrometric notch of 25 µm: **(a)** L1; **(b)** L2; **(c)** L3; **(d1)** the coordinate system of the orientation map; **(d2)** the sheet reference frame; **(d3)** the fixed specimen reference frame for the orientation analyses; and **(d4)** the IPF colorkey map.

5. Data Analyses

Table 10 lists the integral values I of each fiber for each sheet. The data show that L2 has the highest I_{Cube}, L3 the highest I_{γ}, and L1 has the highest I_{α}.

Table 10. Values of fiber integrals.

	$I_{\alpha} \cdot 10^3$	$I_{\gamma} \cdot 10^3$	I_{Cube}
L1	1.53	2.15	0.0041
L2	0.88	2.92	174.34
L3	0.72	4.4	27.87

The fitting parameters for Function (7) were obtained (Table 11) by solving a nonlinear least-squares problem [21,24] using various types of functions and employing a Trust Region Algorithm [24] via the MATLAB R2024a Curve Fitting Toolbox.

Table 11. Values of fitting parameters of Function (7).

		a	**b·10^{-5}**	**c**	**d**
L1	RD	4.4580	3.1488	−4.4923	−0.0011
	TD	4.4537	3.2769	−4.4916	−0.001
L2	RD	3.9032	3.0893	−3.9533	−0.0018
	TD	3.8210	3.1811	−3.8840	−0.0018
L3	RD	4.3769	2.9879	−4.4113	−0.0013
	TD	4.0893	3.3129	−4.1344	−0.0013

The goodness of fit for the experimental data was evaluated using the sum of squared errors (SSE) [24] and the root mean squared errors (RMSE) [25] to determine if the Trust Region Algorithm successfully converged to the optimal minimum. Additionally, the coefficient of determination $\overline{R}^2$ was calculated to assess the correlation between the experimental data and the fitting curve [26], with particular emphasis on the adjusted $\overline{R}^2$. This adjusted $\overline{R}^2$ accounts for the number of fitting parameters and helps identify potential overfitting [27].

The low values of SSE and RMSE, along with very high values of $\overline{R}^2$ (Table 12), confirm a good fit between the experimental and calculated data. Thus, Function (7) accurately describes the magnetic energy loss behavior of the material along the rolling direction (RD) and transverse direction (TD) and is suitable for studying the material's magnetic anisotropy as a function of the intensity of the applied external field.

Table 12. Values of goodness of fitting and correlation parameters of Function (7).

		SSE	**RMSE**	$\overline{R}^2$
L1	RD	0.5406	0.066	0.9990
	TD	0.2903	0.0484	0.9995
L2	RD	0.3934	0.0563	0.9990
	TD	0.6769	0.0739	0.9982
L3	RD	0.5388	0.0659	0.9989
	TD	0.5805	0.0684	0.9987

Figure 8 reports the fitted curves $P_{s,\,fit}$ of Figure 3, while Figure 9 reports the fitted curves of Figure 4.

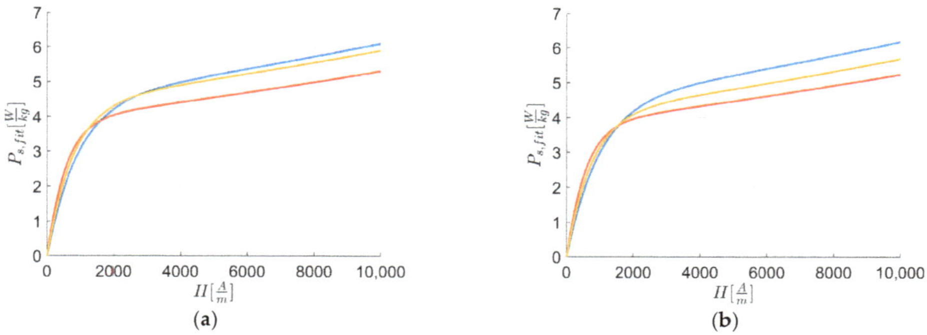

Figure 8. Fitted curves of experimental data of magnetic energy loss $P_{s,fit}$ vs. external applied magnetic field H along RD (**a**) and TD (**b**). The colors are related to the behavior of: L1—blue; L2—orange; and L3—yellow.

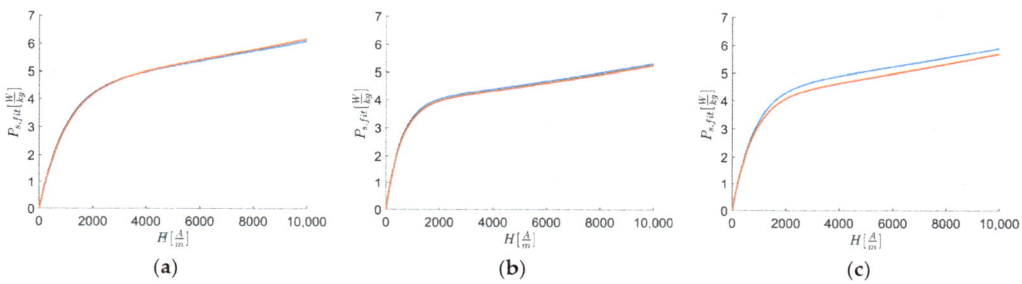

Figure 9. Fitted curves of experimental data of magnetic energy loss $P_{s,fit}$ vs. external applied magnetic field H along RD (blue) and TD (orange). The figures are related to: (**a**) L1; (**b**) L2; and (**c**) L3.

Based on the above, Function (7) can be reformulated as follows:

$$P_{s,fit}(H) = ae^{b\,H} + ce^{d\,H} \tag{12}$$

The linear coefficient values of Function (8) are reported in Table 13 for all the energy loss fitted curves, indicating that $m_{RD} > m_{TD}$ for all the sheets.

Table 13. Linear coefficient of Function (8).

	$m_{lin,1}$	
	RD	**TD**
L1	0.0049	0.0048
L2	0.0072	0.0071
L3	0.0058	0.0057

Table 14 provides the SSE, RMSE, and $\overline{R}^2$ values for $H \geq H_{ref}$, where H_{ref} is the field strength at which the experimental curves become linear. The SSE and RMSE values indicate that the linear function (10) accurately fits the data provided by Function (7) for $H \geq H_{ref}$, while the $\overline{R}^2$ values confirm a strong correlation between the experimental and fitted data.

Table 14. Values of SSE, RMSE, and $\overline{R}^2$ for each sheet for fitted magnetic energy loss with Function (5).

		SSE·10^{-4}	RMSE	$\overline{R}^2$
L1	RD	1.22	0.0078	0.9997
	TD	1.78	0.0094	0.9996
L2	RD	2.92	0.0121	0.9994
	TD	3.2	0.0127	0.9990
L3	RD	0.54	0.0061	0.9998
	TD	2.05	0.01	0.9995

The procedure for finding the best line approximation is the same as applied to Function (7). The values of linear coefficients for the fitting linear Function (10) are reported in Table 15, showing that $m_{RD} < m_{TD}$ for all the sheets.

Table 15. Values of Equation (11).

	$m_{lin,2}$	
	RD·10^{-4}	TD·10^{-4}
L1	1.85	1.95
L2	1.5	1.52
L3	1.65	1.74

6. Discussion

Based on the results of the first magnetization curves along the rolling direction (RD) and transverse direction (TD) (Figures 5 and 6), and considering Equations (1) and (3) along with Tables 3, 4 and 8–10, the following conclusions can be drawn for $H \leq H_k$ where domain wall motion is the predominant mechanism:

$$E_p = E_{p,\alpha} + E_{p,\gamma} + E_{p,Cube} \tag{13}$$

where

$$E_{p,Cube} = -M_s H; E_{p,\alpha} = -M_s H cos(45°); E_{p,\gamma} = -M_s H cos(54.7°).$$

Thus,

$$E_{p,Cube} > E_{p,\alpha} > E_{p,\gamma}$$

and

$$J_{Cube}(H_k) > J_\alpha(H_k) > J_\gamma(H_k).$$

At H_k, considering the integral of textures along RD and TD:

$$I_\gamma + I_\alpha + I_{Cube}(L_1) < I_\gamma + I_\alpha + I_{Cube}(L_2) < I_\gamma + I_\alpha + I_{Cube}(L_3).$$

Hence,

$$E_p(L_1) < E_p(L_2) < E_p(L_3).$$

However,

$$E_{p,Cube} > E_{p,\alpha} > E_{p,\gamma}, I_{Cube}(L_2) > I_{Cube}(L_3) > I_{Cube}(L_1).$$

Thus,

$$J_{L_2}(H_k) > J_{L_3}(H_k) > J_{L_3}(H_k)$$

Despite $I_\gamma + I_\alpha(L_3) > I_\gamma + I_\alpha(L_1)$, the significant difference $I_{Cube}(L_3) \gg I_{Cube}(L_1)$ explains the results.

These observations are supported by polarization differences along RD and TD for $H = H_k$.

Figure 5 points out that the first magnetization curve of L2 consistently shows higher polarization values than L1 and L3, with L3 being higher than L1, and L1 having the lowest polarization values among the curves. Table 9 shows these differences for $H = H_k$, and the texture integral values from Table 10 confirm the results from the previous equations.

Figure 6 highlights that for $H = H_k$ the polarization values are consistently lower along TD than RD. Table 8 illustrates these differences for $H = H_k$, and the texture integral values from Table 10 further confirm the results from the previous equations.

For intensities of external applied field lower than H_k for each sheet, denoted as H_{low}, the following expression is used:

$$J \propto E_t \rightarrow \Delta J \propto \Delta E_t.$$

If

$$E_{p,RD}(H_{low}) = E_{p,TD}(H_{low}) \text{and } \Delta J > 0$$

Then,

$$E_{a,RD}(H_{low}) < E_{a,TD}(H_{low})$$

Thus,

$$J_{RD}(H_{low}) > J_{TD}(H_{low}).$$

These observations are supported by polarization differences along RD and TD for $H < H_k$.

From Figure 6, it is evident that from very low values of external applied field intensity, the polarization values are consistently lower along TD than RD. The texture integral values from Table 10 verify the results from the previous equations.

These observations demonstrate that higher Cube texture integral values correspond to higher magnetization at the onset of the "knee region", indicating that less energy is required for magnetization with this texture and that magnetocrystalline anisotropy also plays a role at relatively low magnetic fields.

For $H \rightarrow H_{end}$, considering Equation (4), Tables 3, 4, 7 and 10, along with Figures 5 and 6, where E_p is approximately constant for all sheets, the following expression is used:

$$E_a = E_{a,\alpha} + E_{a,\gamma} + E_{a,Cube} \tag{14}$$

Along RD:

The following is given:

$$E_{a,\gamma} \cong 0, \text{ then } E_{a,RD} = E_{a,\alpha} + E_{a,Cube}.$$

Since

$$I_\alpha + I_{Cube}(L_3) < I_\alpha + I_{Cube}(L_2) < I_\alpha + I_{Cube}(L_1).$$

Thus,

$$E_{a,RD}(L_1) > E_{a,RD}(L_2) > E_{a,RD}(L_3).$$

Therefore,

$$J_{L_3}(H_{end}) > J_{L_2}(H_{end}) > J_{L_1}(H_{end}).$$

These results are confirmed by polarization values at maximum external fields along RD. Figure 5a highlights that for $H \rightarrow H_{end}$, the polarization values are the highest for L3, lower for L2, and the lowest for L1. Table 7 shows these differences for $H \rightarrow H_{end}$, and the texture integral values from Table 10 prove the results from the previous equations, confirming that lower *Cube* and α texture integral values correspond to higher magnetization at maximum magnetic field intensity.

Along TD:

Given that $E_{a,Cube} \cong 0$,

$$E_{a,TD} = E_{a,\alpha} + E_{a,\gamma}.$$

Since

$$I_\alpha + I_\gamma(L_1) < I_\alpha + I_\gamma(L_2) < I_\alpha + I_\gamma(L_3).$$

Thus,

$$E_{a,TD}(L_3) > E_{a,TD}(L_2) > E_{a,TD}(L_1).$$

Therefore,

$$J_{L_1}(H_{end}) > J_{L_2}(H_{end}) > J_{L_3}(H_{end}).$$

These results are confirmed by polarization values at maximum external fields along TD. Figure 5b points out that for $H \to H_{end}$ along TD, the polarization values are the highest for L1, lower for L2, and the lowest for L3. Table 7 shows these differences for $H \to H_{end}$, and the texture integral values from Table 10 support the results from the previous equations, confirming that lower Cube and γ texture integral values correspond to higher magnetization at the maximum field intensity.

Hence, anisotropy energy is lower at high magnetic field intensities, and the material exhibits less isotropic behavior.

According to [1,5,7], increasing silicon content in Fe–Si alloys generally increases electrical resistivity and should decrease energy loss. However, experimental data show the highest energy loss for L1 (1.879% Si), the lowest for L2 (1.718% Si), and intermediate for L3 (1.971% Si) at H_{end} along RD and TD (Figure 1). This discrepancy suggests that the magnetic behavior of Fe–Si alloys should also consider the effects of N, S, Al, and Mn (Table 5). Specifically, elements like AlN can increase magnetic energy loss. Given the low and nearly constant S content, the presence of AlN is significant: L1, with the highest Al and N content (0.403% Al, 0.0086% N), exhibits the worst magnetic behavior; L3, with low Al and intermediate N (0.353% Al, 0.0042% N), shows intermediate behavior; and L2, with intermediate Al and the lowest N (0.397% Al, 0.003% N), demonstrates the best magnetic behavior.

Examining the initial part of the loss curve (Figures 8 and 9) and using Equation (1), Equation (3), and Equation (8), the following can be derived: $P_{s,\,fit} \propto J(H)$.

Therefore,

$$\Delta P_{s,fit} \propto \Delta J(H) \text{ and } \lim_{H\to 0} \frac{\Delta J(H)}{\Delta H} \cong \mu_i' = cost;$$

The following is given:

$$\lim_{H\to 0} \frac{\Delta P_{s,fit}(H)}{\Delta H} = m_{lin,1} = cost.$$

Thus,

$\mu_i' \propto m_{lin,1}$, with μ_i' being the initial differential permeability.

Assuming homogeneous defect distribution and nearly constant electrical resistivity, at $H \to 0$, the following can be derived:

$$\mu'_{L1} \neq \mu'_{L2} \neq \mu'_{L3} \text{ and } E_p = cost.$$

Thus,

$$E_a \neq 0.$$

From Figure 8, it is evident that along RD and TD for $H \leq H_k$ the magnetic loss curve of L2 consistently shows higher magnetic loss values than L1 and L3, with L3 being higher than L1, and L1 having the lowest magnetic loss values among the curves. By linearizing the first magnetization curves (Figures 4 and 5) with an equation similar to Equation (8) from $H \to 0$ up to H_k, to avoid instrumental acquisition errors at very low fields, it is observed that higher $J(H_k)$ correlates with a higher loss rate, as indicated by Equation (9) and supported by Table 13. Based on linear coefficients (Table 15) and the previous observations of the first magnetization curve and loss rate for $H < H_k$, anisotropy energy significantly influences magnetic energy loss behavior.

From Figure 9, it is evident that for $H \geq H_{ref}$ the magnetic loss curve of L1 consistently shows higher magnetic loss values along TD than RD compared to other sheets, while the others always have lower values along TD than RD. Table 15 shows these differences for rates of loss for $H \geq H_{ref}$, and together with Table 6 and the texture integral values from Table 10, it is confirmed that texture energies impact L1 more than other sheets for $H \geq H_{ref}$ along RD and TD.

The results presented represent a novel method for describing magnetic behavior in the vectorial space of the crystal unit cell, involving saturation magnetization, external applied field vectors, and associated energies for typical fiber texture orientations in rolled materials. This approach offers a new perspective compared to semi-empirical methods. Further research is needed to explore the contributions of different orientations to the magnetization and energy loss of materials.

7. Conclusions

The aim of this paper is to evaluate the effect of material texture on the magnetization and magnetic energy loss of Fe–Si alloys by considering the definitions of potential and magnetocrystalline anisotropy magnetization energy in magnetic materials. This study integrates the orientation distribution function along each fiber and models energy loss as a function of the externally applied field.

The results highlight the following key points:

1. Impact of texture on energy loss: the texture of the material significantly influences the first magnetization curve and the initial energy loss. According to Table 13, the rate of variation in energy loss between the rolling direction (RD) and the transverse direction (TD) is 2.04% for each specimen. Specifically, along the RD, the energy loss rates are as follows: 46.93% between L1 and L2; 15.51% between L1 and L3; and 19.44% between L2 and L3. Along the TD, the rates are: 32.4% between L1 and L2; 15.79% between L1 and L3; and 19.72% between L2 and L3.
2. Variation in energy loss with polarization: although the maximum polarizations achieved by L3 and L2 are similar along RD and TD, their energy loss behavior differs significantly with the applied external field. Table 7 shows that the polarization difference between L2 and L3 is 0.17% along RD and 0.67% along TD. The variation in energy loss rates between L2 and L3 is 9.09% along RD and 12.64% along TD. Additionally, the polarization difference between RD and TD for the same specimen is 0.0648 T for L2 and 0.0795 T for L3 (Table 7). The corresponding variation in energy losses between RD and TD is 0.0639 W/kg for L2 and 0.2057 W/kg for L3 (Table 6). This results in a 1.38% variation in energy loss relative to the maximum polarization for L2 and a 61.35% variation for L3, indicating that the first magnetization curve does not directly influence the energy loss behavior.
3. Modeling energy loss with double exponential curve: The influence of texture on energy loss can be effectively studied using a double exponential curve, which provides insight into the contributions of different textures.
4. Vectorial space analysis: describing saturation magnetization, external applied field vectors, and the related energies in the vectorial space of the crystal unit cell—based on a typical fiber texture set of orientations for the rolled material—proves useful for analyzing both magnetization and magnetic energy losses.
5. These findings underscore the complex interplay between material texture and magnetic behavior, offering a detailed understanding of how different textures affect magnetization and energy loss in Fe–Si alloys.

Author Contributions: D.C.: conceptualization, software, investigation, and writing—original draft. A.M.: methodology, formal analysis, and writing—review and editing. L.C.: resources and data curation. A.F.: resources and data curation. All authors have read and agreed to the published version of the manuscript.

Funding: The research was financed by: The European Union—NextGenerationEU (National Sustainable Mobility Center CN00000023, Italian Ministry of University and Research Decree No. 1033—17/06/2022, Spoke 11—Innovative Materials and Lightweighting); Regione Emilia-Romagna—POR FESR 2021–2027 ACTION 1.1.2 Call For Strategic Industrial Research Projects Aimed At The Priority Areas Of The Smart Specialisation Strategy DGR 2097/2022"—ID 37993 PG/2023/303085 CUP E27G22000330003 (project MAGMOVE—Sustainable magnetic materials in high-efficiency electric motors for the Green transition). The opinions expressed are those of the authors only and should not be considered as representative of the European Union or the European Commission's official position. Neither the European Union nor the European Commission can be held responsible for them.

Institutional Review Board Statement: Not applicable.

Informed Consent Statement: Not applicable.

Data Availability Statement: The raw data supporting the conclusions of this article will be made available by the authors on request.

Conflicts of Interest: The author Alessandro Ferraiuolo is employed by the company Marcegaglia Ravenna SpA and discloses its conflict of interest as outlined by the MDPI guidelines. The remaining authors declare that the research was conducted in the absence of any commercial or financial relationships that could be construed as a potential conflict of interest.

References

1. Fiorillo, F. *Measurement and Characterization of Magnetic Materials*; Elsevier Series in Electromagnetism; Elsevier: Amsterdam, The Netherlands; Heidelberg, Germany, 2004.
2. Tumanski, S. *Handbook of Magnetic Measurements*, 1st ed.; CRC Press: Boca Raton, FL, USA, 2016.
3. Dietrich, D.W. Magnetically Soft Materials. In *Properties and Selection: Nonferrous Alloys and Special-Purpose Materials*; ASM Handbook Committee, A. c. di; ASM International: Materials Park, OH, USA, 1990; pp. 761–781. [CrossRef]
4. Davis, J.R.; A. c. di. Magnetically Soft Materials. In *Metals Handbook Desk Edition*, 2nd ed.; ASM International: Materials Park, OH, USA, 1998; pp. 639–644. [CrossRef]
5. Cullity, B.D.; Graham, C.D. *Introduction to Magnetic Materials*, 1st ed.; Wiley: Hoboken, NJ, USA, 2008. [CrossRef]
6. Elgamli, E.; Anayi, F. Advancements in Electrical Steels: A Comprehensive Review of Microstructure, Loss Analysis, Magnetic Properties, Alloying Elements, and the Influence of Coatings. *Appl. Sci.* **2023**, *13*, 10283. [CrossRef]
7. Chikazumi, S. *Physics of Magnetism*; Wiley Series on the Science and Technology of Materials; Wiley: New York, NY, USA, 1964.
8. Bertotti, G. *Hysteresis in Magnetism: For Physicists, Materials Scientists, and Engineers*; Electromagnetism Series; Academic Press: San Diego, CA, USA, 1998.
9. Landgraf, F.J.G.; Yonamine, T.; Emura, M.; Cunha, M.A. Modelling the angular dependence of magnetic properties of a fully processed non-oriented electrical steel. *J. Magn. Magn. Mater.* **2003**, *254–255*, 328–330. [CrossRef]
10. Yonamine, T.; Landgraf, F.J.G. Correlation between magnetic properties and crystallographic texture of silicon steel. *J. Magn. Magn. Mater.* **2004**, *272–276*, E565–E566. [CrossRef]
11. Barros, J.; Schneider, J.; Verbeken, K.; Houbaert, Y. On the correlation between microstructure and magnetic losses in electrical steel. *J. Magn. Magn. Mater.* **2008**, *320*, 2490–2493. [CrossRef]
12. Gomes, E.; Schneider, J.; Verbeken, K.; Barros, J.; Houbaert, Y. Correlation Between Microstructure, Texture, and Magnetic Induction in Nonoriented Electrical Steels. *IEEE Trans. Magn.* **2010**, *46*, 310–313. [CrossRef]
13. Qin, J.; Yang, P.; Mao, W.; Ye, F. Effect of texture and grain size on the magnetic flux density and core loss of cold-rolled high silicon steel sheets. *J. Magn. Magn. Mater.* **2015**, *393*, 537–543. [CrossRef]
14. Leuning, N.; Steentjes, S.; Hameyer, K. Effect of grain size and magnetic texture on iron-loss components in NO electrical steel at different frequencies. *J. Magn. Magn. Mater.* **2019**, *469*, 373–382. [CrossRef]
15. Bunge, H.-J. *Texture Analysis in Materials Science: Mathematical Methods*, English ed.; Butterworths: London, UK; Boston, MA, USA, 1982.
16. Kocks, U.F.; Tomé, C.N.; Wenk, H.-R.; Beaudoin, A.J.; Mecking, H.; A. c. di. *Texture and Anisotropy: Preferred Orientations in Polycrystals and Their Effect on Materials Properties*; First Paperback Edition (with Corrections); Cambridge University Press: Cambridge, UK, 2000.
17. Engler, O. *Introduction To Texture Analysis: Macrotexture, Microtexture, and Orientation Mapping*, 2nd ed.; Place of Publication Not Identified; CRC Press: Boca Raton, FL, USA, 2017.
18. Schwartz, A.J. *Electron Backscatter Diffraction in Materials Science*, 2nd ed.; Springer: New York, NY, USA, 2009.

19. Humphreys, F.J.; Hatherly, M. *Recrystallization and Related Annealing Phenomena*, 2nd ed.; Elsevier: Amsterdam, The Netherlands, Boston, MA, USA, 2004.
20. Field, D.P. Textured Structures. In *Metallography and Microstructures*; Vander Voort, G.F., A. c. di, Eds.; ASM International: Materials Park, OH, USA, 2004; pp. 215–226. [CrossRef]
21. Nicolas, P.K.E. *Applied Parameter Estimation for Chemical Engineers*; CRC Press: Boca Raton, FL, USA, 2019.
22. *MTEX (v. 5.10)*. [Online]. Disponibile Su. Available online: https://mtex-toolbox.github.io/ (accessed on 3 September 2023).
23. Hielscher, R.; Silbermann, C.B.; Schmidl, E.; Ihlemann, J. Denoising of crystal orientation maps. *J. Appl. Crystallogr.* **2019**, *52*, 984–996. [CrossRef]
24. Nocedal, J.; Wright, S.J. *Numerical Optimization*, 2nd ed.; Springer Series in Operations Research; Springer: New York, NY, USA, 2006.
25. Introduction to Least-Squares Fitting. [Online]. Disponibile Su. Available online: https://it.mathworks.com/help/curvefit/least-squares-fitting.html (accessed on 12 June 2024).
26. Olofsson, P.; Andersson, M. *Probability, Statistics, and Stochastic Processes*, 2nd ed.; Wiley: Hoboken, NJ, USA, 2012.
27. Curve Fitting Onramp. [Online]. Disponibile Su. Available online: https://matlabacademy.mathworks.com/it/details/curve-fitting-onramp/orcf (accessed on 12 June 2024).

materials

MDPI

Article

Cavity-Tuned Exciton Dynamics in Transition Metal Dichalcogenides Monolayers

Kaijun Shen [1], Kewei Sun [2], Maxim F. Gelin [2] and Yang Zhao [1,*]

[1] School of Materials Science and Engineering, Nanyang Technological University, Singapore 639798, Singapore; shen0193@e.ntu.edu.sg

[2] School of Science, Hangzhou Dianzi University, Hangzhou 310018, China; skw79724@163.com (K.S); maxim.gelin@gmail.com (M.F.G.)

* Correspondence: yzhao@ntu.edu.sg; Tel.: +65-65137990

Abstract: A fully quantum, numerically accurate methodology is presented for the simulation of the exciton dynamics and time-resolved fluorescence of cavity-tuned two-dimensional (2D) materials at finite temperatures. This approach was specifically applied to a monolayer WSe_2 system. Our methodology enabled us to identify the dynamical and spectroscopic signatures of polaronic and polaritonic effects and to elucidate their characteristic timescales across a range of exciton–cavity couplings. The approach employed can be extended to simulation of various cavity-tuned 2D materials, specifically for exploring finite temperature nonlinear spectroscopic signals.

Keywords: variational method; coherent states; Davydov Ansatz; transition metal dichalcogenides; nanocavity; exciton dynamics; time-resolved fluorescence

1. Introduction

Monolayer transition metal dichalcogenides (TMDs) are celebrated for their broken inversion symmetry, resulting in a non-zero Berry curvature and robust valley–spin interactions. Such valley–spin interactions enable selective valley polarization through optical or electromagnetic means [1,2]. The reduction in the layer number results in the transformation of TMDs from indirect to direct band gap semiconductors [3,4]. This layer-number-dependent symmetry variation also allows for the targeted manipulation of valley properties, establishing TMDs as a pivotal platform for valleytronic device development. In tungsten-based monolayers, low-temperature photoluminescence studies have shown that dark exciton states [5–8], which are momentum-forbidden and significantly influence the optical and electronic behavior of TMDs [9–12], exhibit lower energy levels than their bright counterparts.

Additionally, the interaction of two-dimensional (2D) materials with light, crucial for devices such as nanolasers [13,14], can be enhanced by coupling these materials with optical cavities or through intrinsic polaritonic resonances [15,16]. The strong interaction between light and matter within optical cavities leads to the creation of quasiparticles termed polaritons. In organic microcavities, when the vibronic coupling between excitons and phonons is comparable in strength to the exciton–photon coupling, multiple polariton branches have been experimentally observed [17,18] alongside phonon-assisted relaxation processes [19,20]. Recently, these phenomena have been directly utilized to investigate polaritonic nonlinearities at the single-photon level [21]. These entities underpin a wide array of both quantum and classical behaviors across various materials and spectral regions, such as polariton lasing [22], polariton-induced blockade [23,24], and the Bose–Einstein condensation of polaritons [25]. Recent developments in 2D TMDs interfaced with photonic architectures have paved the way for detailed studies of exciton–polaritons on an atomic level. The valley-selected property of TMDs excitons, when utilized in hybrid photonic systems, has enabled the creation of valley-polarized exciton–polaritons [26]. Moreover, excitons located in the K and K′ valleys are interconnected through time reversal symmetry. Consequently, removing valley degeneracy in these systems might facilitate the study of topological exciton–polaritons that exhibit broken time reversal symmetry at visible frequencies [27]. Another compelling dimension of exciton behavior in 2D TMDs emerges when two monolayers are angularly misaligned and superposed, creating bilayer heterostructures [28,29]. The resulting moiré patterns lead to moiré excitons within individual layers, which show a pronounced increase in nonlinearity due to exciton blockade effects. This significant interaction between moiré excitons and photonic modes in cavities can produce polaritons with marked nonlinearity [30], forming an excellent platform for studying highly interactive exciton–polaritons. Such integration facilitates the development of compact, on-chip devices [31] by coupling excitons in TMDs with photonic modes within planar nanocavities [32,33], paving the way for novel light–matter interaction platforms.

Photoluminescence (PL) spectroscopy remains a fundamental technique for probing quantum phenomena and complex interactions in TMD monolayers [34]. At a low temperature, PL reveals the presence of bound excitonic complexes, such as biexcitons and trions [5,35–37]. The significant role of phonon-assisted recombination in shaping the PL emission characteristics can be highlighted, particularly within cavity-controlled materials, leading to asymmetrical line shapes and enhanced sidebands—a phenomenon termed cavity-coupled PL emission [31]. Time-resolved fluorescence (TRF) spectroscopy [38–40] further allows the examination of these interactions over brief intervals, providing deeper insights into exciton dynamics and relaxation processes.

Despite the utility of conventional theoretical approaches such as quantum master equations [31] and cluster expansion methods [5], these often fall short in scenarios involving strong exciton–photon or exciton–phonon interactions. However, the multiple Davydov Ansatz (mDA) technique has recently emerged as a robust, numerically precise method [41,42], applied to a diverse array of multidimensional challenges, including analyzing Landau–Zener transitions [43,44], ultrafast dynamics at conical intersections (CIs) [45,46], and hole–magnon dynamics in an antiferromagnet [47,48]. The mDA method is a variational method to solve the many-body time-dependent Schrödinger equation, where the wave function for bosonic degrees of freedom is expanded into multiple coherent states for each boson mode in a multispecies boson bath of arbitrary spectral density functions. If the multiplicity is high enough, the so-obtained solutions converge to an exact solution. Here, we employ the mDA framework, complemented by thermofield dynamics (TFDs) representation [49,50], to offer a comprehensive, microscopically accurate simulation method for excitonic dynamics and spectroscopic responses of cavity-tuned single-layer WSe2 at finite temperature and multiple exciton–cavity (EC) coupling strengths.

In Section 2, we establish a microscopic model of the WSe_2 monolayer embedded in a microcavity and introduce the methodologies adopted. Section 3 is devoted to a comprehensive discussion of calculated exciton populations and TRF spectra for a range of EC coupling strengths from weak to strong. Conclusions are offered in Section 4.

2. Microscopic Model and Methodologies

PL and TRF profiles of WSe_2 monolayers are predominantly determined by KK, KQ, and KK$'$ excitons, of which the first K represents hole momentum in the valence band, and the second symbol indicates the electron momentum in the conduction band, corresponding to momenta at $1\mathbf{\Gamma}$, $2\mathbf{\Lambda}$, and $3\mathbf{K}$ respectively. Excitons of higher energy than the above states have a negligible influence on the dynamic properties and optical signals at room temperature and below. The phonon-mediated PL is attributed to the KQ and KK$'$ intervalley dark states, which is in contrast to the KK intravalley excitons that are optically bright and yield direct PL emission. The contrasting dynamics of direct versus indirect PL are detailed in Figure 1. Following this qualitative picture, the exciton–polariton dynamics of the cavity-modified WSe_2 monolayer can be described by a Hamiltonian containing three excitonic states, a cavity mode, and several dominant phonon modes [5,31,51,52]:

$$\begin{aligned} H &= \sum_{(i\mathbf{Q}_{\|})} E_{(i\mathbf{Q}_{\|})} X^{\dagger}_{(i\mathbf{Q}_{\|})} X_{(i\mathbf{Q}_{\|})} + \hbar\omega_{\sigma+} c^{\dagger}_{\sigma+} c_{\sigma+} \\ &+ M_{\sigma+}[c^{\dagger}_{\sigma+} X_{(1\Gamma)} + c_{\sigma+} X^{\dagger}_{(1\Gamma)}] + \sum_{\alpha\mathbf{q}_{\|}} \hbar\Omega_{\alpha,\mathbf{q}_{\|}} b^{\dagger}_{\alpha,\mathbf{q}_{\|}} b_{\alpha,\mathbf{q}_{\|}} \\ &+ \sum_{ij\alpha\mathbf{Q}_{\|}\mathbf{q}_{\|}} D^{ij}_{\alpha\mathbf{q}_{\|}} X^{\dagger}_{(j\mathbf{Q}_{\|}+\mathbf{q}_{\|})} X_{(i\mathbf{Q}_{\|})} (b^{\dagger}_{\alpha,-\mathbf{q}_{\|}} + b_{\alpha,\mathbf{q}_{\|}}). \end{aligned} \tag{1}$$

Here, operators $X^{(\dagger)}_{(i\mathbf{Q}_{\|})}, c^{(\dagger)}_{\sigma+}, b^{(\dagger)}_{\alpha,\mathbf{q}_{\|}}$ $(X_{(i\mathbf{Q}_{\|})}, c_{\sigma+}, b_{\alpha,\mathbf{q}_{\|}})$ create (annihilate), respectively, exciton states $(i\mathbf{Q}_{\|})$, circularly polarized photon states σ_+ and phonon modes α with momentum $\mathbf{q}_{\|}$, and frequency $\Omega_{\alpha,\mathbf{q}_{\|}}$. $E_{(i\mathbf{Q}_{\|})}$ is the excitonic dispersion relation, where the symbol $\|$ denotes the momentum's in-plane component. The strength of coupling between the exciton and cavity photon modes is quantified by the coefficient M_{σ_+}. The tensor $D^{ij}_{\alpha\mathbf{q}_{\|}}$, which maps the coupling of excitons with phonons, can be computed through the density functional theory (DFT) formalism (mean effective deformation potential approximation) [7,53]:

$$D^{ij}_{\alpha\mathbf{q}_{\|}} = \sum_{\mathbf{k}} \Phi^{i*}_{\mathbf{k}} (\Phi^{j}_{\mathbf{k}+\mathbf{q}_{1\|}} g^{c\mathbf{k}(ij)}_{\alpha\mathbf{q}_{\|}} - \Phi^{j}_{\mathbf{k}-\mathbf{q}_{2\|}} g^{v\mathbf{k}(ij)}_{\alpha\mathbf{q}_{\|}}). \tag{2}$$

where $\Phi^{i}_{\mathbf{k}}$ is the excitonic wave function in the momentum space, which can be obtained by solving the Wannier equation. $g^{c/v\mathbf{k}ij}_{\alpha\mathbf{q}_{\|}}$ are the electron–phonon interaction coefficients within the conduction and valence bands [54,55], where α and $\mathbf{q}_{\|}$ define the phonon modes scattered between the excitonic band index i and j.

$$\begin{aligned} g^{c/v\mathbf{k}(ij)}_{\alpha\mathbf{q}_{\|}} &= \sqrt{\frac{\hbar}{2\mathcal{M}\Omega_{\alpha,\mathbf{q}_{\|}}}} \langle j,\mathbf{k}+\mathbf{q}_{\|}|\Delta V^{\alpha}_{\mathbf{q}_{\|}}|i,\mathbf{k}\rangle \\ &= \sqrt{\frac{\hbar}{2\mathcal{M}\Omega_{\alpha,\mathbf{q}_{\|}}}} \int_{\mathrm{UC}} \psi^{*}_{j,\mathbf{k}+\mathbf{q}_{\|}}(\mathbf{r}) \Delta V^{\alpha}_{\mathbf{q}_{\|}}(\mathbf{r}) \psi_{i,\mathbf{k}}(\mathbf{r})\, d^3r \end{aligned} \tag{3}$$

are the electron–phonon matrix elements in conduction/valence band involving integrals over the unit cell (UC). $\mathcal{M}$ encapsulates the total mass of atoms within a unit cell, while $|i,\mathbf{k}\rangle$ describes the Bloch eigenstates $\psi_{i,\mathbf{k}}(\mathbf{r})$ characterized by wave vector $\mathbf{k}$ and an excitonic band index i. The potential perturbation, $\Delta V^{\alpha}_{\mathbf{q}_{\|}}$, is derived using density functional perturbation theory. The first-order deformation potential, or the acoustic deformation potential, is expressed as $\langle j,\mathbf{k}+\mathbf{q}_{\|}|\Delta V^{\alpha}_{\mathbf{q}_{\|}}|i,\mathbf{k}\rangle/|\mathbf{q}_{\|}|$. It can be shown that acoustic phonons at the $\mathbf{\Gamma}$ do not contribute to the electron–phonon coupling. The zero-order deformation potential, also known as the optical deformation potential, is defined as $\langle j,\mathbf{k}+\mathbf{q}_{\|}|\Delta V^{\alpha}_{\mathbf{q}_{\|}}|i,\mathbf{k}\rangle$ [54]. Detailed parameters for this potential are documented in Ref. [56]. In this work, we neglect cavity energy losses, which can be realized in fabricated photonic crystal nanobeam (PhCnB) cavities with a notably high-quality factor [57]. For instance, the exciton–photon coupling of a thin excitonic material positioned at the center of a high-quality, symmetric Fabry–Pérot cavity near the cavity resonance can be fine-tuned by adjusting the mirror reflectivity and the cavity length [58–60].

To treat finite temperature effects with the TFD method, we introduce the extended Hamiltonian [61–63]:

$$\overline{H} = H - \tilde{H}_{\mathrm{ph}} = H - \sum_{\alpha\mathbf{q}_{\|}} \Omega_{\alpha,\mathbf{q}_{\|}} \tilde{b}^{\dagger}_{\alpha,\mathbf{q}_{\|}} \tilde{b}_{\alpha,\mathbf{q}_{\|}} \tag{4}$$

where

$$\tilde{H}_{\text{ph}} = \sum_{\alpha\mathbf{q}_{\|}} \Omega_{\alpha,\mathbf{q}_{\|}} \tilde{b}^{\dagger}_{\alpha,\mathbf{q}_{\|}} \tilde{b}_{\alpha,\mathbf{q}_{\|}} \tag{5}$$

is the phonon Hamiltonian acting in the fictitious "tilde" vibrational space. Having performed the thermal Bogoliubov transformation, we arrive at the TFD Hamiltonian [49,50]:

$$\begin{aligned}
\overline{H}_{\theta} &= e^{iG}\overline{H}e^{-iG} \\
&= \sum_{(i\mathbf{Q}_{\|})} E_{(i\mathbf{Q}_{\|})} X^{\dagger}_{(i\mathbf{Q}_{\|})} X_{(i\mathbf{Q}_{\|})} + \omega_{\sigma+} c^{\dagger}_{\sigma+} c_{\sigma+} \\
&+ M_{\sigma+} [c^{\dagger}_{\sigma+} X_{(1\Gamma)} + c_{\sigma+} X^{\dagger}_{(1\Gamma)}] + \sum_{\alpha\mathbf{q}_{\|}} \Omega_{\alpha,\mathbf{q}_{\|}} (b^{\dagger}_{\alpha,\mathbf{q}_{\|}} b_{\alpha,\mathbf{q}_{\|}} - \tilde{b}^{\dagger}_{\alpha,\mathbf{q}_{\|}} \tilde{b}_{\alpha,\mathbf{q}_{\|}}) \\
&+ \sum_{ij\alpha\mathbf{Q}_{\|}\mathbf{q}_{\|}} D^{ij}_{\alpha\mathbf{q}_{\|}} X^{\dagger}_{(j\mathbf{Q}_{\|}+\mathbf{q}_{\|})} X_{(i\mathbf{Q}_{\|})} [\cosh(\theta_{\alpha,\mathbf{q}_{\|}})(b_{\alpha,\mathbf{q}_{\|}} + b^{\dagger}_{\alpha,-\mathbf{q}_{\|}}) + \sinh(\theta_{\alpha,\mathbf{q}_{\|}})(\tilde{b}_{\alpha,-\mathbf{q}_{\|}} + \tilde{b}^{\dagger}_{\alpha,\mathbf{q}_{\|}})]
\end{aligned} \tag{6}$$

in which

$$G = G^{\dagger} = -i \sum_{\alpha\mathbf{q}_{\|}} \theta_{\alpha,\mathbf{q}_{\|}} (b_{\alpha,\mathbf{q}_{\|}} \tilde{b}_{\alpha,\mathbf{q}_{\|}} - b^{\dagger}_{\alpha,\mathbf{q}_{\|}} \tilde{b}^{\dagger}_{\alpha,\mathbf{q}_{\|}}) \tag{7}$$

is the Bogoliubov operator, and

$$\theta_{\alpha,\mathbf{q}_{\|}} = \operatorname{arctanh}(e^{-\beta\Omega_{\alpha,\mathbf{q}_{\|}}/2}) \tag{8}$$

are the mixing angles, which account for the influence of the temperature on the exciton–phonon couplings.

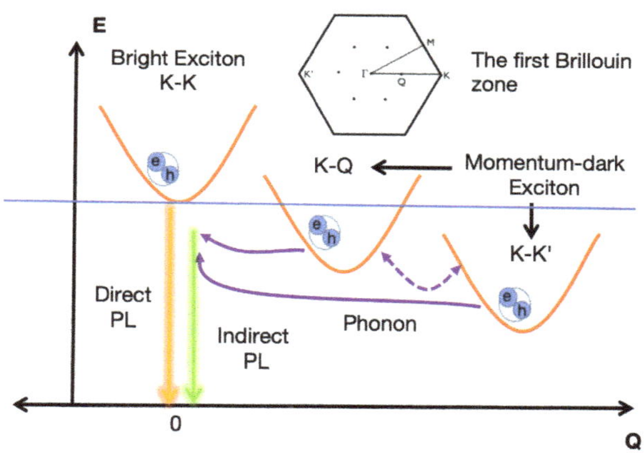

Figure 1. A graphical representation of the direct and phonon-assisted PL emission processes. The momentum–dark excitons located at the K − Q and K − K′ points undergo radiative decay via absorbing or emitting a phonon, thus contributing to the indirect PL emissions. The inset in the upper middle part illustrates the distribution of conduction electrons across various valleys within the first Brillouin zone.

The Hamiltonian of Equation (6) commutes with the number operator

$$N_{\rm ex} = c^{\dagger}_{\sigma_+} c_{\sigma_+} + \sum_{(iQ_{\parallel})} X^{\dagger}_{(iQ_{\parallel})} X_{(iQ_{\parallel})} \tag{9}$$

and conserves the number of excitons. In this work, we consider the dynamics within the singly excited excitonic manifold, where $\langle N_{\rm ex}\rangle = 1$. In this manifold, the solution of the time-dependent Schrödinger equation with the TFD Hamiltonian of Equation (6) can be represented in terms of the mDA wave function of multiplicity M:

$$|D_2^{\rm M}(t)\rangle = \sum_{i=0} |i\rangle \sum_{m=1}^{M} B_{mi}(t) e^{\sum_l (u_{ml}(t) b_l^{\dagger} - {\rm H.C.})} |\mathbf{0}\rangle \times e^{\sum_q (\tilde{u}_{mq}(t) \tilde{b}_q^{\dagger} - {\rm H.C.})} |\tilde{\mathbf{0}}\rangle \tag{10}$$

Here, the indices $i = 0,1,2,3$ denote the photon state, KK, KQ, and KK$'$ excitons, respectively. $B_{mi}(t)$ represents the time-dependent exciton amplitude, where the indices m stand for the mth coherent state superposition (with a total of M superpositions). The operators $b_l^{\dagger}(b_l)$ and $\tilde{b}_l^{\dagger}(b_l)$ are the physical and tilde phonon's creation (annihilation) operators, respectively. The variables $u_{ml}(t)$ and $\tilde{u}_{ml}(t)$ capture displacements of the physical and tilde phonons in mth coherent state. For the calculation of exciton dynamics, the initial population is solely in the photon state, and the initial elements of $u_{ml}(t)$ and $\tilde{u}_{ml}(t)$ are random numbers of the order 10^{-4}. The parameters $B_{mi}(t)$, $u_{ml}(t)$, and $\tilde{u}_{ml}(t)$ are determined through the equations of motion obtained following the variational principle [41,46,64], as detailed in Appendix A. All observables studied in the present work were evaluated in $|D_2^{\rm M}(t)\rangle$, as described in Appendix B.

To establish a realistic model, ab initio input parameters entering the Hamiltonian include the electronic band structure [65], phonon dispersion [56], dielectric constants [66] and electron–phonon coupling elements [56]. In our calculations, we assumed that $E_{(1\Gamma)} = 1.724$ eV [67,68], $E_{(2\Lambda)} = 1.69$ eV, and $E_{(3\mathbf{K})} = 1.678$ eV, where the spectral exciton separations between KK, KQ, and KK$'$ excitons were taken from ref. [5]. The exciton energies and wave functions were derived in the effective mass approximation by solving the Wannier equation [69–71]. This approach leads to an exciton dispersion, which is comparable to that from DFT calculations based on the Bethe–Salpeter equation [72–74]. To amplify cavity-prompted phenomena, a single cavity mode [5,75] was taken to be in resonance with the luminescent KK exciton, where $\hbar\omega_{\sigma_+} = 1.724$ eV. The dielectric constant [5,66] was set to be 4.5 for the hBN-encapsulated TMDs monolayers. For the exciton–cavity couplings, we only consider the interactions between the bright KK exciton and photons, as there is no transition dipole moment of momentum–dark states. Within the photonic crystal cavities, the coupling between excitons and photons is modifiable within a range from 4 meV to 14 meV, as documented by Rosser et al. [31]; for our calculations, $M_{\sigma_+} = 0,4,8,12$ meV [76–78] were used. Table 1 lists the employed phonon frequencies [5,56], illustrating that both longitudinal acoustic (LA) and transverse acoustic (TA) phonon branches exhibit a linear dispersion nearing the long-wavelength limit, aligning their frequencies at zero when $q = 0$. The optical phonon branches with significant exciton interactions include the homopolar (A_1) modes, characterized by out-of-plane vibrations, and the in-plane longitudinal (LO) and transversal (TO) modes.

The Hamiltonian of Equation (6) incorporates 23 phonon modes, three excitonic states (KK, KQ, KK$'$), and a single photonic mode. The number of physical phonon modes (i.e., 23) surpasses 13, as outlined in Table 1, which is due to the inclusion of $\pm$ modes $b_{\alpha,\pm\mathbf{q}_{\parallel}}$ with $\alpha = {\rm KQ}, {\rm KK}'$. The total number of phonon modes (i.e., 46) comes about by the duplication of the number of physical phonons owing to the inclusion of the tilde modes $\tilde{b}_{\alpha,\mathbf{q}_{\parallel}}$, crucial for addressing temperature effects within the TFD framework. The model omits holes near the K point due to their KK$'$ symmetry and excludes spin-forbidden dark states, which are irrelevant for the ultrafast dynamics under study. This model effectively captures coherent interactions among excitons, phonons, and photons, and it incorporates environmental dephasing as proposed, e.g., in Ref. [5].

The impact of the temperature on the exciton dynamics and TRF spectra has been comprehensively investigated in Ref. [55]. Here, we set the temperature to 75 K and focused on a detailed study of how the strength of the EC coupling affects the dynamical and spectroscopic observables. All our numerical results are proven to be convergent for the multiplicity $M = 48$ of the mDA wave function of Equation (10).

Table 1. Phonon frequencies [56] (in units of meV).

Mode	Γ	Λ	K
TA	0	11.6	15.6
LA	0	14.3	18
TO	30.5	27.3	26.7
LO	30.8	32.5	31.5
A_1	30.8	30.4	31

3. Results and Discussion

3.1. Exciton Populations

Figure 2 provides a comprehensive analysis of how varying ECs influence the dynamics of specific photonic and excitonic modes. In the absence of EC coupling in panel (a), the photon mode population persists, reflecting the lack of energy exchange between photons and excitons. For $EC = 4$ meV (red lines), the photon mode population decreases, and the excitonic mode population increases almost linearly with time, exhibiting the short-time ballistic transport on the timescale of ~400 fs. As the EC coupling strength increases, additional oscillations and nonmonotonic behaviors emerge, as depicted in panels (b)–(d). This is indicative of exciton–polariton formation resulting from intensified exciton–photon interactions within the cavity. It is the value of $EC = 8$ meV (yellow lines) that marks the changeover to the oscillatory strong EC coupling regime. The onset of this regime is manifested through the nonmonotonic evolution of the cavity mode population in panel (a), which reaches a minimum (almost zero) at $t \approx 350$ fs (when $EC = 8$ meV) and $t \approx 200$ fs (when $EC = 12$ meV) followed by the increase at longer times. Irrespective of the EC coupling strength, the excitonic states are populated sequentially [55] at short times: the KK state is populated first, followed by the KQ and KK′ states. Enhanced exciton–cavity coupling accelerates the transfer of photon population to excitons, as evidenced by the purple lines for $EC = 12$ meV, reaching their minima (a) and maxima (b)–(d) more swiftly. At an EC of 12 meV both bright KK and dark KQ excitons acquire rapid population growth initially at 100 fs and 200 fs, respectively, followed by noticeable oscillations. Remarkably, an increased EC consistently sustains a higher (though nonmonotonic as a function of time) population level in the dark KK′ exciton due to phonon-assisted transfer when the polaritonic state is formed. This observation, which demonstrates robustness of the cavity-induced population of the KK′ excitons within a wide range of EC coupling strengths, may become important for practical applications.

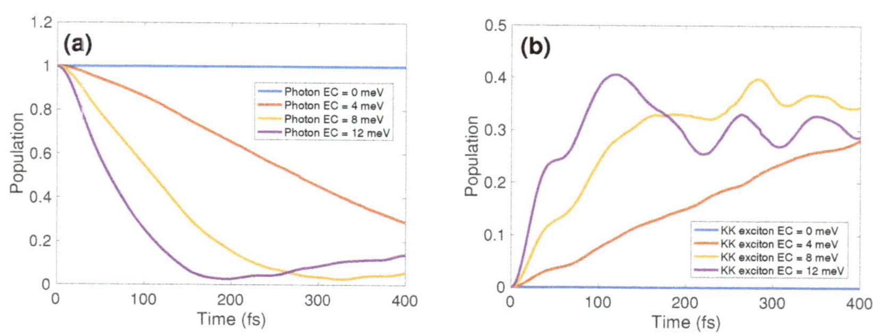

Figure 2. *Cont.*

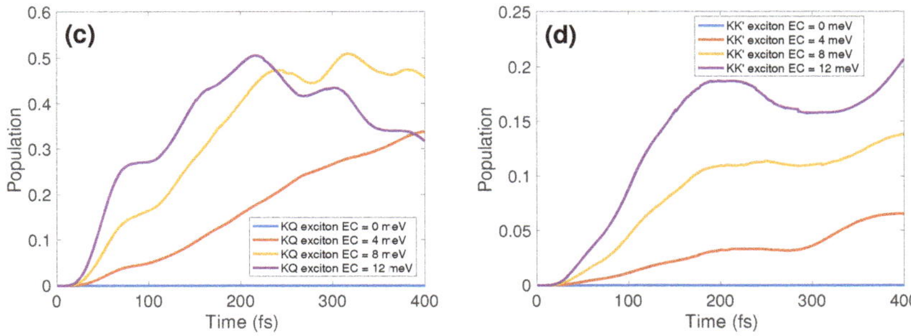

Figure 2. Population plots of (**a**) photons, (**b**) KK excitons, (**c**) KQ excitons, and (**d**) KK′ excitons for different EC coupling strengths.

3.2. TRF

To delve deeper into the dynamics of ultrafast exciton–polariton formation in WSe_2 monolayers, we computed the TRF spectra $S(\omega,t)$ presuming instantaneous excitation of the system by the pump pulse. These spectra, which describe the rate of emission of photons of frequency ω at time t [64,79,80], are evaluated with the mDA wave function, as described in Appendix B. For obtaining the spectral shape of $S(\omega,t)$, we used the overdamped harmonic oscillator lineshape function of Equation (A9) with the Stokes shift parameter $\lambda = 5$ meV and the memory rate parameter $\Lambda = 30$ meV.

A 3D view of the TRF spectra $S(\omega,t)$ is presented in Figure 3 for different ECs from zero (a) through 4 eV (b) and 8 eV (c) to 12 eV (d). The spectra are characterized by a high initial (at $t = 0$) peak revealing the bright KK exciton (direct PL), the intensity of which drops substantially on the timescale of several dozens femtoseconds. Rapid fluorescence attenuation of the bright KK exciton is culminated in panel (d), which is characterized by an approximately 90% loss in intensity within the initial 100 fs. This behavior is indicative of swift internal conversion processes at the (cavity-induced) CIs [45,46,81]. In contrast, the absence of a cavity ($EC = 0$, panel (a)) sees the fluorescence intensity of direct PL at about 50% in the first 100 fs. From panels (a) to (d), the tendency of fast fluorescence decay with the EC coupling agrees with the enhanced population of the dark states (KQ and KK′) indicated in Figure 2c,d.

In the time domain, $S(\omega,t)$ exhibits oscillatory behavior, which mirrors cavity-mediated exciton–phonon wave packet motion. This motion is especially pronounced for $EC = 0$ (panel a). However, it is not strictly periodic, since the spectral maxima are separated by the intermittent time interval of about ∼80–100 fs. Interestingly, these intervals are substantially shorter than the period corresponding to the fastest phonon mode ($2\pi/\Omega_{LO,\Lambda} = 127$ fs. This is a clear indication of the exciton–vibrational–photon coupling, which results in Rabi-like oscillations. As EC coupling strengths increase, the wave packet evolution retains oscillatory features, which, however, become progressively more erratic and spread over a wider range in the frequency domain.

A detailed examination of the early stages of the spectral evolution is provided by the inset of Figure 3, which elucidates the wave packet dynamics influenced by the intrinsic and cavity-induced CIs. The spectral features primarily cluster around $\omega = 1.724$ eV, aligning with the bright KK exciton energy. Notably, this observation confirms that the lower-energy KQ and KK′ excitons, which are intrinsically dark, do not substantially contribute to the emission at these early times (cf. Ref. [82]). This scenario also illustrates the influence of polaritonic effects in Figure 3d: the fundamental excitonic KK state splits into two distinct polaritonic states, with a separation roughly equal to $2M_{\sigma_-} = 24$ meV, thereby widening the spectrum. Additionally, interactions with phonon modes (polaron effects) further broaden the spectrum, extending it toward the blue side. Inversely, decreasing ECs causes a narrowing of $S(\omega,t)$ along the ω axis, which produces a more symmetric spectral shape and leads to a slower fluorescence quenching.

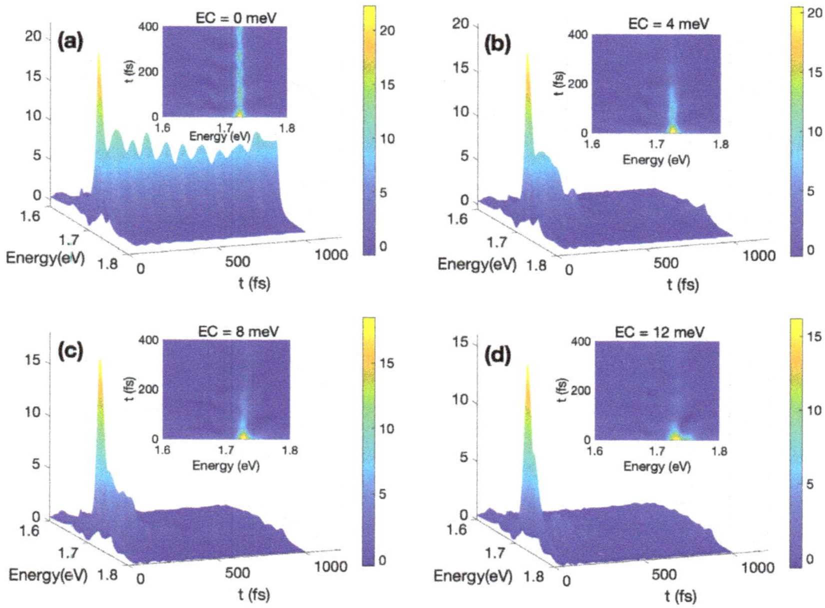

Figure 3. A 3D view of TRF spectra $S(\omega, t)$ within the first 1000 fs: (**a**) $EC = 0$ meV, (**b**) $EC = 4$ meV, (**c**) $EC = 8$ meV, and (**d**) $EC = 12$ meV. Inset figures are 2D view of TRF spectra $S(\omega, t)$ within the first 400 fs.

For obtaining a detailed view of the spectral features, it is worthwhile to inspect $S(\omega, t)$ at specific values of t. TRF spectra at $t = 0$, which can be interpreted as absorption spectra [79], are shown in panel (a) of Figure 4, accompanied by relaxed fluorescence spectra at $t = 1000$ fs in panel (b). At $t = 0$, the spectra for all EC couplings exhibit similar shapes. Yet, larger EC coupling strengths reduce the intensity of the main peak associated with direct PL, as well as inducing a blue shift and spectral broadening. The red wing peak at $\omega \approx E - E_{\text{bright}} = -43$ meV in Figure 4a is related to indirect PL, while the blue wing shoulder located around $\omega \approx E - E_{\text{bright}} = 30$ meV represents a vibronic peak. This implies a modification (enhancement) of the Huang–Rhys factor [83], and a similar spectral redistribution is observed in surface plasmon polaritons [84]. The observed enhancement of the vibronic peak intensity, as outlined in Ref. [85], stems from the close-to-resonance conditions. If the phonon frequency $\Omega_{\alpha,\mathbf{q}_{\|}}$ (25–30 meV) from the energy gap between the prominent KK peak and the adjacent higher energy peak matches the separation Δ between the red wing peak and the central peak, the resonance condition is satisfied. This relationship is quantified as $\Delta \approx k\Omega_{\alpha,\mathbf{q}_{\|}}$, where $k = 1, 2, \ldots$. As EC couplings increase from 0 to 12 meV, Δ nearly doubles the frequency of the ($\mathbf{\Lambda}$ and $\mathbf{K}$) phonon modes.

The relaxed TRF spectrum at $t = 1000$ fs in Figure 4b shows drastic changes in lineshape as a function of the EC coupling strength. The intensity of the central direct PL peak (EC = 0) diminishes significantly due to the EC-enhanced population transfer to the dark excitons. In the strong coupling regime ($EC = 8$ meV and $EC = 12$ meV), the peak splits into two sub-peaks separated by 25 meV, as shown in the inset. These polaritonic features are not seen in the absorption spectrum. Therefore, TRF spectra deliver valuable information on the photoinduced processes in WSe_2 monolayers, which cannot be extracted from the absorption spectra.

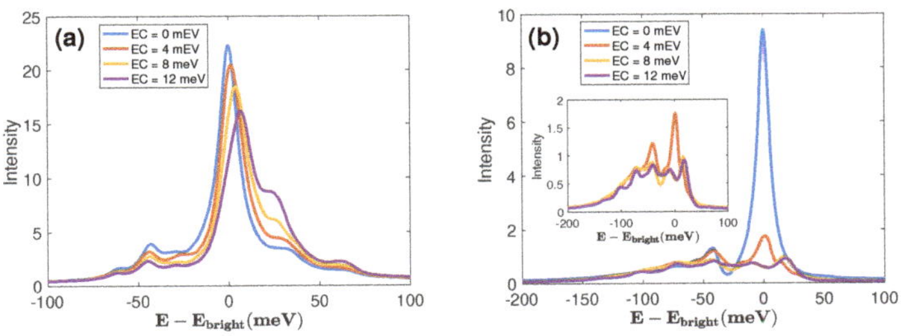

Figure 4. TRF spectra $S(\omega, t)$ at $t = 0$ (**a**) and 1000 fs (**b**) for four EC coupling strengths, as indicated in the panels.

Figure 5 presents the TRF spectra $S(\omega, t)$ at intermediate times for four EC coupling strengths. For $EC = 0$ [panel (a)], the main dynamical events in the evolution of $S(\omega, t)$ are over before $t = 250$ fs (see Figure 3a). Hence, the spectra at 500 fs and 1000 fs in Figure 5a are almost identical, suggesting that the steady state fluorescence is reached quicker for weaker EC coupling. Only the intensity of the KK peak goes out slowly and remains much higher than that of the adjacent vibronic peaks. Additionally, the KK peak position stays unchanged in panels (a) and (b), contrasting with the significant shifts of the peak positions in panels (c) and (d). These shifts can be attributed to the polaritonic effects, which are manifested in significantly richer peak structures that spread over a substantially broader spectral domain. This indicates that the widths of relaxed TFR spectra can be used for estimating the EC coupling strengths. Qualitatively, the spectra in the strong EC coupling regime (panel d) are similar to those calculated in Ref. [31] using the Lindblad master equation for cavity-assisted TMDs monolayers. Nonetheless, the inter-peak separations and spectral widths in our spectra are significantly larger than those in Ref. [31] due to the stronger exciton–phonon and exciton–photon interactions.

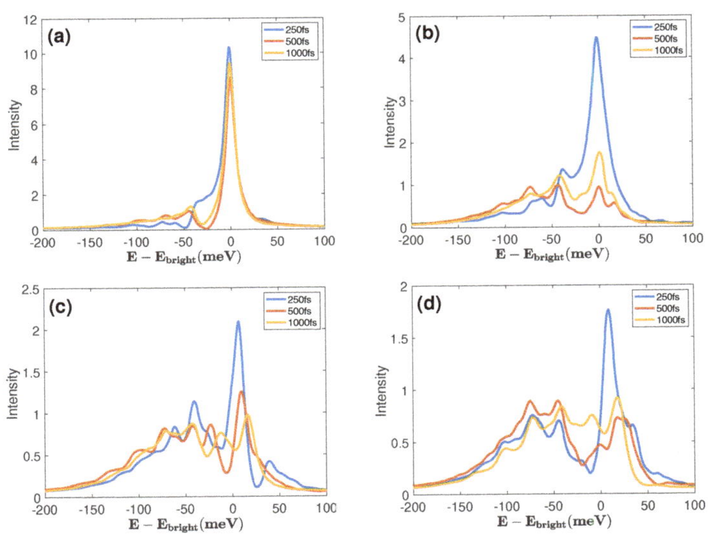

Figure 5. TRF spectra $S(\omega, t)$ at 250 fs, 500 fs, and 1000 fs for (**a**) $EC = 0$ meV, (**b**) $EC = 4$ meV, (**c**) $EC = 8$ meV, and (**d**) $EC = 12$ meV.

4. Conclusions

By integrating the mDA method [41,42] with the TFD framework [49,50], we performed numerically accurate simulations of the exciton dynamics and TRF spectra of cavity-modulated WSe_2 monolayers at 75 K for EC coupling strengths ranging from 0 (no polaritonic effects) to 12 meV (strong polaritonic effects). The efficient computational method employed in the present work can be applied to simulations of various cavity-tuned 2D materials at finite temperatures. More specifically, our approach can be extended to possibly account for the inversion of the band ordering (i.e., the lower polariton mode pushed below the WSe_2 dark state) and phonon-assisted polariton relaxation [86]. Furthermore, anti-Stokes PL [87] activated in nanocavity-integrated WSe_2 monolayers can be simulated by incorporating resonant excitation of a dark exciton [87] in our Hamiltonian.

Our approach identified distinct spectroscopic signatures and characteristic timescales of polaronic and polaritonic effects in WSe_2 monolayers at various EC interactions. Notably, our investigation highlighted the crucial role of multidimensional CIs in governing the dynamics of strongly coupled excitons, photons, and phonons. It comes as no surprise that the EC coupling is an important parameter governing energy transfer in TMDs. What is more significant for applications and less evident is that we found that several aspects of the polaritonic transport in WSe_2 monolayers do not require finetuning of EC interactions. For example, increasing EC coupling strengths enhances the total population of the dark interval excitons KQ and KK' on the timescale of up to 200 fs, and it amplifies populations of the KK' excitons at times longer than 400 fs. This robustness in the cavity-mediated energy transport may be instrumental for engineering 2D materials.

Supplementary Materials: The following supporting information can be downloaded at: https://www.mdpi.com/article/10.3390/ma17164127/s1.

Author Contributions: Conceptualization, K.S. (Kaijun Shen), M.F.G., and Y.Z.; methodology, K.S. (Kewei Sun), K.S. (Kaijun Shen), and Y.Z.; software, K.S. (Kaijun Shen); validation, K.S. (Kaijun Shen), M.F.G., and K.S. (Kewei Sun); formal analysis, K.S. (Kaijun Shen) and M.F.G.; investigation, K.S. (Kaijun Shen) and M.F.G.; resources, K.S. (Kaijun Shen); data curation, K.S. (Kaijun Shen); writing—original draft preparation, K.S. (Kaijun Shen); writing—review and editing, M.F.G., K.S. (Kewei Sun), and Y.Z.; visualization, K.S. (Kaijun Shen); supervision, Y.Z.; project administration, Y.Z.; funding acquisition, Y.Z. All authors have read and agreed to the published version of the manuscript.

Funding: This research was funded by the Singapore Ministry of Education Academic Research Fund Tier 1 (Grant No. RG87/20) and the National Natural Science Foundation of China (Grant No. 22373028).

Institutional Review Board Statement: Not applicable.

Informed Consent Statement: Not applicable.

Data Availability Statement: The data presented in this study are available upon request from the corresponding author.

Conflicts of Interest: The authors declare no conflicts of interest.

Abbreviations

The following abbreviations are used in this manuscript:

TMDs	Transition metal dichalcogenides
2D	Two-dimensional
PL	Photoluminescence
TRF	Time-resolved fluorescence
mDA	Multiple Davydov Ansatz
CIs	Conical intersections
TFDs	Thermofield dynamics
EC	Exciton–cavity

Appendix A. Equations of Motion for the mDA Parameters

The time evolution of the parameters $B_{mi}(t)$, $u_{ml}(t)$, and $\tilde{u}_{mq}(t)$ specifying the mDA wave function (10) are determined using the variational method [41,46,64]:

$$\frac{d}{dt}\frac{\partial \mathcal{L}}{\partial \dot{\xi}_j^*} - \frac{\partial \mathcal{L}}{\partial \xi_j^*} = 0, \tag{A1}$$

where the Lagrangian $\mathcal{L}$ is defined as

$$\mathcal{L} = \frac{i}{2}\left[\langle \mathrm{D}_2^{\mathrm{M}}(t)|\dot{\mathrm{D}}_2^{\mathrm{M}}(t)\rangle - \langle \dot{\mathrm{D}}_2^{\mathrm{M}}(t)|\mathrm{D}_2^{\mathrm{M}}(t)\rangle\right] - \langle \mathrm{D}_2^{\mathrm{M}}(t)|H_{\mathrm{T}}|\mathrm{D}_2^{\mathrm{M}}(t)\rangle. \tag{A2}$$

The multi-D_2 Ansatz converges to reproduce the accurate wave function if the multiplicity M is large enough. The multiplicity $M = 48$ adopted in our simulations yields the converged results.

The Hamiltonian structured within the framework of the multi-D_2 ansatz is expressed as

$$\begin{aligned}
L_H &= \langle D_2^{\mathrm{M}}(t)|\overline{H}_\theta|D_2^{\mathrm{M}}(t)\rangle \\
&= \sum_{i=0}^{3}\sum_{m}^{M}\sum_{m'}^{M} E_i B_{mi}^* B_{m'i} R_{mm'} + \sum_{i=0}^{3}\sum_{m}^{M}\sum_{m'}^{M} B_{mi}^* B_{m'i}\sum_{l}^{23}\Omega_l(u_{ml}^* u_{m'l} - \tilde{u}_{ml}^*\tilde{u}_{m'l})R_{mm'} \\
&+ \sum_{m}^{M}\sum_{m'}^{M}(M_{\sigma_+} B_{m0}^* B_{m'1} + M_{\sigma_+}^* B_{m1}^* B_{m'0})R_{mm'} + \sum_{m}^{M}\sum_{m'}^{M}\{\sum_{l=1}^{3} D_{1l}^{(1)} B_{m1}^* B_{m'1}\cosh(\theta_l)(u_{ml}^* + u_{m'l})R_{mm'} \\
&+ \sum_{l=1}^{3} D_{1l}^{(1)} B_{m1}^* B_{m'1}\sinh(\theta_l)(\tilde{u}_{ml}^* + \tilde{u}_{m'l})R_{mm'} + \sum_{l=4}^{8} D_l^{(2)} B_{m1}^* B_{m'2}(u_{ml}^*\cosh(\theta_l) + \tilde{u}_{m'l}\sinh(\theta_l))R_{mm'} \\
&+ \sum_{l=14}^{18} D_l^{(2)} B_{m1}^* B_{m'2}(u_{m'l}\cosh(\theta_l) + \tilde{u}_{ml}^*\sinh(\theta_l))R_{mm'} \\
&+ \sum_{l=9}^{13} D_l^{(3)} B_{m1}^* B_{m'3}(u_{ml}^*\cosh(\theta_l) + \tilde{u}_{m'l}\sinh(\theta_l))R_{mm'} \\
&+ \sum_{l=19}^{23} D_l^{(3)} B_{m1}^* B_{m'3}(u_{m'l}\cosh(\theta_l) + \tilde{u}_{ml}^*\sinh(\theta_l))R_{mm'}\} \\
&+ \sum_{m}^{M}\sum_{m'}^{M}\{\sum_{l=1}^{3} D_{2l}^{(1)} B_{m2}^* B_{m'2}\cosh(\theta_l)(u_{ml}^* + u_{m'l})R_{mm'} + \sum_{l=1}^{3} D_{2l}^{(1)} B_{m2}^* B_{m'2}\sinh(\theta_l)(\tilde{u}_{ml}^* + \tilde{u}_{m'l})R_{mm'} \\
&+ \sum_{l=4}^{8} D_l^{(2)} B_{m2}^* B_{m'1}(u_{ml}^*\cosh(\theta_l) + \tilde{u}_{m'l}\sinh(\theta_l))R_{mm'} \\
&+ \sum_{l=14}^{18} D_l^{(2)} B_{m2}^* B_{m'1}(u_{m'l}\cosh(\theta_l) + \tilde{u}_{ml}^*\sinh(\theta_l))R_{mm'}\} \\
&+ \sum_{m}^{M}\sum_{m'}^{M}\{\sum_{l=1}^{3} D_{3l}^{(1)} B_{m3}^* B_{m'3}\cosh(\theta_l)(u_{ml}^* + u_{m'l})R_{mm'} + \sum_{l=1}^{3} D_{3l}^{(1)} B_{m3}^* B_{m'3}\sinh(\theta_l)(\tilde{u}_{ml}^* + \tilde{u}_{m'l})R_{mm'} \\
&+ \sum_{l=9}^{13} D_l^{(3)} B_{m3}^* B_{m'1}(u_{ml}^*\cosh(\theta_l) + \tilde{u}_{m'l}\sinh(\theta_l))R_{mm'} \\
&+ \sum_{l=19}^{23} D_l^{(3)} B_{m3}^* B_{m'1}(u_{m'l}\cosh(\theta_l) + \tilde{u}_{ml}^*\sinh(\theta_l))R_{mm'}
\end{aligned} \tag{A3}$$

Here, $D_{il}^{(1)}$ specifies the diagonal exciton–phonon coupling matrix elements for the exciton state $|i\rangle$ interacting with phonon mode l. The terms $D_l^{(2)}$ and $D_l^{(3)}$ represent the exciton–phonon couplings between KK exciton states and KQ (KK′) exciton states with phonon mode l, respectively. Consequently, the governing equations for $B_{m'0}$ are formulated as $B_{m'1}$, $B_{m'2}$, and $B_{m'3}$, which are formulated as

$$\begin{aligned} & i\sum_{m'}^{M}[\dot{B}_{m'0} + B_{m'0}\sum_{l} u^*_{ml}\dot{u}_{m'l} + B_{m'0}\sum_{l}\tilde{u}^*_{ml}\dot{\tilde{u}}_{m'l}]R_{mm'} \\ & = \sum_{m'}^{M} E_0 B_{m'0}R_{mm'} + \sum_{m'}^{M} B_{m'0}\sum_{l}^{23}\Omega_l(u^*_{ml}u_{m'l} - \tilde{u}^*_{ml}\tilde{u}_{m'l})R_{mm'} \\ & + \sum_{m'}^{M} M_{\sigma_+} B_{m'1}R_{mm'} \end{aligned} \tag{A4}$$

Other variational parameters, such as $B_{m'1}$, $B_{m'2}$, $B_{m'3}$, $u_{m'l}$, and $\tilde{u}_{m'l}$, are described in the Supplementary Materials.

Appendix B. Observables in Terms of Multi-D$_2$ Wave Functions

Populations of the photonic and excitonic states can be evaluated in terms of the multi-D$_2$ wave functions (10) as the expectation values of the projection operators:

$$P_i(t) = \langle D_2^{\mathrm{M}}(t)|i\rangle\langle i|D_2^{\mathrm{M}}(t)\rangle \tag{A5}$$

where $i = 0, 1, 2, 3$ correspond to the photon state, KK, KQ, and KK$'$ excitons, respectively.

TRF spectra excited by a short (Dirac delta function) pump pulse can be computed using the following expression [64,79,80]:

$$S(\omega, t) \sim \mathrm{Re}\int_0^{\infty} \mathrm{d}t_3 R_1^{DA}(t_3, t, 0)R_1^{g(t)}(t_3, t, 0)\exp[i\omega t_3]. \tag{A6}$$

Here, $R_1^{DA}(t_3, t, 0)$ is the third-order response function evaluated using the mDA approach as follows:

$$\begin{aligned} R_1^{DA}(t_3, t_2, t_1) = & \sum_{m,m'}^{M}\sum_{i_0,i_1,i_2,i_3}^{3} \mu^*_{i_2}\mu_{i_3}\mu^*_{i_0}\mu_{i_1} B^*_{m'i_1i_0}(t_2)B_{mi_2i_3}(t_3+t_2+t_1) \\ & \times e^{\sum_q u^*_{m'qi_0}(t_2)u_{mqi_3}(t_3+t_2+t_1)e^{i\Omega_q t} + \sum_l \tilde{u}^*_{m'li_0}(t_2)\tilde{u}_{mli_3}(t_3+t_2+t_1)e^{i\Omega_l t}} \\ & \times e^{-\frac{1}{2}\sum_q(|u_{m'qi_0}(t_2)|^2 + |u_{mqi_3}(t_3+t_2+t_1)|^2) - \frac{1}{2}\sum_l(|\tilde{u}_{m'li_0}(t_2)|^2 + |\tilde{u}_{mli_3}(t_3+t_2+t_1)|^2)} \end{aligned} \tag{A7}$$

Here μ_i are the transition dipole moment coupling coefficients, where $\mu_0 = \mu_1 = 1$, $\mu_2 = \mu_3 = 0$. $B_{m'i_1i_0}(t)$ indicates the probability at time t for the exciton to occupy the state $|i_1\rangle$ relative to the initial state $|i_0\rangle$ with a given multiplicity M, while $u_{m'qi_0}(t)$ and $\tilde{u}_{m'li_0}(t)$ denote displacements of the 'physical' and 'tilde' phonon modes.

$R_1^{g(t)}(t_3, t, 0)$ is the response function representing environmental factors, which are not included into the Hamiltonian of Equation (6). It assumes the form

$$R_1^{g(t)}(t_3, t_2, t_1) = \exp[-g(t_1) - g^*(t_2) - g^*(t_3) + g(t_1 + t_2) + g^*(t_2 + t_3) - g(t_1 + t_2 + t_3)] \tag{A8}$$

and can be computed using the overdamped harmonic oscillator lineshape functions [64, 79]:

$$g(t) = g'(t) + ig''(t) \tag{A9}$$

$$g''(t) = -(\lambda/\Lambda)[\exp(-\Lambda t) + \Lambda t - 1] \tag{A10}$$

$$g'(t) = (\lambda/\Lambda)\cot(\hbar\beta\Lambda/2)[\exp(-\Lambda t) + \Lambda t - 1] + \frac{4\lambda\Lambda}{\hbar\beta}\sum_{n=1}^{\infty}\frac{\exp(-\upsilon_n t) + \upsilon_n t - 1}{\upsilon_n(\upsilon_n{}^2 - \Lambda^2)} \tag{A11}$$

$$\upsilon_n = \frac{2\pi}{\hbar\beta}n. \tag{A12}$$

Here, λ is the Stokes shift, and Λ^{-1} represents the memory time of the environment.

References

1. Xie, L.; Cui, X.D. Manipulating spin-polarized photocurrents in 2D transition metal dichalcogenides. *Proc. Natl. Acad. Sci. USA* **2016**, *113*, 3746-3750. [CrossRef] [PubMed]
2. Ye, Y.; Xiao, J.; Wang, H.; Ye, Z.; Zhu, H.; Zhao, M.; Wang, Y.; Zhao, J.; Yin, X.; Zhang, X. Electrical generation and control of the valley carriers in a monolayer transition metal dichalcogenide. *Nat. Nanotech.* **2016**, *11*, 598-602. [CrossRef] [PubMed]
3. Mak, K. F.; Lee, C.; Hone, J.; Shan, J.; Heinz, T.F. Atomically Thin MoS_2: A New Direct-Gap Semiconductor. *Phys. Rev. Lett.* **2010**, *105*, 136805. [CrossRef] [PubMed]
4. Splendiani, A.; Sun, L.; Zhang, Y.; Li, T.; Kim, J.; Chim, C.Y.; Galli, G.; Wang, F. Emerging Photoluminescence in Monolayer MoS_2. *Nano Lett.* **2010**, *10*, 1271–1275. [CrossRef]
5. Brem, S.; Ekman, A.; Christiansen, D.; Katsch, F.; Selig, M.; Robert, C.; Marie, X.; Urbaszek, B.; Knorr, A.; Malic, E. Phonon-Assisted Photoluminescence from Indirect Excitons in Monolayers of Transition-Metal Dichalcogenides. *Nano Lett.* **2020**, *20*, 2849–2856. [CrossRef] [PubMed]
6. Berghäuser, G.; Steinleitner, P.; Merkl, P.; Huber, R.; Knorr, A.; Malic, E. Mapping of the dark exciton landscape in transition metal dichalcogenides. *Phys. Rev. B* **2018**, *98*, 020301. [CrossRef]
7. Selig, M.; Berghäuser, G.; Richter, M.; Bratschitsch, R.; Knorr, A.; Malic, E. Dark and bright exciton formation, thermalization, and photoluminescence in monolayer transition metal dichalcogenides. *2D Mater.* **2018**, *5*, 035017. [CrossRef]
8. Echeverry, J. P.; Urbaszek, B.; Amand, T.; Marie, X.; Gerber, I.C. Splitting between bright and dark excitons in transition metal dichalcogenide monolayers. *Phys. Rev. B* **2016**, *93*, 121107. [CrossRef]
9. Rosati, R.; Schmidt, R.; Brem, S.; Perea-Causín, R.; Niehues, I.; Kern, J.; Preuß, J.A.; Schneider, R.; Michaelis de Vasconcellos, S.; Bratschitsch, R.; Malic, E. Dark exciton anti-funneling in atomically thin semiconductors. *Nat. Commun.* **2021**, *12*, 7221. [CrossRef]
10. Gelly, R.J.; Renaud, D.; Liao, X.; Pingault, B.; Bogdanovic, S.; Scuri, G.; Watanabe, K.; Taniguchi, T.; Urbaszek, B.; Park, H.; Lončar, M. Probing dark exciton navigation through a local strain landscape in a WSe_2 monolayer. *Nat. Commun.* **2022**, *13*, 232. [CrossRef]
11. Madeo, J.; Man, M.K.L.; Sahoo, C.; Campbell, M.; Pareek, V.; Wong, E.L.; Al-Mahboob, A.; Chan, N.S.; Karmakar, A.; Mariserla, B.M.K.; Li, X.; Heinz, T.F.; Cao, T.; Dani, K.M. Directly visualizing the momentum-forbidden dark excitons and their dynamics in atomically thin semiconductors. *Science* **2020**, *370*, 1199–1204. [CrossRef]
12. Rosati, R.; Wagner, K.; Brem, S.; Perea-Causín, R.; Ziegler, J. D.; Zipfel, J.; Taniguchi, T.; Watanabe, K.; Chernikov, A.; Malic, E. Non-equilibrium diffusion of dark excitons in atomically thin semiconductors. *Nanoscale* **2021**, *13*, 19966–19972. [CrossRef]
13. Xia, F.; Wang, H.; Xiao, D.; Dubey, M.; Ramasubramaniam, A. Two-dimensional material nanophotonics. *Nat. Photon* **2014**, *8*, 899–907. [CrossRef]
14. Zhao, L.; Shang, Q.; Li, M.; Liang, Y.; Li, C.; Zhang, Q. Strong exciton-photon interaction and lasing of two-dimensional transition metal dichalcogenide semiconductors. *Nano Res.* **2021**, *14*, 1937–1954. [CrossRef]
15. Liu, M.; Yin, X.; Ulin-Avila, E.; Geng, B.; Zentgraf, T.; Ju, L.; Wang, F.; Zhang, X. A graphene-based broadband optical modulator. *Nature* **2011**, *474*, 64–67. [CrossRef] [PubMed]
16. Koppens, F.H.L.; Chang, D.E.; Garcia de Abajo, F.J. Graphene plasmonics: A platform for strong light-matter interactions. *Nano Lett.* **2011**, *11*, 3370–3377. [CrossRef]
17. Holmes, R.J.; Forrest, S.R. Strong exciton-photon coupling and exciton hybridization in a thermally evaporated polycrystalline film of an organic small molecule. *Phys. Rev. Lett.* **2004**, *93*, 186404. [CrossRef]
18. Fontanesi, L.; Mazza, L.; La Rocca, G.C. Organic-based microcavities with vibronic progressions: Linear spectroscopy. *Phys. Rev. B* **2009**, *80*, 235313. [CrossRef]
19. Coles, D.M.; Michetti, P.; Clark, C.; Tsoi, W.C.; Adawi, A.M.; Kim, J.S.; Lidzey, D.G. Vibrationally Assisted Polariton-Relaxation Processes in Strongly Coupled Organic-Semiconductor Microcavities. *Adv. Funct. Mater.* **2011**, *21*, 3691–3696. [CrossRef]
20. Virgili, T.; Coles, D.; Adawi, A.M.; Clark, C.; Michetti, P.; Rajendran, S.K.; Brida, D.; Polli, D.; Cerullo, G.; Lidzey, D.G. Ultrafast polariton relaxation dynamics in an organic semiconductor microcavity. *Phys. Rev. B* **2011**, *83*, 245309. [CrossRef]

21. Zasedatelev, A.V.; Baranikov, A.V.; Sannikov, D.; Urbonas, D.; Scafirimuto, F.; Shishkov, V.Y.; Andrianov, E.S.; Lozovik, Y.E.; Scherf, U.; Stöferle, T.; Mahrt, R.F. Single-photon nonlinearity at room temperature. *Nature* **2021**, *597*, 493–497. [CrossRef] [PubMed]
22. Deng, H.; Weihs, G.; Snoke, D.; Bloch, J.; Yamamoto, Y. Polariton lasing vs. photon lasing in a semiconductor microcavity. *Proc. Natl. Acad. Sci. USA* **2003**, *100*, 15318. [CrossRef] [PubMed]
23. Delteil, A.; Fink, T.; Schade, A.; Höfling, S.; Schneider, C.; İmamoğlu, A. Towards polariton blockade of confined exciton-polaritons. *Nat. Mater.* **2019**, *18*, 219–222. [CrossRef]
24. Munoz-Matutano, G.; Wood, A.; Johnsson, M.; Vidal, X.; Baragiola, B.Q.; Reinhard, A.; Lemaître, A.; Bloch, J.; Amo, A.; Nogues, G.; Besga, B.; Richard, M.; Garcia, J. M. Emergence of quantum correlations from interacting fibre-cavity polaritons. *Nat. Mater.* **2019**, *18*, 213–218. [CrossRef]
25. Byrnes, T.; Kim, N.Y.; Yamamoto, Y. Exciton-polariton condensates. *Nat. Phys.* **2014**, *10*, 803. [CrossRef]
26. Chen, Y.J.; Cain, J.D.; Stanev, T.K.; Dravid, V.P.; Stern, N.P. Valley-polarized exciton-polaritons in a monolayer semiconductor. *Nat. Photon* **2017**, *11*, 431–435. [CrossRef]
27. He, L.; Wu, J.; Jin, J.; Mele, E. J.; Zhen, B. Polaritonic Chern insulators in monolayer semiconductors. *Phys. Rev. Lett.* **2023**, *130*, 043801. [CrossRef]
28. Seyler, K.L.; Rivera, P.; Yu, H.; Wilson, N.P.; Ray, E.L.; Mandrus, D.G.; Yan, J.; Yao, W.; Xu, X. Signatures of moire-trapped valley excitons in $mose_2/wse_2$ heterobilayers. *Nature* **2019**, *567*, 66. [CrossRef]
29. Jin, C.; Regan, E.C.; Yan, A.; Utama, M.I.B.; Wang, D.; Zhao, S.; Wang, F. Observation of moire excitons in WSe2/WS2 heterostructure superlattices. *Nature* **2019**, *567*, 76. [CrossRef]
30. Zhang, L.; Wu, F.; Hou, S.; Zhang, Z.; Chou, Y.-H.; Watanabe, K.; Taniguchi, T.; Forrest, S.R.; Deng, H. Van der Waals heterostructure polaritons with moire-induced nonlinearity. *Nature* **2021**, *591*, 61. [CrossRef]
31. Rosser, D.; Fryett, T.; Ryou, A.; Saxena, A.; Majumdar, A. Exciton-phonon interactions in nanocavity-integrated monolayer transition metal dichalcogenides. *NPJ 2D Mater. Appl.* **2020**, *4*, 20. [CrossRef]
32. Walther, V.; Johne, R.; Pohl, T. Giant optical nonlinearities from Rydberg excitons in semiconductor microcavities. *Nat. Commun.* **2018**, *9*, 1309. [CrossRef] [PubMed]
33. Ryou, A.; Rosser, D.; Saxena, A.; Fryett, T.; Majumdar, A. Strong photon antibunching in weakly nonlinear two-dimensional exciton-polaritons. *Phys. Rev. B* **2018**, *97*, 16042. [CrossRef]
34. Kira, M.; Jahnke, F.; Hoyer, W.; Koch, S.W. Quantum Theory of Spontaneous Emission and Coherent Effects in Semiconductor Microstructures. *Progr. Quant. Electr.* **1999**, *23*, 189–279. [CrossRef]
35. Mak, K. F.; He, K.; Lee, C.; Lee, G.H.; Hone, J.; Heinz, T.F.; Shan, J. Tightly bound trions in monolayer MoS_2. *Nat. Mater.* **2013**, *12*, 207–211. [CrossRef] [PubMed]
36. Ye, Z.; Waldecker, L.; Ma, E.Y.; Rhodes, D.; Antony, A.; Kim, B.; Zhang, X.-X.; Deng, M.; Jiang, Y.; Lu, Z.; et al. Efficient generation of neutral and charged biexcitons in encapsulated WSe_2monolayers. *Nat. Commun.* **2018**, *9*, 3718. [CrossRef] [PubMed]
37. Huang, J.; Hoang, T.; Mikkelsen, M. Probing the origin of excitonic states in monolayer WSe_2. *Sci Rep.* **2016**, *6*, 22414. [CrossRef] [PubMed]
38. Hyeon-Deuk, K.; Tanimura, Y.; Cho, M. Ultrafast exciton transfers in DNA and its nonlinear optical spectroscopy. *J. Chem. Phys.* **2008**, *128*, 135102. [CrossRef] [PubMed]
39. Ahn, T.-S.; Muller, A.M.; Al-Kaysi, R.O.; Spano, F.C.; Norton, J.E.; Beljonne, D.; Bredas, J.L.; Bardeen, C.J. Experimental and theoretical study of temperature dependent exciton delocalization and relaxation in anthracene thin films. *J. Chem. Phys.* **2008**, *128*, 054505. [CrossRef] [PubMed]
40. Nakamura, R.; Hamada, N.; Ichida, H.; Tokunaga, F.; Kanematsu, Y. Coherent oscillations in ultrafast fluorescence of photoactive yellow protein. *J. Chem. Phys.* **2007**, *127*, 215102. [CrossRef]
41. Zhao, Y.; Sun, K.W.; Chen, L.P.; Gelin, M.F. The hierarchy of Davydov's Ansätze and its applications. *WIREs Comput. Mol. Sci.* **2021**, *12*, e1589. [CrossRef]
42. Zhao, Y. The hierarchy of Davydov's Ansaetze: From guesswork to numerically "exact" many-body wave functions. *J. Chem. Phys.* **2023**, *158*, 080901. [CrossRef]
43. Zheng, F.; Shen, Y.; Sun, K.; Zhao, Y. Photon-assisted Landau-Zener transitions in a periodically driven Rabi dimer coupled to a dissipative mode. *J. Chem. Phys.* **2021**, *154*, 044102. [CrossRef] [PubMed]
44. Huang, Z.; Zheng, F.; Zhang, Y.; Wei, Y.; Zhao, Y. Dissipative dynamics in a tunable Rabi dimer with periodic harmonic driving. *J. Chem. Phys.* **2019**, *150*, 184116. [CrossRef] [PubMed]
45. Sun, K. W.; Xu, Q.; Chen, L. P.; Gelin, M.; Zhao, Y. Temperature effects on singlet fission dynamics mediated by a conical intersection. *J. Chem. Phys.* **2020**, *153*, 194106. [CrossRef]
46. Sun, K.; Liu, X.; Hu, W.; Zhang, M.; Long, G.; Zhao, Y. Singlet fission dynamics and optical spectra of pentacene and its derivatives. *Phys. Chem. Chem. Phys.* **2021**, *23*, 12654-12667. [CrossRef]
47. Shen, K.; Sun, K.; Gelin, M.F.; Zhao, Y. Finite-Temperature Hole-Magnon Dynamics in an Antiferromagnet. *J. Phys. Chem. Lett.* **2024**, *15*, 447-453. [CrossRef]
48. Shen, K.; Gelin, M.F.; Sun, K.; Zhao, Y. Dynamics of a Magnetic Polaron in an Antiferromagnet. *Materials* **2024**, *17*, 469. [CrossRef]
49. Chen, L.; Zhao, Y. Finite temperature dynamics of a Holstein polaron: The thermo-field dynamics approach. *J. Chem. Phys.* **2017**, *147*, 214102. [CrossRef]

50. Borrelli, R.; Gelin, M.F. Finite temperature quantum dynamics of complex systems: Integrating thermo-field theories and tensor-train methods. *WIREs Comput. Mol. Sci.* **2021**, *11*, e1539. [CrossRef]
51. Ivanov, A.; Haug, H. Self-consistent theory of the biexciton optical nonlinearity. *Phys. Rev. B* **1993**, *48*, 1490. [CrossRef] [PubMed]
52. Katsch, F.; Selig, M.; Carmele, A.; Knorr, A. Theory of Exciton-Exciton Interactions in Monolayer Transition Metal Dichalcogenides. *Phys. Status Solidi B* **2018**, *255*, 1800185. [CrossRef]
53. Brem, S.; Selig, M.; Berghaeuser, G.; Malic, E. Exciton Relaxation Cascade in two-dimensional Transition Metal Dichalcogenides. *Sci. Rep.* **2018**, *8*, 8238. [CrossRef] [PubMed]
54. Li, Z.; Graziosi, P.; Neophytou, N. Deformation potential extraction and computationally efficient mobility calculations in silicon from first principles. *Phys. Rev. B* **2021**, *104*, 195201. [CrossRef]
55. Sun, K.; Shen, K.; Gelin, M.F.; Zhao, Y. Exciton Dynamics and Time-Resolved Fluorescence in Nanocavity-Integrated Monolayers of Transition-Metal Dichalcogenides. *J. Phys. Chem. Lett.* **2023**, *14*, 221-229. [CrossRef]
56. Jin, Z.; Li, X.; Mullen, J. T.; Kim, K. W. Intrinsic transport properties of electrons and holes in monolayer transition-metal dichalcogenides. *Phys. Rev. B Condens. Matter Mater. Phys.* **2014**, *90*, 045422. [CrossRef]
57. Deotare, P.B.; McCutcheon, M.W.; Frank, I.W.; Khan, M.; Loncar, M. High quality factor photonic crystal nanobeam cavities. *Appl. Phys. Lett.* **2009**, *94*, 121106. [CrossRef]
58. Fitzgerald, J.M.; Thompson, J.J.P.; Malic, E. Twist angle tuning of moiré exciton polaritons in van der Waals heterostructures. *Nano Lett.* **2022**, *22*, 4468. [CrossRef] [PubMed]
59. Panzarini, G.; Andreani, L.C.; Armitage, A.; Baxter, D.; Skolnick, M.S.; Astratov, V.N.; Roberts, J.S.; Kavokin, A.V.; Vladimirova, M.R.; Kaliteevski, M.A. Exciton-light coupling in single and coupled semiconductor microcavities: Polariton dispersion and polarization splitting. *Phys. Rev. B* **1999**, *59*, 5082. [CrossRef]
60. Fitzgerald, J.M.; Rosati, R.; Ferreira, B.; Shan, H.; Schneider, C.; Malic, E. Circumventing the polariton bottleneck via dark excitons in 2D semiconductors. *arXiv*, **2024**, arXiv:2401.03825.
61. Takahashi, Y.; Umezawa, H. Thermal field dynamics. *Int. J. Mod. Phys. B.* **1996**, *10*, 1755. [CrossRef]
62. Suzuki, M. Density Matrix Formalism, Double-Space and Thermo Field Dynamics in Non-Equilibrium Dissipative Systems. *Int. J. Mod. Phys. B.* **1991**, *05*, 1821. [CrossRef]
63. Suzuki, M. Thermo field dynamics of quantum spin systems. *J. Stat. Phys.* **1986**, *42*, 1047. [CrossRef]
64. Mukamel, S. *Principles of Nonlinear Optical Spectroscopy*; Oxford University Press: New York, NY, USA, 1995.
65. Kormanyos, A.; Burkard, G.; Gmitra, M.; Fabian, J.; Zolyomi, V.; Drummond, N. D.; Falko, V. k· p theory for two-dimensional transition metal dichalcogenide semiconductors. *2D Mater.* **2015**, *2*, 022001. [CrossRef]
66. Laturia, A.; Van de Put, M.L.; Vandenberghe, W.G. Dielectric properties of hexagonal boron nitride and transition metal dichalcogenides: From monolayer to bulk. *NPJ 2D Mater. Appl.* **2018**, *2*, 6. [CrossRef]
67. Chen, S.Y.; Goldstein, T.; Tong, J.; Taniguchi, T.; Watanabe, K.; Yan, J. Superior valley polarization and coherence of 2s excitons in monolayer WSe2. *Phys. Rev. Lett.* **2018**, *120*, 046402. [CrossRef]
68. Li, Z.; Wang, T.; Lu, Z.; Jin, C.; Chen, Y.; Meng, Y.; Lian, Z.; Taniguchi, T.; Watanabe, K.; Zhang, S.; Smirnov, D. Revealing the biexciton and trion-exciton complexes in BN encapsulated WSe2. *Nat. Commun.* **2018**, *9*, 3719. [CrossRef] [PubMed]
69. Koch, S.; Kira, M.; Khitrova, G.; Gibbs, H. Semiconductor excitons in new light. *Nat. Mater.* **2006**, *5*, 523-531. [CrossRef]
70. Berghäuser, G.; Malic, E. Analytical approach to excitonic properties of MoS_2. *Phys. Rev. B* **2014**, *89*, 125309. [CrossRef]
71. Wang, G.; Chernikov, A.; Glazov, M.M.; Heinz, T.F.; Marie, X.; Am, ; T.; Urbaszek, B. Colloquium: Excitons in atomically thin transition metal dichalcogenides. *Rev. Modern Phys.* **2018**, *90*, 021001. [CrossRef]
72. Wu, F.; Qu, F.; MacDonald, A.H. Exciton band structure of monolayer MoS_2. *Phys. Rev. B* **2015**, *91*, 075310. [CrossRef]
73. Qiu, D.Y.; Cao, T.; Louie, S.G. Nonanalyticity, valley quantum phases, and lightlike exciton dispersion in monolayer transition metal dichalcogenides: Theory and first-principles calculations. *Phys. Rev. Lett.* **2015**, *115*, 176801. [CrossRef]
74. Berkelbach, T.C.; Hybertsen, M.S.; Reichman, D.R. Theory of neutral and charged excitons in monolayer transition metal dichalcogenides. *Phys. Rev. B Condens. Matter Mater. Phys.* **2013**, *88*, 045318. [CrossRef]
75. Mandal, A.; Taylor, M.A.; Weight, B.M.; Koessler, E.R.; Li, X.; Huo, P. Theoretical advances in polariton chemistry and molecular cavity quantum electrodynamics. *Chem. Rev.* **2023**, *123*, 9786-9879. [CrossRef] [PubMed]
76. Zhang, L.; Gogna, R.; Burg, W.; Tutuc, E.; Deng, H. Photonic-crystal exciton-polaritons in monolayer semiconductors. *Nat. Commun.* **2018**, *9*, 713. [CrossRef] [PubMed]
77. Dufferwiel, S., Schwarz, S., Withers, F., Trichet, A.A.P., Li, F., Sich, M., Del Pozo-Zamudio, O., Clark, C., Nalitov, A., Solnyshkov, D.D., Malpuech, G. Exciton–polaritons in van der Waals heterostructures embedded in tunable microcavities. *Nat. Commun.* **2015**, *6*, 8579. [CrossRef]
78. Kravtsov, V.; Khestanova, E.; Benimetskiy, F.A.; Ivanova, T.; Samusev, A.K.; Sinev, I.S.; Pidgayko, D.; Mozharov, A.M.; Mukhin, I.S.; Lozhkin, M.S.; et al. Nonlinear polaritons in a monolayer semiconductor coupled to optical bound states in the continuum. *Light Sci. Appl.* **2020**, *9*, 56. [CrossRef] [PubMed]
79. Zhao, Y.; Knox, R. S. A Brownian Oscillator Approach to the Kennard-Stepanov Relation. *J. Phys. Chem. A* **2000**, *104*, 7751–7761. [CrossRef]
80. Gelin, M.F.; Pisliakov, A. V.; Domcke, W. Time and Frequency Gated Spontaneous Emission as a Tool for Studying Vibrational Dynamics in the Excited State. *Phys. Rev. A* **2002**, *65*, 062507. [CrossRef]

81. Domcke, W.; Yarkony, D.; Köppel, H. *Conical Intersections: Electronic Structure, Dynamics and Spectroscopy*; World Scientific: Singapore, 2004.
82. Chen, L.; Gelin, M.F.; Domcke, W.; Zhao, Y. Mapping of Wave Packet Dynamics at Conical Intersections by Time- and Frequency-Resolved Fluorescence Spectroscopy: A Computational Study. *J. Phys. Chem. Lett.* **2019**, *10*, 5873–5880. [CrossRef]
83. Gale, J.; Garcia-Vidal, F. J.; Feist, J. Cavity-induced modifications of molecular structure in the strong-coupling regime. *Phys. Rev. X* **2015**, *5*, 041022.
84. Deshmukh, R.; Marques, P.; Panda, A.; Sfeir, M. Y.; Forrest, S.R.; Menon, V. M. Modifying the spectral weights of vibronic transitions via strong coupling to surface plasmons. *ACS Photonics* **2019**, *7*, 43–48. [CrossRef]
85. Gelin, M.F.; Chen, L.; Borrelli, R.; Thyrhaug, E. Generalized Huang-Rhys factors for molecular aggregates. *Chem. Phys.* **2020**, *528*, 110495. [CrossRef]
86. Shan, H.; Iorsh, I.; Han, B.; Rupprecht, C.; Knopf, H.; Eilenberger, F.; Esmann, M.; Yumigeta, K.; Watanabe, K.; Taniguchi, T.; Klembt, S. Brightening of a dark monolayer semiconductor via strong light-matter coupling in a cavity. *Nat. Commun.* **2022**, *13*, 3001. [CrossRef]
87. Mueller, N.S.; Arul, R.; Kang, G.; Saunders, A.P.; Johnson, A.C.; Sánchez-Iglesias, A.; Hu, S.; Jakob, L.A.; Bar-David, J.; de Nijs, B.; Liz-Marzán, L.M. Photoluminescence upconversion in monolayer WSe_2 activated by plasmonic cavities through resonant excitation of dark excitons. *Nat. Commun.* **2023**, *14*, 5726. [CrossRef] [PubMed]

materials

MDPI

Article

Thermal Relaxation in Janus Transition Metal Dichalcogenide Bilayers

Aristotelis P. Sgouros [1,2,*], Fotios I. Michos [3], Michail M. Sigalas [3] and George Kalosakas [3,*]

[1] School of Chemical Engineering, National Technical University of Athens (NTUA), GR-15780 Athens, Greece
[2] Theoretical and Physical Chemistry Institute, National Hellenic Research Foundation, Vass. Constantinou 48, GR-11635 Athens, Greece
[3] Department of Materials Science, University of Patras, GR-26504 Patras, Greece; fotismixos@gmail.com (F.I.M.); sigalas@upatras.gr (M.M.S.)
* Correspondence: asgouros@eie.gr (A.P.S.); georgek@upatras.gr (G.K.)

Abstract: In this work, we employ molecular dynamics simulations with semi-empirical interatomic potentials to explore heat dissipation in Janus transition metal dichalcogenides (JTMDs). The middle atomic layer is composed of either molybdenum (Mo) or tungsten (W) atoms, and the top and bottom atomic layers consist of sulfur (S) and selenium (Se) atoms, respectively. Various nanomaterials have been investigated, including both pristine JTMDs and nanostructures incorporating inner triangular regions with a composition distinct from the outer bulk material. At the beginning of our simulations, a temperature gradient across the system is imposed by heating the central region to a high temperature while the surrounding area remains at room temperature. Once a steady state is reached, characterized by a constant energy flux, the temperature control in the central region is switched off. The heat attenuation is investigated by monitoring the characteristic relaxation time (τ_{av}) of the local temperature at the central region toward thermal equilibrium. We find that SMoSe JTMDs exhibit thermal attenuation similar to conventional TMDs (τ_{av}~10–15 ps). On the contrary, SWSe JTMDs feature relaxation times up to two times as high (τ_{av}~14–28 ps). Forming triangular lateral heterostructures in their surfaces leads to a significant slowdown in heat attenuation by up to about an order of magnitude (τ_{av}~100 ps).

Keywords: thermal transport; thermal conductivity; heterostructures; phonons; two-dimensional materials; out-of-equilibrium dynamics

1. Introduction

Two-dimensional (2D) materials with a few-atoms thickness have captured significant attention due to their unique properties arising from their planar structure and large relative surface area, giving rise to exceptional electronic, mechanical, thermal, optical, and chemical properties [1–9]. The electrical conductivity of 2D materials varies widely, encompassing insulating materials from hexagonal boron nitride (h-BN), to semiconducting transition metal dichalcogenides (TMDs) [10,11], to conducting transition metal carbides/nitrides (MXenes) [4,12].

The electronic properties of TMD materials and the corresponding band gaps (BGs) can be tuned by adjusting their thickness [13–16], by varying the stoichiometry of their alloys [17–21], by imposing strain [5,22,23], and through functionalization [24–26].

For example, monolayer TMDs with composition MX_2, where M is a transition-metal atom and X is a chalcogen atom (i.e., such as MoS_2, $MoSe_2$, WS_2, and WSe_2), exhibit a direct BG ranging between 1.1 and 2.0 eV at the K point of the Brillouin zone [22,23,27]. Increasing the thickness of TMDs, on the other hand, has been shown to yield indirect band gaps [13,14,16]. The aforementioned properties make the family of atomic thickness TMDs useful in various applications such as fast photodetection, photonic applications (including optical limiting and optical switching applications), and nanoelectromechanical devices (NEMS) such as pressure sensors and more [28–30].

Moreover, the electronic properties of these materials can be tuned by adjusting their stoichiometric composition. For example, photoluminescence characterization in conjunction with density functional theory calculations by Chen et al. [17] demonstrate that tunable band gaps can be obtained in $Mo_{1-x}W_xS_2$ monolayer alloys (where x ranges from 0 to 1) by alloying two materials with different BGs (WS_2 and MoS_2). The optical band gap of $Mo_{1-x}W_xS_2$ has also been measured by Pelaez-Fernandez et al. [21] by means of non-penetrative low-loss EELS analyses, in agreement with Chen et al. [17]. Li et al. [18], on the other hand, reported the simultaneous growth of $MoS_{2x}Se_{2(1-x)}$ triangular nanosheets with precise composition tunability. High-quality in-plane heterojunctions can be formed between monolayer $MoSe_2$ and WSe_2 [19], while band gaps varying between 1.82 eV (680 nm) to 1.64 eV (755 nm) have been found in bilayer $MoS_{2x}Se_{2(1-x)}$ alloys along single triangular nanosheets [20].

The past decade has witnessed a surge in applications exploiting the unique properties and characteristics of TMDs, such as in biomedical devices [31–35], photovoltaic measurements with p-n junctions and the controlled implantation of charge donors and acceptors in silicon crystals to form p–n junctions [36], p-n heterojunctions with MoS_2 in n-type semiconductors [37], high-performance light-emitting transistors and field-effect transistors based on MoS_2 and WS_2 [38,39], and efficient solar cells based on the excellent properties of MoS_2 [40,41].

Recently, a new class of hybrid TMDs has been discovered, the Janus transition metal dichalcogenides (JTMDs), featuring unique structures with distinct chalcogen atoms on their top and bottom layers [42–46]. Janus particles [47] constitute a class of colloidal particles that exhibit multiple distinct physical properties on their surfaces. The same concept has been transferred recently to characterize 2D materials with different chemical constitution/functionalization on their opposing sides. JTMDs are considered Janus in the sense that they break the vertical mirror symmetry [42–46] and incorporate distinct chalcogen atoms on each side.

JTMDs have been successfully synthesized via a modified chemical vapor deposition method [43,45,48], controlled sulfurization [46], and plasma stripping following thermal annealing [46]. Their asymmetric atomic structure offers unique functionalities due to the intrinsic electric field and grants them novel properties like Rashba spin–orbit coupling and piezoelectricity, making them ideal for potential applications in electronics, piezoelectrics, sensors and actuators, and beyond [44,45].

Besides electronic properties, tuning the chemical constitution of the (J)TMD nanomaterials and their hybrid heterostructures has a profound effect on the vibrational properties and other phonon-related phenomena [49–57]. Since the symmetry is broken in these structures, the thermal relaxation cannot be described sufficiently by the Peierls–Boltzmann transport equation [58,59]. Due to their relatively large electron band gaps in TMD-based materials, the thermal conductivity is primarily governed by phonon transport.

As previously demonstrated by molecular dynamics simulations of pristine TMDs, an inverse correlation exists between the thermal conductivity (TC) [52] and the width of the phonon band gap (PBG) [50]. In detail, the TC values of few-layer MoS_2, $MoSe_2$, WSe_2, and WS_2 have been estimated to be 18.0, 15.5, 14.2, and 11.6 W/mK, and the corresponding phonon band gaps are 0.5, 1.0, 0.9, and 2.4 THz^{-1}; see Figure 3 in reference [52]. The thermal conductivity of TMDs such as MoS_2 and $MoSe_2$ is the same along the armchair and zigzag directions [60]. In addition, the TC has been shown [61] to decrease significantly along the normal direction in multilayer TMDs; a similar response is expected for JTMDs as well.

Introducing defects has been shown to affect the phonon spectra upon suppressing existing phonon frequencies and forming new ones [50,62,63]. Localized vacancies in 2D TMDs, created by removing metal or chalcogen atoms, suppress phonon propagation and induce localized phonon states, significantly reducing thermal conductivity [62–64]. On the contrary, the formation of substitutional defects affects thermal conductivity depending on a complex interplay between concentration, spatial distribution (e.g., ordered patterns [52] and sharpness of the interfaces [55]), and the type of substituted atoms. For example, the phonon spectrum mismatch and the variation in thermal conductivity at periodic heterostructure media have been shown to suppress heat dissipation, resulting in a substantial decrease in TC at high densities of heterostructure patterns of up to an order of magnitude (down to ca. 2.5 W/mK) [52]. On the other hand, the rate of thermal relaxation in triangular TMD heterostructures exhibits a complicated behavior on the resolution (sharpness) of the heterostructure interface [55].

Despite existing efforts to characterize the thermal properties of JTMDs [57,65,66], the thermal behavior of hybrid JTMD heterostructures remains currently unexplored. The present work expands upon previous research on traditional TMDs [50,52,55] and deals with investigating the thermal relaxation in various bilayer JTMD heterostructures with diverse compositions regarding the type of transition metal (Mo or W) and chalcogen atom (S or Se). Both pristine and hybrid JTMD lateral heterostructures are considered here, the latter featuring triangular regions with different chemical constitution than the surrounding bulk region.

The outer part of the investigated bilayer nanosheets is kept at an ambient temperature (300 K), while the central part is thermally excited at an elevated temperature, as shown in Figure 1a. Upon removing the heat source at the central part of the material, the local temperature steadily approaches room temperature. By tracking the rate of temperature decrease, we quantify the characteristic relaxation times by fitting with the Kohlrausch–Williams–Watts stretched exponential function [67]. Similar to the cases of low-dimensional nonlinear systems [68,69], laser-excited graphene [70], and various glassy materials [71,72], we find nonexponential relaxation phenomena.

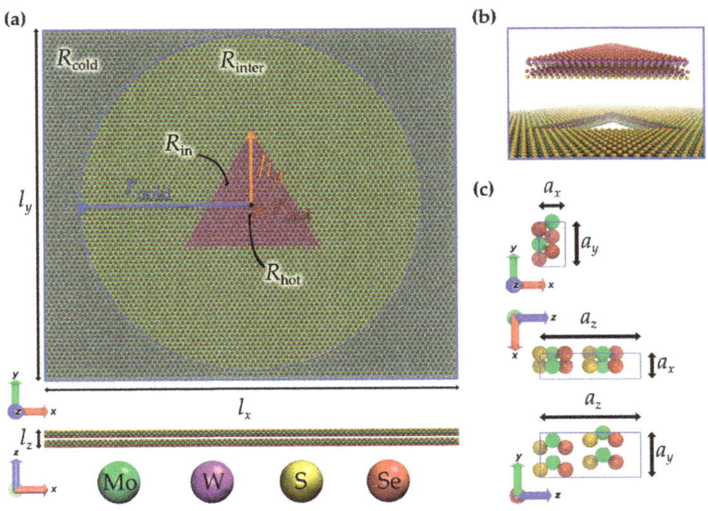

Figure 1. (**a**) Atomistic illustration of a bilayer JTMD heterostructure with $(n_x, n_y) = (80, 40)$ replications of the orthogonal unit cell along the x- and y-axes, respectively. A top view (**top**) and a side view (**bottom**) are shown, where atoms of different species are depicted by spheres of a particular color. The chemical constitution within the inner triangular region (R_{in}) with vertex-to-centroid distance $h_d = 5$ nm differs from the composition outside of this region ($R_{out} = R_{inter} + R_{cold}$). The temperature is initially fixed to T_{hot} in a circular region R_{hot} with radius r_{hot}. In the region R_{cold}, outside a circle with radius r_{cold}, the temperature is permanently fixed to T_{cold}. (**b**) Schematic representations of the atoms within regions R_{in} (**top**) and R_{out} (**bottom**). (**c**) Illustrations of the orthogonal 12-atom unit cell in the P63 mmc (2H) space group along the $-z$, y, and x directions.

Given the substantial interest in Janus TMDs, this research addresses a critical knowledge gap in their thermal relaxation properties. Notably, we quantify similarities and differences in thermal relaxation between traditional and Janus TMDs, highlighting the influence of inhomogeneous interfaces in these triangular heterostructures. Pristine JTMDs with S and Se outer and Mo middle layers (SMoSe) feature a similar response toward equilibrium to conventional transition metal dichalcogenides. On the contrary, SWSe JTMDs exhibit characteristic relaxation times about twice as large. Engineering lateral heterostructures of JTMDs delays heat dissipation considerably due to phonon spectrum mismatch between the different materials and to the variation in thermal conductivity at the interfaces [50,52,73,74], as was discussed previously. The characteristic relaxation times are of the order of tens or even hundreds of picoseconds for the heterostructures examined here depending on their chemical constitution.

2. Materials and Methods

Molecular dynamics simulations were performed with the open-source-code LAMMPS (Large-scale Atomic/Molecular Massively Parallel Simulator). The equations of motion are integrated numerically by employing the velocity-Verlet algorithm [75] with a time step of 1 femtosecond (fs). Interatomic interactions are modeled using harmonic bonds and bond-bending angles, Lennard-Jones dispersive interactions, and Coulomb electrostatic potentials using the parameters from reference [61] (set 9 parameters in Table 3 therein). Long-range interactions are calculated analytically below a cutoff distance of $r_c = 1$ nm. Beyond this distance, they are estimated with Ewald's summation method [76] using the particle–particle/particle–mesh (PPPM) solver [77] in LAMMPS [78]. The relative error in forces was 10^{-5}. According to the results presented in Ref. [50], the vibrational spectra of MoS_2, $MoSe_2$, WS_2, and WSe_2 using this force field [61] conform with density functional theory (DFT) calculations and are dominated by the corresponding atomic masses.

Figure 1a depicts an atomistic illustration of a JTMD bilayer heterostructure featuring a triangular region R_{in} with a vertex-to-centroid distance $h_d = 5$ nm having a different chemical composition from the surrounding region; see Figure 1a,b. The structure was generated by replicating the orthogonal 12-atom unit cell (Figure 1c) $n_x = 80$ and $n_y = 40$ times along the x-axis and y-axis, respectively. Notably, the individual atomic layers within the Janus-like monolayers incorporate different chalcogens at the top and bottom atomic layers.

The aim of the numerical simulations performed here is to investigate the thermal relaxation of irradiated JTMDs. To this end, we track the evolution of the local temperature at the central, initially heated region as it tends toward its equilibrium value after the removal of the radiation source.

The simulation begins by assigning initial velocities to all atoms within the simulation box. The velocities of the atoms in the R_{in} and $R_{out} = R_{inter} + R_{cold}$ regions were sampled from Boltzmann distributions corresponding to the temperatures T_{hot} and T_{cold}, respectively. Following this, the system underwent equilibration upon maintaining temperature control in a spatially dependent manner. The temperature of the region R_{hot} (inside the red circle in Figure 1a) was maintained at $T_{hot} = 370$ K. The temperature of the region R_{cold} (outside the blue circle Figure 1a) was maintained at an ambient temperature ($T_{cold} = 300$ K). The intermediate region (R_{inter}), as well as the remaining part of R_{in} apart from its central part R_{hot}, was not thermostated, allowing for natural heat transfer from R_{hot} to R_{cold}.

In particular, the specimens were simulated for extended time intervals (at least 1 ns) in a hybrid statistical ensemble with a constant number of atoms (N), atmospheric lateral pressure ($P = 1$ atm), and partial temperature control. At the beginning of this phase, the system underwent equilibration for 0.1 nanoseconds (ns) using the velocity rescale algorithm. Then, temperatures/pressures were controlled using the Nosé–Hoover thermostat/barostat, respectively [79–81]; the corresponding relaxation times were 0.1 ps/1.0 ps.

After reaching a steady state, where a constant energy flux was established, the configurations underwent additional simulation under the same ensemble. Every 300,000 steps (0.3 ns), a snapshot of the system's trajectory was captured. These snapshots represent different randomized configurations corresponding to the same steady state and were used as starting points for the temperature relaxation experiments. At least 20 such different realizations were considered in our simulations. In particular, for each one of these snapshots i (corresponding to the ith realization), the thermostating at the central region was removed, resulting in a decline of the local temperature $T_i(t)$ of region R_{hot} over time toward the equilibrium value T_{cold}.

The time-dependent local temperature was averaged over all different realizations:

$$\langle T(t)\rangle = \frac{1}{i_{max}} \sum_{i=1,i_{max}} T_i(t) \tag{1}$$

with $T_i(t)$ being the evolution of the temperature of the inner region from the ith realization and i_{max} representing the number of realizations. The thermal relaxation was quantified by fitting a reduced temperature to a Kohlrausch–Williams–Watts stretched exponential function [67]:

$$\widetilde{T}(t) = \exp\left(-[t/t_K]^{\beta}\right) \tag{2}$$

where the two fitting parameters t_K and β correspond to the characteristic time scale and the stretched exponent, respectively, while

$$\widetilde{T}(t) = \frac{\langle T(t)\rangle - T_{cold}}{T_{hot} - T_{cold}} \tag{3}$$

represents the reduced temperature. The mean relaxation time is obtained analytically through direct integration of Equation (2):

$$\tau_{av} = \int_0^{\infty} \widetilde{T}(t)dt = t_K \frac{\Gamma(1/\beta)}{\beta} \tag{4}$$

where Γ is the gamma function. The uncertainty of τ_{av} was determined based on the standard deviation of the fitting parameters t_K and β from the diagonal of the covariance matrix [82] resulting upon fitting Equation (2) and applying the error propagation theory:

$$\delta\tau_{av} = \sqrt{\left(-\frac{\Gamma(1/\beta)}{\beta^2} t_K \delta\beta - \Psi(1/\beta)\Gamma(1/\beta)\frac{1}{\beta^3} t_K \delta\beta\right)^2 + \left(\frac{\Gamma(1/\beta)}{\beta}\delta t_K\right)^2} \tag{5}$$

where $\Psi = \frac{d}{dx}\ln(\Gamma(x))$ is the digamma function [55].

The visualizations of the atomistic configurations were realized with the open-source programs VMD Molecular Graphics Viewer v1.9.4 [83] and the Advanced Image Editor Photopea [84]. The plots were derived with Veusz v3.6.2 [85].

3. Results and Discussion

3.1. Pristine and Janus Spatially Homogeneous TMDs

In this subsection, the thermal relaxation of spatially homogeneous pristine and Janus TMD materials is presented. Spatial homogeneity is established across the xy plane of the 2D nanostructures. The material composition of these TMDs plays a crucial role in dissipating thermal energy and the corresponding thermal relaxation.

Figure 2a depicts the variation of reduced temperature (calculated through Equations (1) and (3)) during the relaxation toward the equilibrium of pristine TMD materials. Figure 2b presents the corresponding variations for JTMDs using the notation XMY and XMY(s). Here, X and Y denote the chalcogen atoms at the top and bottom atomic layers of each sheet, while M represents the metal atom at the middle atomic layer. The trailing (s) in the notation indicates the relative stacking order in the bilayer; in particular, it denotes whether the individual sheets forming the bilayer exhibit mirror symmetry across their midpoint along the z-axis.

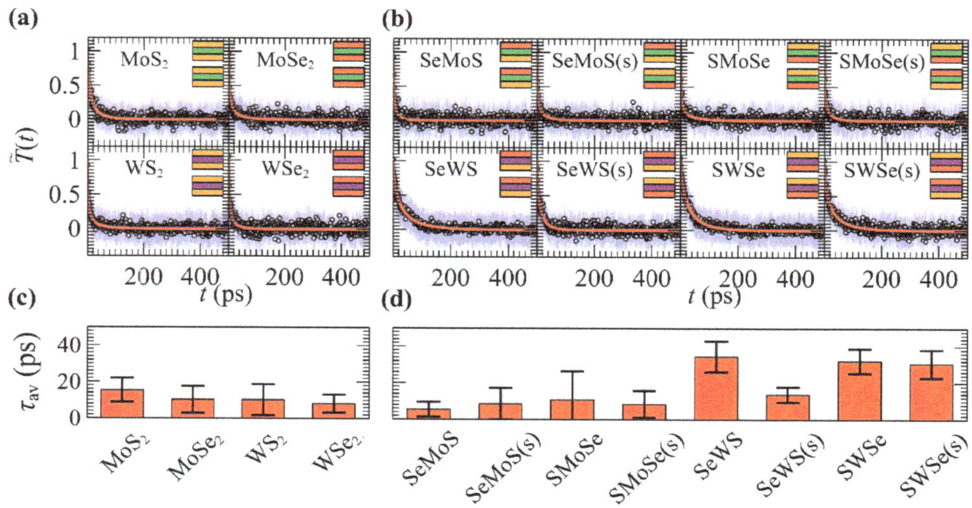

Figure 2. Evolution of the average reduced temperature (circles) through Equations (1) and (3) during the course of the relaxation simulations of (**a**) pristine TMDs and (**b**) JTMDs; blue error bars correspond to the standard deviation. Red lines indicate fittings with Equation (2). The presence of trailing "(s)" in the notation indicates mirror symmetry of the individual sheets that comprise the bilayer along the xy plane. The insets depict side-view schematics of the corresponding bilayer structures following the color code: Mo (green), W (purple), S (orange), and Se (red). (**c**,**d**) The mean relaxation times (τ_{av}) as obtained from the fitting of the reduced temperature shown in (**a**,**b**), respectively, with a stretched exponential function (Equation (2)). Error bars depict the standard deviations from Equation (5).

The mean relaxation times τ_{av} corresponding to the cases presented in (a) and (b) are shown in Figure 2c,d. τ_{av} is calculated through the second equality of Equation (4), where the parameters t_K and β are obtained by fitting the average reduced temperature with the stretched exponential of Equation (2). Error bars represent the standard deviation resulting from Equation (5).

According to Figure 2c, pristine TMDs exhibit very fast characteristic relaxation times, around 8–16 picoseconds (ps), in accordance with the findings of reference [55]. Interestingly, JTMDs containing Mo metal atoms and S/Se chalcogens exhibit a similar response to pristine TMDs, yielding relaxation times within the same range and uncertainty, e.g., compare the relaxation times of pristine MoS_2 and $MoSe_2$ in Figure 2c with the JTMDs relaxation times of the various combinations of Mo with Se/S (first four bars in Figure 2d). On the contrary, JTMDs with W metal atoms display significantly slower relaxation. For instance, heat relaxation is most delayed, (ca. 15–30 ps) in the SeWS and SWSe bilayers (Figure 2d), characterized by the Se and S elements on the outer surface and the W element on the middle layer. Note that the SeWS and SWSe configurations, as well as the SeMoS and SMoSe ones, are equivalent due to their layer arrangement (e.g., compare the first and third columns in the insets of Figure 2b), feature the same relaxation times within the uncertainty of the error bars. On the contrary, the relative orientation of the monolayers appears to affect the relaxation time in a single case, e.g., compare the SeWS and SeWs(s) structures.

3.2. *JTMDs Lateral Heterostructures*

Here, the thermal relaxation of lateral JTMD heterostructures is considered. These materials are no longer spatially homogeneous across the xy plane of the 2D nanostructure. In Figure 3, the time dependence of the average reduced temperature during the numerical simulation of JTMD heterostructures is depicted. The outer region ($R_{out} = R_{inter} + R_{cold}$;

see Figure 1a) comprises Janus SWSe in these structures. The composition of the three distinct atomic layers (the first corresponds to the upper layer, second to the middle, and third to the lower layer) in the inner region (R_{in}) of the heterostructure is represented by the side-view schematics of the bilayer in the insets of Figure 3, with Mo (green), W (purple), S (orange), and Se (red) color codes. Figure 4 presents the mean relaxation times τ_{av} corresponding to the cases shown in Figure 3, with the error bars representing the standard deviations from Equation (5).

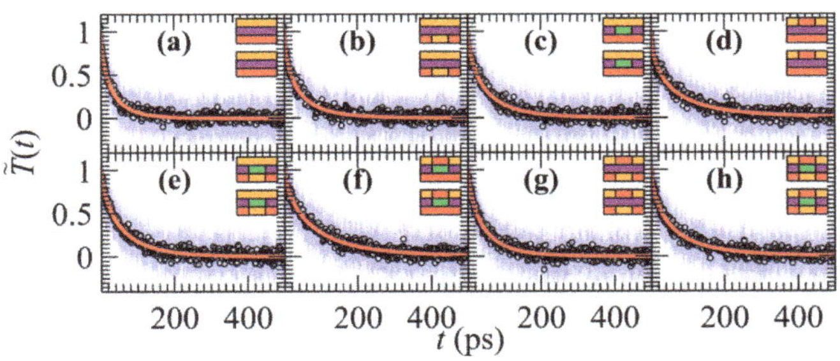

Figure 3. Evolution of the average reduced temperature (circles) through Equations (1) and (3) during the relaxation of bilayer JTMDs lateral heterostructures with pristine and Janus triangular regions; blue error bars correspond to the standard deviations. Red lines indicate fittings with Equation (2). The outer region (R_{out} = R_{inter} + R_{cold}) is composed of SWSe in all cases. The composition of the first, second, and third atomic layers in the inner region of the heterostructure (R_{in}) in panels (**a**–**h**) is illustrated by the side-view schematics of the bilayers in the insets following the color code: Mo (green), W (purple), S (orange), and Se (red).

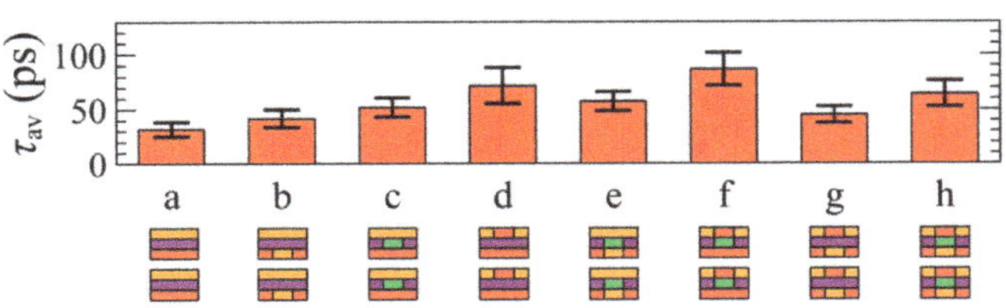

Figure 4. The mean relaxation times for cases (**a**–**h**) shown in Figure 3. Error bars depict the standard deviations from Equation (5). The composition of the first, second, and third atomic layers in the inner region of the heterostructure (R_{in}) is illustrated by the side-view schematics of the bilayers in the insets following the color code: Mo (green), W (purple), S (orange), and Se (red).

It is noted that any substitution on the upper, middle, or lower layer results in an increase in the mean relaxation time. For instance, the spatially homogeneous material without any substitution (label a in Figure 4) exhibits a relaxation time of 30–40 ps, while the next shown heterostructure (label b) has a relaxation time slightly higher (ca. 50 ps). Furthermore, the relaxation time of the heterostructure labeled f is nearly 2.5 times longer than that of the structure without any substitutions. However, performing more substitutions does not necessarily result in longer mean relaxation times. For example, the heterostructure h does not exhibit longer τ_{av}, nor does the g material feature longer relaxation times than the d lateral heterostructure.

Finally, JTMD lateral heterostructures obtained by similar substitutions in pristine transition metal dichalcogenides are considered. The obtained thermal relaxation results are shown in Figure 5. The outer region (R_{out} = R_{inter} + R_{cold}) consists of pristine WS_2 now. The composition of the first, second, and third layers in the inner region (R_{in}) is denoted by the insets based on the same color code as explained above. Figure 6 presents the mean relaxation times τ_{av} for the dynamics shown in Figure 5, with error bars indicating the standard deviations. Again, it is evident that altering atoms in any of the three atomic layers in the inner region of the heterostructure leads to a significant increase in relaxation time. As an example, the relaxation time of the pristine WS_2 (label a in Figure 6) is approximately 10 ps, whereas for the heterostructure labeled c (indicating the presence of Mo atoms in the middle layer of R_{in}), the relaxation time is around 50 ps, which is five times longer. Furthermore, the lateral heterostructures within this category labeled as g or h demonstrate the longest relaxation times, about one order of magnitude longer than that of the pristine structure.

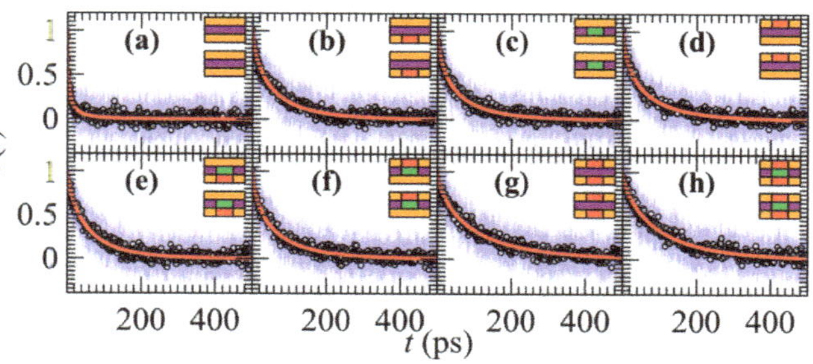

Figure 5. Evolution of the average reduced temperature (circles) through Equations (1) and (3) during the relaxation of bilayer JTMD lateral heterostructures with pristine and Janus triangular regions; blue error bars correspond to the standard deviations. Red lines indicate fittings with Equation (2). The outer region ($R_{out} = R_{inter} + R_{cold}$) is composed of WS_2 in all cases. The composition of the first, second, and third atomic layers in the inner region of the heterostructure (R_{in}) in panels (**a–h**) is illustrated by the side-view schematics of the bilayers in the insets following the color code: Mo (green), W (purple), S (orange), and Se (red).

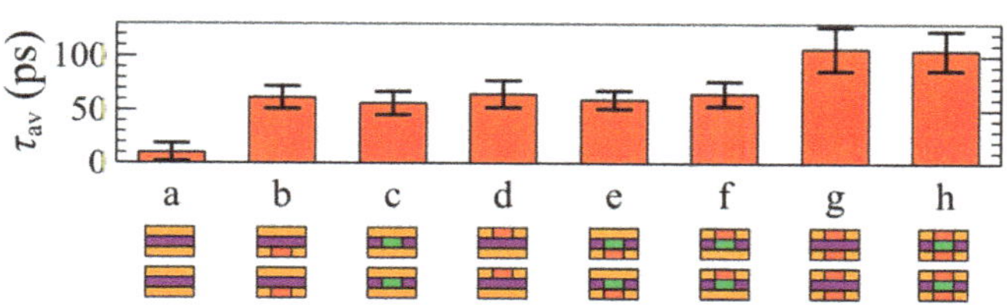

Figure 6. The mean relaxation times for cases (**a–h**) shown in Figure 5. Error bars depict the standard deviations from Equation (5). The composition of the first, second, and third atomic layers in the inner region of the heterostructure (R_{in}) is illustrated by the side-view schematics of the bilayers in the insets following the color code: Mo (green), W (purple), S (orange), and Se (red).

As our results demonstrate, the relaxation of thermal excitations is substantially influenced by the material composition of the JTMD nanostructures. Lateral heterostructures of Janus transition metal dichalcogenides can increase the thermal relaxation times by an order of magnitude as compared to that of spatially homogeneous pristine TMDs. Altering atoms in more than one layer does not necessarily imply longer relaxation times than in structures with fewer layer substitutions. In general, the values of thermal relaxation times in JTMD lateral heterostructures fall in the same regime as the corresponding ones in conventional TMD heterostructures [55].

4. Conclusions

This study investigates the in-plane thermal relaxation of various Janus transition metal dichalcogenide bilayers with triangular heterostructures using molecular dynamics simulations. Our numerical simulations explore the effect of composition (transition-metal and chalcogen atoms) on thermal behavior in both pristine JTMD and hybrid JTMD lateral heterostructures.

The samples were subjected to non-equilibrium molecular dynamics simulations within a hybrid statistical ensemble. This ensemble allows for spatially dependent temperature control, resulting in a temperature gradient across the material. A central circular region was designated as the heat source and was maintained at a higher temperature. Conversely, the outer region of the material acted as a heat sink and was held at a cooler (room) temperature. The intermediate region remained uncoupled from the thermostats, allowing for natural heat transfer from the "hot" center toward the "cooler" edges. Once a steady-state condition was achieved (characterized by a constant energy flux across the material), the thermostat controlling the central region was deactivated. This sudden removal of the heat source resulted in a gradual decrease in the local temperature at the central hot region of the JTMD bilayer. The characteristic relaxation time was determined upon fitting the time dependence of the local temperature at the center of the structure with the Kohlrausch–Williams–Watts stretched exponential function.

The findings demonstrate that engineering lateral heterostructures in JTMDs significantly delays heat dissipation compared to pristine JTMDs. This delay is attributed to the mismatch in vibrational frequencies and variations in thermal conductivity between the different materials within the heterostructure. The characteristic relaxation times for the examined heterostructures can be enhanced by up to an order of magnitude, depending on their specific chemical composition. The thermal properties of bilayer TMDs closely resemble those of their few-layer counterparts while monolayer TMDs exhibit distinct, enhanced relaxation behavior [55]; we anticipate an analogous response for JTMDs as well.

These results highlight the potential for manipulating thermal properties in JTMDs through structural engineering, paving the way for advanced applications in heat management, nanophononics, thermal devices, and optoelectronics. For instance, forming JTMD heterostructures with low thermal conductivity can enhance the efficiency of potential thermoelectric devices. This is because the figure of merit (ZT) improves when thermal conductivity decreases while maintaining electrical conductivity. Synthesizing JTMDs with tailor-made thermo-mechanical properties offers potential for applications in nuclear reactor fuel materials by enabling controlled heat distribution during the fuel cycle [86]. Controlling phonon transport along the JTMD surface holds promise for developing advanced interconnect components (resonators, waveguides, and switches) in quantum computing applications and next-gen phononic transistors [62,87].

Author Contributions: Conceptualization, G.K., A.P.S. and M.M.S.; methodology, A.P.S.; software, A.P.S.; validation, A.P.S., G.K., F.I.M. and M.M.S.; formal analysis, A.P.S.; investigation, A.P.S., G.K., M.M.S. and F.I.M.; resources, A.P.S. and G.K.; data curation, A.P.S. and F.I.M.; writing—original draft preparation, F.I.M., A.P.S., G.K. and M.M.S.; writing—review and editing, A.P.S., F.I.M., G.K. and M.M.S.; visualization, A.P.S. All authors have read and agreed to the published version of the manuscript.

Funding: This research received no external funding.

Data Availability Statement: The raw data supporting the conclusions of this article will be made available by the authors upon request.

Acknowledgments: The computational part of this study was conducted on the METROPOLIS HPC Cluster, operated by the Department of Physics at the University of Crete.

Conflicts of Interest: The authors declare no conflicts of interest.

References

1. Akinwande, D.; Brennan, C.J.; Bunch, J.S.; Egberts, P.; Felts, J.R.; Gao, H.; Huang, R.; Kim, J.-S.S.; Li, T.; Li, Y.; et al. A Review on Mechanics and Mechanical Properties of 2D Materials—Graphene and Beyond. *Extrem. Mech. Lett.* **2017**, *13*, 42–77. [CrossRef]
2. Jie, W.; Yang, Z.; Bai, G.; Hao, J. Luminescence in 2D Materials and van Der Waals Heterostructures. *Adv. Opt. Mater.* **2018**, *6*, 1701296. [CrossRef]

3. Huang, S.; Liu, Y.; Zhao, Y.; Ren, Z.; Guo, C.F. Flexible Electronics: Stretchable Electrodes and Their Future. *Adv. Funct. Mater.* **2018**, *29*, 1805924. [CrossRef]
4. Gogotsi, Y.; Anasori, B. The Rise of MXenes. *ACS Nano* **2019**, *13*, 8491–8494. [CrossRef]
5. Blundo, E.; Felici, M.; Yildirim, T.; Pettinari, G.; Tedeschi, D.; Miriametro, A.; Liu, B.; Ma, W.; Lu, Y.; Polimeni, A. Evidence of the Direct-to-Indirect Band Gap Transition in Strained Two-Dimensional WS_2, MoS_2, and WSe_2. *Phys. Rev. Res.* **2020**, *2*, 12024. [CrossRef]
6. Kalosakas, G.; Lathiotakis, N.N.; Papagelis, K. Width Dependent Elastic Properties of Graphene Nanoribbons. *Materials* **2021**, *14*, 5042. [CrossRef] [PubMed]
7. Shanmugam, V.; Mensah, R.A.; Babu, K.; Gawusu, S.; Chanda, A.; Tu, Y.; Neisiany, R.E.; Försth, M.; Sas, G.; Das, O. A Review of the Synthesis, Properties, and Applications of 2D Materials. *Part. Part. Syst. Charact.* **2022**, *39*, 2200031. [CrossRef]
8. Kalosakas, G.; Lathiotakis, N.N.; Papagelis, K. Uniaxially Strained Graphene: Structural Characteristics and G-Mode Splitting. *Materials* **2022**, *15*, 67. [CrossRef]
9. Sgouros, A.P.; Drougkas, E.; Kallivokas, S.V.; Theodorou, D.N. Buckling Kinetics of Graphene Membranes under Uniaxial Compression. *Phys. Rev. E* **2024**, *109*, 23001. [CrossRef]
10. Ko, T.-J.; Wang, M.; Yoo, C.; Okogbue, E.; Islam, M.A.; Li, H.; Shawkat, M.S.; Han, S.S.; Oh, K.H.; Jung, Y. Large-Area 2D TMD Layers for Mechanically Reconfigurable Electronic Devices. *J. Phys. D Appl. Phys.* **2020**, *53*, 313002. [CrossRef]
11. Luo, Y.K.; Xu, J.; Zhu, T.; Wu, G.; McCormick, E.J.; Zhan, W.; Neupane, M.R.; Kawakami, R.K. Opto-Valleytronic Spin Injection in Monolayer MoS_2/Few-Layer Graphene Hybrid Spin Valves. *Nano Lett.* **2017**, *17*, 3877–3883. [CrossRef]
12. Anasori, B.; Gogotsi, Y. *2D Metal Carbides and Nitrides (MXenes): Structure, Properties and Applications*, 1st ed.; Springer: Cham, Switzerland, 2019; ISBN 978-3-030-19026-2.
13. Mak, K.F.; Lee, C.; Hone, J.; Shan, J.; Heinz, T.F. Atomically Thin MoS_2: A New Direct-Gap Semiconductor. *Phys. Rev. Lett.* **2010**, *105*, 136805. [CrossRef]
14. Sun, Y.; Wang, D.; Shuai, Z. Indirect-to-Direct Band Gap Crossover in Few-Layer Transition Metal Dichalcogenides: A Theoretical Prediction. *J. Phys. Chem. C* **2016**, *120*, 21866–21870. [CrossRef]
15. Hsu, C.; Frisenda, R.; Schmidt, R.; Arora, A.; de Vasconcellos, S.M.; Bratschitsch, R.; van der Zant, H.S.J.; Castellanos-Gomez, A. Thickness-Dependent Refractive Index of 1L, 2L, and 3L MoS_2, $MoSe_2$, WS_2, and WSe_2. *Adv. Opt. Mater.* **2019**, *7*, 1900239. [CrossRef]
16. Splendiani, A.; Sun, L.; Zhang, Y.; Li, T.; Kim, J.; Chim, C.-Y.; Galli, G.; Wang, F. Emerging Photoluminescence in Monolayer MoS_2. *Nano Lett.* **2010**, *10*, 1271–1275. [CrossRef] [PubMed]
17. Chen, Y.; Xi, J.; Dumcenco, D.O.; Liu, Z.; Suenaga, K.; Wang, D.; Shuai, Z.; Huang, Y.-S.; Xie, L. Tunable Band Gap Photoluminescence from Atomically Thin Transition-Metal Dichalcogenide Alloys. *ACS Nano* **2013**, *7*, 4610–4616. [CrossRef]
18. Li, H.; Duan, X.; Wu, X.; Zhuang, X.; Zhou, H.; Zhang, Q.; Zhu, X.; Hu, W.; Ren, P.; Guo, P.; et al. Growth of Alloy $MoS_{2x}Se_{2(1-x)}$ Nanosheets with Fully Tunable Chemical Compositions and Optical Properties. *J. Am. Chem. Soc.* **2014**, *136*, 3756–3759. [CrossRef]
19. Duan, X.; Wang, C.; Fan, Z.; Hao, G.; Kou, L.; Halim, U.; Li, H.; Wu, X.; Wang, Y.; Jiang, J.; et al. Synthesis of $WS_{2x}Se_{2-2x}$ Alloy Nanosheets with Composition-Tunable Electronic Properties. *Nano Lett.* **2016**, *16*, 264–269. [CrossRef]
20. Huang, C.; Wu, S.; Sanchez, A.M.; Peters, J.J.P.; Beanland, R.; Ross, J.S.; Rivera, P.; Yao, W.; Cobden, D.H.; Xu, X. Lateral Heterojunctions within Monolayer $MoSe_2$–WSe_2 Semiconductors. *Nat. Mater.* **2014**, *13*, 1096–1101. [CrossRef]
21. Pelaez-Fernandez, M.; Lin, Y.C.; Suenaga, K.; Arenal, R. Optoelectronic Properties of Atomically Thin $Mo_xW_{(1-x)}S_2$ Nanoflakes Probed by Spatially-Resolved Monochromated Eels. *Nanomaterials* **2021**, *11*, 3218. [CrossRef]
22. Deng, S.; Li, L.; Li, M. Stability of Direct Band Gap under Mechanical Strains for Monolayer MoS_2, $MoSe_2$, WS_2 and WSe_2. *Phys. E Low-Dimens. Syst. Nanostruct.* **2018**, *101*, 44–49. [CrossRef]
23. Chang, C.-H.; Fan, X.; Lin, S.-H.; Kuo, J.-L. Orbital Analysis of Electronic Structure and Phonon Dispersion in MoS_2, $MoSe_2$, WS_2, and WSe_2 Monolayers under Strain. *Phys. Rev. B* **2013**, *88*, 195420. [CrossRef]
24. Stergiou, A.; Tagmatarchis, N. Molecular Functionalization of Two-Dimensional MoS_2 Nanosheets. *Chem.-Eur. J.* **2018**, *24*, 18246–18257. [CrossRef] [PubMed]
25. Sideri, I.K.; Arenal, R.; Tagmatarchis, N. Covalently Functionalized MoS_2 with Dithiolenes. *ACS Mater. Lett.* **2020**, *2*, 832–837. [CrossRef]
26. Vallan, L.; Canton-Vitoria, R.; Gobeze, H.B.; Jang, Y.; Arenal, R.; Benito, A.M.; Maser, W.K.; D'Souza, F.; Tagmatarchis, N. Interfacing Transition Metal Dichalcogenides with Carbon Nanodots for Managing Photoinduced Energy and Charge-Transfer Processes. *J. Am. Chem. Soc.* **2018**, *140*, 13488–13496. [CrossRef]
27. Ross, J.S.; Wu, S.; Yu, H.; Ghimire, N.J.; Jones, A.M.; Aivazian, G.; Yan, J.; Mandrus, D.G.; Xiao, D.; Yao, W.; et al. Electrical Control of Neutral and Charged Excitons in a Monolayer Semiconductor. *Nat. Commun.* **2013**, *4*, 1474. [CrossRef] [PubMed]
28. Mak, K.F.; McGill, K.L.; Park, J.; McEuen, P.L. The Valley Hall Effect in MoS_2 Transistors. *Science* **2014**, *344*, 1489–1492. [CrossRef]
29. Chang, Y.-H.; Zhang, W.; Zhu, Y.; Han, Y.; Pu, J.; Chang, J.-K.; Hsu, W.-T.; Huang, J.-K.; Hsu, C.-L.; Chiu, M.-H.; et al. Monolayer MoSe2 Grown by Chemical Vapor Deposition for Fast Photodetection. *ACS Nano* **2014**, *8*, 8582–8590. [CrossRef] [PubMed]
30. Yu, F.; Liu, Q.; Gan, X.; Hu, M.; Zhang, T.; Li, C.; Kang, F.; Terrones, M.; Lv, R. Ultrasensitive Pressure Detection of Few-Layer MoS_2. *Adv. Mater.* **2017**, *29*, 1603266. [CrossRef]
31. Zhu, C.; Zeng, Z.; Li, H.; Li, F.; Fan, C.; Zhang, H. Single-Layer MoS_2-Based Nanoprobes for Homogeneous Detection of Biomolecules. *J. Am. Chem. Soc.* **2013**, *135*, 5998–6001. [CrossRef]

32. Loan, P.T.K.; Zhang, W.; Lin, C.-T.; Wei, K.-H.; Li, L.-J.; Chen, C.-H. Graphene/MoS_2 Heterostructures for Ultrasensitive Detection of DNA Hybridisation. *Adv. Mater.* **2014**, *26*, 4838–4844. [CrossRef] [PubMed]
33. Zhang, Y.; Zheng, B.; Zhu, C.; Zhang, X.; Tan, C.; Li, H.; Chen, B.; Yang, J.; Chen, J.; Huang, Y.; et al. Single-Layer Transition Metal Dichalcogenide Nanosheet-Based Nanosensors for Rapid, Sensitive, and Multiplexed Detection of DNA. *Adv. Mater.* **2015**, *27*, 935–939. [CrossRef]
34. Cheng, L.; Yuan, C.; Shen, S.; Yi, X.; Gong, H.; Yang, K.; Liu, Z. Bottom-Up Synthesis of Metal-Ion-Doped WS_2 Nanoflakes for Cancer Theranostics. *ACS Nano* **2015**, *9*, 11090–11101. [CrossRef]
35. Wang, S.; Chen, Y.; Li, X.; Gao, W.; Zhang, L.; Liu, J.; Zheng, Y.; Chen, H.; Shi, J. Injectable 2D MoS_2-Integrated Drug Delivering Implant for Highly Efficient NIR-Triggered Synergistic Tumor Hyperthermia. *Adv. Mater.* **2015**, *27*, 7117–7122. [CrossRef] [PubMed]
36. Liang, Q.; Zhang, Q.; Zhao, X.; Liu, M.; Wee, A.T.S. Defect Engineering of Two-Dimensional Transition-Metal Dichalcogenides: Applications, Challenges, and Opportunities. *ACS Nano* **2021**, *15*, 2165–2181. [CrossRef] [PubMed]
37. Jariwala, D.; Howell, S.L.; Chen, K.-S.; Kang, J.; Sangwan, V.K.; Filippone, S.A.; Turrisi, R.; Marks, T.J.; Lauhon, L.J.; Hersam, M.C. Hybrid, Gate-Tunable, van Der Waals p–n Heterojunctions from Pentacene and MoS_2. *Nano Lett.* **2016**, *16*, 497–503. [CrossRef]
38. Ross, J.S.; Klement, P.; Jones, A.M.; Ghimire, N.J.; Yan, J.; Mandrus, D.G.; Taniguchi, T.; Watanabe, K.; Kitamura, K.; Yao, W.; et al. Electrically Tunable Excitonic Light-Emitting Diodes Based on Monolayer WSe_2 p-n Junctions. *Nat. Nanotechnol.* **2014**, *9*, 268–272. [CrossRef]
39. Yang, W.; Shang, J.; Wang, J.; Shen, X.; Cao, B.; Peimyoo, N.; Zou, C.; Chen, Y.; Wang, Y.; Cong, C.; et al. Electrically Tunable Valley-Light Emitting Diode (VLED) Based on CVD-Grown Monolayer WS_2. *Nano Lett.* **2016**, *16*, 1560–1567. [CrossRef]
40. Lin, S.; Li, X.; Wang, P.; Xu, Z.; Zhang, S.; Zhong, H.; Wu, Z.; Xu, W.; Chen, H. Interface Designed MoS_2/GaAs Heterostructure Solar Cell with Sandwich Stacked Hexagonal Boron Nitride. *Sci. Rep.* **2015**, *5*, 15103. [CrossRef]
41. Wang, Y.; Xu, N.; Li, D.; Zhu, J. Thermal Properties of Two Dimensional Layered Materials. *Adv. Funct. Mater.* **2017**, *27*, 1604134. [CrossRef]
42. Jang, C.W.; Lee, W.J.; Kim, J.K.; Park, S.M.; Kim, S.; Choi, S.-H. Growth of Two-Dimensional Janus MoSSe by a Single in Situ Process without Initial or Follow-up Treatments. *NPG Asia Mater.* **2022**, *14*, 15. [CrossRef]
43. Zhang, J.; Jia, S.; Kholmanov, I.; Dong, L.; Er, D.; Chen, W.; Guo, H.; Jin, Z.; Shenoy, V.B.; Shi, L.; et al. Janus Monolayer Transition-Metal Dichalcogenides. *ACS Nano* **2017**, *11*, 8192–8198. [CrossRef]
44. Tang, X.; Kou, L. 2D Janus Transition Metal Dichalcogenides: Properties and Applications. *Phys. Status Solidi Basic Res.* **2022**, *259*, 2100562. [CrossRef]
45. Li, R.; Cheng, Y.; Huang, W. Recent Progress of Janus 2D Transition Metal Chalcogenides: From Theory to Experiments. *Small* **2018**, *14*, 1802091. [CrossRef] [PubMed]
46. Trivedi, D.B.; Turgut, G.; Qin, Y.; Sayyad, M.Y.; Hajra, D.; Howell, M.; Liu, L.; Yang, S.; Patoary, N.H.; Li, H.; et al. Room-Temperature Synthesis of 2D Janus Crystals and Their Heterostructures. *Adv. Mater.* **2020**, *32*, 2006320. [CrossRef]
47. De Gennes, P.-G. Soft Matter (Nobel Lecture). *Angew. Chem. Int. Ed. Engl.* **1992**, *31*, 842–845. [CrossRef]
48. Lu, A.-Y.; Zhu, H.; Xiao, J.; Chuu, C.-P.; Han, Y.; Chiu, M.-H.; Cheng, C.-C.; Yang, C.-W.; Wei, K.-H.; Yang, Y.; et al. Janus Monolayers of Transition Metal Dichalcogenides. *Nat. Nanotechnol.* **2017**, *12*, 744–749. [CrossRef] [PubMed]
49. Zhang, J.; Hong, Y.; Wang, X.; Yue, Y.; Xie, D.; Jiang, J.; Xiong, Y.; Li, P. Phonon Thermal Properties of Transition-Metal Dichalcogenides MoS_2 and $MoSe_2$ Heterostructure. *J. Phys. Chem. C* **2017**, *121*, 10336–10344. [CrossRef]
50. Konstantopoulou, A.; Sgouros, A.P.; Sigalas, M.M. Computational Study of Phononic Resonators and Waveguides in Monolayer Transition Metal Dichalcogenides. *Phys. Chem. Chem. Phys.* **2017**, *19*, 8082–8090. [CrossRef]
51. Jin, C.; Ma, E.Y.; Karni, O.; Regan, E.C.; Wang, F.; Heinz, T.F. Ultrafast Dynamics in van Der Waals Heterostructures. *Nat. Nanotechnol.* **2018**, *13*, 994–1003. [CrossRef]
52. Sgouros, A.P.; Konstantopoulou, A.; Kalosakas, G.; Sigalas, M.M. Temperature Profiles and Thermal Conductivities of Nanostructured Transition Metal Dichalcogenides. *Int. J. Heat Mass Transf.* **2019**, *140*, 579–586. [CrossRef]
53. Zhang, W.; Srivastava, A.; Li, X.; Zhang, L. Chiral Phonons in the Indirect Optical Transition of a MoS_2/WS_2 Heterostructure. *Phys. Rev. B* **2020**, *102*, 174301. [CrossRef]
54. Parzefall, P.; Holler, J.; Scheuck, M.; Beer, A.; Lin, K.-Q.; Peng, B.; Monserrat, B.; Nagler, P.; Kempf, M.; Korn, T.; et al. Moiré Phonons in Twisted $MoSe_2$–WSe_2 Heterobilayers and Their Correlation with Interlayer Excitons. *2D Materials* **2021**, *8*, 35030. [CrossRef]
55. Kanistras, N.; Sgouros, A.P.; Kalosakas, G.; Sigalas, M.M. Delayed Thermal Relaxation in Lateral Heterostructures of Transition-Metal Dichalcogenides. *J. Phys. Chem. C* **2022**, *126*, 6815–6824. [CrossRef]
56. Bharadwaj, S.; Ramasubramaniam, A.; Ram-Mohan, L.R. Lateral Transition-Metal Dichalcogenide Heterostructures for High Efficiency Thermoelectric Devices. *Nanoscale* **2022**, *14*, 11750–11759. [CrossRef]
57. Miao, T.; Xiang, M.; Chen, D.; An, M.; Ma, W. Thermal Transport Characteristics of Two-Dimensional t-$PdTe_2$ and Its Janus Structures. *Int. J. Heat Mass Transf.* **2022**, *183*, 122099. [CrossRef]
58. Lindsay, L.; Hua, C.; Ruan, X.L.; Lee, S. Survey of Ab Initio Phonon Thermal Transport. *Mater. Today Phys.* **2018**, *7*, 106–120. [CrossRef]
59. Seyf, H.R.; Gordiz, K.; DeAngelis, F.; Henry, A. Using Green-Kubo Modal Analysis (GKMA) and Interface Conductance Modal Analysis (ICMA) to Study Phonon Transport with Molecular Dynamics. *J. Appl. Phys.* **2019**, *125*, 81101. [CrossRef]

60. Hong, Y.; Zhang, J.; Zeng, X.C. Thermal Conductivity of Monolayer $MoSe_2$ and MoS_2. *J. Phys. Chem. C* **2016**, *120*, 26067–26075. [CrossRef]
61. Varshney, V.; Patnaik, S.S.; Muratore, C.; Roy, A.K.; Voevodin, A.A.; Farmer, B.L. MD Simulations of Molybdenum Disulphide (MoS_2): Force-Field Parameterization and Thermal Transport Behavior. *Comput. Mater. Sci.* **2010**, *48*, 101–108. [CrossRef]
62. Sgouros, A.P.; Neupane, M.R.; Sigalas, M.M.; Aravantinos-Zafiris, N.; Lake, R.K. Nanoscale Phononic Interconnects in THz Frequencies. *Phys. Chem. Chem. Phys.* **2014**, *16*, 23355–23364. [CrossRef]
63. Cavallini, M.; Gentili, D. Atomic Vacancies in Transition Metal Dichalcogenides: Properties, Fabrication, and Limits. *Chempluschem* **2022**, *87*, e202100562. [CrossRef]
64. Ding, Z.; Pei, Q.-X.; Jiang, J.-W.; Zhang, Y.-W. Manipulating the Thermal Conductivity of Monolayer MoS_2 via Lattice Defect and Strain Engineering. *J. Phys. Chem. C* **2015**, *119*, 16358–16365. [CrossRef]
65. Liu, H.; Yang, L.; Zhao, Y.; Sun, S.; Wei, X. Exploring Optoelectronic and Thermal Characteristics of Janus MoSeTe and WSeTe Monolayers under Mechanical Strain. *Mater. Today Commun.* **2024**, *40*, 109570. [CrossRef]
66. Wang, C.; Chen, Y.-X.; Gao, G.; Xu, K.; Shao, H. Theoretical Investigations of Janus WSeTe Monolayer and Related van Der Waals Heterostructures with Promising Thermoelectric Performance. *Appl. Surf. Sci.* **2022**, *593*, 153402. [CrossRef]
67. Kohlrausch, R. Ueber Das Dellmann'sche Elektrometer. *Ann. Phys.* **1847**, *148*, 353–405. [CrossRef]
68. Tsironis, G.P.; Aubry, S. Slow Relaxation Phenomena Induced by Breathers in Nonlinear Lattices. *Phys. Rev. Lett.* **1996**, *77*, 5225–5228. [CrossRef]
69. Kalosakas, G.; Rasmussen, K.O.; Bishop, A.R. Non-Exponential Decay of Base-Pair Opening Fluctuations in DNA. *Chem. Phys. Lett.* **2006**, *432*, 291–295. [CrossRef]
70. Krylow, S.; Hernandez, F.V.; Bauerhenne, B.; Garcia, M.E. Ultrafast Structural Relaxation Dynamics of Laser-Excited Graphene: Ab Initio Molecular Dynamics Simulations Including Electron-Phonon Interactions. *Phys. Rev. B* **2020**, *101*, 205428. [CrossRef]
71. Ngai, K.L. Dynamic and Thermodynamic Properties of Glass-Forming Substances. *J. Non-Cryst. Solids* **2000**, *275*, 7–51. [CrossRef]
72. Lunkenheimer, P.; Schneider, U.; Brand, R.; Loid, A. Glassy Dynamics. *Contemp. Phys.* **2000**, *41*, 15–36. [CrossRef]
73. Zhang, Z.; Xie, Y.; Ouyang, Y.; Chen, Y. A Systematic Investigation of Thermal Conductivities of Transition Metal Dichalcogenides. *Int. J. Heat Mass Transf.* **2017**, *108*, 417–422. [CrossRef]
74. Zhang, X.; Sun, D.; Li, Y.; Lee, G.H.; Cui, X.; Chenet, D.; You, Y.; Heinz, T.F.; Hone, J.C. Measurement of Lateral and Interfacial Thermal Conductivity of Single- and Bilayer MoS_2 and $MoSe_2$ Using Refined Optothermal Raman Technique. *ACS Appl. Mater. Interfaces* **2015**, *7*, 25923–25929. [CrossRef]
75. Swope, W.C.; Andersen, H.C.; Berens, P.H.; Wilson, K.R. A Computer Simulation Method for the Calculation of Equilibrium Constants for the Formation of Physical Clusters of Molecules: Application to Small Water Clusters. *J. Chem. Phys.* **1982**, *76*, 637–649. [CrossRef]
76. In 'T Veld, P.J.; Ismail, A.E.; Grest, G.S. Application of Ewald Summations to Long-Range Dispersion Forces. *J. Chem. Phys.* **2007**, *127*, 144711. [CrossRef]
77. Hockney, R.W.; Eastwood, J.W. *Computer Simulation Using Particles*; Hilger, A., Ed.; CRC Press: Bristol, UK, 1988; ISBN 9780852743928.
78. Thompson, A.P.; Aktulga, H.M.; Berger, R.; Bolintineanu, D.S.; Brown, W.M.; Crozier, P.S.; in 't Veld, P.J.; Kohlmeyer, A.; Moore, S.G.; Nguyen, T.D.; et al. LAMMPS—A Flexible Simulation Tool for Particle-Based Materials Modeling at the Atomic, Meso, and Continuum Scales. *Comput. Phys. Commun.* **2022**, *271*, 108171. [CrossRef]
79. Nosé, S. A Unified Formulation of the Constant Temperature Molecular Dynamics Methods. *J. Chem. Phys.* **1984**, *81*, 511–519. [CrossRef]
80. Hoover, W.G. Canonical Dynamics: Equilibrium Phase-Space Distributions. *Phys. Rev. A* **1985**, *31*, 1695–1697. [CrossRef]
81. Martyna, G.J.; Tobias, D.J.; Klein, M.L. Constant Pressure Molecular Dynamics Algorithms. *J. Chem. Phys.* **1994**, *101*, 4177–4189. [CrossRef]
82. Virtanen, P.; Gommers, R.; Oliphant, T.E.; Haberland, M.; Reddy, T.; Cournapeau, D.; Burovski, E.; Peterson, P.; Weckesser, W.; Bright, J.; et al. SciPy 1.0: Fundamental Algorithms for Scientific Computing in Python. *Nat. Methods* **2020**, *17*, 261–272. [CrossRef]
83. Humphrey, W.; Dalke, A.; Schulten, K. VMD: Visual Molecular Dynamics. *J. Mol. Graph.* **1996**, *14*, 33–38. [CrossRef] [PubMed]
84. Kuckir, I. Photopea. Available online: https://www.photopea.com/ (accessed on 16 July 2024).
85. Sanders, J. Veusz-3.6.2. Available online: https://veusz.github.io/ (accessed on 16 July 2024).
86. Kuzmin, A.V.; Yurkov, M.M. Thermal Conductivity Coefficient UO_2 of Theoretical Density and Regular Stoichiometry. *MATEC Web Conf.* **2017**, *92*, 01050. [CrossRef]
87. Qiao, H.; Dumur, É.; Andersson, G.; Yan, H.; Chou, M.-H.; Grebel, J.; Conner, C.R.; Joshi, Y.J.; Miller, J.M.; Povey, R.G.; et al. Splitting Phonons: Building a Platform for Linear Mechanical Quantum Computing. *Science* **2023**, *380*, 1030–1033. [CrossRef]

materials

MDPI

Article

Enhanced Raman Scattering in CVD-Grown MoS_2/Ag Nanoparticle Hybrids

Dionysios M. Maratos [1], Antonios Michail [2,3], Alkeos Stamatelatos [4,†], Spyridon Grammatikopoulos [4,5], Dimitris Anestopoulos [3], Vassilis Tangoulis [1], Konstantinos Papagelis [3,6], John Parthenios [3,*] and Panagiotis Poulopoulos [4,*]

1 Department of Chemistry, University of Patras, 26504 Patras, Greece; up1070376@upatras.gr (D.M.M.); vtango@upatras.gr (V.T.)
2 Department of Physics, University of Patras, 26504 Patras, Greece; michailantonios@gmail.com
3 Institute of Chemical Engineering Sciences, Foundation for Research and Technology-Hellas, 26504 Patras, Greece; anestopd@gmail.com (D.A.); kpapag@physics.auth.gr (K.P.)
4 Department of Materials Science, University of Patras, 26504 Patras, Greece; spiridongramma@upatras.gr (S.G.)
5 Department of Mechanical Engineering, University of Peloponnese, M. Alexandrou 1, 26334 Patras, Greece
6 School of Physics, Department of Solid State Physics, Aristotle University of Thessaloniki, 54124 Thessaloniki, Greece
* Correspondence: jparthen@iceht.forth.gr (J.P.); poulop@upatras.gr (P.P.)
† Current address: Department of Physics, University of Warwick, Coventry CV4 7AL, UK.

Abstract: Surface-Enhanced Raman Spectroscopy (SERS) is a powerful, non-destructive technique for enhancing molecular spectra, first discovered in 1974. This study investigates the enhancement of Raman signals from single- and few-layer molybdenum disulfide (MoS_2) when interacting with silver nanoparticles. We synthesized a MoS_2 membrane primarily consisting of monolayers and bilayers through a wet chemical vapor deposition method using metal salts. The silver nanoparticles were either directly grown on the MoS_2 membrane or placed beneath it. Raman measurements revealed a significant increase in signal intensity from the MoS_2 membrane on the silver nanoparticles, attributed to localized surface plasmon resonances that facilitate SERS. Our results indicate that dichalcogenide/plasmonic systems have promising applications in the semiconductor industry.

Keywords: SERS; LSPRs; MoS_2; Ag nanoparticles

1. Introduction

Raman spectroscopy serves as an effective, non-destructive technique for material characterization by probing molecular vibrations and phonon modes within crystalline structures. However, in field testing conditions, the method's analytical capabilities are diminished due to its low sensitivity, which arises from signal attenuation and the inherently small extinction cross-sections of the analytes. The phenomenon of Surface-Enhanced Raman Spectroscopy (SERS) shows great potential as a technique to enhance signal quality and sensitivity, making it more easily applicable and potentially capable of achieving even single-molecule detection [1]. The mechanisms responsible for signal enhancement in SERS involve both chemical and electromagnetic effects, which are often challenging to distinguish from one another. The electromagnetic mechanism, which is the dominant and most well-understood in SERS, occurs when a laser beam irradiates a metallic surface with free electrons. This laser frequency resonates with the collective oscillation frequency of the conduction band electrons, enhancing the Raman signal [2]. Also, in specific areas known as "hot spots", a strong local field amplification is produced around the metallic interface due to the light entrapment, which causes an oscillating dipole to be formed around the noble metal nanoparticles (NPs), thus enhancing the intensity of the outcoming radiation [3]. The interface chemistry between the NPs and MoS_2 may also play an important role in the modification of the contact resistance, which is crucial for applications [4,5]. However, this subject goes beyond the scope of the present manuscript.

Given the affinity between substrate and analyte, electromagnetic enhancement is a tunable parameter for refining the SERS method by controlling the thin films' NP growth parameters. However, the self-assembly process used to grow the nanoparticles leads to an inhomogeneous distribution on the substrate. The lack of uniform production methods of thin films leads to randomly placed hot spots, while a partial oxidation of silver plasmonic NPs may further affect the stability of the system.

The chemical mechanism, which is less well understood, is observed primarily on the surfaces of adsorbed molecules. It primarily involves charge transfer mechanisms where the wavelength of the excitation resonates with the metal–ligand complex charge transfer electronic states. Charge transfer between the metal surface and the absorbed molecule is considered to be a resonant effect that can enhance the Raman signal by roughly 10^4. This enhancement is dependent on the application of an electrical potential. When the energy gap between the Fermi and the electronic level of the noble metal matches the energy of the laser, photon/electron transfer occurs, leading to a pronounced Raman signal [6]. The enhancement depends on the nature of the material, composition, geometry, and aggregation state, as these factors directly influence plasmon oscillation. Noble metal nanoparticles are known to exhibit localized surface plasmon resonances (LSPRs) in the visible spectrum, making them widely used in SERS applications [7]. Other factors affecting the phenomenon include the manufacturing conditions, the orientation of the adsorbate, and the surface roughness.

While a major limitation of SERS in quantitative analysis is its poor reproducibility, recent advancements in the development of relatively homogeneous substrates have addressed this issue. Another challenge during SERS is the potential interference from fluorescence, which can obscure the specific spectra of the materials being analyzed. It has been demonstrated that MoS_2, a two-dimensional chalcogenide material, can generate SERS phenomena when in nanosheet form [8]. MoS_2 is increasingly valued for its ideal properties in optoelectronic and biosensing applications [9], and is suitable for stabilizing metal nanoparticles to form composites. Noble metal NPs also possess unique properties that make them optimal for these applications, suggesting that combining MoS_2 with noble metals could enhance their capabilities. We have previously shown that that highly crystalline single- and few-layer MoS_2 can be synthesized via chemical vapor deposition using sodium molybdate (Na_2MoO_4) as a wet precursor, employing a novel, controllable, and eco-friendly method [10]. The produced MoS_2 nanosheets were transferred on a silver thin film, which were prepared via physical vapor deposition and post-annealed to form nanoparticles. The MoS_2 on Ag NP nanocomposites exhibited enhanced SERS features due to an increased quantity of the hot spots. As mentioned earlier, in the electromagnetic mechanism of SERS, the laser frequency matches the oscillation frequency of conduction band electrons, enhancing the Raman signal. In "hot spots", where strong local field amplification occurs around the metallic interface, the signal is further boosted. Our goal is to demonstrate that SERS in monolayer and bilayer MoS_2 is activated at these hot spots, which are characterized by higher strain and charge accumulation. To achieve this, we elaborate a Raman-spectroscopy-based method to measure strain and doping in MoS_2 on Ag NP nanocomposites.

The Raman spectrum of multilayer MoS_2 features two prominent Raman peaks at 383 cm^{-1} and 408 cm^{-1}, corresponding to the E_{2g} and A_{1g} vibrational modes, respectively [11]. The E_{2g} mode is a doubly degenerate in-plane vibration, and the vibrations of sulfur (S) and molybdenum (Mo) atoms are in anti-phase, whereas the A_{1g} mode is a non-degenerate out-of-plane vibration involving only vertical vibrations of the sulfur atoms relative to the layer [12]. As the number of layers decreases, the A_{1g} mode softens due to weakened interlayer coupling, whereas the E_{2g} mode stiffens because of enhanced dielectric screening [12,13]. It is a common practice to identify the number of layers by measuring the frequency difference $\Delta\omega$ between the A_{1g} and E_{2g} modes [13,14]. The reduction in symmetry changes the phonon assignment from E′ and A′ in even-layered MoS_2, to E_{2g} and A′ in odd-layered MoS_2 such as monolayer MoS_2 [12,15]. To simplify notation, we will use the monolayer assignments E′ and A′ regardless of the number of layers.

2. Materials and Methods

Synthesis of the 2D MoS_2 flakes was carried out through chemical vapor deposition (CVD), by the reaction of sodium molybdate (Sigma-Aldrich, Darmstadt, Germany. 331058) with sulfur (Sigma Aldrich n.84683). More specifically, a sodium molybdate aqueous solution was spin-coated for 120 s at 2000 rotations per minute (rpm) onto quartz and Si/SiO_2 wafers, which were previously cleaned with organic solvents and treated using an oxygen plasma procedure. The coated wafer was then placed at the center of a quartz tube positioned in a high-temperature furnace with a diameter of 1 inch. The crucible containing the sulfur powder was placed at the edge of the quartz tube in an independent zone heated by a heating mantle, outside the furnace, from the side where the purging nitrogen gas flows at a rate of 50 sccm. The temperature of the furnace zone was then increased with a steady ramp of 13 °C/min until it stabilized at 800 °C. Afterwards, the independent zone was heated to 250 °C in order for the sulfur to melt, with a rate of 60 °C, and then stabilized for 15 min. The sulfur vapors were carried through the quartz tube by the nitrogen gas flow and ended up at the fume hood. After 15 min, the furnace was shut down and the quartz tube was immediately removed for cooling at room temperature. The whole procedure was carried out with the quartz tube shut at both edges with flanges that allowed nitrogen to flow in one direction under atmospheric pressure.

Silver thin films with nominal thickness t varying between 10 and 15 nm were deposited on quartz glass substrates at room temperature (RT) or at 420 °C using direct current (DC) magnetron sputtering. The base pressure of the medium vacuum sputter coater device (custom-modified Balzers Union model SCD040, Oerlikon Balzers, Balzers, Liechtenstein) was 1.5×10^{-2} mbar. During deposition, the pressure increased to 5×10^{-2} mbar due to the insertion of sputtering gas (argon) through a fine valve. The deposition rate was kept at about 0.35 nm/s. The fabrication of the silver NPs was achieved by annealing the thin films, which were deposited at RT in a furnace with air at 460 °C. A muffle furnace was used for the annealing (model Linn elektronik VMK 22, Escenfelden, Germany). For deposition at 420 °C, silver NPs were formed directly during the deposition time due to the elevated temperature of the sample holder. Response to optical and ultraviolet light was recorded at RT by means of a Perkin-Elmer λ-35 UV-Vis spectrometer (PerkinElmer, Akron, OH, USA) working at the wavelength range of 200–1100 nm. The UV-Vis spectra were recorded in transmission geometry.

The experimental procedure involved the fabrication of three different samples in total.

Sample 1 (MoS_2 on Ag NPs): Initially, silver thin films with t = 10 nm were deposited for 30 sec on quartz substrate at RT and subsequently annealed to produce NPs. MoS_2 2D membrane was produced by CVD on a 300 nm SiO_2 and transferred onto the silver NPs lying on the quartz substrate NPs by means of the wet transfer method. In particular, a 200 nm PMMA thin film (Micro Chem PMMA 495K A3, NEWTON, MA, USA) was spin-coated on the as-grown MoS_2 and the stack was then immersed into triply deionized water (ρ = 18.2 MΩ cm), obtained through a tabletop ultrapure water production system (Synergy UV, Merck KGaA, Darmstadt, Germany). The edges of the film were lightly poked with tweezers to allow water to penetrate between the PMMA and SiO_2. Due to the hydrophilic nature of SiO_2, the MoS_2 film was separated from the SiO_2 substrate but remained attached to the PMMA film. The PMMA/MoS_2 stack was then placed onto the Ag nanoparticle-coated quartz substrate. Finally, the PMMA film was removed by exposing the sample to hot acetone vapors [16].

Sample 2 (Ag NPs on MoS_2): This sample consisted of an initial layer of MoS_2, which was produced by CVD, transferred onto a quartz substrate followed by the deposition of silver via sputtering. The MoS_2/quartz substrate was kept at 420 °C during the silver deposition for 30 sec. This process is adequate for silver NPs to be formed on substrates such as glass or quartz or MoS_2 because the surface energy of those substrates is much smaller than that of Ag [17,18].

Sample 3 (MoS_2 on annealed Ag NPs): Ag NPs were formed on a quartz substrate by sputtering with the substrate temperature at 420 °C for 45 sec. MoS_2 crystals were again grown by CVD separately on a Si/SiO_2 wafer and then transferred via the PDMS stamping method [19]. More specifically, a PDMS stamp was attached to the Si/SiO_2 wafer containing the as-grown MoS_2. The stack was again immersed in water and the stamp was slowly released from the substrate. Similar to the PMMA-based transfer used in sample 1, the triply deionized water interpenetrated the MoS membrane/SiO_2 interface, resulting in the detachment of the MoS_2 membrane from the SiO_2 substrate. Finally, the PDMS/MoS_2 stack was gently stamped onto the quartz substrate covered with the Ag NPs. PDMS was then slowly removed, releasing the MoS_2 membrane onto the Ag NPs.

Raman spectra were collected in the backscattering geometry using a Renishaw InVia 2000 spectrometer equipped with a 2400 groove per mm grating, providing a spectral resolution of nearly 2.5 cm^{-1} with spectral accuracy better than 0.1 cm^{-1}. In order to avoid unintentional heating of the sample, the excitation power was kept below 200 μW in all measurements. The Raman spectra were fitted with Lorentzian peaks using the software WiRe 4.2 by Renishaw, Gloucestershire, UK.

Surface topography images of the sample were acquired by the FSM Nanoview 2000 (Suzhou FlyingMan Precision Instruments Co., Ltd., Suzhou, China) operating in tapping mode. The cantilever used was beam-shaped silicon supplied by Budget sensors with a nominal force constant of 48 N/m and a tip radius of ~10 nm. Image processing was carried out using open-source software: WSxM version 3.1 (Nanotec Electronica, Madrid, Spain) and Gwyddion version 2.46 (CMI, Brno, Czech Republic).

3. Results

3.1. First Sample: MoS_2 Wet-Transferred on Ag NPs Sputtered and Annealed on Quartz

In Figure 1, we show the optical density (absorbance) of MoS_2 on quartz. The selection of quartz as a substrate allows for spectral recording at least up to 5 eV, as quartz is practically transparent across the entire range. To ensure precise results, we first recorded the transparency T of the films and then normalized this by dividing it by the transparency T_S of the substrate [20]. Finally, we used the formula absorbance = $-\log(T/T_S)$. The absorbance spectrum of Figure 1 shows a doublet of maxima A and B. This doublet is due to the excitonic transitions in K point of the Brillouin zone of few-layered MoS_2 [21,22]. On the other hand, the broader maximum at about 2.8 eV, consisting of the doublet C and D, is due to excitonic transitions in regions with a high density of states within Brillouin zone, e.g., P and Q points [23]. The direct band gap of MoS_2 is about 2.1 eV [24]. In Figure 1, we also show the absorbance of MoS_2 on quartz covered by silver nanoparticles (NPs). For direct comparison, we include the absorbance spectra of Ag NPs on quartz before the deposition of MoS_2. The Ag film before annealing had a thickness of 10 nm. After annealing, self-organized Ag NPs were formed on the quartz surface to minimize the surface energy of the system [25]. The spectrum of Ag NPs consists of an intense peak at about 2.6 eV and a smaller one, sometimes appearing as a doublet, at about 3.5 eV. The first peak is the main localized surface plasmon resonance (LSPR) of Ag NPs. The second is a higher-order mode. The origin of the second peak is discussed in Ref. [17]. Interestingly, the spectrum of MoS_2 combined with Ag NPs shows a superposition of the MoS_2 and Ag NP characteristics. The characteristic doublet peaks A and B of MoS_2 remain at their original positions. The MoS_2 peak at 2.8 eV overlaps with the main LSPR peak of Ag NPs, which broadens without a significant shift. In contrast, the higher-order LSPR peak at 3.5 eV of the Ag NPs nearly fades away. This indicates an interaction between MoS_2 and Ag NPs that affects the overall absorbance of the system. One would generally expect the LSPR of Ag to exhibit a notable red shift due to the increase in the refractive index as air is replaced by MoS_2 [26]. The absence of this red shift may suggest inadequate coverage of Ag by

MoS_2. This observation will be further explored in the next paragraph when discussing the Raman signal.

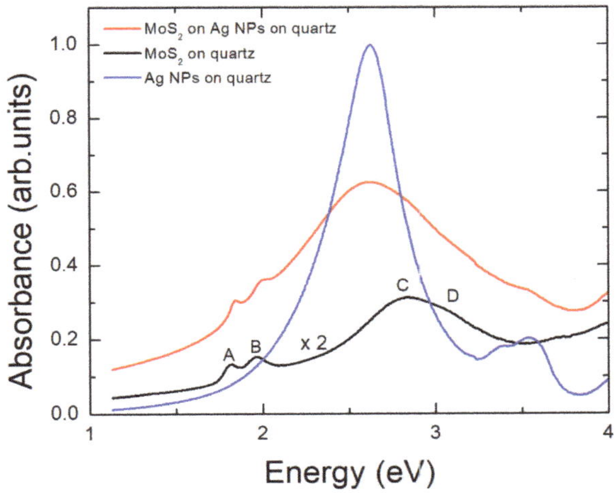

Figure 1. UV-Vis absorption spectra of the various samples on quartz substrate. The absorption spectrum of MoS_2 on quartz substrate (black line) has been multiplied by 2 for better visibility.

In Figure 2, we present a characteristic Raman spectrum of monolayer MoS_2 on bare quartz (black solid line). The spectrum features two prominent peaks, namely, the E′ mode, located around 383 cm^{-1}, which corresponds to the in-plane vibrations of sulfur atoms, and the A′, located around 407 cm^{-1}, which corresponds to the out-of-plane vibrations of molybdenum atoms. The difference $\Delta\omega$ = Pos(A′) − Pos(E′), where Pos(A′) and Pos (E′) are the peak positions of the A′ and E′ Raman bands, respectively, is commonly used to determine the thickness of the MoS_2 crystal [13]. As the layer thickness decreases to a single layer crystal, the in-plane mode shifts closer to 386 cm^{-1}, while the out-of-plane mode shifts towards 404 cm^{-1} [27]. The observed peak differences in the spectra are consistently less than 20 cm^{-1}, confirming the monolayer thickness of the analyzed regions. The total Raman signal, TRS, (TRS = I(E′) + I(A′), where I(E′), I(A′) are the peak heights of E′ and A′ Raman bands, respectively, was found to be 10 times more intense compared to areas with no signal amplification and 300 times more intense compared to the MoS_2 on bare quartz (Figure 2). Based on the above, we conducted a detailed Raman mapping, recording 1435 spectra at 2.0 μm intervals within the yellow rectangular area shown in Figure 2a.

In Figure 3a, a contour plot of $\Delta\omega$ for the Raman mapping region is presented in order to identify the crystal thickness. Blue and teal regions indicate monolayer MoS_2, with $\Delta\omega$ ranging from approximately 17.5 to 20 cm^{-1}. The green area, which covers the majority of our mapping window, corresponds to bilayer MoS_2, with $\Delta\omega$ values between 20 and 22 cm^{-1}. The yellowish and red areas represent thicker layers of MoS_2, with $\Delta\omega$ values exceeding 22 cm^{-1}. By comparing Figures 2a and 3a, points 1, 2, and 3 lie within the monolayer region.

Figure 3b presents the total Raman signal (TRS) of the probed area. As established in previous studies [28], a thicker sample generally yields a higher TRS, so one might expect a contour plot similar to that of $\Delta\omega$ (Figure 3a). However, the TRS contour plot in Figure 3b reveals lower signal intensities in regions with bilayer and trilayer MoS_2, particularly in the distinct diagonal dark area extending from the top left edge to the bottom right. Additionally, variations in TRS are also observable within the monolayer regions spanning from the top left edge to the bottom right in Figure 3b. By comparing Figures 2a and 3b, it can be observed that point 1 is located in a region where monolayer

MoS_2 is attached directly to the substrate without any Ag nanoparticles beneath it. Point 2, although situated on the monolayer with Ag nanoparticles, exhibits a low TRS signal because it is positioned within a dark TRS spot in an otherwise high TRS area (Figure 3b). Finally, point 3 is positioned on the monolayer with Ag nanoparticles and corresponds to a region with a high TRS signal, identified as a SERS hot spot.

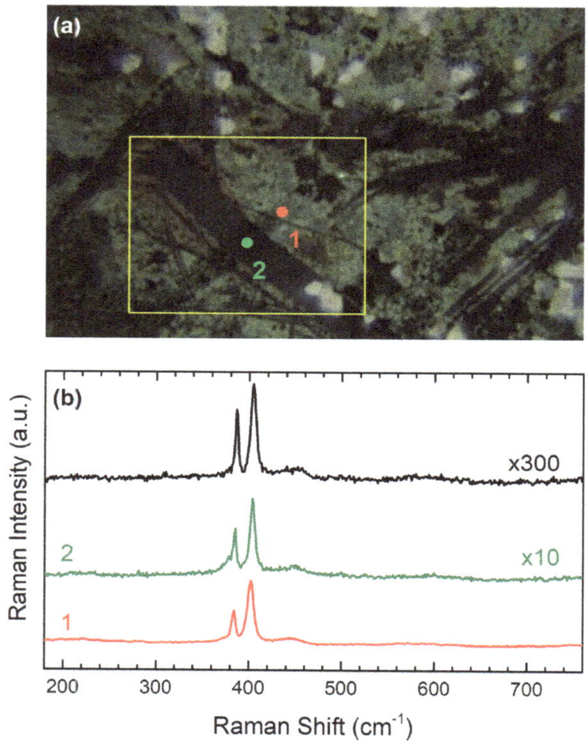

Figure 2. (**a**) Optical microscopy image of sample 1 (inset: schematic of its structure). The yellow line rectangle delineates the Raman mapping window of 80 × 68 μm². Points 1 and 2 lie within the monolayer region as shown in Figure 3b. (**b**) Raman spectra of monolayer MoS_2 recorded from the spots shown in (**a**) in comparison with monolayer MoS_2 on quartz (black line).

We attributed signal amplification (high TRS) in monolayer to the SERS effect. This effect arises from the concentration of electrons due to plasmonic phenomena occurring when the laser beam interacts with the silver nanoparticles, leading to localized amplification of the Raman signal. Using the methodology from our previous work [29], we analyzed the collected Raman data to determine the strain and doping states of single-layer MoS_2 as influenced by the interaction with the nanoparticles in the mapped region. The results are presented in Figure 3c,d, which display the strain and doping values, respectively.

By correlating the areas with TRS amplification (Figure 3b) with the strain (Figure 3c) and doping (Figure 3d) maps, we deduced that the amplified TRS on the monolayer is located on areas with higher charge density as well as strain. Similar observations were made for the bilayer areas, which cover most of the mapping window (green areas in Figure 3a). It is interesting to note that higher strain is expected in regions with higher mechanical interaction between the Ag nanoparticles and the overlaying MoS_2 sheet, as will be demonstrated in the AFM topography images (Figure 4).

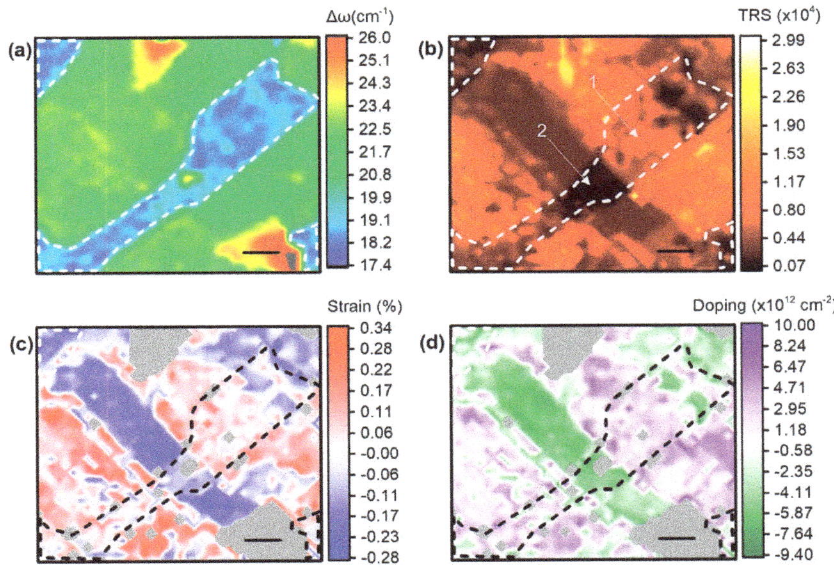

Figure 3. Heat maps of (**a**) peak difference, Δω, (**b**) total Raman signal (TRS) in arbitrary units, (**c**) strain, and (**d**) charge density in sample 1. The dashed area denotes the 1 L area of the mapping window. The numbered arrows in Figure 3b point to the spots shown in Figure 2a.

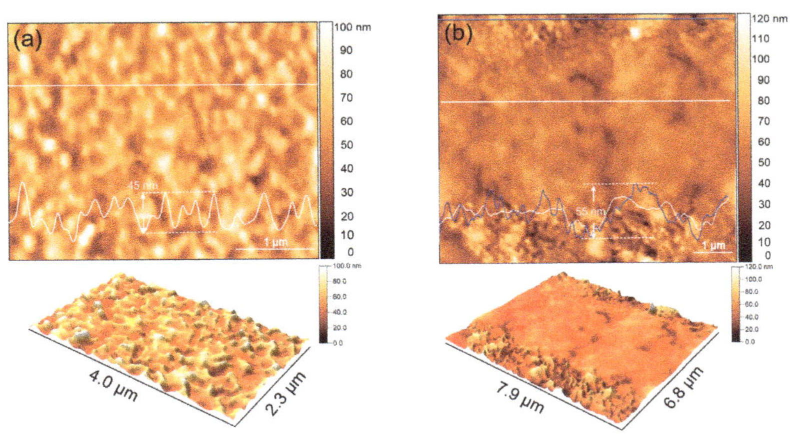

Figure 4. AFM topography images of monolayer MoS2 onto Ag nanoparticles. (**a—upper** panel) Wrinkling of the MoS2 membrane induced by Ag nanoparticles. (**a—lower** panel) The 3D image of the percolating network of wrinkles corresponding to (**a**). (**b—upper** panel) Delamination with increased membrane thickness or incomplete conformation occurs in the mid region. (**b—lower** panel) The 3D image of (**b—upper** panel) clearly showing the delaminated area. Characteristic line profiles are also included in the topography images.

Figure 4a,b depict the formation of wrinkles in monolayer MoS2 induced by the presence of Ag nanoparticles, as visualized through AFM topography images. Similar to graphene, the wrinkling morphology of the MoS2 monolayer is determined by the density and size of the Ag nanoparticles [30]. In areas with low nanoparticle density, the monolayer MoS2 adheres closely to the substrate, with detachment only occurring in small regions around the nanoparticles. As the nanoparticle density increases, wrinkles begin to form, linking adjacent nanoparticles. In the analyzed samples (1 and 3), the nanoparticle density results in a percolating network of wrinkles across the sample (Figure 4a). The conformation of the monolayer on the nanoparticles induces strain and excess charges in the MoS2 membrane [31]. However, when the MoS_2 membrane thickness is increased, or if the membrane does not fully conform to the nanoparticles due to the transfer process, global delamination from the substrate is observed (Figure 4b). This delamination leads to strain relaxation and a decrease in charge density within the membrane. The roughness of the line profiles in Figure 4a,b (upper panel) can indicate how well the membrane conforms to or delaminates from the nanoparticles. Specifically, the roughness in well-conformed areas is greater than 6 nm, while in delaminated areas, it is less than 1.5 nm. This difference is clearly shown by comparing the profile lines in the upper panel of Figure 4b.

3.2. Sample 2: Ag NPs Grown in Situ on Top of CVD-Produced MoS_2 Membrane Transferred on Quartz

In Figure 5, we show the absorbance spectra for the second sample. The spectrum of MoS_2 on quartz for this sample is similar to that of sample 1. Notably, the spectrum of MoS_2 on quartz covered by Ag nanoparticles (NPs) reveals significant features. Given that MoS_2 has a lower surface energy than quartz or glass, one would expect Ag NPs to self-organize directly on the heated substrate. The thermodynamic process governing this self-organization is detailed in Ref. [32]. The characteristic features A, B, C, and D of the MoS_2 spectrum are preserved, indicating that the short in situ annealing process does not cause extensive damage to the MoS_2 structure. It is important to note that the optical beam covers more than 1 mm^2 of the sample surface, resulting in a signal that is an average over this area. The main LSPR of Ag shows a considerable red shift. For comparison, Figure 5 includes the spectrum of sample 1, which only exhibits a broadening of the LSPR due to increased inhomogeneity, without any red shift relative to the LSPR of Ag NPs on quartz shown in Figure 1. The fact that the LSPR of sample 2 shows a prominent red shift, which is depicted in Figure 5, indicates a full adhesion of MoS_2 with the Ag NPs. The higher refractive index of MoS_2 compared to SiO_2 contributes to this red shift. We will revisit this observation in the next paragraph when discussing the Raman signal. As we will detail, the Raman signal indicates that single-layer MoS_2 areas are degraded during the growth process, while thicker layers remain intact, which is why the main features of MoS_2 in Figure 5 are preserved.

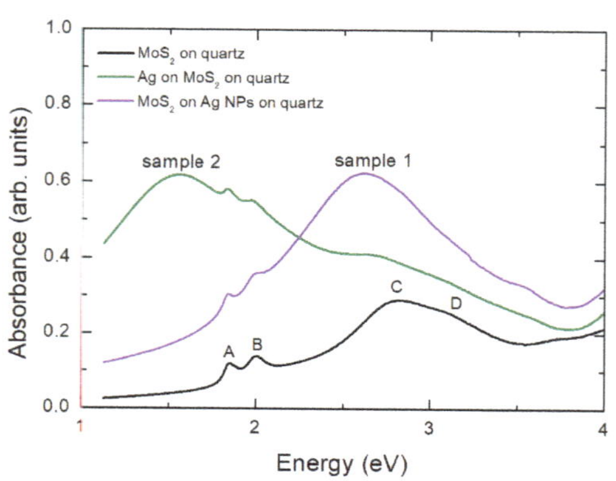

Figure 5. UV-Vis absorption spectra of various samples, as indicated, on quartz substrate.

In Figure 6 spectra from a multi-layered ($N \geq 3$), bilayer and monolayer region are compared. The regions are marked as points 1, 2, and 3 in Figure 6a, respectively, and the corresponding spectra are presented in Figure 6b. Trilayer and multilayer areas ($\Delta\omega \geq 25$ cm^{-1})

are present with a strong signal (point 1). Bilayer areas ($20 < \Delta\omega \leq 23$ cm^{-1}) are significantly degraded (point 2), while monolayer regions ($\Delta\omega \leq 20$ cm^{-1}) seem to be totally extinct (point 3). Unfortunately, a TRS comparison for SERS location could not be carried out in this case due to the absence of single-layer areas.

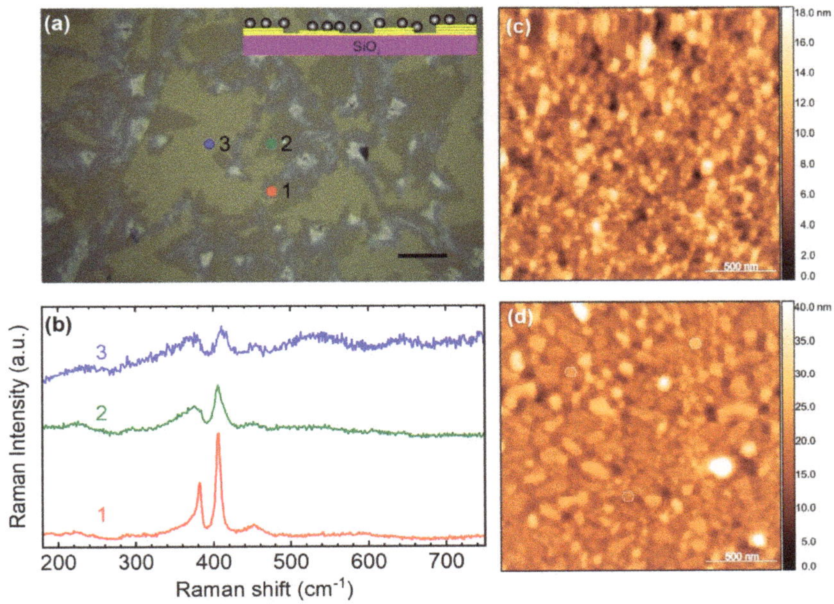

Figure 6. (**a**) Optical microscopy image of sample 2 (inset: schematic of its structure). (**b**) Representative Raman spectra from (1) multilayer, (2) bilayer, and (3) region where the single-layer MoS_2 was destroyed by heat during the annealing process. (**c**,**d**) AFM images of high-density Ag nanoparticles on MoS_2 sheet. Ag nanoparticles on MoS_2, of small lateral dimensions on thinner membrane regions like point 1 (**c**) and larger sizes, occasionally hexagonal shapes on thicker regions like point 2 (**d**).

However, it is interesting to assess the grown nanoparticles on locations with varying thicknesses of MoS_2 membrane. Figure 6c,d show AFM images of Ag nanoparticles grown on the MoS_2 membrane in sample 2. The nanoparticles exhibit high density and display noticeable variations in shape and size. In Figure 6c, the flat nanoparticles, with random shapes and small lateral dimensions, are observed on thinner regions of the MoS_2 membrane located on point 1 (Figure 6a). In contrast, Figure 6d shows nanoparticles grown on thicker regions of the membrane located on point 2 (Figure 6a), where most of them are faceted, with some adopting a hexagonal shape. These observations closely align with the findings of Deng Y. et al. [33].

3.3. *Sample 3: CVD-Grown MoS_2 Membrane Transferred onto Ag NPs Formed in Situ on Quartz*

The rectangular area in Figure 7a is the region of interest. The white aggregates on the left and right are multilayer and bulky crystals [10]. Due to low optical contrast on quartz, MoS_2 monolayers and few layers are not visible. However, the $\Delta\omega$ map in Figure 7 clearly identifies monolayer and bilayer regions along with some thicker islands. The signal on the areas covered with NPs is significantly enhanced with a median increase of 10x, as shown in Figure 7. This enhancement might be attributed to the in situ annealing process conducted in a vacuum chamber, which likely resulted in better uniformity of the nanoparticles, or it could be due to the MoS_2 transfer method using PDMS instead of PMMA, as described in the Section 2.

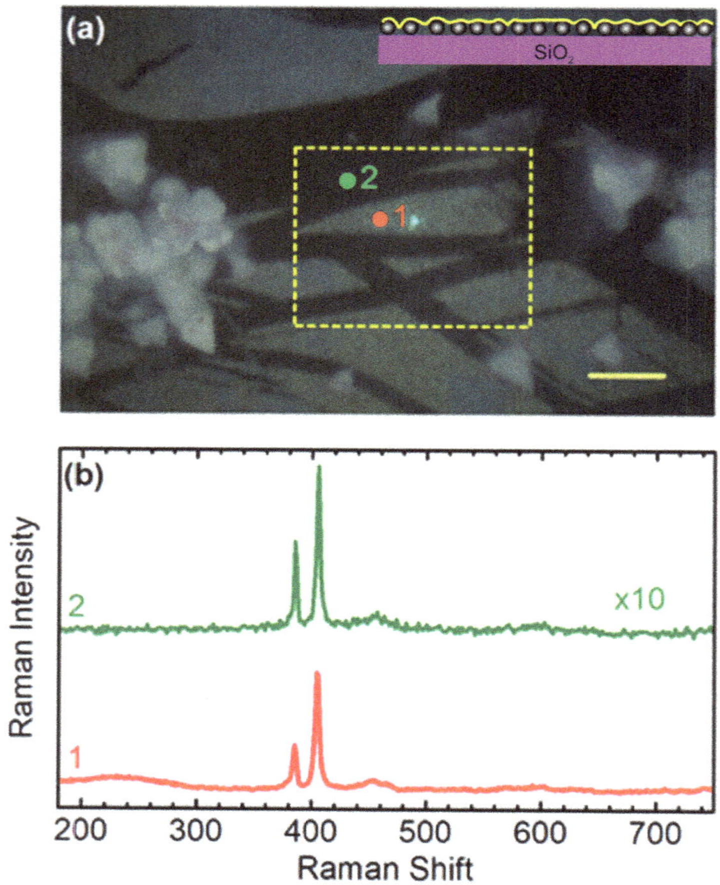

Figure 7. (**a**) Optical microscopy image of sample 3 (inset is a schematic showing its structure). The yellow dashed line rectangle delineates the Raman mapping window of 31.5 × 25.5 μm^2. (**b**) Representative Raman spectra of MoS_2 from locations on sample 3 shown in (**a**). 1 (Red) corresponds to areas with underlying NPs and 2 (green) to areas without NPs.

Similar to sample 1, we conducted a detailed Raman mapping, capturing 396 spectra at 1.5 μm intervals within the yellow dashed rectangular area shown in Figure 7a. Figure 8a presents the contour map of the $\Delta\omega$ intensity across the mapped region, illustrating the crystal thickness. The teal regions correspond to monolayers with $\Delta\omega$ values ranging from about 18.5 to 20 cm^{-1}, while the green regions, which dominate the mapping window, represent bilayer areas with $\Delta\omega$ values ranging from about 20 to 22 cm^{-1}. The TRS contour map in Figure 8b reveals a clear enhancement of the Raman signal in areas where the substrate is covered with Ag nanoparticles. The teal areas in Figure 8a, with a $\Delta\omega$ of ~19.5 cm^{-1}, exhibit strong TRS, suggesting the presence of an SERS phenomenon. The data in Figs. 8c and 8d indicate the presence of strain and doping features in the area of interest, further supporting the idea of interactions with a conformally coated MoS_2 sheet.

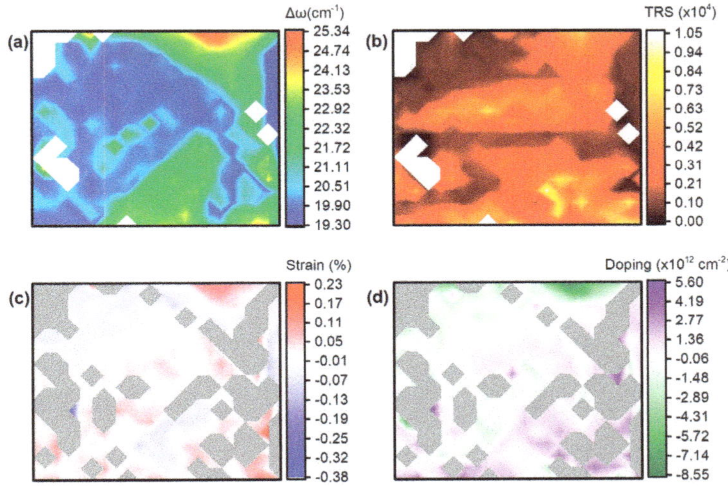

Figure 8. Contour plots of (**a**) peak difference, Δω, (**b**) total Raman signal, TRS in arbitrary units, (**c**) strain and (**d**) charge density in sample 3.

4. Conclusions

Thin MoS_2 flakes studied using Raman spectroscopy show an increase in Raman signal intensity as the crystal thickness increases. However, MoS_2 placed on top of Ag nanoparticles show a significantly enhanced signal. In certain areas, referred to as "hot spots", the Raman signal can be nearly an order of magnitude larger. This enhancement is attributed to the localized surface plasmon resonances exhibited by Ag nanoparticles in the visible spectrum, which interact with the laser radiation and the overlaying MoS_2 sheet. The greatest enhancement occurs in areas where the MoS_2 sheet covers larger nanoparticles, resulting in direct contact and notable strain due to the geometrical placement and transfer process. Thus, this work demonstrates the SERS effect facilitated by the LSPRs of noble metal nanoparticles. Our findings show that dichalcogenide/plasmonic systems hold potential for applications in the semiconductor industry.

Author Contributions: Conceptualization, J.P. and K.P.; methodology, A.M. and A.S.; software, D.M.M. and A.M.; formal analysis, A.S. and A.M.; investigation, D.A., A.S., A.M. and D.M.M.; resources, P.P., J.P. and K.P.; data curation, A.M. and D.M.M.; writing—original draft preparation, D.M.M., J.P. and P.P.; writing—review and editing, K.P., A.S. and V.T.; visualization, S.G.; supervision, J.P., P.P., K.P. and V.T.; project administration, S.G. and J.P. All authors have read and agreed to the published version of the manuscript.

Funding: This research received no external funding.

Institutional Review Board Statement: Not applicable.

Informed Consent Statement: Not applicable.

Data Availability Statement: Data are contained within the article.

Conflicts of Interest: The authors declare no conflicts of interest.

References

1. Demirel, G.; Usta, H.; Yilmaz, M.; Celik, M.; Alidagi, H.A.; Buyukserin, F. Surface-enhanced Raman spectroscopy (SERS): An adventure from plasmonic metals to organic semiconductors as SERS platforms. *J. Mater. Chem. C* **2018**, *6*, 5314–5335. [CrossRef]
2. Ding, S.-Y.; You, E.-M.; Tian, Z.-Q.; Moskovits, M. Electromagnetic theories of surface-enhanced Raman spectroscopy. *Chem. Soc. Rev.* **2017**, *46*, 4042–4076. [CrossRef] [PubMed]

3. Chulhai, D.V.; Hu, Z.; Moore, J.E.; Chen, X.; Jensen, L. Theory of linear and nonlinear surface-enhanced vibrational spectroscopies. *Annu. Rev. Phys. Chem.* **2016**, *67*, 541–564. [CrossRef] [PubMed]
4. Schauble, K.; Zakhidov, D.; Yalon, E.; Deshmukh, S.; Grady, R.W.; Cooley, K.A.; McClellan, C.J.; Vaziri, S.; Passarello, D.; Mohney, S.E. Uncovering the effects of metal contacts on monolayer MoS. *ACS Nano* **2020**, *14*, 14798–14808. [CrossRef]
5. Wang, X.; Kim, S.Y.; Wallace, R.M. Interface chemistry and band alignment study of Ni and Ag contacts on MoS_2. *ACS Appl. Mater. Interfaces* **2021**, *13*, 15802–15810. [CrossRef]
6. Aranda, D.; Valdivia, S.; Avila, F.J.; Soto, J.; Otero, J.C.; López–Tocón, I. Charge transfer at the nanoscale and the role of the out–of–plane vibrations in the selection rules of surface–enhanced Raman scattering. *Phys. Chem. Chem. Phys.* **2018**, *20*, 29430–29439. [CrossRef] [PubMed]
7. Li, W.; Zhao, X.; Yi, Z.; Glushenkov, A.M.; Kong, L. Plasmonic substrates for surface enhanced Raman scattering. *Anal. Chim. Acta* **2017**, *984*, 19–41. [CrossRef]
8. Ling, X.; Fang, W.; Lee, Y.-H.; Araujo, P.T.; Zhang, X.; Rodriguez-Nieva, J.F.; Lin, Y.; Zhang, J.; Kong, J.; Dresselhaus, M.S. Raman Enhancement Effect on Two-Dimensional Layered Materials: Graphene, H-BN and MoS_2. *Nano Lett.* **2014**, *14*, 3033–3040. [CrossRef]
9. Li, Y.; Zhang, J.; Huang, D.; Sun, H.; Fan, F.; Feng, J.; Wang, Z.; Ning, C.Z. Room-temperature continuous-wave lasing from monolayer molybdenum ditelluride integrated with a silicon nanobeam cavity. *Nat. Nanotechnol.* **2017**, *12*, 987–992. [CrossRef]
10. Michail, A.; Parthenios, J.; Anestopoulos, D.; Galiotis, C.; Christian, M.; Ortolani, L.; Morandi, V.; Papagelis, K. Controllable, eco-friendly, synthesis of highly crystalline 2D- MoS_2 and clarification of the role of growth-induced strain. *2D Mater.* **2018**, *5*, 035035. [CrossRef]
11. Verble, J.L.; Wieting, T.J. Lattice Mode Degeneracy in MoS_2 and Other Layer Compounds. *Phys. Rev. Lett.* **1970**, *25*, 362. [CrossRef]
12. Molina-Sanchez, A.; Wirtz, L. Phonons in single-layer and few-layer MoS_2 and WS_2. *Phys. Rev. B* **2011**, *84*, 155413. [CrossRef]
13. Lee, C.; Yan, H.; Brus, L.E.; Heinz, T.F.; Hone, J.; Ryu, S. Anomalous Lattice Vibrations of Single-and Few-Layer MoS2. *ACS Nano* **2010**, *4*, 2695–2700. [CrossRef] [PubMed]
14. Li, H.; Zhang, Q.; Ray, C.C. Ray Yap, Beng Kang Tay, Teo Hang Tong Edwin, Aurelien Olivier, Dominique Baillargeat. *Adv. Funct. Mater.* **2012**, *22*, 1385–1390. [CrossRef]
15. Zhang, X.; Han, W.P.; Wu, J.B.; Milana, S.; Lu, Y.; Li, Q.Q.; Ferrari, A.C.; Tan, P.H. Raman spectroscopy of shear and layer breathing modes in multilayer MoS_2. *Phys. Rev. B* **2013**, *87*, 115413. [CrossRef]
16. Sharma, M.; Singh, A.; Aggarwal, P.; Singh, R. Large-Area Transfer of 2D TMDCs Assisted by a Water-Soluble Layer for Potential Device Applications. *ACS Omega* **2022**, *7*, 11731–11741. [CrossRef]
17. Chronis, A.G.; Stamatelatos, A.; Grammatikopoulos, S.; Sigalas, M.M.; Karoutsos, V.; Maratos, D.M.; Lysandrou, S.P.; Trachylis, D.; Politis, C.; Poulopoulos, P. Microstructure and Plasmonic Behavior of Self-Assembled Silver Nanoparticles and Nanorings. *J. Appl. Phys.* **2019**, *125*, 023106. [CrossRef]
18. Tsarmpopoulou, M.; Ntemogiannis, D.; Stamatelatos, A.; Geralis, D.; Karoutsos, V.; Sigalas, M.; Poulopoulos, P.; Grammatikopoulos, S. Silver Nanoparticles' Localized Surface Plasmon Resonances Emerged at Polymeric Environments by Theory and Experiment. *Micro* **2024**, *4*, 318–333. [CrossRef]
19. Watson, A.J.; Lu, W.; Guimarães, M.H.D.; Stöhr, M. Transfer of large-scale two-dimensional semiconductors: Challenges and developments. *2d Mater.* **2021**, *8*, 032001. [CrossRef]
20. Poulopoulos, P.; Baskoutas, S.; Pappas, S.D.; Garoufalis, C.S.; Droulias, S.A.; Zamani, A.; Kapaklis, V. Intense Quantum Confinement Effects in Cu_2O Thin Films. *J. Phys. Chem. C* **2011**, *115*, 14839–14843. [CrossRef]
21. Ghayeb Zamharir, S.; Karimzadeh, R.; Aboutalebi, S.H. Laser-assisted tunable optical nonlinearity in liquid-phase exfoliated MoS_2 dispersion. *Appl. Phys. A* **2018**, *124*, 692. [CrossRef]
22. Eda, G.; Yamaguchi, H.; Voiry, D.; Fujita, T.; Chen, M.; Chhowalla, M. Photoluminescence from Chemically Exfoliated MoS_2. *Nano Lett.* **2011**, *11*, 5111–5116. [CrossRef] [PubMed]
23. Beal, A.R.; Knights, J.C.; Liang, W.Y. Transmission spectra of some transition metal dichalcogenides. II. Group VIA: Trigonal prismatic coordination. *J. Phys. Solid State Phys.* **1972**, *5*, 3540–3551. [CrossRef]
24. Ryou, J.; Kim, Y.-S.; Santosh, K.C.; Cho, K. Monolayer MoS_2 bandgap modulation by dielectric environments and tunable bandgap transistors. *Sci. Rep.* **2016**, *6*, 29184. [CrossRef] [PubMed]
25. Sousanis, A.; Poulopoulos, P.; Karoutsos, V.; Trachylis, D.; Politis, C. Giant enhancement of small photoluminescent signals on glass surfaces covered by self-assembled silver nanorings. *J. Nanosci. Nanotechnol.* **2017**, *17*, 1428–1433. [CrossRef]
26. Ntemogiannis, D.; Tsarmpopoulou, M.; Stamatelatos, A.; Grammatikopoulos, S.; Karoutsos, V.; Anyfantis, D.I.; Barnasas, A.; Alexopoulos, V.; Giantzelidis, K.; Ndoj, E.A.; et al. ZnO Matrices as a Platform for Tunable Localized Surface Plasmon Resonances of Silver Nanoparticles. *Coatings* **2024**, *14*, 69. [CrossRef]
27. Li, H.; Zhang, Q.; Yap, C.C.R.; Tay, B.K.; Edwin, T.H.T.; Olivier, A.; Baillargeat, D. From Bulk to Monolayer MoS_2: Evolution of Raman Scattering. *Adv. Funct. Mater.* **2012**, *22*, 1385–1390. [CrossRef]
28. Van Velson, N.; Zobeiri, H.; Wang, X. Thickness-Dependent Raman Scattering from Thin-Film Systems. *J. Phys. Chem. C* **2023**, *127*, 2995–3004. [CrossRef]
29. Michail, A.; Delikoukos, N.; Parthenios, J.; Galiotis, C.; Papagelis, K. Optical detection of strain and doping inhomogeneities in single layer MoS_2. *Appl. Phys. Lett.* **2016**, *108*, 173102. [CrossRef]

30. Yamamoto, M.; Pierre-Louis, O.; Huang, J.; Fuhrer, M.S.; Einstein, T.L.; Cullen, W.G. "The Princess and the Pea" at the Nanoscale: Wrinkling and Delamination of Graphene on Nanoparticles. *Phys. Rev. X* **2012**, *2*, 041018. [CrossRef]
31. Kayal, A.; Dey, S.; Harikrishnan; Nadarajan, R.; Chattopadhyay, S.; Mitra, J. Mobility Enhancement in CVD-Grown Monolayer MoS_2 Via Patterned Substrate-Induced Nonuniform Straining. *Nano Lett.* **2023**, *23*, 6629–6636. [CrossRef] [PubMed]
32. Grammatikopoulos, S.; Pappas, S.D.; Dracopoulos, V.; Poulopoulos, P.; Fumagalli, P.; Velgakis, M.J.; Politis, C. Self-assembled au nanoparticles on heated corning glass by dc magnetron sputtering: Size-dependent surface plasmon resonance tuning. *J. Nanoparticle Res.* **2013**, *15*, 1446. [CrossRef]
33. Deng, Y.; Chen, M.; Zhang, J.; Wang, Z.; Huang, W.; Zhao, Y.; Nshimiyimana, J.P.; Hu, X.; Chi, X.; Hou, G.; et al. Thickness-dependent morphologies of Ag on n-layer MoS_2 and its surface-enhanced Raman scattering. *Nano Res.* **2016**, *9*, 1682–1688. [CrossRef]

materials

MDPI

Article

A DFT Computational Study of Type-I Clathrates A_8Sn_{46-x} (A = Cs or NH_4, x = 0 or 2)

Nikolaos Kelaidis, Emmanuel Klontzas and Andreas Kaltzoglou *

Theoretical and Physical Chemistry Institute, National Hellenic Research Foundation, 11635 Athens, Greece; nkelaidis@eie.gr (N.K.); klontzas@eie.gr (E.K.)
* Correspondence: akaltzoglou@eie.gr

Abstract: Semiconducting clathrates have attracted considerable interest in the field of thermoelectric materials. We report here a computational study on the crystal structure, the enthalpy of formation, and the physical properties of the following type-I clathrates: (a) experimentally studied Cs_8Sn_{44} and hypothetical Cs_8Sn_{46} and (b) hypothetical $(NH_4)_8Sn_{46-x}$ (x = 0 or 2). The ab initio VASP calculations for the nominal stoichiometries include the geometry optimization of the initial structural models, enthalpies of formation, and the electronic and phonon density of states. Comparison of the chemical bonding of the structural models is performed via the electron localization function. The results show that the presence and distribution of defects in the Sn framework for both Cs_8Sn_{46-x} and $(NH_4)_8Sn_{46-x}$ systems significantly alters the formation energy and its electrical properties, ranging from metallic to semiconducting behavior. In particular, one defect per six-membered Sn ring in a 3D spiro-network is the thermodynamically preferred configuration that results in the Cs_8Sn_{44} and $(NH_4)_8Sn_{44}$ stoichiometries with narrow-band gap semiconducting behavior. Moreover, the rotation of the ammonium cation in the polyhedral cavities is an interesting feature that may promote the use of ammonium or other small molecular cations as guests in clathrates for thermoelectric applications; this is due to the decrease in the lattice thermal conductivity.

Keywords: clathrates; host-guest interactions; Zintl-Klemm concept; DFT calculations; phonon scattering

1. Introduction

"Semiconducting" or "intermetallic" clathrates are a distinct class of inclusion compounds. In contrast to clathrate hydrates or organic clathrates that incorporate water or organic molecules as host framework, respectively, semiconducting clathrates mainly incorporate moderately electronegative (p-block and d-block) elements as the host framework (atoms X) and highly electropositive or electronegative elements as guests (atoms A) [1,2]. Various clathrate structural types are known that combining polyhedra into a 3D network, with type-I A_8X_{46} (Figure 1) and type-II $A_{24}X_{136}$ as the most common. The covalently, four-bonded X atoms form the following polyhedra: (a) pentagonal dodecahedron (pdod, $[5^{12}]$) with 20 vertices, (b) tetrakaidecahedron (tkad, $[5^{12}6^2]$) with 24 vertices, (c) pentakaidecahedron (pkad, $[5^{12}6^3]$) with 26 vertices, and (d) hexakaidecahedron (hkad, $[5^{12}6^4]$) with 28 vertices. The symbols in brackets denote the number of pentagonal and hexagonal faces of the polyhedron. The formation of different clathrate types depends mainly on the guest size. These compounds are often non-stoichiometric on either the X or A site, and mixed occupancies between different atom types occur. Even "empty" clathrates exist as thermodynamically metastable allotropic forms of host elements, e.g., type-II Ge_{136} [3]. In general, these open framework structures show a thermodynamic tendency towards dense states. The guest atom or ion must be small enough to "fit" in the rigid covalent cage without stereochemical hindrance, but, on the other hand, should not be too small and bound very loosely or even "escape" from it. In terms of cavity size, the small pdod made of Si atoms (e.g., in type-I Ba_8Si_{46}) has a $d(Si\text{-}Si)_{max}$ of ca. 6.8 Å, whereas the large hkad made of Sn atoms (e.g., in type-II $K_{0.9}Ba_{15.1}Ga_{31}Sn_{105}$) has a d(Sn-Sn)$_{max}$ of ca. 9.1 Å. In this context, clathrate compounds do not usually exist with small cations such as A = H, Li Na, Be, and Mg but do exist with A = K, Rb, Cs, Ca, Sr, and Ba [4].

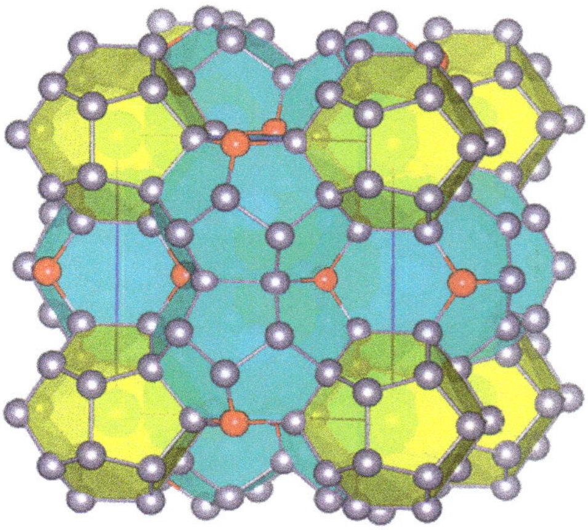

Figure 1. Crystal structure of the ideal clathrate-I Cs_8Sn_{46}. Columns of tkad (light blue) run orthogonally to one another and the voids are filled with pdod (yellow). Cs atoms reside in the center of the polyhedra. Sn atoms that bridge the pdod are in red to denote their experimentally determined partial occupancy, which leads to the actual Cs_8Sn_{44} composition. Unit cell edges are shown in blue.

With regard to the physical properties, the majority of clathrates are either intrinsic or extrinsic semiconductors and their stoichiometry follows the electron-counting rules of the Zintl–Klemm formalism [5] for polyanionic (more common) and polycationic (less common) frameworks. The underoccupancy of crystallographic sites can be described as Zintl defects [6,7]. Overall, many classes of compounds that follow the Zintl–Klemm concept have been considered as promising candidates for thermoelectric applications [8,9]. In particular, the thermoelectric efficiency of the clathrates has been proposed by G. Slack in terms of the Phonon Glass-Electron Crystal concept (PGEC) [10]. The basic hypothesis is that a semiconducting host framework has high Seebeck coefficient and electrical conductivity (like in crystals), while the extreme thermal motion—"rattling"–of the loosely-bound guest atoms efficiently scatters the heat-carrying phonons and reduces the lattice contribution of thermal conductivity (like in glasses) [11–14]. Moreover, small molecular cations, such as methylammonium and formamidinium, exhibit an off-centering from the framework cavity, which hinders phonon propagation, hence reducing the lattice thermal conductivity. This effect was first seen in halide perovskite compounds for application in state-of-the-art third-generation solar cells [15].

With regard to A_8Sn_{46-x} system, it is well established that the Sn framework contains two defects in order to fulfill the Zintl–Klemm concept according to the electron count $(Cs^+)_8(3b\text{-}Sn^-)_8(4b\text{-}Sn^0)_{36}$ (Figure 1). The tin clathrates were originally found to crystallize in the cubic space group *Pm-3n* and the partially occupied positions were distributed along a 4_2 screw axis [16]. These framework sites are part of six-membered rings of the tkad and do not belong to the pdod. A Mulliken population analysis based on topological charge stabilization showed that the 6*c* framework sites (6 out of 46 for the space group *Pm-3n*, No. 223) possess the lowest electron density and therefore it is easier to remove a Sn atom from there than from any other position [17]. Other theoretical studies for Cs_8Sn_{44} [18] indicated that one vacancy per six-membered ring is energetically favored over two vacancies per ring. This implies that the compound may exhibit an even higher ordering of the defects. Such a superstructure was experimentally established for Rb_8Sn_{44} [19], which crystallizes at room temperature with a 2 × 2 × 2 unit cell in the space group *Ia-3d* (No.

230); the partially occupied sites (3 out of 46) are distributed along a 4_1 screw axis. The same defect distribution and space group are found for the type-I clathrate Ba_8Ge_{43}, where the three Ge vacancies are fully ordered after annealing and fast cooling of the material [6]. Eventually, the conditions under which the two modifications of Rb_8Sn_{44} and Cs_8Sn_{44} exist were clarified based on an enantiotropic, order-disorder phase transition at 353 K and 363 K, respectively [20,21]. In particular, below the transition temperature, the high-order, lower-symmetry (space group *Ia-3d*) occurs, whereas above the transition temperature, the low-order, high-symmetry configuration (space group *Pm-3n*) occurs. The thermoelectric properties (electrical resistivity, thermal conductivity, and Seebeck coefficient) of both Rb_8Sn_{44} and Cs_8Sn_{44} are significantly affected by the phase transition [21–23]. Moreover, multitemperature Rietveld refinements in Rb_8Sn_{44} have shown strong dynamic disorder effect of the Rb cation, as it fades almost linearly to zero with decreasing temperature to 0 K [21].

From a practical point of view, the synthesis of the A_8Sn_{44} (A = Rb, Cs) compounds can only be performed in very inert conditions, namely in Nb or Ta crucibles and under high-purity Argon atmosphere, due to the reducing and pyrophoric nature of the alkali metals. This limits the materials preparation to small quantities (*ca.* 1 g per batch) and has a high cost. Here, we perform a computational study on the exact distribution of defects in Cs_8Sn_{44} and we also study, for the first time, the incorporation of the ammonium cation in type-I Sn-based clathrates. On one hand, the ammonium cation is very light ($M_r(NH_4^+) = 18.04$) compared to other commonly used cations, such as Cs ($A_r(Cs^+) = 132.91$). On the other hand, it can rotate in the center of the polyhedra cavities. In terms of 'effective ionic radius', the ammonium cation equals to ca. 1.5 Å [24,25], which is similar to the ionic radius of Cs at 1.7 Å. This combination of low weight and large size of the guest species may affect the phonon scattering of the system and ultimately reduce the lattice contribution to the thermal conductivity of the clathrate materials.

2. Materials and Methods

2.1. DFT Calculations

The preparation of all structural models for the DFT calculations as well as the illustration of the crystal structures and the electron localization function plots (see Section 2.3) was performed using the version 3.9 VESTA software [26]. For Cs_8Sn_{46-x}, the experimentally determined structure for Cs_8Sn_{44} at room temperature from synchrotron diffraction data [21] was used as starting point. Symmetry reduction from space group *Ia-3d* (No. 230) to space group *P*1 (No. 1) was performed to avoid any underoccupied Sn sites and the following four models were studied: (a) Cs_8Sn_{46} with no defects, where the atomic coordinates are equal to the type-I clathrate archetype symmetry (space group *Pm-3n*, No. 223), (b) Cs_8Sn_{44} model A, where the defect sites are located in neighboring six-membered Sn rings (4 Sn atoms in between) and form a non-crystallographic helical axis (similar to 2_1) across the tkad chains in one of the three crystal axes (there are only half the defect sites in the other two crystal axes), (c) Cs_8Sn_{44} model B, where the defect sites are located in five-membered rings (two Sn atoms in between) of both pdod and tkad in one of the three crystal axes, and (d) Cs_8Sn_{44} model C, where the defects are located in six-membered Sn rings of the tkad (three Sn atoms in between) without forming a helical axis as in model A, but forming a spiro-network between interconnected tkad in all three crystal axes (Figure 2).

For $(NH_4)_8Sn_{46-x}$, the same four models were used as in the case of Cs_8Sn_{46-x} with regard to the framework atoms. As for the ammonium cation, its initial coordinates were chosen to form a regular tetrahedron (as in the NH_4Cl crystal structure [27]) where the N atom replaces the Cs atom. The orientation of the ammonium cation in the pdod is such that the N–H–Sn angles of the closest neighbors are in the range 110–160° and no 5-fold rotational axis of the pdod coincides with a 3-fold rotational axis of the ammonium cation. The orientation of the ammonium cation in the tkad is such that a 3-fold rotational axis of the ammonium cation coincides with the 6-fold rotational axis of the tkad, and this results in two sets of N–H–Sn angles at 125° (for the six-membered rings) and 158° (for the five-membered rings) (Figure 3a).

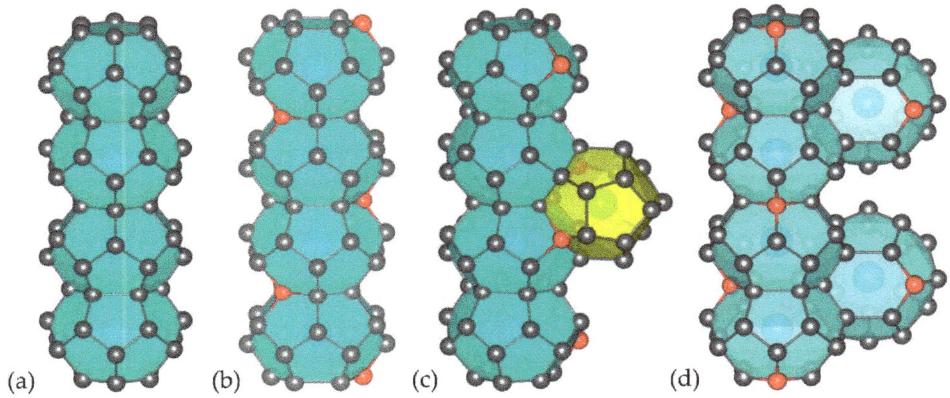

Figure 2. Initial structural models: (**a**) Cs_8Sn_{46} without framework defects, (**b**) model A, (**c**) model B, and (**d**) model C for Cs_8Sn_{44} with different distribution of the defect sites (red spheres).

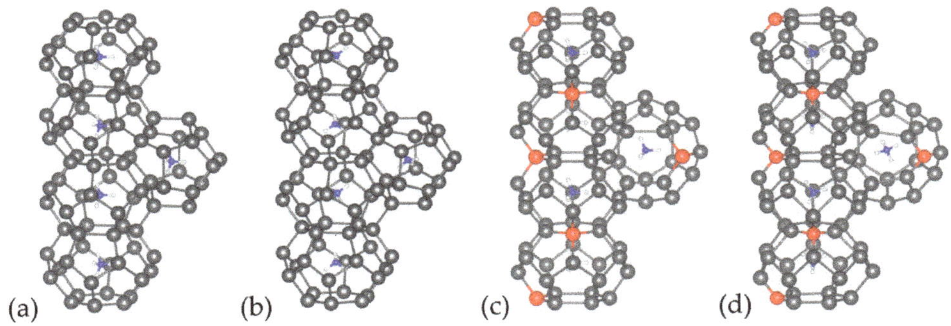

Figure 3. Structural models ammonium-containing clathrates: (**a**) before and (**b**) after relaxation for $(NH_4)_8Sn_{46}$, (**c**) before and (**d**) after relaxation for $(NH_4)_8Sn_{44}$ (model C). Red spheres indicate vacant sites within a tkad.

The DFT calculations were performed on the Vienna Ab initio Simulation Package (VASP, vs. 5.4.4.) [28–30], which employs the plane-wave pseudopotential method. We applied the Perbew–Burke–Ernzerhof functional revised for solids (PBESOL) in order to describe the exchange-correlation interactions and we utilized the soft pseudopotentials [31,32] provided by VASP. The cut-off energy was set at 520 eV and a k-point grid of 4 × 4 × 4 was applied for the sampling of the Brillouin zone after convergence tests. The self-consistent field (SCF) criterium for our calculations is set at 1.0×10^{-6} eV/atom for the energy tolerance and 0.02 eV/Å for the force tolerance. The conjugate gradient algorithm was used for the geometry optimization of the compounds and all degrees of freedom (positions of ions, cell volume, and shape) were allowed to change. The formation energy of a compound is calculated by calculating the total energies of the compounds and their constituent elements using the formula:

$$\Delta H_f = E_{compound} - \sum_i n_i \cdot \mu_i \quad (1)$$

where E_{compound} is the total energy calculated by DFT and μ_i the chemical potential of each element, calculated by DFT by calculating total energy (per atom) in its standard, most stable state. The Sn vacancy formation energy of compound A_8Sn_{44} is also calculated (per pair of Sn atoms) by employing the following formula, where error correction terms have been omitted:

$$E_{V_{Sn}} = E_{A_8Sn_{44}} - \left(E_{A_8Sn_{46}} - 2\mu\ E_{Sn}\right) \quad (2)$$

2.2. Electronic Properties Calculation Method

To gain an understanding of the electronic structure of the Cs and ammonium-containing clathrates, we calculate the band structures, density of states (DOS), and electron localization function (ELF) of the structural models under investigation. For the DOS calculations, a denser Monkhorst–Pack grid of 8 × 8 × 8 was applied. The band structure was calculated along the high symmetry points in the Brillouin zone. As for the geometry optimization, we employed the modified for solids Perdew–Burke–Ernzerhof exchange-correlation functional within the Generalized Gradient Approximation (GGA-PBEsol). Band structure and DOS plots are drawn with the SUMO software [33].

2.3. Electron Localization Function Method

The Electron Localization Function (ELF) was introduced by Becke and Edgcombe [34] in 1990 to provide additional insight into the nature of bonding between atoms and the charge distribution in the system. It is a dimensionless indicator of the localization of electrons, correlated to the probability density of a same-spin electron pair, and ranges from 0 to 1. The regions with high ELF values that are close to unity are regions of high electron localization, usually an indication of a strong covalent bond. Regions of lower ELF are regions of lower electron localization and can be found in ionic bonds. The value of 0.5 is of a free electron like delocalized gas, and lower values denote a minimal electron density concentration [35,36]. An excellent review paper on this and examples of its applications can be found in reference [37].

2.4. Phonon Calculations Method

Phonons were calculated by DFT using the finite displacement method (FDM) with the aid of the Phonopy [38,39] software. Phonopy introduces a systematic set of atomic displacements in the equilibrium structure, and DFT calculations were performed on each displaced structure to obtain the forces on the atoms. After the individual DFT calculations were completed, Phonopy was used to compute the force constants and the vibrational properties and stability of the material. For these calculations, the optimized structures were used. Due to the size of our unit cell of our models and the number of atoms, no supercell was used. The DFT conditions are kept the same with the geometry optimization; the GGA-PBEsol was applied, with a cutoff energy of 520 eV and a 4 × 4 × 4 Monkhorst-Pack k-point grid, with the exception of the energy convergence criterion, which was set to 1×10^{-8} eV.

3. Results

3.1. Structural Optimization and Calculation of Formation Enthalpies

Structural optimizations were performed for all structural models (Table 1). The selection of the PBEsol functional with respect to other functionals (LDA, PBE) was based on a better agreement with experimental results for the case of the experimentally available Cs- compounds, namely Cs_8Sn_{44}, consistent with the findings in Ref. [7], where it is also observed that PBEsol leads to more accurate vacancy formation energies in Si/Ge based clathrates. The calculated vacancy formation energies of the Cs_8Sn_{44} and $(NH_4)_8Sn_{44}$ configurations are related to the stability of Sn vacancies within the lattice. Model A and C exhibit negative vacancy formation energies (energies shown are for the formation of two V_{Sn} per cell), suggesting that these structures are energetically favorable, whereas model B shows a slightly positive vacancy formation energy, indicating a less favorable configuration. The A_8Sn_{46} models are more stable than the corresponding B models, but significantly less stable than A and C models, the latter being the most stable configuration overall.

Table 1. Chemical composition, closest distance between framework vacancies, and lattice parameters for the models of Cs_8Sn_{46-x} and $(NH_4)_8Sn_{46-x}$ as well as the calculated enthalpies of formation based on DFT calculations. For reference, the experimentally determined lattice for Cs_8Sn_{44} is $a = b = c = 12.107$ Å [21], namely half of the 2 × 2 × 2 supercell (space group *Ia-3d*, No. 230).

Compound	Closest Distance between Vacancies (Å)	a (Å)	b (Å)	c (Å)	Volume (Å^3)	Vacancy $E_{formation}$ (eV)	Compound $E_{formation}$ (eV)
Cs_8Sn_{46}	–	12.497	12.497	12.497	1951.75	–	−8.948
Cs_8Sn_{44} (model A)	8.154	12.100	12.135	12.137	1782.13	−0.411	−9.359
Cs_8Sn_{44} (model B)	6.165	12.176	12.101	12.122	1785.98	0.125	−8.832
Cs_8Sn_{44} (model C)	7.407	12.157	12.069	12.157	1783.71	−0.555	−9.503
$(NH_4)_8Sn_{46}$	–	12.384	12.406	12.406	1906.062	–	−6.689
$(NH_4)_8Sn_{44}$ (model A)	8.154	12.016	12.044	12.051	1743.993	−0.5063	−7.195
$(NH_4)_8Sn_{44}$ (model B)	6.165	12.001	11.982	12.255	1760.361	0.3164	−6.372
$(NH_4)_8Sn_{44}$ (model C)	7.407	12.083	11.957	12.075	1744.555	−0.9322	−7.621

The Sn vacancies cause the neighboring Sn atoms to move in order to reduce the strain and charge imbalances induced by the defects and minimize the total energy of the system. This is in agreement with the crystallographic [20] and the spectroscopic (Sn Moessbauer [21] and multi-temperature Raman [40]) results for A_8Sn_{44}, excluding the possibility of having two defects within a six-membered ring even for the high-temperature modification (space group *Pm-3n*). Such a case would accumulate large negative charge between the bonding Sn atoms (the remaining four of the defect hexagon), which renders it unstable (similar to our model B). The higher thermodynamic stability of model C over model A is attributed to the more isotropic distribution of the defects in the unit cell, as model A accumulates higher negative charge in only one of the crystal axes. Overall, no other possibility for a distance greater than four atoms between the defects is available for the type-I A_8M_{44} clathrates. In principle, the distance between two defects (or two atoms) within a cubic cell cannot exceed $(3^{0.5} \times a)/2 = 10.4$ Å.

With regard to the displacement of the guest atoms from the center of the polyhedra, for Cs_8Sn_{44} model C, Cs in pdod is displaced by 0.029 Å, whereas in the tkad it is displaced by 0.035 Å. This small displacement for the Cs cation is expected due to its relatively large size. For reference, a similar study on the dislocation or even migration of Li and Na guests from the center of the polyhedra in type-I clathrates showed much smaller mobility for Na cations compared to Li cations due to the smaller ionic radius of the latter [41]. In $(NH_4)_8Sn_{44}$ model C, the deviation of the N atoms from the center of the pdod is 0.02 Å and 0.17 Å from the center of the tkad. This increase in the off-centering in the tkad compared to the pdod for both compounds is attributed to the asymmetry of the tkad imposed by the presence of one defect site, which attracts the cationic guest due to its opposite charge. In contrast, the pdod remains almost regular and neutral in terms of Sn charge. With regard to the guest-host interactions in the $(NH_4)_8Sn_{46-x}$ system, the nearest Sn–H distances are at ca. 2.9 Å for $(NH_4)_8Sn_{46}$ and at ca. 2.75 Å for $(NH_4)_8Sn_{44}$ model C, which renders the interactions mainly ionic, rather than covalent (see Sections 3.2.2 and 3.3 for more details). The Sn–Sn distances range from 2.80 Å (4-bonded) to 2.99 Å (3-bonded) for $(NH_4)_8Sn_{44}$

model C. This elongation of the Sn–Sn bonds next to the defect sites is clearly visible in Figure 3d. With regard to the orientation of the ammonium group within the pdod and the tkad, significant differences are observed both before and after the structure relaxation as the tetrahedron rotates significantly. For model C after relaxation, even though there is no direct dual relationship between the regular dodecahedron and the tetrahedron (both are Platonic solids) [42], two neighboring ammonium group within a column of tkad exhibit a symmetry similar to the reflection plane along with a ca. 6 ° rotation.

3.2. Electronic Properties

3.2.1. Band-Structure and Density of States for Cs_8Sn_{46-x}

We have calculated the band structures and DOS of the optimized Cs_8Sn_{46} and C_8Sn_{44} (model C) (Figure 4). In Figure 4a, the band-structure of Cs_8Sn_{46} is shown, plotted along the k point path of L(0.5,0.5,0.5), Γ(0,0,0), X(0.5,0,0), M(0.5,0.5,0), and Γ. The metallic structure of Cs_8Sn_{46} is evident as the Fermi level (dashed line) is in the conduction band, which is in alignment with previous studies [12]. The partial density of states (Figure 4b) of Cs_8Sn_{46} illustrates the contribution of each element to the total DOS. The total DOS is dominated by the Sn atoms, while the contribution of Cs atoms is significantly smaller. Both atoms contribute with the *p* orbitals above and below the pseudo-gap, while the d orbitals of Cs are more prominent above the pseudo-gap. There are few overlapping peaks between Cs and Sn, with some small peaks just above the Fermi level and deep in the valence band around −3 eV. This indicates minimal hybridization between Cs and Sn orbitals, suggesting that the Cs-Sn bond is not strongly covalent, leaning more towards an ionic character. The Cs atoms can act as electron donors, contributing to the conductivity, which is also evident from the Cs contribution in the band structure (Figure 4a).

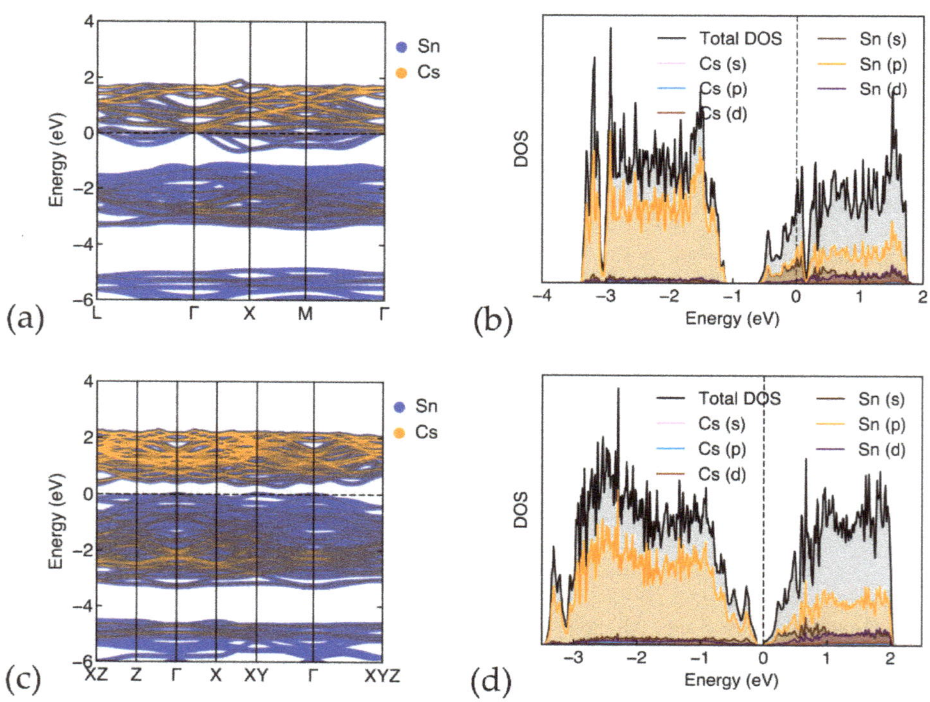

Figure 4. (**a**) Band structure and (**b**) DOS for Cs_8Sn_{46}, (**c**) Band structure and (**d**) DOS for Cs_8Sn_{44} (model C). Fermi level is set at zero.

The band structure of the defect Cs_8Sn_{44} (model C) is shown in Figure 4c. The band structure is plotted along the path of high symmetry points of the Brillouin zone XZ(0.5,0,0.5), Z(0,0,0.5), Γ(0,0,0), X(0.5,0,0), XY(0.5,0.5,0), and Γ. The band structure shows a semiconducting character for this structure, with the presence of a band gap calculated at 0.28 eV. The band structure also shows that this material is an indirect band gap, where the valence band maximum and the conduction band minimum appears at XY and between XY and Γ respectively. In this work, we will be presenting model C as it is calculated to be the most energetically stable. Model A has similar electronic properties (shape of band structure and DOS) but shows an even lower band gap. Examining the partial DOS calculations of this structure (Figure 4d), we observe similarities with the Cs_8Sn_{46} structure. Here, the main contributions arise from the Sn atoms, with Sn (p) orbitals being more prominent. Nevertheless, we observe a flatter (broader) distribution of states along the energy range than in the case of Cs_8Sn_{44}, without significant peaks.

There is an increased correlation between the Sn (p) and Cs (p and d) orbitals and there appears to be a greater hybridization in Cs_8Sn_{44} compared to Cs_8Sn_{46}. Despite this, the overall picture is not that different, with even fewer peaks present, leading to a weak or negligible covalent interaction between Cs and Sn and pointing towards an ionic character of the Cs-Sn bond. Additionally, the bands of Cs_8Sn_{44} are flatter (broader), which indicate less dispersion of electronic states and different electronic properties, such as reduced metallic character or changes in electrical conductivity, which is consistent with the fact that calculations predict this structure to be a semiconductor [43].

3.2.2. Band-Structure and Density of States for $(NH_4)_8Sn_{46-x}$

In Figure 5, the band structure and density-of-states of $(NH_4)_8Sn_{46-x}$ (x = 0,2) are shown. In Figure 5a, the calculated band structure of $(NH_4)_8Sn_{46}$ is presented. As in the case with Cs guest atoms, the band structure reveals a metallic character, with the Fermi energy level situated in the conduction band. In Figure 5b, the total and partial DOS are shown for this structure. We observe a similar picture to that of the Cs_8Sn_{46} structure, mainly in the sense the DOS is dominated by framework atoms, with the Sn (p) orbitals being the most prominent, while the guest atoms, N and H, contribute less than 2% of the total DOS. Both nitrogen s and p states show small and almost equal contributions, with a peak at the Fermi level. Hydrogen s and p states display minor contributions and are more prominent in the conduction band region. The are no significant overlapping peaks in the PDOS for Sn and N or H below the Fermi level, except for the peaks at 3 eV below the Fermi level and a peak at the Fermi level. When examining the band structure of $(NH_4)_8Sn_{44}$, we observe semiconducting behavior with a small band gap of approx. 0.18 eV and the Fermi level at the top of the valence band. Similar to the case of Cs_8Sn_{44}, this material is an indirect band gap, where the valence band maximum and the conduction band minimum appears at XY and Z, respectively. The DOS of $(NH_4)_8Sn_{44}$ are shown in Figure 5d. In comparison with the DOS of $(NH_4)_8Sn_{46}$, we observe a flatter image with less prominent peaks. As in the case of Cs_8Sn_{46-x}, we observe a non-covalent character for guest–host interactions and the transition from metal to semiconductor when x changes from 0 to 2. The similarities between the Sn based clathrates with Cs and NH_4 clathrates are encouraging for the potential thermoelectric properties of the latter. Further details on the partial DOS for both A_8Sn_{46} and $(NH_4)_8Sn_{44}$ (model C) (A = Cs, NH_4) are found in Supplementary Materials (Figures S1–S4).

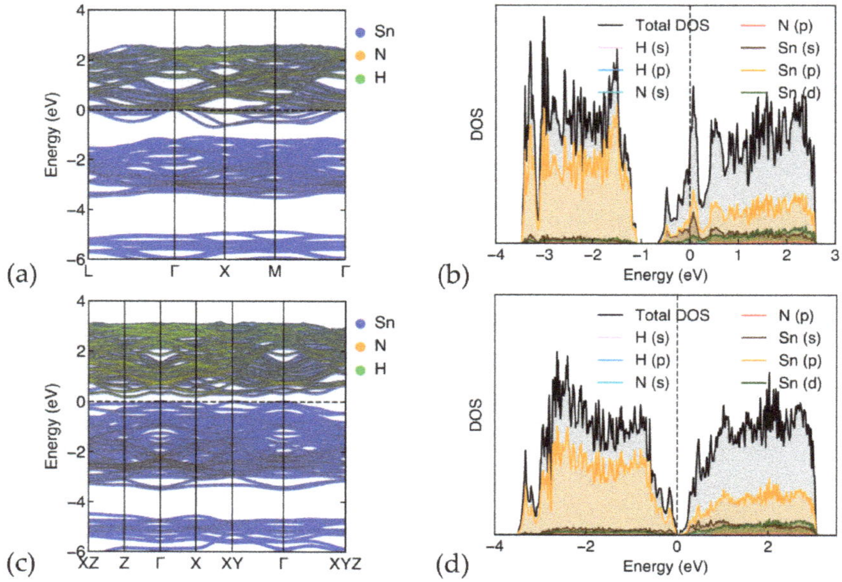

Figure 5. (**a**) Band structure and (**b**) DOS for $(NH_4)_8Sn_{46}$, (**c**) Band structure and (**d**) DOS for $(NH_4)_8Sn_{44}$ (model C). Fermi level is set at zero.

3.2.3. Electron Localization Function

In order to provide further insights regarding the chemical bonding of the examined systems, ELF analysis was conducted. The corresponding 2D and 3D ELF plots for A_8Sn_{44} (A = Cs, NH_4) are illustrated in Figures 6 and 7, respectively. We focused on the analysis of the A_8Sn_{44} structures since they are more interesting due to the Sn vacancies in their framework. The ELF is dimensionless and takes values in the range of 0 and 1 depending on the magnitude of the localization of electrons. For ELF = 1 (red), the electrons are strongly localized, while values 0.5 (green) and 0 (blue) define regions with delocalized electrons and low electron density, respectively. The ELF analysis for the Cs_8Sn_{44} clathrate is presented in Figure 6. In Figure 6a, we present a 2D ELF map along the (001) plane of model A, along with the corresponding 3D ELF map (Figure 6b).

In the case of the three Sn atoms belonging to a six-membered ring in the bottom of the 2D map, it can be noticed that there is a strong electron localization between the Sn atoms, confirming the covalent nature of the bonds between Sn atoms. Each of these Sn atoms forms four covalent bonds with their neighboring Sn atoms in the framework. Interestingly, in the case of the Sn atoms belonging to a hypothetical six-membered ring with a missing Sn atom, as depicted in the opposite site of the 2D map, there is a region with strong electron localization, which is located in the position of vacancy and can be attributed to the formation of electron pairs from the neighboring undercoordinated Sn atoms. The ELF value in the region between the Cs guest atoms and the Sn atoms of the clathrate framework is very low, confirming the ionic nature of the interaction between the guest atoms and the host material. We have also plotted the corresponding 2D and 3D ELF maps for the model C of Cs_8Sn_{44} (Figure 6c,d). In contrast to the 2D ELF map of model A, it can be observed that identical ELF values were found for the region between the Sn atoms forming the half of six-membered rings, since the (001) plane of the model C contain completed six-membered rings without Sn vacancies.

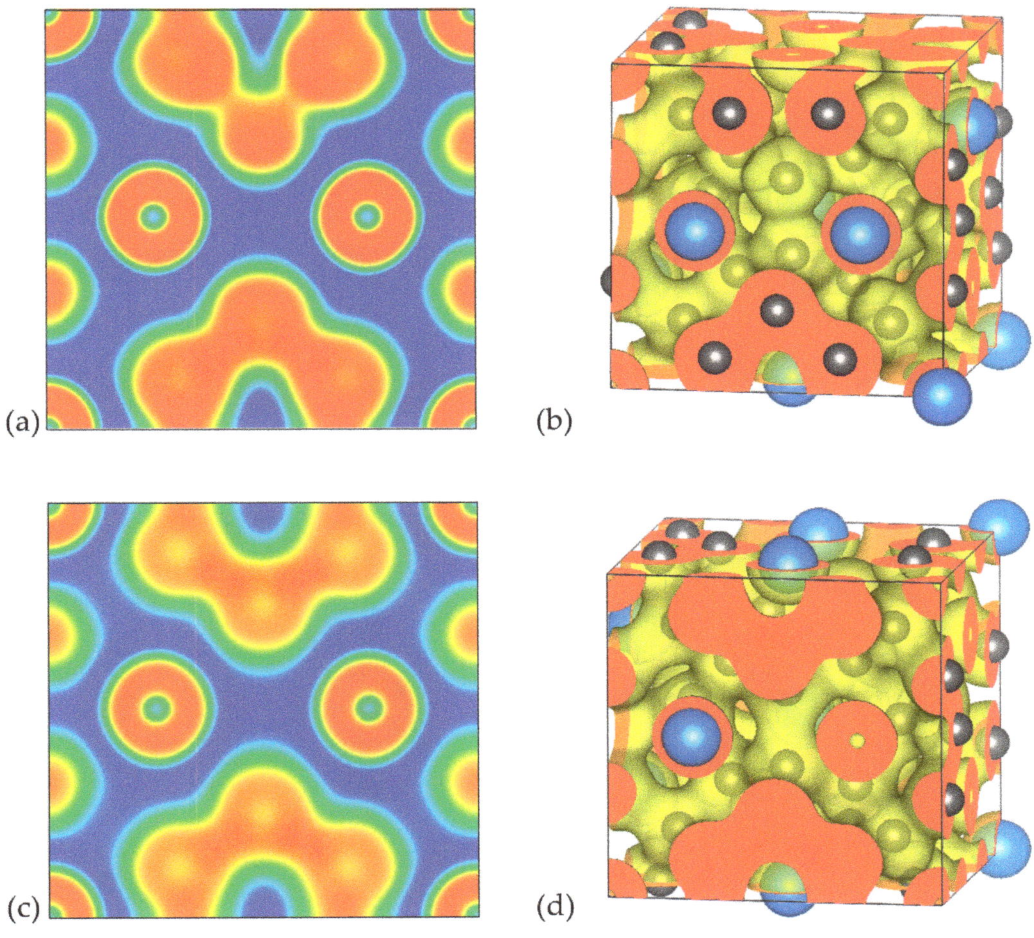

Figure 6. Electron localization function plots for Cs_8Sn_{44} (**a**) 2D ELF map of the (001) plane isosurface (n = 0.01) and (**b**) 3D ELF map isosurface (n = 0.5) for model A, (**c**) 2D ELF map of the (001) plane isosurface (n = 0.01) and (**d**) 3D ELF map isosurface (n = 0.5) for model C. The colors blue, green and red refer to ELF values of 0, 0.5 and 1, respectively.

The 2D and 3D ELF plots for $(NH_4)_8Sn_{44}$ for models A and C are illustrated in Figure 7. The 2D ELF maps shown in Figure 7a,c correspond to the same (001) plane as for Cs clathrates. Qualitatively, the obtained 2D ELF maps for the corresponding models for Cs and NH_4 clathrates share the same features regarding the regions of electron localization in the neighbor of Sn atoms and the ionic character of the interaction between guest cations and the host framework (blue region around NH_4 cations). High values of ELF are observed in NH_4^+, indicating the N-H covalent bonds existing within the guest cation. A slight distinction in the 2D ELF maps between model A for Cs and NH_4 clathrates can be observed if we compare the illustrations in Figures 6a and 7a. In the case of the NH_4 clathrate, lower ELF values are observed in the region of the Sn vacancy, which can be attributed to the weaker ability of charge transfer from the ammonium cation with respect to Cs.

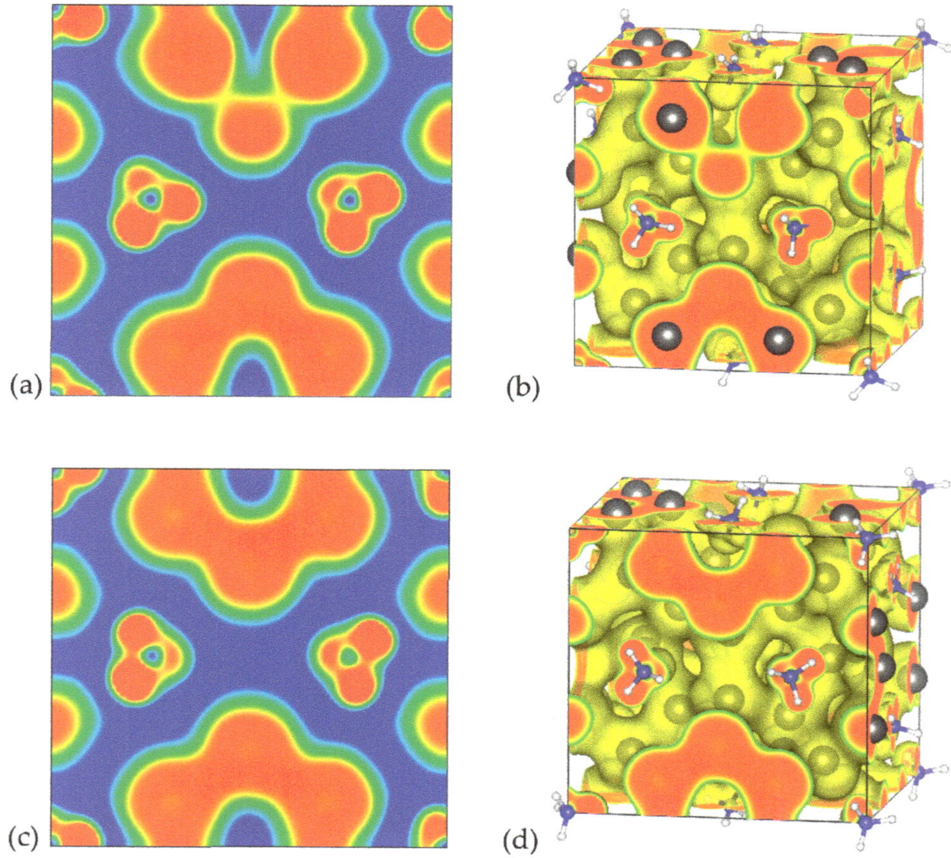

Figure 7. Electron localization function plots for $(NH_4)_8Sn_{44}$. (**a**) 2D ELF map of the (001) plane isosurface (n = 0.01) and (**b**) 3D ELF map isosurface (n = 0.5) for the model A, (**c**) 2D ELF map of the (001) plane isosurface (n = 0.01) and (**d**) 3D ELF map isosurface (n = 0.5) for model C. The colors blue, green and red refer to ELF values of 0, 0.5 and 1, respectively.

3.3. *Phonon Calculations*

We have conducted DFT calculations to investigate the phonon properties of the examined structures. We present the phonon spectrum, including the phonon band structure and phonon density of states (pDOS) of Cs_8Sn_{46-x} (x = 0, 2). In Figure 8, we present the calculations for the phonon band structure and phonon DOS of the Cs_8Sn_{46} compound and compare them with the phonon band structure and phonon DOS of the Cs_8Sn_{44} structure (model C). The phonon bands are in good agreement with previous experimental and computational studies [22,44,45]. Additionally, we investigate the theoretical properties of the proposed $(NH_4)_8Sn_{44}$ compound. The absence of imaginary frequencies in the phonon band structure confirms that the investigated materials are dynamically stable at 0 K. In the case of Cs_8Sn_{46} (Figure 8a,b), we observe that several flat bands are introduced both in the acoustic and optical regions of the band structure, and that these are in good agreement with previous computational and experimental findings. As can be seen in pDOS plot, three distinct phonon bands resulting from Cs appear at 23 cm^{-1}, 35 cm^{-1}, and 50 cm^{-1}, which can be attributed to the phonon modes of Cs atoms in the two different Sn cages of the material. It can be also seen that there are two regions with high density of states located above the acoustic modes and another near the top of the optical modes. For Cs_8Sn_{44} (Figure 8c,d), the overall appearance of the phonon band structure and the pDOS are quite similar to the Cs_8Sn_{46} with some distinctions. In the band structure diagram, we observe that the optical modes of Cs_8Sn_{44} reach larger frequencies with respect to Cs_8Sn_{46} analogue. As seen in the pDOS plot, these optical modes are related to the Sn atoms of the framework and their appearance can be attributed to the existence of three-coordinated Sn atoms in the framework. The number of localized bands in the band diagram are further increased for the same reason. For the phonon modes related to the Cs atoms, we can observe a shift of the low frequency peak in higher frequencies and a broadening of the two peaks in the region of 30 cm^{-1} to 60 cm^{-1} owing to the small distortions of the Sn cages and the reduction of framework symmetry due to the creation of Sn vacancies.

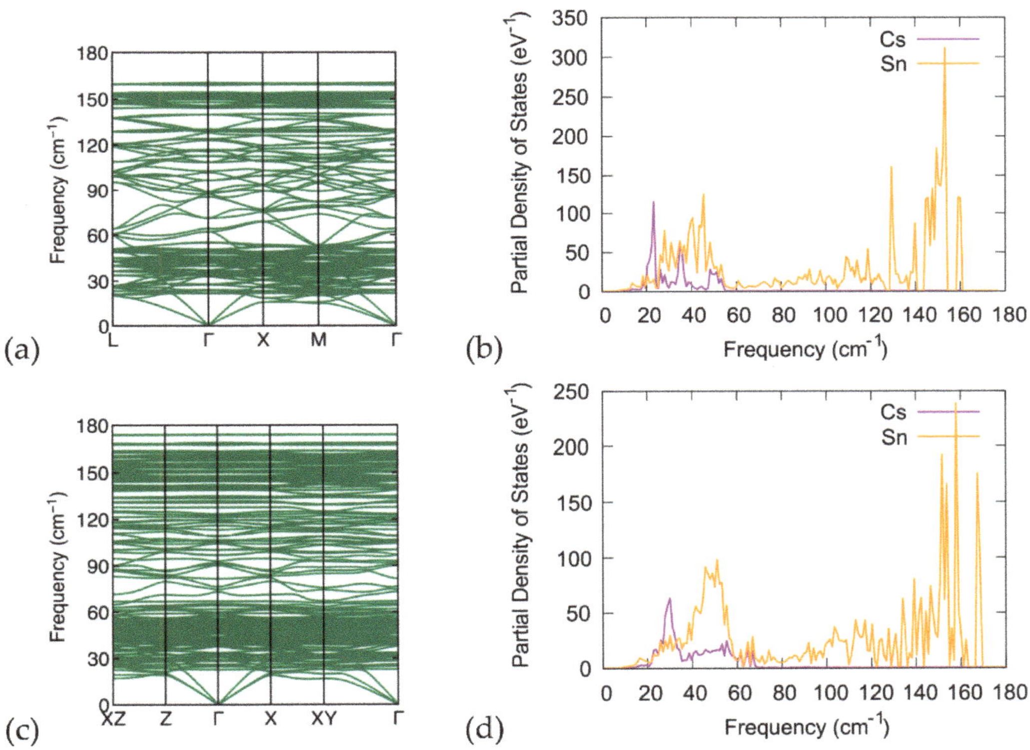

Figure 8. (**a**) Calculated phonon band structure along the L-Γ-X-M-Γ path for the Cs_8Sn_{46}, (**b**) partial phonon density of states for Cs and Sn of the Cs_8Sn_{46} structure, (**c**) calculated phonon band structure along the XZ-Z-Γ-X-XY-Γ path, and (**d**) partial phonon density of states for Cs and Sn of the Cs_8Sn_{44} (model C).

We also present the $(NH_4)_8Sn_{44}$ (model C) calculations of the phonon plots (Figure 9a) and phonon density of states (Figure 9b). The NH_4 cation produces increased phonon bands compared to Cs_8Sn_{44}, which is expected as a result of the large number of rotational and stretching modes of the ammonium cation (e.g., in NH_4Cl [46]). Apart from the low frequency modes, vibrational modes of the ammonium cation at 1380 cm^{-1} and ca. 3000 cm^{-1} are found.

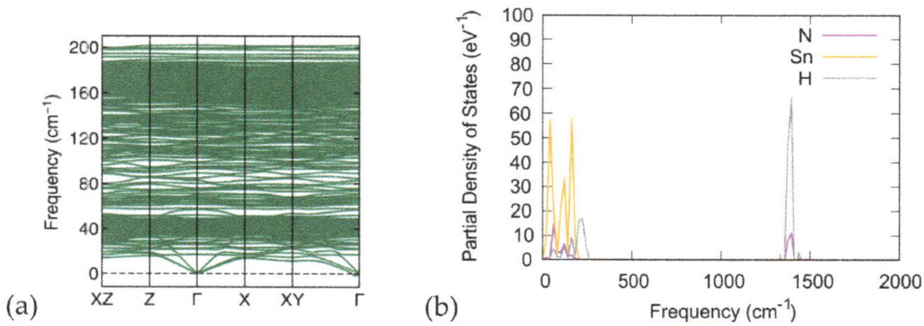

Figure 9. (**a**) Calculated phonon band structure along the XZ-Z-Γ-X-XY-Γ path and (**b**) partial phonon density of states for N, H, and Sn atoms for the $(NH_4)_8Sn_{44}$ (model C).

4. Discussion

Our computational study shows that the most energetically favored distribution of the Sn defects in the crystal structure of Cs_8Sn_{44} is model C, with one defect per six-membered ring in the 3D spiro-network of Sn framework atoms. This is in good agreement with the literature for A_8Sn_{44} [20]. It is worth noting that a fully ordered configuration has not been experimentally confirmed even at temperatures close to 0 K via synchrotron powder diffraction [21]. This analysis also led us to a reliable initial model for the hypothetical $(NH_4)_8Sn_{44}$. As with Cs_8Sn_{44}, the $(NH_4)_8Sn_{44}$ model C is the most stable and shows semiconducting behavior.

The use of ammonium cations instead of cesium may trigger a new synthetic approach based on reactive precursors and ammonium salts. The only experimental outcome relevant to our current study concerning the presence of a molecular guest in semiconducting clathrates is the clathrate-I $Na_{5.5}(H_2)_{2.15}Si_{46}$, which is stable at ambient temperature and pressure [47] and is prepared by mixing the Zintl phase NaSi and NH_4Br under dynamic vacuum at 573 K. The presence of H_2 seems to stabilize the framework, as no 'empty' clathrate-I Si_{46} can be formed under these or any other conditions. Except for H_2, no neutral guests have been reported so far in semiconducting clathrates. This is attributed to the limited chemical affinity of the neutral molecules with an inorganic framework. Only one theoretical study on the use of Xe gas as guest has been performed, and that study suggests the incorporation under specific conditions in type-II Sn clathrate $Xe_{24}[Sn_{136}]$ [48]. On the other hand, ammonia molecules as neutral guests are known in clathrates hydrates and they stabilize the framework, e.g., type-II tetrahydrofuran + NH_3 water clathrate [49].

With regard to thermoelectric efficiency, there are only limited experimental data on the A_8Sn_{44} system that show a low performance compared to the current record in type-I clathrates, namely $Ba_8Ga_{16}Ge_{30}$ that reaches a ZT of ca. 1 at 800 K [4]. The current analysis on the distribution of defects is of significance for the elucidation of the phonon scattering mechanism in this system.

5. Conclusions

The current work describes in detail the defect distribution in the Cs_8Sn_{44} type-I clathrates and shows how it hinders the presence of the Sn defects exclusively in spiro-connected six-membered rings as the most stable configuration. It also suggests that the incorporation of ammonium as a light, molecular cationic guest in the polyhedral cavities is thermodynamically possible. Both Cs_8Sn_{44} and $(NH_4)_8Sn_{44}$ materials behave as narrow band gap semiconductors. Due to the strong vibration of the ammonium cation in the framework cavities, phonon scattering is achieved. Given the high cost and hazards when handling alkali metals, this study may also trigger the exploration of new members in this class of compounds with molecular cations and ultimately improve the thermoelectric properties in A_8Sn_{44} clathrates. New synthetic attempts with the use of reactive precursors are currently in progress in our research group.

Supplementary Materials: The following supporting information can be downloaded at: https://www.mdpi.com/article/10.3390/ma17184595/s1, Figure S1: partial DOS of Cs_8Sn_{46}.; Figure S2: partial DOS of Cs_8Sn_{44} (model C); Figure S3: partial DOS of $(NH_4)_8Sn_{46}$, Figure S4: partial DOS of $(NH_4)_8Sn_{44}$ (model C).

Author Contributions: Conceptualization, E.K. and A.K.; methodology, N.K. and E.K.; software, N.K. and E.K.; validation, N.K., E.K. and A.K.; writing—original draft preparation, N.K., E.K. and A.K.; writing—review and editing, N.K., E.K. and A.K.; supervision, A.K.; project administration, A.K.; funding acquisition, A.K. All authors have read and agreed to the published version of the manuscript.

Funding: The research project is implemented in the framework of H.F.R.I call "Basic research Financing (Horizontal support of all Sciences)" under the National Recovery and Resilience Plan "Greece 2.0" funded by the European Union–NextGenerationEU (H.F.R.I. Project Number: 14730).

Institutional Review Board Statement: Not applicable.

Informed Consent Statement: Not applicable.

Data Availability Statement: Dataset available on request from the authors.

Acknowledgments: This work was supported by computational time granted from the National Infrastructures for Research and Technology S.A. (GRNET S.A.) in the National HPC facility ARIS -under project ID pa240203.

Conflicts of Interest: The authors declare no conflicts of interest.

References

1. Rogl, P. Chapter 32. Formation and Crystal Chemistry of Clathrates. In *Thermoelectrics Handbook Macro to Nano*; CRC Taylor & Francis: Boca Raton, FL, USA, 2006.
2. Konvir, K.; Shevelkov, A. Semiconducting clathrates: Synthesis, structure and properties. *Russ. Chem. Rev.* **2004**, *73*, 923–938. [CrossRef]
3. Guloy, A.M.; Ramlau, R.; Tang, Z.; Schnelle, W.; Baitinger, M.; Grin, Y. A guest-free germanium clathrate. *Nature* **2006**, *443*, 320–323. [CrossRef] [PubMed]
4. Freer, R.; Ekren, D.; Ghosh, T.; Biswas, K.; Qiu, P.; Wan, S.; Chen, L.; Han, S.; Fu, C.; Zhu, T.; et al. Key properties of inorganic thermoelectric materials—Tables (version 1). *J. Phys. Energy* **2022**, *4*, 02200. [CrossRef]
5. Corbett, J. Polyatomic Zintl Anions of the Post-Transition Elements. *Chem. Rev.* **1985**, *85*, 383–397. [CrossRef]
6. Baitinger, M.; Böhme, B.; Wagner, F.R.; Schwarz, U. Zintl Defects in Intermetallic Clathrates. *Z. Anorg. Allg. Chem.* **2020**, *646*, 1034–1041. [CrossRef]
7. Bhattacharya, A.; Carbogno, C.; Böhme, B.; Baitinger, M.; Grin, Y.; Scheffler, M. Formation of Vacancies in Si- and Ge-based Clathrates: Role of Electron Localization and Symmetry Breaking. *Phys. Rev. Lett.* **2017**, *118*, 236401. [CrossRef]
8. Kauzlarich, S.M.; Brown, S.R.; Snyder, G.J. Zintl phases for thermoelectric devices. *Dalton Trans.* **2007**, 2099–2107. [CrossRef]
9. Yang, L.; Chen, Z.-G.; Dargusch, M.S.; Zou, J. High Performance Thermoelectric Materials: Progress and Their Applications. *Adv. Energy Mater.* **2017**, *8*, 1701797. [CrossRef]
10. Slack, G. Chapter 34. New Materials and Performance Limits for Thermoelectric Cooling. In *Handbook of Thermoelectrics*; CRC Press: Boca Raton, FL, USA, 1995.
11. Iversen, B.; Palmqvist, A.E.C.; Cox, D.E.; Nolas, G.S.; Stucky, G.D.; Blake, N.P.; Metiu, H. Why are Clathrates Good Candidates for Thermoelectric Materials? *J. Solid State Chem.* **2000**, *149*, 455–458. [CrossRef]
12. Myles, C.; Dong, J.; Sankey, O. Rattling guest atoms in Si, Ge, and Sn-based type-II clathrate materials. *Phys. Status Solidi B* **2003**, *239*, 26–34. [CrossRef]
13. Wu, J.; Xu, J.; Tanigaki, K. Universal van der Waals-type interactions in rattler containing cage materials. *arXiv* **2018**, arXiv:1802.09228. [CrossRef]
14. Liu, Z.; Zhang, W.; Gao, W.; Mori, T. A material catalogue with glass-like thermal conductivity mediated by crystallographic occupancy for thermoelectric application. *Energy Environ. Sci.* **2021**, *14*, 3579–3587. [CrossRef]
15. Laurita, G.; Fabini, D.H.; Stoumpos, C.C.; Kanatzidis, M.G.; Seshadri, R. Chemical tuning of dynamic cation off-centering in the cubic phases of hybrid tin and lead halide perovskites. *Chem. Sci.* **2017**, *8*, 5628–5635. [CrossRef] [PubMed]
16. von Schnering, H.G.; Kröner, R.; Baitinger, M.; Peters, K.; Nesper, R.; Grin, Y. Crystal structure of the defect clathrate Cs_8Sn_{44}. *Z. Kristallogr. NCS* **2000**, *215*, 205–206.
17. Miller, G.J. Chapter 1. Structure and Bonding at the Zintl Border. In *Chemistry, Structure and Bonding of Zintl Phases and Ions*; VCH Publishers, Inc.: Hoboken, NJ, USA, 1996.

18. Myles, C.; Dong, J.; Sankey, O. Structural and electronic properties of tin clathrate materials. *Phys. Rev. B* **2001**, *64*, 165202. [CrossRef]
19. Dubois, F.; Fässler, T. Ordering of Vacancies in Type-I Tin Clathrate: Superstructure of $Rb_8Sn_{44}\square_2$. *J. Am. Chem. Soc.* **2005**, *127*, 3264. [CrossRef]
20. Kaltzoglou, A.; Hoffmann, S.; Fässler, T. Order-disorder phase transition in type-I clathrate Cs_8Sn_{44}. *Eur. J. Inorg. Chem.* **2007**, *2007*, 4162–4167. [CrossRef]
21. Kaltzoglou, A.; Fässler, T.; Christensen, M.; Johnsen, S.; Iversen, B.; Presniakov, I.; Sobolev, A.; Shevelkov, A. Effects of the order-disorder phase transition on the physical properties of A_8Sn_{44} (A = Rb, Cs). *J. Mater. Chem.* **2008**, *18*, 5630–5637. [CrossRef]
22. Christensen, S.; Bjerg, L.; Kaltzoglou, A.; Juranyi, F.; Fässler, T.; Unruh, T.; Christensen, M. Guest host interaction and low energy host structure dynamics in tin clathrates. *J. Appl. Phys.* **2013**, *113*, 084902. [CrossRef]
23. Candolfi, C.; Koza, M.M.; Aydemir, U.; Carrillo-Cabrera, W.; Grin, Y.; Steglich, F.; Baitinger, M. Vibrational dynamics of the type-I clathrates A_8Sn_{44} (A = Cs, Rb, K) from lattice-dynamics calculations, inelastic neutron scattering, and specific heat measurements. *J. Appl. Phys.* **2020**, *127*, 145104. [CrossRef]
24. Kieslich, G.; Sun, S.; Cheetham, A.K. Solid-state principles applied to organic–inorganic perovskites: New tricks for an old dog. *Chem. Sci.* **2014**, *5*, 4712–4715. [CrossRef]
25. Sidey, V. On the effective ionic radii for ammonium. *Acta Crystallogr.* **2016**, *B72*, 626–633. [CrossRef] [PubMed]
26. Momma, K.; Izumi, F. VESTA 3 for Three-Dimensional Visualization of Crystal, Volumetric and Morphology Data. *J. Appl. Crystallogr.* **2011**, *44*, 1272–1276. [CrossRef]
27. Vainshtein, B.K. Refinement of the Structure of the Group NH4 in the Structure of Ammonium Chloride. *Tr. Instituta Krist. Akad. Nauk. SSSR* **1956**, *12*, 18–24.
28. Kresse, G.; Hafner, J. Ab initio molecular dynamics for liquid metals. *Phys. Rev. B* **1993**, *47*, 558. [CrossRef]
29. Kresse, G.; Furthmüller, J. Efficiency of ab-initio total energy calculations for metals and semiconductors using a plane-wave basis set. *Comput. Mater. Sci.* **1996**, *6*, 15–50. [CrossRef]
30. Kresse, G.; Furthmüller, J. Efficient iterative schemes for ab initio total-energy calculations using a plane-wave basis set. *Phys. Rev. B* **1996**, *54*, 11169. [CrossRef]
31. Blöchl, P.E. Projector augmented-wave method. *Phys. Rev. B* **1994**, *50*, 17953. [CrossRef]
32. Kresse, I.G.; Joubert, D. From ultrasoft pseudopotentials to the projector augmented-wave method. *Phys. Rev. B* **1999**, *59*, 1758. [CrossRef]
33. Ganose, A.M.; Jackson, A.J.; Scanlon, D.O. sumo: Command-line tools for plotting and analysis of periodic ab initio calculations. *J. Open Source Softw.* **2018**, *3*, 717. [CrossRef]
34. Becke, A.D.; Edgecombe, K.E. A simple measure of electron localization in atomic and molecular systems. *J. Chem. Phys.* **1990**, *92*, 5397–5403. [CrossRef]
35. Fuentealba, P.; Chamorro, E.; Santos, J.C. Chapter 5. Understanding and using the electron localization function. *Theor. Asp. Chem. React.* **2007**, *19*, 57–85. [CrossRef]
36. Burdett, J.K.; McCormick, T.A. Electron Localization in Molecules and Solids: The Meaning of ELF. *J. Phys. Chem. A* **1998**, *102*, 6366–6372. [CrossRef]
37. Savin, A.; Nesper, R.; Wengert, S.; Fässler, T.F. ELF: The Electron Localization Function. *Angew. Chem. Int. Ed.* **1997**, *36*, 1808–1832. [CrossRef]
38. Togo, A.; Chaput, L.; Tadano, T.; Tanaka, I. Implementation strategies in phonopy and phono3py. *J. Phys. Condens. Matter* **2023**, *35*, 353001. [CrossRef]
39. Togo, A. First-principles Phonon Calculations with Phonopy and Phono3py. *J. Phys. Soc. Jpn.* **2023**, *92*, 012001. [CrossRef]
40. Shimizu, H.; Imai, T.; Kume, T.; Sasaki, S.; Kaltzoglou, A.; Fässler, T. Raman spectroscopy study of type-I clathrates A_8Sn_{44} (A = Rb, Cs) and $Rb_8Hg_4Sn_{42}$. *Chem. Phys. Lett.* **2008**, *464*, 54–57. [CrossRef]
41. Dopilka, A.; Peng, X.; Chan, C.K. Ab Initio Investigation of Li and Na Migration in Guest-Free, Type I Clathrates. *J. Phys. Chem. C* **2019**, *123*, 22812–22822. [CrossRef]
42. Alvarez, S. Polyhedra in (inorganic) chemistry. *Dalton Trans.* **2005**, 2209–2233. [CrossRef]
43. Egbele, P.O.; Shoko, E.; Joubert, D.P. Structural and thermoelectric properties of the type-I Sn clathrates Cs8Sn46-n(n=0,2) from Density Functional Theory (DFT). *MRS Adv.* **2018**, *3*, 1031–1038. [CrossRef]
44. Nolas, G.S.; Chakoumakos, B.C.; Mahieu, B.; Long, G.J.; Weakley, T.J.R. Structural Characterization and Thermal Conductivity of Type-I Tin Clathrates. *Chem. Mater.* **2000**, *12*, 1947–1953. [CrossRef]
45. Tse, J.S.; Li, Z.; Uehara, K. Phonon band structures and resonant scattering in Na_8Si_{46} and Cs_8Sn_{44} clathrates. *Europhys. Lett.* **2001**, *56*, 261. [CrossRef]
46. Lu, M.; Tian, F.; Zhou, Q.; Cui, T. First principle studies of ammonium chloride under high pressure. *RSC Adv.* **2021**, *11*, 5149–5155. [CrossRef] [PubMed]
47. Neiner, D.; Okamoto, N.L.; Condron, C.L.; Ramasse, Q.M.; Yu, P.; Browning, N.D.; Kauzlarich, S.M. Hydrogen Encapsulation in a Silicon Clathrate Type I Structure: $Na_{5.5}(H_2)_{2.15}Si_{46}$: Synthesis and Characterization. *J. Am. Chem. Soc.* **2007**, *129*, 13857–13862. [CrossRef]

48. Karttunen, A.J.; Fassler, T.F. Semiconducting Clathrates Meet Gas Hydrates: $Xe_{24}[Sn_{136}]$. *Chem. Eur. J.* **2014**, *20*, 6693–6698. [CrossRef]
49. Shin, K.; Kumar, R.; Udachin, K.A.; Alavi, S.; Ripmeester, J.A. Ammonia clathrate hydrates as new solid phases for Titan, Enceladus, and other planetary systems. *Proc. Natl. Acad. Sci. USA* **2012**, *109*, 14785–14790. [CrossRef]

materials

MDPI

Article

Improvement of the Stability of Quantum-Dot Light Emitting Diodes Using Inorganic HfO_x Hole Transport Layer

Jung Min Yun [1,2], Min Ho Park [1,2], Yu Bin Kim [1,2], Min Jung Choi [1,2], Seunghwan Kim [3,4], Yeonjin Yi [4], Soohyung Park [3,5] and Seong Jun Kang [1,2,*]

1 Department of Advanced Materials Engineering for Information and Electronics, Kyung Hee University, Yongin 17104, Republic of Korea; msd426@khu.ac.kr (J.M.Y.); triolight@khu.ac.kr (M.H.P.); bebeto@khu.ac.kr (Y.B.K.); okcmj@khu.ac.kr (M.J.C.)
2 Integrated Education Program for Frontier Materials (BK21 Four), Kyung Hee University, Yongin 17104, Republic of Korea
3 Advanced Analysis and Data Center, Korea Institute of Science and Technology (KIST), Seoul 02792, Republic of Korea; clam1127@kist.re.kr (S.K.); soohyung.park@kist.re.kr (S.P.)
4 Department of Physics, Yonsei University, Seoul 03722, Republic of Korea; yeonjin@yonsei.ac.kr
5 Division of Nanoscience & Technology, KIST School, University of Science and Technology (UST), Seoul 02792, Republic of Korea
* Correspondence: junkang@khu.ac.kr; Tel.: +82-31-201-3324

Abstract: One of the major challenges in QLED research is improving the stability of the devices. In this study, we fabricated all inorganic quantum-dot light emitting diodes (QLEDs) using hafnium oxide (HfO_x) as the hole transport layer (HTL), a material commonly used for insulator. Oxygen vacancies in HfO_x create defect states below the Fermi level, providing a pathway for hole injection. The concentration of these oxygen vacancies can be controlled by the annealing temperature. We optimized the all-inorganic QLEDs with HfO_x as the HTL by changing the annealing temperature. The optimized QLEDs with HfO_x as the HTL showed a maximum luminance and current efficiency of 66,258 cd/m^2 and 9.7 cd/A, respectively. The fabricated all-inorganic QLEDs exhibited remarkable stability, particularly when compared to devices using organic materials for the HTL. Under extended storage in ambient conditions, the all-inorganic device demonstrated a significantly enhanced operating lifetime (T_{50}) of 5.5 h, which is 11 times longer than that of QLEDs using an organic HTL. These results indicate that the all-inorganic QLEDs structure, with ITO/MoO_3/HfO_x/QDs/ZnMgO/Al, exhibits superior stability compared to organic-inorganic hybrid QLEDs.

Keywords: QLEDs; all-inorganic device; quantum dots; stable; oxygen vacancies; optoelectronics; solution process

1. Introduction

Colloidal quantum dot (QD) light-emitting diodes have been extensively researched over the past few decades as a next-generation display technology. Unlike other LEDs that use two-dimensional (2D) materials, organics, or perovskites in the emitting layer (EML), QLEDs that use QDs offer advantages such as easy band gap tuning for color customization, high stability, brightness, high color purity, and low cost fabrication [1–6].

In QLED research, significant efforts have been made not only to enhance emission performance, such as luminance and efficiency, but also to improve device stability. Factors contributing to improved QLED stability include charge balance optimization, QD ligand exchange, and layer modification [4,7,8]. Among these factors, layer modification arises from the weakness of organic materials to moisture and oxygen. This has led to extensive research on all-inorganic QLEDs [9–12]. A review of the research done so far indicates that various metal oxides, such as V_2O_5, MoO_3, and WO_3, are used as hole injection layers (HIL), while ZnO, ZnMgO, and SnO_2 are commonly employed as electron transport layers (ETL), with these materials being optimized for solution processing [7,13–17].

However, the inorganic materials used for the hole transport layer (HTL) are limited, and films are fabricated through complex processes. For example, commonly used inorganic HTLs such as NiO and Al_2O_3 films are typically produced via physical deposition methods, and doping processes are often employed to enhance performance. In some cases, additional layers are inserted or further film engineering is performed to improve hole injection and optimize the device [11,14,18–21].

In this study, we fabricated all-inorganic QLEDs using HfO_x thin films, produced via solution processing, as the hole transport layer (HTL). Although HfO_x is widely used as an insulating layer and is not typically used as a charge transport layer (CTL) in LEDs, we used the defect states generated by oxygen vacancies within the HfO_x film. These defect states form below the Fermi level, and by controlling the annealing temperature after spin coating, we optimized the defect states to suit the intended device [22–24]. The optimized device achieved a maximum luminance of 66,258 cd/m^2, a current efficiency of 9.7 cd/A, and an external quantum efficiency (EQE) of 2.4%. Additionally, the device exhibited an 11-fold increase in lifetime compared to a device using TFB as the HTL, along with superior stability when stored in ambient conditions. These findings demonstrate that solution-processed HfO_x HTL films can be used to create highly stable all-inorganic QLEDs.

2. Materials and Method

2.1. Solution Synthesis

The HfO_x solution was prepared by dissolving 0.01 M of $HfCl_4$ (Sigma Aldrich, St. Louis, MO, USA) in ethylene glycol monomethyl ether. The solution was stirred for 1 h at room temperature. The MoO_3 solution was prepared by dissolving 0.05 wt% of MoO_3 nanoparticle powder (Sigma Aldrich, St. Louis, MO, USA) in deionized water. The solution of CdSe/ZnS Qds (Finelab, Deajeon, Republic of Korea) was prepared by diluting it in a toluene solution to a concentration of 20 mg/mL. The 15% Mg doped ZnMgO solution (Nanofix, Suwon, Republic of Korea) was prepared by diluting it in ethyl alcohol solution at a 1:1 ratio.

2.2. Fabrication of the Device

The patterned ITO glass was ultrasonicated with deionized water (DI water), acetone, and isopropyl alcohol (IPA) for 15 min, respectively. Subsequently, the substate was exposed to UV-Ozone for 15 min to make the surface hydrophilic, to remove residual organics, and to increase the work function, followed by four cycles of spin coating and annealing. First, the MoO_3 solution was spin-coated on the substrates at 3000 rpm for 30 s, followed by annealing at 120 °C for 10 min. Next, for the all-inorganic QLEDs, the HfO_x solution was spin-coated at 3000 rpm for 30 s, followed by pre-annealing at 120 °C for 5 min to evaporate the solvent. Subsequently, three different devices were annealing at 150 °C, 250 °C, and 350 °C, respectively, for 45 min each. For the control device, a solution of TFB was spin-coated at 3000 rpm for 30 s and annealed at 180 °C for 30 min. Then, CdSe/ZnS QDs was spin coated onto the HTL layer at 2000 rpm for 30 s and annealed at 90 °C for 10 min. The optical band gap of the QDs was determined from UV-Vis measurements, and the photoluminescence (PL) spectra are shown in Figure S2a,b. The QDs layer was covered with a ZnMgO layer, which was spin-coated at 2000 rpm for 60 s and annealed at 90 °C for 10 min. Finally, a 130 nm aluminium cathode was deposited via thermal evaporation at a rate of 3 Å/s under a pressure of 8×10^{-6} torr.

2.3. Characterization and Measurement of Films and Devices

The absorption spectra were measured using UV-Vis spectroscopy (V-570, JASCO, Easton, MD, USA). The PL characteristics were measured with a PL spectrometer (Ocean Optics, USB 2000+, Rochester, NY, USA) using a 365 nm light source. The EL characteristics of the QLEDs were measured with a OLED I-V-L test system (M6100, Mcscience Inc., Suwon, Republic of Korea). The device structure was analyzed using TEM (ThermoFisher Talos, Waltham, MA, USA). Interfacial properties were investigated through XPS and UPS measurements (ThermoFisher NEXSA, Waltham, MA, USA) using Al Kα (1486.8 eV) and He–I (21.22 eV) sources.

3. Results and Discussion

Figure 1a displays the device structure. ITO was used as the anode, MoO_3 as the hole injection layer (HIL), HfO_x as the hole transport layer (HTL), QDs as the emitting layer (EML), ZnMgO as the electron transport layer (ETL), and Al as the cathode. Figure S1a is a cross-section HR-TEM image of the device. The thickness of the HfO_x layer was found to be approximately 21 nm. Due to the thin HfO_x film, measuring the Hf 4f peak on the ITO substrate via XPS is challenging because it overlaps with the In 4d peak [25]. Therefore, thin films of HfO_x were fabricated on a bare Si substrate under the same conditions as device fabrication, with varying annealing temperatures. Figure 1b–d shows the XPS O 1s spectra of the HfO_x film at different annealing temperatures. The O 1s peak at ~530.4 eV indicates metal-oxide bonding, while the peak at ~531.5 eV represents oxygen vacancies. The ratio of metal oxide to oxygen vacancies decreases from 1:0.59 at 150 °C to 1:0.31 at 350 °C. Since oxygen vacancies create defect states below the Fermi level in the HfO_x bandgap, the annealing temperature influences the defect states of HfO_x [26]. Figure S1b–d show the XPS Hf 4f spectra of the HfO_x films at different annealing temperatures. The Hf $4f_{7/2}$ peak is located at 16.9 eV and the Hf $4f_{5/2}$ peak is at 18.6 eV for all temperatures, consistent with reference values [22]. The XPS Cl 2p spectra of the HfO_x film were only measured for the film annealed at 150 °C, as shown in Figure S2d. The presence of residual Cl is due to the incomplete decomposition of $HfCl_4$ below 200 °C [24]. These XPS spectra indicate that the annealing temperature of the HfO_x films changes the film stoichiometry and the ratio of metal oxide to oxygen vacancies. In particular, the variation in oxygen vacancies is expected to influence the defect states of the HfO_x film, thereby affecting hole injection in QLEDs.

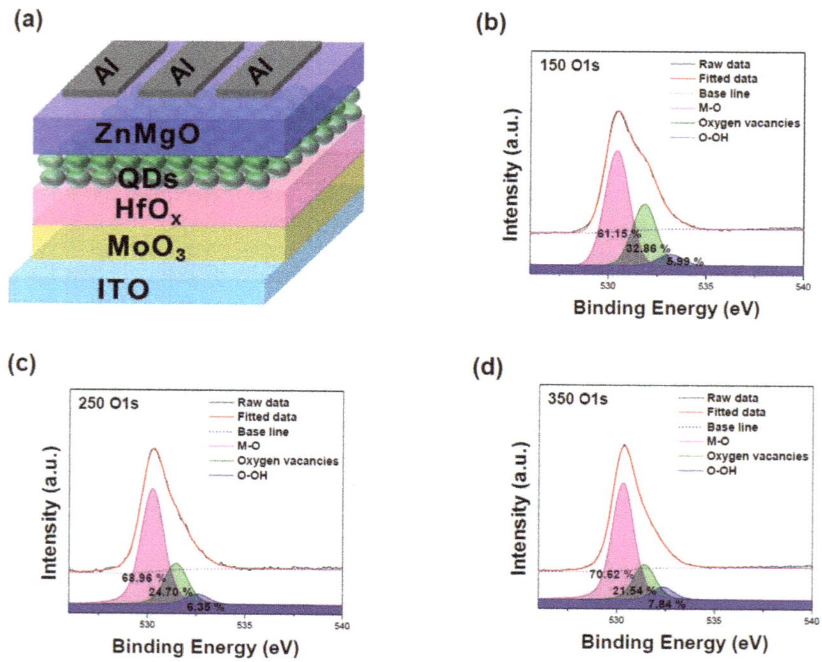

Figure 1. (**a**) Schematic illustration of the device. XPS measurement for O 1s of (**b**) MoO_3/HfO_x (150 °C), (**c**) MoO_3/HfO_x (250 °C), and (**d**) MoO_3/HfO_x (350 °C).

Figure 2 shows the UPS measurements and the band structure of the optimized device for the following films: ITO, ITO/MoO_3, ITO/MoO_3/HfO_x (150 °C), ITO/MoO_3/HfO_x (250 °C), ITO/MoO_3/HfO_x (350 °C), ITO/MoO_3/HfO_x (150 °C)/QDs, ITO/MoO_3/HfO_x (250 °C)/QDs, ITO/MoO_3/HfO_x (350 °C)/QDs, and ITO/MoO_3/HfO_x (250 °C)/QDs/ZnMgO. The UPS-determined work function (WF), valence band maximum (VBM) region, and secondary electron cutoff (SEC) region are shown in Figure 2a,b. Figure 2c presents the near valance region of each HfO_x film showing the different position of defect states, which form below the Fermi level due to oxygen vacancies, and vary with the annealing temperature. The defect states of HfO_x annealed at 150 °C, 250 °C, and 350 °C are located 1.20 eV, 2.09 eV, and 1.25 eV below the Fermi level, respectively. Figure 2d shows that the UPS intensity near the defect states of HfO_x varies with the annealing temperature. The intensity changes with the annealing temperature, with the HfO_x film annealed at 250 °C exhibiting the highest intensity, indicating the greatest concentration of defect states. Figure 2e is a schematic band diagram of the device using UPS data from the HfO_x film annealed at 250 °C. MoO_3, when used as an HIL, is known for facilitating hole transport via electron extraction from the CBM due to its deep-lying electronic states [27,28]. The defect states of HfO_x are well aligned between the CBM of MoO_3 and the VBM of the QDs, making it effective for hole injection into the EML. Additionally, HfO_x is an insulator with a high CBM—it effectively blocks electrons from the QDs, which helps maintain charge balance [2]. UPS measurements of the device films revealed that the position and concentration of defect states in HfO_x shift according to the annealing temperature due to changes in oxygen vacancy concentration. Analysis of the schematic band diagram derived from UPS data illustrates the band structure and defect states of HfO_x, confirming its advantages as an HTL in QLEDs.

To confirm the effect of the aforementioned results on QLEDs, we measured the density–voltage–luminance (J–V–L) characteristics of QLEDs with HfO_x annealed at different temperatures and density–voltage (J-V) measurements on hole-only devices (HODs) with an ITO/MoO_3/HfO_x/Al structure. Figure 3a,b show the luminance, current efficiency, and EQE of each device. Although the turn-on voltage was the same at 2.7 V for all three devices, the device annealed at 250 °C exhibited the highest performance in all aspects, with maximum values of 66,528 cd/m^2 for luminance, 9.7 cd/A for current efficiency, and 2.4% for EQE, respectively. The structure of the HODs is shown in Figure 3c. Figure 3d shows the current density–voltage (J–V) characteristics of the hole-only device (HOD) at different annealing temperatures of HfO_x. Four distinct regions are observed: ohmic contact, trap-partially-filled space charge limited currents (T-SCLC), trap-filled-limited (TFL) region, and Child's law of space charge limited currents (SCLC). The device annealed at 250 °C shows the TFL region at the lowest voltage, indicating the highest hole mobility and the lowest trap concentration [29–31]. Figure S2c–e compared the EL intensity peaks of QLEDs with HfO_x annealed at different temperatures. The comparison at every 1 V from 3 V to 7 V shows that the EL peak positions are the same under all conditions. These data indicate that the differences in defect states of the HfO_x film affect hole injection, leading to variations in device performance.

All-inorganic QLEDs are known to have superior stability compared to QLEDs using organic materials. Figure 4 shows the stability comparison between QLEDs using TFB as the HIL and other QLEDs. The measured QLEDs were fabricated under the same conditions except for the HTL. Figure 4a shows the lifetime data. The lifetimes of QLEDs using TFB and HfO_x as the HTL were measured to be 0.43 h and 5.5 h, respectively. Additionally, Figure S2f shows the evolution of EL spectra during the operational lifetime measurement. These data demonstrate that the device with HfO_x exhibited a lifetime approximately 11 times longer, with no significant changes in the EL spectrum until T_{50}. Lifetime is a representative measurement to estimate the stability of QLEDs, but it shows degradation occurring over a relatively short period [32]. Figure 4b displays the degradation of QLEDs over a comparatively longer period in an ambient environment. After fabricating the devices, they were stored in ambient conditions, and luminance values were measured at 5 V every two days. Each measurement was taken from five devices, and the results are presented with error bars. It was observed that the luminance of the device using TFB as the HTL decreased more significantly with time compared to the device using HfO_x. Through these stability data, it was confirmed that the all-inorganic QLED using HfO_x as the HTL has superior stability compared to the device using TFB.

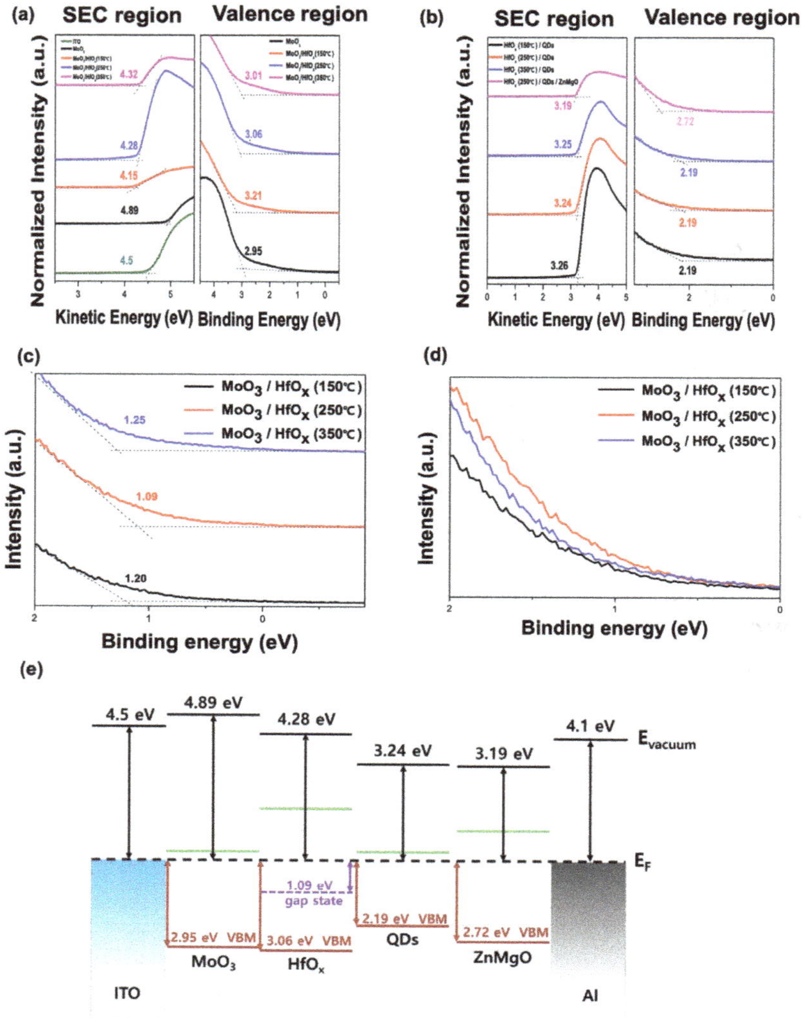

Figure 2. SEC and valence region spectra of (**a**) ITO, ITO/MoO_3, ITO/MoO_3/HfO_x (150 °C), ITO/MoO_3/HfO_x (250 °C), ITO/MoO_3/HfO_x (350 °C) films, and (**b**) ITO/MoO_3/HfO_x (150 °C)/QDs, ITO/MoO_3/HfO_x (250 °C)/QDs, ITO/MoO_3/HfO_x (350 °C)/QDs, ITO/MoO_3/HfO_x (250 °C)/QDs/ZnMgO films. (**c**) Near valance region of each HfO_x film showing the different positions of defect states. (**d**) Near valance region of each HfO_x film showing the different intensity (**e**) schematic energy level diagram of QLEDs (250 °C).

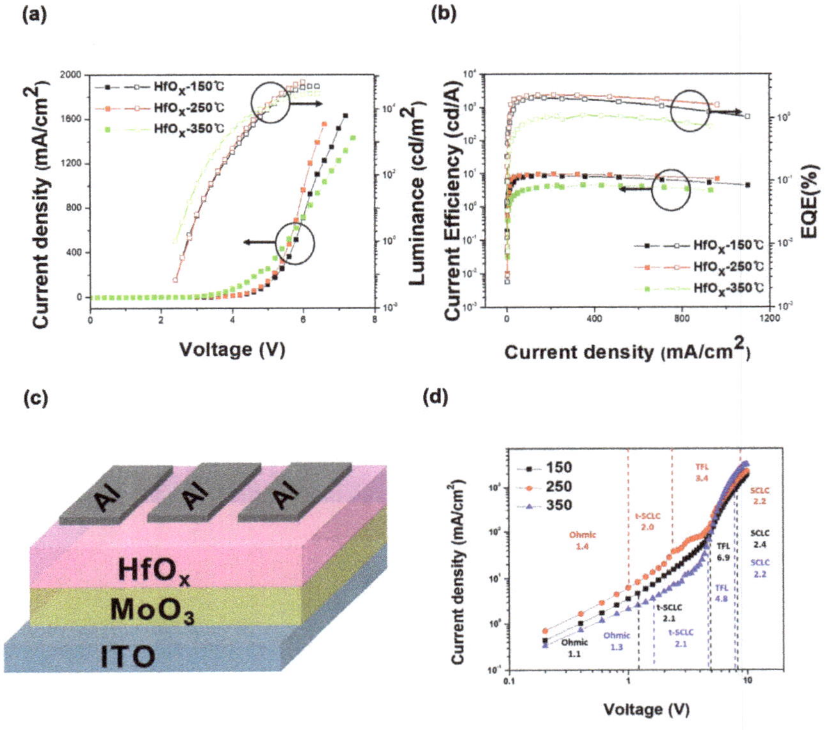

Figure 3. (**a**) J–V–L characteristics and (**b**) current efficiency–current density plot of QLED devices. (**c**) Schematic illustration of hole-only device. (**d**) J–V characteristics of HOD devices.

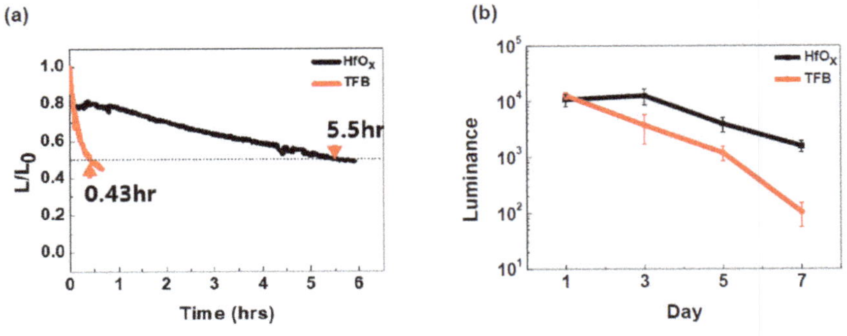

Figure 4. (**a**) Lifetime data of QLED devices using HfO_x and TFB for HTL. (**b**) Luminance value measured data every two days of QLED devices using HfO_x and TFB for HTL.

4. Conclusions

In this study, we developed a highly stable oxide-based HfO_x HTL, instead of the conventional organic TFB HTL. XPS and UPS analyses of the HfO_x film revealed the presence of oxygen vacancies, which resulted in the formation of defect states in the solution-processed film. Additionally, we observed that the oxygen vacancies concentration and defect states changed with the annealing temperature during the film formation process. Optimizing these characteristics, we successfully fabricated high-stability QLEDs using HfO_x as the HTL. The optimized device, achieved by adjusting the annealing temperature of HfO_x, exhibited maximum values of 66,528 cd/m^2 for luminance, 9.7 cd/A for current efficiency, and 2.4% for EQE. Additionally, the device exhibited an 11-fold increase in lifetime and greater stability in ambient conditions compared to the TFB-used device. Our results suggest that we have developed a novel HTL, which is crucial for all-inorganic QLED research, offering a useful and simple method for fabricating highly stable QLEDs.

Supplementary Materials: The following supporting information can be downloaded at: https://www.mdpi.com/article/10.3390/ma17194739/s1, Figure S1. (a) TEM Image of QLED device. XPS measurement for Hf 4f of (b) MoO_3/HfOx (150 °C), (c) MoO_3/HfOx (250 °C), and (d) MoO_3/HfOx (350 °C). (e) XPS measurement for Cl 2p of MoO_3/HfOx (150 °C). Figure S2. (a) The optical band gap of CdSe/ZnS green QDs. (b) The Photoluminescence spectra of CdSe/ZnS green QDs. Normalized EL intensity of devices with HfOx annealed at (c) 150 °C (d) 250 °C, (e) 350 °C. (f) EL intensity spectra during the operational lifetime. Figure S3. Thermal stability of the fabricated all-inorganic QLED devices. Figure S4. The PL intensity variation under different HfOx conditions.

Author Contributions: J.M.Y. and S.J.K. planned the experiment. J.M.Y. and M.H.P. fabricated the QLEDs and conducted the experimental analysis. J.M.Y., Y.B.K., S.K., Y.Y., S.P. and M.J.C. examined the data obtained from measurements. J.M.Y. drafted the manuscript, and it was confirmed and edited by and S.J.K. All authors have read and agreed to the published version of the manuscript.

Funding: This study was supported by a research project grant from the National Research Foundation of Korea (NRF-2021R1A2C1009442 and 2020R1A6A1A03048004).

Institutional Review Board Statement: Not applicable.

Informed Consent Statement: Not applicable.

Data Availability Statement: The authors declare that all data supporting the findings of this study are available within the paper and its Supplementary Information.

Conflicts of Interest: The authors declare no conflict of interest.

References

1. Jang, E.; Jang, H. Quantum dot light-emitting diodes. *Chem. Rev.* **2023**, *123*, 4663–4692. [CrossRef] [PubMed]
2. Jeong, J.H.; Kim, M.G.; Ma, J.H.; Park, M.H.; Ha, H.J.; Kang, S.J.; Maeng, M.-J.; Kim, Y.D.; Park, Y.; Kang, S.J. Improving the Performance of Solution−Processed Quantum Dot Light−Emitting Diodes via a HfOx Interfacial Layer. *Materials* **2022**, *15*, 8977. [CrossRef] [PubMed]
3. Ji, K.; Anaya, M.; Abfalterer, A.; Stranks, S.D. Halide perovskite light-emitting diode technologies. *Adv. Opt. Mater.* **2021**, *9*, 2002128. [CrossRef]
4. Moon, H.; Lee, C.; Lee, W.; Kim, J.; Chae, H. Stability of quantum dots, quantum dot films, and quantum dot light-emitting diodes for display applications. *Adv. Mater.* **2019**, *31*, 1804294. [CrossRef] [PubMed]
5. Sudheendran Swayamprabha, S.; Dubey, D.K.; Shahnawaz; Yadav, R.A.K.; Nagar, M.R.; Sharma, A.; Tung, F.C.; Jou, J.H. Approaches for long lifetime organic light emitting diodes. *Adv. Sci.* **2021**, *8*, 2002254. [CrossRef]
6. Zhang, L.; Sun, C.; He, T.; Jiang, Y.; Wei, J.; Huang, Y.; Yuan, M. High-performance quasi-2D perovskite light-emitting diodes: From materials to devices. *Light: Sci. Appl.* **2021**, *10*, 61. [CrossRef]
7. Chen, Z.; Li, H.; Yuan, C.; Gao, P.; Su, Q.; Chen, S. Color Revolution: Prospects and Challenges of Quantum-Dot Light-Emitting Diode Display Technologies. *Small Methods* **2024**, *8*, 2300359. [CrossRef]
8. Vasudevan, D.; Gaddam, R.R.; Trinchi, A.; Cole, I. Core–shell quantum dots: Properties and applications. *J. Alloys Compd.* **2015**, *636*, 395–404. [CrossRef]
9. Cao, F.; Wu, Q.; Sui, Y.; Wang, S.; Dou, Y.; Hua, W.; Kong, L.; Wang, L.; Zhang, J.; Jiang, T. All-Inorganic Quantum Dot Light-Emitting Diodes with Suppressed Luminance Quenching Enabled by Chloride Passivated Tungsten Phosphate Hole Transport Layers. *Small* **2021**, *17*, 2100030. [CrossRef]
10. Wang, L.; Pan, J.; Qian, J.; Liu, C.; Zhang, W.; Akram, J.; Lei, W.; Chen, J. Performance enhancement of all-inorganic quantum dot light-emitting diodes via surface modification of nickel oxide nanoparticles hole transport layer. *ACS Appl. Electron. Mater.* **2019**, *1*, 2096–2102. [CrossRef]
11. Wang, T.; Zhu, B.; Wang, S.; Yuan, Q.; Zhang, H.; Kang, Z.; Wang, R.; Zhang, H.; Ji, W. Influence of shell thickness on the performance of NiO-based all-inorganic quantum dot light-emitting diodes. *ACS Appl. Mater. Interfaces* **2018**, *10*, 14894–14900. [CrossRef] [PubMed]
12. Yang, X.; Zhang, Z.-H.; Ding, T.; Wang, N.; Chen, G.; Dang, C.; Demir, H.V.; Sun, X.W. High-efficiency all-inorganic full-colour quantum dot light-emitting diodes. *Nano Energy* **2018**, *46*, 229–233. [CrossRef]

13. Chen, M.; Chen, X.; Ma, W.; Sun, X.; Wu, L.; Lin, X.; Yang, Y.; Li, R.; Shen, D.; Chen, Y. Highly stable SnO_2-based quantum-dot light-emitting diodes with the conventional device structure. *ACS Nano* **2022**, *16*, 9631–9639. [CrossRef] [PubMed]
14. Ji, W.; Lv, Y.; Jing, P.; Zhang, H.; Wang, J.; Zhang, H.; Zhao, J. Highly efficient and low turn-on voltage quantum dot light-emitting diodes by using a stepwise hole-transport layer. *ACS Appl. Mater. Interfaces* **2015**, *7*, 15955–15960. [CrossRef]
15. Kim, H.S.; Lee, D.; Kim, B.; Hwang, B.; Kim, C.K. Improved performance of quantum dot light-emitting diodes by introducing WO3 hole injection layers. *Mol. Cryst. Liq. Cryst.* **2022**, *735*, 51–60. [CrossRef]
16. Kim, T.Y.; Park, S.; Kim, B.J.; Heo, S.B.; Yu, J.H.; Shin, J.S.; Hong, J.-A.; Kim, B.-S.; Kim, Y.D.; Park, Y. Dual-functional quantum-dots light emitting diodes based on solution processable vanadium oxide hole injection layer. *Sci. Rep.* **2021**, *11*, 1700. [CrossRef]
17. Nie, L.; Fan, J.; Li, Y.; Xiang, C.; Zhang, T. Direct Optical Patterning of MoO_3 Nanoparticles and Their Application as a Hole Injection Layer for Solution-Processed Quantum Dot Light-Emitting Diodes. *ACS Appl. Nano Mater.* **2024**, *7*, 9499–9506. [CrossRef]
18. Sun, Y.; Chen, W.; Wu, Y.; He, Z.; Zhang, S.; Chen, S. A low-temperature-annealed and UV-ozone-enhanced combustion derived nickel oxide hole injection layer for flexible quantum dot light-emitting diodes. *Nanoscale* **2019**, *11*, 1021–1028. [CrossRef]
19. Wang, F.; Wang, Z.; Zhu, X.; Bai, Y.; Yang, Y.; Hu, S.; Liu, Y.; You, B.; Wang, J.; Li, Y. Highly efficient and super stable full-color quantum dots light-emitting diodes with solution-processed all-inorganic charge transport layers. *Small* **2021**, *17*, 2007363. [CrossRef]
20. Ji, W.; Liu, S.; Zhang, H.; Wang, R.; Xie, W.; Zhang, H. Ultrasonic spray processed, highly efficient all-inorganic quantum-dot light-emitting diodes. *ACS Photonics* **2017**, *4*, 1271–1278. [CrossRef]
21. Ji, W.; Shen, H.; Zhang, H.; Kang, Z.; Zhang, H. Over 800% efficiency enhancement of all-inorganic quantum-dot light emitting diodes with an ultrathin alumina passivating layer. *Nanoscale* **2018**, *10*, 11103–11109. [CrossRef] [PubMed]
22. Hadacek, N.; Nosov, A.; Ranno, L.; Strobel, P.; Galéra, R.-M. Magnetic properties of HfO2 thin films. *J. Phys. Condens. Matter* **2007**, *19*, 486206. [CrossRef]
23. Kaiser, N.; Vogel, T.; Zintler, A.; Petzold, S.; Arzumanov, A.; Piros, E.; Eilhardt, R.; Molina-Luna, L.; Alff, L. Defect-stabilized substoichiometric polymorphs of hafnium oxide with semiconducting properties. *ACS Appl. Mater. Interfaces* **2021**, *14*, 1290–1303. [CrossRef] [PubMed]
24. Yoo, Y.B.; Park, J.H.; Lee, K.H.; Lee, H.W.; Song, K.M.; Lee, S.J.; Baik, H.K. Solution-processed high-k HfO_2 gate dielectric processed under softening temperature of polymer substrates. *J. Mater. Chem. C* **2013**, *1*, 1651–1658. [CrossRef]
25. Krishna, D.N.G.; Philip, J. Review on surface-characterization applications of X-ray photoelectron spectroscopy (XPS): Recent developments and challenges. *Appl. Surf. Sci. Adv.* **2022**, *12*, 100332. [CrossRef]
26. Yun, H.W.; Lee, J.; Kim, R.N.; Ji, S.H.; Ryu, S.O.; Kim, W.-B. Dipole formation and electrical properties of high-k/SiO_2 interface according to the density of SiO_2 interfacial layer. *Curr. Appl. Phys.* **2022**, *37*, 45–51. [CrossRef]
27. Yang, J.-H.; Jang, G.-P.; Kim, S.-Y.; Chae, Y.-B.; Lee, K.-H.; Moon, D.-G.; Kim, C.-K. Highly Efficient All-Solution-Processed Quantum Dot Light-Emitting Diodes Using MoOx Nanoparticle Hole Injection Layer. *Nanomaterials* **2023**, *13*, 2324. [CrossRef]
28. Kröger, M.; Hamwi, S.; Meyer, J.; Riedl, T.; Kowalsky, W.; Kahn, A. Role of the deep-lying electronic states of MoO_3 in the enhancement of hole-injection in organic thin films. *Appl. Phys. Lett.* **2009**, *95*, 123301. [CrossRef]
29. Hildebrandt, E.; Kurian, J.; Müller, M.M.; Schroeder, T.; Kleebe, H.-J.; Alff, L. Controlled oxygen vacancy induced p-type conductivity in HfO_{2-x} thin films. *Appl. Phys. Lett.* **2011**, *99*, 112902. [CrossRef]
30. Ha, H.J.; Kim, M.G.; Ma, J.H.; Jeong, J.H.; Park, M.H.; Kang, S.J.; Kim, W.; Park, S.; Kang, S.J. Enhanced performance of flexible quantum dot light-emitting diodes using a low-temperature processed PTAA hole transport layer. *Sci. Rep.* **2023**, *13*, 3780. [CrossRef]
31. Haneef, H.F.; Zeidell, A.M.; Jurchescu, O.D. Charge carrier traps in organic semiconductors: A review on the underlying physics and impact on electronic devices. *J. Mater. Chem. C* **2020**, *8*, 759–787. [CrossRef]
32. Padmasali, A.N.; Kini, S.G. A lifetime performance analysis of LED luminaires under real-operation profiles. *IEEE Trans. Electron Devices* **2019**, *67*, 146–153. [CrossRef]

materials

MDPI

Article

Multiferroic and Phonon Properties of the Double Perovskite Pr_2FeAlO_6

Angel T. Apostolov [1], Iliana N. Apostolova [2] and Julia M. Wesselinowa [3,*]

1 Department of Physics, Faculty of Hydrotechnics, University of Architecture, Civil Engineering and Geodesy, Hristo Smirnenski Blvd. 1, 1046 Sofia, Bulgaria; angelapos@abv.bg
2 Faculty of Forest Industry, University of Forestry, Kl. Ohridsky Blvd. 10, 1756 Sofia, Bulgaria; inaapos@abv.bg
3 Faculty of Physics, Sofia University "St. Kliment Ohridski", J. Bouchier Blvd. 5, 1164 Sofia, Bulgaria
* Correspondence: julia@phys.uni-sofia.bg

Abstract: With the help of a microscopic model and Green's function technique, we studied the multiferroic and phonon properties of the recently reported new multiferroic Pr_2FeAlO_6 (PFAO) compound, which belongs to the double perovskite $A_2BB'O_6$ family. The magnetization decreases with the increase in temperature and disappears at the ferromagnetic Curie temperature T_C^{FM}. The polarization increases with the application of an external magnetic field, indicating strong magneto-electric coupling and confirming the multiferroic behavior of PFAO. In the curves of dependence of the phonon energy and their damping with respect to temperature, a kink is observed at T_C^{FM}. This is due to the strong anharmonic spin–phonon interactions, which play a crucial role below T_C^{FM} and are frequently observed in other double perovskite compounds. Above T_C^{FM}, only anharmonic phonon–phonon coupling remains. The phonon mode is controlled by an external magnetic field.

Keywords: Pr_2FeAlO_6; magnetoelectric effect; phonon energy; microscopic model; Green's function theory

1. Introduction

Multiferroic materials, which possess interacted ferroelectric and ferromagnetic order parameters, are promising candidates for multifunctional devices that can be controlled both electrically and magnetically [1]. However, multiferroics with significant magnetoelectric coupling are relatively rare. Recently, Liu et al. [2] studied the multiferroic and magnetodielectric effects in Pr_2FeAlO_6 (PFAO). Its crystal structure is similar to the earlier reported nanopolycrystalline Pr_2FeCrO_6 [3]. PFAO is a type I multiferroics and belongs to the family of double perovskites $A_2BB'O_6$, where A is a rare earth element and B/B′ are transition metals. It exhibits significant magnetoresistive and electronic properties, a high Curie temperature, and a strong spin polarization [4,5]. In such materials, electric polarization can be further enhanced by the formation of square bonds between B or B′ ions and through the presence of oxygen vacancies. For example, introducing non-magnetic Al ions into the B-site, as shown in Y_2FeAlO_6, increases polarization [6]. Double perovskites feature a doubled unit cell compared with the conventional perovskite structure (ABO_3), with alternating BO_6 and $B'O_6$ octahedra forming two interpenetrating sublattices. The wide range of possible A and B/B′ ion combinations in double perovskites allows for various configurations that support the coexistence of ferromagnetism and ferroelectricity. Many multiferroic double perovskite compounds, such as La_2NiMnO_6, Dy_2NiMnO_6, Sm_2NiMnO_6, Pr_2FeCrO_6, and Y_2FeAlO_6, have been investigated, both theoretically and experimentally [7–11]. These compounds exhibit various remarkable properties, suggesting potential applications. The coupling and changing of multiple properties may also enable their use in multifunctional devices that intersect various technological domains [5].

There are not many studies on the phonon properties of these double perovskites. Raman spectroscopy measurements as a function of temperature were performed on Y_2NiMnO_6 and La_2CoMnO_6 ceramics by Filho et al. [12], Silva et al. [13], and Iliev et al. [14]. The Raman spectra of PFAO and La_2CoMnO_6 were observed by Liu et al. [2] and Harbi et al. [15]. Raman studies of La_2CoMnO_6 thin film grown on $NdGaO_3$ at high temperatures were observed by Kumar et al. [16]. Spin–phonon coupling was detected in Tb_2NiMnO_6, although it was found to be weaker than in La_2NiMnO_6 and Pr2NiMnO6 [13,17,18].

In this study, we will explore the electric, magnetic, and phonon properties of the recently discovered multiferroic double perovskite compound PFAO using Green's function theory and a microscopic model.

2. Model and Method

We propose a model to describe the properties of PFAO. As PFAO is a multiferroic compound, the Hamiltonian must include the following three terms for the magnetic system, electrical system, and the one that connects them:

$$H = H_m + H_e + H_{me}. \tag{1}$$

The Hamiltonian for the magnetic properties, including the spin–phonon coupling, which is observed experimentally in PFAO, can be formulated as follows:

$$H_m = H_{sp} + H_{ph} + H_{sp-ph}. \tag{2}$$

The Hamiltonian H_{sp} for the magnetic subsystem is described by the Heisenberg model, which is extended taking into account the super-exchange interactions between Fe and Pr ions:

$$\begin{aligned} H_{sp} &= -\sum_{i,j} J_{ij}^{Fe-Fe} \mathbf{S}_i^{Fe} \cdot \mathbf{S}_j^{Fe} - \sum_{i,j} J_{ij}^{Fe-Pr} \mathbf{S}_i^{Fe} \cdot \mathbf{S}_j^{Pr} \\ &- \sum_i D_i (S_i^{zFe})^2 - \sum_i D_i (S_i^{zPr})^2 - g\mu_B \mathbf{h} \cdot \sum_i \mathbf{S}_i^{Fe}. \end{aligned} \tag{3}$$

The Heisenberg spin operator $\mathbf{S}_i$ refers to the localized spins at site i. The exchange interaction constants J_{ij}^{Fe-Fe} include contributions from both nearest-neighbor interactions $J_1 > 0$ and next-nearest-neighbor interactions $J_2 < 0$ between Fe–Fe ions, and $J_{ij}^{Fe-Pr} > 0$ describes the super-exchange interaction of Fe^{3+}-O-Pr^{3+}. The last interaction is responsible for the ferromagnetic behavior of PFAO due to the much higher anisotropy of the Pr ion compared with that of the Fe ion. The Al^{3+} ion is non-magnetic. The term D_i denotes the single-ion anisotropy constant and h is the external magnetic field.

The Hamiltonian H_{ph}, which is used to study the phonon properties in PFAO, accounts for lattice vibrations, incorporating anharmonic phonon–phonon interactions described by the constants B and A:

$$\begin{aligned} H_{ph} &= \frac{1}{2!} \sum_i \omega_{0i} a_i a_i^+ + \frac{1}{3!} \sum_{i,j,r} B(i,j,r) Q_i Q_j Q_r \\ &+ \frac{1}{4!} \sum_{i,j,r,s} A(i,j,r,s) Q_i Q_j Q_r Q_s. \end{aligned} \tag{4}$$

In this formulation, the frequency of the lattice mode is denoted by ω_{0i}, while Q_i represents the normal coordinate. Q_i can be expressed in terms of the phonon creation operator a^+ and the annihilation operator a:

$$Q_i = (2\omega_{0i})^{-1/2} (a_i + a_i^+). \tag{5}$$

The Hamiltonian describing spin–phonon coupling is expressed as follows:

$$H_{sp-ph} = \frac{1}{2}\sum_{i,j} F(i,j)Q_i S_j^z - \frac{1}{4}\sum_{i,j,r} R(i,j,r)Q_i Q_j S_r^z + h.c. \tag{6}$$

The amplitudes for the interacting phonons and spin excitations are represented by parameters F and R, corresponding to first-order and second-order interactions, respectively.

The spontaneous magnetization, $M = \langle S^z \rangle$, is calculated using the following equation:

$$M = \langle S^z \rangle = \frac{1}{N}\sum_{ij}\Big[(S+0.5)\coth[(S+0.5)\beta E_{mij})] - 0.5\coth(0.5\beta E_{mij})\Big]. \tag{7}$$

Here, S denotes the spin value, $\beta = 1/k_B T$, and E_{mij} represents the spin excitations derived from the spin Green's function $G_{ij} = \ll S_i^+; S_j^- \gg$ using the method of Tserkovnikov [19]. We will provide a brief description. After a formal integration of the equation of motion for the retarding two-time GF

$$G_{ij}(t) = \langle\langle S_i^+(t); S_j^- \rangle\rangle \tag{8}$$

one obtains

$$G_{ij}(t) = -i\theta(t)\langle [S_i^+; S_j^-] \rangle \exp(-iE_{ij}(t)t), \tag{9}$$

$$\begin{aligned} E_{ij}(t) = E_{ij} &- \frac{i}{t}\int_0^t dt' t' \Bigg(\frac{\langle [j_i(t); j_j^+(t')] \rangle}{\langle [S_i^+(t); S_j^-(t')] \rangle} \\ &- \frac{\langle [j_i(t); S_j^-(t')] \rangle \langle [S_i^+(t); j_j^+(t')] \rangle}{\langle [S_i^+(t); S_j^-(t')] \rangle^2} \Bigg). \end{aligned} \tag{10}$$

Here, $j_i(t) = \langle [S_i^+(t), H_{interaction}] \rangle$. E_{ij} is the excitation energy:

$$E_{ij} = \frac{\langle [[S_i^+, H]; S_j^-] \rangle}{\langle [S_i^+; S_j^-] \rangle} \tag{11}$$

From the time-dependent term in Equation (10) we can calculate the damping effects.

The ferroelectric subsystem is described by the transverse Ising model:

$$H_e = -\Omega\sum_i B_i^x - \frac{1}{2}\sum_{ij} J'_{ij} B_i^z B_j^z. \tag{12}$$

In this context, B_i^x and B_i^z denote the pseudo-spin operators, while J'_{ij} represents the pseudo-spin exchange interaction. The dynamics of the ferroelectric component are governed by the first term connected with a flipping rate Ω and operator B^x. A new coordinate system is introduced by rotating the original one used in Equation (12) by an angle θ in the xz plane, such that $\langle B^{x'} \rangle = 0$ in this rotated system.

The spontaneous polarization P is calculated from Green's function:

$$G_{ij} = \ll B_i^+; B_j^- \gg \tag{13}$$

as

$$P = \frac{1}{2N^2}\sum_{ij} \tanh\frac{E_{fij}}{2k_B T}. \tag{14}$$

E_{fij} is the pseudo-spin wave energy.

The term describing the magnetoelectric coupling between the two order parameters is given by the following:

$$H_{me} = -g\sum_{ijkl} B_i^z B_j^z \mathbf{S}_k \cdot \mathbf{S}_l. \tag{15}$$

We assume a quadratic magnetoelectric interaction in the spin and pseudo-spin operators with a small value for the magnetoelectric constant g due to the large difference between the phase transition temperatures of the ferroelectric and ferromagnetic subsystems, $T_C^{FE} >> T_C^{FM}$, observed in PFAO.

To study the phonon properties, we define Green's function

$$G_{ij}(t) = \langle\langle a_i(t); a_j^+ \rangle\rangle. \tag{16}$$

Using Hamiltonian (2), we derive the expression for the phonon energy, which is renormalized by the anharmonic interactions between spin–phonon and phonon–phonon:

$$\begin{aligned} \omega(\mathbf{k})^2 &= \omega_0^2 - 2\omega_0 \frac{1}{N'}\sum_{\mathbf{q}}\Big(R(\mathbf{k},\mathbf{q})\langle S^z\rangle^2 \\ &- \frac{1}{2N'}\sum_{\mathbf{q}} A(\mathbf{k},\mathbf{q})(2\bar{N}(\mathbf{q})+1) - B(\mathbf{k})\langle Q_{\mathbf{k}}\rangle\Big), \end{aligned} \tag{17}$$

with

$$\langle Q_{\mathbf{k}}\rangle = \frac{F_{\mathbf{k}}\langle S^z\rangle^2 - \frac{1}{N'}\sum_{\mathbf{q}} B_{\mathbf{k},\mathbf{q}}(2\bar{N}_{\mathbf{q}}+1)}{\omega_0 - R_{\mathbf{k}}\langle S^z\rangle^2 + \frac{1}{N'}\sum_{\mathbf{q}} A_{\mathbf{k},\mathbf{q}}(2\bar{N}_{\mathbf{q}}+1)}. \tag{18}$$

The phonon correlation function $\bar{N} = \langle a^+ a^- \rangle$ is determined through the Spectral theorem. Additionally, phonon damping is evaluated employing the method of Tserkovnikov [19].

3. Numerical Results and Discussion

For the numerical evaluation of the properties of PFAO, the following model parameters are used: J_1^{Fe-Fe} = 55 K, $J_2^{Fe-Fe} = -115$ K [20], J^{Fe-Pr} = 47 K, D^{Fe} = 0.2 K, J' = 235 K, Ω = 20 K, g = 10 K, F = 23 cm^{-1}, R = 18 cm^{-1}, A = 6.61 cm^{-1}, B = −2.94 cm^{-1}, S = 2.5 for the magnetic spins, S = 0.5 for the pseudo-spins.

We will provide a short description for some of the observed parameters. From the expression $J = 3k_B T_C / zS(S+1)$, T_C is the calculated interaction constants, z denotes the number of nearest neighbors, S is the spin value, and T_C is the Curie temperature. The value for Ω is observed from the ferroelectric energy at very high temperatures: $2\Omega = E_f$.

Please note that because this is the first theoretical paper describing the multiferroic properties of PFAO, we only made a qualitative comparison with the first and only experimental work of Liu et al. [2]. Thus, the curves in Figures 1–4 are theoretically calculated. The points are the calculated values that are connected through lines.

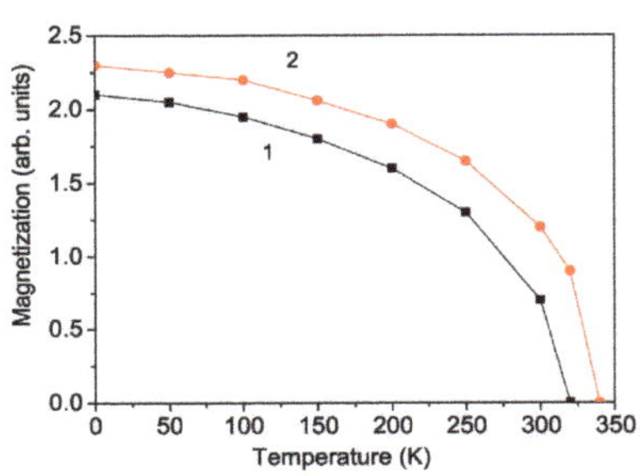

Figure 1. Temperature dependence of the magnetization M in PFAO for different g values: (1) 10 and (2) 15 K.

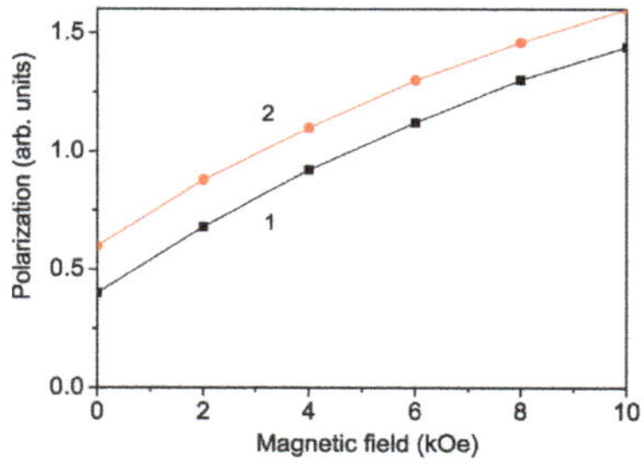

Figure 2. Magnetic field dependence of the polarization P in PFAO for different T values: (1) 300 and (2) 150 K.

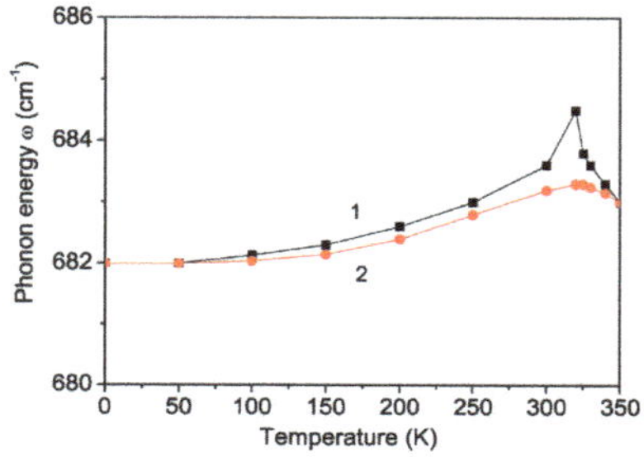

Figure 3. Temperature dependence of the phonon mode ω = 682 cm^{-1} in PFAO for different magnetic fields h: (1) 0 and (2) 50 kOe.

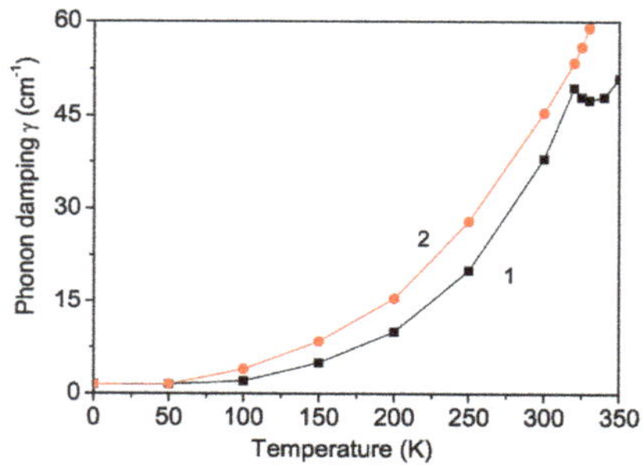

Figure 4. Temperature dependence of the damping of the phonon mode ω = 682 cm^{-1} in PFAO for different magnetic fields h: (1) 0 and (2) 50 kOe.

3.1. *Temperature Dependence of the Magnetization in PFAO*

Firstly, in order to test our model, using Equation (7), we investigated the magnetization M of the PFAO compound. It must be noted that the spin–phonon and magnetoelectric interactions lead to a renormalization of the exchange spin-interaction constant J, which is modified to J_{eff}:

$$J_{eff} = J = 2F^2/(\omega_0 - MR) + 2gP^2\cos^2\theta. \tag{19}$$

Figure 1 presents $M(T)$. It is evident that PFAO demonstrates the ferromagnetic behavior, with M decreasing as the temperature rises and reaching zero at T_C^{FM} = 320 K. It is the transition temperature from the ferromagnetic to paramagnetic phases, as reported by Liu et al. [2]. This ferromagnetic behavior is primarily attributed to the super-exchange interaction between the Fe and Pr ions (Fe^{3+}-O-Pr^{3+}). Moreover, magnetization increases with a higher magnetoelectric coupling constant g (see Figure 1, curve 2).

3.2. Magnetic Field Dependence of the Polarization in PFAO

The polarization is analyzed using Equation (14). A key characteristic of multiferroic materials is the impact of an external magnetic field on their electric properties (and similarly, the influence of an external electric field on their magnetic properties). To explore the multiferroic nature of PFAO, we calculated the magnetic field dependence of the polarization P. This behavior arises from the magnetoelectric coupling g, which renormalizes the pseudo-spin exchange interaction constant J' to J'_{eff}:

$$J'_{eff} = J' + 2g(\langle S^- S^+ \rangle + \langle S^z S^z \rangle)/\cos\theta. \tag{20}$$

The polarization P enhances with the increase in the external magnetic field h, as is illustrated in Figure 2, which indicates the multiferroic behavior of the double perovskite compound PFAO. In the double perovskite compound under consideration, the structural units responsible for the occurrence of magnetization and polarization are different, as well as the temperatures, and the corresponding phase transitions differ significantly. This determines the nature of the magnetoelectric interaction, which is quadratic in terms of polarization and magnetization. Thus, the effective pseudo-spin arrangement is renormalized by the magnetoelectric coupling, and this renormalization is proportional to $\langle S^z \rangle^2$. The experimentally found and theoretically proven ferromagnetic behavior of the system assumes an increase in magnetization with an increase in the magnetic field h. This leads to an increase in J'_{eff}, which, at a fixed temperature will lead to an increase in polarization with an increase in the magnetic field. This is in disagreement with the result of Liu et al. [2]. They reported a reduction in polarization and dielectric constant of PFAO with an increase in magnetic field h. Let us emphasize that we observed an increase in the dielectric constant with an increase in the magnetic field in the double-perovskite La_2NiMnO_6 [21]. This increase is confirmed by many experimental data. Additionally, we observed that the polarization P increased as the temperature T decreased (see Figure 2, curve 2).

3.3. Temperature and Magnetic Field Dependence of the Phonon Energy in PFAO

Now, we will investigate the temperature dependence of the A_{1g} phonon mode with a frequency of ω = 682 cm^{-1}. Reference [22] calculated the temperature behavior of the phonon energy in hexagonal bulk multiferroics for different R, which can be either positive ($R > 0$) or negative ($R < 0$). It was found that for $R > 0$, the phonon energy increased with temperature below T_C^{FM}, whereas for $R < 0$, it decreased with the rising temperature. For PFAO, we assumed a positive anharmonic spin–phonon constant ($R > 0$). The result is depicted in Figure 3, curve 1. It shows a softening below T_C^{FM}. It is noteworthy that Filho et al. [12] reported a similar decrease in temperature dependence of the phonon mode $\omega(T)$ = 652 cm^{-1} below the ferromagnetic phase transition temperature T_C^{FM} = 100 K in the double perovskite Y_2NiMnO_6. A similar behavior was noted in several double perovskites (La_2NiMnO_6, A_2CoMnO_6, where A = La, Pr, Nd) [23,24], respectively, confirming the presence of spin–phonon coupling in these materials. Additionally, a distinct kink in the $\omega(T)$ curve at T_C^{FM} = 320 K was observed (see Figure 3, curve 1), which indicates a strong spin–phonon interaction R. This finding suggests that the stretching vibrational mode of the oxygen octahedra plays a role in stabilizing the magnetic interaction. The combination of ferromagnetic ordering and spin–phonon coupling below T_C^{FM} leads to a decrease in the phonon mode with decreasing temperature T. Bhatti et al. [25] reported a similar kink at T_C^{FM} = 172 K in the temperature dependence of both phonon energy and damping in the double perovskite manganite Pr_2CoMnO_6.

Nair et al. [17] proposed a correlation between the size of the rare earth ion and the strength of the spin–phonon coupling, indicating that a decrease in the rare earth ionic radius resulted in reduced spin-phonon coupling. Consequently, they reported that the spin–phonon coupling observed in Tb_2NiMnO_6 was weaker than that in La_2NiMnO_6 and Pr_2NiMnO_6. However, Iliev et al. [14] demonstrated that the strong spin–phonon coupling in Y_2NiMnO_6 suggested that rare earth ionic size was not the sole factor influencing the spin–phonon coupling, as this effect occurred in Y_2NiMnO_6 despite the smaller ionic radius of Y compared with Tb. We propose that the exchange interactions between the two transition metal ions, as well as those between the transition metal and rare earth ions, are also crucial for determining the strength of the spin–phonon interaction. This topic will be explored in a future publication. Comparing the properties of these materials poses challenges due to their diverse physical characteristics, including various magnetic orderings such as ferromagnetic, antiferromagnetic, or ferrimagnetic [5]. For instance, Bi_2NiMnO_6 is a ferromagnet, but La_2NiMnO_6 is a ferromagnetic semiconductor. Furthermore, the double perovskites can exhibit ferroelectric, antiferroelectric, or relaxor ferroelectric properties.

The kink shown in Figure 3 is not present when $R = 0$. Below T_C^{FM}, the temperature dependence of the phonon mode is primarily influenced by spin–phonon interactions, while above T_C^{FM}, where $\langle S^z \rangle$ terms are absent, only anharmonic phonon–phonon interactions remain.

Furthermore, a characteristic of multiferroics is that their properties can be controlled by external fields. Applying an external magnetic field h to PFAO causes the phonon energy $\omega(T)$ to decrease, which eliminates the kink at T_C^{FM}. The result is demonstrated in Figure 3, curve 2.

3.4. Temperature and Magnetic Field Dependence of Phonon Damping in PFAO

The phonon damping γ, represented by the full width at the half maximum (FWHM) of the Raman peaks, enhances with an increase in temperature. It is important to note that γ rises with T for both $R > 0$ and $R < 0$, as it is proportional to R^2. Figure 4, curve 1, illustrates the temperature dependence of phonon damping without the magnetic field, $h = 0$. $\gamma(T)$ increases with an increase in temperature T and also shows a kink at T_C^{FM}, although this kink is smaller than that observed in the phonon energy $\omega(T)$. The phonon damping γ increases, i.e., the width of the Raman line becomes larger, and the kink disappears with the application of a magnetic field h (see Figure 4, curve 2).

Unfortunately, there are currently no experimental data available for $\omega(T,h)$ and $\gamma(T,h)$ in PFAO. We hope that future experimental and theoretical works related to phonon results will confirm our results.

4. Conclusions

Using Green's function technique and a microscopic model, we have examined the multiferroic and phonon properties of the newly discovered multiferroic double perovskite PFAO. The magnetization M diminishes as the temperature increases, vanishing at the ferromagnetic Curie temperature T_C^{FM}. M increases with a higher magnetoelectric coupling constant g. The application of an external magnetic field enhances the polarization, demonstrating the multiferroic nature of PFAO. The temperature dependence of the phonon mode at $\omega = 682$ cm^{-1} and its damping reveals a kink at T_C^{FM}, attributed to the anharmonic spin–phonon coupling, which is significant below T_C^{FM}. Above T_C^{FM}, only the anharmonic phonon–phonon interactions persist. The application of a magnetic field causes the kinks in both the phonon energy and damping to disappear.

Author Contributions: Conceptualization, J.M.W.; Methodology, A.T.A.; Software, I.N.A.; Formal analysis, A.T.A.; Investigation, A.T.A., I.N.A. and J.M.W.; Writing—original draft, J.M.W. All authors have read and agreed to the published version of the manuscript.

Funding: This research received no external funding.

Institutional Review Board Statement: Not applicable.

Informed Consent Statement: Not applicable.

Data Availability Statement: The raw data supporting the conclusions of this article will be made available by the authors on request.

Conflicts of Interest: The authors declare no conflicts of interest.

References

1. Fiebig, M. Revival of the magnetoelectric effect. *J. Phys. D* **2005**, *38*, R123. [CrossRef]
2. Liu, S.; Xiang, F.; Cheng, Y.; Luo, Y.; Sun, J. Multiferroic and Magnetodielectric Effects in Multiferroic Pr_2FeAlO_6 Double Perovskite. *Nanomater* **2022**, *12*, 3011. [CrossRef]
3. Ravi, S. Multiferroism in Pr_2FeCrO_6 perovskite. *J. Rare Earth.* **2018**, *36*, 1175–1178. [CrossRef]
4. Vasala, S.; Karppinen, M. $A_2B'B''O_6$ perovskites: A review. *Prog. Solid State Chem.* **2015**, *43*, 1–36. [CrossRef]
5. Hossain, A.; Bandyopadhyay, P.; Roy, S. An overview of double perovskites $A_2B'B''O_6$ with small ions at A site: Synthesis, structure and magnetic properties. *J. Alloys Compd.* **2018**, *740*, 414–427. [CrossRef]
6. Song, B.; Shen, J.; Zhao, H.; Kumar, A.; Xu, Q.; Zhai, Y.; Li, Q. A new room-temperature multiferroic material: Y_2FeAlO_6. *Ceram. Int.* **2021**, *47*, 10873–10879. [CrossRef]
7. Das, R.; Choudhary, R.N.P. Dielectric relaxation and magneto-electric characteristics of lead-free double perovskite: Sm_2NiMnO_6. *J. Adv. Ceram.* **2019**, *8*, 174–185. [CrossRef]
8. Wu, S.-Q.; Cheng, S.; Lu, L.; Liu, M.; Jin, X.-W.; Cheng, S.-D.; Mi, S.-B. B-Site ordering and strain-induced phase transition in double-perovskite La_2NiMnO_6 Films. *Sci. Rep.* **2018**, *8*, 2516. [CrossRef] [PubMed]
9. Masud, M.G.; Dey, K.; Ghosh, A.; Majumdar, S.; Giri, S. Occurrence of magnetoelectric effect correlated to the Dy order in Dy_2NiMnO_6 double perovskite. *J. Appl. Phys.* **2015**, *118*, 064104. [CrossRef]
10. Das, N.; Singh, S.; Joshi, A.G.; Thirumal, M.; Reddy, V.R.; Gupta, L.C.; Ganguli, A.K. Pr2FeCrO6: A type I multiferroic. *Inorg. Chem.* **2017**, *56*, 12712–12718. [CrossRef]
11. Zhao, H.; Song, B.; Shen, J.; Li, Q. Enhancement of Magnetoelectric Coupling in Rare Earth Substituted Y_2FeAlO_6 Ceramics. *J. Supercond. Novel Magn.* **2023**, *36*, 1133–1142. [CrossRef]
12. Filho, R.B.M.; Ayala, A.P.; Paschoal, C.W.A. Spin-phonon coupling in Y_2NiMnO_6 double perovskite probed by Raman spectroscopy. *Appl. Phys. Lett.* **2013**, *102*, 192902. [CrossRef]
13. Silva, R.X.; Santos, C.C.; Reichlova, H.; Marti, X.; Paniago, R.; Paschoal, C.W.A. Short-range antiferromagnetic interaction and spin-phonon coupling in La2CoMnO6 double perovskite. *Vibr. Spectr.* **2022**, *123*, 103426. [CrossRef]
14. Iliev, M.N.; Guo, K.; Gupta, A. Raman spectroscopy evidence of strong spin-phonon coupling in epitaxial thin films of the double perovskite La2NiMnO6. *Appl. Phys. Lett.* **2007**, *90*, 151914. [CrossRef]
15. Harbi, A.; Azouaoui, A.; Benmokhtar, S.; Moutaabbid, M. Synthesis, Characterization, and DFT Investigation of the Role of Oxygen Vacancy on the Structural, Magnetic, and Electronic Properties of La_2CoMnO_6. *J. Supercond. Novel Magn.* **2022**, *35*, 1–8. [CrossRef]
16. Kumar, D.; Kumar, S.; Sathe, V.G. Raman studies of ordered double perovskite thin film at high temperatures. *AIP Conf. Proc.* **2015**, *1665*, 140030. [CrossRef]
17. Nair, H.S.; Swain, D.; Adiga, H.N.S.; Narayana, C.; Elzabeth, S. Griffiths phase-like behavior and spin-phonon coupling in double perovskite Tb_2NiMnO_6. *J. Appl. Phys.* **2011**, *110*, 123919. [CrossRef]
18. Truong, K.D.; Singh, M.P.; Jandl, S.; Fournier, P. Investigation of phonon behavior in Pr_2NiMnO_6 by micro-Raman spectroscopy. *J. Phys. Condens. Matter* **2011**, *23*, 052202. [CrossRef] [PubMed]
19. Tserkovnikov, Y.A. Decoupling of chains of equations for two-time Green's functions. *Theor. Math. Phys.* **1971**, *7*, 511. [CrossRef]
20. Apostolov, A.T.; Apostolova, I.N.; Wesselinowa, J.M. Magnetic field effect on the dielectric properties of rare earth doped multiferroic $BiFeO_3$. *J. Magn. Magn. Mater.* **2020**, *513*, 167101. [CrossRef]
21. Apostolov, A.T.; Apostolova, I.N.; Wesselinowa, J.M. The magnetoelectric effect in thin films of ferromagnetic semiconductor La_2NiMnO_6. *Phys. Status Solidi B* **2014**, *251*, 1219–1224. [CrossRef]
22. Wesselinowa, J.M.; Kovachev, S. Theoretical study of the phonon spectra of hexagonal multiferroics $RMnO_3$. *J. Phys. Condens. Matter* **2007**, *19*, 386218. [CrossRef]
23. Nasir, M.; Ahmed, A.; Park, H.J.; Sen, S. Evidence of spin-phonon coupling in La2NiMnO6 double perovskite. *J. Korean Crystal GrowthCrystal Techn.* **2021**, *31*, 112–115. [CrossRef]
24. Kumar, D.; Kumar, S.; Sathe, V.G. Spin–phonon coupling in ordered double perovskites A_2CoMnO_6 (A=La, Pr, Nd) probed by micro-Raman spectroscopy. *Solid State Commun.* **2014**, *194*, 59–64. [CrossRef]
25. Bhatti, I.N.; Bhatti, I.N. Spin-phonon coupling and dielectric spectroscopy in nano-crystalline Pr2CoMnO6 double perovskite manganite. *Physica B* **2021**, *610*, 412943. [CrossRef]

materials

MDPI

Review

Advances in Powder-Filled Mold Processes: A Comprehensive Review and Outlook

Pengyu Bai [1,2], Shuhua Yang [1,*], Yalin Yan [3,4], Dongliang Wang [2,3,4,*] and Yanwei Ma [2,3,4]

1 School of Material Science and Engineering, University of Jinan, Jinan 250022, China
2 Institute of Electrical Engineering and Advanced Electromagnetic Drive Technology, Qilu Zhongke, Jinan 250013, China
3 Key Laboratory of Applied Superconductivity, Institute of Electrical Engineering, Chinese Academy of Sciences, Beijing 100190, China
4 University of Chinese Academy of Sciences, Beijing 100049, China
* Correspondence: yangshuhua78@163.com (S.Y.); dongliangwang@mail.iee.ac.cn (D.W.)

Abstract: Powder molding technology is a versatile process widely used in the pharmaceutical, ceramic, chemical, food, and powder metallurgy industries. The powder-filling mold process is a key link in powder compression molding, and the uniformity and consistency of powder filling directly affect the final quality of powder products. Powder filling of molds is a more complex flow process. This paper first reviews the methods used to test powder flow characteristics and comments on their applicability to the mold-filling process, provides an in-depth discussion of four different filling techniques, focusing on the flow behavior of the powder during the filling process, and analyzes the effects of powder characteristics and process parameters on the filling effect. By reviewing the latest advances and identifying the key challenges, a valuable reference is provided for the mold-filling process.

Keywords: die filling; powder flow; uniform fill; powder-filling process

1. Introduction

Powder-pressing molding technology is a universal powder processing technique widely applied across various industries, such as pharmaceuticals [1–4], ceramics [5–8], chemicals [9–11], food [12–14], and powder metallurgy [15,16]. As global manufacturing innovation and processing scale continue to expand, the application of powder molding technology across industries is becoming increasingly widespread. The demand for product performance is also rising, with conventional methods of producing powder products often resulting in defects such as voids [17–19], cracks [20,21], and uneven textures [22]. Therefore, meeting the requirements for uniformity, densification, high precision, and strong shape adaptability has become the direction of development for powder molding technology. The powder pressing process generally includes three stages [23]: mold filling, pressing, and ejection, where the powder mixture is filled into the mold cavity either through gravity or other external forces, compressed under high pressure into a continuous dense structure, and finally ejected from the mold. These three stages have a significant impact on the final performance of the powder compact [24], especially the process of powder filling into the mold cavity.

In 1998, Demetry et al. used tactile sensors to measure the density distribution of powder filling in the mold before and after pressing, discovering that the post-pressing density distribution was consistent with that before pressing, indicating that the pressing process could not eliminate the adverse factors during mold filling [25]. In 2008, Korachkin et al. studied the impact of unevenness during the filling process on the results of pressing, showing that unevenness during the mold-filling process significantly increased variation in pressing stress and density gradient [26]. Therefore, the powder-filling process (mold filling) is key to powder-pressing molding. The densification and uniformity of powder filling directly affect the subsequent pressing process and, consequently, the final quality and performance of the powder products. The powder loading process involves filling a certain amount of powder into the mold, ensuring that the filled powder is densely and reasonably distributed within the mold cavity uniformly. The majority of actual powder production processes adopt a free-filling method [27–31], where a certain amount of powder is poured into the mold, and the powder loosely fills the cavity solely under the action of gravity coupled with the effect of the geometric properties of the powder; there are often a large number of voids between adjacent powders, with the loose packing density being only 30–65% of the theoretical density. The size and shape of the voids affect the density and hardness of the green body, leading to defects such as cracks and dimensional deformation during subsequent processing. The presence of defects affects the strength [32], hardness [33], toughness [34], ductility [35], dimensional stability [36], and other mechanical properties of the finished product. The complexity of powder mold filling stems from the interaction of multiple variables, including the physical properties of the powder [37,38], environmental conditions [39,40], filling methods [41,42], etc. In addition, the design of the mold is critical, and the shape [37], size [43], and surface quality of the mold [44] can significantly affect the flow of powder and filling efficiency. Accurate mold design can reduce friction in the powder-filling process to improve mobility and achieve uniform filling. Therefore, revealing the densification and uniformity mechanism of powder filling, understanding the flow process of mold-filling powder, exploring the impact of powder characteristic parameters and powder-filling process parameters on powder filling, and providing reference data for improving and optimizing the powder-filling process is of crucial importance for enhancing the performance of powder products.

In recent years, along with the development of materials science and engineering technology, the research of powder mold filling has gradually moved toward intelligence and automation, combined with computer simulation and emulation, which can help us better understand the dynamic changes in the filling process and then formulate the best filling strategy [42,45–47]. In addition, the continuous emergence of new materials, such as composite materials [48], high entropy alloys [49], and functional gradient materials [50], further stimulate the research needs of powder mold technology; in the face of how to ensure the performance of new materials under the premise of achieving high-efficiency filling, researchers are actively exploring new filling methods and technical means. Although there has been much research focusing on various aspects of powder mold filling, there is relatively little literature that systematically integrates and summarizes the results of this research. In order to fill this gap, this paper aims to provide a comprehensive review of the related theories and technologies of powder mold filling, discuss its influencing factors in detail, analyze the key problems and challenges in the field at present, and look forward to the direction of future development in order to provide theoretical guidance and practical references on the filling of powder molds to support the development of high-performance materials.

In this paper, the research background of the powder-filling process is briefly introduced in Section 1; Section 2 outlines the flow characteristics of powder during the mold-filling process and reviews the assistance of different flowability testing methods in the mold-filling process; Section 3 focuses on the flow behavior of powder under different filling methods; and, finally, Section 4 concludes with an outlook on the powder-filling process.

2. Powder Flowability in Mold-Filling Process

During the powder-filling process, powder flowability is an important evaluation parameter in industrial production [51–57]. To achieve consistent and uniform filling outcomes, freely flowing powders are preferred, as poor flowability can lead to product uniformity issues and unnecessary economic losses in industrial production. Therefore, understanding the flow characteristics of powder is crucial.

Powder flowability is not an inherent property of the powder particles but a derivative characteristic closely related to the powder's physical properties and flow environment parameters, such as powder particle size [58,59], size distribution [60], shape [58], moisture content [61,62], and interparticle forces [63,64]. Generally, powders with larger particle sizes and narrower size distributions exhibit better flow properties, displaying lower interaction behaviors when larger particles contact each other. As particle size decreases, van der Waals forces begin to overcome gravity, leading to higher adhesion and cohesion between adjacent particles and resulting in particle agglomeration. Typically, spherical particles have smaller contact areas, smoother morphologies, and are more prone to sliding past each other, thus exhibiting better flowability. The impact of moisture on powder flow characteristics depends on the moisture content; on the one hand, when the moisture content is low, the water molecules between particles can reduce interparticle friction, acting as a lubricant and eliminating static electricity and positively affecting powder flow. On the other hand, as moisture content increases, the water adsorbed on the particle surfaces forms liquid bridges and generates capillary forces, causing the powder fill volume to increase and small particles to agglomerate into irregularly shaped larger particles, worsening powder flowability. The aforementioned parameters ultimately affect interparticle forces, such as electrostatic forces, capillary forces, van der Waals forces, cohesion, and friction, which interact with each other and ultimately determine the powder's flow characteristics [65].

As can be seen in the above discussion, assessing powder flowability is a complex issue, and a single measurement parameter cannot accurately evaluate powder flow properties. Although various methods for evaluating powder flowability exist in the industry, finding a universal characterization method suitable for all applications is challenging. Powder flowability rankings can change across different evaluation systems, and linking test results to actual powder flow behavior is very challenging. The mold-filling process is an instantaneous process, and the flow behavior of powder is quite complex. To better study the scientific issues during the mold-filling process, Wu et al. developed a model system consisting of a fixed mold and a moving feeding mechanism to simulate the powder-filling process. This system comprises transparent molds and boots, using a high-speed video system to observe and record the filling process, as shown in Figure 1 [37]. In this system, the filling ratio of the cavity is defined as:

$$\delta = \frac{m_x}{M}, \tag{1}$$

where m_x is the mass of powder filled into the cavity and M is the mass of powder when the cavity is fully filled. The filling ratio is used to evaluate mold-filling efficiency, and the concept of "critical filling speed" is introduced. The relationship between the filling ratio and the charging boot's movement speed is given by the formula:

$$\delta = \left(\frac{v_c}{v_s}\right)^n, \tag{2}$$

where v_c is the critical movement speed of the charging boot, v_s is the actual movement speed of the charging boot, and n is the exponent [66]. The critical filling speed is used to evaluate powder flow performance, indicating the maximum filling speed that allows the powder particles to fully fill the cavity. When the filling speed exceeds the critical speed, the powder cannot completely fill the cavity; when the filling speed is less than the critical speed, the cavity is fully filled (i.e., the higher the critical filling speed, the better the flowability of the powder particles [37]).

Over the past few decades, various innovative methods for evaluating powder flow have been developed. Therefore, understanding the different evaluation methods for powder flowability and their interrelationships, as well as making reasonable use of powder flow characterization techniques, is crucial. Table 1 outlines different evaluation methods and their scopes of application, reviewing their assistance in the powder-filling process. Despite the existence of multiple testing methods to evaluate the flowability of powder during the mold-filling process, there are still some limitations that need to be further explored and improved to provide more comprehensive flowability data.

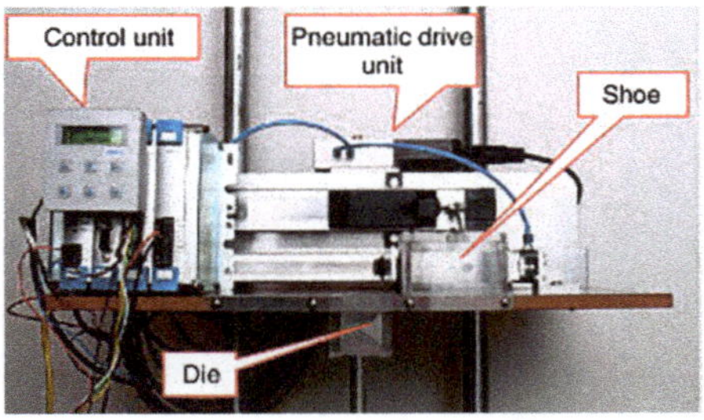

Figure 1. A model die-filling system [67].

Table 1. Comparison of different powder flow test methods.

	Test Methods	Measurement Parameter	Evaluation Criterion	Pros and Cons	Applicability for Mold Filling
Traditional Methods	Angle of Repose [68–70]	The three-dimensional angle formed by the height and radius of a conical pile of powder	The larger the angle of repose, the worse the flowability	Pros: Broad measurement range, can be used to test powders from free-flowing to very non-flowing Cons: Results are not repeatable, highly influenced by external factors	Does not provide insight into the dynamic flow process, only roughly indicates flow characteristics, not suitable for evaluating powder flow in molds
	Carr Index and Hausner Ratio [71–73]	Tapped density, bulk density	The smaller these values, the better the flowability	Pros: Simple, quick, and direct measurement Cons: Highly influenced by external factors	Lacks the dynamic flow process of powder, lacks predictability, not suitable for mold filling
	Orifice Flow Rate [74]	Flow rate through an orifice	The greater the orifice flow rate, the better the flowability	Pros: Powder container shape and orifice are flexible Cons: Difficult to measure non-flowing powders	This test is more for basic flowability assessment of free-flowing powders and is of limited practical help for mold filling
Modern Methods	FT4 Powder Rheometer [75–78]	Flow function, wall friction, compressibility index	Higher flow function values, smaller wall friction angles, and lower compressibility index indicate better flowability	Pros: Wide testing range, repeatable measurements, high sensitivity Cons: Complex operation, powder samples require pre-treatment	Can simulate the dynamic flowability of powder in actual operations and provide relevant parameters, but its direct predictive capability for mold filling remains limited
	RST Ring Shear Tester [79,80]	Flow function, wall friction force, bulk density, compressibility	Higher flow function values, smaller wall friction angles, and lower compressibility index indicate better flowability	Pros: High accuracy, good repeatability Cons: Not very suitable for viscous powders	Can accurately assess powder flowability and densification, offering some guidance for the mold-filling process, but also has certain limitations
	PFT Brookfield Powder Flow Tester [80–82]	Flow function, wall friction, compressibility index	Higher flow function values, smaller wall friction angles, and lower compressibility index indicate better flowability	Pros: Simple operation, reduces operator involvement in the testing process Cons: Shear stress values are fixed and not adjustable	Provides flowability parameters of reference value for mold filling, but the fixed shear stress of the equipment may limit its application across different powder types

3. Powder-Filling Methods

3.1. Gravity Filling

In the actual production process, powder filling is mostly done through gravity filling, where the powder mixture is poured into the mold from a feeder located above. The powder mixture is deposited into the mold cavity solely under the influence of gravity. This method of filling has been extensively studied over the past two decades [41,67,83–85].

Gravity filling typically utilizes a hopper for feeding. Research on powder flow within hoppers is widespread. Broadly speaking, powder flow within hoppers can be divided into two modes—"mass flow" and "funnel flow", as illustrated in Figure 2a,b [46,86]. In mass flow, all powder moves simultaneously, adhering to the "first-in, first-out" principle. In contrast, during funnel flow discharge, the powder is divided into two regions: a central flowing region and a stagnant region near the walls [37]. Powder from the central flowing region leaves the hopper first, followed by powder from the stagnant region. This occurs because the powder in the central flowing zone experiences the same pressure at the top and bottom, allowing for free fall under gravity. The powder in the stagnant region is influenced by the angle of the hopper and wall friction and exits subsequently [87,88]. Mass flow exhibits stability and lacks segregation, representing an almost ideal flow mode. On the other hand, funnel flow results in uneven powder distribution and is prone to segregation and separation [46,89,90]. Moreover, the discharge process from the hopper is not merely a singular flow mode, but a transition from mass flow to funnel flow occurs [46,91–93], indicating the existence of a flow transition zone.

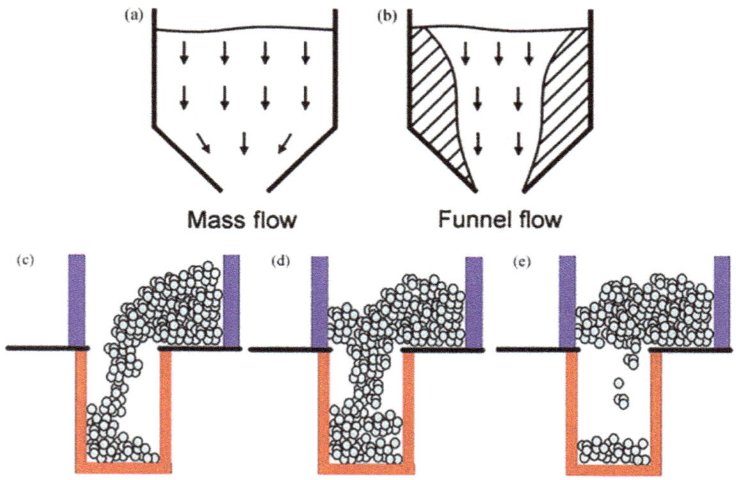

Figure 2. Powder flow patterns in the hopper: (**a**) mass flow, and (**b**) funnel flow [86]; and types of powder flow in gravity filling: (**c**) nose flow, (**d**) bulk flow, and (**e**) intermittent flow [39].

The analysis above suggests that the flow behavior in funnel flow mode is more complex, sparking research interest. When stagnant zones exist in funnel flow, the flow process resembles that of a shoe feeder system [37]. Utilizing shoe feeder systems, numerous scholars have systematically studied the flow behavior of powder during gravity filling. Wu et al. discovered that during the movement of the shoe feeder, powder can exhibit two flow modes: nose flow and bulk flow [23,37], as shown in Figure 2c–e [39]. When the shoe feeder, carrying a fixed mass of powder, slides over the mold, powder moves toward the rear end of the shoe due to friction between the powder and the shoe bottom, and the shoe feeder's initial acceleration, forming a nose-like profile. Powder from the upper region of the shoe feeder can deposit into the mold cavity through this profile, termed "nose flow", as depicted in Figure 2c. In scenarios of high speed or smaller mold openings, the nose profile swiftly slides through the mold opening, allowing the powder to fall freely from the shoe feeder bottom into the mold cavity. If this flow is continuous, it is termed "bulk flow", as shown in Figure 2d. During gravity filling, nose flow and bulk flow are the primary modes of powder movement. Nose flow predominates at lower shoe feeder speeds, while bulk flow takes precedence at higher speeds [39]. They also found that nose flow rapidly expels air from the mold cavity, achieving better filling results, whereas bulk flow, due to slower air expulsion, results in lower overall powder fill rates due to the interaction between residual air and powder particles [37]. Additionally, for certain viscous materials, powder flow tends to be discrete and random, with powder often depositing into the mold cavity in large aggregations or clusters of small particles, termed "intermittent flow", as shown in Figure 2e [39,94].

Powder-filling efficiency during gravity filling largely depends on powder characteristics and process conditions. Generally, smaller particle sizes, more complex particle shapes, and greater interparticle cohesion tend to lead to arching phenomena, poorer powder flowability, and worse filling outcomes. Wu et al. found through shoe feeder systems that the fill rate decreases with decreasing particle average size and increasing irregularity of particle shape [37]. Mills et al. also confirmed this, noting that larger-particle powders have higher critical speeds and better filling outcomes under gravity filling [38]. Mehrabi et al. studied the powder fill rate of Ti_6Al_4V metal powders produced by two different manufacturing processes—hydrogenated dehydrogenated irregular powder (HDH) (Figure 3a) and gas atomized spherical powder (GA) (Figure 3b)—and found irregular-shaped powder (HDH) had lower fill fractions in all regions compared to spherical powder (GA) due to the tendency of irregular HDH particles to interlock, resulting in lower packing rates. However, for both samples, fill fractions increased from the central region to the walls, since the friction between powder particles and the walls was less than the interparticle friction, leading to higher packing rates near the walls and lower rates in the center (Figure 3c) [95]. Moreover, numerous studies indicate that powder flow entering the mold interacts with airflow within the mold cavity, ultimately determining powder stacking outcomes. Wu [37], Schneider [39], Guo [40,96], and others have explored the effect of airflow on mold-filling outcomes. Guo et al. compared the powder-filling outcomes of dispersed and mixed particles in air and vacuum environments in a fixed boot, using a coupled discrete element method (DEM) and computational fluid dynamics (CFD) approach, finding that air significantly impacts filling outcomes, especially for smaller or lighter powder particles. For single-dispersed particles, as shown in Figure 3d,e, powder particles exhibit better flowability in a vacuum environment at the same moment, because air tends to slow down powder flow due to adhesion forces acting on the powder particles. Figure 3f demonstrates the variation in deposited particle mass in the mold over time during filling in both vacuum and air environments. The study found that during filling in a vacuum, the filling speed remains constant for most of the time until a deceleration phase is observed, by which point the mold is almost entirely filled. In air, powder flow can be categorized into two phases: an early slow flow state and a later rapid flow state, and it is evident that filling in the air is much slower than in a vacuum [40,96]. For poly-dispersed particles, when filling in air, due to the influence of air, lighter particles fall slower than denser particles, resulting in a layer of light particles at the top of the powder layer, which, after filling, enriches in the middle of the top layer of the powder. Meanwhile, in vacuum filling, such aggregation of light particles is not observed and both types of powder particles fall at the same speed and are evenly distributed [96]. Therefore, how to more quickly expel air from the cavity during the mold-filling process becomes crucial, especially for molds with complex shapes. However, the study also found that when the particle size and density are sufficiently large, the impact of airflow can be negligible. Moreover, some scholars have proposed that airflow during the mold-filling process also has certain advantages. The airflow acts as a lubricant between powder particles, making them more likely to slide past each other, thereby improving powder flowability [97,98].

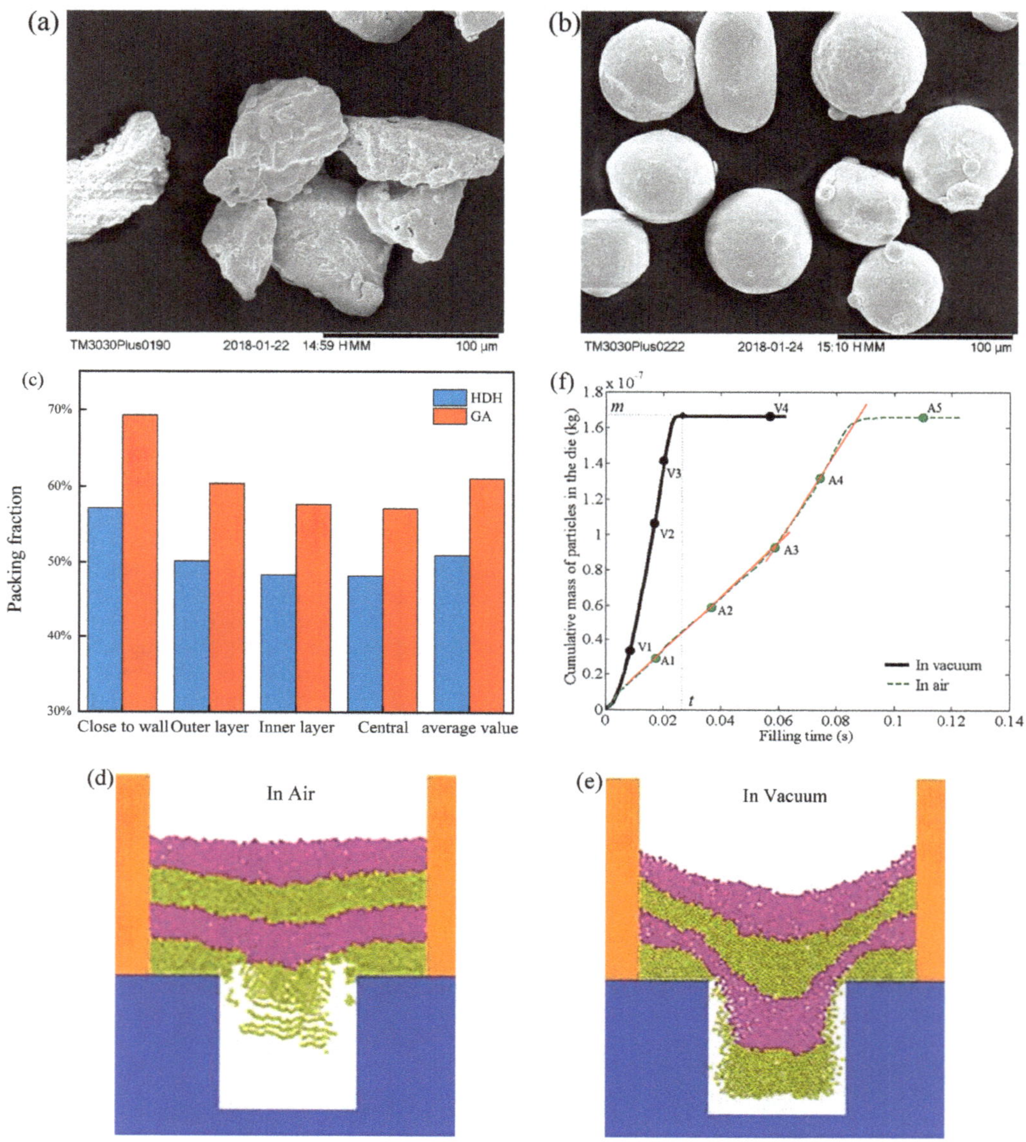

Figure 3. SEM images of (**a**) hydride-dehydration (HDH), and (**b**) gas atomization (GA) of Ti_6Al_4V samples [95]. (**c**) Filling fraction variation of HDH and GA powders along the radial region. Adapted from [95]. Powder flow patterns in air (**d**) and vacuum (**e**) at the same moment during the moldfilling process, where yellow and purple are the same particles with different colors in order to observe the macroscopic flow patterns. (**f**) Curves of the mass of particles deposited into the mold with mold-filling time during mold filling in vacuum and air [40].

Achieving a consistent and uniform density distribution of powder in the mold-filling process is challenging. Apart from the inherent characteristics of the powder affecting the filling, the dimensions and shape of the mold also significantly impact the powder fill density, prompting many meaningful explorations. Rice and Tengzelius conducted powder-filling experiments using circular molds of different diameters and found that fill density decreased with decreasing mold diameter, and when the mold diameter was sufficiently large, the fill density remained constant [99]. Bocchini et al. studied the impact of mold width on fill density using copper powder and found that fill density increased with increasing mold width, as smaller molds are not conducive to particle rearrangement [100]. Ready et al. later also confirmed that fill density decreased as the mold aspect ratio decreased [101]. Guillot et al. found that deeper molds resulted in lower fill densities. For molds with complex shapes, the density variation is even more complicated [43]. Wu et al. used a simulated shoe system to study the effect of simple and complex molds on the powder-filling process, where the complex mold included steps of varying width. The study showed that the powder-filling process is very sensitive to the geometry of the mold cavity. When using stepped molds, the fill rate was always slow, and the fill density was lower in narrower areas. For the moving simulation shoe, the position of the steps on either side differed, further altering the powder flow pattern, as shown in Figure 4. When the narrow section is downstream of the moving shoe direction, the chimney effect of the powder is stronger, producing noticeable turbulence. When the narrow section is upstream, the chimney effect is quickly suppressed, with no significant turbulence, making the filling process gentler. Additionally, they found that when using simple molds for filling, wider molds had higher critical speeds [37] (critical shoe speed can be used as an indicator of filling efficiency; the higher the critical shoe speed, the better the flowability of the powder). However, in actual applications, product shapes are often complex and varied. Precisely designing mold shapes and dimensions to accommodate uniform powder distribution within the mold according to actual needs presents a significant challenge.

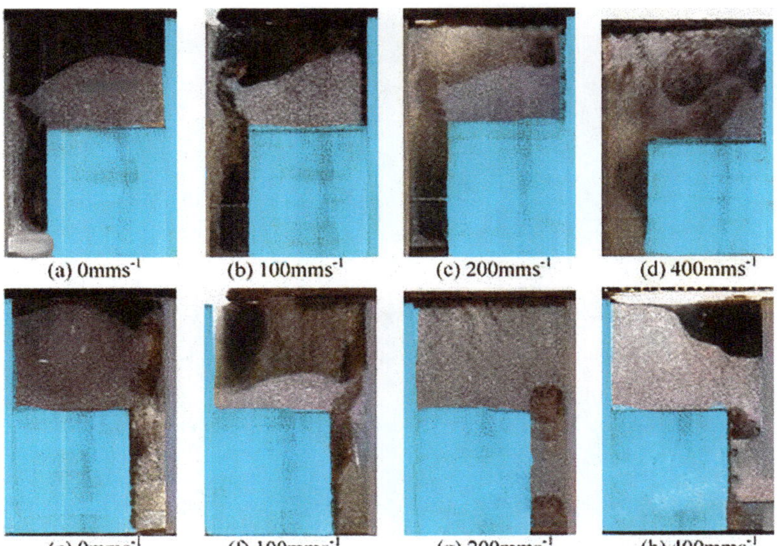

Figure 4. Powder flow states into a complex stepped mold at different feeding boot speeds, with the (**a–d**) steps on the right side and the (**e–h**) steps on the left side [37].

Over the past few decades, gravity-filling methods have been widely utilized across various industries due to their ease of implementation, low cost, and strong adaptability. However, ensuring the uniformity of powder within molds using gravity filling can be challenging, and the efficiency of mold filling is often low. Accordingly, the optimization of gravity-filling mold designs with respect to filling speed [41,102], filling time [41], and filling accuracy [103], with the objective of achieving higher filling density and minimizing powder segregation within the mold, represents a pivotal challenge. In this regard, utilizing computer modeling for computational analysis has shown great potential, For gravity filling of the material to be determined, the filling process was simulated by coupling the discrete element method (DEM) and computational fluid dynamics (CFD) using software such as ANSYS Fluent 2023 R1, OpenFOAM 10, and COMSOL Multiphysics 6.0 [104–109], and the effect of different process conditions on filling uniformity and product quality was evaluated by using different settings (e.g., filling speed, filling height). At the same time, actively exploring and innovating new powder-filling methods presents a promising new direction.

3.2. Suction Filling

Due to reliance solely on gravity for powder filling into mold cavities, the filling rate is relatively low, and the impact of air on powder-filling effectiveness is significant. As more powder fills the mold, the air pressure within the cavity increases, severely hindering the efficiency of powder filling [40,96]. To improve the powder-filling outcome, Jackson et al. devised a suction-based forced powder-filling method. In this approach, a movable piston is installed within the mold cavity, and the downward movement of the piston applies a suction force on the powder to accomplish the filling process, as illustrated in Figure 5 [67]. Studies have shown that compared to gravity-based filling, suction filling can significantly increase the mold-filling rate by approximately 2.5 times [110].

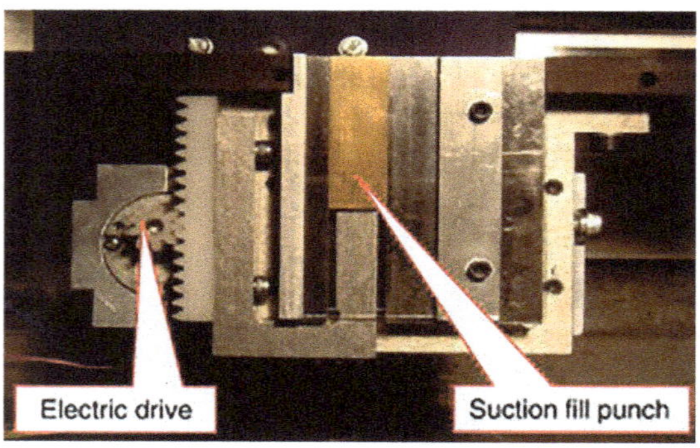

Figure 5. Model of the pumping filling schematic diagram of the system [67].

During the suction-filling process, initially, the absence of residual air in the mold cavity does not hinder the deposition of powder into the mold. On the other hand, due to the suction effect, a significant pressure difference is created above and below the powder bed, thereby accelerating the powder-filling rate. Wu and Guo et al. simulated the flow process of powder in the suction-filling mode. The simulation results revealed that the downward movement of the piston generates a low-pressure environment beneath the powder bed (inside the mold cavity), with the pressure inside the cavity gradually decreasing from top to bottom, as shown in Figure 6a–c. This creates a pressure gradient over the powder bed, further promoting powder flow, while the pressure distribution remains relatively uniform in the horizontal direction [45].

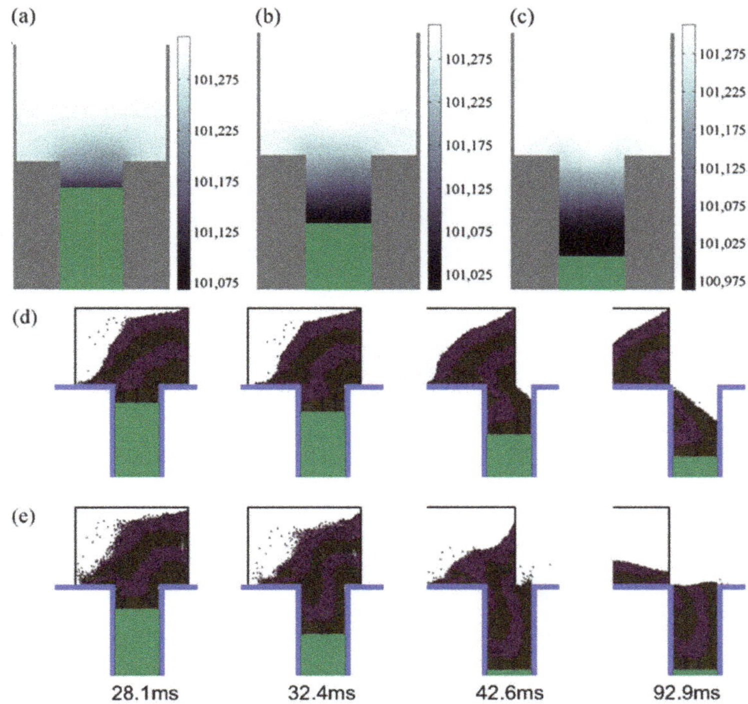

Figure 6. Air pressure distribution at different moments when suction filling is carried out at a piston speed of 100 mm/s: (**a**) t = 8.5 ms, (**b**) t = 21.3 ms, and (**c**) t = 31.5 ms. Powder flow state at different piston speeds when suction filling is carried out at a feed shoe speed of 140 mm/s: (**d**) piston speed of 100 mm/s, and (**e**) 276 mm/s piston speed [45].

The suction-filling process typically involves simultaneous powder feeding and suction. Therefore, for the suction-filling system, the relationship between the feeding speed and suction speed determines the powder-filling behavior. In initial experimental studies, the piston could only fall freely under the influence of gravity, uncontrollably. Subsequently, Sinka et al. improved and upgraded this mode by driving the piston with an electric device to conduct deeper studies on suction-filling conditions. When the suction speed is set to a constant 100 mm/s, whether in air or vacuum, the critical speed for suction filling is much higher than that for gravity filling, and a high filling rate is maintained when the shoe feeder speed exceeds the critical speed [67]. Wu et al. also studied the effect of different suction speeds on the filling process using a moving shoe feeder. When the shoe feeder moves at a constant speed and the suction filling is performed at a lower suction speed of 100 mm/s, the piston has not reached the bottom of the mold cavity by the time the shoe feeder completely passes the mold opening, resulting in incomplete filling of powder particles, as shown in Figure 6d. However, by increasing the suction speed to 276 mm/s, the piston can reach the bottom before the shoe feeder completely passes the mold opening, achieving complete powder filling, as shown in Figure 6e [45]. Zakhvatayeva et al. later proposed the existence of a critical ratio between the feeding speed and suction speed, i.e.,

$$V_{CR} = \frac{V_f}{V_s} \quad (3)$$

where V_f is the feeding speed, and Vs is the suction speed, meaning when suction filling operates at this critical speed ratio, the piston exactly reaches the bottom of the mold cavity as the shoe feeder fully passes the mold opening, resulting in complete powder filling. Thus, suction filling can be categorized into two different powder flow states: slow filling and rapid filling. When the speed ratio V_R is less than V_{CR}, it is termed "slow filling", where the piston reaches the bottom of the mold cavity before the shoe feeder fully passes the mold opening. Conversely, when V_R is greater than V_{CR}, it is "rapid filling", where the piston has not reached the bottom by the time the shoe feeder fully passes the mold opening. Generally, in the slow-filling state, complete powder filling can be achieved, while in the rapid-filling state, the powder-filling rate is lower, resulting in incomplete filling [57,111]. In summary, by evaluating the ratio of filling speed to suction speed and precisely controlling the suction speed and filling speed, various powder-filling behaviors can be scientifically predicted, providing strong support for industrial manufacturing.

It is noteworthy that the permeability of the powder plays a crucial role in suction filling. Mills et al. reported the effects of particle size and density under suction filling, conducting experiments with microcrystalline cellulose of different particle sizes and densities. The results showed that, compared to gravity filling, suction filling achieves a higher critical speed and filling rate, especially for powders with smaller particle sizes and lower bulk densities. This is because the principle of suction filling differs from that of gravity filling, where powder permeability plays a key role. Small and irregular powder particles generally have lower permeability, making it easier for small particles to fill the gaps between each other. During suction, the applied suction force significantly affects the powder, whereas large spherical particles typically have high permeability, allowing air to pass more easily through the gaps between particles, with less impact from the suction force [38].

Moreover, during the suction-filling process, the airflow within the mold cavity significantly differs from that in gravity filling. As depicted in Figure 7, in gravity filling, air can only be expelled through the tiny gaps between powder particles or system gaps. As the powder fills up, the gaps between particles gradually reduce, making it difficult to expel air, and the air pressure inside the cavity gradually increases, eventually leading to incomplete powder filling or uneven density distribution. In suction filling, the downward movement of the piston creates a negative pressure state in the mold cavity, promoting powder filling. During this stage, due to the pressure effect, air enters the mold cavity through system gaps or powder gaps. However, once the piston moves to the lowest end of the mold cavity, ending the suction process, powder then fills the mold under the influence of gravity, and air can only be expelled through the gaps, leading to a corresponding increase in cavity air pressure, thereby inhibiting further powder filling. Nonetheless, due to the negative pressure effect during the initial suction process, the filling outcome is usually better than pure gravity filling [38,41]. To improve this, mold design can be optimized by adding exhaust channels or using mold materials with good ventilation, such as special metals or porous ceramics, to further enhance airflow. Additionally, a vacuum-assisted exhaust system can be used during the filling process to reduce air retention, lower air pressure, and thus improve the filling outcome.

In conclusion, the powder suction-filling process is an efficient and precise powder-filling technique. By creating a negative pressure, it enables rapid and uniform filling of the mold cavity, reducing the formation of bubbles and voids, thereby enhancing the density and uniformity of the product. This process is applicable to a variety of powder materials, and is especially suitable for powders with small particle sizes, high viscosity, and poor permeability. Through continuous technological innovation and process optimization, integrating intelligent control systems and online monitoring technologies, and real-time prediction of powder flow behavior, the suction-filling process continues to show great development potential for the future.

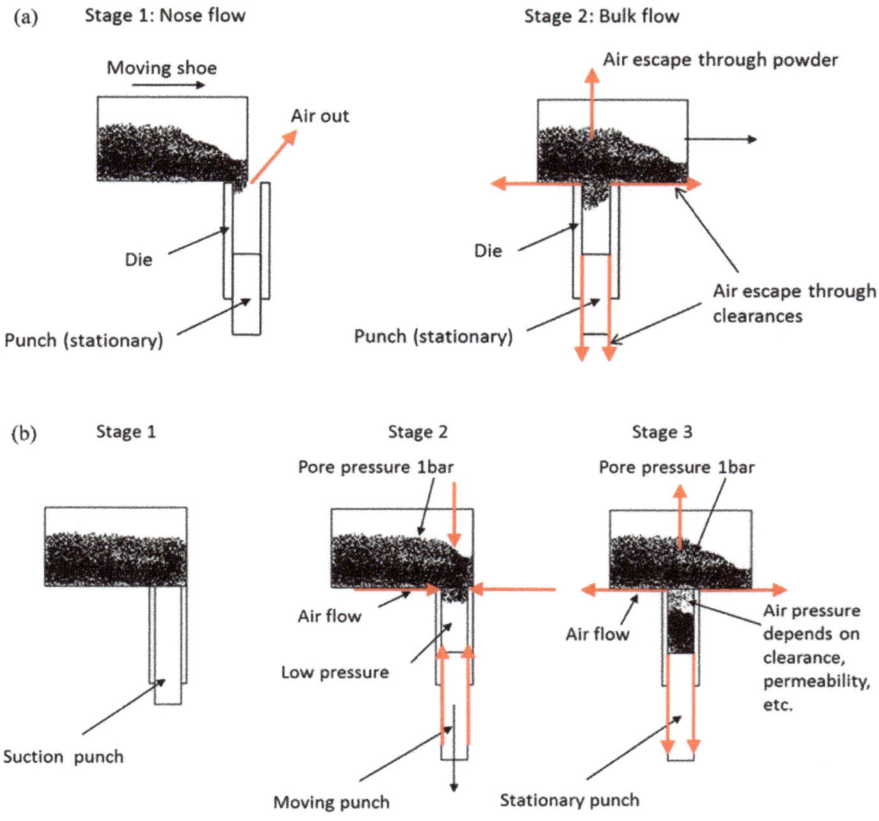

Figure 7. Schematic diagram of airflow mechanism during gravity filling (**a**) and suction filling (**b**) [41].

3.3. Vibration Filling

In most technical applications, there is a need for densely packed powder. However, traditional filling methods such as gravity filling often result in the phenomenon of arching between powder particles, leading to larger internal voids and lower fill densities. Additionally, the presence of particles of different shapes and sizes in the powder can cause uneven distribution and, consequently, segregation. Common methods to increase densification include vibration, tapping, and hammering [112], with vibration being a standard densification technique. During the powder-filling process, applying a certain degree of external vibration to the piled powder particles within the mold, under the combined action of gravity and external vibration, facilitates particle displacement and rearrangement. Thus, smaller particles can fill the gaps between larger particles, increasing the powder's fill density and ensuring a uniform distribution.

Optimizing vibration parameters is a critical step in enhancing the powder-filling outcome. To investigate the effect of different vibration conditions on the fill result, researchers have conducted extensive experimental studies and numerical simulations on the powder-filling process under vibration conditions. Evans et al. examined the impact of powder particle characteristics and vibration parameters on fill density during vibration filling. The study showed that applying external vibration could improve the fill density of powder, and changes in vibration parameters correspondingly altered the fill density. Figure 8a shows the theoretical fill density variation over time under the same frequency but different amplitudes, indicating that fill density increases with amplitude [113]. Knight et al. observed changes in the fill density of spherical particles within a cylindrical tube under vibration conditions and found that the powder-fill fraction increased with vibration intensity and eventually stabilized [114]. Fan et al. explored the effects of vibration time and frequency on the filling and mixing effect of Si powder, finding that powder-fill density increased with vibration time and remained constant after reaching a certain value. Furthermore, they discovered that, with an increase in vibration frequency, the time to reach a constant fill density decreased, as shown in Figure 8b, which displays the relationship between vibration time and powder-fill density at different vibration frequencies. Initially, the consolidation velocity of the powder is rapid, but it gradually slows over time, eventually ceasing, at which point the fill density stabilizes and reaches its maximum value. This is because when the powder is freely piled, interparticle arching is common, and the powder has many internal voids. In the early stages of vibration, under the combined action of gravity and external vibration, interparticle friction and collision cause displacement and rearrangement, breaking the arching effect present in loose packing and rapidly filling the internal gaps, leading to a quick consolidation velocity initially. As time progresses, the internal voids of the powder decrease, limiting particle displacement, thus gradually reducing the consolidation velocity until the interparticle gaps are essentially filled, and the fill density gradually stabilizes. Figure 8c shows the powder-fill density at different vibration frequencies over the same vibration time, indicating that, with an increase in vibration frequency, the fill density first increases and then decreases [115]. Jaggannagari et al. reported similar trends in elongated prismatic containers during particle vibration filling. At lower vibration frequencies, the interparticle forces are weaker, leading to reduced ability to fill gaps and break arching effects, resulting in lower fill density. As the vibration frequency increases, the capacity to destroy arching and fill gaps strengthens, further increasing the fill density. However, beyond a certain frequency, a noticeable "boiling" phenomenon occurs in the powder at the top of the mold cavity during vibration, causing particle stratification, and the higher the frequency, the more severe the "boiling" and stratification phenomena, leading to a decrease in fill density [116]. Hence, mid- to low-frequency vibration seems to be a good choice, providing enough energy for particle flow while avoiding the density reduction caused by excessive fluidization, thus achieving optimal filling results.

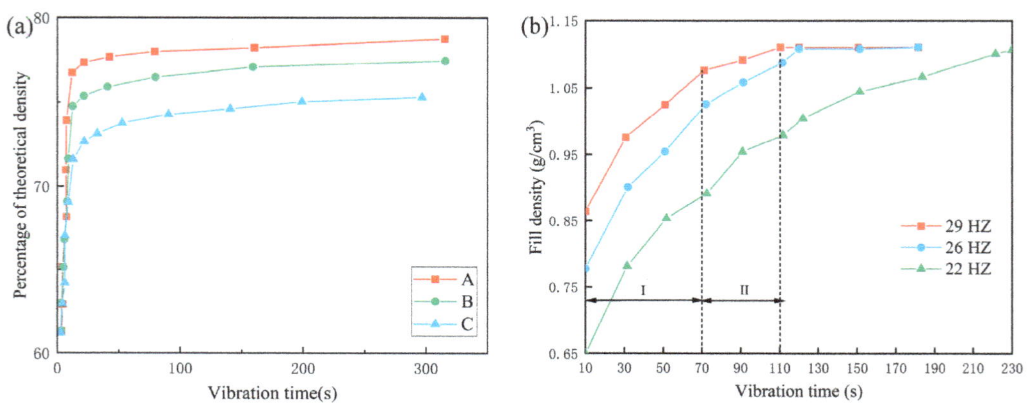

Figure 8. *Cont.*

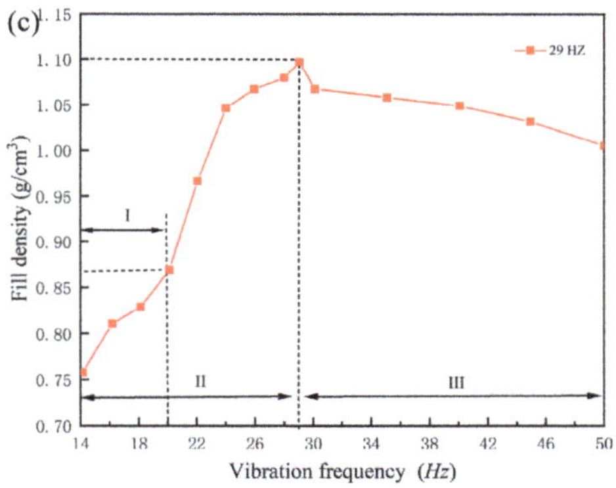

Figure 8. (**a**) Curves of the theoretical filling density of powder with time under different amplitudes (the amplitudes of A, B, and C in the curves are 2.6 mm, 2.0 mm, and 1.3 mm, respectively). Adapted from [113]. (**b**) Curve of filling density of Si powder with time under different vibration frequencies. (**c**) Variation curves of Si powder-filling density with vibration frequency under the same vibration time. Adapted from [115].

During the vibration-filling process, the particle system exhibits many fluid-like behaviors, such as the Brazil nut effect [117–120], particle convection [121–123], surface standing waves [124,125], arching, and separation [122,126], among others.

In the transport process of Brazil nuts, after long-term vibration during transit, the larger Brazil nuts are found at the top of the storage bin, while smaller nuts have moved to the bottom, a phenomenon known as the Brazil nut effect [117–120]. Conversely, under specific vibration conditions, particle shapes, and container shapes, a reverse Brazil nut effect may occur [127]. In previous studies, both phenomena have been explained using convection [128,129] and geometric mechanisms [119,130]. In 1831, Faraday first observed convection phenomena in a vibratory particle system within a cylindrical container: particles in the center of the container moved upward, while those along the container walls moved downward [131]. Over the following decades, researchers conducted extensive and in-depth studies on particle convection, with Gallas et al. discovering different types of convection cells using molecular dynamic methods [132]. Knight et al. studied convection in cylindrical containers with glass particles, finding that the upward flow velocity along the axial direction decays exponentially from top to bottom [114]. Hu et al. reported the effect of vibration frequency on the convection modes of particle systems under vertical vibration, identifying four different convection modes as the vibration frequency varied. When the frequency $f < 25$ Hz, the convection mode is as shown in Figure 9a, where convection particles begin to rise from the center of the container, then disperse and move downward along the container walls, forming nearly symmetrical double convection cells. As f increases from 25 Hz to 40 Hz, a noticeable particle mound forms on the mixture surface as shown in Figure 9b, with convection particles rising but deviating from the center of the container, then dispersing and moving downward along the container walls. With further increases in frequency, the deviation of convection particles from the central axis increases. When 40 Hz $< f <$ 55 Hz, convection particles first rise from one side of the container and then move downward on the opposite side, forming a single convection cell as shown in Figure 9c. Figure 9d displays the convection mode when $f > 55$ Hz, with convection particles first rising along the container's central axis, then moving downward along the container walls, forming double convection cells. However, this mode differs from Figure 9a, featuring a protrusion in the center of the powder particle top (whereas the top in Figure 9a is relatively flat [133]). Meanwhile, Hu et al. found that vibration acceleration had little effect on the system's convection mode. Zhang et al. also confirmed this conclusion, discussing the convection modes of particles under various vibration parameters, suggesting that vibration frequency has a significant influence on convection modes and directions, while vibration acceleration mainly affects the size and strength of convection [134]. Therefore, convection modes can be controlled by adjusting vibration frequency. Particle convection behavior has both advantages and disadvantages during the vibration-filling process. On one hand, convection behavior can promote uniform mixing of different particles, reduce void rates, accelerate the filling process, and improve filling efficiency. On the other hand, convection phenomena may lead to uneven local density distribution and accelerate equipment wear. Thus, whether convection behavior is beneficial should be evaluated based on actual application scenarios. Generally, applications that benefit from high-density filling and are insensitive to particle breakage, such as powder metallurgy and building materials, may find convection behavior advantageous. Applications that require high uniformity and particle integrity, such as pharmaceutical powders and food processing, may find convection behavior detrimental, and production adjustments in vibration frequency and amplitude can mitigate the negative impacts of convection.

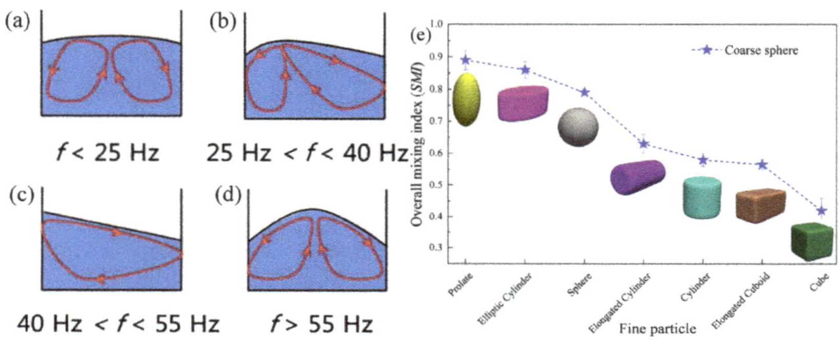

Figure 9. (**a–d**) Schematic diagrams of four convection modes under vertical vibration with different vibration frequencies [133] and (**e**) overall mixing index (SMI) of seven different shapes of particles [135].

In industrial applications, most particles are not standard spheres, and results from spherical particle systems studies may mislead non-spherical particle filling. Hence, research on non-spherical particles is increasingly crucial. Chen et al. observed the convection behavior of ellipsoidal particles under vertical vibration, revealing that due to differences in shape and surface characteristics between ellipsoidal and spherical particles, similar convection modes correspond to different vibration parameters. Moreover, under vibration conditions, the alignment direction of ellipsoidal particles tends to become consistent [74]. Yogi et al. investigated the impact of non-spherical particles in binary mixtures on convective vibrational mixing using the discrete element method (DEM). They found that elongated particles mixed best, while cubic particles mixed worst under vibration filling, as illustrated in Figure 9e. Additionally, they calculated the average velocity of coarse particles in binary mixtures, with the overall trend similar to SMI (Ellipsoids > Spheres > Cylinders > Cubes), and higher average velocity of coarse spherical particles indicating higher convection intensity and better mixing outcomes [135]. The mobility of non-spherical particles is usually not as good as that of spherical particles, and the flow process is prone to clogging and uneven filling, which may lead to uneven pressure distribution inside the mold during the compaction process, thus affecting the compaction effect [106]. Currently, it is difficult to accurately predict the flow and stacking behavior of non-spherical particles due to their more complex geometry, which further increases the difficulty of physical and mathematical modeling. Therefore, compared to our understanding of spherical particles, our understanding of the mechanisms of non-spherical particle arrangement and the impact of irregular shapes on convection is not comprehensive, requiring further exploration.

To date, most studies on vibration-induced powder particle behavior have focused on vertical vibrations, extensively exploring separation, segregation, and the impact of various parameters on mixtures of vibratory powder particles through experimental and numerical studies. However, research on vibrations in other directions is scarce. Liffman et al. discovered different convection modes from vertical vibrations through horizontal vibration [136]. Under horizontal vibration, Chung et al. also observed the Brazil nut effect within rectangular plates [137]. Further, Li et al. studied the impact of one-dimensional, two-dimensional, and three-dimensional vibration modes on the stacking of ellipsoidal particles, finding that applied vibration significantly increased particle packing density. Vibration effectively eliminated large voids and overcame particle interlocking, forming dense packing. Moreover, compared to one-dimensional and two-dimensional vibrations, three-dimensional vibration achieved higher packing density, as shown in Figure 10. This is mainly because one-dimensional vibration only provides energy for particles in the vertical direction, with horizontal movement restricted. Similarly, two-dimensional vibration only allows particles to move horizontally, not vertically. However, three-dimensional vibration provides energy for particles in both vertical and horizontal directions, thus obtaining a denser packing effect [138]. The study also found that batch feeding achieved higher packing density than full feeding, as full feeding might form more arch structures during filling, resulting in lower packing density. In contrast, batch feeding allows particles more opportunities to move and rearrange, forming more stable positions, especially when vibration is applied.

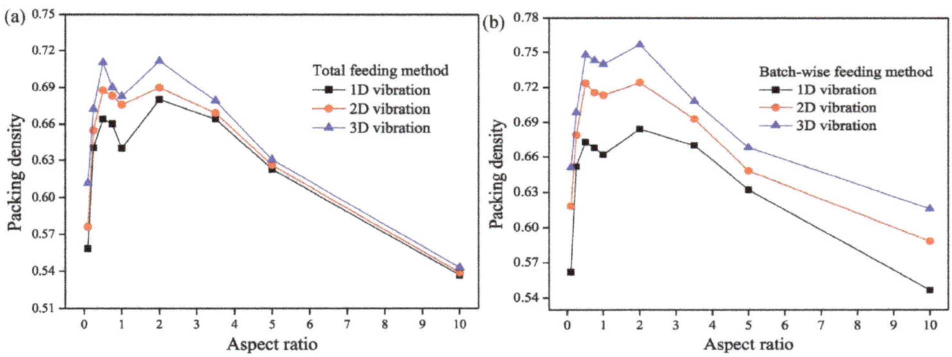

Figure 10. Effects of vibration modes on the stacking density of (**a**) total feeding, and (**b**) batch feeding [86].

Additionally, container size, shape, and wall friction significantly impact the vibrational mixing effect of the powder. Larger container sizes provide more free space for powder particles to move, reducing interparticle interference. During vibration filling, powder particles can move and diffuse more easily, resulting in a more uniform powder distribution. Zhang et al. used the discrete element method (DEM) to simulate the impact of wall friction on the mixing process of binary spherical particles in a vertically vibrated container. The study found that particles near the wall easily moved downward due to wall friction, with particles moving down along the side walls returning to the top from the middle position, forming convection cells. With an increased wall friction coefficient, the macroscopic convection phenomenon became more pronounced. Macroscopic convection caused by wall friction can promote particle flow mixing but may also cause large particles to aggregate at the center of the convection cell, leading to segregation. This phenomenon is particularly evident under different wall friction conditions (as shown in Figure 11). The study showed that, in the first phase of mixing, the variation in the mixing index became more significant with increasing wall friction coefficient, as convection is induced by wall friction [134]. Therefore, the higher the friction coefficient, the more pronounced the convection effect and the clearer the mixing outcome. However, in the third phase of mixing, the mixing index gradually stabilizes, and the degree of segregation increases with the wall friction coefficient. Therefore, to achieve rapid and uniform mixing of particles, it is necessary to consider the wall friction coefficient and other operating conditions comprehensively. An appropriate wall friction coefficient can promote convective mixing while avoiding excessive particle segregation. Hsiau et al. studied the impact of wall friction and container geometry on particle vibration processes, discovering similar convective effects. They also found that the container side wall's inclination angle significantly affects the direction and flow rate of convection cells. When the side wall's inclination angle reaches a certain critical value, the direction of the convection cells reverses, meaning particles no longer circulate in the original direction but move and mix in a new direction. As the inclination angle increases, the flow rate of the convection cells gradually decreases [139]. This is because a larger inclination angle increases resistance to particle displacement, restricting free particle flow. Therefore, designing an appropriate side wall inclination angle based on particle characteristics and mixing requirements to promote convective mixing and reduce segregation is one of the key steps to optimizing powder-vibration-filling processes.

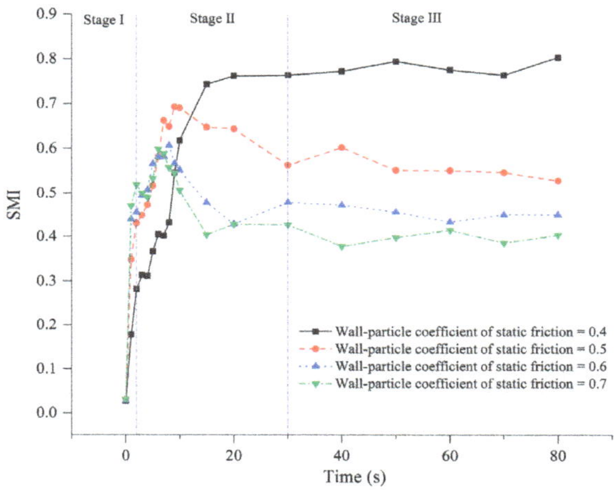

Figure 11. Mixing index versus time curves under different wall friction conditions [134].

3.4. Rotary Filling

Rotary filling is extensively adopted within the chemical, food, and pharmaceutical industries, holding an especially vital position in the pharmaceutical sector. Uniform mold filling is crucial, as the quantity of powder in the die determines the weight of the tablet, thereby influencing the drug content and the mechanical-physical properties of the tablet [38,140]. Powder rotary filling is typically an active process, where powder flows into the feeding frame through a hopper and is then pushed into the mold by a paddle feeder, a process we refer to as "an active mold-filling process".

Accurate mold filling constitutes a complex process, completed under the combined influence of gravity, forced feeding, suction, and rotation [141]. Hence, researchers have extensively studied powder rotary filling to optimize this process. Mendez et al. utilized a combination of a feed frame and a die plate to simulate the powder-filling process, exploring the impact of the operating parameters of the feed frame on the powder-filling process. The feed frame consists of a box with inlets and outlets and several paddles, where powder is poured from the top into the feed frame and then forced out by rotating paddles through the bottom slits, pushing it into the mold. The study found that the weight of powder in the mold increased with the speed of the feed frame and decreased with the speed of the turret and that the powder weight also depended on powder flow characteristics; for example, viscous powders resulted in reduced tablet mass [142], a phenomenon similarly observed by Peeters [143]. For powders with poor flowability, a higher feed frame speed might be necessary for effective filling, while powders with good flowability could achieve uniform filling at lower speeds. However, rotation of the paddles generates shear stress within the powder particles, potentially causing particle abrasion or excessive lubrication, leading to material loss and quality degradation. Ketterhagen et al. evaluated the impact of paddle wheel shapes on mold filling, modeling three different paddle wheel designs. Their study indicated that paddle wheels with larger hubs were preferable, as larger-diameter hubs reduced the amount of powder lingering in the feed frame, lowering the likelihood of severe powder abrasion [144]. Grymonpré et al. designed four different feed frame paddles and evaluated the mold-filling effect, eventually identifying an optimal paddle type [145]. Thus, for different types of powder and process requirements, it is necessary to flexibly adjust paddle designs to ensure optimal filling outcomes. Additionally, the selection of materials for the feed frame and paddles is paramount, as wear-resistant materials can reduce powder particle abrasion and extend equipment lifespan.

In recent years, numerical simulation, particularly the discrete element method (DEM), has played a significant role in analyzing the mechanical behavior of powder during rotary filling and optimizing design. Mateo-Ortiz and Méndez et al. utilized DEM to analyze forces applied to particles during the rotary-filling process, finding that particle translational motion is primarily governed by gravity, tangential forces, and normal forces. The farther the particles from the paddle, the smaller the forces applied to them, because the energy provided by the paddles decreases with distance due to particle collisions, particle friction, and paddle friction [146]. Siegmann et al. reported on different feed frame designs' flow mixing effects on powder particles using DEM, indicating that optimal geometry shapes and process parameters can be flexibly adjusted based on different powder characteristics [147]. Hildebrandt et al. evaluated the impact of paddle wheel shapes (circular and square) in a three-chamber feed frame on forced filling using DEM. They found that paddle wheel shape significantly affects the filling outcome; cylindrical paddles achieved higher fill quality under low turret-high paddle wheel speeds, while rectangular paddles performed better under high turret-low paddle wheel speeds [148]. An appropriate combination of paddle wheel speeds can yield better filling outcomes, providing a theoretical basis for optimizing paddle wheel design and process parameters.

Despite significant progress in powder-filling technology, many areas warrant further exploration and research, such as developing adjustable feed frames that can flexibly adjust inlet sizes and paddle arrangements based on powder characteristics and process requirements; developing more wear-resistant materials for manufacturing feed frames and paddles; and incorporating machine learning technology to further enhance the efficiency and precision of numerical simulations. In the future, as research in various fields continues to progress, combined with high-precision numerical simulations, more comprehensive and efficient solutions can be provided for optimizing the powder-filling process, expanding the application breadth of powder-filling technology.

Different filling techniques are suitable for different powder characteristics and application demands; these powder-filling technologies each have advantages and limitations. Table 2 summarizes the working principles, applicable powder types, and the advantages and disadvantages of different filling technologies. By choosing the appropriate filling technique, production efficiency, and product quality can be effectively improved, meeting the diversified needs of various industries. With the continuous progress of science and technology, brand-new powder-filling methods are emerging, such as laser powder melting filling [149], ultrasonic-assisted filling [150,151], and so on. These brand-new methods show more possibilities in the field of powder filling, helping to improve production efficiency, product quality, and material utilization. With the continuous development of technology, more innovative filling methods may appear in the future, further promoting industrial upgrading.

Table 2. Advantages and disadvantages of different filling techniques and the type of powder to which they are applied.

Filling Process	Principle	Applicable Powders	Advantages	Disadvantages
Gravity filling	Powder naturally flows under the action of gravity	Powders with good flowability, larger particles	Simple operation, low energy consumption, low maintenance cost	Limited filling accuracy, susceptible to environmental influences
Suction filling	Uses vacuum or negative pressure to draw powder into the mold	Fine particles, powders with poor flowability, such as pharmaceutical powders	Precise control of filling volume, strong adaptability	Complex equipment, high maintenance cost, high energy consumption
Vibration filling	Uses vibration to rearrange powder in the mold	Suitable for powders requiring high fill density and uniformity, such as metal and ceramic powders	High fill density, good uniformity, wide applicability	Complex equipment structure, high maintenance cost, noise
Rotary filling	Rotary motion pushes powder into the mold	Suitable for a variety of powders, such as pharmaceutical and food powders	Strong adaptability, uniform filling, high production efficiency	Complex equipment structure, rotary parts prone to wear, high energy consumption

4. Conclusions and Future Perspectives

The mold-filling process is a critical step in powder pressing and molding, where the uniformity and consistency of powder filling directly impact subsequent processing steps, thus affecting the final quality and performance of powder products. Powder filling into molds is a complex flow process, and accurately assessing powder flowability with appropriate methods has significant implications for product design and process optimization. This article discusses four main mold-filling methods, detailing the flow behavior of powder under different filling mechanisms and analyzing the impact of powder characteristics and processing parameters on filling outcomes. For example, under gravity filling, larger and denser particles achieve a higher mold-filling rate compared to finer particles; for powders with poor flowability, suction filling often yields better results. Therefore, in practical applications, it is essential to flexibly design filling equipment and optimize processing parameters tailored to different powder characteristics and process requirements to better control the filling process.

With the continuous development of new materials and the upgrading of filling equipment, computer modeling still shows enormous potential in assisting the evaluation of process parameters and the design of filling systems. Through computer modeling, the flow and filling effects of powder under different conditions can be simulated and predicted, thus optimizing design schemes, reducing experimental costs, and enhancing production efficiency. Therefore, computer modeling remains an important direction for the future development of powder-filling technology, playing an increasingly significant role in assisting the evaluation and optimization of filling processes. Through ongoing research and innovation, powder-filling technology will continue to advance, providing more efficient and precise solutions across various industries.

Author Contributions: Conceptualization, P.B. and D.W.; methodology, P.B.; formal analysis, P.B. and Y.Y.; investigation, P.B.; resources, Y.M.; writing—original draft preparation, P.B.; writing—review and editing, P.B., S.Y., and D.W.; supervision, S.Y. and D.W.; project administration, D.W. and Y.M.; funding acquisition, D.W. and Y.M. All authors have read and agreed to the published version of the manuscript.

Funding: This research was funded by the National Natural Science Foundation of China (Grant Nos. 52172275, 52107031, 52377032, and 52377033), Beijing Municipal Natural Science Foundation (Grant No. 3222061), and the International Partnership Program of Chinese Academy of Sciences (Grant No. 116GJHZ2023005MI).

Institutional Review Board Statement: Not applicable.

Informed Consent Statement: Not applicable.

Data Availability Statement: No new data were created or analyzed in this study.

Conflicts of Interest: The authors declare no conflicts of interest.

References

1. Zhao, J.; Burt, H.; Miller, R. The Gurnham equation in characterizing the compressibility of pharmaceutical materials. *Int. J. Pharm.* **2006**, *317*, 109–113. [CrossRef] [PubMed]
2. Han, L.H.; Elliott, J.A.; Bentham, A.C.; Mills, A.; Amidon, G.E.; Hancock, B.C. A modified Drucker-Prager Cap model for die compaction simulation of pharmaceutical powders. *Int. J. Solids Struct.* **2008**, *45*, 3088–3106. [CrossRef]
3. Ojha, V.K.; Schiano, S.; Wu, C.-Y.; Snášel, V.; Abraham, A. Predictive modeling of die filling of the pharmaceutical granules using the flexible neural tree. *Neural Comput. Appl.* **2016**, *29*, 467–481. [CrossRef]
4. Shi, G.; Lin, L.; Liu, Y.; Chen, G.; Luo, Y.; Wu, Y.; Li, H. Pharmaceutical application of multivariate modelling techniques: A review on the manufacturing of tablets. *RSC Adv.* **2021**, *11*, 8323–8345. [CrossRef]
5. Antipov, A.B.; Grigoriev, A.N.; Zayats, S.V.; Ivanov, V.V.; Kaul, A.R. Low-field magnetoresistance of $La_{0.7}Sr_{0.3}MnO_3$ nanoceramics processed by different precursor approaches and powder compaction techniques. *Mendeleev Commun.* **2004**, *14*, 148–150. [CrossRef]
6. Wu, Y.; Bandyopadhyay, A.; Bose, S. Processing of alumina and zirconia nano-powders and compacts. *Mater. Sci. Eng. A* **2004**, *380*, 349–355. [CrossRef]
7. Chen, P.; Kim, G.-Y.; Ni, J. Investigations in the compaction and sintering of large ceramic parts. *J. Mater. Process. Technol.* **2007**, *190*, 243–250. [CrossRef]
8. Kumagai, T. Isostatic compaction behavior of yttria-stabilized tetragonal zirconia polycrystal powder granules. *Powder Technol.* **2018**, *329*, 345–352. [CrossRef]
9. Singh, A.; Singh, J.; Dhami, H.S.; Sinha, M.K.; Viswanathan, K.; Kumar, R. Quantifying the influences of coal fly ash on rheological and compaction behaviour of iron powder and mild steel machining chips. *Adv. Powder Technol.* **2022**, *33*, 103819. [CrossRef]
10. Yaowiwat, N.; Madmusa, N.; Yimsuwan, K. Potential of Thai aromatic fruit (Artocarpus species) seed as an alternative natural starch for compact powder. *Int. J. Biol. Macromol.* **2023**, *242*, 124940. [CrossRef]
11. Murtaza, M.; Zhang, J.; Yang, C.; Su, C.; Wu, H. Durability of high strength self-compacting concrete with fly ash, coal gangue powder, cement kiln dust, and recycled concrete powder. *Constr. Build. Mater.* **2024**, *449*, 138345. [CrossRef]
12. Weinekötter, R. Compact and efficient continuous mixing processes for production of food and pharmaceutical powders. *Trends Food Sci. Technol.* **2009**, *20*, S48–S50. [CrossRef]
13. Pai, D.A.; Okos, M.R. Predicting the density and tensile strength of viscoelastic soy powder compacts. *J. Food Eng.* **2013**, *116*, 184–194. [CrossRef]
14. Mitchell, W.R.; Forny, L.; Althaus, T.; Dopfer, D.; Niederreiter, G.; Palzer, S. Compaction of food powders: The influence of material properties and process parameters on product structure, strength, and dissolution. *Chem. Eng. Sci.* **2017**, *167*, 29–41. [CrossRef]
15. Schmidt, R.; Pilz, S.; Lindemann, I.; Damm, C.; Hufenbach, J.; Helth, A.; Geissler, D.; Henss, A.; Rohnke, M.; Calin, M.; et al. Powder metallurgical processing of low modulus β-type Ti-45Nb to bulk and macro-porous compacts. *Powder Technol.* **2017**, *322*, 393–401. [CrossRef]
16. Verma, P.; Saha, R.; Chaira, D. Waste steel scrap to nanostructured powder and superior compact through powder metallurgy: Powder generation, processing and characterization. *Powder Technol.* **2018**, *326*, 159–167. [CrossRef]
17. Uematsu, K. Processing defects in ceramic powders and powder compacts. *Adv. Powder Technol.* **2014**, *25*, 154–162. [CrossRef]
18. Soldati, R.; Zanelli, C.; Guarini, G.; Piancastelli, A.; Melandri, C.; Fazio, S.; Bignozzi, M.C.; Dondi, M. Pore evolution and compaction behaviour of spray-dried bodies for porcelain stoneware slabs. *J. Eur. Ceram. Soc.* **2018**, *38*, 4127–4136. [CrossRef]
19. Okuma, G.; Osada, T.; Minagawa, H.; Arai, Y.; Inoue, R.; Kakisawa, H.; Shimoda, K.; Takeuchi, A.; Uesugi, M.; Tanaka, S.; et al. Heterogeneities and defects in powder compacts and sintered alumina bodies visualized by using the synchrotron X-ray CT. *J. Eur. Ceram. Soc.* **2023**, *43*, 486–492. [CrossRef]
20. McDonald, S.A.; Motazedian, F.; Cocks, A.C.F.; Withers, P.J. Shear cracking in an Al powder compact studied by X-ray microtomography. *Mater. Sci. Eng. A* **2009**, *508*, 64–70. [CrossRef]
21. Melnikov, A.; Sivagurunathan, K.; Guo, X.; Tolev, J.; Mandelis, A.; Ly, K.; Lawcock, R. Non-destructive thermal-wave-radar imaging of manufactured green powder metallurgy compact flaws (cracks). *NDT E Int.* **2017**, *86*, 140–152. [CrossRef]
22. Jonsén, P.; Häggblad, H.-Å. Modelling and numerical investigation of the residual stress state in a green metal powder body. *Powder Technol.* **2005**, *155*, 196–208. [CrossRef]
23. Wu, C.Y.; Cocks, A.C.F. Flow behaviour of powders during die filling. *Powder Metall.* **2004**, *47*, 127–136. [CrossRef]
24. Coube, O.; Cocks, A.C.F.; Wu, C.Y. Experimental and numerical study of die filling, powder transfer and die compaction. *Powder Metall.* **2005**, *48*, 68–76. [CrossRef]

25. Demetry, C.; Souto, F.S.; Ryden, B.C.; Roy, J.M. Tactile sensing of density uniformity in powder beds after die filling. *Powder Technol.* **1998**, *99*, 119–124. [CrossRef]
26. Korachkin, D.; Gethin, D.T.; Lewis, R.; Tweed, J. Effect of die filling on powder compaction. *Int. J. Powder Metall.* **2008**, *44*, 22–34. [CrossRef]
27. Wu, C.Y.; Hung, W.L.; Miguélez-Morán, A.M.; Gururajan, B.; Seville, J.P.K. Roller compaction of moist pharmaceutical powders. *Int. J. Pharm.* **2010**, *391*, 90–97. [CrossRef]
28. Ammarcha, C.; Gatumel, C.; Dirion, J.L.; Cabassud, M.; Mizonov, V.; Berthiaux, H. Predicting bulk powder flow dynamics in a continuous mixer operating in transitory regimes. *Adv. Powder Technol.* **2012**, *23*, 787–800. [CrossRef]
29. Bianchini, A.; Farnish, R.J.; Gambuti, M.; Saccani, C. Predictive model for corn flour discharge rates through an aerated hopper. *Powder Technol.* **2016**, *298*, 84–89. [CrossRef]
30. Bostijn, N.; Dhondt, J.; Ryckaert, A.; Szabó, E.; Dhondt, W.; Van Snick, B.; Vanhoorne, V.; Vervaet, C.; De Beer, T. A multivariate approach to predict the volumetric and gravimetric feeding behavior of a low feed rate feeder based on raw material properties. *Int. J. Pharm.* **2019**, *557*, 342–353. [CrossRef]
31. Allenspach, C.; Timmins, P.; Lumay, G.; Holman, J.; Minko, T. Loss-in-weight feeding, powder flow and electrostatic evaluation for direct compression hydroxypropyl methylcellulose (HPMC) to support continuous manufacturing. *Int. J. Pharm.* **2021**, *596*, 120259. [CrossRef] [PubMed]
32. Kulkarni, H.; Dabhade, V.V. Green machining of powder-metallurgy-steels (PMS): An overview. *J. Manuf. Process.* **2019**, *44*, 1–18. [CrossRef]
33. Khudyakov, O.; Vashchenko, S.; Baiul, K.; Semenov, Y.; Krot, P. Development of a versatile method for predicting the density of monocomponent dry fine materials compacts based on comparative study of compression factors. *Powder Technol.* **2024**, *445*, 120104. [CrossRef]
34. Romero, C.; Yang, F.; Bolzoni, L. Fatigue and fracture properties of Ti alloys from powder-based processes—A review. *Int. J. Fatigue* **2018**, *117*, 407–419. [CrossRef]
35. Rofman, O.V.; Mikhaylovskaya, A.V.; Kotov, A.D.; Mochugovskiy, A.G.; Mohamed, A.K.; Cheverikin, V.V.; Short, M.P. AA2024/SiC metal matrix composites simultaneously improve ductility and cracking resistance during elevated temperature deformation. *Mater. Sci. Eng. A* **2020**, *790*, 139697. [CrossRef]
36. Shu, G.J.; Hwang, K.S.; Pan, Y.T. Improvements in sintered density and dimensional stability of powder injection-molded 316L compacts by adjusting the alloying compositions. *Acta Mater.* **2006**, *54*, 1335–1342. [CrossRef]
37. Wu, C.Y.; Dihoru, L.; Cocks, A.C.F. The flow of powder into simple and stepped dies. *Powder Technol.* **2003**, *134*, 24–39. [CrossRef]
38. Mills, L.A.; Sinka, I.C. Effect of particle size and density on the die fill of powders. *Eur. J. Pharm. Biopharm.* **2013**, *84*, 642–652. [CrossRef]
39. Schneider, L.C.R.; Sinka, I.C.; Cocks, A.C.F. Characterisation of the flow behaviour of pharmaceutical powders using a model die–shoe filling system. *Powder Technol.* **2007**, *173*, 59–71. [CrossRef]
40. Guo, Y.; Kafui, K.D.; Wu, C.Y.; Thornton, C.; Seville, J.P.K. A coupled DEM/CFD analysis of the effect of air on powder flow during die filling. *AIChE J.* **2009**, *55*, 49–62. [CrossRef]
41. Baserinia, R.; Sinka, I.C. Powder die filling under gravity and suction fill mechanisms. *Int. J. Pharm.* **2019**, *563*, 135–155. [CrossRef] [PubMed]
42. Bobba, D.; Tabei, S.A.; Cherukuri, H.P.; Pando, M.A. DEM simulation of particle segregation in filling of vibratory dies. *Adv. Powder Technol.* **2020**, *31*, 3474–3484. [CrossRef]
43. Chtourou, H.; Gakwaya, A.; Guillot, M. Modeling of the metal powder compaction process using the cap model. Part II: Numerical implementation and practical applications. *Int. J. Solids Struct.* **2002**, *39*, 1077–1096. [CrossRef]
44. Shima, S.; Sakamoto, Y.; Kotera, H. Simulation of rubber isostatic pressing and shape optimization of rubber mold. *Int. J. Mech. Sci.* **2002**, *44*, 1603–1623. [CrossRef]
45. Wu, C.Y.; Guo, Y. Numerical modelling of suction filling using DEM/CFD. *Chem. Eng. Sci.* **2012**, *73*, 231–238. [CrossRef]
46. Zhang, Y.; Jia, F.; Zeng, Y.; Han, Y.; Xiao, Y. DEM study in the critical height of flow mechanism transition in a conical silo. *Powder Technol.* **2018**, *331*, 98–106. [CrossRef]
47. Zheng, C.; Yost, E.; Muliadi, A.R.; Govender, N.; Zhang, L.; Wu, C.-Y. DEM analysis of the influence of stirrer design on die filling with forced powder feeding. *Particuology* **2024**, *88*, 107–115. [CrossRef]
48. Yang, B.; Ouyang, J.; Li, X. Simulation of fiber reinforced composite materials mold filling process and mechanical properties analysis. *J. Non-Newton. Fluid Mech.* **2011**, *166*, 1129–1136. [CrossRef]
49. Ragunath, S.; Radhika, N.; Saleh, B. Advancements and future prospects of additive manufacturing in high-entropy alloy applications. *J. Alloys Compd.* **2024**, *997*, 174859. [CrossRef]
50. Huang, J.; Liu, G.; Yu, X.; Wu, H.; Huang, Y.; Yu, S.; Fan, D. Microstructure regulation of titanium alloy functionally gradient materials fabricated by alternating current assisted wire arc additive manufacturing. *Mater. Des.* **2022**, *218*, 110731. [CrossRef]
51. Van den Eynde, M.; Verbelen, L.; Van Puyvelde, P. Assessing polymer powder flow for the application of laser sintering. *Powder Technol.* **2015**, *286*, 151–155. [CrossRef]
52. Polizzi, M.A.; Franchville, J.; Hilden, J.L. Assessment and predictive modeling of pharmaceutical powder flow behavior in small-scale hoppers. *Powder Technol.* **2016**, *294*, 30–42. [CrossRef]

53. Legoix, L.; Gatumel, C.; Milhé, M.; Berthiaux, H. Characterizing powders in order to determine their flow behavior in a mixer: From small scale observations to macroscopic in-mixer rheology for powders of various flowabilities. *Powder Technol.* **2017**, *322*, 314–331. [CrossRef]
54. Schiano, S.; Chen, L.; Wu, C.-Y. The effect of dry granulation on flow behaviour of pharmaceutical powders during die filling. *Powder Technol.* **2018**, *337*, 78–83. [CrossRef]
55. Baserinia, R.; Brockbank, K.; Dattani, R. Correlating polyamide powder flowability to mechanical properties of parts fabricated by additive manufacturing. *Powder Technol.* **2022**, *398*, 117147. [CrossRef]
56. Yim, S.; Wang, H.; Aoyagi, K.; Yamanaka, K.; Chiba, A. Role of gravity magnitude on flowability and powder spreading in the powder bed fusion additive manufacturing process: Towards additive manufacturing in space. *Addit. Manuf.* **2024**, *94*, 104441. [CrossRef]
57. Ismail, O.; Zheng, C.; Chamberlain, T.; Zakhvatayeva, A.; Hare, C.; Yost, E.; Muliadi, A.R.; Wu, C.-Y. An experimental study on flow behaviour of pharmaceutical powders during suction filling. *Int. J. Pharm.* **2024**, *662*, 124527. [CrossRef]
58. Fu, X.; Huck, D.; Makein, L.; Armstrong, B.; Willen, U.; Freeman, T. Effect of particle shape and size on flow properties of lactose powders. *Particuology* **2012**, *10*, 203–208. [CrossRef]
59. Liang, C.; Xie, X.; Xu, P.; Chen, X.; Zhao, C.; Wu, X. Investigation of influence of coal properties on dense-phase pneumatic conveying at high pressure. *Particuology* **2012**, *10*, 310–316. [CrossRef]
60. Yu, W.; Muteki, K.; Zhang, L.; Kim, G. Prediction of Bulk Powder Flow Performance Using Comprehensive Particle Size and Particle Shape Distributions. *J. Pharm. Sci.* **2011**, *100*, 284–293. [CrossRef]
61. Landi, G.; Barletta, D.; Poletto, M. Modelling and experiments on the effect of air humidity on the flow properties of glass powders. *Powder Technol.* **2011**, *207*, 437–443. [CrossRef]
62. Blanco, D.; Antikainen, O.; Räikkönen, H.; Yliruusi, J.; Juppo, A.M. Effect of colloidal silicon dioxide and moisture on powder flow properties: Predicting in-process performance using image-based analysis. *Int. J. Pharm.* **2021**, *597*, 120344. [CrossRef] [PubMed]
63. Tomasetta, I.; Barletta, D.; Poletto, M. Correlation of powder flow properties to interparticle interactions at ambient and high temperatures. *Particuology* **2014**, *12*, 90–99. [CrossRef]
64. Capece, M.; Silva, K.R.; Sunkara, D.; Strong, J.; Gao, P. On the relationship of inter-particle cohesiveness and bulk powder behavior: Flowability of pharmaceutical powders. *Int. J. Pharm.* **2016**, *511*, 178–189. [CrossRef] [PubMed]
65. Shah, D.S.; Moravkar, K.K.; Jha, D.K.; Lonkar, V.; Amin, P.D.; Chalikwar, S.S. A concise summary of powder processing methodologies for flow enhancement. *Heliyon* **2023**, *9*, e16498. [CrossRef]
66. Zhong, W.Z.; Zakhvatayeva, A.; Zhang, L.; Wu, C.Y. Powder flow during linear and rotary die filling. *Int. J. Pharm.* **2021**, *602*, 120654. [CrossRef]
67. Sinka, I.C.; Motazedian, F.; Cocks, A.C.F.; Pitt, K.G. The effect of processing parameters on pharmaceutical tablet properties. *Powder Technol.* **2009**, *189*, 276–284. [CrossRef]
68. Beakawi Al-Hashemi, H.M.; Baghabra Al-Amoudi, O.S. A review on the angle of repose of granular materials. *Powder Technol.* **2018**, *330*, 397–417. [CrossRef]
69. Müller, D.; Fimbinger, E.; Brand, C. Algorithm for the determination of the angle of repose in bulk material analysis. *Powder Technol.* **2021**, *383*, 598–605. [CrossRef]
70. Kalman, H. Quantification of mechanisms governing the angle of repose, angle of tilting, and Hausner ratio to estimate the flowability of particulate materials. *Powder Technol.* **2021**, *382*, 573–593. [CrossRef]
71. Thalberg, K.; Lindholm, D.; Axelsson, A. Comparison of different flowability tests for powders for inhalation. *Powder Technol.* **2004**, *146*, 206–213. [CrossRef]
72. Santomaso, A.; Lazzaro, P.; Canu, P. Powder flowability and density ratios: The impact of granules packing. *Chem. Eng. Sci.* **2003**, *58*, 2857–2874. [CrossRef]
73. Montanari, D.; Agostini, A.; Bonini, M.; Corti, G.; Ventisette, C. The Use of Empirical Methods for Testing Granular Materials in Analogue Modelling. *Materials* **2017**, *10*, 635. [CrossRef] [PubMed]
74. Chen, B.; Wu, P.; Xing, H.; Liu, H.; Li, L.; Wang, L. Convection behavior of ellipsoidal particles in a quasi-two-dimensional bed under vertical vibration. *Powder Technol.* **2020**, *363*, 575–583. [CrossRef]
75. Lindberg, N.O.; Pålsson, M.; Pihl, A.C.; Freeman, R.; Freeman, T.; Zetzener, H.; Enstad, G. Flowability Measurements of Pharmaceutical Powder Mixtures with Poor Flow Using Five Different Techniques. *Drug Dev. Ind. Pharm.* **2004**, *30*, 785–791. [CrossRef]
76. Leturia, M.; Benali, M.; Lagarde, S.; Ronga, I.; Saleh, K. Characterization of flow properties of cohesive powders: A comparative study of traditional and new testing methods. *Powder Technol.* **2014**, *253*, 406–423. [CrossRef]
77. Freeman, R. Measuring the flow properties of consolidated, conditioned and aerated powders—A comparative study using a powder rheometer and a rotational shear cell. *Powder Technol.* **2007**, *174*, 25–33. [CrossRef]
78. Moravkar, K.K.; Shah, D.S.; Magar, A.G.; Bhairav, B.A.; Korde, S.D.; Ranch, K.M.; Chalikwar, S.S. Assessment of pharmaceutical powders flowability and comparative evaluation of lubricants on development of gastro retentive tablets: An application of powder flow tester. *J. Drug Deliv. Sci. Technol.* **2022**, *71*, 103265. [CrossRef]
79. Schulze, D. Round robin test on ring shear testers. *Adv. Powder Technol.* **2011**, *22*, 197–202. [CrossRef]
80. Salehi, H.; Barletta, D.; Poletto, M. A comparison between powder flow property testers. *Particuology* **2017**, *32*, 10–20. [CrossRef]

81. Lopes Neto, J.P.; Meira, A.S.; Nascimento, J.W.B.d. Flow Properties and Pattern Flow Prediction of Food Industrial Powders. *Eng. Agríc.* **2017**, *37*, 627–636. [CrossRef]
82. Wang, Y.; Koynov, S.; Glasser, B.J.; Muzzio, F.J. A method to analyze shear cell data of powders measured under different initial consolidation stresses. *Powder Technol.* **2016**, *294*, 105–112. [CrossRef]
83. Schneider, L.C.R.; Cocks, A.C.F.; Apostolopoulos, A. Ceramic comparison of filling behaviour of metallic, ceramic, hardmetal and magnetic powders. *Powder Metall.* **2005**, *48*, 77–84. [CrossRef]
84. Wu, C.-Y.; Cocks, A.C.F. Numerical and experimental investigations of the flow of powder into a confined space. *Mech. Mater.* **2006**, *38*, 304–324. [CrossRef]
85. Guo, Y.; Wu, C.Y.; Kafui, K.D.; Thornton, C. 3D DEM/CFD analysis of size-induced segregation during die filling. *Powder Technol.* **2011**, *206*, 177–188. [CrossRef]
86. Ketterhagen, W.R.; Curtis, J.S.; Wassgren, C.R.; Hancock, B.C. Predicting the flow mode from hoppers using the discrete element method. *Powder Technol.* **2009**, *195*, 1–10. [CrossRef]
87. Yu, Y.; Saxén, H. Discrete element method simulation of properties of a 3D conical hopper with mono-sized spheres. *Adv. Powder Technol.* **2011**, *22*, 324–331. [CrossRef]
88. Gandia, R.M.; Gomes, F.C.; de Paula, W.C.; Rodriguez, P.J.A. The influence of flow pattern and hopper angle on static and dynamic pressures in slender silos. *Powder Technol.* **2023**, *427*, 118756. [CrossRef]
89. Huang, X.; Zheng, Q.; Liu, D.; Yu, A.; Yan, W. A design method of hopper shape optimization with improved mass flow pattern and reduced particle segregation. *Chem. Eng. Sci.* **2022**, *253*, 117579. [CrossRef]
90. Faqih, A.N.; Alexander, A.W.; Muzzio, F.J.; Tomassone, M.S. A method for predicting hopper flow characteristics of pharmaceutical powders. *Chem. Eng. Sci.* **2007**, *62*, 1536–1542. [CrossRef]
91. Kumar, V.; Arora, R.; Goyal, D.; Pal, D.; Sehgal, S. Investigations of powder flow patterns in hopper model through discrete element method. *Met. Powder Rep.* **2020**, *75*, 40–47. [CrossRef]
92. Rabinovich, E.; Kalman, H.; Peterson, P.F. Granular material flow regime map for planar silos and hoppers. *Powder Technol.* **2021**, *377*, 597–606. [CrossRef]
93. You, M.; Wang, X.; Shi, Y.; Luo, B.; Liang, C.; Liu, D.; Ma, J.; Chen, X. The dynamic evolution of powder flow and wall normal stress in different flow pattern silos. *Int. J. Multiph. Flow* **2024**, *176*, 104844. [CrossRef]
94. Sinka, I.C.; Schneider, L.C.R.; Cocks, A.C.F. Measurement of the flow properties of powders with special reference to die fill. *Int. J. Pharm.* **2004**, *280*, 27–38. [CrossRef]
95. Mehrabi, M.; Hassanpour, A.; Bayly, A. An X-ray microtomography study of particle morphology and the packing behaviour of metal powders during filling, compaction and ball indentation processes. *Powder Technol.* **2021**, *385*, 250–263. [CrossRef]
96. Guo, Y.; Wu, C.Y.; Thornton, C. The effects of air and particle density difference on segregation of powder mixtures during die filling. *Chem. Eng. Sci.* **2011**, *66*, 661–673. [CrossRef]
97. Khonsari, M.M. On the modeling of multi-body interaction problems in tribology. *Wear* **1997**, *207*, 55–62. [CrossRef]
98. Li, H.X. Anti-Friction Effect of Ball-Milled Expanded Graphite/Fe Powders. *Adv. Mater. Res.* **2012**, *496*, 395–398. [CrossRef]
99. Rice, E.R.; Tengzelius, J. Die Filling Characteristics of Metal Powders. *Powder Metall.* **1986**, *29*, 183–194. [CrossRef]
100. Bocchini, G.F. Influence of Small Die Width on Filling and Compacting Densities. *Powder Metall.* **1987**, *30*, 261–266. [CrossRef]
101. Readey, M.J.; Mahoney, F.M. *Compaction of Spray-Dried Ceramic Powders: An Experimental Study of the Factors That Control Green Density*; U. S. Department of Energy Office of Scientific and Technical Information: Oak Ridge, TN, USA, 1995. [CrossRef]
102. Schomberg, A.K.; Kwade, A.; Finke, J.H. Modeling die filling under gravity for different scales of rotary tablet presses. *Powder Technol.* **2023**, *424*, 118550. [CrossRef]
103. Santamaría Alvarez, F.N.; Kleinebudde, P.; Thies, J. Influence of refilling on dosing accuracy of loss-in-weight powder feeders in continuous manufacturing. *Powder Technol.* **2024**, *436*, 119501. [CrossRef]
104. Bierwisch, C.; Kraft, T.; Riedel, H.; Moseler, M. Three-dimensional discrete element models for the granular statics and dynamics of powders in cavity filling. *J. Mech. Phys. Solids* **2009**, *57*, 10–31. [CrossRef]
105. Zheleznyakova, A.L. A versatile 3D computational model based on molecular dynamics for real-time simulation of the automatic powder filling processes. *Powder Technol.* **2021**, *381*, 141–163. [CrossRef]
106. Ma, H.; Zhou, L.; Liu, Z.; Chen, M.; Xia, X.; Zhao, Y. A review of recent development for the CFD-DEM investigations of non-spherical particles. *Powder Technol.* **2022**, *412*, 117972. [CrossRef]
107. Tiscar, J.M.; Boix, J.; Mallol, G.; Pérez, J.A.; Gilabert, F.A. Design of new industrial mould filling systems for the manufacture of ceramic tiles using a discrete element framework. *Powder Technol.* **2022**, *404*, 117446. [CrossRef]
108. Studeník, O.; Isoz, M.; Kotouč Šourek, M.; Kočí, P. OpenHFDIB-DEM: An extension to OpenFOAM for CFD-DEM simulations with arbitrary particle shapes. *SoftwareX* **2024**, *27*, 101871. [CrossRef]
109. Yu, A.B. Powder processing—Models and simulations. In *Encyclopedia of Condensed Matter Physics*; Academic Press: Cambridge, MA, USA, 2024; pp. 679–693.
110. Jackson, S.; Sinka, I.C.; Cocks, A.C.F. The effect of suction during die fill on a rotary tablet press. *Eur. J. Pharm. Biopharm.* **2007**, *65*, 253–256. [CrossRef]
111. Zakhvatayeva, A.; Hare, C.; Wu, C.Y. Suction filling of pharmaceutical powders. *Powder Technol.* **2019**, *355*, 438–448. [CrossRef]
112. Ratnaswamy, V.; Rosato, A.D.; Blackmore, D.; Tricoche, X.; Ching, N.; Zuo, L. Evolution of solids fraction surfaces in tapping: Simulation and dynamical systems analysis. *Granul. Matter* **2012**, *14*, 163–168. [CrossRef]

113. Evans, P.E.; Millman, R.S. The vibratory packing of powders. *Powder Metall.* **1964**, *7*, 50–63. [CrossRef]
114. Knight, J.B.; Fandrich, C.G.; Lau, C.N.; Jaeger, H.M.; Nagel, S.R. Density relaxation in a vibrated granular material. *Phys. Rev. E* **1995**, *51*, 3957–3963. [CrossRef] [PubMed]
115. Fan, S.; Xu, Y.; Zhang, L. Effects of vibration frequency and time on packed density of powder. *Refract. Mater.* **2005**, *39*, 123–125, 129.
116. Jaggannagari, S.R.; Desu, R.K.; Reimann, J.; Gan, Y.; Moscardini, M.; Annabattula, R.K. DEM simulations of vibrated sphere packings in slender prismatic containers. *Powder Technol.* **2021**, *393*, 31–59. [CrossRef]
117. Rosato, A.; Strandburg, K.J.; Prinz, F.; Swendsen, R.H. Why the Brazil nuts are on top: Size segregation of particulate matter by shaking. *Phys. Rev. Lett.* **1987**, *58*, 1038–1040. [CrossRef]
118. Jullien, R.; Meakin, P.; Pavlovitch, A. Three-dimensional model for particle-size segregation by shaking. *Phys. Rev. Lett.* **1992**, *69*, 640–643. [CrossRef]
119. Hong, D.C.; Quinn, P.V.; Luding, S. Reverse Brazil Nut Problem: Competition between Percolation and Condensation. *Phys. Rev. Lett.* **2001**, *86*, 3423–3426. [CrossRef]
120. Sanders, D.A.; Swift, M.R.; Bowley, R.M.; King, P.J. Are Brazil Nuts Attractive? *Phys. Rev. Lett.* **2004**, *93*, 208002. [CrossRef] [PubMed]
121. Tatemoto, Y.; Mawatari, Y.; Yasukawa, T.; Noda, K. Numerical simulation of particle motion in vibrated fluidized bed. *Chem. Eng. Sci.* **2004**, *59*, 437–447. [CrossRef]
122. Saez, A.; Vivanco, F.; Melo, F. Size segregation, convection, and arching effect. *Phys. Rev. E* **2005**, *72*, 021307. [CrossRef]
123. Majid, M.; Walzel, P. Convection and segregation in vertically vibrated granular beds. *Powder Technol.* **2009**, *192*, 311–317. [CrossRef]
124. Aoki, K.M.; Akiyama, T. Spontaneous Wave Pattern Formation in Vibrated Granular Materials. *Phys. Rev. Lett.* **1996**, *77*, 4166–4169. [CrossRef] [PubMed]
125. Butzhammer, L.; Völkel, S.; Rehberg, I.; Huang, K. Pattern formation in wet granular matter under vertical vibrations. *Phys. Rev. E* **2015**, *92*, 012202. [CrossRef]
126. Liu, C.; Wu, P.; Wang, L. Particle climbing along a vibrating tube: A vibrating tube that acts as a pump for lifting granular materials from a silo. *Soft Matter* **2013**, *9*, 4762–4766. [CrossRef]
127. Shinbrot, T.; Muzzio, F.J. Reverse Buoyancy in Shaken Granular Beds. *Phys. Rev. Lett.* **1998**, *81*, 4365–4368. [CrossRef]
128. Knight, J.B.; Jaeger, H.M.; Nagel, S.R. Vibration-induced size separation in granular media: The convection connection. *Phys. Rev. Lett.* **1993**, *70*, 3728–3731. [CrossRef]
129. Hsiau, S.S.; Chen, C.H. Granular convection cells in a vertical shaker. *Powder Technol.* **2000**, *111*, 210–217. [CrossRef]
130. Trujillo, L.; Alam, M. Segregation in a fluidized binary granular mixture: Competition between buoyancy and geometric forces. *Europhys. Lett.* **2003**, *64*, 190–196. [CrossRef]
131. Faraday, M. On a peculiar class of acoustical figures; and on certain forms assumed by groups of particles upon vibrating elastic surfaces. *R. Soc.* **1831**, *121*, 299–340. [CrossRef]
132. Gallas, J.A.C.; Herrmann, H.J.; Sokołowski, S. Convection cells in vibrating granular media. *Phys. Rev. Lett.* **1992**, *69*, 1371–1374. [CrossRef]
133. Hu, K.; Xie, Z.-A.; Wu, P.; Sun, J.; Li, L.; Jia, C.; Zhang, S.; Liu, C.; Wang, L. Convecting particle diffusion in a binary particle system under vertical vibration. *Soft Matter* **2014**, *10*, 4348–4359. [CrossRef] [PubMed]
134. Zhang, S.; Wang, X. Effect of vibration parameters and wall friction on the mixing characteristics of binary particles in a vertical vibrating container subject to cohesive forces. *Powder Technol.* **2023**, *413*, 118078. [CrossRef]
135. Yogi, J.; Dubey, P.; Khatoon, S.; Iyer, K.; Kumar, S.; Anand, A. Effect of non-sphericity of a narrow-sized binary mixture on mixing in convective vibrated packed bed using Discrete Element Method. *Powder Technol.* **2023**, *426*, 118600. [CrossRef]
136. Liffman, K.; Metcalfe, G.; Cleary, P. Granular Convection and Transport due to Horizontal Shaking. *Phys. Rev. Lett.* **1997**, *79*, 4574–4576. [CrossRef]
137. Chung, F.F.; Liaw, S.-S.; Ju, C.-Y. Brazil nut effect in a rectangular plate under horizontal vibration. *Granul. Matter* **2009**, *11*, 79–86. [CrossRef]
138. Li, C.X.; Zou, R.P.; Pinson, D.; Yu, A.B.; Zhou, Z.Y. An experimental study of packing of ellipsoids under vibrations. *Powder Technol.* **2020**, *361*, 45–51. [CrossRef]
139. Hsiau, S.S.; Wang, P.C.; Tai, C.H. Convection cells and segregation in a vibrated granular bed. *AIChE J.* **2002**, *48*, 1430–1438. [CrossRef]
140. Van Snick, B.; Grymonpré, W.; Dhondt, J.; Pandelaere, K.; Di Pretoro, G.; Remon, J.P.; De Beer, T.; Vervaet, C.; Vanhoorne, V. Impact of blend properties on die filling during tableting. *Int. J. Pharm.* **2018**, *549*, 476–488. [CrossRef] [PubMed]
141. Goh, H.P.; Heng, P.W.S.; Liew, C.V. Understanding effects of process parameters and forced feeding on die filling. *Eur. J. Pharm. Sci.* **2018**, *122*, 105–115. [CrossRef]
142. Mendez, R.; Muzzio, F.; Velazquez, C. Study of the effects of feed frames on powder blend properties during the filling of tablet press dies. *Powder Technol.* **2010**, *200*, 105–116. [CrossRef]
143. Peeters, E.; De Beer, T.; Vervaet, C.; Remon, J.-P. Reduction of tablet weight variability by optimizing paddle speed in the forced feeder of a high-speed rotary tablet press. *Drug Dev. Ind. Pharm.* **2014**, *41*, 530–539. [CrossRef] [PubMed]

144. Ketterhagen, W.R. Simulation of powder flow in a lab-scale tablet press feed frame: Effects of design and operating parameters on measures of tablet quality. *Powder Technol.* **2015**, *275*, 361–374. [CrossRef]
145. Grymonpré, W.; Vanhoorne, V.; Van Snick, B.; Blahova Prudilova, B.; Detobel, F.; Remon, J.P.; De Beer, T.; Vervaet, C. Optimizing feed frame design and tableting process parameters to increase die-filling uniformity on a high-speed rotary tablet press. *Int. J. Pharm.* **2018**, *548*, 54–61. [CrossRef] [PubMed]
146. Mateo-Ortiz, D.; Méndez, R. Relationship between residence time distribution and forces applied by paddles on powder attrition during the die filling process. *Powder Technol.* **2015**, *278*, 111–117. [CrossRef]
147. Siegmann, E.; Forgber, T.; Toson, P.; Martinetz, M.C.; Kureck, H.; Brinz, T.; Manz, S.; Grass, T.; Khinast, J. Powder flow and mixing in different tablet press feed frames. *Adv. Powder Technol.* **2020**, *31*, 770–781. [CrossRef]
148. Hildebrandt, C.; Gopireddy, S.R.; Scherließ, R.; Urbanetz, N.A. A DEM approach to assess the influence of the paddle wheel shape on force feeding during pharmaceutical tableting. *Adv. Powder Technol.* **2020**, *31*, 755–769. [CrossRef]
149. Sarkar, D.; Kapil, A.; Sharma, A. Advances in computational modeling for laser powder bed fusion additive manufacturing: A comprehensive review of finite element techniques and strategies. *Addit. Manuf.* **2024**, *85*, 104157. [CrossRef]
150. Wang, A.; Zou, P.; Xu, J.; Wang, W.; Li, B. Investigation on reduction in compaction of ultrasonic vibration-assisted powder feeding for additive manufacturing. *Powder Technol.* **2024**, *435*, 119414. [CrossRef]
151. George, M.; Antony, A.; Jobin, M.V.; Amarnath, K.S.; Bayins, A. Ultrasonic induced additive powder deposition manufacturing. *Mater. Today Proc.* **2023**, *72*, 2978–2984. [CrossRef]

materials

MDPI

Article

Improving Visible Light Photocatalysis Using Optical Defects in CoO_x-TiO_2 Photonic Crystals

Alexia Toumazatou [1], Elias Sakellis [1,2] and Vlassis Likodimos [1,*]

[1] Section of Condensed Matter Physics, Department of Physics, National and Kapodistrian University of Athens, University Campus, 15784 Athens, Greece; alextoum@phys.uoa.gr (A.T.); e_sakel@phys.uoa.gr (E.S.)
[2] Institute of Nanoscience and Nanotechnology, National Center for Scientific Research "Demokritos", Agia Paraskevi, 15341 Athens, Greece
* Correspondence: vlikodimos@phys.uoa.gr; Tel.: +30-2107276824

Abstract: The rational design of photonic crystal photocatalysts has attracted significant interest in order to improve their light harvesting and photocatalytic performances. In this work, an advanced approach to enhance slow light propagation and visible light photocatalysis is demonstrated for the first time by integrating a planar defect into CoO_x-TiO_2 inverse opals. Trilayer photonic crystal films were fabricated through the successive deposition of an inverse opal TiO_2 underlayer, a thin titania interlayer, and a photonic top layer, whose visible light activation was implemented through surface modification with CoO_x nanoscale complexes. Optical measurements showed the formation of "donor"-like localized states within the photonic band gap, which reduced the Bragg reflection and expanded the slow photon spectral range. The optimization of CoO_x loading and photonic band gap tuning resulted in a markedly improved photocatalytic performance for salicylic acid degradation and photocurrent generation compared to the additive effects of the constituent monolayers, indicative of light localization in the defect layer. The electrochemical impedance results showed reduced recombination kinetics, corroborating that the introduction of an optical defect into inverse opal photocatalysts provides a versatile and effective strategy for boosting the photonic amplification effects in visible light photocatalysis by evading the constraints imposed by narrow slow photon spectral regions.

Keywords: photonic crystals; titanium dioxide; cobalt oxides; inverse opals; photocatalysis; visible light; optical defects; slow photons; multilayers

1. Introduction

Tailoring the morphology of semiconductor photocatalysts in the form of three-dimensional (3D) photonic crystal (PC) architectures has been attracting attention as an advanced approach that may selectively improve light harvesting at frequencies of weak electronic absorption by using slow photon effects, i.e., slow light propagation at the photonic band gap's (PBG) edges [1,2]. Significant progress has been achieved in slow photon-assisted visible light-activated (VLA) photocatalysis through defect engineering [3–5] and mostly the heterostructuring of benchmark wide band gap semiconductor photocatalysts such as TiO_2 with suitable semiconducting [6–10], plasmonic [11–13], and graphene [14–16] nanomaterials or by PC fabrication using narrow band gap semiconductors that are intrinsically responsive to visible light [17–21]. In all cases, photonic amplification was based on the spectral matching of the red or blue [22,23] slow photon frequency regions to the weak electronic absorption of the photocatalyst, while avoiding detrimental Bragg reflection losses. PBG/slow photon engineering is commonly implemented by varying the PC periodicity, e.g., the macropore void diameter in the case of inverse opal structures [24,25] and the angle of light incidence [26,27] or, more recently, by the compositional tuning of heterostructured WO_3/TiO_2 inverse opal films that provides an additional degree of freedom for the control of the PC lattice parameter beyond that of templating spheres [28].

In addition, PC structures have been utilized as inert photonic supports for photocatalytic layers, acting as dielectric mirrors, whose Bragg reflection is tuned to the catalysts' absorption edge leading to enhanced light harvesting independently of the photocatalytic process [29–32]. Moreover, PBG engineering of CoO_x-TiO_2 bilayer films consisting of an inverse opal substrate and a benchmark mesoporous titania overlayer showed that PCs can be very effective as both photocatalytic layers and Bragg mirrors in multilayer photonic films [33], while a significantly enhanced performance on photoelectrochemical water splitting was reported for bilayer $BiVO_4$-WO_3 photoanodes deposited upon inverse opal WO_3 underlayers [34]. Very recently, multi-spectral slow photon tuning to the photocatalyst electronic absorption was demonstrated for bilayers consisting of TiO_2-$BiVO_4$ inverse opals with two distinct PBGs leading to enhanced light trapping for VLA photocatalysis [35]. A Fabry–Pérot-like resonator was previously implemented by Eftekhari et al. [36] using a trilayer stack of 3D TiO_2 PCs consisting of identical top and bottom inverse opal layers and a middle one with larger pores, whose red and blue PBG edges overlapped with the TiO_2 absorption edge. The trilayer film enabled light trapping between the top and bottom PC layers, leading to a significantly improved photocatalytic activity under UVA light and corroborating the unique prospects of slow photon management for photocatalytic applications in the absence of visible light activation.

An alternative approach to manipulate slow light propagation in PC structures was previously proposed based on the introduction of planar "defects" that disrupt their periodicity and create localized states within the PBG [37]. Optical defect engineering in PCs has been keenly pursued since the early days of the field of photonics for optoelectronic applications [38,39], while planar defects have been reported for polymeric colloidal opals [40] and SiO_2 inverse opals [41,42] by the deposition of an intermediate layer with the aim of recovering optical transmission within the forbidden PBG spectral range. Additional slow photon regions can therefore be created around the optical defect state, which reduce the group velocity, similar to the band bending at the edges of the photonic band gap (PBG). However, the incorporation of optical defects into PC photocatalysts to expand the relatively narrow slow light spectral range, which limits photonic amplification effects, has not been explored until now. There have been only a few reports addressing this challenge, primarily focusing on engineering multiple slow photon spectral regions within multilayer PC structures with different lattice constants [35,36].

This work demonstrates for the first time that introducing an optical defect within monolithic CoO_x-TiO_2 inverse opals is an effective strategy to enhance visible light harvesting and photocatalytic performance by reducing Bragg reflection losses and expanding the slow photon spectral regions. Composite PC films were fabricated by the successive deposition of an inverse opal TiO_2 substrate, a thin mesoporous layer of the benchmark Aeroxide P25 titania catalyst acting as planar defect, and a TiO_2 inverse opal top layer identical to the bottom one. Visible light activation of the trilayer PC films was performed through their surface modification with CoO_x "molecular" complexes at various stages of the multilayer deposition in order to optimize visible light absorption. The optimization of CoO_x loading in combination with PBG engineering resulted in a greatly improved photocatalytic performance of the CoO_x-TiO_2 trilayer PCs, which was evaluated based on salicylic acid (SA) degradation as well as photoelectrochemical measurements under visible light. Marked enhancements of the SA degradation rate and photocurrent density as well as reduced recombination kinetics were observed, indicating the unique potential of optical defect engineering in augmenting the reactivity of PC photocatalysts.

2. Materials and Methods

2.1. Photonic Film Fabrication

TiO_2 inverse opal underlayers were deposited on glass substrates in a single-step process via the evaporation-induced co-assembly of 0.1% w/v monodisperse poly(methyl methacrylate) (PMMA) colloidal spheres with a hydrolyzed Ti alkoxide precursor based on a titanium(IV) bis(ammonium lactato)dihydroxide (TiBALDH) 50 wt.% aqueous solution, followed by calcination at 500 °C. PMMA templating spheres with 406 and 499 nm diameters were selected to deposit the photonic substrates, labeled as PC406 and PC499, leading to well-defined PBGs (in water) that approached the spectral region of electronic absorption of CoO_x oxides [43]. A thin nanocrystalline TiO_2 layer of about 200 nm was then deposited by spin coating a paste made from the benchmark Aeroxide® P25 titania catalyst at 2000 rpm for 60 s on the PC substrate, followed by drying at 120 °C and calcination at 450 °C. Finally, a TiO_2 inverse opal upper layer was deposited on the P25/PC bilayer films under the same conditions as the original substrate, PC406 or PC499, leading to PC406/P25/PC406 and PC499/P25/PC499 trilayer structures, as depicted in Figure 1.

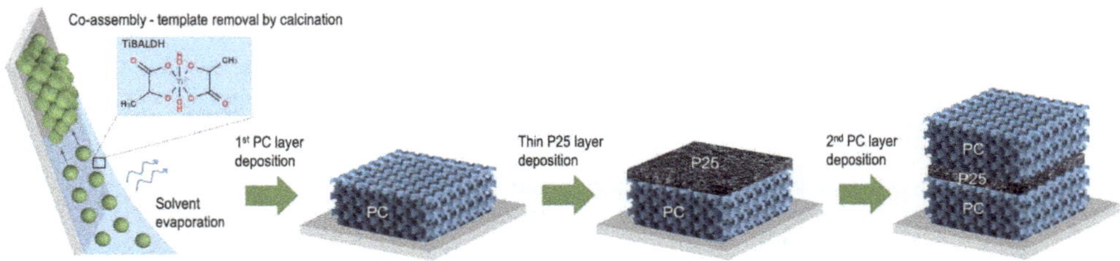

Figure 1. Schematic illustration of the PC/P25/PC trilayer deposition process.

A surface modification using nanoscale Co oxides was performed via the chemisorption of cobalt(II) acetylacetonate dihydrate $Co(acac)_2(H_2O)_2$ on the titania films after immersion in 100 mL of a 10^{-3} M solution (3:17 v/v EtOH/n-hexane) for 24 h followed by thorough washing with the same solvent, drying, and post-calcination in air at 500 °C for 1 h [43]. The CoO_x modification was performed using the chemisorption–calcination method at various stages of the films' deposition to achieve the optimal absorption of visible light. In particular, the trilayer films CoO_x-PC406/P25/PC406, PC406/CoO_x-P25/PC406, CoO_x-PC406/CoO_x-P25/PC406, and PC499/CoO_x-P25/PC499 were deposited as well as the reference films CoO_x-PC406, CoO_x-PC499, and CoO_x-P25.

2.2. Materials Characterization

The films' morphology and phase composition were studied using a scanning electron microscope (SEM, Quanta Inspect, FEI, Eindhoven, Netherlands) and a micro-Raman microscope (inVia Reflex, Renishaw, London, UK) with a ×50 objective at 785 nm excitation and a low power density (0.1 mW/μm^2). The optical properties were investigated by specular and diffuse reflectance spectroscopy on a Cary 60 UV-Vis spectrometer equipped with fiber-optic diffuse reflectance and specular reflectance accessories, using a Halon powder standard and a UV-enhanced Al mirror for baseline measurements, respectively. Photoelectrochemical experiments were performed in a three-electrode configuration on a CS350 potentiostat/galvanostat using the photonic films deposited on cleaned fluorine-doped tin oxide (FTO) transparent conductive glass (7 ohms/sq Pilkington) substrates as working electrodes, Ag/AgCl as a reference electrode, and a Pt foil as a counter electrode in 0.5 M NaOH aqueous electrolyte. The measurements were carried out in the dark and under visible light illumination provided by a 300 W Xe lamp with a 400 nm long-pass filter at 100 mW/cm^2. Electrochemical impedance spectroscopy (EIS) was carried out at open-circuit voltage (V_{OC}) in the frequency range of 10^4–10^{-1} Hz with an ac amplitude of 10 mV. Mott–Schottky measurements were carried out at 1 kHz and a scan rate of 10 mV/s at each potential.

2.3. Photocatalytic Evaluation

The photocatalytic activity of trilayer PC and reference films was evaluated on the aqueous phase degradation of SA under visible light. Films with an area of 2 cm^2 were placed horizontally to vials containing aqueous SA solutions (3 mL, 25 μM), which were stirred for 60 min in the dark to reach adsorption–desorption equilibrium, while adsorption was enhanced by stabilizing the solution's pH at 3 using diluted HCl. UV–Vis irradiation was provided by a 150 W Xe lamp along with a long-pass edge filter with a 305 nm cut-on and a heat-reflective mirror, while visible light was selected using an additional 400 nm long-pass filter. The horizontal beam was directed onto the film surface by a UV-enhanced Al mirror with an incident power density on the film surface of 76 mW/cm^2. Aliquots of 0.5 mL were periodically withdrawn from the SA solution and analyzed in a 10 mm quartz micro cell in the Cary 60 spectrophotometer. The photocatalytic experiments were performed in triplicate, and standard errors were calculated for the mean kinetic constants.

3. Results and Discussion

3.1. Morphological, Structural, and Optical Properties

SEM images of the co-assembled PC406 and PC499 photonic substrates, shown in Figure 2a–c, display the formation of well-organized periodic inverse opal structures with macropore sizes of 255 and 310 nm, respectively. Cross-sectional SEM images of the PC406/CoO_x-P25/PC406 trilayer films at successively higher magnifications, shown in Figure 2d–f, indicate the formation of a thin nanocrystalline P25 layer of about 200 nm thickness, which is inserted between the two photonic layers interrupting the periodicity of the inverse opal structure. Figure 2g,h show specular reflectance (R%) measurements at a 15° incidence angle for the PC406/P25/PC406 and PC499/P25/PC499 trilayer films in comparison to the corresponding monolithic PC406 and PC499 inverse opals (without an intermediate P25 layer).

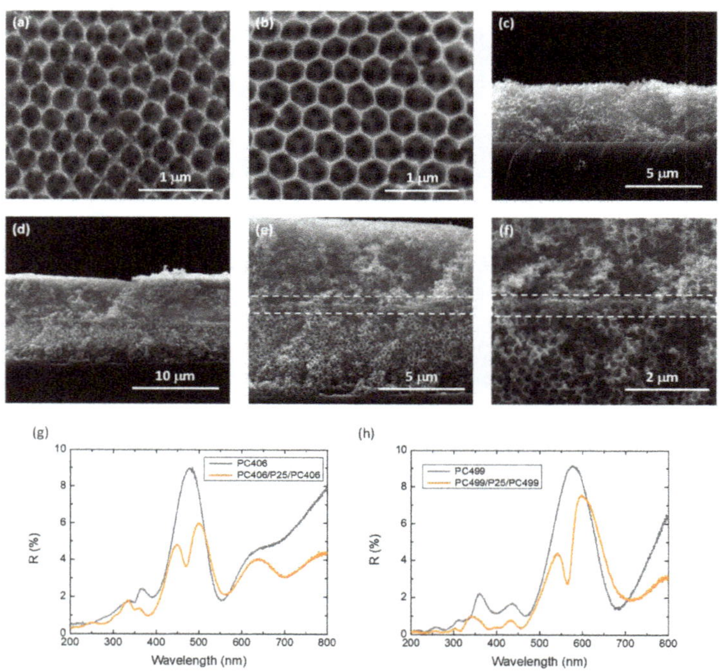

Figure 2. SEM images of the (**a**) PC406 and (**b**,**c**) PC499 inverse opal substrates as well as (**d**–**f**) cross-sections of the PC406/P25/PC406 trilayer photonic films at different magnifications. Specular reflectance (R%) spectra for the (**g**) PC406/P25/PC406 και (**h**) PC499/P25/PC499 in comparison to the monolithic PC406 and PC499 inverse opals.

The incomplete PBGs (stop bands) of the PC406 and PC499 inverse opals were determined from the corresponding Bragg reflections at 486 and 578 nm, respectively. The stop band positions can be approximated using the modified Bragg's law for first-order diffraction from the (111) planes of the fcc inverse opal structure [9]

$$\lambda = 2d_{111}\sqrt{n_{eff}^2 - sin^2\theta} \quad (1)$$

where λ is the stop band wavelength, $d_{111} = \sqrt{2/3}D$ is the spacing between the (111) planes, D is the macropore diameter, and $n_{eff}^2 = n_{sphere}^2 f + n_{TiO_2}^2(1-f)$ shows the volume-weighted average of the sphere's refractive index (n_{sphere}) and titania (n_{TiO_2}), while f is the sphere-filling fraction ($f = 0.74$ for the fcc lattice), and θ is the angle between the incident beam and the [111] direction. Using the experimental PBG wavelengths at $\theta = 15°$ and the sphere diameters for $n_{TiO_2} = 2.55$ and $n_{sphere} = n_{air} = 1.0$, the n values and the titania filling fractions $(1-f)$ were determined in air (Table 1). These values were much smaller than the theoretical one of 0.26 for complete filling of the inverse fcc lattice, supporting the mesoporosity of the nanocrystalline anatase skeleton. Furthermore, using the obtained $(1-f)$ values for $n_{sphere} = n_{H_2O} = 1.33$, the stop bands were calculated at 603 and 731 nm for PC406 and PC499 at 0° incidence in water, where the photocatalytic reactions took place.

Table 1. Structural and optical parameters of the TiO_2 PC substrates.

Film	D (nm)	λ (15°) (nm)	n_{eff} (air)	1−f	n_{eff} (H_2O)	λ (0°) (nm) (air)	λ (0°) (nm) (H_2O)
PC406	255(5)	478	1.18(2)	0.070(9)	1.45(2)	490(13)	603(13)
PC499	310(5)	578	1.17(2)	0.067(9)	1.44(2)	593(16)	731(16)

Importantly, a distinctive local minimum within the Bragg reflection of the monolithic PCs was clearly observed in the corresponding R% spectra for the PC406/P25/PC406 and PC499/P25/PC499 trilayer photonic films (Figure 2g,h). The observed dip within the Bragg R% peak indicates the creation of a localized state that enhances transmittance through the inverse opal PBG, confirming the formation of a structural "defect" that interrupts the periodicity of the PC structure [41]. The observed local minima in the R% spectra of the trilayer photonic films were shifted to lower wavelengths with respect to the Bragg reflection approaching the high-energy PBG edge. Such a spectral shift leads to donor-like optical states similar to the localized electronic states close to the conduction band in n-type semiconductors. This behavior is in close agreement with previous results on high-quality SiO_2 inverse opals fabricated by chemical vapor deposition, incorporating planar defects of a homogeneous SiO_2 layer with thicknesses smaller than the lattice constant ($a = \sqrt{2D}$) of the photonic crystal [41,42]. In this case, besides the decrease in group velocity due to the reduced gradients of the dispersion relation in the spectral region of the planar defect likewise the slow photon effects at the PBG edges, a strong localization of light is also expected [38], which can enhance the photocatalytic process through the mesoporous VLA CoO_x-P25 defect layer.

The structural properties and phase composition of the trilayer photonic films were studied by Raman spectroscopy after CoO_x surface modification through the $Co(acac)_2(H_2O)_2$ chemisorption–calcination treatment at various stages of their fabrication in comparison to the CoO_x-P25/PC406, CoO_x-PC406, and CoO_x-P25 reference films (Figure 3). The monolithic TiO_2 inverse opals showed the characteristic Raman-active phonons of anatase at approximately 147 (E_g), 197 (E_g), 398 (B_{1g}), 518 (A_{1g} + B_{1g}), and 642 cm^{-1} (E_g), confirming their crystallization in the single anatase phase after calcination at 500 °C. The anatase modes, especially the most intense low-frequency E_g band, exhibited appreciable shifts and broadening for the PC films compared to P25, due to the formation of small-size (≤10 nm) anatase nanocrystals in the co-assembled TiO_2 inverse opals using the TiBALDH precursor that cause the breakdown of the $q = 0$ selection rule for Raman scattering in titania nanomaterials [15]. The Raman spectra of the surface-modified titania films showed, in addition to the anatase TiO_2 phonons (the weak peaks of the rutile phase are only discernible in the P25 films), characteristic peaks of the Co_3O_4 spinel phase at 194, 483, 619, and 692 cm^{-1} [44]. The latter bands were significantly enhanced due to the resonance Raman scattering with excitation at 785 nm, which approaches the strong electronic transition of Co_3O_4 at about 750 nm [45]. The Co_3O_4 Raman modes were observed in all multilayer films containing the P25 layer, in contrast to the CoO_x-PC406 films, where these vibrations were not observed, confirming the preferential formation of the Co_3O_4 spinel in the mesoporous CoO_x-P25 films [43]. This was directly observed in the corresponding Raman spectrum of the reference CoO_x-P25 thin film, where the spinel Raman peaks and the parasitic fluorescence of the glass substrate (1200–1600 cm^{-1}) were dominant. The relative intensity of the Co_3O_4 Raman peaks with respect to the TiO_2 Raman mode intensities in the trilayer photonic films depended on the stage of the application of the surface modification.

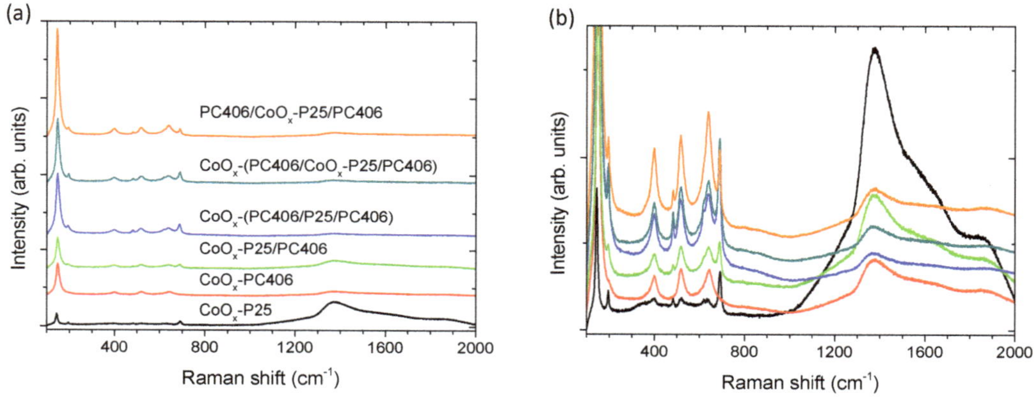

Figure 3. Raman spectra of the trilayer films CoO_x-PC406/P25/PC406, PC406/CoO_x-P25/PC406, and CoO_x-PC406/CoO_x-P25/PC406 in comparison to CoO_x-P25/PC406, CoO_x-PC406 and CoO_x-P25 reference films at 785 nm and different intensity scales (**a**,**b**). Shaded bands indicate the Co_3O_4 Raman modes.

The lowest intensity was observed for the PC406/CoO_x-P25/PC406 films, where the surface modification was performed after the deposition of the P25 interlayer in the photonic substrate, while the highest intensity was detected for the films where the CoO_x deposition was performed after the completion of the trilayer film fabrication. However, it should be pointed out that, apart from the deposition of nanoscopic Co_3O_4, the formation of monovalent oxides CoO and Co_2O_3 is also possible, which show weak Raman scattering [44,46] especially in the photonic layers, as previously reported for CoO_x-TiO_2 P25/PC bilayer films [33].

The influence of CoO_x surface modification on the optical properties of the trilayer films was observed in the diffuse reflectance (DR%) spectra, which are presented in Figure 4. The deposition of CoO_x by a chemisorption–thermal treatment cycle of the $Co(acac)_2(H_2O)_2$ complex following the deposition of the P25 mesoporous layer on the PC406 and PC499 photonic substrates resulted in a significant reduction in DR% over a wide spectral range due to the formation of CoO_x oxides, as shown in Figure 4a,b. The deposition of the TiO_2 photonic top layer on the P25/PC bilayer films to form the trilayer structure resulted in an increase in DR%, which, however, remained lower than that of the original PC substrate. Surface modification of the trilayer films with CoO_x caused an even greater decrease in DR% due to the formation of CoO_x in the upper photonic layer, which became particularly pronounced in the case of CoO_x-PC406/CoO_x-P25/PC406 films, as shown in Figure 4c, where CoO_x was deposited twice.

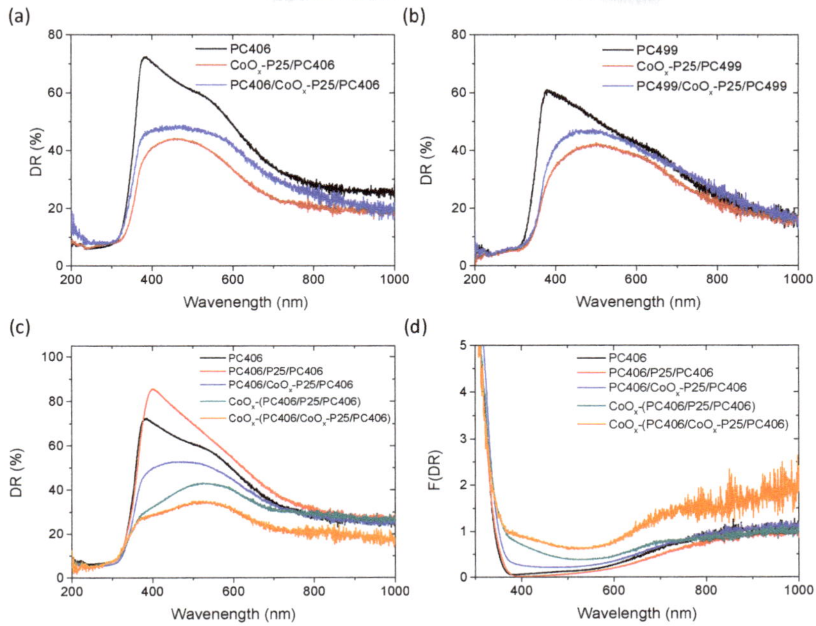

Figure 4. Diffuse (DR%) reflectance spectra for the mono-, bi-, and tri-layer photonic films: (**a**) PC406, CoO_x-P25/PC406, PC406/CoO_x-P25/PC406 and (**b**) PC499, CoO_x-P25/PC499, and PC499/CoO_x-P25/PC499. (**c**) DR% and (**d**) the corresponding Kubelka–Munk F(DR) absorbance spectra of the PC406/CoO_x-P25/PC406, CoO_x-PC406/P25/PC406, and CoO_x-PC406/CoOx-P25/PC406 trilayer films in comparison to the PC406 and PC406/P25/PC406 reference ones.

In all cases, the drop in the DR% after CoO_x modification occurred mainly in the 400-500 nm spectral range and to a lesser extent at wavelengths above 700 nm, which were also detected in the corresponding Kubelka–Munk absorbance spectra (Figure 4d). These absorption bands comply favorably with the electronic transitions at about 2.8 and 1.7 eV for Co_3O_4 [47], corroborating the preferential spinel formation detected by Raman spectroscopy on the P25 interlayer (Figure 3) as well as the single-valence CoO and Co_2O_3 oxides presenting wider band gaps of approximately 2.6 and 1.8 eV, respectively [48].

3.2. Photocatalytic Activity

The photocatalytic performance of the trilayer photonic films compared to their constituent monolithic layers was evaluated in the photodegradation of SA under visible light irradiation ($\lambda > 400$ nm), using aqueous solutions with pH = 3 that favor the chemisorption of SA on TiO_2 and its direct oxidation via photoinduced holes [16]. All the trilayer films showed a strong photocatalytic activity in the degradation of SA, whose concentration (C) was determined from the absorption peak at 300 nm, which decreased gradually with irradiation time, as shown in Figure 5a. In all cases, the variation in $\ln(C/C_0)$ with time t presented a linear dependence, where C_0 was the initial concentration of SA at $t = 0$ after its adsorption on the films' surface in the dark for 30 min, showing that the photocatalytic reaction follows first-order kinetics (Figure 5b). Figure 5c shows the change in the reaction rates $r = kC_0$, which were determined from the kinetic constants k obtained from the slopes of the linear plots $\ln(C/C_0)$ vs t and the corresponding C_0 values. The surface-modified PC406/P25/PC406 films presented significantly higher reaction rates than the sum of the respective rates of the constituent reference CoO_x-P25 and CoO_x-PC406 films, regardless of the processing step in which the CoO_x deposition was carried out.

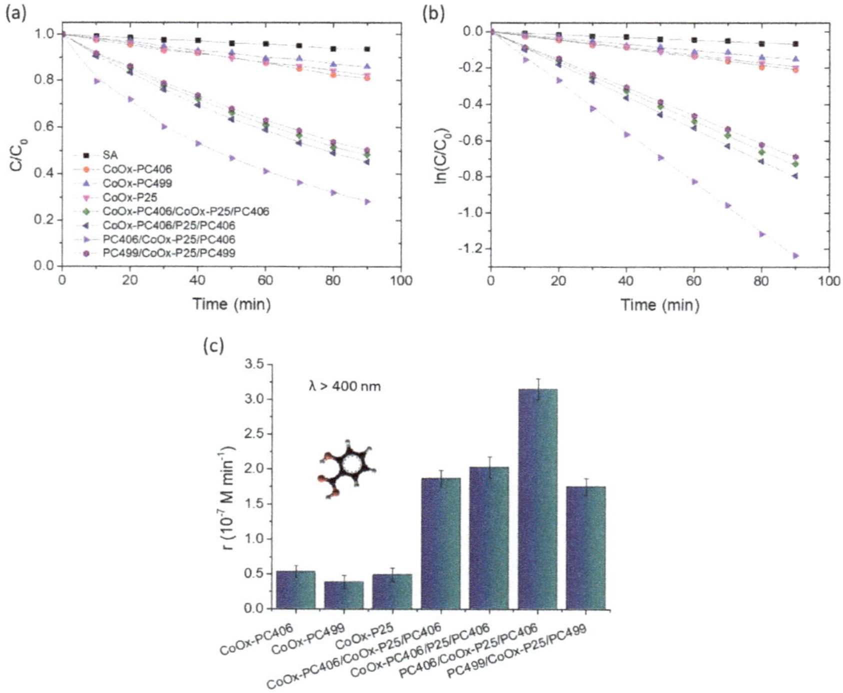

Figure 5. (**a**,**b**) SA photodegradation kinetics and (**c**) reaction rates for the CoO_x-surface modified trilayer photonic films in comparison to the constituent monolithic CoO_x-TiO_2 films under visible light ($\lambda > 400$ nm) irradiation.

This behavior indicates that the trilayer structure can enhance visible light harvesting by extending the slow light spectral region for the CoO_x-PC406 inverse opals, which exhibit the best photocatalytic performance in SA decomposition as the two edges, high and low-energy, of the PBG expected at ~600 nm in water (Table 1) overlap with the electronic absorption bands of CoO_x-TiO_2. It is noteworthy that, among the surface-modified trilayer films, the PC406/CoO_x-P25/PC406 films exhibited the best reaction rate even though they presented the lowest visible light optical absorption (Figure 4c) as the upper PC406 layer is not optically active in the visible range due to the deposition of CoO_x in the intermediate fabrication stage of the trilayer structure, i.e., on the P25/PC406 bilayer film. This behavior shows that the enhancement of visible light collection and photocatalytic activity in the trilayer films is largely caused by the localization of light in the CoO_x-P25 interlayer, which, despite its small thickness, exhibits a high photocatalytic activity. In addition, the strong optical absorption due to the higher CoO_x concentration in the CoO_x-PC406/P25/PC406 and CoO_x-PC406/CoO_x-P25/PC406 films can deteriorate their photonic properties as well as the light transmittance in the trilayer structure. An appreciable photocatalytic performance was also observed for the PC499/CoO_x-P25/PC499 trilayer films, where visible light localization in the PBG region of PC499 in water is expected at ~730 nm (Table 1), which overlaps with the electronic absorption region of CoO_x oxides and particularly the nanoscale Co_3O_4 clusters formed in the P25 mesoporous interlayer that absorb strongly at 450 and 750 nm [45,47]. This indicates that Bragg reflection losses in the PC499 layers and also the contribution of the most active CoO_x-PC406 photonic substrate are important to the overall photocatalytic performance.

It should be noted that the introduction of a mesoporous P25 nanocrystalline interlayer with a surface area of about 40 m^2 g^{-1} [49] is not expected to appreciably influence the pore structure of the monolithic PC406 and PC499 inverse opal films with surface areas of about 64 and 68 m^2 g^{-1}, respectively, arising mainly from the wall mesoporosity of the co-assembled inverse opal films [43] and, thus, the pore structure of the trilayer films. Although there are no reports in the literature on the performance of PC photocatalysts with optical defects, a general comparison can be made by examining the enhancement factor (EF) of the observed reaction rates in water pollutant degradation, relative to those of bilayer PCs, where amplification is pursued by Bragg reflection and multi-spectral slow photons via double PBG formation [33,35]. The EF, calculated from the ratio of the reaction rate for the best-performing PC406/CoO_x-P25/PC406 film to the sum of the r values for the individual CoO_x-PC406 support, the CoO_x-P25 interlayer, and the bare PC406 top layer, reached 2.67. This is comparable to, yet slightly higher than, the EF of 2.45 obtained for the CoO_x-P25/PC406 bilayer film in the VLA SA degradation with a thicker P25 top layer [33], and the EF of 2.2 recently reported from the photocatalytic degradation of the rhodamine B dye using $BiVO_4$-decorated TiO_2 bilayer inverse opals compared to the corresponding monolayer PC films [35]. These results suggest that introducing an optical defect through the insertion of a mesoporous layer in inverse opal photocatalysts offers a versatile and competent strategy for enhancing visible light harvesting through slow photons in spectral regions of the weak electronic absorption.

3.3. Charge Separation

The photocatalytic responses of PC406/CoO_x-P25/PC406 and PC499/CoO_x-P25/PC499 were further studied through photoelectrochemical measurements on films deposited on conductive FTO substrates. Figure 6 shows the photocurrent density produced during photocatalytic water splitting in the presence of mono-, bi-, and tri-layer films as a function of time under chopped visible light illumination (>420 nm) in a 0.5 M NaOH electrolyte at 0.2 V vs. Ag/AgCl as well as the Nyquist plots of the electrochemical impedance under visible light. Upon illumination, the photocurrent density increased instantly to the highest value that corresponds to the current density of photogenerated holes reaching the semiconductor–electrolyte interface. Subsequently, due to surface electron–hole recombination, the photocurrent density decreased exponentially to a steady-state value, which corresponded to the flux of the holes in the electrolyte without undergoing recombination with electrons at the surface [50]. The negative photocurrent overshoot feature that appears when switching the light off has been associated with the back flow of electrons that recombine with the remaining holes at the surface. An increase in both instantaneous (hole current) and steady-state photocurrent density values was observed for the trilayer photonic films, with a higher value in the case of PC406/CoO_x-P25/PC406 confirming the enhanced visible light harvesting and the improved charge separation for the specific inverse opal diameter in the trilayer structure. The same conclusion is obtained from the corresponding Nyquist diagrams, which show that the charge transfer resistance at the interface of the photoelectrode with the liquid electrolyte, which corresponds to the curvature of the graph of the imaginary part of the impedance –Z″ with respect to the real Z′, is significantly reduced in the trilayer films, regardless of the inverse opal diameter.

In addition, capacitance measurements ($C = -1/Z''\omega$) were performed as a function of the applied potential V at a constant frequency of 1000 Hz for the defect P25 interlayer (Figure 6e). Neglecting the contribution of the Helmholtz layer (C_H), the inverse capacitance at the semiconductor–electrolyte interface for n-type semiconductors such as TiO_2 is approximated using the inverse capacitance ($1/C_{sc} \gg 1/C_H$) in the space charge region (depletion zone) according to the following relation

$$\frac{1}{C^2} = \left(\frac{1}{C_{sc}} + \frac{1}{C_H}\right)^2 \approx \frac{1}{{C_{sc}}^2} = \frac{2}{eA^2\varepsilon\varepsilon_0 N_D}\left(V - V_{fb} - \frac{kT}{e}\right) \quad (2)$$

where e is the electron charge, ε is the dielectric constant, A is the geometric surface of the semiconductor electrode, ε_0 is the vacuum permeability, and k is Boltzmann constant. N_D is the concentration of donors of the semiconductor and V_{fb} is the flatband potential, which is defined as the potential that neutralizes the bending of the semiconductor energy bands when equilibrium is reached at the semiconductor–electrolyte interface through the equation of the Fermi energy with the electrochemical potential of the electrolyte.

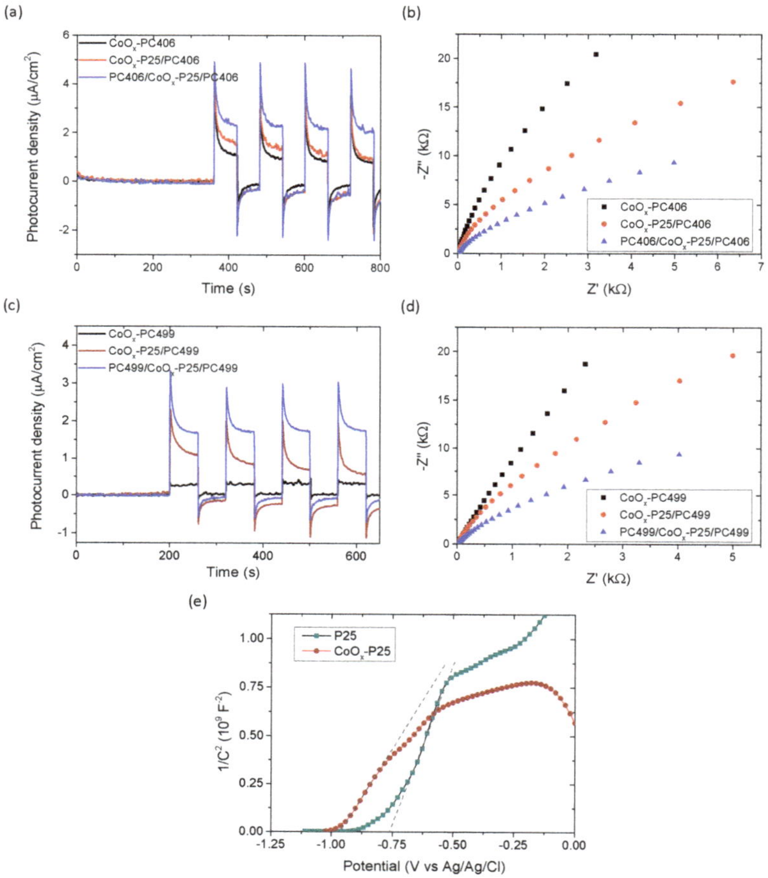

Figure 6. Transient photocurrent density and EIS Nyquist plots in 0.5 M NaOH electrolyte under visible light illumination for the mono-, bi-, and tri-layer films (**a**), (**b**) CoO_x-PC406, CoO_x-P25/PC406, PC406/CoO_x-P25/PC406, and (**c**), (**d**) CoO_x-PC499, CoO_x-P25/PC499, PC499/CoO_x-P25/PC499. (**e**) Mott–Schottky plots for the P25 και CoO_x-P25 films at 1000 Hz.

The linear part of the plot of $1/C^2$vs. V (Mott–Schottky plot) with the applied potential was used to determine N_D and V_{fb} from the slope and the intercept with the potential axis [51]. Figure 6e shows the Mott-Schottky diagrams of the P25 and CoO_x-P25 films, which form the intermediate photocatalytic layer (planar defect) for the localization of visible light in the trilayer structure. A shift in V_{fb} to negative potentials from −0.78 to −0.99 V vs. Ag/AgCl was observed as well as an increase in the donor concentration from 4.1×10^{19} cm^{-3} to 7.2×10^{19} cm^{-3} in the CoO_x-P25 film compared to the unmodified mesoporous titania P25 film. This behavior can be attributed to the formation of a surface band in the CoO_x/TiO_2 composite system that shifts the valence band of TiO_2 to lower potentials [52,53] as well as the formation of p-n Co_3O_4-TiO_2 heterojunctions through the preferential formation of Co_3O_4 spinel nanoparticles, a p-type semiconductor, on the surface of the modified P25 films [54].

4. Conclusions

An alternative structural approach to enhance visible light harvesting and the photocatalytic performance of PC photocatalysts has been implemented for the first time by the introduction of a planar defect in visible light activated CoO_x-TiO_2 inverse opal films in order to alleviate the major limitation of slow light propagation within narrow spectral ranges at the PBG edges. The deposition of a mesoporous titania layer within monolithic inverse opals was studied as a means to interrupt the PC periodicity and create localized defect states within the PBG. To this end, trilayer photonic films were fabricated by the successive deposition of an inverse opal TiO_2 underlayer, a thin mesoporous titania interlayer, and a photonic top layer identical to the original one. Visible light activation of the trilayer photonic films was carried out by their surface modification with low-concentration CoO_x "molecular" complexes at various stages of the film fabrication in order to optimize visible light absorption. Optical measurements showed the formation of "donor"-like states within the trilayer PBG, which reduced the Bragg reflection and broadened the slow light spectral regions. The optimization of CoO_x loading in combination with PBG engineering resulted in an improved photocatalytic performance of the CoO_x-TiO_2 trilayer PCs that were evaluated using SA degradation and photoelectrochemical measurements under visible light. Significant enhancements in the SA degradation rate and photocurrent density along with reduced electron–hole recombination were observed for the optimal trilayer PCs in comparison to the additive effect of the constituent layers and trilayers with variable CoO_x loading, indicative of light localization in the middle defect layer. Optical defect engineering within the PBG of PC photocatalysts is, accordingly, concluded to be a promising strategy to enhance slow photon amplification effects and reduce Bragg reflection losses, which can be further combined with suitable compositional modifications for visible light photocatalysis.

Author Contributions: Conceptualization, V.L.; methodology, V.L. and A.T.; investigation, A.T. and E.S.; resources, V.L.; formal analysis, A.T. and E.S.; writing—original draft preparation, A.T. and V.L.; writing—review and editing, V.L. and A.T.; supervision, V.L. All authors have read and agreed to the published version of the manuscript.

Funding: The research work was supported by the Hellenic Foundation for Research and Innovation (H.F.R.I.) under the "First Call for H.F.R.I. Research Projects to support Faculty members and Researchers and the procurement of high-cost research equipment grant" (Project Number: 543).

Institutional Review Board Statement: Not applicable.

Informed Consent Statement: Not applicable.

Data Availability Statement: The original contributions presented in this study are included in the article. Further inquiries can be directed to the corresponding author.

Conflicts of Interest: The authors declare no conflicts of interest.

References

1. Chen, J.I.L.; von Freymann, G.; Choi, S.Y.; Kitaev, V.; Ozin, G.A. Amplified photochemistry with slow photons. *Adv. Mater.* **2006**, *18*, 1915–1919. [CrossRef]
2. Curti, M.; Schneider, J.; Bahnemann, D.W.; Mendive, C.B. Inverse opal photonic crystals as a strategy to improve photocatalysis: Underexplored questions. *J. Phys. Chem. Lett.* **2015**, *6*, 3903–3910. [CrossRef] [PubMed]
3. Cai, J.M.; Wu, M.Q.; Wang, Y.T.; Zhang, H.; Meng, M.; Tian, Y.; Li, X.A.; Zhang, J.; Zheng, L.R.; Gong, J.L. Synergetic enhancement of light harvesting and charge separation over surface-disorder engineered TiO2 photonic crystals. *Chem* **2017**, *2*, 877–892. [CrossRef]
4. Wang, Y.; Peng, C.; Jiang, T.; Zhang, J.; Jiang, Z.; Li, X. Construction of defect-engineered three-dimensionally ordered macroporous WO_3 for efficient photocatalytic water oxidation reaction. *J. Mater. Chem. A* **2020**, *9*, 3036–3043. [CrossRef]

5. Feng, L.; Wang, F.; Luo, H.; Xu, Z.; Zhao, T.; Zhu, J.; Qin, Y. Thermal vacuum de-oxygen fabrication of new catalytic pigments: $SiO_2@TiO_{2-x}$ amorphous photonic crystals for formaldehyde removal. *J. Mater. Chem. B* **2023**, *11*, 1533–1544. [CrossRef]
6. Zhao, H.; Li, C.-F.; Hu, Z.-Y.; Liu, J.; Li, Y.; Hu, J.; Van Tendeloo, G.; Chen, L.-H.; Su, B.-L. Size effect of bifunctional gold in hierarchical titanium oxide-gold-cadmium sulfide with slow photon effect for unprecedented visible-light hydrogen production. *J. Colloid Interface Sci.* **2021**, *604*, 131–140. [CrossRef]
7. Pylarinou, M.; Toumazatou, A.; Sakellis, E.; Xenogiannopoulou, E.; Gardelis, S.; Boukos, N.; Dimoulas, A.; Likodimos, V. Visible light trapping against charge recombination in FeO_x–TiO_2 photonic crystal photocatalysts. *Materials* **2021**, *14*, 7117. [CrossRef]
8. Madanu, T.L.; Mouchet, S.R.; Deparis, O.; Liu, J.; Li, Y.; Su, B.-L. Tuning and transferring slow photons from TiO_2 photonic crystals to $BiVO_4$ nanoparticles for unprecedented visible light photocatalysis. *J. Colloid Interface Sci.* **2022**, *634*, 290–299. [CrossRef]
9. Loukopoulos, S.; Sakellis, E.; Kostakis, M.G.; Gerokonstantis, D.-T.; Tsipas, P.; Gardelis, S.; Kontos, A.G.; Katsaros, F.K.; Sideratou, Z.; Romanos, G.E.; et al. Co-assembled MoS_2–TiO_2 inverse opal photocatalysts for visible light-activated pharmaceutical photodegradation. *ACS Omega* **2023**, *8*, 33639–33650. [CrossRef]
10. Hedrich, C.; James, N.T.; Maragno, L.G.; De Lima, V.; González, S.Y.G.; Blick, R.H.; Zierold, R.; Furlan, K.P. Enhanced photocatalytic properties and photoinduced crystallization of TiO_2–Fe_2O_3 inverse opals fabricated by atomic layer deposition. *ACS Appl. Mater. Interfaces* **2024**, *16*, 46964–46974. [CrossRef]
11. Collins, G.; Lonergan, A.; McNulty, D.; Glynn, C.; Buckley, D.; Hu, C.; O'Dwyer, C. Semiconducting metal oxide photonic crystal plasmonic photocatalysts. *Adv. Mater. Interfaces* **2020**, *7*, 1901805. [CrossRef]
12. Raja-Mogan, T.; Lehoux, A.; Takashima, M.; Kowalska, E.; Ohtani, B. Slow photon-induced enhancement of photocatalytic activity of gold nanoparticle-incorporated titania inverse opal. *Chem. Lett.* **2021**, *50*, 711–713. [CrossRef]
13. Temerov, F.; Pham, K.; Juuti, P.; Mäkelä, J.M.; Grachova, E.V.; Kumar, S.; Eslava, S.; Saarinen, J.J. Silver-decorated TiO_2 inverse opal structure for visible light-induced photocatalytic degradation of organic pollutants and hydrogen evolution. *ACS Appl. Mater. Interfaces* **2020**, *12*, 41200–41210. [CrossRef] [PubMed]
14. Boppella, R.; Kochuveedu, S.T.; Kim, H.; Jeong, M.J.; Mota, F.M.; Park, J.H.; Kim, D.H. Plasmon-sensitized graphene/TiO_2 inverse opal nanostructures with enhanced charge collection efficiency for water splitting. *ACS Appl. Mater. Interfaces* **2017**, *9*, 7075–7083. [CrossRef]
15. Diamantopoulou, A.; Sakellis, E.; Gardelis, S.; Tsoutsou, D.; Glenis, S.; Boukos, N.; Dimoulas, A.; Likodimos, V. Advanced photocatalysts based on reduced nanographene oxide–TiO_2 photonic crystal films. *Materials* **2019**, *12*, 2518. [CrossRef]
16. Apostolaki, M.-A.; Toumazatou, A.; Antoniadou, M.; Sakellis, E.; Xenogiannopoulou, E.; Gardelis, S.; Boukos, N.; Falaras, P.; Dimoulas, A.; Likodimos, V. Graphene quantum dot-TiO_2 photonic crystal films for photocatalytic applications. *Nanomaterials* **2020**, *10*, 2566. [CrossRef]
17. Oh, Y.; Yang, W.; Tan, J.; Lee, H.; Park, J.; Moon, J. Boosting visible light harvesting in p-type ternary oxides for solar-to-hydrogen conversion using inverse opal structure. *Adv. Funct. Mater.* **2019**, *29*, 1900194. [CrossRef]
18. Zhu, H.; Zhang, Y.; Zhu, J.; Li, Y.; Jiang, S.; Wu, N.; Wei, Y.; Zhou, J.; Song, Y. Crack-free hematite inverse opal photo-anodes for enhancing photo-electrochemical water splitting. *J. Mater. Chem. A* **2020**, *8*, 22929–22937. [CrossRef]
19. Lin, Q.; Liang, S.; Wang, J.; Zhang, R.; Wang, X. Cadmium sulfide 3D photonic crystal with hierarchically ordered macropores for highly efficient photocatalytic hydrogen generation. *Inorg. Chem.* **2022**, *61*, 2920–2928. [CrossRef]
20. Pylarinou, M.; Sakellis, E.; Tsipas, P.; Romanos, G.E.; Gardelis, S.; Dimoulas, A.; Likodimos, V. Mo-$BiVO_4$/Ca-$BiVO_4$ homojunction nanostructure-based inverse opals for photoelectrocatalytic pharmaceutical degradation under visible light. *ACS Appl. Nano Mater.* **2023**, *6*, 6759–6771. [CrossRef]
21. Dong, Y.; Ai, F.; Sun-Waterhouse, D.; Murai, K.-I.; Moriga, T.; Waterhouse, G.I.N. Optical and photocatalytic properties of three-dimensionally ordered macroporous Ta_2O_5 and Ta_3N_5 inverse opals. *Chem. Mater.* **2023**, *35*, 8281–8300. [CrossRef]
22. Deparis, O.; Mouchet, S.R.; Su, B.-L. Light harvesting in photonic crystals revisited: Why do slow photons at the blue edge enhance absorption? *PCCP* **2015**, *17*, 30525–30532. [CrossRef] [PubMed]
23. Zhang, J.; Cai, X.; Fu, X.; Teng, D.; Murtaza, G.; Meng, Z.; Jia, Z.; Qiu, L. Slow light effect enhances the photocatalytic effect of inverse opal TiO_2-based photonic nanocrystals. *ACS Appl. Nano Mater.* **2024**, *7*, 15376–15386. [CrossRef]
24. Liu, J.; Jin, J.; Li, Y.; Huang, H.-W.; Wang, C.; Wu, M.; Chen, L.-H.; Su, B.-L. Tracing the slow photon effect in a ZnO inverse opal film for photocatalytic activity enhancement. *J. Mater. Chem. A* **2014**, *2*, 5051–5059. [CrossRef]
25. Curti, M.; Mendive, C.B.; Grela, M.A.; Bahnemann, D.W. Stopband tuning of TiO_2 inverse opals for slow photon absorption. *Mater. Res. Bull.* **2017**, *91*, 155–165. [CrossRef]
26. Curti, M.; Zvitco, G.; Grela, M.A.; Mendive, C.B. Angle dependence in slow photon photocatalysis using TiO_2 inverse opals. *Chem. Phys.* **2018**, *502*, 33–38. [CrossRef]
27. Wang, L.; Mogan, T.R.; Wang, K.; Takashima, M.; Ohtani, B.; Kowalska, E. Fabrication and characterization of inverse-opal titania films for enhancement of photocatalytic activity. *ChemEngineering* **2022**, *6*, 33. [CrossRef]
28. Apostolaki, M.-A.; Sakellis, E.; Tsipas, P.; Giannouri, M.; Gardelis, S.; Boukos, N.; Dimoulas, A.; Likodimos, V. Three-phase co-assembly of compositionally tunable WO_3/TiO_2 inverse opal photoelectrodes. *Appl. Surf. Sci.* **2022**, *613*, 155919. [CrossRef]
29. Li, P.; Chen, S.-L.; Wang, A.-J.; Wang, Y. Probing photon localization effect between titania and photonic crystals on enhanced photocatalytic activity of titania film. *Chem. Eng. J.* **2015**, *284*, 305–314. [CrossRef]
30. Zhang, R.; Zeng, F.; Pang, F.; Ge, J. Substantial enhancement toward the photocatalytic activity of CdS quantum dots by photonic crystal-supporting films. *ACS Appl. Mater. Interfaces* **2018**, *10*, 42241–42248. [CrossRef]

31. Xie, C.; Fan, T.; Wang, A.; Chen, S.-L. Enhanced visible-light photocatalytic activity of a TiO_2 membrane-assisted with n-doped carbon quantum dots and SiO_2 opal photonic crystal. *Ind. Eng. Chem. Res.* **2018**, *58*, 120–127. [CrossRef]
32. Lan, D.; Sheng, W.; Fu, Q.; Ge, J. Enhancement of CO_2 photoreduction efficiency by supporting blue TiO_2 with photonic crystal substrate. *Nano Res.* **2023**, *16*, 9310–9317. [CrossRef]
33. Loukopoulos, S.; Toumazatou, A.; Sakellis, E.; Xenogiannopoulou, E.; Boukos, N.; Dimoulas, A.; Likodimos, V. Heterostructured CoO_x–TiO_2 mesoporous/photonic crystal bilayer films for enhanced visible-light harvesting and photocatalysis. *Materials* **2020**, *13*, 4305. [CrossRef] [PubMed]
34. Taga, Y.; Pan, Z.; Katayama, K.; Sohn, W.Y. $BiVO_4$-Dotted WO_3 Photoanode with an inverse opal underlayer for photoelectrochemical water splitting. *ACS Appl. Energy Mater.* **2022**, *5*, 5750–5755. [CrossRef]
35. Madanu, T.L.; Chaabane, L.; Mouchet, S.R.; Deparis, O.; Su, B.-L. Manipulating multi-spectral slow photons in bilayer inverse opal TiO_2@$BiVO_4$ composites for highly enhanced visible light photocatalysis. *J. Colloid Interface Sci.* **2023**, *647*, 233–245. [CrossRef]
36. Eftekhari, E.; Broisson, P.; Aravindakshan, N.; Wu, Z.; Cole, I.S.; Li, X.; Zhao, D.; Li, Q. Sandwich-structured TiO_2 inverse opal circulates slow photons for tremendous improvement in solar energy conversion efficiency. *J. Mater. Chem. A* **2017**, *5*, 12803–12810. [CrossRef]
37. Galisteo-López, J.F.; Galli, M.; Andreani, L.C.; Mihi, A.; Pozas, R.; Ocaña, M.; Míguez, H. Phase delay and group velocity determination at a planar defect state in three dimensional photonic crystals. *Appl. Phys. Lett.* **2007**, *90*, 101113. [CrossRef]
38. Yablonovitch, E.; Gmitter, T.J.; Meade, R.D.; Rappe, A.M.; Brommer, K.D.; Joannopoulos, J.D. Donor and acceptor modes in photonic band structure. *Phys. Rev. Lett.* **1991**, *67*, 3380–3383. [CrossRef]
39. Karathanos, V.; Modinos, A.; Stefanou, N. Planar defects in photonic crystals. *J. Condens. Matter Phys.* **1994**, *6*, 6257–6264. [CrossRef]
40. Pozas, R.; Mihi, A.; Ocaña, M.; Míguez, H. Building nanocrystalline planar defects within self-assembled photonic crystals by spin-coating. *Adv. Mater* **2006**, *18*, 1183–1187. [CrossRef]
41. Palacios-Lidón, E.; Galisteo-López, J.F.; Juárez, B.H.; López, C. Engineered planar defects embedded in opals. *Adv. Mater* **2004**, *16*, 341–345. [CrossRef]
42. Tétreault, N.; Mihi, A.; Míguez, H.; Rodríguez, I.; Ozin, G.A.; Meseguer, F.; Kitaev, V. Dielectric planar defects in colloidal photonic crystal films. *Adv. Mater* **2004**, *16*, 346–349. [CrossRef]
43. Toumazatou, A.; Antoniadou, M.; Sakellis, E.; Tsoutsou, D.; Gardelis, S.; Romanos, G.E.; Ioannidis, N.; Boukos, N.; Dimoulas, A.; Falaras, P.; et al. Boosting visible light harvesting and charge separation in surface modified TiO_2 photonic crystal catalysts with CoO_x nanoclusters. *Mater. Adv.* **2020**, *1*, 2310–2322. [CrossRef]
44. Rivas-Murias, B.; Salgueiriño, V. Thermodynamic CoO–Co_3O_4 crossover using Raman spectroscopy in magnetic octahedron-shaped nanocrystals. *J. Raman Spectrosc.* **2017**, *48*, 837–841. [CrossRef]
45. Qiao, L.; Xiao, H.Y.; Meyer, H.M.; Sun, J.N.; Rouleau, C.M.; Puretzky, A.A.; Geohegan, D.B.; Ivanov, I.N.; Yoon, M.; Weber, W.J.; et al. Nature of the band gap and origin of the electro-/photo-activity of Co_3O_4. *J. Mater. Chem. C* **2013**, *1*, 4628–4633. [CrossRef]
46. Li, Y.; Qiu, W.; Qin, F.; Fang, H.; Hadjiev, V.G.; Litvinov, D.; Bao, J. Identification of cobalt oxides with Raman scattering and Fourier Transform infrared spectroscopy. *J. Phys. Chem. C* **2016**, *120*, 4511–4516. [CrossRef]
47. Singh, V.; Kosa, M.; Majhi, K.; Major, D.T. Putting DFT to the test: A first-principles study of electronic, magnetic, and optical properties of CO_3O_4. *J. Chem. Theory Comput.* **2014**, *11*, 64–72. [CrossRef]
48. Singh, V.; Major, D.T. Electronic structure and bonding in Co-based single and mixed valence oxides: A quantum chemical perspective. *Inorg. Chem.* **2016**, *55*, 3307–3315. [CrossRef]
49. Kontos, A.I.; Kontos, A.G.; Tsoukleris, D.S.; Bernard, M.-C.; Spyrellis, N.; Falaras, P. Nanostructured TiO_2 films for DSSCS prepared by combining doctor-blade and sol–gel techniques. *J. Mater. Process. Technol.* **2008**, *196*, 243–248. [CrossRef]
50. Dunn, H.K.; Feckl, J.M.; Müller, A.; Fattakhova-Rohlfing, D.; Morehead, S.G.; Roos, J.; Peter, L.M.; Scheu, C.; Bein, T. Tin doping speeds up hole transfer during light-driven water oxidation at hematite photoanodes. *PCCP* **2014**, *16*, 24610–24620. [CrossRef]
51. Hankin, A.; Bedoya-Lora, F.E.; Alexander, J.C.; Regoutz, A.; Kelsall, G.H. Flat band potential determination: Avoiding the pitfalls. *J. Mater. Chem. A* **2019**, *7*, 26162–26176. [CrossRef]
52. Jin, Q.; Yamamoto, H.; Yamamoto, K.; Fujishima, M.; Tada, H. Simultaneous induction of high level thermal and visible-light catalytic activities to titanium (IV oxide by surface modification with cobalt(III) oxide clusters. *PCCP* **2013**, *15*, 20313. [CrossRef] [PubMed]
53. Tada, H.; Jin, Q.; Iwaszuk, A.; Nolan, M. Molecular-scale transition metal oxide nanocluster surface-modified titanium dioxide as solar-activated environmental catalysts. *J. Phys. Chem. C* **2014**, *118*, 12077–12086. [CrossRef]
54. Lang, D.; Cheng, F.; Xiang, Q. Enhancement of photocatalytic H_2 production activity of CdS nanorods by cobalt-based cocatalyst modification. *Catal. Sci. Technol.* **2016**, *6*, 6207–6216. [CrossRef]

materials

Article

$Lu_3Al_5O_{12}:Ce^{3+}$ Fluorescent Ceramic with Deep Traps: Thermoluminescence and Photostimulable Luminescence Properties

Junwei Zhang [1], Miao Zhao [2], Qiao Hu [2], Renjie Jiang [2], Hao Ruan [2,*] and Hui Lin [1]

[1] Engineering Research Center of Optical Instrument and System, Ministry of Education and Shanghai Key Lab of Modern Optical System, University of Shanghai for Science and Technology, No. 516 Jungong Road, Shanghai 200093, China; junweizhang@siom.ac.cn (J.Z.); linh8112@163.com (H.L.)

[2] Aerospace Laser Technology and System Department, Wangzhijiang Innovation Center for Laser, Shanghai Institute of Optics and Fine Mechanics, Chinese Academy of Sciences, Shanghai 201800, China; miaozhao@siom.ac.cn (M.Z.); huqiao@siom.ac.cn (Q.H.); jiangrj@shanghaitech.edu.cn (R.J.)

* Correspondence: ruanhao@siom.ac.cn

Abstract: Electron-trapping materials have attracted a lot of attention in the field of optical data storage. However, the lack of suitable trap levels has hindered its development and application in the field of optical data storage. Herein, $Lu_3Al_5O_{12}:Ce^{3+}$ fluorescent ceramics were developed as the optical storage medium, and high-temperature vacuum sintering induced the formation of deep traps (1.36 eV). The matrix based on the garnet-structured material ensures excellent rewritability. By analyzing the thermoluminescence and photostimulable luminescence, it is found that the transition of electrons provided by Ce^{3+} between the conduction band and trap levels offers the possibility for optical data storage. As evidence of its application, the optical information encoding using 254 nm light and decoding using a light stimulus and thermal stimulus were applied. These findings are expected to provide candidate material for novel optical storage technology, and further promote the development of advanced information storage technology.

Keywords: $Lu_3Al_5O_{12}$; electron-trapping materials; optical data storage; photostimulable luminescence

1. Introduction

With the rapid development of society, the amount of global data increase exponentially. The huge amount of data puts forward challenges for information storage technology. Optical data storage (ODS) is widely used in information storage because of its outstanding performance, environmentally friendly nature, high capacity, and long lifetime [1–3]. However, caused by the limitation of optical diffraction resolution, the application of traditional ODS technologies is facing considerable challenges. The emergence of super-resolution recording [4] and information multiplexing technologies is expected to solve these problems [5–9]. To better apply the two technologies, a new optical storage medium needs to be developed.

Since Lindmayer first proposed the concept of electron-trapping materials (ETMs) in 1986 [10], they have been widely used in data storage [11–13], emergency display [14], and information processing [15]. ETMs, which can retain excitation energy and release it as light emission upon external thermal and optical stimulations, are also referred to as photostimulable luminescence (PSL) materials. PSL materials have great application prospects in the field of ODS due to its all-optical operation mode [16]. Some sulfide-based PSL materials have been developed in the previous period, but most of them focus on optical computing operations [17–20]; there are few studies on ODS performance, and the problems of sulfide pollution and poor thermal stability limit its applications. The emergence of ceramic phosphors solves the above problems and exhibits good ODS properties, such as rewritability and high speed of reading/writing. Some researchers have prepared organic–inorganic composite fluorescent films. Zhuang et al. proposed two PSL materials $(Sr_{1-x}Ba_x)Si_2O_2N_2$:Eu/Yb, Dy and $NaMgF_3$:Tb that can complete the writing and reading of information by 405 nm and 980 nm laser/heating, which allows for information multiplexing; Deng et al. prepared Y_2GeO_5:Pr, Tb PSL materials that can record information in a bit by bit mode by a 515 nm laser, which demonstrated the potential of PSL materials in the field of high-density ODS [21–23]; Lin et al. prepared phosphor-in-glass $BaSi_2O_5$:Eu, Nd for data storage, and established the relationship between trap and PSL performance [24]. As a result of the poor stability and uniformity of these materials, they are not ideal materials for data storage. Cerium-doped lutetium aluminum garnet ($Lu_3Al_5O_{12}$:Ce) ceramic is a promising ceramic material with excellent optical and thermal properties, good chemical stability, and high mechanical strength, which has been widely used in the fields of fluorescent materials, laser materials, and scintillation materials [25–29]. The above characteristics are very suitable as optical storage matrix materials.

However, the lack of appropriate depth of traps limits the application of new information storage materials. The trap levels of ETMs should be considered. Generally, the trap depth is between 1 and 1.5 eV, which can not only ensure the stable storage of information, but also be read out by near-infrared light [24,30]. Ion doping is a typical modulation method to construct traps. Xie et al. reported a series of $Y_3Al_{5-x}Ga_xO_{12}$:Ce^{3+}, V^{3+} PSL materials with a tunable trap depth (1.2–1.6 eV). According to the vacuum referred binding energy (VRBE) diagram, the effect of Ga^{3+} doping on the formation of deep traps was analyzed [31]. In the typical spinel-structured system, there are lots of anti-site defects of Ga^{3+} and Mg^{2+}, and the doping of Bi^{3+}/Cr^{3+} can induce the formation of deep traps [32,33].

In this work, $Lu_3Al_5O_{12}$:Ce^{3+} PSL materials were developed by vacuum sintering technology. Its spectral properties and crystal structure were studied in detail, capable of emitting 512 nm green light, and SEM shows that the grain size distribution is between 10 and 12 μm. In particular, the origin of the trap and the mechanism of optical storage were analyzed by thermoluminescence curves. Then, its rewritability and information retention ability were tested. It also exhibited the writing/reading process of the optical information by a 254 nm light and 980 nm laser/heating.

2. Experiments

$Lu_{3-x}Al_5O_{12}$:xCe (x = 0.011, 0.013 and 0.015) (abbreviated as LuAG:xCe) materials were prepared by the vacuum sintering technique. The required chemical regents include Lu_2O_3 (4N), Al_2O_3 (4N), CeO_2 (4N), MgO, and TEOs. Firstly, raw materials were mixed with anhydrous alcohol according to the stoichiometric ratio and ground in a planet-type ball mill at 230 rpm for 12 h. The ball-milled mixtures were rinsed with anhydrous alcohol and dried in an oven at 80 °C for 6 h. Then, the dried samples were thoroughly ground. After grinding, the dried powders were sieved through a 200-mesh screen. Prior to sintering, the green bodies were shaped from powders under a uniaxial press of 30 MPa and were cold isostatic pressed (CIP) at 210 MPa. The obtained samples were pressed and pre-sintered at 800 °C for 6 h and high-temperature sintered at 1850 °C for 6 h; the preparation flow chart of fluorescent ceramics as shown in Figure S1.

X-ray diffraction (XRD) patterns were obtained by an X-ray diffractometer, and the test range is 2θ = 10–90° (Bruker D8 ADVANCE, Karlsruhe, Germany). The steady photoluminescence (PL) and photoluminescence excitation (PLE) were tested by a fluorescence spectrophotometer (Hitachi F-4600, Tokyo, Japan). The morphologies of the ceramics were measured using a scanning electron microscope (SEM, Verios G4 UC, Waltham, MA, USA). Thermoluminescence (TL) was tested by a TOSL-3DS meter (Guangzhou Rongfan, Guangzhou, China). Luminescence decay curves and Photostimulated luminescence (PSL) spectra were measured by a Time-resolved spectrometer (Edinburgh Instruments, FLSP-920, Livingston, UK).

3. Results and Analysis

3.1. Structure and Optical Properties

As shown in Figure 1a, we tested the XRD spectrum of the $Lu_{3-x}Al_5O_{12}$:xCe (x = 0.011, 0.013 and 0.015), and the diffraction peaks are in good agreement with the standard XRD pattern of $Lu_3Al_5O_{12}$ (NO. 73-1368); the space group is Ia–3d (NO. 230), and no obvious impurity peaks are observed. This indicates that Ce^{3+} is doped into the $Lu_3Al_5O_{12}$ lattice, forming a solid solution. Different concentrations of Ce^{3+} doping have little effect on the lattice structure of LuAG, and the peak positions are basically consistent [29,34]. The calculated lattice parameters are a = b = c =11.8903 Å, α = β = γ = 90°. The XRD Rietveld refinement of the $Lu_{2.987}Al_5O_{12}$:$0.013Ce^{3+}$ is presented in Figure 1b; the reliability parameters Rwp, Rp, and χ^2 were calculated at 10.73%, 8.10%, and 10.04. Almost all the peaks are well-indexed with the $Lu_3Al_5O_{12}$ material, which indicates the results are reliable. Figure 1c presents the PLE (λ_{em} = 512 nm) and PL (λ_{ex} = 437 nm) of the LuAG:xCe^{3+} material. The excitation spectrums have two peaks located at 350 nm and 450 nm, which matches the 4f→5d transition of Ce^{3+}. The PL peak of LuAG:Ce^{3+} is located at 512 nm, which is equivalent to the transitions of $5d_1 \rightarrow {}^2F_{5/2}$ and $5d_1 \rightarrow {}^2F_{7/2}$. With the increase in the Ce^{3+} concentration, the PL intensity decreases first and then increases in a small range, which may be related to the difference in the crystallinity of ceramics. The SEM micrograph of the surface region of LuAG:0.013Ce^{3+} ceramics are displayed in Figure 1d. The samples have a dense structure, few micropores and defects, and the grain size was almost the same at 10–12 μm. From the EDS mapping in Figure S2, it can be seen that the elements of Lu, Al, O, and Ce are evenly distributed. This result confirmed that high-quality ceramics have been prepared and provides a good foundation for optical information storage.

Trap levels are closely related to the information storage properties of ETMs; they are generally analyzed by the thermoluminescence (TL) curve. TL involves the process in which storage optical energy by irradiation with light is re-emitted by heating. As an effective tool to study the trap levels, the TL curve provide lots of information about the traps. The peak position is related to the trap depth, and the peak intensity is related to the number of traps. Therefore, the TL curves of the LuAG:xCe^{3+} materials were measured at 1 K s^{-1} in Figure 2a. There are two different depths of trap (T_1 and T_2). Compared with the pure LuAG, the TL intensity of LuAG:0.013Ce^{3+} is significantly improved. The calculation of the depth is as shown in Figure 2b. The specific calculation method is shown in the following formula [35,36]:

$$\beta E / k_B T_m^2 = s\exp(-E / k_B T_m) \quad (1)$$

where E (eV), β (K/s), and k_B are the trap depth, heating rate, and Boltzmann constant; T_m (K) is the temperature corresponding to the maximum vertical axis; and s (s^{-1}) is the frequency factor. In Figure 2b, By plotting $\ln(T_m^2/\beta)$ against $1/(k_B T_m)$, the trap depth of the TL curve can be determined from the slope of the straight line. When x = 0.013, the trap depths are 1.36 eV and 0.89 eV, respectively, which are larger than those of most typical ETMs [37]. Next, the information retention ability of the sample was tested, which was expressed as the integral area under the TL curves. As shown in Figure 2c, the TL curve of the LuAG:0.013Ce^{3+} materials were tested for different decay times after being irradiated by 254 nm light. With the decay time increasing, the signal intensity attenuates and gradually flattens out. After a delay for 168 h, the sample still has about 40% integrated intensity, and the stability of information storage is slightly lower than that of the YAG: Ce crystal [38]. This indicates that the sample has a relatively stable information storage capacity.

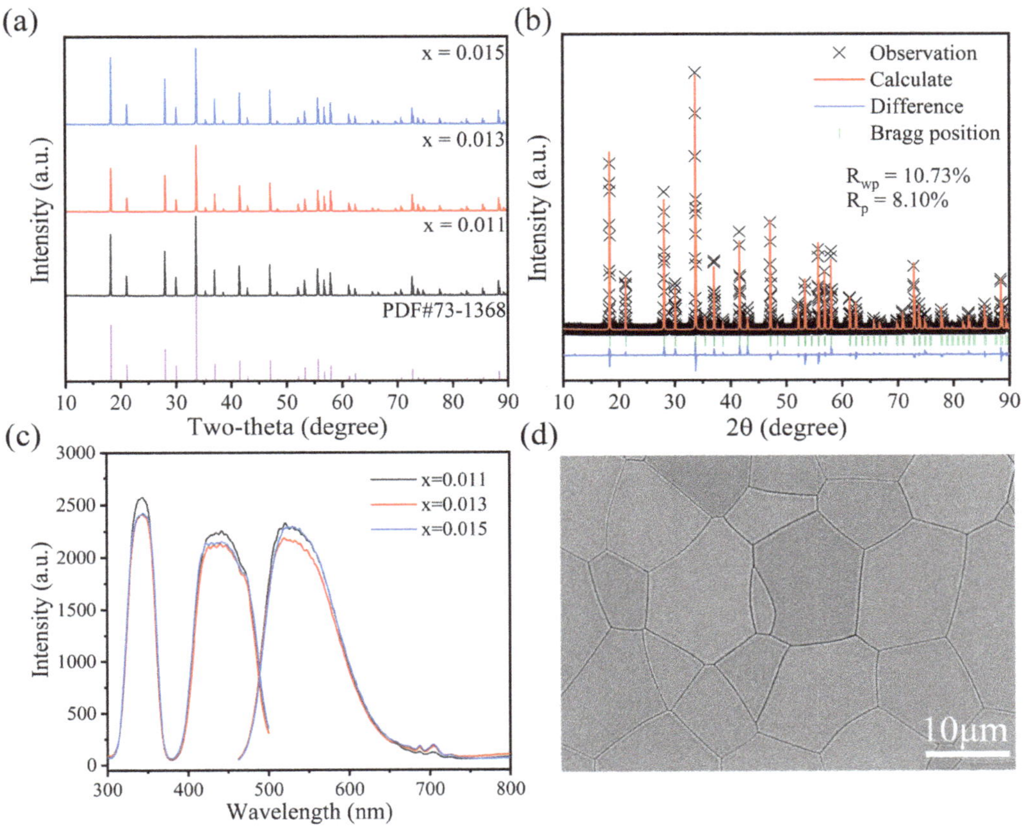

Figure 1. (**a**) XRD patterns of LuAG:xCe (x = 0.011, 0.013 and 0.015). (**b**) Rietveld refinement of LuAG:0.013Ce. (**c**) PLE and PL spectra of LuAG:xCe (λ_{em} = 512 nm, λ_{ex} = 437 nm). (**d**) SEM image of LuAG:0.013Ce^{3+}.

Rewritability is one of the most important properties for electron-trapping materials. Therefore, the rewritability of the LuAG:Ce material was verified. This cycle includes write-in, readout, and erasing. Firstly, UV light is used to excite the ceramic for 1 min; then, the ceramic is heated to 673 K after 30 s, and the data were recorded during heating. Finally, the sample is cooled down to room temperature. The area integral of the TL curve in each cycle is taken as the optical signal intensity, as shown in Figure 2d. Obviously, after 25 cycles, the signal intensity remains basically unchanged, and the size fluctuated around 7%, which was basically the same as the literature results [21]. This is due to the electron capture and de-trapping mechanism. In addition, it is proving that the sample has a higher physical and chemical stability than sulfide and glass phosphors [17,18,24].

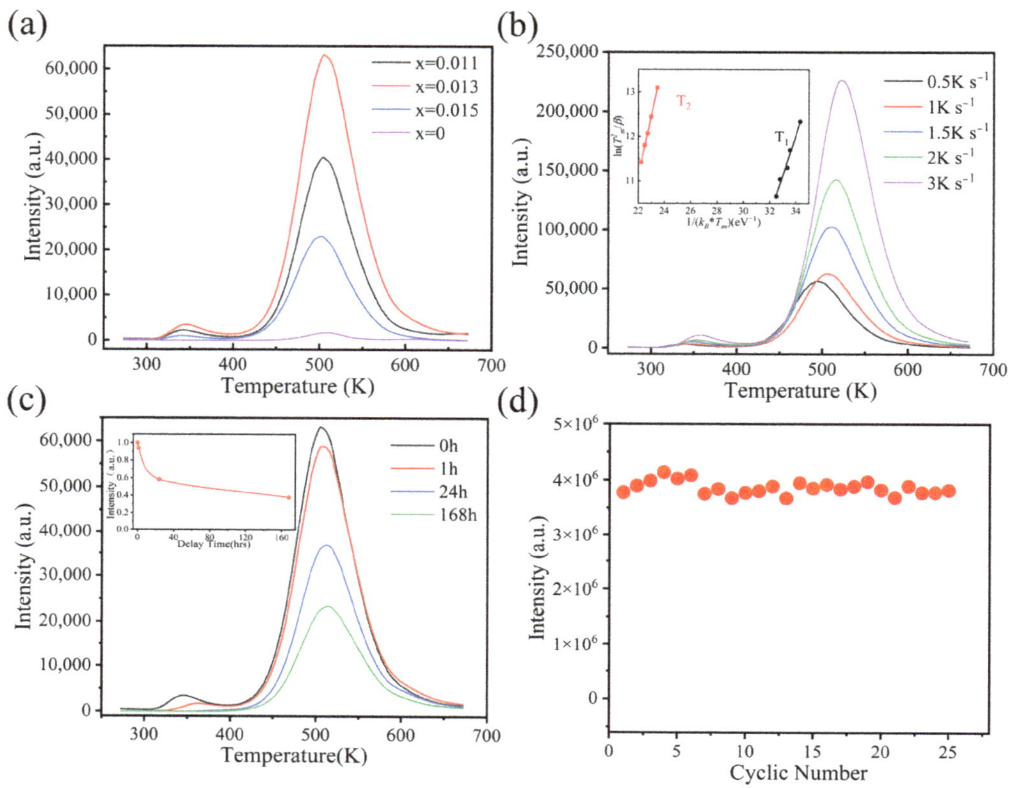

Figure 2. (**a**) TL curves of LuAG:xCe^{3+}. (**b**) TL curves with different heating rate of LuAG:0.013Ce^{3+} (Inset: Heating rate plots of LuAG:0.013Ce^{3+}). (**c**) TL curves of LuAG:0.013Ce^{3+} for different delay time (Inset: Integral area of TL curves as function of delay time). (**d**) The rewritability test of LuAG:0.013Ce^{3+}.

3.2. Analysis of Trap and Optical Storage Principle

From the analysis of the TL glow curve, we know that there are two traps of different depths in LuAG:Ce. This may be due to the difference in the attraction of the traps to the carriers, which makes the carrier trapped in different positions around the traps. The principle of information storage is based on the optical quantum effect of the trap level on electron capture and release. Therefore, it is necessary to study the transfer process of electrons clearly. In LuAG:Ce, Ce^{3+} is not only the luminescence center [39], but also provides electrons; this is because Ce^{3+} is inclined to lose electrons and become Ce^{4+}, due to its special external electron configuration [40,41]. These electrons are captured by traps and stored stably. This explains the significant enhancement of the TL intensity of LuAG with Ce^{3+} doping.

For the nature of the trap level, it is generally considered to have oxygen vacancies [42,43], and few $Lu_{Al,16a}$ anti-site defects (ADs) [44]. Herein, the samples were prepared in a high-temperature vacuum environment, which is more likely to generate defects. To study the origin of defects, the TL intensity of the sample was tested after high-temperature annealing in air, and the TL curves are displayed in Figure 3a. Compared with the unannealed sample, the TL intensity of the sample after air annealing almost disappeared; this is because oxygen vacancies are filled by oxygen under the high-temperature and oxygen-rich atmosphere [45,46]. The above results indicate that the trap levels in the material are related to oxygen vacancies.

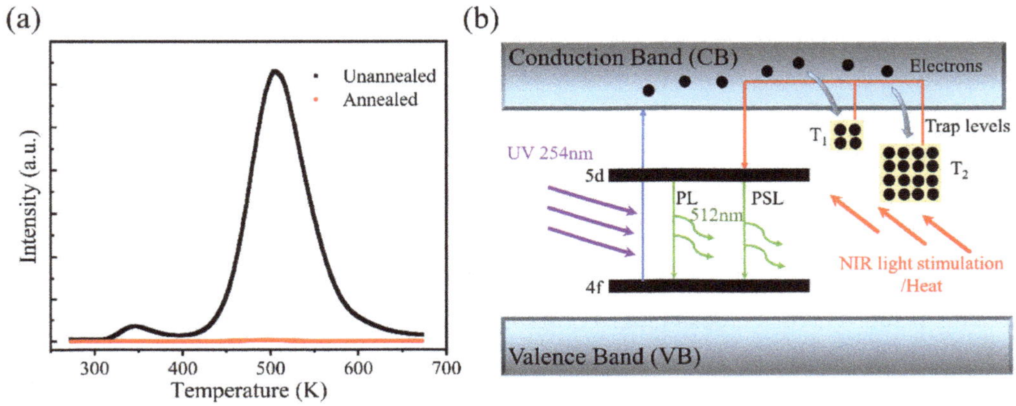

Figure 3. (**a**) TL curves of LuAG:Ce before and after annealing at 1000 °C. (**b**) A diagram of the information storage principle.

On the basis of the above experiments, an information storage principle of the LuAG:Ce material was proposed, as shown in Figure 3b. After the ceramic is irradiated, the electrons on the Ce^{3+} ground state 4f are excited to the 5d energy level and continue to the conduction band. Some activated electrons in the conduction band return to the 5d energy level and transition to the 4f ground state with 512 nm. The other electrons can transfer to the conduction band (CB) and are captured by trap levels. Electrons are released from the trap when stimulated by heating or 980 nm and transfer to the Ce^{3+} excited state energy level and ground state, resulting in the green PSL (readout).

3.3. *Performance of ODS*

An experimental scheme of the optical storage and readout is shown in Figure 4a. The 254 nm UV light is irradiated to the sample through the mask. The electrons in the material are excited and transitioned to the CB, and then captured by the deep trap level and stabilized for a period of time, that is, the letters "USST" are recorded into the storage medium, and the data are read by heating the sample. Figure 4b was photographed in sunlight (i) and 254 nm light (ii). As shown in Figure 4b(iii), LuAG:Ce was covered with a mask, then irradiated with a 254 nm UV lamp for 5 min, and heated to 250 °C; the letters "U, S, S, and T" were read out, and the unirradiated area was black.

Furthermore, the reading of information by NIR photostimulation (980 nm laser) was demonstrated, as shown in Figure 5a. The luminescence decay curve was recorded at room temperature. The sample produced persistent luminescent (PersL) in the dark due to the shallow traps. After 300 s, we used the 980 nm laser (0.7 W) in a periodically on/off mode to excite the sample and recorded the intensity change. A bright emission from the sample was observed after turning on the 980 nm laser, and the intensity was much higher than the PersL. With the increase in laser irradiation time, the intensity gradually decreases, since the electrons in the trap are gradually released by the 980 nm laser. The PSL emission spectra by using the 980 nm laser in different power modes were tested, and as shown in Figure 5b, the emission peak position is basically consistent with Figure 1c. In addition, with the increase in laser power, the emission intensity first increases and then decreases, which corresponds with the "readout" and "erasing" of the 980 nm laser. The above results show that it is possible to readout information at a high speed by the 980 nm laser, which is of great significance to information storage.

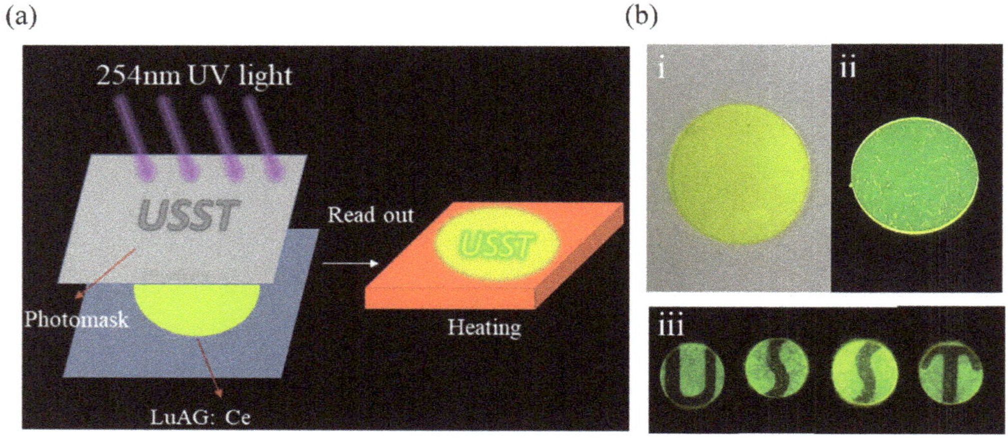

Figure 4. (**a**) Diagram of the optical storage application. (**b**) Photos of LuAG:Ce under (**i**) sunlight; (**ii**) under 254 nm light; (**iii**) the sample was covered with photomask "U", "S", "S", and "T" for 5 min and heated to 250 °C.

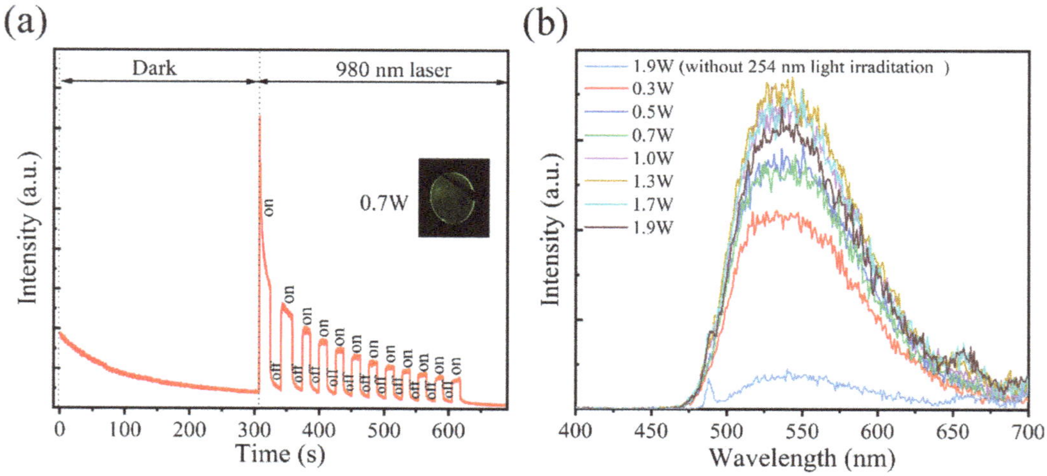

Figure 5. Sample was excited by 254 nm light for 15 min. (**a**) Luminescence decay curve of the sample. (**b**) Photostimulated luminescence (PSL) curves of the ceramic by 980 nm laser in different power settings.

4. Conclusions

In conclusion, we prepared $Lu_{3-x}Al_5O_{12}$:xCe (x = 0.011, 0.013 and 0.015) fluorescent ceramics by the vacuum sintering technique, and its crystal structure, trap level, and optical storage characteristics were systematically studied. The TL intensity is significantly increased by the selective introduction of Ce^{3+}; it acts as a luminescence center. The lutetium aluminum garnet-structured ceramics ensure good cycle stability (25 times). The nature of the trap was discussed, and is it believed that the oxygen vacancies generated by the vacuum sintering environment capture electrons and lead to PSL and TL. Finally, the thermally stimulated readout and PSL properties of the sample were verified, demonstrating the potential of LuAG:Ce in field of optical storage. Currently, LuAG:Ce's short information storage time and heating-based readout method limit its ODS application. With these results, we hope to promote the development of new storage technology by optimizing material design and combining advanced optical technologies.

Supplementary Materials: The following supporting information can be downloaded at: https://www.mdpi.com/article/10.3390/ma18010063/s1, Figure S1: Flowchart of the preparation of LuAG:Ce ceramics; Figure S2: Flowchart of the preparation of LuAG:Ce ceramics.

Author Contributions: Conceptualization, H.R.; methodology, M.Z. and J.Z.; software, Q.H.; validation, J.Z.; formal analysis, R.J.; investigation, M.Z. and J.Z.; resources, H.R. and H.L.; data curation, Q.H. and J.Z.; writing—original draft preparation, J.Z.; writing—review and editing, H.R. and H.L.; visualization, R.J. and Q.H.; supervision, H.R. and H.L.; project administration, H.R.; funding acquisition, H.R. and H.L. All authors have read and agreed to the published version of the manuscript.

Funding: This work was supported by the National Key Research and Development Program of China (2021YFB2802001).

Institutional Review Board Statement: Not applicable.

Informed Consent Statement: Not applicable.

Data Availability Statement: The original contributions presented in this study are included in the article/Supplementary Materials. Further inquiries can be directed to the corresponding author.

Conflicts of Interest: The authors declare no conflicts of interest.

References

1. Gu, M.; Li, X.; Cao, Y. Optical storage arrays: A perspective for future big data storage. *Light Sci. Appl.* **2014**, *3*, 11. [CrossRef]
2. Gu, M.; Zhang, Q.; Lamon, S. Nanomaterials for optical data storage. *Nat. Rev. Mater.* **2016**, *1*, 14. [CrossRef]
3. Zhao, M.; Wen, J.; Hu, Q.; Wei, X.; Zhong, Y.; Ruan, H.; Gu, M. A 3D nanoscale optical disk memory with petabit capacity. *Nature* **2024**, *626*, 22. [CrossRef] [PubMed]
4. Hell, S.W.; Wichmann, J. Breaking the diffraction resolution limit by stimulated emission: Stimulated-emission-depletion fluorescence microscopy. *Opt. Lett.* **1994**, *19*, 780–782. [CrossRef] [PubMed]
5. Bozinovic, N.; Yue, Y.; Ren, Y.; Tur, M.; Kristensen, P.; Huang, H.; Willner, A.E.; Ramachandran, S. Terabit-Scale Orbital Angular Momentum Mode Division Multiplexing in Fibers. *Science* **2013**, *340*, 1545–1548. [CrossRef] [PubMed]
6. Huang, L.; Chen, X.; Mühlenbernd, H.; Zhang, H.; Chen, S.; Bai, B.; Tan, Q.; Jin, G.; Cheah, K.W.; Qiu, C.; et al. Three-dimensional optical holography using a plasmonic metasurface. *Nat. Commun.* **2013**, *4*, 8. [CrossRef]
7. Cumpston, B.H.; Ananthavel, S.P.; Barlow, S.; Dyer, D.L.; Ehrlich, J.E.; Erskine, L.L.; Heikal, A.A.; Kuebler, S.M.; Lee, I.Y.S.; McCord-Maughon, D.; et al. Two-photon polymerization initiators for three-dimensional optical data storage and microfabrication. *Nature* **1999**, *398*, 51–54. [CrossRef]
8. Kawata, S.; Kawata, Y. Three-dimensional optical data storage using photochromic materials. *Chem. Rev.* **2000**, *100*, 1777–1788. [CrossRef] [PubMed]
9. Parthenopoulos, D.A.; Rentzepis, P.M. Three-dimensional optical storage memory. *Science* **1989**, *245*, 843–845. [CrossRef] [PubMed]
10. Lindmayer, J. A new erasable optical memory. *Solid State Technol.* **1988**, *31*, 135–138.
11. Zhuang, Y.; Chen, D.; Chen, W.; Zhang, W.; Su, X.; Deng, R.; An, Z.; Chen, H.; Xie, R. X-ray-charged bright persistent luminescence in $NaYF_4$:Ln^{3+}@$NaYF_4$ nanoparticles for multidimensional optical information storage. *Light Sci. Appl.* **2021**, *10*, 10. [CrossRef] [PubMed]

12. Li, P.; Hua, Y.; Ye, R.; Cai, M.; Xu, S.; Zhang, J. $SrAl_2O_4$ crystallite embedded inorganic medium with super-long persistent luminescence, thermoluminescence, and photostimulable luminescence for smart optical information storage. *Photonics Res.* **2022**, *10*, 381–388. [CrossRef]
13. Lin, S.; Lin, H.; Ma, C.; Cheng, Y.; Ye, S.; Lin, F.; Li, R.; Xu, J.; Wang, Y. High-security-level multi-dimensional optical storage medium: Nanostructured glass embedded with $LiGa_5O_8$: Mn^{2+} with photostimulated luminescence. *Light Sci. Appl.* **2020**, *9*, 10. [CrossRef] [PubMed]
14. Ueda, J.; Kuroishi, K.; Tanabe, S. Bright persistent ceramic phosphors of Ce^{3+}-Cr^{3+}-codoped garnet able to store by blue light. *Appl. Phys. Lett.* **2014**, *104*, 4. [CrossRef]
15. Bian, H.; Qin, X.; Wu, Y.; Yi, Z.; Liu, S.; Wang, Y.; Brites, C.D.S.; Carlos, L.D.; Liu, X. Multimodal Tuning of Synaptic Plasticity Using Persistent Luminescent Memitters. *Adv. Mater.* **2022**, *34*, 7. [CrossRef] [PubMed]
16. Lv, X.; Liang, Y.; Zhang, Y.; Chen, D.; Shan, X.; Wang, X.-J. Deep-trap ultraviolet persistent phosphor for advanced optical storage application in bright environments. *Light Sci. Appl.* **2024**, *13*, 253. [CrossRef] [PubMed]
17. Jutamulia, S.; Storti, G.M.; Lindmayer, J.; Seiderman, W. Use of electron trapping materials in optical signal processing. 1: Parallel Boolean logic. *Appl. Opt.* **1990**, *29*, 4806–4811. [CrossRef] [PubMed]
18. Wu, J.; Newman, D.; Viney, I.V.F. Study on relationship of luminescence in CaS:Eu,Sm and dopants concentration. *J. Lumin.* **2002**, *99*, 237–245. [CrossRef]
19. Pashaie, R.; Farhat, N.H. Optical realization of bioinspired spiking neurons in the electron trapping material thin film. *Appl. Opt.* **2007**, *46*, 8411–8418. [CrossRef]
20. Sun, J.; Liu, Z.; Du, H. Effect of trivalent rare earth ions doping on the fluorescence properties of electron trapping materials $SrS:Eu^{2+}$. *J. Rare Earths* **2011**, *29*, 101–104. [CrossRef]
21. Zhuang, Y.; Wang, L.; Lv, Y.; Zhou, T.; Xie, R. Optical Data Storage and Multicolor Emission Readout on Flexible Films Using Deep-Trap Persistent Luminescence Materials. *Adv. Funct. Mater.* **2018**, *28*, 9. [CrossRef]
22. Wang, Y.; Chen, D.; Zhuang, Y.; Chen, W.; Long, H.; Chen, H.; Xie, R. $NaMgF_3$:Tb^{3+}@$NaMgF_3$ Nanoparticles Containing Deep Traps for Optical Information Storage. *Adv. Opt. Mater.* **2021**, *9*, 8. [CrossRef]
23. Deng, M.; Liu, Q.; Zhang, Y.; Wang, C.; Guo, X.; Zhou, Z.; Xu, X. Novel Co-Doped Y_2GeO_5:Pr^{3+},Tb^{3+}: Deep Trap Level Formation and Analog Binary Optical Storage with Submicron Information Points. *Adv. Opt. Mater.* **2021**, *9*, 10. [CrossRef]
24. Lin, S.; Lin, H.; Huang, Q.; Cheng, Y.; Xu, J.; Wang, J.; Xiang, X.; Wang, C.; Zhang, L.; Wang, Y. A Photostimulated $BaSi_2O_5$:Eu^{2+},Nd^{3+} Phosphor-in-Glass for Erasable-Rewritable Optical Storage Medium. *Laser Photonics Rev.* **2019**, *13*, 11. [CrossRef]
25. Fu, J.; Feng, S.; Genevois, C.; Veron, E.; Yang, Y.; Wang, H.; Ma, Z.; Bai, L.; Xu, W.; Fan, R.; et al. Green-emissive Ce^{3+}:$Lu_3Al_5O_{12}$-Al_2O_3 nanoceramics elaborated via glass crystallization for high-power laser lighting applications. *J. Mater. Chem. C* **2024**, *12*, 7188–7196. [CrossRef]
26. Hao, W.; Wang, X.-J.; Guo, J.; Liu, J.; Li, S.; Xu, X. Fabrication of LuAG:Ce-Al_2O_3 eutectics via laser-heated pedestal growth technique for high-power laser-driven lighting. *J. Am. Ceram. Soc.* **2024**, *107*, 8291–8298. [CrossRef]
27. Ling, J.; Zhang, Y.; Zhang, H.; Xu, W.; Zhou, Y.; Hong, M. Ultrathin green-emitting LuAG:Ce-Al_2O_3 composite fine-grained ceramics for high-brightness chip-scale packaging LEDs. *J. Eur. Ceram. Soc.* **2025**, *45*, 116948. [CrossRef]
28. Zhu, D.; Cai, J.; Beitlerova, A.; Kucerkova, R.; Hu, C.; Zhou, Z.; Chewpraditkul, W.; Nikl, M.; Li, J. Stoichiometric modulation and scintillation property characterization of Ce,Ca:LuAG ceramics. *J. Am. Ceram. Soc.* **2024**, *107*, 6844–6852. [CrossRef]
29. Ding, H.; Liu, Z.; Hu, P.; Liu, Y.; Sun, P.; Luo, Z.; Chen, X.; Jiang, H.; Jiang, J. High Efficiency Green-Emitting LuAG:Ce Ceramic Phosphors for Laser Diode Lighting. *Adv. Opt. Mater.* **2021**, *9*, 2002141. [CrossRef]
30. Zhuang, Y.; Chen, D.; Xie, R. Persistent Luminescent Materials with Deep Traps for Optical Information Storage. *Laser Optoelectron. Prog.* **2021**, *58*, 1516001-1. [CrossRef]
31. Li, W.; Zhuang, Y.; Zheng, P.; Zhou, T.-L.; Xu, J.; Ueda, J.; Tanabe, S.; Wang, L.; Xie, R.-J. Tailoring Trap Depth and Emission Wavelength in $Y_3Al_{5-x}GaxO_{12}$:Ce^{3+},V^{3+} Phosphor-in-Glass Films for Optical Information Storage. *ACS Appl. Mater. Interfaces* **2018**, *10*, 27150–27159. [CrossRef] [PubMed]
32. Tang, Y.; Deng, M.; Wang, M.; Liu, X.; Zhou, Z.; Wang, J.; Liu, Q. Bismuth-Activated Persistent Phosphors. *Adv. Opt. Mater.* **2023**, *11*, 2201827. [CrossRef]
33. Tang, Y.; Deng, M.; Zhou, Z.; Xu, X.; Wang, J.; Liu, Q. Bi-activated $MgGa_2O_4$ phosphors with rich defect energy levels: Spectral property and optical storage applications. *Ceram. Int.* **2022**, *48*, 19141–19149. [CrossRef]
34. Wen, Q.; Wang, Y.; Zhao, C.; Xu, L.; Wang, X.; Xu, Y.; Lin, S.; Liang, X.; Liu, J.; Xiang, W. Ultrahigh Power Density LuAG:Ce Green Converters For High-Luminance Laser-Driven Solid State Lighting. *Laser Photonics Rev.* **2023**, *17*, 2200909. [CrossRef]
35. Bos, A.J.J.; Dorenbos, P.; Bessière, A.; Lecointre, A.; Bedu, M.; Bettinelli, M.; Piccinelli, F. Study of TL glow curves of YPO_4 double doped with lanthanide ions. *Radiat. Meas.* **2011**, *46*, 1410–1416. [CrossRef]

36. Bos, A.J.J. High sensitivity thermoluminescence dosimetry. *Nucl. Instrum. Meth. B* **2001**, *184*, 3–28. [CrossRef]
37. Yuan, L.; Jin, Y.; Su, Y.; Wu, H.; Hu, Y.; Yang, S. Optically Stimulated Luminescence Phosphors: Principles, Applications, and Prospects. *Laser Photonics Rev.* **2020**, *14*, 34. [CrossRef]
38. Hu, Z.; Guan, J.; Zheng, S.; Chai, Z.; Wu, S.; Liu, D.; Su, J.; Shi, F.; Duan, C.; Wang, Y.; et al. Burn After Read: A Rewritable Multiplexing Optical Information Storage and Encryption Method. *Laser Photonics Rev.* **2024**, *18*, 2301024. [CrossRef]
39. Lai, X.; Fang, Z.; Zhang, J.; Wang, B.; Zhu, W.; Zhang, R. Structure and luminescence properties of Ce3+-activated $BaLu_2Al_2Ga_2SiO_{12}$ persistent phosphors for optical information storage. *Opt. Mater.* **2021**, *120*, 7. [CrossRef]
40. Li, L.; Li, T.; Hu, Y.; Cai, C.; Li, Y.; Zhang, X.; Liang, B.; Yang, Y.; Qiu, J. Mechanism of the trivalent lanthanides' persistent luminescence in wide bandgap materials. *Light Sci. Appl.* **2022**, *11*, 8. [CrossRef]
41. Gallo, S.; Veronese, I.; Vedda, A.; Fasoli, M. Evidence of Optically Stimulated Luminescence in $Lu_3Al_5O_{12}$:Ce. *Phys. Status Solidi A* **2019**, *216*, 1900103. [CrossRef]
42. Trojan-Piegza, J.; Niittykoski, J.; Holsa, J.; Zych, E. Thermoluminescence and kinetics of persistent luminescence of vacuum-sintered Tb^{3+}-doped and Tb^{3+}, Ca^{2+}-codoped LU_2O_3 materials. *Chem. Mater.* **2008**, *20*, 2252–2261. [CrossRef]
43. Wang, C.; Deng, M.; Zhang, Y.; Liu, Q.; Zhou, Z.; Xu, X.; He, H. Multicomponent garnet phosphor $(LuYGd)(Al_4Ga)O_{12}$:Ce^{3+},V^{3+}: Trap modulation via reductive gas annealing and an optical information storage property. *Opt. Mater. Express* **2021**, *11*, 2256–2267. [CrossRef]
44. Hu, C.; Liu, S.; Shi, Y.; Kou, H.; Li, J.; Pan, Y.; Feng, X.; Liu, Q. Antisite defects in nonstoichiometric $Lu_3Al_5O_{12}$:Ce ceramic scintillators. *Phys. Status Solidi B* **2015**, *252*, 1993–1999. [CrossRef]
45. Zhou, T.; Zhang, L.; Selim, F.A.; Sun, R.; Wong, C.; Chen, H.; Zhang, Q. Annealing induced discoloration of transparent YAG ceramics using divalent additives in solid-state reaction sintering. *J. Eur. Ceram. Soc.* **2017**, *37*, 4123–4128. [CrossRef]
46. Chen, X.; Hu, Z.; Dai, J.; Chen, H.; Shi, Y.; Kou, H.; Wang, T.; Vedda, A.; Beitlerova, A.; Kucerkova, R.; et al. The influence of air annealing on the microstructure and scintillation properties of Ce,Mg:LuAG ceramics. *J. Am. Ceram. Soc.* **2019**, *102*, 1805–1813. [CrossRef]

materials

MDPI

Article

Thermal Optimization of Edge-Emitting Lasers Arrays

Robert P. Sarzała *, Dominika Dąbrówka and Maciej Dems *

Institute of Physics, Lodz University of Technology, ul. Wólczańska 217/221, 90-003 Łódź, Poland; dominika.dabrowka@dokt.p.lodz.pl

* Correspondence: robert.sarzala@p.lodz.pl (R.P.S.); maciej.dems@p.lodz.pl (M.D.)

Abstract: This paper presents a novel approach to address the issue of uneven temperature distribution in one-dimensional laser arrays, specifically in gallium nitride edge-emitting lasers emitting green light of 540 nm. The results were obtained using heat flow numerical analysis, which included an optimization method specifically developed for this type of array. It was demonstrated that thermal optimization of a one-dimensional edge-emitting laser array can be achieved by adjusting the placement of the emitters within the array and the size of the top gold contact, without changing the overall dimensions of the device. The proposed design alterations ensure an even temperature distribution across the array without the need for a complex and expensive cooling systems. The proposed optimization method can be applied to arrays made from various material systems, including nitrides, arsenides, and phosphides.

Keywords: nitride edge-emitting laser arrays; thermal optimization; computer simulation; GaN

1. Introduction

In recent years, nitride-based semiconductor lasers emitting green light have garnered significant interest due to their potential applications in various fields. These lasers are utilized in laser displays, lighting systems, projection systems, and fiber-optic telecommunications. There is also a demand for compact green lasers suitable for small (micro- or pico-scale) portable projectors, semiconductor lighting, medical equipment, and military applications, as the green light beam exhibits weak dependence on external conditions [1–6]. The development of arrays based on these lasers would enable achieving higher emitted powers than those provided by single emitters, significantly expanding the range of applications for these devices. However, currently, there are no commercially available green light sources (>500 nm), let alone their arrays [1–3]. For laser emitters based on nitride materials, commercial production has so far only mastered edge-emitting lasers (EELs) and their one-dimensional arrays emitting wavelengths from violet to blue–green light [7–12].

Laser arrays, or semiconductor laser systems, are key components of many modern technologies. They consist of multiple individual emitters placed in close proximity in an orderly configuration. Despite their widespread use across various fields, several challenges can limit their performance and application. One of the main challenges with laser arrays is effective thermal management [7–9,13]. The simultaneous operation of multiple emitters generates significant heat, which can lead to excessive heating of the device. This, in turn, may alter the operating parameters of the array, reduce its efficiency, and shorten its lifespan. In particular, an increase in temperature degrades material properties, such as reducing thermal and electrical conductivity. Moreover, temperature rise alters the band gap of the materials forming the active region of the laser, leading to a shift in the gain spectrum of the active region and a reduction in its value, negatively affecting the emitted power. The significant amounts of heat generated also contribute to uneven temperature distribution across the individual emitters in the array, particularly within their active regions. This leads to uneven operation of the emitters, which not only switch on and off asynchronously but also emit light at different wavelengths, thereby degrading the output beam quality of the entire array [14]. These issues are closely related to the effect of emitter mutual heating (thermal crosstalk), which occurs primarily within the array itself—rather than at the solder/heatsink interface or within the housing—critically impacting the performance of laser bars [15]. Therefore, precise management of heat flow between individual emitters in laser arrays is crucial for further improving their efficiency.

The thermal crosstalk effect in emitter arrays has been analyzed since the 1970s [16–18]. Scientists have analyzed it in one-dimensional Distributed Feedback Lasers arrays [19], arrays made of arsenide materials [15,20–22], and in arrays made of nitride materials [23–25]. They mainly focused on investigating the impact of the emitter spacing [19,26], the substrate dimensions [22,25,26], the material of the heat sink [19], and the dimensions of selected layers forming a single emitter or the entire array [20,22,25]. Other analyzed factors include the impact of additional heat spreaders, such as diamond-based ones [24], on temperature rise within the array.

Temperature equalization in the active regions of individual emitters of the array and maintaining it at the lowest possible level are crucial for improving its efficiency and longevity. Therefore, scientists are intensively working on developing appropriate methods and solutions, as well as optimizing the array design. One of the simplest solutions is to significantly increase the distance between individual emitters. Unfortunately, this results in a loss of the array's compactness. Specialized systems are also used to individually control the power supply of each emitter, but this solution complicates the device's construction and increases its cost. Considerable attention has been devoted to improving commonly used solutions for temperature equalization and reduction, such as stackable micro-channel coolers and impingement-type heat exchangers. Over the years, a number of innovations have been developed that have improved cooling efficiency [27–34]. Studies have also been conducted on the impact of solder materials on the thermal performance and reliability of arrays [35–38]. Another approach to addressing the thermal issues of the array was presented in a 2018 paper, which analyzed the thermal properties of mini-laser arrays based on GaN, consisting of five laser bars in a TO-9 package [26]. It was observed that the temperature difference between evenly spaced bars exceeded 5 K, but by modifying the width of the bars and the spacing between them, this difference was reduced to less than 0.4 K. The outer bars were widened, while the remaining three were narrowed, while maintaining the same total width of the bars. Increasing the volume of the AlN substrate and changing its shape to a trapezoidal form further optimized the temperature distribution. In the present work, we propose a much simpler solution to this problem. The control of self-heating in the emitters can be achieved by properly selecting the arrangement of the individual emitters in combination with the appropriate width of the gold contact on the top of the device, as will be demonstrated in this paper. This approach eliminates the need for changes in bar widths, which could potentially affect optical modes.

The proposed solution is demonstrated through a case study of a multi-element laser bar made from edge-emitting lasers that emit green light at a wavelength of 540 nm. The main goal of the work was to search for design solutions that could equalize and reduce the temperature in the aforementioned array while maintaining its width. The solutions were sought by utilizing the self-heating effect of the emitters in the array. It was noticed that this generally unfavorable phenomenon could, under certain conditions, contribute to the equalization of temperatures in all emitters in the array. By appropriately adjusting the distance between selected emitters, it is possible to raise the temperature in some emitters while lowering it in others, effectively reducing the temperature differences. However, it turned out that simply adjusting the emitter positions is insufficient for full temperature equalization. Therefore, an additional solution was sought by modifying the width of the upper gold contact, which allowed us to adjust the temperature of the outermost emitters. The proposed approach ensures that all emitters in the array, especially their active regions, maintain the same temperature.

Although this work is purely theoretical and numerical, it focuses on the practical implementation of efficient high-power laser arrays. All known technological limitations of gallium nitride light emitters have been carefully considered. Therefore, the results obtained in this study can serve as a foundation for the design of new devices. The proposed solutions can be extended to arrays made from various material systems, including arsenides, phosphides, and antimonides.

2. Materials and Methods

2.1. Structure

The thermal optimization was conducted for edge-emitting lasers with a ridge waveguide. Each emitter was designed to emit green light at 540 nm [39]. The structure is based on a gallium nitride (GaN) substrate, 2500 μm in width and 50 μm in thickness. Directly above it is an n-type layer of $Al_{0.83}In_{0.17}N$, composition-matched to the underlying GaN. These two layers function as the n-type cladding layers. Next is an n-type waveguide layer of $In_{0.08}Ga_{0.92}N$. The active region consists of three 2.7 nm $In_{0.29}Ga_{0.71}N$ quantum wells, surrounded by four 10 nm GaN barriers. Directly above the active region is a p-$Al_{0.20}Ga_{0.80}N$ electron blocking layer (EBL) and a p-type waveguide made of $In_{0.08}Ga_{0.92}N$. The p-type cladding layer, consisting of a p-GaN layer and a ZnO layer, is located above the waveguide. The ZnO layer has already been used in lasers as an optical confinement layer [40–42]. A ridge waveguide with a width of 2 μm was fabricated in these layers. A gold contact is situated on top of the emitter, insulated from the structure by a silicon dioxide layer except at a small area atop the ridge waveguide. A schematic and a three-dimensional visualization of the edge-emitting laser array are shown in Figure 1, and detailed design data of the analyzed structure is provided in Table 1. In Figure 1a, the individual layers of the array and the most important dimensions are labeled. The width of the top gold contact is denoted as W_{Au} and its thickness as d_{Au}.

The emitters forming the optimized array are primarily composed of materials from the III-N group, including binary GaN and ternary compounds such as AlGaN, InGaN, and AlInN. For these materials, a temperature-dependent thermal conductivity relationship has been applied, expressed using the temperature coefficient δ as follows:

$$k(T) = k_{300\mathrm{K}} \left(\frac{T}{300\,\mathrm{K}} \right)^{\delta}, \tag{1}$$

where $k_{300\mathrm{K}}$ is the thermal conductivity at room temperature [W/(m·K)], and T is the temperature [K]. The relationship given in Equation (1) is commonly found in the literature and is applied to nitride materials. For the mentioned layers, the temperature coefficient δ is -1.4 [23,43].

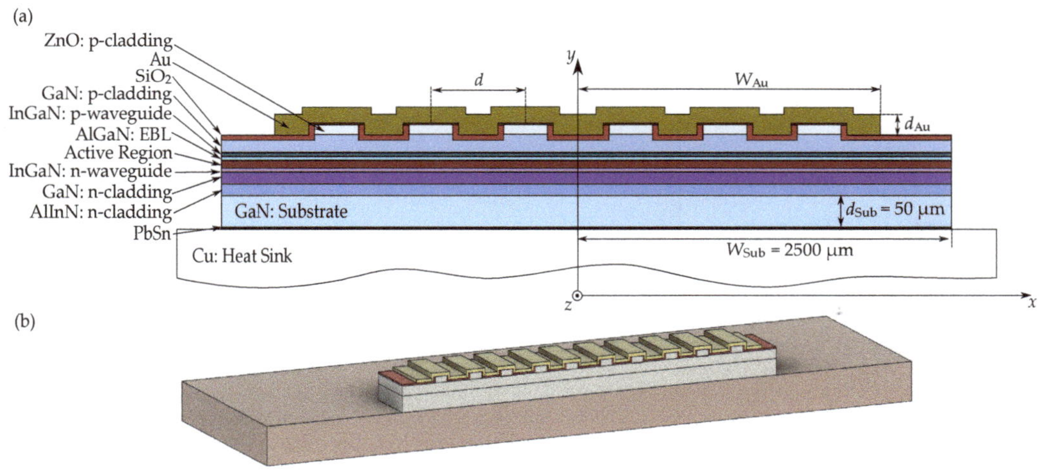

Figure 1. Schematic (**a**) of the laser bar based on edge-emitting nitride lasers, with marked layers that make up the structure, as well as key parameters: d_{Au}—thickness of the top gold contact, W_{Au}—width of the top gold contact, d_{Sub}—thickness of the substrate, W_{Sub}—width of the substrate. Three-dimensional visualization (**b**) of a 10-element laser bar. The drawings are not to scale.

Table 1. Design data of the modeled edge-emitting lasers and thermal parameters of materials used in simulations: $k_{300\,K}$, $k_{400\,K}$, and $k_{500\,K}$ are thermal conductivities at 300 K, 400 K, and 500 K, respectively. QW stands for quantum well and EBL for electron blocking layer.

Layer		Material	Thickness (μm)	Doping (cm^{-3})	k_{300K} (W/(m·K))		k_{400K} (W/(m·K))		k_{500K} (W/(m·K))	
p-side contact		Au	1.00	-	317.10		310.70		304.30	
Insulator		SiO_2	0.25	-	1.29		1.45		1.56	
p-cladding		ZnO p-GaN	0.28 0.28	- Mg: 2×10^{19}	50.00 92.00		50.00 61.50		50.00 45.00	
p-waveguide		p-$In_{0.08}Ga_{0.92}N$	0.045	Mg: 1×10^{19}	23.00		15.37		11.25	
EBL		p-$Al_{0.20}Ga_{0.80}N$	0.01	Mg: 5×10^{19}	13.00		8.69		6.36	
Active region	3 × QW 4 × barrier	$In_{0.29}Ga_{0.71}N$ GaN	0.0027 0.01	Si-doped -	k_x 48.46	k_y 8.39	k_x 32.39	k_y 5.61	k_x 23.70	k_y 4.10
n-waveguide		n-$In_{0.08}Ga_{0.92}N$	0.01	Si: 5×10^{18}	28.00		18.72		13.70	
n-cladding		n-GaN n-$Al_{0.83}In_{0.17}N$	0.35 0.50	Si: 2×10^{18} Si: 5×10^{18}	50.00 94.87		50.00 3.26		50.00 2.38	
Substrate		n-GaN	50.00	Si: 2×10^{18}	166.00		110.97		81.19	
Solder		PbSn	1.00	-	50.00		50.00		50.00	
Heat sink		Cu	5000	-	400.80		392.47		386.13	

The thermal conductivities at room temperature for the binary material GaN were taken from [43,44]. For the ternary materials, i.e., $Al_xGa_{1-x}N$ and $In_yGa_{1-y}N$, the thermal conductivities were calculated based on the known values of thermal conductivity at room temperature for binary materials using linear interpolation with a curvature coefficient c, which indicates the deviation from linearity [23]. For calculating the thermal conductivity of $Al_xGa_{1-x}N$, AlN and GaN were used [23,45], and for $In_yGa_{1-y}N$, InN and GaN were used [23,46]. For the AlInN material, the data from [47] were used as a reference.

Since the active region consists of seven layers with small thicknesses, an equivalent thermal conductivity was calculated for it. This structural simplification does not affect the accuracy of the results but significantly reduces memory requirements and computation time, which is particularly important for calculations involving laser arrays. A standard approach was used, in which the equivalent thermal conductivities in the parallel direction k_x (along the x-axis) and in the perpendicular direction k_y (along the y-axis) to the layer were calculated for the layered materials. These conductivities were calculated using the following formulas [48]:

$$k_x = \frac{\sum_i^N d_i k_i}{d_{\text{tot}}}, \tag{2}$$

$$k_y = \frac{d_{\text{tot}}}{\sum_i^N \frac{d_i}{k_i}}, \tag{3}$$

where N is the number of combined layers, d_i is the thickness of the i-th combined layer, k_i is the thermal conductivity of the i-th combined layer, and d_{tot} is the total thickness of the combined layers. For the combined layers, the base thermal conductivities are 2.00 W/(m·K) for the quantum wells made of n-$In_{0.29}Ga_{0.71}N$ and 61.00 W/(m·K) for the GaN barriers. A temperature-dependent thermal conductivity relationship, as described by Equation (1), was also applied for the active region.

The simulated device comprises not only semiconductor materials of the III–N group but also other materials such as gold (Au), silicon dioxide (SiO_2), zinc oxide (ZnO), copper (Cu), and a tin–lead alloy (PbSn), which serves as a thermal contact layer between the copper heatsink and the GaN substrate. In the calculations, temperature-dependent effects were not considered for some materials, as the temperature variations within the analyzed range were small and did not significantly affect the final results. The thermal conductivity data were taken from [49,50] for Au, SiO_2, PbSn, and Cu. For ZnO material, significant discrepancies exist in the literature regarding the reported thermal conductivity values [51]. Furthermore, its heat management can be significantly impacted by the morphology of the ZnO layer. The value used in the calculations was selected based on studies describing the application of a ZnO layer as a cladding layer [40–42].

Table 1 presents the thermal conductivities of the materials forming the analyzed structure, determined for three temperatures: 300 K, 400 K, and 500 K.

It must be noted that the design of the nitride laser diode emitting green light that we used is just one example. It shows very interesting properties, which is why we chose it. Nevertheless, the proposed optimization method can be applied to other types of laser arrays, including those emitting light of different wavelengths, made from various materials, such as arsenides, phosphides, or antimonides.

2.2. Numerical Model

The heat flow and temperature distribution in the analyzed structures were modeled using the Finite Element Method (FEM) to solve the heat flow equation:

$$\nabla \cdot (\lambda_T \nabla T) = -g, \tag{4}$$

where T is the temperature, λ_T is the thermal conductivity, and g is the heat source density. It is assumed that all the heat is generated in the active region. While this is not entirely accurate, carrier recombination in the active region contributes to the vast majority of heating in correctly manufactured devices. Additionally, other heat sources, such as contacts and high-resistive layers, are located in close proximity to the active region (relative to the overall array dimensions), and their impact can be considered within the active region heat source density. The device temperature at the bottom of the 5 mm thick heatsink is assumed to be maintained at 300 K.

Calculations were performed using the multi-physics modeling software PLaSK, developed by the authors [52] and available as open source [53]. It is a comprehensive tool for the numerical analysis of a broad range of physical phenomena in photonic devices. Although it is primarily designed for simulating semiconductor lasers, its range of applications includes transistors, light-emitting diodes, photodetectors, and more. Due to its modular nature, it is possible to perform computations of virtually any physical phenomenon in micro-scale structures.

Since its advent, PLaSK has been used in numerous scientific publications. Investigated phenomena included thermal, electrical, and optical analysis in edge-emitting lasers [39,54], advanced vertical-cavity surface-emitting lasers [55,56], photonic-crystal lasers [57], photodetectors [58], and high-contrast grating lasers [59]. Simulation results obtained with it have shown very good agreement with experimental data [59–62], proving the reliability and accuracy of the software.

In this work, we utilized the PLaSK module for the analysis of heat dissipation using the Finite Element Method. It employed rectangular bilinear elements, with a total mesh size of approximately 100,000 elements per emitter. Due to the mirror symmetry of the entire array, only half of it was modeled, with adiabatic boundary conditions applied along the central plane. However, due to the temperature dependence of the thermal conductivity throughout the structure, an iterative approach was necessary to solve the heat flow equation. Consequently, the total computation time varied from seconds for small arrays (fewer than 10 emitters) to 20–30 min for larger arrays comprising around 20 emitters.

PLaSK by default implements such an iterative process and considers the temperature dependence of the material parameters within its default material database, which is compiled from multiple available sources. Additionally, it allows for refinement of these parameters for the particular structure under investigation. For the analysis presented in this article, we used the relations shown in Table 1 and Equation (1).

The optimization procedure was performed with the SciPy [63] package. The method used was Sequential Least Squares Programming (SLSQP) [64], with constraints ensuring that no emitters overlap. The figure of merit was the standard deviation of the temperature in the active regions of all emitters. To prevent undesired solutions where emitters combine in pairs (each pair comprising two emitters very close to each other and mutually heating), which significantly increased both the maximum and average temperature of the entire array, an additional constraint was introduced. We required that the distances between consecutive emitters increase progressively towards the center of the array. Thus, the emitters near the array edges were closest to each other, while those in the center were the most distant. This approach not only reduced the temperature difference between all the array emitters but also slightly decreased the maximum temperature of the entire device.

3. Results and Discussion

As mentioned earlier, in edge-emitting lasers, heat is primarily generated in the active region, where laser radiation is emitted. Consequently, this region experiences the highest temperature increase, with the generated heat spreading outward to external components, primarily towards the heat sink. In our study [25], we demonstrated that if the thickness of the top metallization layer (mainly composed of gold) exceeds 1 μm, this contact can function as a heat spreader, significantly reducing the maximum temperature increase in the active region. The efficiency of heat dissipation through the top contact can be enhanced by introducing layers with the highest possible thermal conductivity between the active region and the contact. For the laser structure analyzed in this study, heat dissipation efficiency can be improved by replacing the cladding layer material, made of ITO, with ZnO. The use of a ZnO layer facilitates effective heat transfer to the gold contact, spreading the heat over a wider area of the structure. This, in turn, promotes more efficient heat removal from the laser's interior. An example of heat distribution in an edge-emitting laser is shown in Figure 2.

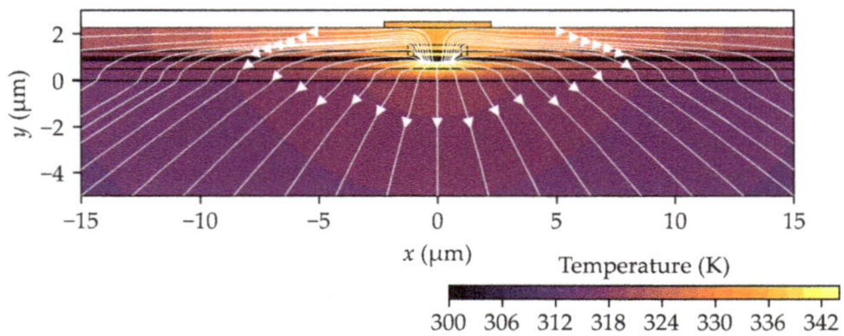

Figure 2. Example of the temperature distribution with directions of heat flow generated in the active region marked by white arrows.

Nitride materials, such as GaN, InGaN, InAlN, and AlN, exhibit significant shifts in emission wavelength as a function of temperature, which impact the stability and performance of optoelectronic devices like LEDs and edge-emitting lasers. The $d\lambda/dT$ values for these materials typically range from 0.1 nm/K to 0.4 nm/K, depending on the device structure and operating conditions [65–69]. For the structure under consideration, this is illustrated in Figure 3, which shows the calculated material gain in the laser's active region as a function of wavelength for various temperatures. As the temperature increases, a reduction in gain is observed, accompanied by a red shift in the gain peak relative to the target emission wavelength. Consequently, temperature affects other parameters, such as threshold current and efficiency, emphasizing the importance of effective thermal management in these devices, especially under extreme conditions, where wavelength shifts can reach up to 0.5 nm/K.

For a 20-element non-optimized array operating at 1 W per emitter, the spectral non-uniformity is 10.4 nm, assuming a wavelength shift of 0.4 nm/K resulting from a maximum temperature difference of 26 K between the emitters. However, with optimization, the spectral non-uniformity is significantly reduced to an extremely low value of 0.02 nm.

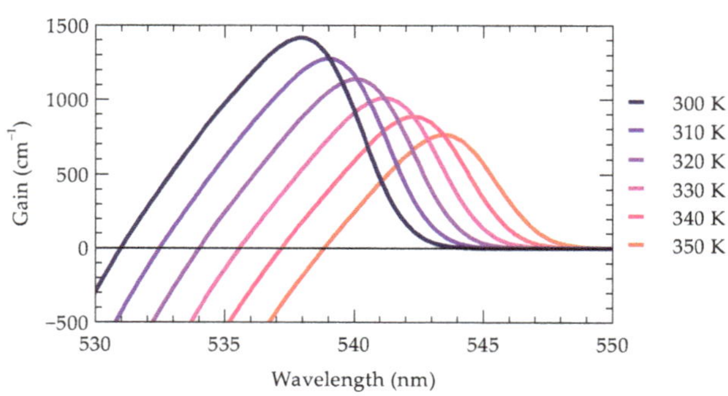

Figure 3. The effect of temperature on the gain spectrum of the active region in a nitride-based edge-emitting green laser.

In laser arrays, temperatures are even higher than in single emitters. This is primarily due to the previously mentioned thermal crosstalk effect, which involves mutual heating among the emitters within the array. The self-heating of individual emitters in the array intensifies as the number of emitters in the laser bar increases and/or the distance between them decreases. With an increasing number of emitters, not only does the maximum temperature in the array rise (see Figure 4a,b), but greater temperature unevenness is also observed among the individual emitters in the array (see Figure 4b,c). The differences in maximum temperature increases between emitters, particularly between the central and edge emitters in the array, can reach several tens of percent. For the analyzed 10-element laser bar, a temperature difference of 12.5 K (11%) is observed between the central and edge emitters; doubling the number of emitters in the array doubles this temperature difference. Such significant temperature variations between emitters lead to differences in material gain and emitted wavelengths among the lasers, resulting in uneven operation across the array.

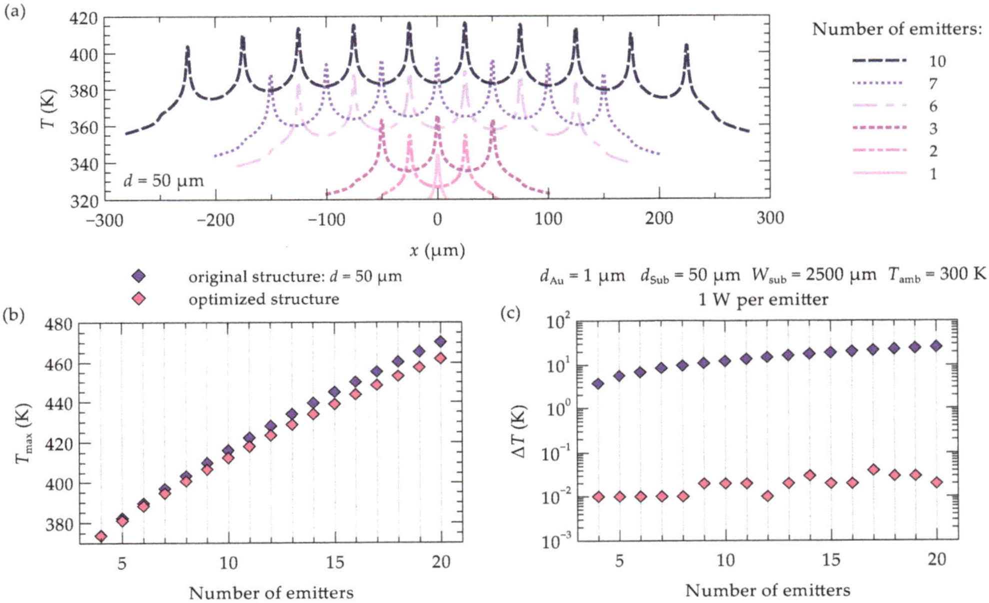

Figure 4. Temperature profiles (**a**) in the active region depending on the number of evenly spaced emitters in a one-dimensional array. Maximum temperature (**b**) in the laser bar and maximum temperature differences (**c**) between emitters in the array as a function of the number of emitters. In graphs (**b**,**c**), results for arrays with evenly spaced emitters are shown in purple, while results for laser bars with optimally adjusted distances between emitters and tailored gold width are shown in pink.

The thermal crosstalk phenomenon, although generally detrimental to heat dissipation, can be utilized to equalize the maximum temperatures across individual emitters in a laser array. The proposed solution involves unevenly spacing the emitters within the array. This approach imposes a constraint: the positions of the outermost emitters remain fixed to ensure that the overall device size is identical to that of an array with evenly spaced emitters. Furthermore, selecting an appropriate width (W_{Au}) for the gold layer on top of the array (see Figure 1a) is crucial. This layer serves as the p-side electrical contact and acts as a heat spreader, facilitating lateral heat distribution within the laser bar. The developed optimization method enables the determination of emitter spacing and gold contact width that result in nearly identical temperatures in the active regions of the array emitters (see Figures 4c and 5). In optimized multi-emitter arrays, these temperature differences do not exceed 0.02 K, and in larger arrays, they are no more than 0.05 K. The effect of these modifications on the temperature profiles in the emitters' active regions is shown in Figure 5, which compares the temperature profiles before (dashed line) and after (solid line) optimization for two arrays: (a) a 10-emitter array and (b) a 20-emitter array. Additionally, Figure 6 illustrates the temperature distribution in a 10-emitter laser bar before and after optimization.

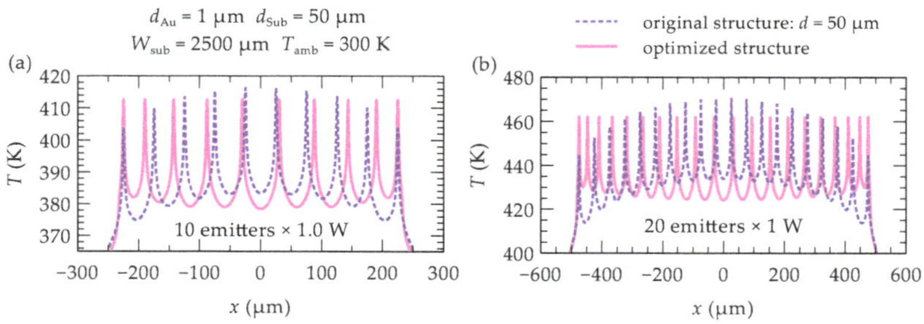

Figure 5. Temperature profile in the active region of the laser bar with (**a**) 10 emitters and (**b**) 20 emitters, each with a heat source power of 1 W, before (dashed line) and after (solid line) optimization.

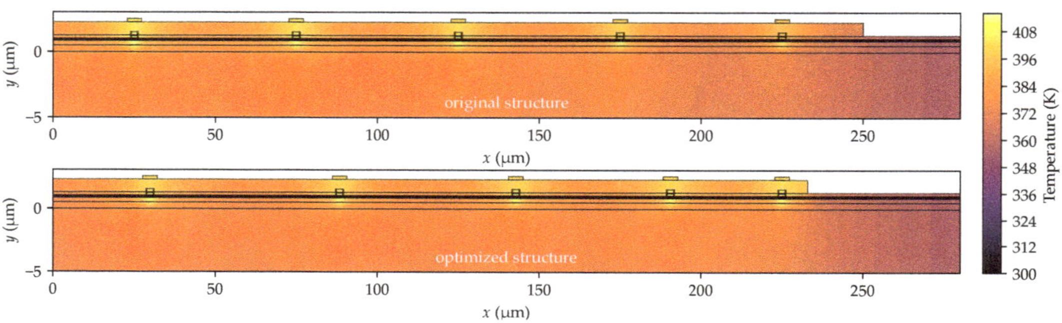

Figure 6. Temperature distribution in a 10-element one-dimensional laser bar before and after optimization. Diagram not to scale. Because of the mirror symmetry of the structure, only half of it is shown.

Temperature equalization across all emitters is possible because moving the outer emitters closer raises their temperatures, while increasing the spacing between central emitters improves heat dissipation from the center of the array. This approach lowers the temperature in the central part and raises it on the sides of the array, resulting in a more uniform temperature distribution. The temperature increase in the outer emitters is also influenced by a 17 μm reduction in the width of the gold contact layer (compared to the array with evenly spaced emitters—see Figure 6). This narrowing limits the surface area over which heat spreads and restricts its flow to the heat sink, thereby increasing the temperature of the outer emitters.

Proper selection of the gold contact layer width allows not only for temperature equalization across the outermost emitters but also for a reduction in the maximum temperature across the entire array (see Figures 4b and 5). This effect is particularly pronounced in arrays with a larger number of emitters. For instance, in a 20-emitter array, the maximum temperature is reduced by 2% compared to the unoptimized structure, while in a smaller array (e.g., 10 emitters), the reduction is approximately 1%.

The recommended dimensions of the gold layer, determined through simulations, are specified with a precision of up to thousandths of a micrometer. However, considering technological capabilities, the impact of manufacturing precision for this layer on the effectiveness of the proposed optimization was investigated. A manufacturing error range of ±3 μm for the gold layer was considered (see Figure 7a). The results, shown as a graph of the maximum temperature difference (ΔT) between the emitters in the array as a function of the gold layer's manufacturing precision, are presented in Figure 7b. Making the gold contact wider than the dimensions obtained during optimization reduces the temperature of the outer emitters due to better heat dissipation from the structure. However, this also increases the temperature unevenness between emitters. In the case of a gold layer that is 3 µm wider in the analyzed 10-emitter array, the temperature difference between the central and outermost emitters is 1.1 K. On the other hand, if the gold layer is made 3 µm narrower, the temperature difference increases to 1.3 K. A smaller amount of gold limits and hinders the heat spread from the outer emitters, thereby raising their temperature. This effect is most noticeable in the two outer emitters, but the manufacturing precision of the gold layer does not significantly impact the maximum temperatures in the central emitters, as shown in Figure 7c. Compared to the temperature differences between emitters before optimization (12.5 K), the changes within the considered range are relatively small and are unlikely to have a significant impact.

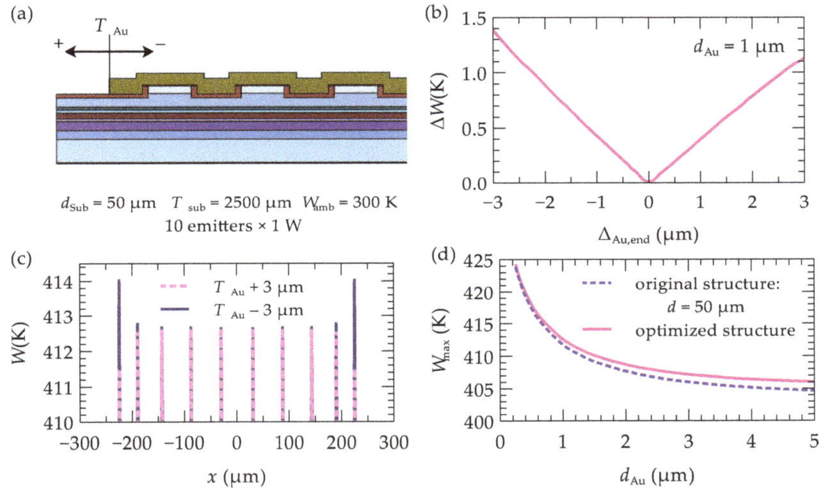

Figure 7. Impact of manufacturing precision of the gold width: (**a**) schematic of the examined precision, (**b**) maximum temperature difference between the emitters of the optimized 10-element array as a function of the dimensional deviation from the recommended gold width, (**c**) temperature profile section of the optimized array with gold layers 3 µm longer and shorter, and (**d**) impact of the gold contact thickness on the maximum temperature in the array.

As mentioned earlier, the thickness of the gold contact can significantly affect the temperature within the device. A thicker gold layer causes the heat to spread more widely, allowing it to flow more efficiently toward the heat sink at the bottom of the array. According to Fourier's law, a wider heat flux (i.e., lower flux density) at the same vertical distance to the heat sink results in a smaller temperature gradient within the structure. As a result, the temperature inside the array decreases, as shown in Figure 7d. However, beyond a certain thickness of the gold layer, no further significant temperature reductions are observed.

The presented optimization results were obtained under the assumption that each emitter dissipates 1 W of power. However, varying this value does not lead to significant changes. In other words, the optimized structure shows robustness for a large range of powering. The impact of the dissipated heat in individual emitters of the array on the maximum temperature differences (ΔT) between the emitters is shown in Figure 8. Reducing the dissipated power in the laser leads to slight temperature changes in the emitters, but the temperature unevenness does not exceed 0.4 K. On the other hand, at higher powers, a faster increase in temperature unevenness between the emitters is observed. Increasing the power from 1.0 W to 1.5 W results in a maximum temperature difference of 1.5 K (purple line in Figure 8). Therefore, it is most appropriate to perform optimization for the maximum assumed power at which the emitters will operate. An example of performing optimization for higher power and applying it to lower powers is shown in Figure 8 (pink line). With this approach, the maximum temperature difference between emitters increases, but this increase remains relatively small and does not exceed 1.6 K for all lower power values compared to the power for which the optimization was performed.

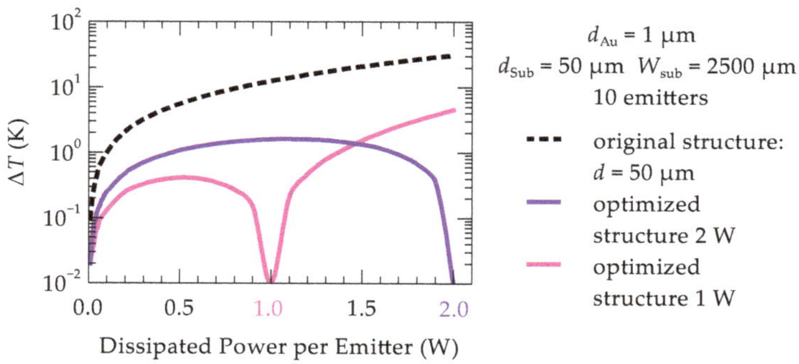

Figure 8. Temperature difference between the central and outermost emitter in the 10-element optimized array as a function of the dissipated power in each heat source: while maintaining the emitter arrangement obtained during the optimization of the 10-emitter array with 1 W power per emitter (pink line), for the optimal arrangements obtained for 2 W power (purple line), and the difference in the unoptimized structure (dashed line).

4. Conclusions

This paper presents an innovative method for equalizing the temperatures of individual emitters in an array, demonstrated using a one-dimensional array of edge-emitting lasers emitting green light. Temperature equalization was achieved by optimizing the arrangement of the inner emitters and adjusting the width of the array's upper gold contact, without altering the position of the outermost emitters. The proposed solution enables uniform temperature distribution in each emitter of the array without the need for specialized power supply or cooling systems. The introduced modifications not only equalize the temperature but also contribute to a slight reduction in the overall temperature. For a 10-emitter array, the maximum temperature increase is 116.30 K, with a temperature increase difference between emitters of 11% relative to the maximum increase. Using the proposed modifications, the overall temperature increase was reduced by 3%, and the difference in temperature increase between emitters was minimized to nearly 0% of the maximum increase before optimization (0.01 K). The proposed method is effective for both small arrays and larger arrays, such as those with 20 emitters. In the case of a 20-emitter array, the temperature increase is reduced by nearly 5%. Prior to optimization, the temperature increase difference between emitters reached 15% of the maximum increase, but after optimization, this difference was reduced to nearly 0% (0.02 K).

Determining the appropriate geometric parameters is made possible through a developed and customized computational optimization method tailored to the analyzed array. The phenomenon of thermal crosstalk, which typically causes unfavorable thermal effects, as demonstrated by the presented results, can also be harnessed to improve the operating conditions of the laser array, thereby enhancing its performance parameters. Furthermore, improving the thermal conditions of arrays solely through structural modifications can be crucial for achieving low-cost, high-power devices. It is worth noting that the proposed optimization method is applicable to dies made from any material system, such as nitrides, arsenides, or phosphides.

Author Contributions: Conceptualization, M.D. and R.P.S.; methodology, M.D.; software, M.D.; validation, R.P.S. and D.D.; formal analysis, M.D., R.P.S. and D.D.; investigation, D.D., R.P.S. and

M.D.; writing—original draft preparation, D.D.; writing—review and editing, M.D.; visualization, D.D. and M.D.; supervision, R.P.S. and M.D.; funding acquisition, D.D. and R.P.S. All authors have read and agreed to the published version of the manuscript.

Funding: This research was conducted as part of the research project no. 2024/53/N/ST7/02024 financed by the National Science Centre, Poland.

Institutional Review Board Statement: Not applicable.

Informed Consent Statement: Not applicable.

Data Availability Statement: The original contributions presented in this study are included in the article. Further inquiries can be directed to the corresponding authors.

Acknowledgments: This work was completed while the second author was a Doctoral Candidate in the Interdisciplinary Doctoral School at Lodz University of Technology, Poland. This research was funded in part by National Science Centre, Poland: 2024/53/N/ST7/02024. For the purpose of open access, the author has applied a CC-BY public copyright licence to any Author Accepted Manuscript (AAM) version arising from this submission.

Conflicts of Interest: The authors declare no conflicts of interest.

References

1. Jiang, L.; Liu, J.; Tian, A.; Cheng, Y.; Li, Z.; Zhang, L.; Zhang, S.; Li, D.; Ikeda, M.; Yang, H. GaN-based green laser diodes. *J. Semicond.* **2016**, *37*, 111001. [CrossRef]
2. Usman, M.; Munsif, M.; Mushtaq, U.; Anwar, A.R.; Muhammad, N. Green gap in GaN-based light-emitting diodes: In perspective. *Crit. Rev. Solid State Mater. Sci.* **2021**, *46*, 450–467. [CrossRef]
3. Sizov, D.S.; Bhat, R.; en Zah, C. Challenges and approaches of fabricating GaN-based green lasers. In Proceedings of the 2011 International Semiconductor Device Research Symposium (ISDRS), College Park, MD, USA, 7–9 December 2011; pp. 1–2. [CrossRef]
4. Sizov, D.; Bhat, R.; Zah, C.E. Gallium indium nitride-based green lasers. *J. Light. Technol.* **2012**, *30*, 679–699. [CrossRef]
5. Nakatsu, Y.; Nagao, Y.; Kozuru, K.; Hirao, T.; Okahisa, E.; Masui, S.; Yanamoto, T.; ichi Nagahama, S. High-efficiency blue and green laser diodes for laser displays. In Proceedings of the SPIE, San Francisco, CA, USA, 1 March 2019; Volume 10918, p. 109181D. [CrossRef]
6. Tian, A.; Hu, L.; Zhang, L.; Liu, J.; Yang, H. Design and growth of GaN-based blue and green laser diodes. *Sci. China Mater.* **2020**, *63*, 1348–1363. [CrossRef]
7. Samonji, K.; Yoshida, S.; Hagino, H.; Yamanaka, K.; Takigawa, S. 6.3 W InGaN laser diode array with highly efficient wide-striped emitters. In Proceedings of the IEEE Photonic Society 24th Annual Meeting, Arlington, VA, USA, 9–13 October 2011; pp. 507–508.
8. Suzuki, N.; Morimoto, K. 10-W CW blue-violet diode laser array on the micro-channel cooler. In Proceedings of the SPIE, San Francisco, CA, USA, 8 February 2012; Volume 8241, p. 82410J. [CrossRef]
9. Perlin, P.; Marona, L.; Holc, K.; Wisniewski, P.; Suski, T.; Leszczynski, M.; Czernecki, R.; Najda, S.; Zajac, M.; Kucharski, R. InGaN Laser Diode Mini-Arrays. *Appl. Phys. Express* **2011**, *4*, 62103. [CrossRef]
10. Goto, S.; Ohta, M.; Yabuki, Y.; Hoshina, Y.; Naganuma, K.; Tamamura, K.; Hashizu, T.; Ikeda, M. Super high-power AlGaInN-based laser diodes with a single broad-area stripe emitter fabricated on a GaN substrate. *Phys. Status Solidi (A)* **2003**, *200*, 122–125. [CrossRef]
11. Najda, S.P.; Perlin, P.; Suski, T.; Marona, L.; Boćkowski, M.; Leszczyński, M.; Wisniewski, P.; Czernecki, R.; Kucharski, R.; Targowski, G. AlGaInN laser diode bar and array technology for high-power and individual addressable applications. In Proceedings of the SPIE, Brussels, Belgium, 28 April 2016; Volume 9892, p. 98920Z. [CrossRef]
12. Samonji, K.; Yoshida, S.; Hagino, H.; Yamanaka, K.; Takigawa, S. High-power operation of a wide-striped InGaN laser diode array. In Proceedings of the SPIE, San Francisco, CA, USA, 8 February 2012; Volume 8277, p. 82771K. [CrossRef]
13. Jawulski, K.; Kuc, M.; Sarzala, R.P. Simplified thermal analysis of impact of diamond heat spreader on InGaN laser diode arrays. In Proceedings of the SPIE, Szczecin, Poland, 22 January 2013; Volume 8702, p. 87020D. [CrossRef]
14. Kozlowska, A.; Malag, A. Investigations of transient thermal properties of conductively cooled diode laser arrays operating under quasicontinuous-wave conditions. *Microelectron. Reliab.* **2006**, *46*, 2079–2084. [CrossRef]
15. Qiao, Y.; Feng, S.; Xiong, C.; Ma, X.; Zhu, H.; Guo, C.; Wei, G. The thermal properties of AlGaAs/GaAs laser diode bars analyzed by the transient thermal technique. *Solid-State Electron.* **2013**, *79*, 192–195. [CrossRef]
16. Laff, R.A.; Comerford, L.D.; Crow, J.D.; Brady, M.J. Thermal performance and limitations of silicon–substrate packaged GaAs laser arrays. *Appl. Opt.* **1978**, *17*, 778–784. [CrossRef] [PubMed]
17. Garmire, E.; Tavis, M. Heatsink requirements for coherent operation of high-power semiconductor laser arrays. *IEEE J. Quantum Electron.* **1984**, *20*, 1277–1283. [CrossRef]

18. Murata, S.; Nishimura, K. Improvement in Thermal Properties of a Multi-Beam Laser Diode Array. *Jpn. J. Appl. Phys.* **1989**, *28*, 165. [CrossRef]
19. Sato, K.; Murakami, M. Experimental investigation of thermal crosstalk in a distributed feedback laser array. *IEEE Photonics Technol. Lett.* **1991**, *3*, 501–503. [CrossRef]
20. Osinski, M.; Nakwaski, W. Thermal analysis of closely-packed two-dimensional etched-well surface-emitting laser arrays. *IEEE J. Sel. Top. Quantum Electron.* **1995**, *1*, 681–696. [CrossRef]
21. Wipiejewski, T.; Young, D.B.; Thibeault, B.J.; Coldren, L.A. Thermal crosstalk in 4 × 4 vertical-cavity surface-emitting laser arrays. *IEEE Photonics Technol. Lett.* **1996**, *8*, 980–982. [CrossRef]
22. Nakwaski, W.; Osinski, M. Erratum: Thermal resistance of top-surface-emitting vertical-cavity semiconductor lasers and monolithic two-dimensional arrays. *Electron. Lett.* **1992**, *28*, 572–574. [CrossRef]
23. Kuc, M.; Sarzała, R.P.; Nakwaski, W. Thermal crosstalk in arrays of III-N-based Lasers. *Mater. Sci. Eng. B* **2013**, *178*, 1395–1402. [CrossRef]
24. Sarzała, R.P.; Sokół, A.K.; Kuc, M.; Nakwaski, W. How to enhance a room-temperature operation of diode lasers and their arrays. *Opt. Appl.* **2016**, *46*, 213–226.
25. Dąbrówka, D.; Sarzała, R.; Wasiak, M.; Kafar, A.; Perlin, P.; Saba, K. Thermal analysis of a two-dimensional array with surface light emission based on nitride EEL lasers. *Opto-Electron. Rev.* **2022**, *30*, e144115. [CrossRef]
26. Hu, J.J.; Zhang, S.M.; Li, D.Y.; Zhang, F.; Feng, M.X.; Wen, P.Y.; Liu, J.P.; Zhang, L.Q.; Yang, H. Thermal analysis of GaN-based laser diode mini-array*. *Chin. Phys. B* **2018**, *27*, 094208. [CrossRef]
27. Vetrovec, J. Improved cooling for high-power laser diodes. In Proceedings of the SPIE, San Jose, CA, USA, 20 February 2008; Volume 6876, p. 687603. [CrossRef]
28. Kozłowska, A.; Teodorczyk, M.; Dąbrowska-Tumańska, E.; Chmielewski, M.; Podniesiński, D.; Maląg, A. Novel micro-channel cooler for high-power diode laser arrays. *Photonics Lett. Pol.* **2014**, *6*, 23–25. lett. pl.v6i1.435 [CrossRef]
29. Dix, J.; Jokar, A.; Martinsen, R. Enhanced microchannel cooling for high-power semiconductor diode lasers. In Proceedings of the SPIE, San Jose, CA, USA, 13 February 2008; Volume 6876, p. 687606. [CrossRef]
30. Pais, M.; Chang, M.; Morgan, M.; Chow, L. Spray Cooling of a High Power Laser Diode. In Proceedings of the Aerospace Atlantic Conference & Exposition. SAE International, Dayton, OH, USA, 18–22 April 1994. [CrossRef]
31. Benett, W.J.; Freitas, B.L.; Ciarlo, D.R.; Beach, R.J.; Sutton, S.B.; Emanuel, M.A.; Solarz, R.W. Microchannel-cooled heatsinks for high-average-power laser diode arrays. In Proceedings of the SPIE, Tromsö, Norway, 30 June 1993; Volume 1865, pp. 144–153. [CrossRef]
32. Krause, V.K.; Treusch, H.G.; Loosen, P.; Kimpel, T.; Biesenbach, J.; Koesters, A.; Robert, F.; Oestreicher, H.; Marchiano, M.; DeOdorico, B. Microchannel coolers for high-power laser diodes in copper technology. In Proceedings of the SPIE, Bergen, Norway, 13–15 June 1994; Volume 2148, pp. 351–358. [CrossRef]
33. Huddle, J.J.; Chow, L.C.; Lei, S.; Marcos, A.; Rini, D.P.; Lindauer, S.J.; Bass, M.; Delfyett, P.J. Thermal management of diode laser arrays. In Proceedings of the Sixteenth Annual IEEE Semiconductor Thermal Measurement and Management Symposium (Cat. No.00CH37068), San Jose, CA, USA, 23–23 March 2000; pp. 154–160. [CrossRef]
34. Wang, J.; Kang, L.; Zhang, P.; Nie, Z.; Li, X.; Xiong, L.; Liu, X. High-power semiconductor laser array packaged on microchannel cooler using gold-tin soldering technology. In Proceedings of the SPIE, San Francisco, CA, USA, 8 February 2012; Volume 8241, p. 82410H. [CrossRef]
35. Liu, X.; Davis, R.W.; Hughes, L.C.; Rasmussen, M.H.; Bhat, R.; Zah, C.E.; Stradling, J. A study on the reliability of indium solder die bonding of high power semiconductor lasers. *J. Appl. Phys.* **2006**, *100*, 013104. [CrossRef]
36. Barwolff, A.; Tomm, J.; Muller, R.; Weiß, S.; Hutter, M.; Oppermann, H.; Reichl, H. Spectroscopic measurement of mounting-induced strain in optoelectronic devices. *IEEE Trans. Adv. Packag.* **2000**, *23*, 170–175. [CrossRef]
37. Mizuishi, K. Some aspects of bonding-solder deterioration observed in long-lived semiconductor lasers: Solder migration and whisker growth. *J. Appl. Phys.* **1984**, *55*, 289–295. [CrossRef]
38. Dhamdhere, A.R.; Malshe, A.P.; Schmidt, W.F.; Brown, W.D. Investigation of reliability issues in high power laser diode bar packages. *Microelectron. Reliab.* **2003**, *43*, 287–295. [CrossRef]
39. Kuc, M.; Piskorski, Ł.; Dems, M.; Wasiak, M.; Sokół, A.K.; Sarzała, R.P.; Czyszanowski, T. Numerical Investigation of the Impact of ITO, AlInN, Plasmonic GaN and Top Gold Metalization on Semipolar Green EELs. *Materials* **2020**, *13*, 1444. [CrossRef]
40. Myzaferi, A.; Reading, A.H.; Farrell, R.M.; Cohen, D.A.; Nakamura, S.; DenBaars, S.P. Semipolar III-nitride laser diodes with zinc oxide cladding. *Opt. Express* **2017**, *25*, 16922–16930. [CrossRef] [PubMed]
41. Myzaferi, A.; Mughal, A.J.; Cohen, D.A.; Farrell, R.M.; Nakamura, S.; Speck, J.S.; DenBaars, S.P. Zinc oxide clad limited area epitaxy semipolar III-nitride laser diodes. *Opt. Express* **2018**, *26*, 12490–12498. [CrossRef]
42. Reading, A.H.; Richardson, J.J.; Pan, C.C.; Nakamura, S.; DenBaars, S.P. High efficiency white LEDs with single-crystal ZnO current spreading layers deposited by aqueous solution epitaxy. *Opt. Express* **2012**, *20*, A13–A19. [CrossRef] [PubMed]
43. Mion, C.; Muth, J.; Preble, E.; Hanser, D. Accurate dependence of gallium nitride thermal conductivity on dislocation density. *Appl. Phys. Lett.* **2006**, *89*, 92123. [CrossRef]

44. Oshima, Y.; Yoshida, T.; Eri, T.; Shibata, M.; Mishima, T. Thermal and electrical properties of high-quality freestanding GaN wafers with high carrier concentration. *Phys. Status Solidi C* **2007**, *4*, 2215–2218. [CrossRef]
45. Daly, B.C.; Maris, H.J.; Nurmikko, A.V.; Kuball, M.; Han, J. Optical pump-and-probe measurement of the thermal conductivity of nitride thin films. *J. Appl. Phys.* **2002**, *92*, 3820–3824. [CrossRef]
46. Pantha, B.N.; Dahal, R.; Li, J.; Lin, J.Y.; Jiang, H.X.; Pomrenke, G. Thermoelectric properties of InxGa1-xN alloys. *Appl. Phys. Lett.* **2008**, *92*, 042112. [CrossRef]
47. Zhang, J.; Tong, H.; Liu, G.; Herbsommer, J.A.; Huang, G.S.; Tansu, N. Thermoelectric properties of MOCVD-grown AlInN alloys with various compositions. In Proceedings of the SPIE, San Francisco, CA, USA, 3 March 2011; Volume 7939, p. 79390O. [CrossRef]
48. Osinski, M.; Nakwaski, W.; Leal, A. Effective thermal conductivity analysis of vertical-cavity top-surface-emitting lasers with semiconducting Bragg mirrors. In Proceedings of the SPIE, Los Angeles, CA, USA, 2 June 1994; Volume 2147, pp. 85–96. [CrossRef]
49. Lide, D.R. (Ed.) *CRC Handbook of Chemistry and Physics*; CRC Press: Boca Raton, FL, USA, 2004; Volume 85.
50. Shackelford, J.F.; Alexander, W. (Eds.) *CRC Materials Science and Engineering Handbook*; CRC Press: Boca Raton, FL, USA, 2000. [CrossRef]
51. Wu, X.; Lee, J.; Varshney, V.; Wohlwend, J.L.; Roy, A.K.; Luo, T. Thermal Conductivity of Wurtzite Zinc-Oxide from First-Principles Lattice Dynamics–a Comparative Study with Gallium Nitride. *Sci. Rep.* **2016**, *6*, 22504. [CrossRef] [PubMed]
52. Dems, M.; Beling, P.; Gębski, M.; Piskorski, Ł.; Walczak, J.; Kuc, M.; Frasunkiewicz, L.; Michał, W.; Sarzała, R.; Czyszanowski, T. VCSEL Modeling with Self-Consistent Models: From Simple Approximations to Comprehensive Numerical Analysis. In Proceedings of the SPIE, San Francisco, CA, USA, 4 March 2015; Volume 9381, p. 93810K. [CrossRef]
53. PLaSK: A Comprehensive Tool for Numerical Analysis of a Broad Range of Physical Phenomena in Photonic Devices. 2024. Available online: https://plask.app (accessed on 19 December 2024).
54. Czyszanowski, T.; Stańczyk, S.; Kafar, A.; Perlin, P. Optical Optimization of InGaN/GaN Edge-Emitting Lasers with Reduced AlGaN Cladding Thickness. *Jpn. J. Appl. Phys.* **2014**, *53*, 032701. [CrossRef]
55. Janczak, M.; Sarzała, R.P.; Dems, M.; Kolek, A.; Bugajski, M.; Nakwaski, W.; Czyszanowski, T. Threshold Performance of Pulse-Operating Quantum-Cascade Vertical-Cavity Surface-Emitting Lasers. *Opt. Express* **2022**, *30*, 45054. [CrossRef] [PubMed]
56. Więckowska, M.; Dems, M. Increasing Single-Mode Power in VCSELs With Antiresonant Oxide Island. *IEEE Photonics J.* **2024**, *16*, 1–7. [CrossRef]
57. Dems, M.; Chung, I.S.; Nyakas, P.; Bischoff, S.; Panajotov, K. Numerical Methods for Modeling Photonic-Crystal VCSELs. *Opt. Express* **2010**, *18*, 16042. [CrossRef] [PubMed]
58. Janaszek, A.; Wróbel, P.; Dems, M.; Ceylan, O.; Gurbuz, Y.; Kubiszyn, Ł.; Piotrowski, J.; Kotyński, R. Rigorous Optical Modelling of Long-Wavelength Infrared Photodetector with 2D Subwavelength Hole Array in Gold Film. *Opto-Electron. Rev.* **2024**, *32*, e148831. [CrossRef]
59. Ekielski, M.; Głowadzka, W.; Bogdanowicz, K.; Rygała, M.; Mikulicz, M.; Śpiewak, P.; Kowalski, M.; Gębski, M.; Motyka, M.; Szerling, A.; et al. Monolithic High Contrast Grating Integrated with Metal: Infrared Electrode with Exceptionally High Conductivity and Transmission. *Adv. Funct. Mater.* **2023**, *34*, 2312392. [CrossRef]
60. Pruszyńska-Karbownik, E.; Jandura, D.; Dems, M.; Zinkiewicz, Ł.; Broda, A.; Gębski, M.; Muszalski, J.; Pudiš, D.; Suffczyński, J.; Czyszanowski, T. Concept of Inverted Refractive-Index-Contrast Grating Mirror and Exemplary Fabrication by 3D Laser Micro-Printing. *Nanophotonics* **2023**, *12*, 3579–3588. [CrossRef]
61. Mikulicz, M.; Rygała, M.; Smołka, T.; Janczak, M.; Badura, M.; Łozińska, A.; Wolf, A.; Emmerling, M.; Ściana, B.; Höfling, S.; et al. Enhancement of Quantum Cascade Laser Intersubband Transitions via Coupling to Resonant Discrete Photonic Modes of Subwavelength Gratings. *Opt. Express* **2023**, *31*, 26898–26909. [CrossRef] [PubMed]
62. Schade, A.; Bader, A.; Huber, T.; Kuhn, S.; Czyszanowski, T.; Pfenning, A.; Rygała, M.; Smołka, T.; Motyka, M.; Sęk, G.; et al. Monolithic High Contrast Grating on GaSb/AlAsSb Based Epitaxial Structures for Mid-Infrared Wavelength Applications. *Opt. Express* **2023**, *31*, 16025–16034. [CrossRef] [PubMed]
63. Virtanen, P.; Gommers, R.; Oliphant, T.E.; Haberland, M.; Reddy, T.; Cournapeau, D.; Burovski, E.; Peterson, P.; Weckesser, W.; Bright, J.; et al. SciPy 1.0: Fundamental Algorithms for Scientific Computing in Python. *Nat. Methods* **2020**, *17*, 261–272. [CrossRef]
64. Kraft, D. *A Software Package for Sequential Quadratic Programming*; Tech. Rep. DFVLR-FB 88-28; DLR German Aerospace Center—Institute for Flight Mechanics: Koln, Germany, 1988.
65. Ocaya, R.O. Thermal tuning of light-emitting diode wavelength as an implication of the Varshni equation. *Measurement* **2020**, *162*, 107910. [CrossRef]
66. Madhusoodhanan, S.; Sabbar, A.; Al-kabi, S.; Atcitty, S.; Kaplar, R.; Dong, B.; Wang, J.; Yu, Q.; Chen, Z. High-Temperature Optical Characterization of Wide Band Gap Light Emitting Diodes and Photodiodes for Future Power Module Application. *Adv. Sci. Technol. Eng. Syst. J.* **2019**, *4*, 17–22. [CrossRef]
67. Wang, G.; Tao, X.; Liu, J.; Jiang, F. Temperature-dependent electroluminescence from InGaN/GaN green light-emitting diodes on silicon with different quantum-well structures. *Semicond. Sci. Technol.* **2014**, *30*, 015018. [CrossRef]

68. Jafar, N.; Jiang, J.; Lu, H.; Qasim, M.; Zhang, H. Recent Research on Indium-Gallium-Nitride-Based Light-Emitting Diodes: Growth Conditions and External Quantum Efficiency. *Crystals* **2023**, *13*, 1623. [CrossRef]
69. Liu, J.; Tam, W.; Wong, H.; Filip, V. Temperature-dependent light-emitting characteristics of InGaN/GaN diodes. *Microelectron. Reliab.* **2009**, *49*, 38–41. [CrossRef]

materials

MDPI

Article

Bilayer TiO_2/Mo-$BiVO_4$ Photoelectrocatalysts for Ibuprofen Degradation

Martha Pylarinou [1], Elias Sakellis [1,2], Spiros Gardelis [1], Vassilis Psycharis [2], Marios G. Kostakis [3], Nikolaos S. Thomaidis [3] and Vlassis Likodimos [1,*]

1 Section of Condensed Matter Physics, Department of Physics, National and Kapodistrian University of Athens, University Campus, 15784 Athens, Greece; pylarin@phys.uoa.gr (M.P.); e_sakel@phys.uoa.gr (E.S.); sgardelis@phys.uoa.gr (S.G.)

2 Institute of Nanoscience and Nanotechnology, National Center for Scientific Research "Demokritos", Agia Paraskevi, 15341 Athens, Greece; v.psycharis@inn.demokritos.gr

3 Laboratory of Analytical Chemistry, Department of Chemistry, National and Kapodistrian University of Athens, University Campus, 15771 Athens, Greece; makostak@chem.uoa.gr (M.G.K.); ntho@chem.uoa.gr (N.S.T.)

* Correspondence: vlikodimos@phys.uoa.gr; Tel.: +30-2107276824

Abstract: Heterojunction formation between $BiVO_4$ nanomaterials and benchmark semiconductor photocatalysts has been keenly pursued as a promising approach to improve charge transport and charge separation via interfacial electron transfer for the photoelectrocatalytic degradation of recalcitrant pharmaceutical pollutants. In this work, a heterostructured TiO_2/Mo-$BiVO_4$ bilayer photoanode was fabricated by the deposition of a mesoporous TiO_2 overlayer using the benchmark P25 titania catalyst on top of Mo-doped $BiVO_4$ inverse opal films as the supporting layer, which intrinsically absorbs visible light below 490 nm, while offering improved charge transport. A porous P25/Mo-$BiVO_4$ bilayer structure was produced from the densification of the inverse opal underlayer after post-thermal annealing, which was evaluated on photocurrent generation in aqueous electrolyte and the photoelectrocatalytic degradation of the refractory anti-inflammatory drug ibuprofen under back-side illumination by visible and UV–Vis light. Significantly enhanced photoelectrochemical performance on both photocurrent density and pharmaceutical degradation was achieved for the bilayer structure with respect to the additive effect of the constituent layers, which was related to the improved light harvesting arising from the backscattering by the mesoporous TiO_2 layer in combination with the favorable charge transfer at the TiO_2/Mo-$BiVO_4$ interface.

Keywords: $BiVO_4$; TiO_2; photocatalytic films; heterojunctions; photoelectrocatalysis; pharmaceuticals; ibuprofen

1. Introduction

Nanostructured bismuth vanadate ($BiVO_4$) has been established as a benchmark n-type semiconductor photoanode because of its exceptional performance in photoelectrochemical (PEC) water splitting, which originates from its favorable physical and chemical properties, including suitable valence band edge potential and visible light activated band gap, accompanied by environmental inertness, earth abundance, and relatively low cost as well as its long-term stability in neutral pH regions compared to other metal oxide PEC materials [1–3]. Major research efforts have been accordingly devoted to elaborate and optimize synthetic methods together with surface, bulk, and interface engineering of $BiVO_4$ in order to alleviate persistent electron-hole recombination losses arising from the poor electron transfer kinetics that is the major drawback of $BiVO_4$ photoelectrodes [4–6]. Despite the marked progress in the improvement of photoelectrochemical performance [7], application of $BiVO_4$ photocatalysts in water purification has lagged behind as the relatively narrow band gap (2.4–2.6 eV) of the photocatalytically most active, monoclinic scheelite (*ms*) $BiVO_4$ phase [8] may impede simultaneous generation of highly reactive oxygen species such as hydroxyl radicals ($^{\bullet}OH$) and superoxide radical anions ($O^{2-\bullet}$), which underlie the non-selective photocatalytic degradation of organic contaminants in the absence of sacrificial scavengers [9,10]. Recently, though, judicious compositional and morphology modifications have rendered $BiVO_4$ an emerging photocatalyst in organics degradation under visible light, applied mostly in the photocatalytic removal of dye pollutants [11,12] and to a lesser extent on colorless water contaminants such as antibiotics and bacteria [13–17]. Advanced application of nanostructured $BiVO_4$ photocatalysts has been further pursued on the photoelectrocatalytic degradation of refractory emerging contaminants such as pharmaceuticals [18], where simultaneous occurrence of both reduction and oxidation processes is not thermodynamically a prerequisite as the counter electrode can assist one of the redox reactions [19]. Metal doping of $BiVO_4$ has been the most competent modification to enhance its poor charge transport/separation efficiency for PEC water splitting with Mo^{6+} and W^{6+} that replace lattice V^{5+} ions being the most efficient dopants [2,3]. Specifically, it has been established that the introduction of Mo-dopants in $BiVO_4$ at low levels, which act as shallow donors, leads to the increase in donor density and n-type conductivity along with electron mobility and electron diffusion length by lowering the small polaron hopping barrier [20]. Heterojunction formation between pristine or doped $BiVO_4$ nanomaterials with other semiconductor photocatalysts [21–25] and metal nanoparticles (NPs) [26–28] has accordingly attracted particular interest in order to improve charge transport and charge separation via interfacial electron transfer for the photoelectrocatalytic degradation of recalcitrant pharmaceuticals.

In addition to compositional engineering, tailoring the morphology of $BiVO_4$ in the form of nanoporous photoanodes has been keenly pursued in order to enhance light harvesting and assist charge separation for photocatalytic applications [29–32]. Among different morphologies, nanocomposite photonic crystals (PCs) have been attracting attention for advanced photoelectric and absorption devices [33–35]. Inverse opals have been the most common three-dimensionally ordered structure, which, besides the selective enhancement of light harvesting via slow photons, offers a macroporous interconnected network that facilitates molecular transport and adsorption during photocatalytic reactions [36]. Nanoparticulate $BiVO_4$ inverse opals have been successfully applied as visible light-activated photocatalysts in powder form [37,38], while Mo-doped $BiVO_4$ inverse opal photoelectrodes have shown excellent performance in photoelectrochemical water splitting [39], especially after surface decoration by Au [40] and Ag [41] plasmonic NPs. Recently, markedly improved activity was reported for Mo-$BiVO_4$/Ca-$BiVO_4$ homojunctions [42] and (Ag, Au) plasmonic-modified Mo-$BiVO_4$ inverse opal films [43] in the visible light photoelectrocatalytic degradation of pharmaceutical emerging contaminants. Heterostructuring of $BiVO_4$ with benchmark TiO_2 photocatalysts has been the subject of particular interest, where post-deposition of $BiVO_4$ NPs has been commonly employed for the visible light sensitization of TiO_2 inverse opals that were applied in the photocatalytic degradation of dye pollutants [44,45]. Likewise, $BiVO_4$ NP decoration has been exploited for the improvement of water splitting over WO_3 inverse opal photoanodes [46] as well as a nanocrystalline coating layer on top of WO_3 [47] and TiO_2 [48] inverse opal photoelectrodes leading to high photocurrent densities. Recently, $BiVO_4$ NPs deposition on single [49] and bilayer [50] TiO_2 inverse opal films resulted in greatly enhanced photocatalytic degradation rates of rhodamine B dye upon slow photon tuning, while significant enhancement of water oxidation was reported for a thin Mo-$BiVO_4$ photoanode when coated with a TiO_2 inverse opal layer with suitable photonic band gap with respect to the $BiVO_4$ absorption edge [51].

In this work, heterostructured TiO_2/Mo-$BiVO_4$ films are demonstrated as efficient photoanodes for the photoelectrocatalytic degradation of recalcitrant pharmaceutical water pollutants. Bilayer photoelectrodes were fabricated by the deposition of a mesoporous TiO_2 overlayer on top of Mo-$BiVO_4$ inverse opal substrates, which intrinsically absorb in the visible range below 490 nm, while offering improved charge transport due to Mo-doping. Post-thermal annealing of the composite structure resulted in a porous P25/Mo-$BiVO_4$ bilayer film after the collapse of the inverse opal bottom layer due to crystallite growth. Comparative PEC evaluation showed that the bilayer photoelectrodes present greatly improved performance with respect to the additive effect of the constituent layers in both photocurrent density and the photoelectrocatalytic degradation of ibuprofen (IBU), a non-steroidal anti-inflammatory drug under back-side illumination by visible and UV–Vis light. Optical and PEC measurements indicate that the rise of photoelectrocatalytic activity of the bilayer films is related to the enhanced light harvesting arising from backscattering by the mesoporous TiO_2 layer in combination with the favorable electron transfer at the TiO_2/Mo-$BiVO_4$ interface.

2. Materials and Methods

2.1. Bilayer Fabrication

A Mo-doped $BiVO_4$ inverse opal layer was initially fabricated using the liquid phase infiltration of a colloidal opal template of 340 nm diameter polystyrene (PS) spheres with the appropriate metal salt precursor and optimized composition. Specifically, the 3% Mo:V molar ratio and the templating sphere size were selected based on our recent work [42], where the Mo-doping levels and the template sphere size were varied in order to optimize the photoelectrocatalytic activity of Mo-$BiVO_4$ on pharmaceutical degradation under visible light. The horizontal colloidal crystallization method was applied in order to rapidly deposit the PS template films without consuming large amounts of colloidal suspension as in conventional self-assembly methods [52]. Specifically, 100 μL of 1.25 wt% monodisperse PS sphere suspension (Microparticles GmbH, Berlin, Germany, standard deviation SD = 10 nm, 2.9% CV, in the form of 5% solids (w/v) colloidal dispersions in deionized water) was deposited on a horizontal fluorine-doped tin oxide (FTO, thickness 2.2 mm, surface resistivity 7 Ω/sq, Sigma Aldrich, St. Louis, MO, USA) conductive glass substrate, which was kept at 20 °C until the solvent evaporated. In that case, inward convective flow as the evaporation rate at the periphery is higher than that at the center results in the self-assembly of PS microspheres to the face-centered cubic (fcc) close-packed lattice. FTO substrates, prior to their use, have been cleaned by ultrasound sonication in Hellmanex™ III (Helma Analytics, Müllheim, Germany) and propanol-2, followed by dipping in water and drying in N_2, while their hydrophilicity was enhanced by dipping in H_2O:HNO_3 (5:1). Then, liquid infiltration of the PS opal template was performed in a complex metal salt precursor based on ammonium vanadate (NH_4VO_3) (ACS, 99.0% min) and bismuth(III) nitrate pentahydrate, $Bi(NO_3)_3 \cdot 5H_2O$ (99.999% trace metals basis) with the addition of ammonium molybdate tetrahydrate $(NH_4)_6Mo_7O_{24} \cdot 4H_2O$ (BioUltra, ≥99.0%) as source of molybdenum (Mo) doping [43]. Specifically, 0.5 mmol NH_4VO_3 was initially mixed with 9 mL of propanol-2 (HPLC, 99.8%) and 0.5 mL of nitric acid (puriss. p.a., 65%)

under continuous stirring for 30 min at room temperature, followed by sonication and the addition of 0.5 mmol citric acid (anhydrous, ACS, 99.5+%) until dissolution. Mo-doping was implemented by substituting part of NH_4VO_3 with $(NH_4)_6Mo_7O_{24}\cdot 4H_2O$ at Mo:V molar ratio of 3%. Another solution was prepared by dissolving 0.5 mmol $Bi(NO_3)_3\cdot 5H_2O$ with 0.5 mL acetic acid (glacial, ACS, 99.7+%). The two solutions were mixed and stirred for 30 min to obtain the Mo-$BiVO_4$ metal salt precursor, wherein the PS opal films were dipped for 5 min, followed by drying for 1 h at 70 °C. The dip-coating process was repeated three times in order to ensure complete precursor infiltration to the opal interstices followed by post-calcination 400 °C in air to remove the PS template and crystallize the amorphous precursor in the Mo-$BiVO_4$ inverse opal structure. Reference films of undoped $BiVO_4$ PC were likewise deposited on FTO substrates without adding the Mo-salt in the precursor mixture. The PC films were labeled as Mo-$BiVO_4$ PC and $BiVO_4$ PC, respectively.

The bilayer photoelectrode was then constructed by the deposition of a mesoporous nanocrystalline TiO_2 layer [53] on the Mo-$BiVO_4$ films, using Aeroxide P25 (Evonik, Essen, Germany) as benchmark TiO_2 photocatalyst due to the mixed phase of anatase and rutile NPs and the concomitant interfacial charge transfer [54]. The deposition of the P25 paste on top of the Mo-$BiVO_4$ PC films was performed by spin coating (Ossila, Sheffield, UK) at 400 rpm for 30 s followed by drying at 120 °C and post-annealing at 450 °C in air for 30 min with rate 5 °C/min, in order to remove organics, leading to P25/Mo-$BiVO_4$ films (Scheme 1). In addition, Mo-$BiVO_4$/P25 bilayer films with reversed deposition sequence i.e., a mesoporous P25 underlayer was first coated on the FTO substrate followed by the Mo-$BiVO_4$ inverse opal deposition, were fabricated in order to determine the optimal bilayer structure. Nanocrystalline Mo-$BiVO_4$ planar films and the corresponding bilayers were also deposited as reference films, applying the same procedure without the colloidal template.

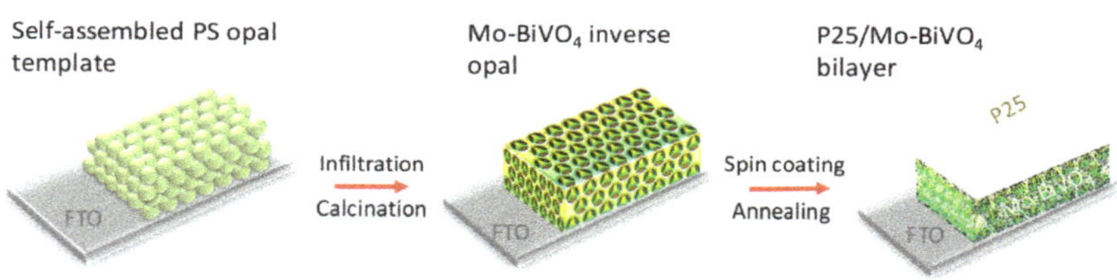

Scheme 1. Fabrication process of P25/Mo-$BiVO_4$ bilayer films.

2.2. *Materials Characterization*

The films' morphology and phase composition were characterized using a scanning electron microscope (SEM) (FEI Quanta Inspect, Eindhoven, The Netherlands) coupled with an energy-dispersive X-ray (EDX) analyzer (EDAX, Mahwah, NJ, USA) and an FEI Talos F200i scanning transmission electron microscope (TEM) (Thermo Fisher Scientific Inc., Waltham, MA, USA) operating at 200 keV, equipped with a windowless energy-dispersive spectroscopy microanalyzer (6T/100 Bruker, Hamburg, Germany). The materials' structural properties were investigated by X-ray powder diffraction (XRD) using a SmartLab Rigaku θ/θ Bragg–Brentano diffractometer (Rigaku, Tokyo, Japan), equipped with a pyrolytic graphite monochromator using Cu Kα radiation. The power conditions were set at 40 kV/35 mA and the aperture slits were set at 2/3°, 2/3°, 0.6°. The continuous step-scanning technique was used at steps of 0.03° with measuring time of 11 s/step and the recorded 2θ range was 15–55°. The films' phase composition was further studied by micro-Raman spectroscopy using a confocal Raman microscope (LabRAM Soleil™, Horiba Scientific, Longjumeau, France) equipped with two automated built-in lasers emitting at 532 and 785 nm. The laser beam was focused on the films' surface using a 100× (NA = 0.9) objective on a high-end microscope with a motorized high-precision X-Y-Z sample stage. The scattered light was filtered by edge Rayleigh rejection filters with cutoffs at 50 cm^{-1} and analyzed by a 330 mm focal length spectrometer, a 1800 gr/mm diffraction grating, and a high-sensitivity UV–Vis CCD detector (Horiba Scientific, Longjumeau, France). Diffuse and specular reflectance measurements were performed using a fiber-optic diffuse and a 15° specular reflectance accessory on a Cary60 UV–Vis spectrometer (Agilent, Santa Clara, CA, USA), respectively. A Halon reference and a UV-enhanced Al mirror were used for background determination.

PEC experiments were performed in a three-electrode configuration on a CS350 potentiostat/galvanostat (Corrtest Instruments, Wuhan, China) using the single and bilayer films on FTO substrates as working electrode, Ag/AgCl as reference, and a Pt foil as counter electrodes in 0.5 M $NaHCO_3$ aqueous electrolyte. The PEC measurements were carried out under dark and back-side illumination using a 300 W Xe lamp as source of UV–Vis and visible light (400 nm cutoff filter, 100 mW/cm^2). Linear sweep voltammetry was performed at a potential scan rate of 10 mV s^{-1}. The applied potential vs. Ag/AgCl was converted to reversible hydrogen electrode (RHE) scale by using the following equation:

$$E_{RHE} = E_{Ag/AgCl} + 0.059 \text{ pH} + E^0{}_{Ag/AgCl}, \quad (1)$$

where $E^0{}_{Ag/AgCl}$ = 0.1976 V is the standard potential for Ag/AgCl at 25 °C. Electrochemical impedance spectroscopy (EIS) was performed at the open-circuit voltage, in the 10^4–10^{-2} Hz frequency range with ac amplitude of 10 mV. Mott–Schottky measurements were carried out at 500 Hz and a scan rate of 10 mV/s at each potential. The flat band potential and donor densities for the PC photoelectrodes were determined from the corresponding Mott–Schottky plots, i.e., the variation of $1/C^2$ vs. applied potential, using the following relation:

$$\frac{1}{C^2} = \frac{2}{eA^2\varepsilon\varepsilon_0 N_D}\left(V - V_{fb} - \frac{kT}{e}\right), \quad (2)$$

where C is the space-charge capacitance, e is the elementary charge and A is the electrode area, N_D is the donor density, ε is the semiconductor permittivity and ε_0 the vacuum permittivity, V_{fb} is the flat band potential, T is the absolute temperature, and k is Boltzman's constant. Incident photon-to-current efficiency (IPCE) measurements were also performed in the same PEC cell in 0.5 M $NaHCO_3$ aqueous electrolyte under back-side illumination at 1.23 V vs. RHE potential using a 1 kW Xe lamp in combination with an Oriel 77200 1/4 monochromator (Newport, Irvine, CA, USA) and a calibrated Si photodiode. IPCE was calculated according to the following relation:

$$\text{IPCE}(\%) = \frac{1239.8[\text{V nm}] \times J[\text{mA/cm}^2]}{\lambda[\text{nm}] \times P[\text{mW/cm}^2]} \times 100 \quad (3)$$

where J is the photocurrent density in mA/cm^2, λ is the wavelength of incident light in nm, and P is the monochromatic light power density (mW/cm^2).

2.3. Photoelectrocatalytic Performance

Photoelectrocatalytic experiments were also carried out in three-electrode configuration under back-side illumination by a 300 W Xe lamp in combination with a 400 nm long-pass filter (100 mW/cm^2) using 36 mL working solution of 0.1 M $NaHCO_3$ supporting electrolyte under mild stirring, comprising 5 mg/L of IBU pollutant at pH = 4. Degradation kinetics was monitored by spectrophotometric detection for small aliquots periodically withdrawn from the reaction cell. The degradation tests were performed in triplicate, and standard errors were calculated for the mean kinetic constants. The IBU concentration was also monitored by high-performance liquid chromatography (HPLC) under UV–Vis light up to 5 h for the P25/Mo-$BiVO_4$ bilayer photocatalyst using an Agilent 1260 Infinity II (Agilent, Santa Clara, CA, USA) HPLC-DAD equipped with a quaternary pump G7111B, multicolumn thermostat G7116A, vial sampler G7129A and DAD HS G7117C. Chromatographic separation was achieved with ZORBAX eclipse plus C8 column (150 mm × 4.6 mm, 5 μm), provided by Agilent. For the separation, the mobile phase consisted of Solution of Monobasic Potassium Phosphate in water and Tetrahydrofuran (56:44, v/v) adjusted in pH 2.0 with phosphoric acid. The column temperature was maintained at room temperature, the injection volume was 20 μL, and the flow rate 1 mL/min.

3. Results

3.1. Morphological, Structural, and Optical Properties

TEM images of the Mo-$BiVO_4$ PC supporting layer that was initially deposited on the FTO glass substrates indicate the formation of an ordered inverse opal structure with macropores of about 200 nm size (Figure 1a), whose chemical composition was confirmed by the corresponding EDX elemental maps of Bi, V, O, and Mo (Figure 1b–e). Subsequently, a mesoporous P25 TiO_2 layer was deposited on top of the Mo-doped $BiVO_4$ PC film (Figure 1f). Post-thermal treatment resulted in the formation of a porous bilayer structure comprising a nanocrystalline Mo-$BiVO_4$ bottom layer produced from the collapse of the inverse opal structure after the second thermal annealing and the resulting crystallite growth and a mesoporous titania overlayer consisting of interconnected P25 aggregates, each layer being about 500 nm thick (Figure 1g,h).

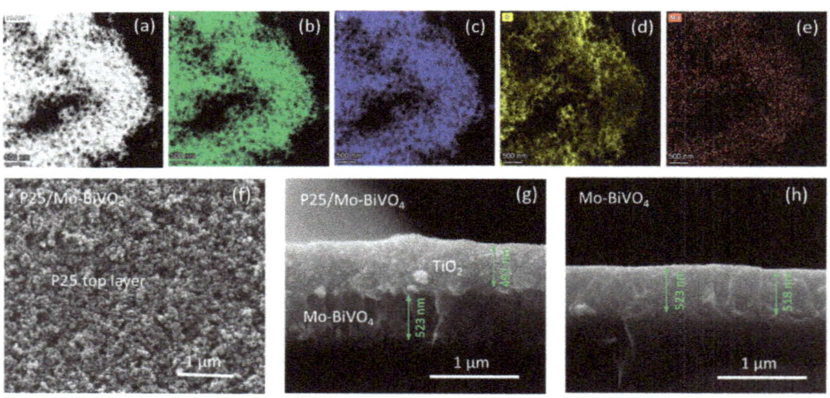

Figure 1. (**a**) TEM image of the Mo-$BiVO_4$ PC underlayer and the corresponding (**b**) Bi, (**c**) V, (**d**) O, and (**e**) Mo EDX elemental maps. (**f**) Top view and (**g**) cross-section SEM image of the P25/Mo-$BiVO_4$ bilayer structure. (**h**) Cross-section SEM image of the bottom Mo-$BiVO_4$ layer.

XRD measurements on the single $BiVO_4$ and Mo-$BiVO_4$ films as well as the P25/Mo-$BiVO_4$ bilayers (Figure 2a) confirmed that the films crystallize in the single *ms* phase (space group I2/a). This can be clearly evidenced by the characteristic splitting of the (002)/(200) (Figure 2c) and the (240)/(042) (Figure 2d) diffraction peaks. However, the observed peak splitting was appreciably reduced and less distinct for Mo-$BiVO_4$.

This verifies that the introduction of Mo^{6+} dopants, which can substitute for V^{5+} cations in the $BiVO_4$ lattice, moderate the monoclinic distortion of the *ms* phase towards the tetragonal scheelite (space group $I4_1/a$) structure [55]. However, a reentrant increase in the splitting and a narrowing of the *ms* diffraction peaks were observed for the P25/Mo-$BiVO_4$ bilayer films (Figure 2c,d), suggesting the growth of Mo-$BiVO_4$ crystallites following the post-thermal annealing that was applied during the deposition of the nanocrystalline P25 titania top layer. The XRD pattern of the individual P25 films deposited on FTO showed the characteristic mixed phase of P25 nanocrystalline titania, consisting of a major anatase (space group $I4_1/amd$) TiO_2 phase along with secondary rutile (space group $P4_2/mnm$) phase [49]. However, deposition of the P25 layer on the macroporous Mo-$BiVO_4$ inverse opal substrate resulted in the slight broadening of the main (101) anatase peak (Figure 2b), which indicates the possibility for a modified interfacial layer between the TiO_2 and Mo-$BiVO_4$ NPs during the post-annealing treatment.

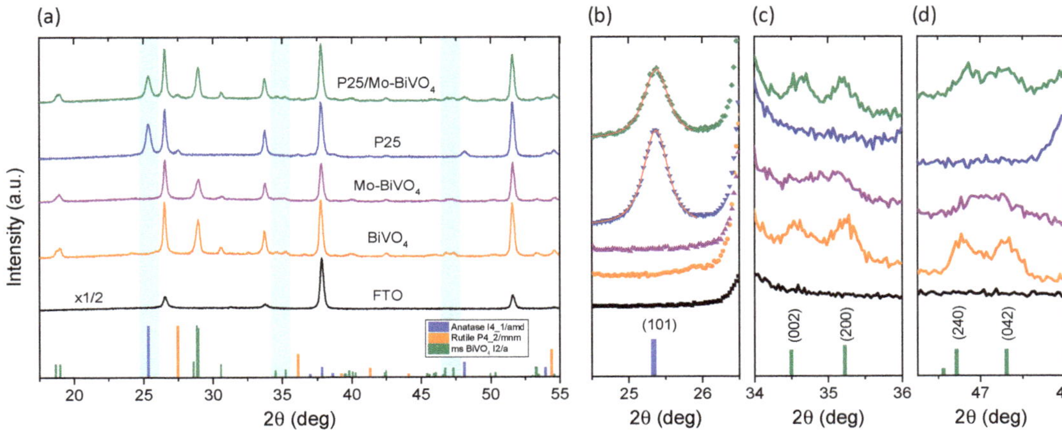

Figure 2. (**a**) XRD patterns of the P25/Mo-$BiVO_4$ bilayer in comparison to the individual Mo-$BiVO_4$ and P25 layer, the undoped $BiVO_4$ PC film, and the FTO substrate. Shaded bands indicate the 2θ regions where the variations of the (**b**) main (101) diffraction peak of the anatase TiO_2 phase of P25, and the splitting of the (**c**) (002)/(200) and (**d**) (240)/(042) diffraction peaks of the *ms* $BiVO_4$ phase, are shown in detail. Red solid lines in (**b**) depict the best fits of the (101) peaks to Voight profiles with full widths at half-maximum of 0.40 and 0.41 deg for the P25 and P25/Mo-$BiVO_4$ bilayer films.

The phase composition of the individual layers and the bilayer structure was further examined by micro-Raman spectroscopy at 785 and 532 nm excitations (Figure 3a,b). The individual $BiVO_4$ films exhibited all the characteristic Raman-active modes of the *ms* phase, without any evidence of polymeric residues or other $BiVO_4$ polymorphic phases under both excitations. Notably, the strong symmetric (ν_s) V-O stretching mode appeared at 828 cm^{-1}, along with the much weaker antisymmetric (ν_{as}) band (~712 cm^{-1}), which was more clearly observed at 532 nm (2.33 eV). The latter energy approaches the band gap of $BiVO_4$ (*vide infra*) leading to enhanced Raman scattering due to near-resonant excitation. The *ms* $BiVO_4$ formation was confirmed by the observation of two distinct bands at 369 and 330 cm^{-1}, corresponding to the symmetric (δ_s) and antisymmetric (δ_{as}) bending modes of the VO_4 tetrahedra, as well as two intense external lattice modes at 213 and 129 cm^{-1} [42,56,57]. The effect of Mo-doping was evident through the shifts of the Raman modes compared to the pristine $BiVO_4$ films. The ν_s and ν_{as} stretching modes, as well as the δ_s and δ_{as} bending ones, shifted closer together in Mo-$BiVO_4$, indicating a reduction in the difference between the two V-O bond lengths and a decrease in the VO_4 tetrahedral deformation of the *ms* phase after Mo-doping, in accordance with previous works [42,55,57]. In addition, a weak shoulder appeared around 880 cm^{-1} after Mo-doping, associated with the MoO_4

stretching vibration [20,55]. A reentrant behavior of the Mo-$BiVO_4$ Raman modes towards those of undoped $BiVO_4$ was observed in the P25/Mo-$BiVO_4$ bilayer, especially under near-resonant excitation at 532 nm, which enhanced the Mo-$BiVO_4$ Raman intensity relative to the TiO_2 top layer's Raman modes (Figure 3b). This suggests that post-annealing during P25 layer deposition improved the crystallinity of the Mo-$BiVO_4$ underlayer consistently with the XRD results. The individual P25 layers exhibited the characteristic Raman-active phonons of the major anatase TiO_2 phase at approximately 144 (E_g), 196 (E_g), 397 (B_{1g}), 517 (A_{1g} + B_{1g}), and 638 cm^{-1} (E_g), while the most intense Raman modes of the minor rutile TiO_2 phase were traced around 445 and 610 cm^{-1} (not discernible in the intensity scale of Figure 3a,b) [58]. In this case, depositing the P25 layer on the Mo-$BiVO_4$ substrate did not significantly affect the most intense, low-frequency anatase E_g mode, except for a slight broadening (left insets of Figure 3a,b), in contrast to the distinctive frequency shifts observed for the *ms* Raman modes due to the improved crystallinity of the Mo-$BiVO_4$ substrate in the bilayer structure. This implies that any modification of the P25 layer in the bilayer structure is confined at the TiO_2/Mo-$BiVO_4$ interface, which contributes minimally to the Raman scattering volume.

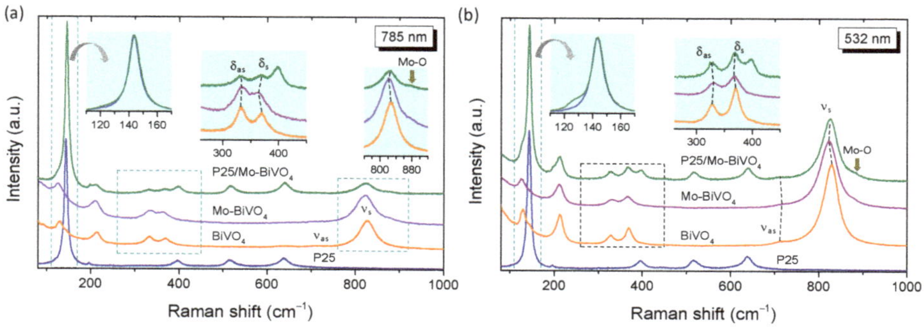

Figure 3. Raman spectra of the P25/Mo-$BiVO_4$ bilayer in comparison to the single $BiVO_4$, Mo-$BiVO_4$ and P25 layers at (**a**) 785 and (**b**) 532 nm excitation wavelengths. Dashed boxes delineate the spectral regions that are shown in detail in the shaded insets. In the left inset for each laser excitation, the anatase E_g modes of the P25 and P25/Mo-$BiVO_4$ films are compared after background removal at normalized intensity scale. Arrows indicate the weak shoulder that is associated with the Mo-O stretching vibration for the Mo-$BiVO_4$ layers.

Figure 4 shows the diffuse reflectance (DR%) and transmittance (T%) spectra of the P25/Mo-$BiVO_4$ bilayer films compared to the individual layers. Both DR% and T% measurements showed the superposition of the absorption edges of the two constituent Mo-$BiVO_4$ and TiO_2 metal oxides, which were clearly resolved in the corresponding absorption spectra obtained from the Kubelka–Munk transforms of the DR% spectra. The derived Tauc plots for indirect band gaps indicated E_g values of 3.28 and 2.59 eV for the individual P25 and Mo-$BiVO_4$ layers, respectively (Figure 4d). The corresponding E_g values for the P25/Mo-$BiVO_4$ bilayer indicated a slight narrowing of E_g to 2.56 eV for Mo-$BiVO_4$ that may be associated with the post-thermal annealing and the enhanced crystallinity in contrast to the identical value for the P25 upper layer. Comparison was also made with nanocrystalline Mo-$BiVO_4$ planar films and the corresponding bilayers, which were deposited on FTO glass substrates using the same procedure without the colloidal template, in order to validate the role of the inverse opal structure in the films' performance. These reference films showed qualitatively similar optical spectra though with considerably reduced DR% and enhanced T% arising from the markedly decreased DR% for the planar Mo-$BiVO_4$ underlayer in comparison to the inverse opal supporting layer (Figure 4a), indicative of lower surface roughness and porosity.

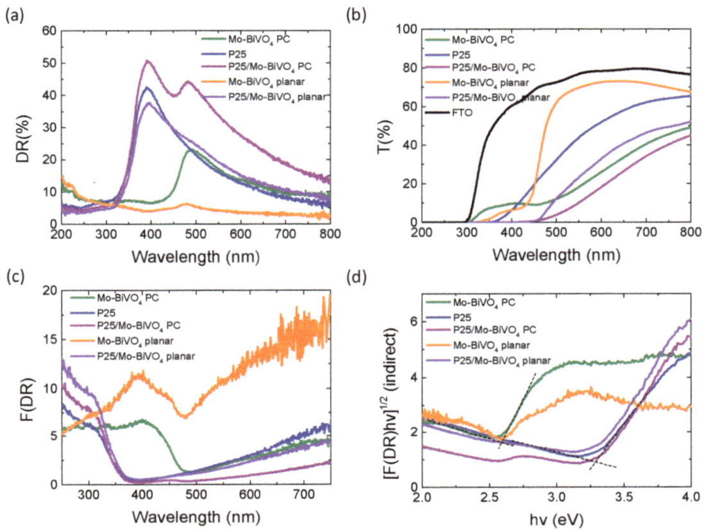

Figure 4. (**a**) Diffuse reflectance (DR%), (**b**) transmittance (T%) and the corresponding (**c**) Kubelka–Munk absorption, and (**d**) Tauc plots of the individual layers and the P25/Mo-$BiVO_4$ bilayers.

3.2. PEC Properties

The PEC performance of the bilayer structures and the individual layers was subsequently investigated by photocurrent density-potential curves in 0.5 M $NaHCO_3$ aqueous electrolyte under visible and UV–Vis back-side illumination (Figure 5a,b). The Mo-$BiVO_4$ films presented increased photocurrent density compared to plain $BiVO_4$, confirming the improved charge transport/separation efficiency arising from Mo-doping [2,3,20,42]. The P25/Mo-$BiVO_4$ bilayer showed much higher photocurrent density than the reverse Mo-$BiVO_4$/P25 bilayer structure under both illumination conditions, indicating greatly improved charge separation efficiency. On the other hand, charge transport and collection efficiency were greatly obstructed when the P25 layer was deposited directly on the FTO substrate, even under UV–Vis illumination, reflecting its poor PEC performance for water oxidation compared to Mo-$BiVO_4$ photoanodes [2].

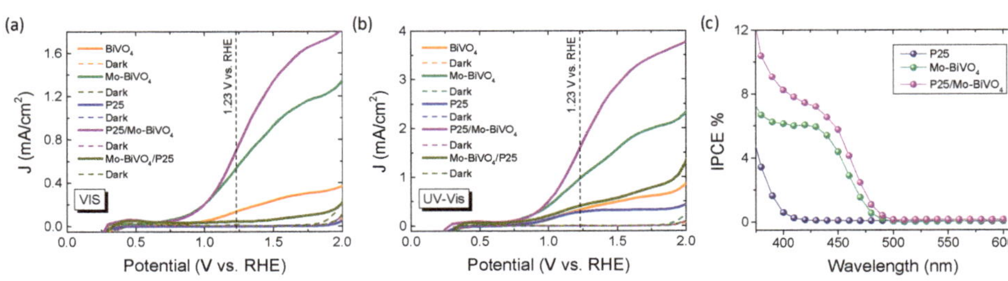

Figure 5. Photocurrent density-potential (J-V) curves for the P25/Mo-$BiVO_4$ and Mo-$BiVO_4$/P25 films in comparison to the individual layers and $BiVO_4$ reference under (**a**) visible and (**b**) UV–Vis back-side illumination in 0.5 M $NaHCO_3$ aqueous electrolyte. (**c**) IPCE spectra at 1.23 V vs. RHE potential for the P25/Mo-$BiVO_4$ bilayer with respect to the P25 and Mo-$BiVO_4$ constituent layers.

The photocurrent density of the Mo-$BiVO_4$/P25 bilayer exceeded the corresponding sums of the individual Mo-$BiVO_4$ and P25 layers by about 31% under both visible and UV–Vis light, indicating improved PEC performance by enhanced light capture and/or charge separation. The former was corroborated by IPCE measurements that showed appreciable enhancement for the P25/Mo-$BiVO_4$ bilayer films, especially below the absorption edge (490 nm) of Mo-$BiVO_4$ down to about 440 nm (Figure 5c). This suggests improved light harvesting by the Mo-$BiVO_4$ layer from the backscattering of light that reached the mesoporous P25 film, which presents significant diffuse reflectance in this spectral range (Figure 4a).

Charge transfer between the metal oxide layers was investigated by Mott–Schottky measurements, which were used for the estimation of the flat band potentials V_{fb} and donor densities N_D of the bilayer film with respect to its individual constituents (Figure 6). The Mo-$BiVO_4$ and P25 films as well as the bilayer P25/Mo-$BiVO_4$ presented positive slopes, indicative of n-type semiconducting behavior with considerably higher donor density N_D for the n-doped Mo-$BiVO_4$ films [42]. The V_{fb} values for the Mo-$BiVO_4$, P25 layers, and P25/Mo-$BiVO_4$ bilayer that were determined from the intercept of the Mott–Schottky plots with the potential axis indicated significant offsets between the individual Mo-$BiVO_4$ and P25 films. Specifically, the obtained V_{fb}, which provide an estimate of the corresponding Fermi level positions, indicate considerable downshift of the energy band positions for Mo-$BiVO_4$ with respect to that of the TiO_2 top layer [59], leading to a staggered alignment between their conduction (CB) and valence (VB) bands before contact, as shown by the tentative band diagram of Figure 6e. When the two layers come into contact, their Fermi levels will align, close to that of Mo-$BiVO_4$ whose V_{fb} is not appreciably shifted in the corresponding Mott–Schottky plots. The formation of a type I heterojunction (straddling band alignment) would be accordingly expected (Figure 6e), where both photogenerated electrons and holes will accumulate in the Mo-$BiVO_4$ layer, which, however, is not favorable for the spatial charge separation between the two metal oxides. In that case, it is very likely that the formation of intermediate and/or defect states at the Mo-$BiVO_4$/TiO_2 interface could mediate hole transfer from $BiVO_4$ to TiO_2 despite the energy barrier between their VBs, as previously reported [3,60].

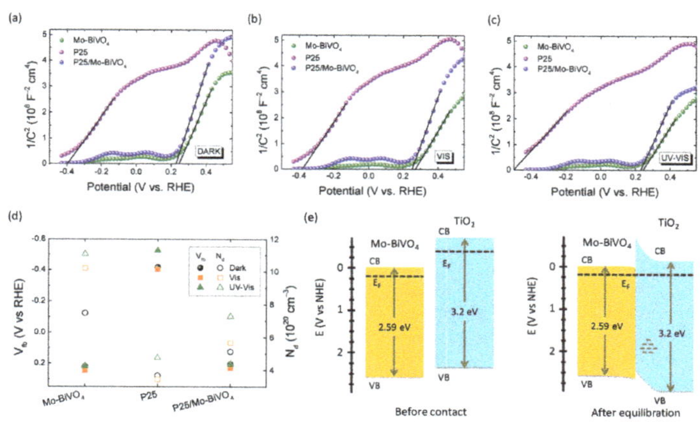

Figure 6. Mott–Schottky plots for the P25/Mo-$BiVO_4$ bilayer in comparison to the Mo-$BiVO_4$ and P25 individual films under (**a**) dark, (**b**) visible, and (**c**) UV–Vis light illumination in 0.5 M $NaHCO_3$ aqueous electrolyte. (**d**) The derived values of V_{fb} and N_D and (**e**) tentative band alignment for the Mo-$BiVO_4$/TiO_2 heterojunction before and after Fermi level equilibration.

In order to investigate charge carrier recombination, EIS experiments were performed under visible and UV–Vis light as well in the dark for the P25/Mo-$BiVO_4$ bilayer compared to the individual layers (Figure 7). The Nyquist plots were fitted to a Randles equivalent circuit consisting of a series resistance R_S, the charge transfer resistance R_{CT} at the electrode/electrolyte interface, and a constant phase element CPE (Table 1). Under light irradiation, the films presented lower R_{CT} compared to dark conditions due to the generation of electron-hole pairs. Visible light irradiation resulted in markedly reduced R_{CT} values for the Mo-$BiVO_4$ and P25/Mo-$BiVO_4$ bilayer in comparison to the P25 films.

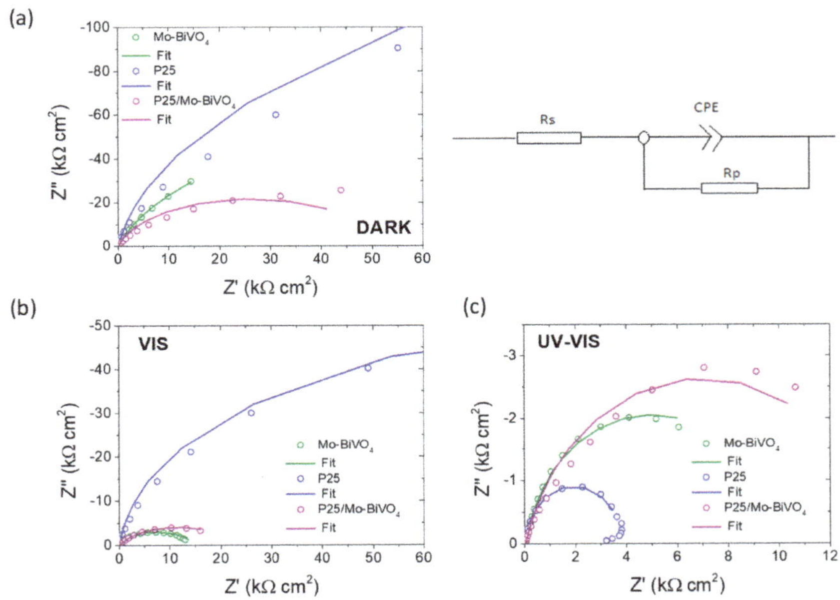

Figure 7. EIS Nyquist plots for the P25/Mo-$BiVO_4$ bilayer in comparison to Mo-$BiVO_4$ and P25 films under (**a**) dark, (**b**) visible, and (**c**) UV–Vis back-side illumination in 0.5 M $NaHCO_3$ aqueous electrolyte. Black lines denote the best fit curves to the Randles circuit shown in (**a**).

Table 1. Parameters obtained by fitting the experimental impedance spectra to a modified Randles circuit for the P25/Mo-$BiVO_4$ bilayer in comparison to Mo-$BiVO_4$ and P25 films.

	Dark		Visible Light (>400 nm)		UV–Vis Light	
Film	R_S (Ω)	R_{CT} (Ω)	R_S (Ω)	R_{CT} (Ω)	R_S (Ω)	R_{CT} (Ω)
Mo-$BiVO_4$	11.9	2.23×10^5	13.1	7292	12.1	4974
P25	9.0	4.62×10^5	20.6	1×10^5	9.1	1909
P25/Mo-$BiVO_4$	4.8	26,034	9.4	11,498	9.4	6979

This variation can be associated with the weak visible light absorption of the mixed-titania phase of P25 nanocatalysts consisting of (10–20%) rutile nanocrystals with a band gap of ~3.0 eV besides the major (80–90%) anatase phase with smaller (20 nm) nanoparticles and a larger band gap of 3.2 eV [54]. Nevertheless, the charge transfer resistance for the single P25 layer was greatly reduced under UV–Vis light in comparison to the Mo-$BiVO_4$

and P25/Mo-$BiVO_4$ bilayer, indicating that under these illumination conditions at open-circuit, charge separation and transport are not major obstacles in the PEC performance. Moreover, the R_{CT} values of the P25/Mo-$BiVO_4$ bilayer do not vary much with respect to the illumination conditions and the single Mo-$BiVO_4$ layer corroborating the proposed charge transfer scheme.

3.3. Photocatalytic Performance

Advanced application of the P25/Mo-$BiVO_4$ bilayer films was performed on the photoelectrocatalytic degradation of the pharmaceutical emerging contaminant ibuprofen (IBU), a non-steroidal anti-inflammatory drug that is widely used as an analgesic and antipyretic [61]. It is the third consumable drug and highly detected in many water bodies [62]. It has been shown that IBU is selectively metabolized in humans and animals and the excretion, which is more toxic than the parent molecule, enters into rivers, lakes, oceans, ground water, etc. Initial photocatalytic tests were performed on IBU degradation in 0.1 M $NaHCO_3$ supporting electrolyte using the bare P25 films at zero bias potential, +0.5 and +1.0 V vs. Ag/AgCl under UV–Vis light (Figure 8). Dark adsorption of IBU on the P25 films was about 5%, while UV–Vis illumination resulted in the continuous decrease in the pollutant concentration (*C*) with time (Figure 8a), which was detected spectrophotometrically by monitoring the main aromatic absorbance band of IBU at 222 nm [63]. The $\ln(C/C_0)$ vs. t plots, where C_0 is the IBU concentration after dark adsorption, varied linearly with time, indicative of pseudo first-order kinetics (Figure 8b). The values of the kinetic constants k obtained from the corresponding linear slopes increased significantly with the applied potential (Figure 8b) that promoted electron-hole separation and pollutant degradation.

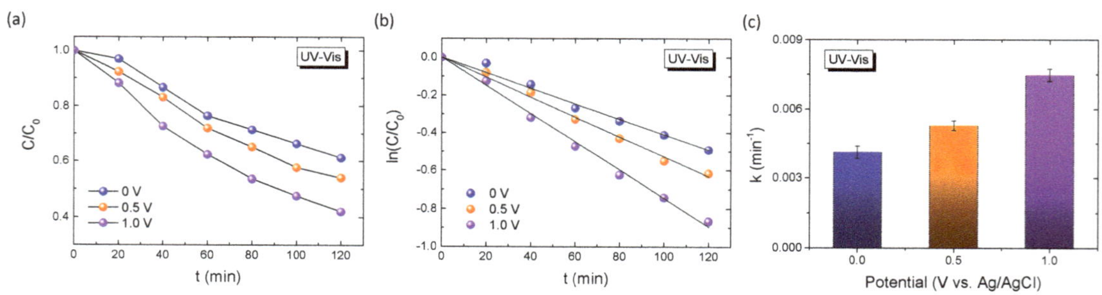

Figure 8. (**a**,**b**) IBU photoelectrocatalytic degradation kinetics and (**c**) corresponding kinetic constants for the P25 films in 0.1 M $NaHCO_3$ supporting electrolyte at bias potentials of 0, +0.5, and +1.0 V vs. Ag/AgCl under UV–Vis illumination.

IBU photodegradation was accordingly performed in 0.1 M $NaHCO_3$ supporting electrolyte at applied potential +1.0 V vs. Ag/AgCl under visible and UV–Vis light. The reaction kinetics monitored by the corresponding IBU absorbance spectra (Figure 9a) followed pseudo first-order kinetics under both illumination conditions (Figure 9b–e). The photocatalytic reaction rates r_{Vis} and $r_{UV\text{-}Vis}$ were calculated as $r = kC_0$, independent of differences in the initial concentration C_0 after dark adsorption. The degradation of IBU was accompanied by the appearance of a new red-shifted absorption band at about 257 nm, indicative of the formation of reaction intermediates, the most prominent being 4-isobutylacetophenone [64]. The corresponding band reached maximum intensity after 80 min of photocatalytic treatment under UV–Vis light (Figure 9a), while illumination for longer times resulted in its continuous reduction signaling the side group oxidation and oxidative opening of the IBU benzene ring towards mineralization [63,64]. The P25 films showed rather weak reactivity due to the inherently low visible light absorbance of both anatase and rutile TiO_2 polymorphic phases [65], which increased considerably under UV–Vis light causing electron-hole excitation in P25 titania. The pristine Mo-$BiVO_4$ photoelectrodes showed smaller differences in their photocatalytic activity under visible and UV–Vis light, reflecting their intrinsic visible light activation below 490 nm. More importantly, the P25/Mo-$BiVO_4$ bilayers presented the highest performance that exceeded the sum of the reaction rate for the individual P25 and Mo-$BiVO_4$ layers by approximately 30 and 28% under visible and UV–Vis light, respectively.

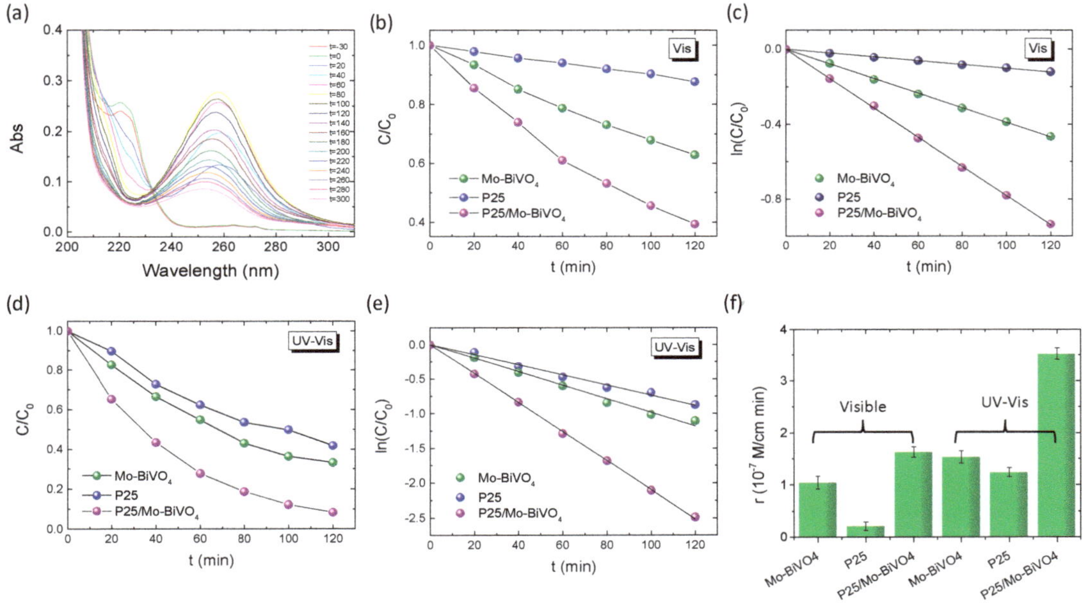

Figure 9. (**a**) IBU absorbance spectra under UV–Vis illumination in the presence of P25/Mo-$BiVO_4$. IBU photodegradation kinetics under (**b**,**c**) visible and (**d**,**e**) UV–Vis light as well as the corresponding (**f**) reaction rates for the individual constituents and the P25/Mo-$BiVO_4$ bilayer photoelectrodes at +1.0 V vs. Ag/AgCl.

In order to validate the bilayer performance, IBU photodegradation was also monitored by HPLC for the P25/Mo-$BiVO_4$ bilayer photoelectrode under UV–Vis light for 5 h (Figure 10). The total run was 15 min and the retention time was 6.3 min for IBU. At 5.4 min, a degradation product of IBU was detected after 1 h, which according to the corresponding UV–Vis spectra (Figure 10a,b) may arise from 4-isobutylacetophenone [64].

The obtained results showed that IBU was completely removed after 2 h (Figure 10e), whereas the degradation of the intermediate reaction product that appeared after 1 h (Figure 10d) reached 94% after 5 h (Figure 10f) under UV–Vis light, supporting the high photocatalytic activity of the bilayer films. Further analytical studies by mass spectrometry techniques would be required, though, in order to identify the transformation products and the reaction pathways during IBU degradation and mineralization [64,66]. Performance comparison with literature results on the photoelectrocatalytic IBU degradation shows that the obtained kinetic constants for the P25/Mo-$BiVO_4$ films are comparable to those reported by highly efficient, optimized metal oxide photoanodes (Table 2).

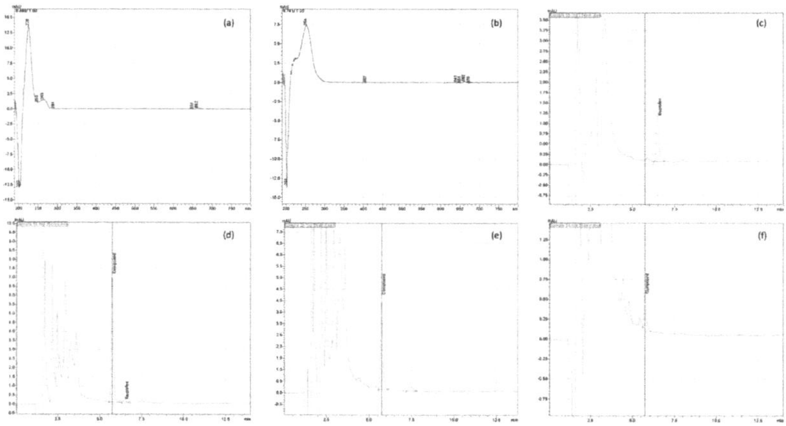

Figure 10. Absorbance spectra of (**a**) IBU and (**b**) the degradation product. HPLC chromatograms of IBU in the presence of P25/Mo-$BiVO_4$ films under UV–Vis light for (**c**) 0, (**d**) 1, (**e**) 2, and (**f**) 5 h.

Table 2. Performance comparison on IBU photoelectrocatalytic degradation by highly efficient metal oxide photoanodes.

Photoanode	Reactor Cell	Light Source	IBU Concentration	% Degradation	Kinetic Constant	Ref.
BiOI-TiO_2 nanotube arrays	Double compartment cell, three electrode configuration, 1.2 V vs. Ag/AgCl	100 W Hg lamp	5 mg/L	100% in 2 h	0.0321 min^{-1}	[67]
g-C_3N_4/Ag/AgCl/$BiVO_4$	Single compartment cell, three electrode configuration, 1.23 V vs. RHE, sewage water	150 W solar simulator	2 ppm	–	4.33% h^{-1} cm^{-2}	[26]
WO_3/$BiVO_4$	Double compartment cell, three electrode configuration, 1.75 V vs. RHE, 1 M NaCl	Solar simulator 350–1800 nm	100 mg/L	96.7% in 2.5 h	–	[25]
(Ag, Au)/Mo-$BiVO_4$	Single compartment cell, three electrode configuration, 1.0 V vs. Ag/AgCl, 0.1 M $NaHCO_3$	300 W Xe lamp >400 nm 100 mW/cm^2	7 mg/L	85% in 4 h	0.0112 min^{-1}	[43]
P25/Mo-$BiVO_4$	Single compartment cell, three electrode configuration, 1.0 V vs. Ag/AgCl, 0.1 M $NaHCO_3$	300 W Xe lamp 100 mW/cm^2	5 mg/L	100% in 2 h	0.0212 min^{-1}	This work

The stability of the P25/Mo-$BiVO_4$ bilayer photocatalyst was evaluated by performing three consecutive IBU photoelectrocatalytic degradation tests under UV–Vis light using the same film. The film was subjected to intermediate photocatalytic cleaning of pollutant residues under UV–Vis illumination for 3 h in DI water (Figure 11a). The derived kinetic constants decreased successively from 0.0210(2), to 0.0202(2) and 0.0193(2) min^{-1}, after the 2nd and 3rd run, respectively, indicating substantial stability and resistance to photocorrosion given the lack of any co-catalyst or surface treatment of the bilayer photoanode [25]. This was corroborated by Raman measurements of the bilayer film before and after the 3rd photocatalytic test, which showed no sign of film degradation (Figure 11b). Further application and validation of the P25/Mo-$BiVO_4$ bilayer photoanodes can be accordingly envisaged in the photoelectrocatalytic degradation of other recalcitrant nonsteroidal anti-inflammatory drugs for water purification, which can be effectively combined with simultaneous hydrogen generation at the cathode under solar light.

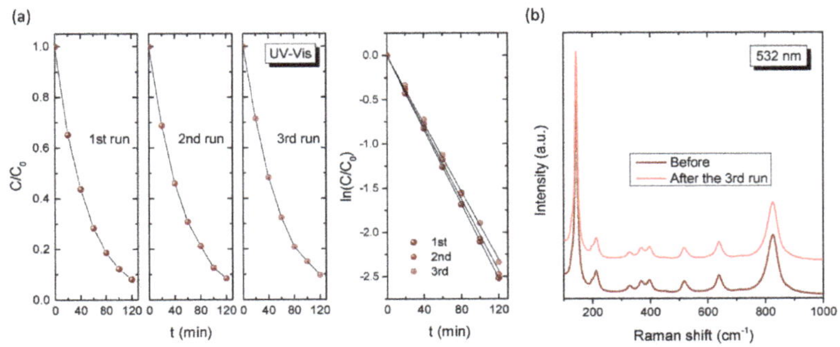

Figure 11. (**a**) Three consecutive IBU photoelectrocatalytic degradation tests using the same P25/Mo-$BiVO_4$ bilayer films under UV–Vis light at +1.0 V vs. Ag/AgCl. (**b**) Raman spectra of the bilayer film before and after the third photocatalytic run at 532 nm.

4. Conclusions

Advanced bilayer photoelectrodes were fabricated utilizing the highly performing Mo-$BiVO_4$ PC films as a supporting layer that absorbs visible light below 490 nm, while providing improved charge transport by means of Mo-doping. Spin coating of a mesoporous P25 titania overlayer resulted in a porous heterostructured P25/Mo-$BiVO_4$ bilayer structure produced from the densification of the inverse opal underlayer after the second thermal annealing. The bilayer film performance was evaluated on the photoelectrocatalytic degradation of the IBU anti-inflammatory drug, a highly refractory water pollutant of emerging concern, under back-side illumination by visible and UV–Vis light. Significantly enhanced photocatalytic performance was achieved for the bilayer structure with respect to the additive effect of the constituent layers, which, according to optical and PEC measurements, can be associated with the improved light harvesting due to backscattering from the P25 TiO_2 layer in combination with the favorable interfacial charge transfer at the TiO_2/Mo-$BiVO_4$ heterojunction.

Author Contributions: Conceptualization, V.L.; methodology, M.P. and V.L.; investigation, M.P., E.S., M.G.K., V.P., S.G. and N.S.T.; validation, M.G.K. and N.S.T.; formal analysis, M.P., E.S., V.P. and S.G.; resources, V.L. and N.S.T.; writing—original draft preparation, M.P. and V.L.; writing—review and editing, V.L. and M.P.; supervision, V.L. All authors have read and agreed to the published version of the manuscript.

Funding: This research work was supported by the Hellenic Foundation for Research and Innovation (H.F.R.I.) under the "First Call for H.F.R.I. Research Projects to support Faculty members and Researchers and the procurement of high-cost research equipment grant" (Project Number: 543).

Institutional Review Board Statement: Not applicable.

Informed Consent Statement: Not applicable.

Data Availability Statement: The original contributions presented in this study are included in the article. Further inquiries can be directed to the corresponding author.

Conflicts of Interest: The authors declare no conflicts of interest.

References

1. Sivula, K.; van de Krol, R. Semiconducting materials for photoelectrochemical energy conversion. *Nat. Rev. Mater.* **2016**, *1*, 15010. [CrossRef]

2. Kim, J.H.; Lee, J.S. Elaborately modified $BiVO_4$ photoanodes for solar water splitting. *Adv. Mater.* **2019**, *31*, 1806938. [CrossRef] [PubMed]
3. Bae, D.; Seger, B.; Vesborg, P.C.K.; Hansen, O.; Chorkendorff, I. Strategies for stable water splitting via protected photoelectrodes. *Chem. Soc. Rev.* **2017**, *46*, 1933–1954. [CrossRef]
4. Lu, Y.; Yang, Y.; Fan, X.; Li, Y.; Zhou, D.; Cai, B.; Wang, L.; Fan, K.; Zhang, K. Boosting charge transport in $BiVO_4$ photoanode for solar water oxidation. *Adv. Mater.* **2022**, *34*, 2108178. [CrossRef]
5. Liu, B.; Wang, X.; Zhang, Y.; Xu, L.; Wang, T.; Xiao, X.; Wang, S.; Wang, L.; Huang, W. A $BiVO_4$ photoanode with a VO_x layer bearing oxygen vacancies offers improved charge transfer and oxygen evolution kinetics in photoelectrochemical water splitting. *Angew. Chem. Int. Ed.* **2023**, *62*, e202217346. [CrossRef] [PubMed]
6. Salih, A.K.; Khan, A.Z.; Drmosh, Q.; Kandiel, T.; Qamar, M.; Jahangir, T.; Ton-That, C.; Yamani, Z. Nanostructured $BiVO_4$ photoanodes fabricated by vanadium-infused interaction for efficient solar water splitting. *ACS Appl. Nano Mater.* **2024**, *7*, 14115–14122. [CrossRef]
7. Gaikwad, M.A.; Suryawanshi, U.P.; Ghorpade, U.V.; Jang, J.S.; Suryawanshi, M.P.; Kim, J.H. Emerging surface, bulk, and interface engineering strategies on $BiVO_4$ for photoelectrochemical water splitting. *Small* **2022**, *18*, 2105084. [CrossRef] [PubMed]
8. Cooper, J.K.; Gul, S.; Toma, F.M.; Chen, L.; Liu, Y.-S.; Guo, J.; Ager, J.W.; Yano, J.; Sharp, I.D. Indirect bandgap and optical properties of monoclinic bismuth vanadate. *J. Phys. Chem. C* **2015**, *119*, 2969–2974. [CrossRef]
9. Castillo, N.C.; Ding, L.; Heel, A.; Graule, T.; Pulgarin, C. On the photocatalytic degradation of phenol and dichloroacetate by $BiVO_4$: The need of a sacrificial electron acceptor. *J. Photochem. Photobiol. A* **2010**, *216*, 221–227. [CrossRef]
10. Saison, T.; Chemin, N.; Chanéac, C.; Durupthy, O.; Mariey, L.; Maugé, F.; Brezová, V.; Jolivet, J.-P. New insights into $BiVO_4$ properties as visible light photocatalyst. *J. Phys. Chem. C* **2015**, *119*, 12967–12977. [CrossRef]
11. Malathi, A.; Madhavan, J.; Ashokkumar, M.; Arunachalam, P. A review on $BiVO_4$ photocatalyst: Activity enhancement methods for solar photocatalytic applications. *Appl. Catal. A* **2018**, *555*, 47–74.
12. Lotfi, S.; Ouardi, M.E.; Ahsaine, H.A.; Assani, A. Recent progress on the synthesis, morphology and photocatalytic dye degradation of $BiVO_4$ photocatalysts: A review. *Catal. Rev.* **2022**, *66*, 214–258. [CrossRef]
13. Regmi, C.; Kshetri, Y.K.; Kim, T.H.; Pandey, R.P.; Ray, S.K.; Lee, S.W. Fabrication of Ni-doped $BiVO_4$ semiconductors with enhanced visible-light photocatalytic performances for wastewater treatment. *Appl. Surf. Sci.* **2017**, *413*, 253–265. [CrossRef]
14. Akbarzadeh, R.; Fung, C.S.L.; Rather, R.A.; Lo, I.M.C. One-pot hydrothermal synthesis of $g\text{-}C_3N_4/Ag/AgCl/BiVO_4$ micro-flower composite for the visible light degradation of ibuprofen. *Chem. Eng. J.* **2018**, *341*, 248–261.
15. Regmi, C.; Kshetri, Y.K.; Pandey, R.P.; Kim, T.-H.; Gyawali, G.; Lee, S.W. Understanding the multifunctionality in Cu-doped $BiVO_4$ semiconductor photocatalyst. *J. Environ. Sci.* **2019**, *75*, 84–97.
16. Algethami, J.S.; Hassan, M.S.; Amna, T.; Alqarni, L.S.; Alhamami, M.A.; Seliem, A.F. Bismuth vanadate decked polyaniline polymeric nanocomposites: The robust photocatalytic destruction of microbial and chemical toxicants. *Materials* **2023**, *16*, 3314. [CrossRef]
17. Singla, S.; Pooja Devi, P.; Basu, S. Revolutionizing the role of solar light responsive $BiVO_4$/BiOBr heterojunction photocatalyst for the photocatalytic deterioration of tetracycline and photoelectrocatalytic water splitting. *Materials* **2023**, *16*, 5661. [CrossRef]
18. Orimolade, B.O.; Arotiba, O.A. Bismuth vanadate in photoelectrocatalytic water treatment systems for the degradation of organics: A review on recent trends. *J. Electroanal. Chem.* **2020**, *878*, 114724.
19. McMichael, S.; Fernández-Ibáñez, P.; Byrne, J.A. A review of photoelectrocatalytic reactors for water and wastewater treatment. *Water* **2021**, *13*, 1198. [CrossRef]
20. Zhang, W.; Wu, F.; Li, J.; Yan, D.; Tao, J.; Ping, Y.; Liu, M. Unconventional relation between charge transport and photocurrent via boosting small polaron hopping for photoelectrochemical water Splitting. *ACS Energy Lett.* **2018**, *3*, 2232–2239. [CrossRef]
21. Liu, J.; Li, J.; Li, Y.; Guo, J.; Xu, S.-M.; Zhang, R.; Shao, M. Photoelectrochemical water splitting coupled with degradation of organic pollutants enhanced by surface and interface engineering of $BiVO_4$ photoanode. *Appl. Catal. B Environ.* **2020**, *278*, 119268. [CrossRef]
22. Orimolade, B.O.; Arotiba, O.A. Towards visible light driven photoelectrocatalysis for water treatment: Application of a $FTO/BiVO_4/Ag_2S$ heterojunction anode for the removal of emerging pharmaceutical pollutants. *Sci. Rep.* **2020**, *10*, 5348. [CrossRef] [PubMed]
23. Prado, T.M.; Silva, F.L.; Grosseli, G.; Fadini, P.S.; Fatibello-Filho, O.; de Moraes, F.C. Using $BiVO_4$/CuO-based photoelectrocatalyzer for 4-nitrophenol degradation. *Materials* **2020**, *13*, 1322. [CrossRef]
24. Kahng, S.; Kim, J.H. Heterojunction photoanode of SnO_2 and Mo-doped $BiVO_4$ for boosting photoelectrochemical performance and tetracycline hydrochloride degradation. *Chemosphere* **2022**, *291*, 132800. [CrossRef] [PubMed]

25. Davies, K.R.; Allan, M.G.; Nagarajan, S.; Townsend, R.; Dunlop, T.; McGettrick, J.D.; Asokan, V.S.; Ananthraj, S.; Watson, T.; Godfrey, A.R.; et al. Solar light-driven simultaneous pharmaceutical pollutant degradation and green hydrogen production using a mesoporous nanoscale $WO_3/BiVO_4$ heterostructure photoanode. *J. Environ. Chem. Eng.* **2023**, *11*, 110256. [CrossRef]
26. Rather, R.A.; Lo, I.M.C. Photoelectrochemical sewage treatment by a multifunctional g-C_3N_4/Ag/AgCl/$BiVO_4$ photoanode for the simultaneous degradation of emerging pollutants and hydrogen production, and the disinfection of *E. coli*. *Water Res.* **2020**, *168*, 115166. [CrossRef] [PubMed]
27. Fan, H.; Yi, G.; Zhang, Z.; Zhang, X.; Li, P.; Zhang, C.; Chen, L.; Zhang, Y.; Sun, Q. Fabrication of Ag particles deposited $BiVO_4$ photoanode for significantly efficient visible-light driven photoelectrocatalytic degradation of β-naphthol. *J. Environ. Chem. Eng.* **2022**, *10*, 107221. [CrossRef]
28. Orimolade, B.O.; Arotiba, O.A. Enhanced photoelectrocatalytic degradation of diclofenac sodium using a system of Ag-$BiVO_4$/BiOI anode and Ag-BiOI cathode. *Sci. Rep.* **2022**, *12*, 4214. [CrossRef] [PubMed]
29. Nair, V.; Perkins, C.L.; Lin, Q.; Law, M. Textured nanoporous Mo:$BiVO_4$ photoanodes with high charge transport and charge transfer quantum efficiencies for oxygen evolution. *Energy Environ. Sci.* **2016**, *9*, 1412–1429. [CrossRef]
30. Mali, S.S.; Park, G.R.; Kim, H.; Kim, H.H.; Patil, J.V.; Hong, C.K. Synthesis of nanoporous Mo:$BiVO_4$ thin film photoanodes using the ultrasonic spray technique for visible-light water splitting. *Nanoscale Adv.* **2019**, *1*, 799–806. [CrossRef] [PubMed]
31. Bera, S.; Lee, S.A.; Lee, W.-J.; Kim, J.-H.; Kim, C.; Kim, H.G.; Khan, H.; Jana, S.; Jang, H.W.; Kwon, S.-H. Hierarchical nanoporous $BiVO_4$ photoanodes with high charge separation and transport efficiency for water oxidation. *ACS Appl. Mater. Interfaces* **2021**, *13*, 14291–14301. [CrossRef] [PubMed]
32. Shim, S.G.; Tan, J.; Lee, H.; Park, J.; Yun, J.; Park, Y.S.; Kim, K.; Lee, J.; Moon, J. Facile morphology control strategy to enhance charge separation efficiency of Mo:$BiVO_4$ photoanodes for efficient photoelectrochemical water splitting. *Chem. Eng. J.* **2022**, *430*, 133061. [CrossRef]
33. Chen, Z.Y.; Cheng, S.B.; Zhang, H.F.; Yi, Z.; Tang, B.; Chen, J.; Zhang, J.G.; Tang, C.J. Ultra wideband absorption absorber based on Dirac semimetallic and graphene metamaterials. *Phys. Lett. A* **2024**, *517*, 129675. [CrossRef]
34. Jiang, B.; Hou, Y.; Wu, J.; Ma, Y.; Gan, X.; Zhao, J. In-fiber photoelectric device based on graphene-coated tilted fiber grating. *Opto-Electron. Sci.* **2023**, *2*, 230012. [CrossRef]
35. Cheng, S.B.; Li, W.X.; Zhang, H.F.; Akhtar, M.N.; Yi, Z.; Zeng, Q.D.; Ma, C.; Sun, T.Y.; Wu, P.H.; Ahmad, S. High sensitivity five band tunable metamaterial absorption device based on block like Dirac semimetals. *Opt. Commun.* **2024**, *569*, 130816. [CrossRef]
36. Likodimos, V. Photonic crystal-assisted visible light activated TiO_2 photocatalysis. *Appl. Catal. B Environ.* **2018**, *230*, 269–303. [CrossRef]
37. Liu, Y.X.; Dai, H.X.; Deng, J.G.; Zhang, L.; Au, C.T. Three-dimensional ordered macroporous bismuth vanadates: PMMA-templating fabrication and excellent visible light-driven photocatalytic performance for phenol degradation. *Nanoscale* **2012**, *4*, 2317–2325. [CrossRef] [PubMed]
38. Zhang, K.; Liu, Y.; Deng, J.; Xie, S.; Lin, H.; Zhao, X.; Yang, J.; Han, Z.; Dai, H. Fe_2O_3/3DOM $BiVO_4$: High-performance photocatalysts for the visible light-driven degradation of 4-nitrophenol. *Appl. Catal. B Environ.* **2017**, *202*, 569–579. [CrossRef]
39. Zhou, M.; Bao, J.; Xu, Y.; Zhang, J.; Xie, J.; Guan, M.; Wang, C.; Wen, L.; Lei, Y.; Xie, Y. Photoelectrodes based upon Mo:$BiVO_4$ inverse opals for photoelectrochemical water splitting. *ACS Nano* **2014**, *8*, 7088–7098. [CrossRef] [PubMed]
40. Zhang, L.W.; Lin, C.Y.; Valev, V.K.; Reisner, E.; Steiner, U.; Baumberg, J.J. Plasmonic enhancement in $BiVO_4$ photonic crystals for efficient water splitting. *Small* **2014**, *10*, 3970–3978. [CrossRef]
41. Fang, L.; Nan, F.; Yang, Y.; Cao, D. Enhanced photoelectrochemical and photocatalytic activity in visible-light-driven Ag/$BiVO_4$ inverse opals. *Appl. Phys. Lett.* **2016**, *108*, 093902.
42. Pylarinou, M.; Sakellis, E.; Tsipas, P.; Romanos, G.E.; Gardelis, S.; Dimoulas, A.; Likodimos, V. Mo-$BiVO_4$/Ca-$BiVO_4$ homojunction nanostructure-based inverse opals for photoelectrocatalytic pharmaceutical degradation under visible light. *ACS Appl. Nano Mater.* **2023**, *6*, 6759–6771.
43. Pylarinou, M.; Sakellis, E.; Tsipas, P.; Gardelis, S.; Psycharis, V.; Dimoulas, A.; Stergiopoulos, T.; Likodimos, V. Light concentration and electron transfer in plasmonic–photonic Ag,Au modified Mo-$BiVO_4$ inverse opal photoelectrocatalysts. *Nanoscale* **2024**, *16*, 10366–10376.
44. Zalfani, M.; Van Der Schueren, B.; Hu, Z.Y.; Rooke, J.; Bourguiga, R.; Wu, M.; Li, Y.; Van Tendeloo, G.; Su, B.L. Novel 3DOM $BiVO_4$/TiO_2 nanocomposites for highly enhanced photocatalytic activity. *J. Mater. Chem. A* **2015**, *3*, 21244–21256.
45. Zalfani, M.; Hu, Z.-Y.; Yu, W.-B.; Mahdouani, M.; Bourguiga, R.; Wu, M.; Li, Y.; Van Tendeloo, G.; Djaoued, Y.; Su, B.-L. $BiVO_4$/3DOM TiO_2 nanocomposites: Effect of $BiVO_4$ as highly efficient visible light sensitizer for highly improved visible light photocatalytic activity in the degradation of dye pollutants. *Appl. Catal. B Environ.* **2017**, *205*, 121–132.
46. Taga, Y.; Pan, Z.; Katayama, K.; Sohn, W.Y. $BiVO_4$—Dotted WO_3 photoanode with an inverse opal underlayer for photoelectrochemical water splitting. *ACS Appl. Energy Mater.* **2022**, *5*, 5750–5755.

47. Ma, M.; Kim, J.K.; Zhang, K.; Shi, X.; Kim, S.J.; Moon, J.H.; Park, J.H. Double-deck inverse opal photoanodes: Efficient light absorption and charge separation in heterojunction. *Chem. Mater.* **2014**, *26*, 5592–5597.
48. Zhang, H.; Cheng, C. Three-Dimensional FTO/TiO_2/$BiVO_4$ composite inverse opals photoanode with excellent photoelectrochemical performance. *ACS Energy Lett.* **2017**, *2*, 813–821.
49. Madanu, T.L.; Mouchet, S.R.; Deparis, O.; Liu, J.; Li, Y.; Su, B.-L. Tuning and transferring slow photons from TiO_2 photonic crystals to $BiVO_4$ nanoparticles for unprecedented visible light photocatalysis. *J. Colloid Interface Sci.* **2023**, *634*, 290–299. [PubMed]
50. Madanu, T.L.; Chaabane, L.; Mouchet, S.R.; Deparis, O.; Su, B.-L. Manipulating multi-spectral slow photons in bilayer inverse opal TiO_2@$BiVO_4$ composites for highly enhanced visible light photocatalysis. *J. Colloid Interface Sci.* **2023**, *647*, 233–245.
51. Al Rammal, M.; Akkaoui, K.; Halaoui, L.I. Multijunction photoanode of Mo:$BiVO_4$ layered with TiO_2 inverse opal and NiBi oxygen evolution catalyst to trap light and enhance water splitting. *J. Phys. Chem. C* **2022**, *126*, 6960–6972. [CrossRef]
52. Yan, Q.; Zhou, Z.; Zhao, X.S. Inward-growing self-assembly of colloidal crystal films on horizontal substrates. *Langmuir* **2005**, *21*, 3158–3164. [CrossRef]
53. Antoniadou, M.; Bouras, P.; Strataki, N.; Lianos, P. Hydrogen and electricity generation by photoelectrochemical decomposition of ethanol over nanocrystalline titania. *Int. J. Hydrogen Energy* **2008**, *33*, 5045–5051. [CrossRef]
54. Likodimos, V.; Chrysi, A.; Calamiotou, M.; Fernández-Rodríguez, C.; Doña-Rodríguez, J.M.; Dionysiou, D.D.; Falaras, P. Microstructure and charge trapping assessment in highly reactive mixed phase TiO_2 photocatalysts. *Appl. Catal. B Environ.* **2016**, *192*, 242–252. [CrossRef]
55. Zhou, D.; Pang, L.-X.; Guo, J.; Wang, H.; Yao, X.; Randall, C. Phase evolution, phase transition, raman spectra, infrared spectra, and microwave dielectric properties of low temperature firing $(K_{0.5x}Bi_{1-0.5x})(Mo_xV_{1-x})O_4$ ceramics with scheelite related structure. *Inorg. Chem.* **2011**, *50*, 12733–12738.
56. Hardcastle, F.D.; Wachs, I.E.; Eckert, H.; Jefferson, D.A. Vanadium(V) Environments in Bismuth Vanadates: A structural investigation using Raman spectroscopy and solid state ^{51}V NMR. *J. Solid State Chem.* **1991**, *90*, 194–210. [CrossRef]
57. Pellicer-Porres, J.; Vázquez-Socorro, D.; López-Moreno, S.; Muñoz, A.; Rodríguez-Hernández, P.; Martínez-García, D.; Achary, S.N.; Rettie, A.J.E.; Buddie Mullins, C. Phase transition systematics in $BiVO_4$ by means of high-pressure–high-temperature Raman experiments. *Phys. Rev. B* **2018**, *98*, 214109. [CrossRef]
58. Papadakis, D.; Diamantopoulou, A.; Pantazopoulos, P.A.; Palles, D.; Sakellis, E.; Boukos, N.; Stefanou, N.; Likodimos, V. Nanographene oxide-TiO_2 photonic films as plasmon-free substrates for surface-enhanced Raman scattering. *Nanoscale* **2019**, *11*, 21542–21553. [CrossRef]
59. Mansfeldova, V.; Zlamalova, M.; Tarabkova, H.; Janda, P.; Vorokhta, M.; Piliai, L.; Kavan, L. Work function of TiO_2 (anatase, rutile, and brookite) single crystals: Effects of the environment. *J. Phys. Chem. C* **2021**, *125*, 1902–1912. [CrossRef]
60. Bae, D.; Kanellos, G.; Wedege, K.; Draževič, E.; Bentien, A.; Smith, A.W. Tailored energy level alignment at MoO_X/GaP interface for solar-driven redox flow battery application. *J. Chem. Phys.* **2020**, *152*, 124710. [CrossRef] [PubMed]
61. Buser, H.-R.; Poiger, T.; Müller, M.D. Occurrence and environmental behavior of the chiral pharmaceutical drug ibuprofen in surface waters and in wastewater. *Environ. Sci. Technol.* **1999**, *33*, 2529–2535. [CrossRef]
62. Huynh, N.C.; Nguyen, T.T.T.; Nguyen, D.T.C.; Tran, T.V. Occurrence, toxicity, impact and removal of selected non-steroidal anti-inflammatory drugs (NSAIDs): A review. *Sci. Total Environ.* **2023**, *898*, 165317. [CrossRef]
63. Choina, J.; Kosslick, H.; Fischer, C.; Flechsig, G.-U.; Frunza, L.; Schulz, A. Photocatalytic decomposition of pharmaceutical ibuprofen pollutions in water over titania catalyst. *Appl. Catal. B Environ.* **2013**, *129*, 589–598. [CrossRef]
64. Miranda, M.O.; Cavalcanti, W.E.C.; Barbosa, F.F.; de Sousa, J.A.; da Silva, F.I.; Pergher, S.B.C.; Braga, T.P. Photocatalytic degradation of ibuprofen using titanium oxide: Insights into the mechanism and preferential attack of radicals. *RSC Adv.* **2021**, *11*, 27720–27733. [CrossRef] [PubMed]
65. Khedr, T.M.; El-Sheikh, S.M.; Hakki, A.; Ismail, A.A.; Badawy, W.A.; Bahnemann, D.W. Highly active non-metals doped mixed-phase TiO_2 for photocatalytic oxidation of ibuprofen under visible light. *J. Photochem. Photobiol. A Chem.* **2017**, *346*, 530–540. [CrossRef]
66. Jallouli, N.; Pastrana-Martínez, L.M.; Ribeiro, A.R.; Moreira, N.F.F.; Faria, J.L.; Hentati, O.; Silva, A.M.T.; Ksibi, M. Heterogeneous photocatalytic degradation of ibuprofen in ultrapure water, municipal and pharmaceutical industry wastewaters using a TiO_2/UV-LED system. *Chem. Eng. J.* **2018**, *334*, 976–984. [CrossRef]
67. Chen, H.; Peng, Y.-P.; Chen, T.-Y.; Chen, K.-F.; Chang, K.-L.; Dang, Z.; Lu, G.-N.; He, H. Enhanced photoelectrochemical degradation of Ibuprofen and generation of hydrogen via BiOI-deposited TiO_2 nanotube arrays. *Sci. Total Environ.* **2018**, *633*, 1198–1205. [CrossRef]

materials

MDPI

Article

Topology Design of Soft Phononic Crystals for Tunable Band Gaps: A Deep Learning Approach

Jingru Li [1,*], Minqi Qian [1], Jingming Yin [1], Wei Lin [2], Zhifu Zhang [1] and Shihao Liu [1]

1 School of Mechanical and Electrical Engineering, Hainan University, Haikou 570228, China; q1961361499@163.com (M.Q.); 22220854060008@hainanu.edu.cn (J.Y.); jeff.zfzhang@foxmail.com (Z.Z.); liushihao1102@126.com (S.L.)

2 Qingdao Innovation and Development Center of Harbin Engineering University, Qingdao 266400, China; linwei23@hrbeu.edu.cn

* Correspondence: ljr@hainanu.edu.cn; Tel.: +86-13591769177

Abstract: The phononic crystals composed of soft materials have received extensive attention owing to the extraordinary behavior when undergoing large deformations, making it possible to provide tunable band gaps actively. However, the inverse designs of them mainly rely on the gradient-driven or gradient-free optimization schemes, which require sensitivity analysis or cause time-consuming, lacking intelligence and flexibility. To this end, a deep learning-based framework composed of a conditional variational autoencoder and multilayer perceptron is proposed to discover the mapping relation from the band gaps to the topology layout applied with prestress. The nonlinear superelastic neo-Hookean model is employed to describe the constitutive characteristics, based on which the band structures are obtained via the transfer matrix method accompanied with Bloch theory. The results show that the proposed data-driven approach can efficiently and rapidly generate multiple candidates applied with predicted prestress. The band gaps are in accord with each other and also consistent with the prescribed targets, verifying the accuracy and flexibility simultaneously. Furthermore, based on the generalization performance, the design space is deeply exploited to obtain desired soft structures whose stop bands are characterized by wider bandwidth, lower location, and enhanced wave attenuation performance.

Keywords: intelligent inverse design; deep learning; topology layout; soft phononic crystals; tunable band gaps

1. Introduction

Phononic crystals (PnCs) [1] or metamaterials (MMs), a typical class of periodic composites, have become one of the hottest research topics in the field of vibration and sound reduction owing to the inherent characteristics of band gaps. In general, they consist of two components; in detail, the component of pass band represents frequency ranges where the wave propagation is allowed, and the component of stop band denotes frequency intervals where the wave propagation is prohibited. The Bragg scattering (BS) [2] and local resonance (LR) [3] are the primary mechanisms for inducing band gaps, based on which increasingly elaborate designs of the composites relating to the geometries, materials, boundaries, and attached mechanisms have been explored to meet various requirements, such as multi-stop bands [4–7], low-frequency stop bands [8–11] and broad bandwidth [12–15]. This facilitates the developments in a wide range of engineering applications, including vibration isolators [16–18], wave filters [19–21], sound insulation [4,22,23], wave guides [24–26], and energy harvesting [27–29]. Motivated by the applicability of band gaps, the tunable band structures endowed with more flexibility to manipulate wave propagation and attenuation are expected to gain more significant scientific and technological value. Compared to ordinary materials, soft materials can undergo large deformation when subjected to external mechanical, electrical, magnetic, or thermal loads, leading to adjusting the constitutive behavior in an active manner, which lays a solid foundation for realizing tunable band gaps [30–36]. Especially, the technique of exerting mechanical prestress or prestretch is a stable tool to measure the external load and has been gradually applied to assemble soft PnCs for regulating wave performance. Based on the nonlinear elasticity theory [37], Huang et al. [38] studied the band structure of one-dimensional (1D) compressible hyperelastic PnCs and validated the efficiency to tune band gaps via applying longitudinal prestress. Afterward the authors [39] included the post-buckling deformations in the design of the PnC plate made of soft materials, where the resonant units were periodically displayed. The results demonstrated remarkable controllability of band gaps with small prestretch or extension that can be realized if strong resonance appears for some particular modes. Wu et al. [40] analyze the axisymmetric guided wave propagation in a pressurized functionally graded (FG) elastomeric hollow cylinder and find that material tailoring along with an adjustment of the pre-stretch and pressure difference can be used to control the propagation of elastic waves. Under the equibiaxial compression, a two-dimensional (2D) acoustic metamaterial was designed by Ning et al. [41], which was composed of resonating elements embedded in the elastomeric matrix. In this study, the numerical results indicated the locally resonant and Bragg scattering band gaps can be simultaneously controlled by the deformation. Wang et al. [42] also applied the soft materials to compose metamaterials for obtaining lower frequency band gaps and demonstrated that the proposed design is able to have a good performance in the design of low-frequency band gaps with large amplitude.

From previous studies, it can be concluded that the tunability of soft PnCs makes them promising candidates as adjustable components of adaptive metamaterials via integrating with other mechanisms, such as the local resonance and inertial amplification mechanisms. As the mentioned mechanisms have been already realized by the applications in practice [43–45], the adaptive metamaterials based on the soft PnCs can be further designed to isolate the vibration and noise of the automobile, pipe system, building, ship engineering, and so on. For instance, the soft PnCs can be conceived as the substrates of sandwich structures [46], whose large deformation is able to control the location of the inner cores. Thus, fulfilling the task of solving low-frequency vibration problems of the marine power system using the tunable band gaps. Also, the combination of different layouts of soft PnCs embedded with sophisticated cavities or local resonant elements is helpful in absorbing sound in water, promoting the engineering application in the field of anechoic coating [47]. As can be seen, the adaptive PnCs made of soft materials exhibit great potential and high efficiency in controlling band gaps. Also, intensive dependency can be observed from the external stimuli, material properties, geometrical parameters, and topological layout. Hence, in order to make full use of the improved tunability level, the design space of soft PnCs should be deeply exploited by the method of the inverse design—a one-fits-all approach with the function of offering solutions with desired performances [48]. Using this approach, on-demand designs can be achieved with respect to prescribed tunable band gaps, which is also a main focus of the present work.

The optimization technique is regarded as a prominent tool to perform the task of inverse design. In recent years, considerable efforts have been devoted to searching for the optimal design of PnCs and MMs to acquire target properties of band gaps using gradient-based or gradient-free optimization algorithms. The size optimization [49], the shape optimization [50], and especially the topology optimization (TO) [51–60], is able to explore the periodic space to a full extent, have drawn intensive interest and been conducted in combination with the band structure analysis methods. Nevertheless, the design space is required to be updated at each iteration during the optimization procedure, which makes the simulations conducted for calculating band gaps recurrent and time-demanding. Also, if the objective is varied, researchers need to restart the entire framework of the optimization project to find the new optimal; that is, an iterative optimization process is required to be completed again. To eliminate the above limitations, the data-driven method that is able to derive the mapping from band gaps to configurations of PnCs in a real-time manner is indispensable. To this end, deep learning (DL), a representative technique of machine learning (ML), which originated from the artificial neural network (ANN), is able to fulfill this task. The term "deep learning" was first proposed by Hinton et al. [61], and remarkable achievements have been obtained in the fields of computer vision [62], natural language processing [63,64], speech recognition [65], decision-making [66], and so on [67] by this flourishing method. In line with the research stream, the DL gradually progressed into the inverse design of artificial structures through two patterns of frameworks: in one approach, the DL technique is integrated with optimization algorithms to optimize the hyperparameters of neural networks instead of manual setting or function as the surrogate model to improve the efficiency of optimal searching [68–73]; in another approach, the DL model is established to produce the sophisticated man-made structures directly with respect to given targets [48,74–78]. The latter approach definitely exhibits more flexibility and instantaneity, but owing to the "one-to-many" phenomenon, uncertain problems appear as a result of solving the inverse problems of PnCs and MMs, as different unit cells can share the same band gaps.

To address the above issue, the DL models designed in an unsupervised learning way, such as the tandem neural network (TNN) [79], the autoencoder (AE) [80], and the generative adversarial network (GAN) [81], are proposed to eliminate the negative effects and have been successfully used in generating feasible PnCs to match target features of band gaps. Through training, the TNN composed of multilayer perceptron (MLP), 1D phononic slabs can be obtained automatically within prescribed band gaps [82]. Then Li et al. [83] focused on the flexural wave attenuation at the nanoscale and combined TNN with the probabilistic distribution to complete the inverse design of phononic nanobeams made up of functionally graded materials, where the results demonstrated that the well-trained DL model can provide multiple feasible designs that possess the same man-made band gaps. The MLP was also directly applied to complete the inverse design of structural parameters of the Y-shaped core sandwich beam for given transmission characteristics based on the band gaps, where the average relative error of outputted structural parameters is below 3% [84]. Lee et al. [78] employed the deep neural network (DNN), TNN, conditional variation autoencoder (CVAE), and conditional GAN (CGAN), respectively, to implement the inverse design framework for a one-dimensional defective PnC as a narrow bandpass filter. The diversity performances were compared among them, indicating that the frameworks using the CVAE and the CGAN can present the best performance by solving the nonunique response-to-design mapping problem. When extended to the 2D space, the topological features of in-plane PnCs were extracted and mapped to the distribution of pass bands and stop bands by establishing the AE and MLE models. The topology layouts were generated within anticipated band gaps as a result [85]. In order to offer controllability and stability, Xiao et al. [86] proposed a topology generator based on the CGAN that was applied to generate candidate topological structures with geometric constraints. And then this approach was combined with the automatic differentiation technique to acquire target band gaps. Moreover, the dispersion curves, which determine the boundaries of band gaps, were regarded as inputs and proactively tailored to achieve diverse configurations of 2D PnCs using CGAN [87]. However, to the best knowledge of the authors, the DL-based inverse designs were mainly concentrated on the desired performance that the unit cells made up of linear elastic hard materials can achieve, the intelligent framework responsible for providing feasible design of soft PnCs characterized by topology and applied mechanical load at an instant speed has not been reported so far. Inspired by this fact, the present work aims at establishing the framework of inverse design based on the DL, and the objective is to generate multiple soft PnCs that can induce customized band gaps and efficiently offer designs corresponding to tunable stop bands within lower frequencies, wider bandwidth, or enhanced attenuation levels. It is worth mentioning that, in contrast to the studies using the TO technique to design soft compressible composites where the optimization projects were performed under constant mechanical load [46,88], the present work regards both the discrete and continuous parameters as the design variables. In other words, the design of a soft PnC is represented by its material distribution in conjunction with the applied prestress. The proposed inverse framework is concentrated on presenting multiple feasible solutions concerning the same given input. Further, the generalization and diversity abilities can also make it possible to acquire on-demand designs. In addition, when composing the dataset for training and testing the DL-based model, the present work adopts the transfer matrix method (TMM) rather than the finite element method (FEM) to calculate the tunable band structure, as the former approach is superior in terms of computation efficiency.

The remainder of this paper is organized as follows: Section 2 presents the formulations of wave motion through 1D soft PnC based on the nonlinear elastic theory and the derivations of TMM modeling to predict tunable band gaps. In Section 3, a DL-based framework for the inverse design is established, where the components are delineated to show how accuracy and diversity can be assured. Section 4 gives the generated results of the topological layout with applied prestress, which corresponds to various on-demand requirements of longitudinal wave propagation and attenuation characteristics. Finally, Section 5 summarizes the conclusions reached in this paper.

2. Model Descriptions and Theoretical Formulations

This section aims to describe the mechanical properties of the 1D soft PnCs using the neo-Hookean model to illustrate the effects induced by the external prestress and establish the model based on the TMM to obtain the tunable band structures of longitudinal waves, which are shown in Sections 2.1 and 2.2, respectively.

2.1. The Constitutive Properties from Nonlinear Elasticity Theory

In this subsection, the soft materials are assumed to be compressible and homogeneous, and the PnC is composed of two soft phases. Following the nonlinear elasticity theory [89], the initial configuration of the considered structure without any deformation is taken as the reference configuration, where each point is marked by a position vector $\mathbf{X}$. When simulated by the external prestrain or prestress, the points on the reference configuration are expected to move to new points, and the position vector on the current configuration is denoted by x. The mapping relationship between x and $\mathbf{X}$ is defined as:

$$\mathrm{x} = \kappa(\mathbf{X}, t) \tag{1}$$

From Equation (1), the deformation gradient $\mathbf{F}$ can be obtained by $\mathbf{F} = \partial \mathrm{x}/\partial \mathbf{X}$, and its determinant $J = \det(\mathbf{F}) > 0$, which is able to determine the local measure of the volume change. Ignoring the dissipative effects of soft compressible elastomers, the first Piola-

Kirchhoff stress tensor **P** is acquired by differentiating the energy density function $\psi(\mathbf{F})$, that is, $\mathbf{P} = \partial\psi(\mathbf{F})/\partial\mathbf{F}$, and the Cauchy stress tensor $\boldsymbol{\sigma}$ is related to the tensor **P**: $\sigma = J^{-1}\mathbf{P}\mathbf{F}^{\mathrm{T}}$. Further, assuming that there are no mechanical body forces, the mechanical momentum balance equation is written as:

$$\mathrm{Div}(\mathbf{P}) = \rho_0 \partial^2 \mathrm{x}/\partial t^2 \tag{2}$$

where, ρ_0 denotes the initial density of the material. If performing static or quasi-static calculations, the inertial part can be omitted, and Equation (2) is correspondingly rewritten as: Div(**P**) = 0. Additionally, the ideal compressible neo-Hookean model is employed to derive the free energy density function:

$$\psi(I_1, I_3) = (I_1 - 3)\mu/2 + (I_3^{-\beta} - 1)\mu/(2\beta) \tag{3}$$

where I_1 = tr(**C**) ($\mathbf{C} = \mathbf{F}^{\mathrm{T}}\mathbf{F}$) and I_3 = det(**C**), which represent the first and third invariants of the right Cauchy strain tensor, respectively. v and μ denote the Poisson ratio and shear modulus of the elastomer phase, and the parameter β is defined as $\beta = v/(1-2v)$. Based on the above expressions, the Cauchy stress tensor $\boldsymbol{\sigma}$ is required to satisfy:

$$J\sigma = 2\mathbf{F}\mathbf{F}^{\mathrm{T}}\partial\psi/\partial I_1 + 2I_3\xi\partial\psi/\partial I_3 \tag{4}$$

where $\boldsymbol{\xi}$ is the Kronecker symbol. Considering the infinitesimal increment in the displacement field, the incremental form of the mechanical momentum balance equation is:

$$\mathrm{Div}(\overline{\mathbf{P}}) = \rho_0 \partial^2 \overline{\mathbf{u}}/\partial t^2 \tag{5}$$

In Equation (5), $\overline{\mathbf{P}}$ is the incremental change of the first Piola-Kirchhoff stress tensor and $\overline{\mathbf{u}}$ is the incremental displacement. The incremental change in the deformation gradient at this time is $\overline{\mathbf{F}} = \mathrm{Grad}(\overline{\mathbf{u}})$. Through linearizing the constitutive law, the incremental change of the first Piola-Kirchhoff stress tensor is obtained by $\overline{P}_{iJ} = C_{0iJkL}\overline{F}_{kL}$ and C_{0iJkL} represent the fourth-order referential elastic moduli tensor, which is determined via $C_{0iJkL} = \partial^2\psi/(\partial F_{iJ}\partial F_{kL})$. Substituting this relation into Equation (5) leads to: $C_{0iJkL}\partial^2\overline{u}_k/(\partial X_J \partial X_L) = \rho_0\partial^2\overline{u}_i/\partial t^2$. After transforming the reference configuration to the current configuration, it can be further updated as follows:

$$C_{ijkl}\partial^2\overline{u}_k/(\partial x_j \partial x_l) = \rho\partial^2\overline{u}_i/\partial t^2 \tag{6}$$

where $\rho = J^{-1}\rho_0$, denotes the density of the material in its current configuration, and $C_{ijkl} = J^{-1}C_{0iJkL}F_{jJ}F_{lL}$, represents the equivalent elastic modulus. According to the neo-Hookean model, the equivalent elastic modulus C_{ijkl} considering instantaneous effects induced by the static finite deformation, can be obtained by:

$$\begin{aligned} C_{ijkl} &= \mu J^{-1}F_{iJ}F_{kL}(\xi_{jl}\xi_{JL} + I_3^{-\beta}F_{Jl}^{-1}F_{Lj}^{-1} + 2\beta I_3^{-\beta}F_{Jj}^{-1}F_{Ll}^{-1}) \\ &= \mu J^{-1}(\xi_{jl}F_{iL}F_{kL} + I_3^{-\beta}\xi_{il}\xi_{kj} + 2\beta I_3^{-\beta}\xi_{ij}\xi_{kl}) \end{aligned} \tag{7}$$

2.2. Dispersion Analysis of Tunable Band Structure

The dynamic characteristics of longitudinal waves are examined in this subsection, based on which the formulations to acquire band structures are derived by applying the TMM. This approach is responsible for providing the labels to compose the dataset.

2.2.1. The Governing Equation of Longitudinal Wave

As the above formulations state, the geometrical and material characteristics can be tuned by the applied mechanical load, in which case the controllable dispersion relations of longitudinal waves are induced by manipulating the applied prestress. As shown in Figure 1, the considered compressible elastomers are denoted by material phase A and phase B. For simplicity, the initial lengths of them denoted by $H^{(A)}$ and $H^{(B)}$ are equal, which is assumed to be 0.5 mm. The 1D soft PnC is subject to fixed equi-biaxial prestretch in the x_1 and x_2 directions, indicating that $\lambda_1 = \lambda_2 = \lambda$. Also, $\lambda = 1$ is adopted in the subsequent project for completing the inverse design. The external prestress is denoted by σ_{33}, which is designed to range from the negative value to the positive value, corresponding to the compressible and stretching effects.

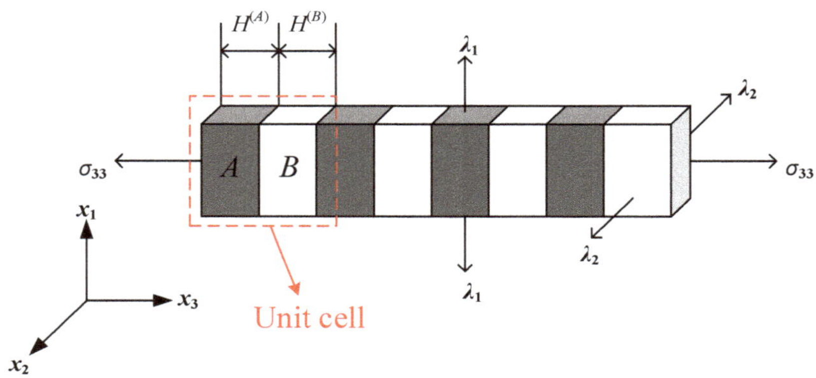

Figure 1. The schematic diagram of finite 1D soft phononic crystals with applied stress σ_{33}. In this work, fixed equi-biaxial prestretch in the x_1 and x_2 directions is employed; that is, $\lambda_1 = \lambda_2 = \lambda$, and $\lambda = 1$ is adopted in this work. The unit cell is manifested by the red dashed rectangle, which is composed of material phases A and B sharing equal volume. The initial length of the unit cell $H = H^{(A)} + H^{(B)}$.

Taking into account the deformation, the length of each phase is redefined as $h^{(p)} = H^{(p)}\lambda_3^{(p)}$ where $\lambda_3^{(p)}$ is the stretch ratio of phase p (p = material phase A or B) is in the direction of x_3, and thus the instantaneous lattice constant h is obtained by $h = h^{(A)} + h^{(B)}$. Obviously, in this case, the deformation gradient $\mathbf{F}$ can be derived as: $\mathbf{F} = \mathrm{diag}\left[\lambda,\ \lambda,\ \lambda_3^{(p)}\right]$. Based on the expressions of the invariants with the free energy density function and the Jacobian, the prestress σ_{33} is able to be formulated as:

$$\sigma_{33} = \mu^{(p)}[(\lambda_3^{(p)})^2 - \lambda^{-4\beta^{(p)}}(\lambda_3^{(p)})^{-2\beta^{(p)}}/(\lambda^2\lambda_3^{(p)}). \tag{8}$$

Focusing on the longitudinal wave propagation, Equation (6) is rewritten as:

$$C_{3333}\partial^2\overline{u}_3/\partial x_3^2 = \rho\partial^2\overline{u}_3/\partial t^2 \tag{9}$$

here, $\overline{u}_3$ is the longitudinal incremental displacement, and C_{3333} is the component of the instantaneous elastic moduli tensor. Employing the expression of $\mathbf{F}$ and Equation (7), the modulus of material phase p-$C_{3333}^{(p)}$ can be formulated as:

$$C_{3333}^{(p)} = \mu^{(p)}[\left(\lambda_3^{(p)}\right)^2 + (2\beta^{(p)}+1)\lambda^{-4\beta^{(p)}}\left(\lambda_3^{(p)}\right)^{-2\beta^{(p)}}]/(\lambda^2\lambda_3^{(p)}) \tag{10}$$

As can be seen, the magnitude of σ_{33} determines the magnitude of $\lambda_3^{(p)}$, meanwhile the variation of $\lambda_3^{(p)}$ plays an important role in controlling the equivalent modulus $C_{3333}^{(p)}$, thereby realizing tuning band gaps of the soft PnCs through controlling σ_{33}.

2.2.2. The Modelling Based on TMM for Band Gaps

Figure 1 describes the unit cell composed of two segments filled with material phases A and B for brevity to discover the relation between the external load and mechanical properties. Here, the unit cell is assumed to consist of multiple segments to make a more general statement to formulate the TMM model focusing on longitudinal waves. The unit cell considered in the finite soft PnCs is illustrated in Figure 2.

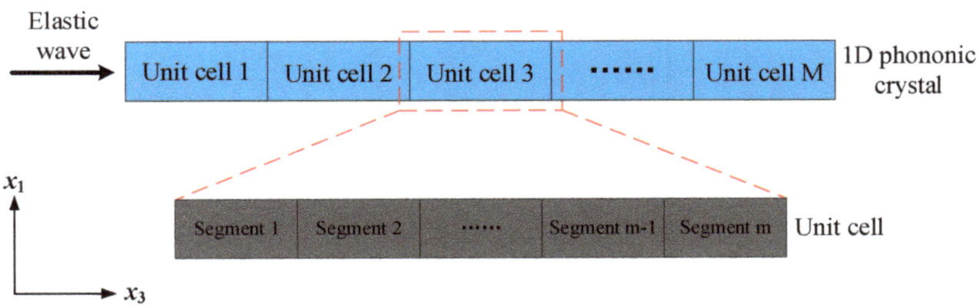

Figure 2. Schematic descriptions of waves propagating through the unit cell. Here, the unit cell is assumed to consist of multiple segments to make a more general statement to formulate the TMM model.

The elastic wave propagates along the longitudinal direction whose characteristics are governed by the following:

$$C_{3333}\partial^2 u(x_3, t)/\partial x_3^2 = \rho \partial^2 u(x_3, t)/\partial t^2 \quad (11)$$

where u represents the displacement in the x_3 direction. Assuming harmonic motion, Equation (11) can be rewritten as:

$$C_{3333}\partial^2 U(x_3)/\partial x_3^2 = -\rho\omega^2 U(x_3) \quad (12)$$

According to Bloch's theory, the periodic boundary condition requires $U(x_3 + h) = e^{iqh}U(x_3)$. Applying this boundary condition to TMM, the eigenvalue equation can be obtained:

$$\left|\mathbf{T} - e^{iqh}\mathbf{I}\right| = 0 \quad (13)$$

where h is the instantaneous length of the unit cell and q is the wavenumber of the Bloch wave. In Equation (13), the matrix $\mathbf{T}$ denotes the transfer matrix, and the detailed derivations can be found in Appendix A. The real part and imaginary part of q are obtained by solving Equation (13), which regards q as the eigenvalue for given frequencies f. In this case, the location, width, and attenuation level of wave band gaps can be examined. In order to validate the proposed TMM model, the material properties are employed from Ref. [88], and the normalized frequency Ω is adopted to eliminate the effects of the unit cell length on the band structure. Also, for brevity's purpose, the prestress is normalized by $\sigma_0 = \sigma_{33}/\mu_A$ and exerted at the two ends of the unit cell. These assumptions remain still unless otherwise specified in the following. Appendix B provides more details about validating the proposed TMM model.

3. Implementation of the Inverse Design Based on DL

The inverse design of the present work is expected to provide multiple feasible topological layouts with different prestresses that can share the same prescribed band gaps. The band gap properties are characterized by the bounding frequencies of interested stop bands in an automatic manner. It is noted that both the topology and the prestress being responsible for controlling wave propagation are required as the outputs here, showing more flexibility and complexity in contrast to the studies using the traditional TO conducted under constant external load. The inputs are assigned as the lower and upper bounding frequencies, and in this work, they are classified into three cases: the lower and upper boundary of the first stop band (Ω_{1l}, Ω_{1u}), that of the second stop band (Ω_{2l}, Ω_{2u}), and the combination of them (Ω_{1l}, Ω_{1u}, Ω_{2l}, Ω_{2u}). The DL-based models that correspond to the above-specified conditions are trained independently for outputting the topology layout and the prestress.

3.1. Description and Acquisition of the Dataset

In order to facilitate investigating the effects of topological distribution on the band gaps of soft PnCs, the unit cell is assumed to be composed of two soft compressible materials, *A* and *B*, as described before and discretized into *N* elements that share equal length of H/N. *H* is designed as 1 mm, denoting the unit cell length at the undeformed state. In order to avoid the emergency of topological layouts manifesting that the unit cell consists of multiple representative units, the number of divisions *N* is taken as a prime number. For example, as shown in Figure 3a, two repeatable elements constitute the unit cell, and thus the actual unit cell length accounts for 0.5 of *H* in such a displayed configuration. Hence, in this case, the absolute frequencies calculated by TMM are twice as large as those derived from the soft PnC shown in Figure 3b. However, the unit cell length *H* participating in normalizing the bounding frequencies is kept still during the whole procedure of training DL models, which causes incorrect inputs to the inverse network. Thus, in the present work, *N* is assumed to be 97, that is, the unit cell of the soft PnC with applied prestress is divided into 97 linear elements.

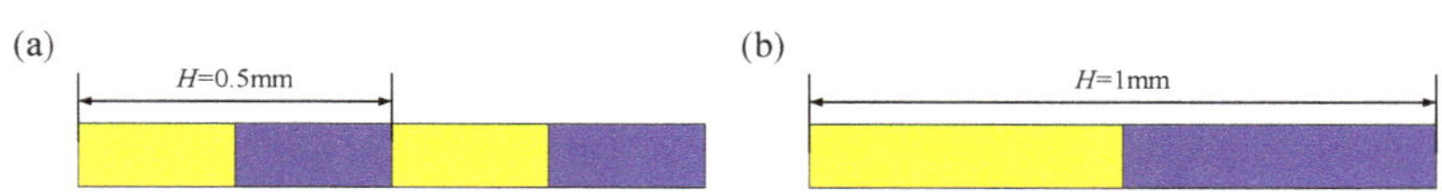

Figure 3. The 1D PnC whose initial unit cell length *H* is assumed as (**a**) 0.5 mm and (**b**) 1 mm, respectively. The analysis of band gaps based on such configurations is conducted to explain the reason for adopting $N = 97$ (where the color blue represents the element filled with phase *A* and yellow represents phase *B*).

Assuming that a_k (k = 1–97) represents the phase of filling material. The division is filled by phase *A* via setting $a_k = 0$ and phase *B* via setting $a_k = 1$. It is evidently seen that 2^{97} distributions can be achieved without considering prestress, exhibiting higher dimensional mapping complexity of the inverse problem. To depict the effects of topology layouts more intuitively, a material distribution is randomly selected, and the sensitivity information to each divided element is derived to indicate the robustness of the interested inverse problem. As the material distribution is described by a set of discrete variables a_k (k = 1–97) whose values can only be 0 or 1, the sensitivity of the bounding frequencies $\mathbf{\Omega}$ = (Ω_{1l}, Ω_{1u}, Ω_{2l}, Ω_{2u}) with respect to a_k is calculated by the following equation:

$$g_k = \left(\mathbf{\Omega}' - \mathbf{\Omega}\right)./\left(\left(a_k' - a_k\right)\mathbf{\Omega}\right) \times 100\% \tag{14}$$

where a_k' represents the reverse of a_k, that is, if $a_k = 0$ then $a_k' = 1$ and vice versa. $\boldsymbol{\Omega}'$ represents the corresponding bounding frequencies calculated based on the material distribution where only a_k is replaced by a_k'. And for offering a fair evaluation, the relative difference is applied as formulated in Equation (14). The results of sensitivity information are shown in Figure 4. It can be seen that the sensitivity to each element is different, and the maximum is about 1%, which also reflects the complexity of implementing the inverse design to generate the topology layouts for on-demand band gaps.

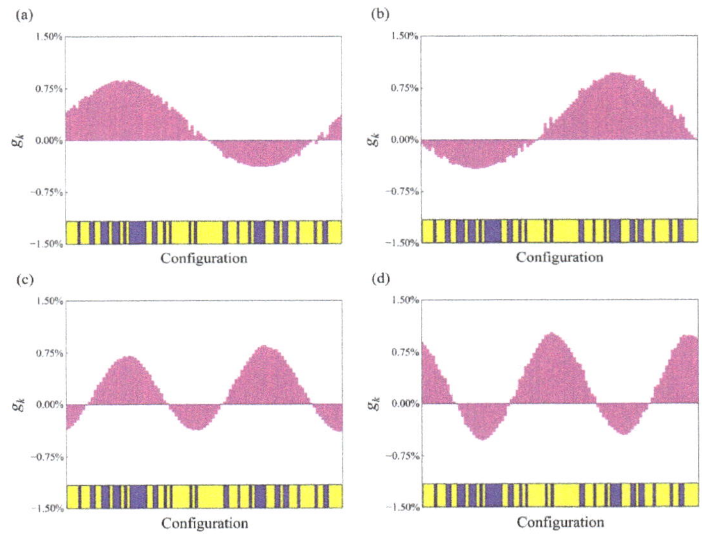

Figure 4. The sensitivity information of the bounding frequencies: (**a**) Ω1l, (**b**) Ω1u, (**c**) Ω2l, and (**d**) Ω2u for a randomly selected material distribution. The configuration is displayed close to the bottom of each plot, and the amplitude of each slender column indicates the sensitivity to the corresponding division (where the color blue represents the element filled with phase A and yellow represents phase B).

Apart from the material distribution, the normalized prestress σ_0 increasing from -1 to 1 is sampled with an interval of 0.1. The discrete variables determining the topological layout are associated with the continuous variable reflecting external load together as the output vector to the inverse design. In total, the amount of 63,000 data samples where the material distributions are randomly generated is taken into account, and the corresponding dispersion relations are calculated by the established TMM model to compose the entire dataset, 70% of which is selected as the training dataset. The remaining 30% is divided into two parts, equally, that are used as the validation dataset and test dataset, respectively.

3.2. The Framework of the Inverse Design

The CVAE that is generative is employed to learn the mapping from the bounding frequencies of stop bands to the soft PnCs described by the material distribution and applied prestress. To enhance the stability of training the CVAE and further acquire optimal soft PnCs for on-demand requirements, the forward DNN model is trained firstly, and subsequently, the task of filtering out the soft PnCs that do not fall within admissible stop bands as well as ranking the generated candidates according to the wave performance is fulfilled. The DNN based on the MLP architecture is proposed; given the prestress and distribution, the forward model is able to predict the band diagram, functioning as a surrogate model of TMM simulations.

Employing the well-trained forward DNN and CVAE concurrently, the proposed DL-based inverse design framework is visualized in Figure 5, consisting of three steps to accelerate the exploration of on-demand soft PnCs. Regarding the band gaps prescribed by special requirements, the established CVAE at the first step allows the generation of multiple classes of topological layouts with predicted prestress. It is worth noting that the target characteristics of band gaps can be specified by researchers, and the training of the CVAE is performed three times independently when the input vector is designed as (Ω_{1l}, Ω_{1u}), (Ω_{2l}, Ω_{2u}), and (Ω_{1l}, Ω_{1u}, Ω_{2l}, Ω_{2u}), respectively. The second step is responsible for filtering out the unsatisfactory soft PnCs among the massive generated candidates with the aid of forward DNN, and in this work, the well-trained inverse model has been performed at a maximum of 5000 times. The error limit of MAPE less than 1% is adopted, and the soft PnCs whose bounding frequencies exceed this criterion will be removed. The expression of MAPE can be found in Equation (15). As can be seen, the inverse design with respect to special requirements of real band diagrams can be solved by accomplishing step 1 and step 2. Nevertheless, the real part of dispersion relations is unable to estimate the attenuation level of elastic waves inside the stop bands. Thus, step 3 is added to the framework, which aims at undertaking the task of ranking the feasible unit cells in terms of wave attenuation performance. The TMM is employed again at this step to calculate the attenuation constant δ, which determines the decaying degree of Bloch waves for given frequencies. Based on the comparison results of δ among the generated feasible PnCs, the topological layout within applied prestress that can possess the maximum attenuation constant is emerging and regarded as the desired solution exhibiting enhanced performance.

$$\mathrm{MAPE} = \frac{1}{n}\left(\left|\left(\Omega' - \Omega\right)./\Omega\right|\right) \times 100\%,\ \mathrm{n} - \text{thedimension of predicted frequencies} \quad (15)$$

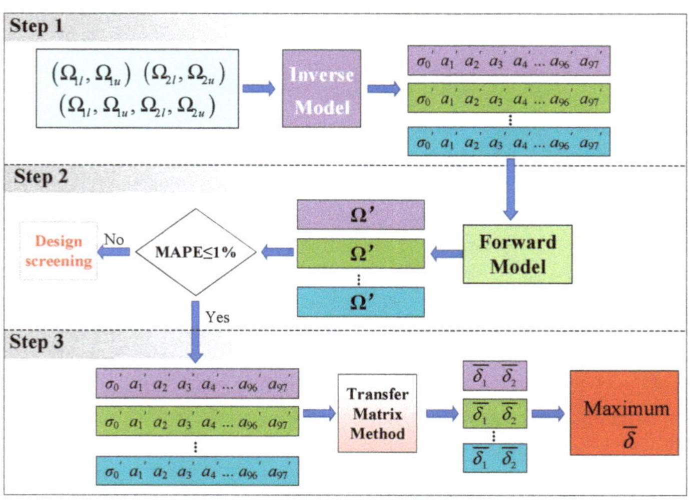

Figure 5. Schematic diagram of the framework of the intelligent inverse design of soft PnCs. The first step illustrates the basic function of the inverse model, that is, outputting different designs to the same

input, while the second step and the third step are conducted after validating the accuracy of the inverse model, which is responsible for achieving on-demand goals of tunable band gaps by utilizing the generalized performance of the inverse model.

3.3. The Architecture of Forward and Inverse Models

Focusing on the forward path, the DL-based model is more likely to be established in a deterministic supervised learning way to predict the nonlinear dynamic behavior of various configurations of soft PnCs with a reasonable trade-off in accuracy. Therefore, the forward model implemented follows the MLP architecture and consists of one input layer, five fully connected layers attached with activation functions (HardSwish), and one output layer. The hidden layers contain 196, 196, 196, 196, and 160 neurons, respectively, and the input is represented by the prestress combined with material distribution, while the output is the bounding frequencies of the first and second stop bands. The optimization algorithm is used with a learning rate of 0.0005 to adjust the hyperparameters automatically to minimize the mean absolute error (MAE) loss function that is formulated as follows:

$$L_{\text{DNN}} = \frac{1}{4m}\sum_{i=1}^{m}\left(\left|\Omega_{1l} - \Omega_{1l}^{'}\right| + \left|\Omega_{1u} - \Omega_{1u}^{'}\right| + \left|\Omega_{2u} - \Omega_{2u}^{'}\right| + \left|\Omega_{2l} - \Omega_{2l}^{'}\right|\right) \quad (16)$$

where m denotes the number of samples in the dataset.

In order to ensure the generation of multiple feasible candidates for the same input of given bounding frequencies, the DL-based model in the inverse path is used in the form of the CVAE consisting of the encoder and decoder parts. Under the behavior condition of target bounding frequencies $\boldsymbol{\Omega}$, the components of design variables, including the continuous prestress exerted at the unit cell and the discrete variables determining the topological layout, are fed into the encoder for mapping them to the latent variables $\boldsymbol{z}$ following the Gaussian distribution characterized by the expectancy $\boldsymbol{\alpha}$ and standard deviation γ. Then the latent variables $\boldsymbol{z}$ are concatenated with targets $\boldsymbol{\Omega}$, which is regarded as the input vector to the decoder for generating soft PnCs in terms of the topology with prestress. The viable and diverse capacities are induced by minimizing the reconstruction error and KL divergence. Reconstruction error L_{CVAE1} measures the difference between the reconstructed and the original samples, while KL divergence is used to evaluate the difference between the distribution of latent variables $\boldsymbol{z}$ and the standard normal distribution. By optimizing the error terms, the CVAE can learn the mapping between the band gaps and topological layouts with applied prestress. It is worth noting that the Sigmoid function is displayed at the end of the decoder as the discrete nature of the variables representing the topological layout needs to be restored. Furthermore, the well-trained forward model is in tandem with the CVAE, which is responsible for predicting the bounding frequencies of the generated soft PnCs.

For brevity purposes, the architectures of both the forward and inverse models are illustrated in Appendix C, from which it can be seen that the error L_{CVAE2} between the predictions by the forward model and targets also takes part in calculating the loss function when training the inverse model. Therefore, the loss function L_{total} of the entire inverse design contains error terms along both the inverse and forward paths, in which case L_{total} can be formulated as:

$$L_{\text{total}} = L_{\text{CVAE1}} + \chi\left(\text{KL} + \sum_{i=1}^{m}\sum_{k=1}^{n}(1/\gamma_{ik}^{2})\right) + 0.5L_{\text{CVAE2}} \quad (17)$$

where:

$$L_{\text{CVAE1}} = \sum_{i=1}^{m} \left| \sigma_{0i} - \sigma'_{0i} \right| + \sum_{i=1}^{m} \sum_{j=1}^{N} \left| a_{ij} - a'_{ij} \right|, \ \text{KL} = \left[-\sum_{i=1}^{m} \sum_{k=1}^{n} \left(1 + \log\left(\gamma_{ik}^2\right) - \gamma_{ik}^2 - \alpha_{ik}^2 \right) \right] / (2mn) \tag{18}$$

$$L_{\text{CVAE2}} = \begin{cases} \left|\Omega_{1l} - \Omega'_{1l}\right| + \left|\Omega_{1u} - \Omega'_{1u}\right|, & \text{Target : } 1^{\text{st}} \text{ BG} \\ \left|\Omega_{2l} - \Omega'_{2l}\right| + \left|\Omega_{2u} - \Omega'_{2u}\right|, & \text{Target : } 2^{\text{nd}} \text{ BG} \\ \left|\Omega_{1l} - \Omega'_{1l}\right| + \left|\Omega_{1u} - \Omega'_{1u}\right| + \left|\Omega_{2l} - \Omega'_{2l}\right| + \left|\Omega_{2u} - \Omega'_{2u}\right|, & \text{Target : } 1^{\text{st}} \text{ and } 2^{\text{nd}} \text{ BGs} \end{cases} \tag{19}$$

In the above equations, m and n represent the number of samples in the training dataset and the number of elements in the variable vector z, respectively. The weight of KL divergence χ is the hyperparameter, and its value is determined via a search process on the validation set. In detail, the optimal value is explored within the specified search spaces using Bayesian single-objective optimization from Optuna, with a total of 200 trials: χ = (0.001, 0.1, step = 0.001). The Nondominated Sorting Genetic Algorithm III (NSGA-III) in Optuna is used for sampling, and finally, this hyperparameter yields the best performance at χ = 0.06. Additionally, during the optimizing process, each trial takes 100 epochs. Regarding the other hyperparameters of the networks, they are determined using heuristic methods.

After assembling the architectures of the forward and inverse models, they are trained in sequence and terminated at 1500 epochs and 1000 epochs, respectively. The accuracy of DL-based models is then examined, and the details are given in Appendix C, from which the superior capacity to generate feasible designs is observed by combining the forward DNN and inverse CVAE. Moreover, Table 1 provides the cost time for training models as well as predicting bounding frequencies or generating soft PnCs. As can be seen, the proposed models exhibit extremely high computation efficiency.

Table 1. The cost time for training and performing DL-based models.

	Forward Model		Inverse Model	
	Training	Predicting (9549 data/single data)	Training	Generating (9549 data/single data)
Time (s)	8771.80	8.10/0.00084	8778.12	33.30/0.0035

4. Extended Investigation on Topology Design for Soft PnC

The accuracy of the inverse network has been proved already. Moreover, to demonstrate the generalized performance and deeply explore the design potential under specified conditions, the band gaps are customized with "on-demand" requirements and then employed as targets. By using the well-trained models, this section determines to motivate the generation of multiple innovative candidates of soft PnCs corresponding to these assigned targets.

4.1. *Inverse Design for Artificial Band Gaps*

In this subsection, three cases termed Case I, Case II, and Case III are taken into account, where the generated unit cells are expected to possess prescribed bounding frequencies of the first stop band, the second stop band, and both simultaneously. To verify the generalized performance, the boundaries are constructed in a man-made manner. Firstly, an averaging operation is applied to the bounding frequencies of four data samples randomly selected from the dataset. Then the numerical values of operated boundaries are reserved to two decimal places to ensure that the formed band gaps are beyond the entire dataset. Performing the mentioned instructions, nine groups of artificial band gaps can be achieved and regarded as on-demand inputs. As a result, Case I, Case II, and Case III are divided into different categories: Case I–I, Case I–II, Case I–III, Case II–I, Case II–II, Case II–III, Case III–I, Case III–II, and Case III–III. The target for each category can be found in Table 2.

Table 2. The boundaries of stop bands that are regarded as on-demand inputs.

Classifications	The Normalized Bounding Frequencies		
	Case I	Case II	Case III
I	1.39~1.49	3.02~3.15	(1.62~1.71, 3.24~3.41)
II	1.05~1.12	2.12~2.25	(1.18~1.26, 2.35~2.54)
III	0.88~0.93	1.83~1.92	(0.72~0.76, 1.44~1.55)

The inherent probabilistic feature of the CVAE allows multiple generations, and here three types of the topology accompanying prestress are provided as the solutions for brevity to the inverse design problem concerning the same input. The topological layouts of the unit cells and the corresponding band diagrams are presented in Figures 6–8, where the stop bands are highlighted by gray regions. As can be seen, the generated unit cells exhibit unique configurations but share overlapping stop bands in accordance with the targets given in Table 2, which demonstrates the diversity and generalized performance of the established framework.

In addition, it can be noticed that the outputted prestress also presents different levels even the same input is assigned, and both compression and stretch effects appear as the prior condition to tune the band diagram among the total 27 groups. Surprisingly, the fourth stop band, located between the fourth and fifth dispersion curves, as found from the plots, has become inconsistent with each other. For example, the fourth stop band in Figure 6a starts at 5.49164 and ends at 6.07446, but the location of that in Figure 6b ranges from 5.31006 to 6.26236. This can be explained by the differences between the topological layouts, which comply with the mechanism of Bragg scattering and indicate the complexity of the considered inverse problem.

4.2. Inverse Design for Broadening and Lowering Band Gaps

By taking advantage of the diversity ability of the present network, the topology with applied prestress that can exhibit lower and wider stop bands is expected to be generated. To deeply explore the potential for providing innovative structures, the bounding frequencies reflecting the location and bandwidth of the demand-satisfying band gaps are achieved by a step-by-step approach. Specifically, initial stop bands are randomly selected out of the entire dataset at first, and then the upper boundary remains still while the lower boundary is continuously reduced by a gradual deduction. As a result, the bandwidth is broadened, and the central position of the stop bands is moving towards lower frequencies. The newly formed band gap in each step is regarded as the input, in which case the well-trained CVAE is required to provide multiple feasible candidates. The accuracy is assured with the aid of the forward network through filtering out the designs that violate the criterion defined by MAPE defined by Equation (15) of being less than 1%. This manipulation for decreasing the lower bounding is terminated until the values of MAPE of all generated structures are larger than 1%. Finally, the inverse design is completed with high accuracy and efficiency in regard to searching for soft PnCs corresponding to lower and wider band gaps. Three cases named Case ①, Case ②, and Case ③ are investigated, respectively, which corresponds to the condition that the above operations are performed on the first and second stop bands simultaneously, the first stop band only, and the second stop band only.

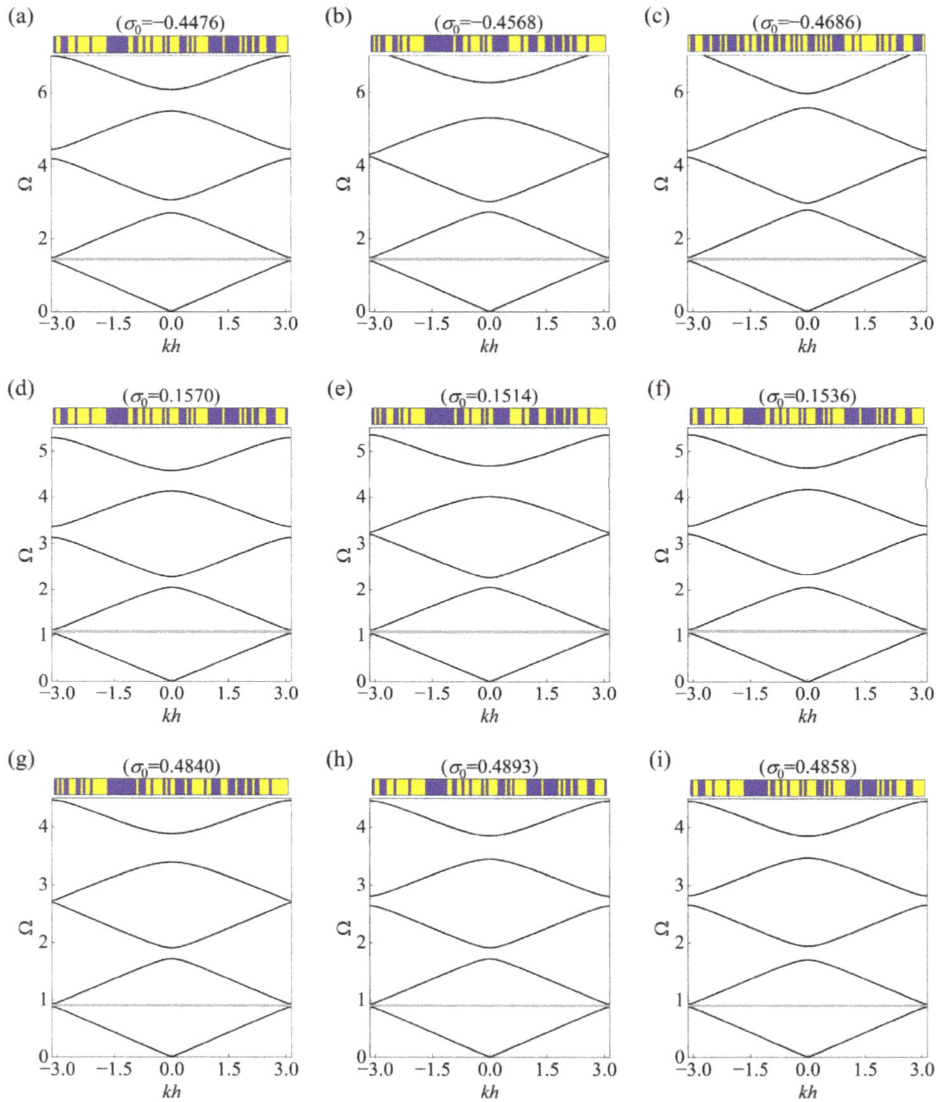

Figure 6. Three solutions of topological layout with prestress for the same input of bounding frequencies specified in Table 2. The continuous curves represent the corresponding real band diagrams in the mentioned cases of (**a–c**): Case I–I, (**d–f**) Case I–II, and (**g–i**) Case I–III. The generated configuration is displayed at the top of each plot, where the color blue represents the element filled with phase *A* and yellow represents phase *B*. Also, the gray regions denote the prescribed stop bands.

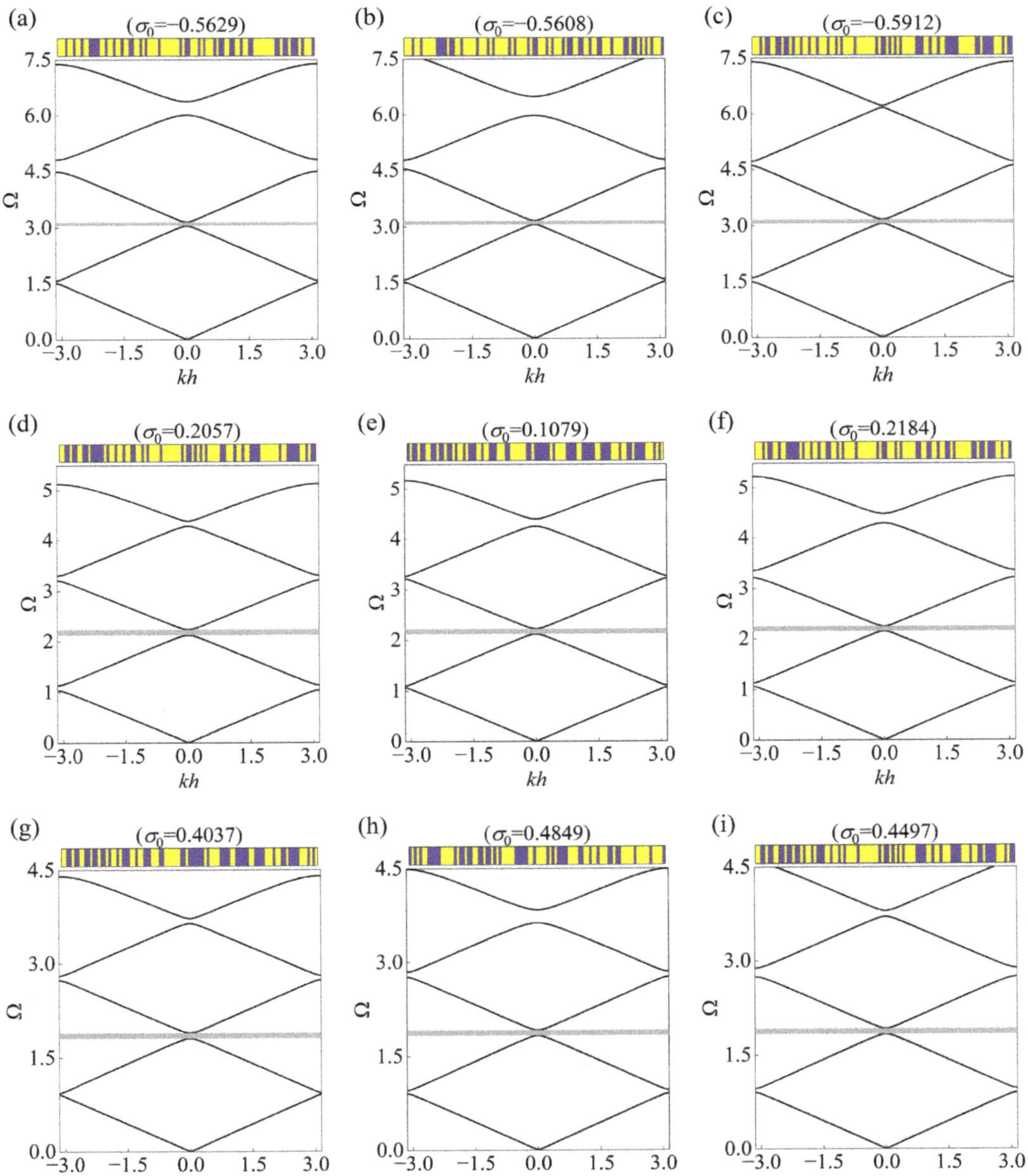

Figure 7. Three solutions of topological layout with prestress for the same input of bounding frequencies specified in Table 2. The continuous curves represent the corresponding real band diagrams in the mentioned cases of (**a–c**): Case II–I, (**d–f**) Case II–II, and (**g–i**) Case II–III. The generated configuration is displayed at the top of each plot, and the gray regions denote the prescribed stop bands.

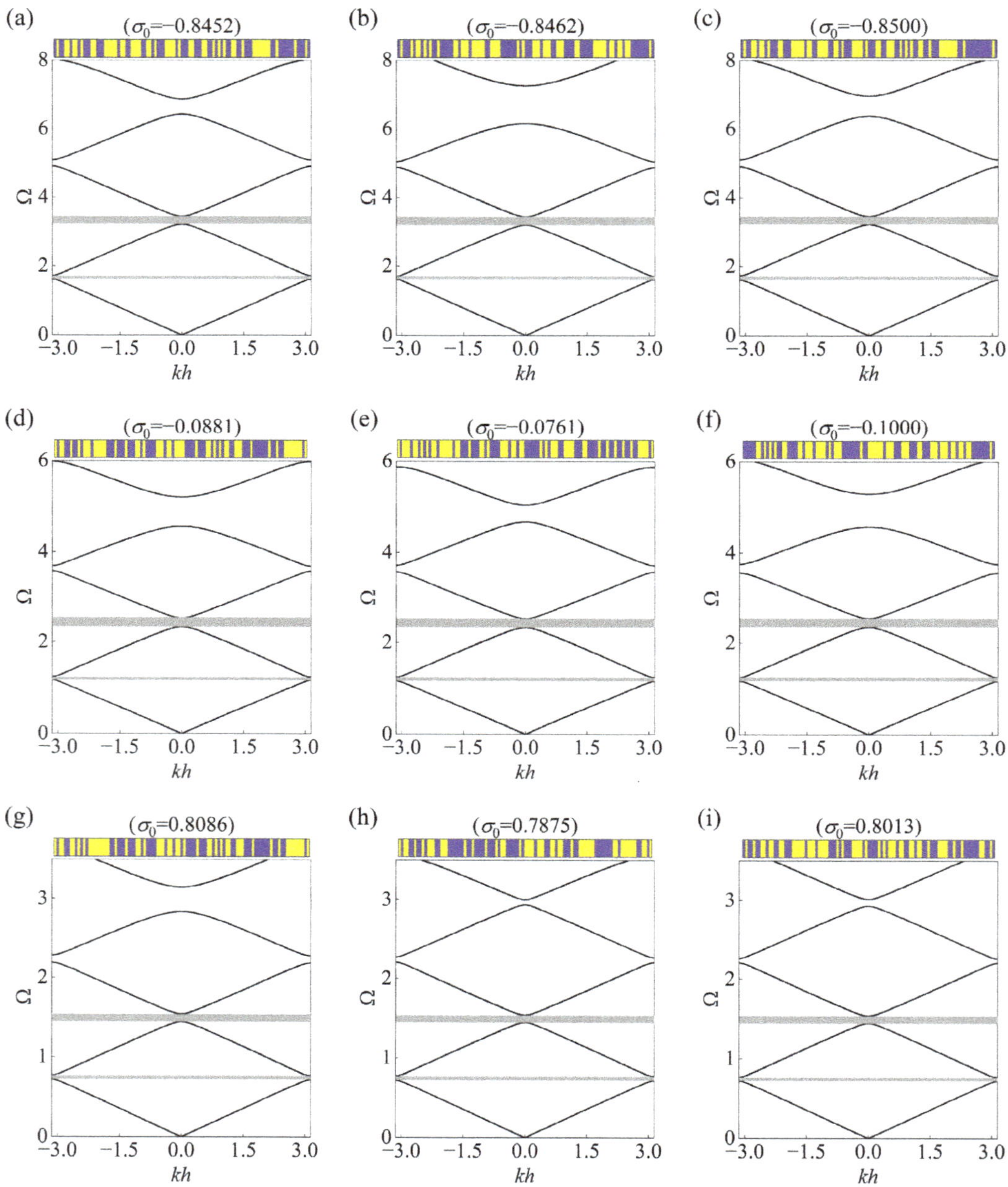

Figure 8. Three solutions of topological layout with prestress for the same input of bounding frequencies specified in Table 2. The continuous curves represent the corresponding real band diagrams in the mentioned cases of (**a–c**): Case III–I, (**d–f**) Case III–II, and (**g–i**) Case III–III. The generated configuration is displayed at the top of each plot, and the gray regions denote the prescribed stop bands.

As shown in Table 3, the initial ranges of the first and second stop bands in Case ① are 1.75–1.80 and 3.50–3.70, the lower bounding frequencies of which are deducted by 0.05 and 0.1 per step separately. For brevity purposes, one presentation of the topological layout with the outputted prestress is illustrated at each step. It is clearly seen that the first and second stop bands continuously shift to lower frequencies and are broadened until they reach the range covering 1.55–1.80 and 3.10–3.70, to which the CVAE cannot provide material distribution with satisfactory accuracy. The final layout corresponds to the first stop band, ranging from 1.60 to 1.80, and the second stop band, from 3.20 to 3.70, whose absolute bandwidth is 4 times and 2.5 times larger than that of the initial stop band, while the relative bandwidth is merely 4.2 times and 2.6 times larger. Moreover, the stop bands can be further expanded in Case ② and Case ③ when employed as the individual target. It is evident from Table 4 to Table 5 that the absolute bandwidth of the first stop band can approach 5 times larger as the upper boundary is confined to 1.80. Meanwhile, the second stop band can be increased to 3 times larger as the upper boundary yields at 3.70.

Table 3. The topological layout accompanying prestress at each step for lowering and broadening the target stop bands in Case ①.

Case ①: (1.75~1.80, 3.50~3.70)	**The Output Design Scheme**
(1.70~1.80, 3.40~3.70)	($\sigma_0 = -1.0521$) 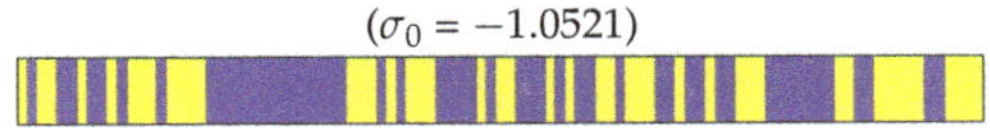
(1.65~1.80, 3.30~3.70)	($\sigma_0 = -0.9405$) 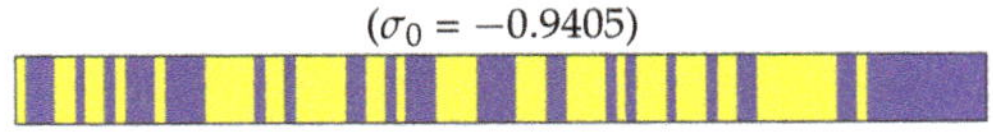
(1.60~1.80, 3.20~3.70)	($\sigma_0 = -0.8957$) 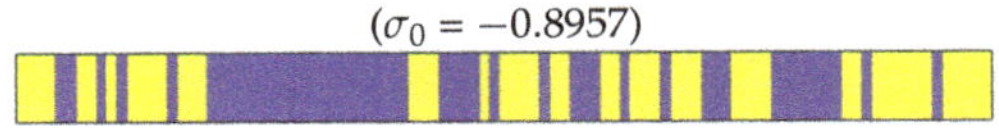
(1.55~1.80, 3.10~3.70)	NaN

Table 4. The topological layout accompanying prestress at each step for lowering and broadening the target stop bands in Case ②.

Case ②: 1.75~1.80	**The Output Design Scheme**
1.70~1.80	($\sigma_0 = -0.9768$) 
1.65~1.80	($\sigma_0 = -0.9228$) 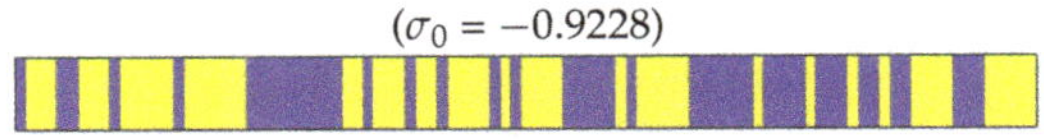
1.60~1.80	($\sigma_0 = -0.8689$) 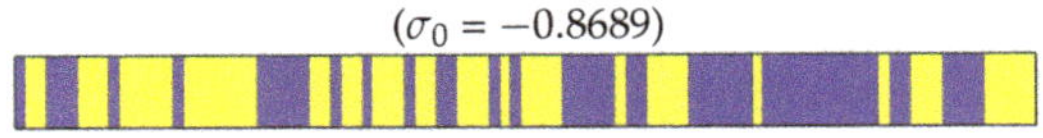
1.55~1.80	($\sigma_0 = -0.8139$) 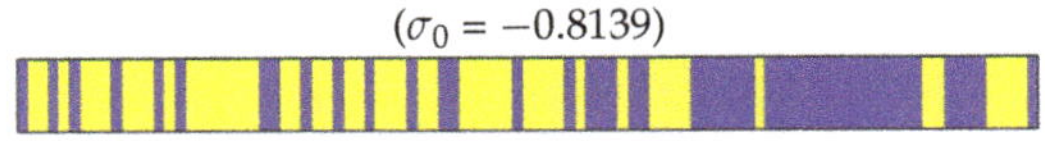
1.50~1.80	NaN

In addition, the predicted compression stress is found to be gradually decreased to lower the band diagram. Figure 9 presents the band structures of the initial design and final designs. From the comparison results, it can be observed that various "on-demand" requirements on the tunable band gaps are realized by the proposed network.

Table 5. The topological layout accompanying prestress at each step for lowering and broadening the target stop bands in Case ③.

Case ③: 3.50~3.70	The Output Design Scheme
3.40~3.70	($\sigma_0 = -0.9395$) 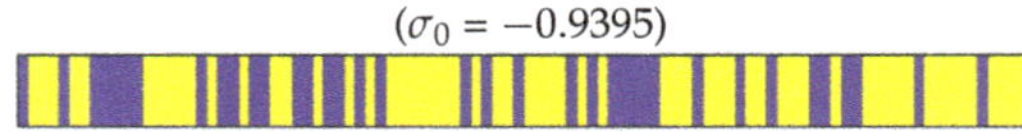
3.30~3.70	($\sigma_0 = -0.8889$) 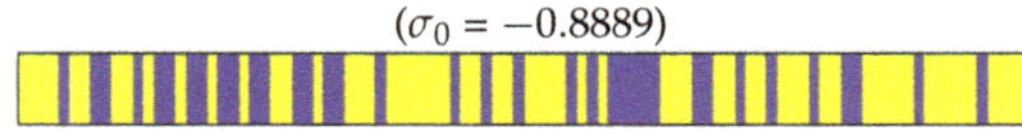
3.20~3.70	($\sigma_0 = -0.8453$) 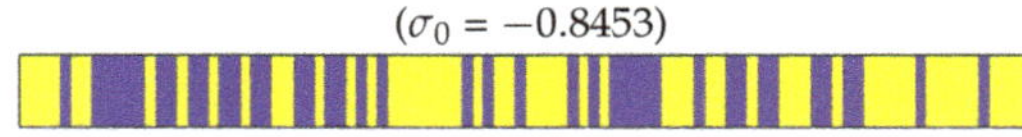
3.10~3.70	($\sigma_0 = -0.7995$) 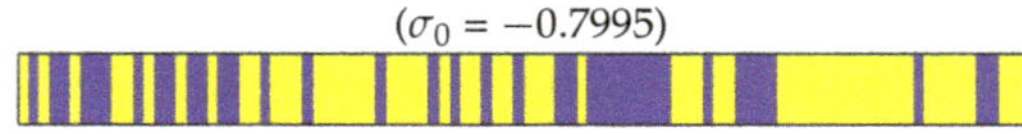
3.00~3.70	NaN

4.3. *Inverse Design for Enhanced Wave Attenuation*

The results of the real band structure shown in Section 4.1 manifest that the unit cells sharing merely the same first and second stop bands can still present distinctive features of dispersion curves, which further suggests the attenuation performance inside the considered stop bands is not completely consistent with each other. To demonstrate this deduction, the attenuation performance of the generated structures in Case I–I, Case I–II, and Case I–III is evaluated by the attenuation constant δ, respectively, via using the TMM model again. As a result, the attenuation peaks of the first stop band are 0.0844, 0.0609, and 0.1056, and those of the second stop band are 0.1355, 0.2262, and 0.2085. Obviously, the bigger the value of δ, the more intensive the degree of prohibiting wave propagation can be achieved. This evident difference between various feasible designs within identical bounding frequencies makes it possible to find the optimal topological layout with improved wave attenuation performance.

The generated unit cells are explored by comparing the average attenuation constant $\overline{\delta}$ over the target stop bands. The candidate possessing the maximization of $\overline{\delta}$ is regarded as the desired solution in this subsection. In order to add the flexibility of this inverse problem, the average attenuation constant is formulated as:

$$\overline{\delta} = n_1\overline{\delta}_1 + n_2\overline{\delta}_2 \tag{20}$$

where $\overline{\delta}_1$ and $\overline{\delta}_2$ represent the mean value of the attenuation constant over the first-stop band and second-stop band, respectively.

Based on Equation (20), three conditions are taken into account that correspond to $(n_1, n_2) = (1, 0)$, $(n_1, n_2) = (0, 1)$, and $(n_1, n_2) = (0.5, 0.5)$. The final design, characterized by the topology layout and applied prestress, is expected to be found under each condition. Here, the target bounding frequencies are derived from Table 2; as stated before, these are randomly selected beyond the training, validation, and test datasets. Now, the considered three conditions are termed Case I-(i), (ii), (iii), Case II-(i), (ii), (iii), and Case III-(i), (ii), (iii). In detail, Cases I, II, and III correspond to the bounding frequencies of the first and second stop bands that are (1.62, 1.71, 3.24, 3.41), (1.18, 1.26, 2.35, 2.54), and (0.72, 0.76, 1.44, 1.55), while the sequence numbers (i), (ii), and (iii) denote the selection of the parameter group of (n_1, n_2) described before.

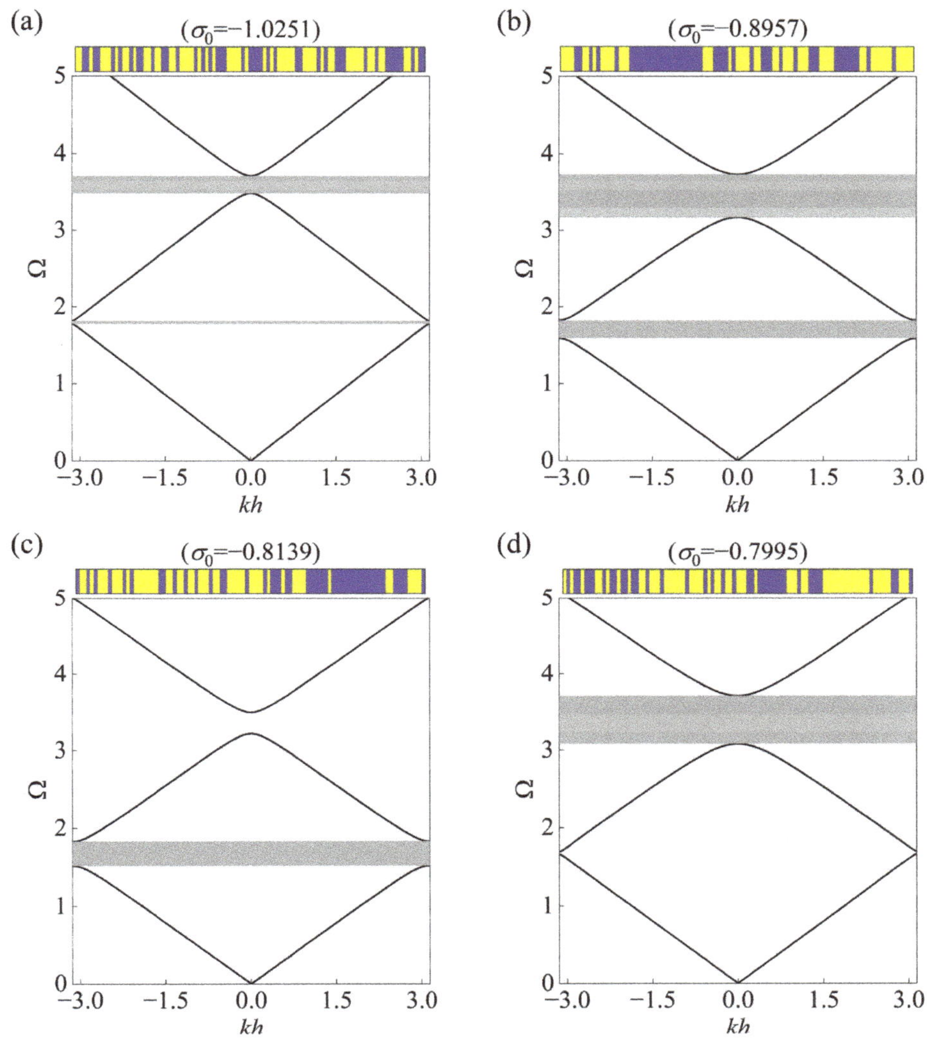

Figure 9. The comparison results between the band diagrams of the (**a**) initial design and offered design at the last step in (**b**) Case ①, (**c**) Case ②, and (**d**) Case ③, where the gray regions denote the stop bands of interested order. The initial and generated configurations with prestress are displayed at the top of each plot.

After running the well-trained inverse model 5000 times in each case, the generated structures that do not satisfy the error limit of being less than 1% of MAPE are filtered out to ensure accuracy. After filtering, there are 4515, 4641, and 3756 feasible designs to be retained in Case I, Case II, and Case III, respectively. By comparing the target values defined in Equation (20) among the feasible designs, the corresponding maximums and minimums of the average attenuation constants are obtained and given in Table 6. As can be seen, whether focusing on the first and second stop bands simultaneously or solely on one stop band, the ability to forbid wave propagation can be significantly improved by searching for the optimal design with a bigger attenuation constant. Correspondingly, the topological layouts with prestress possessing the best attenuation performance among the feasible candidates are given in Tables 7–9. Different weight coefficients of (n_1, n_2) can result in diverse optimal designs. To gain a more straightforward understanding of the improved ability, Figure 10 presents the attenuation diagrams of the unit cells given in these tables. It is noted that the presented design in this subsection does not represent the "global optimal" or "local optimal" through searching for the design space using optimization algorithms. It is achieved by making full use of the potential of the inverse network, which is able to find a satisfactory solution at an incomparable speed.

Table 6. The best and worst attenuation performance over target stop bands among generated feasible designs.

	The Average Attenuation Constant Estimating Attenuation Level					
	Maximum in Case I	**Minimum in Case I**	**Maximum in Case II**	**Minimum in Case II**	**Maximum in Case III**	**Minimum in Case III**
Case i: (n_1, n_2) = (1, 0)	$\overline{\delta}_1 = 0.1244$	$\overline{\delta}_1 = 0.0193$	$\overline{\delta}_1 = 0.1388$	$\overline{\delta}_1 = 0.0261$	$\overline{\delta}_1 = 0.1184$	$\overline{\delta}_1 = 0.0186$
Case ii: (n_1, n_2) = (0, 1)	$\overline{\delta}_2 = 0.2188$	$\overline{\delta}_2 = 0.0296$	$\overline{\delta}_2 = 0.3050$	$\overline{\delta}_2 = 0.0963$	$\overline{\delta}_2 = 0.2656$	$\overline{\delta}_2 = 0.0891$
Case iii: (n_1, n_2) = (0.5, 0.5)	$\overline{\delta} = 0.1441$	$\overline{\delta} = 0.0467$	$\overline{\delta} = 0.2041$	$\overline{\delta} = 0.0816$	$\overline{\delta} = 0.1718$	$\overline{\delta} = 0.0745$

Table 7. The generated unit cells with applied prestress that can produce enhanced attenuation performance over the target band gaps: 1.62–1.71 and 3.24–3.41.

	Design for Improved Wave Attenuation
Case I-(i)	$(\sigma_0 = -0.8496)$ 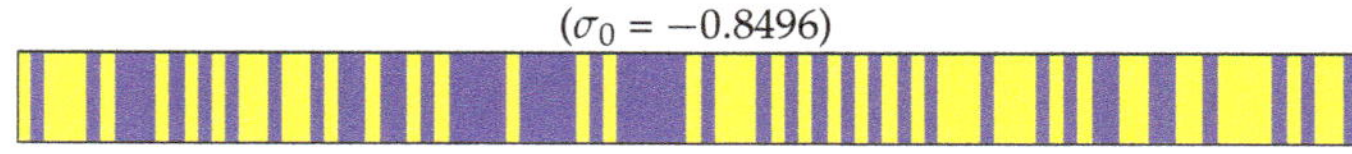
Case I-(ii)	$(\sigma_0 = -0.8413)$ 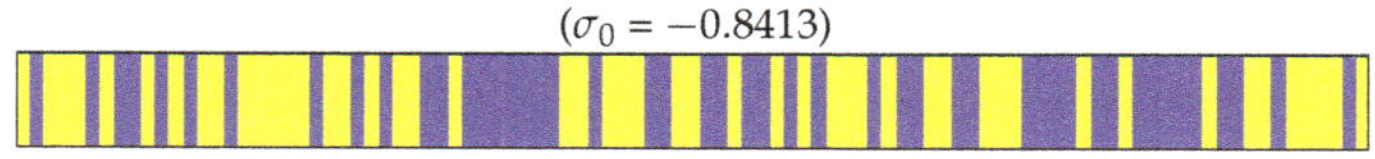
Case I-(iii)	$(\sigma_0 = -0.8413)$ 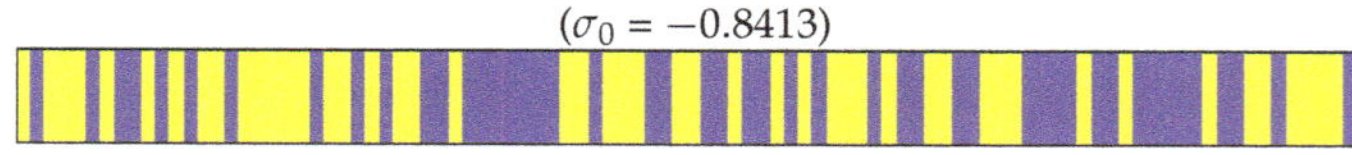

Table 8. The generated unit cells with applied prestress that can produce enhanced attenuation performance over the target band gaps: 1.18–1.26 and 2.35–2.54.

	Design for Improved Wave Attenuation
Case II-(i)	$(\sigma_0 = -0.0936)$ 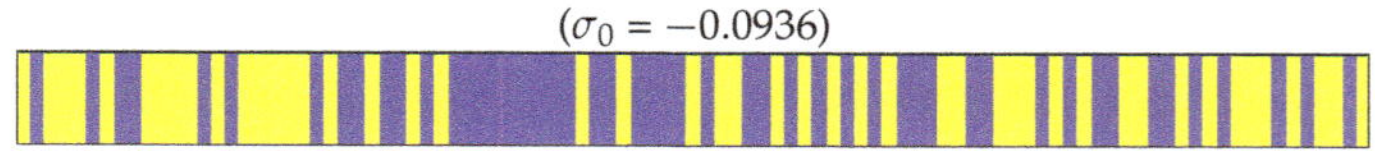
Case II-(ii)	$(\sigma_0 = -0.1137)$ 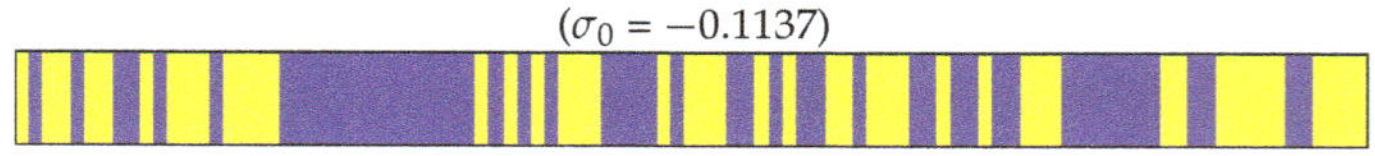
Case II-(iii)	$(\sigma_0 = -0.0892)$ 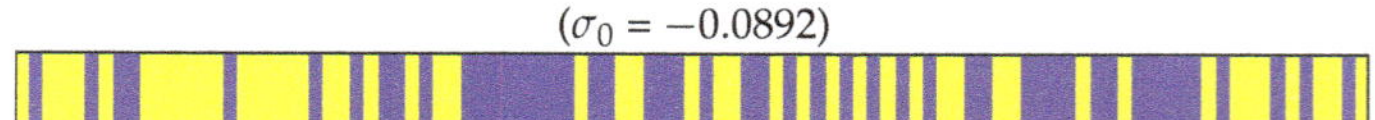

Table 9. The generated unit cells with applied prestress that can produce enhanced attenuation performance over the target band gaps: 0.72–0.76 and 1.44–1.55.

	Design for Improved Wave Attenuation
Case III-(i)	($\sigma_0 = 0.8379$) 
Case III-(ii)	($\sigma_0 = 0.7719$) 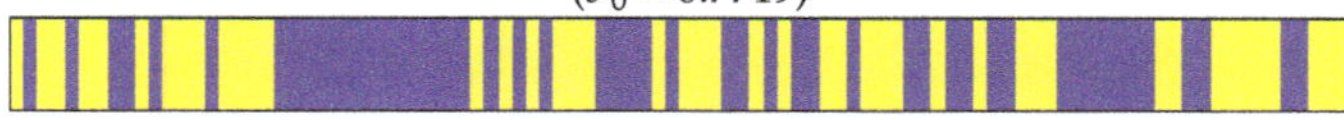
Case III-(iii)	($\sigma_0 = 0.8142$) 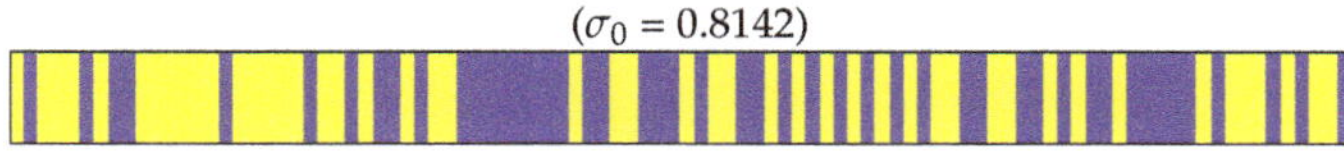

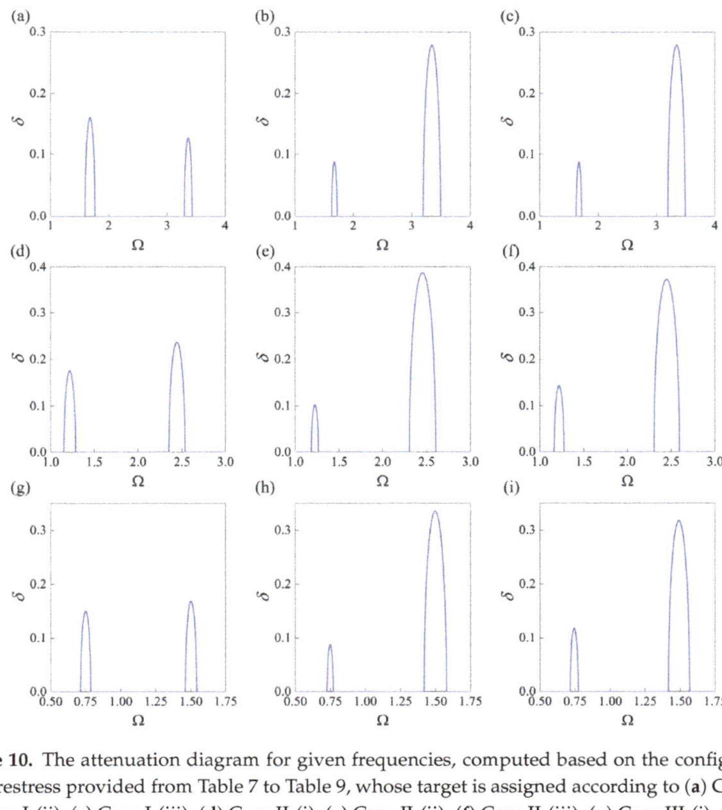

Figure 10. The attenuation diagram for given frequencies, computed based on the configuration and prestress provided from Table 7 to Table 9, whose target is assigned according to (**a**) Case I-(i), (**b**) Case I-(ii), (**c**) Case I-(iii), (**d**) Case II-(i), (**e**) Case II-(ii), (**f**) Case II-(iii), (**g**) Case III-(i), (**h**) Case III-(ii), and (**i**) Case III-(iii).

Further, to explore the technical value of the design solutions, finite soft structures based on the results of Tables 7–9 are taken into consideration to evaluate the vibration insulation performance. Each is constituted by 50 corresponding unit cells with predicted prestress, and the clamped condition (CC) is employed. The harmonic force with unit amplitude displacement is imposed on the left edge, and the vibration isolation is examined by the transmittance TL that is expressed in terms of the displacement ratio: $TL =$

$20\log_{10}(u_r/u_l)$. Here, u_r and u_l represent the displacement at the left side and right side, respectively. The displacement field is acquired by the FEM model. The results of transmittance are shown in Figure 11, where the gray regions denote the stop bands calculated by the TMM. From Figure 11, one can easily observe the location and bandwidth of the attenuation bands with significant insulation are in accord with the stop bands. In addition, as comparing the plots displayed in the same row, the average level for reducing vibration of finite soft PnCs is also found to conform to the objectives assigned in Cases (i), (ii), and (iii), that is, the emphasis is assigned on improving wave attenuation inside the first stop band only, the second stop band only, and the first two stop bands simultaneously, which demonstrates the potential of the proposed framework applied in practical engineering.

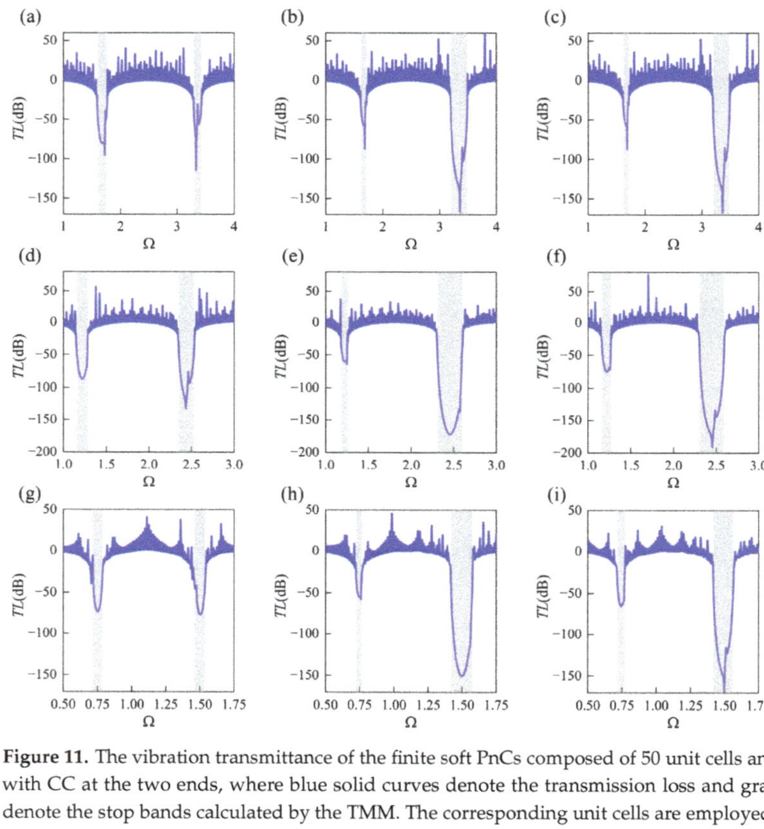

Figure 11. The vibration transmittance of the finite soft PnCs composed of 50 unit cells and applied with CC at the two ends, where blue solid curves denote the transmission loss and gray regions denote the stop bands calculated by the TMM. The corresponding unit cells are employed from the results given in Tables 7–9, within the objective of improving wave attenuation performance under different cases: (**a–c**): Case I-(i) to Case I-(iii), (**d–f**): Case II-(i) to Case II-(iii), and (**g–i**): Case III-(i) to Case III-(iii).

5. Conclusions

In this work, an innovative design method based on the DL is presented to realize two purposes for designing the soft compressible PnCs: one is for outputting the combination of discrete and continuous design variables in terms of the topological layout and external mechanical prestress, and the other is for automatically offering multiple design solutions with respect to the given target of longitudinal wave band gaps at an extremely fast speed. Based on the nonlinear elastic theory and Bloch-Floquet theory, the TMM model is established to extract the complex dispersion relations of the compressible unit cells. In order to assure the accuracy and diversity performance of DL-based models, the inverse framework is composed of CVAE in tandem with pretrained MLP, where the former, displayed in the inverse path, is responsible for generating candidates and the latter, fixed at the forward path, is regarded as the surrogate model for enhancing the stability and filtering out unfeasible designs.

The average value of MAPE is 0.81% over the test dataset, verifying the accuracy of the inverse network. When the target bounding frequencies are randomly selected beyond the entire dataset, the well-trained CAVE can still output various design schemes characterized by different distributions of material phases applied with prestress, which also validates the superior generalized performance. Furthermore, by making use of the outstanding performance in exploiting design space of the proposed method, this work aims to offer topologies with longitudinal prestress that correspond to lower and broader stop bands. In particular, the generated soft, compressible unit cells possess the first stop band and the second stop band, whose widths are at least 2.5 times larger than that of initial designs. In addition, the presented network shows the capacity to efficiently acquire the combination of the topological layout and external prestress that can induce improved attenuation performance. At first, numerous unit cells are generated by the proposed inverse model. Then they are filtered out according to the accuracy criterion. Eventually, the desired solution is obtained via the ranking in terms of the attenuation constant among the feasible generated candidates. Moreover, the attenuation level is also validated by the vibration insulation of finite soft PnCs. Overall, the proposed method can serve as a potent tool for realizing on-demand soft PnCs, enabling researchers to intelligently control vibration and sound characteristics.

Author Contributions: Conceptualization, J.L.; methodology, M.Q.; software, M.Q.; validation, Z.Z.; formal analysis, J.Y.; investigation, M.Q.; resources, J.L.; data curation, M.Q.; writing—original draft preparation, J.L.; writing—review and editing, J.Y.; visualization, S.L.; supervision, W.L.; project administration, W.L.; funding acquisition, J.L. All authors have read and agreed to the published version of the manuscript.

Funding: The present work was supported by the National Natural Science Foundation of China (Grant No. 52365009) and the Natural Science Foundation of Hainan Province (Grant No. 122MS004).

Institutional Review Board Statement: Not applicable.

Informed Consent Statement: Not applicable.

Data Availability Statement: The data presented in this study are available on request from the corresponding author due to privacy.

Conflicts of Interest: The authors declare that they have no known competing financial interests or personal relationships that could have appeared to influence the work reported in this paper.

Abbreviations

The following abbreviations are used in this manuscript:

PnCs	Phononic crystals
MMs	Metamaterials
BS	Bragg scattering
LR	Local resonance
1D	One-dimensional
FG	Functionally graded
2D	Two-dimensional
TO	Topology optimization
DL	Deep learning
ML	Machine learning
ANN	Artificial neural network
TNN	Tandem neural network
AE	Autoencoder
GAN	Generative adversarial network
MLP	Multilayer perceptron
DNN	Deep neural network
CVAE	Conditional variation autoencoder
CGAN	Conditional generative adversarial network
TMM	Transfer matrix method
FEM	Finite element method
MAE	Mean absolute error
NSGA-III	Nondominated Sorting Genetic Algorithm III
CC	Clamped condition

Appendix A. Derivations of the Transfer Matrix Method

For j-th segment of the n-th unit cell, Equation (12) can be written as:

$$U(x_3) = P_j^{n+} e^{ik_j x_3} + P_j^{n-} e^{-ik_j x_3} \tag{A1}$$

where P_j^{n+} and P_j^{n-} are the displacement amplitudes in the x_3 direction, while $k_j = \sqrt{\omega^2 \rho / C_{3333}^{(j)}}$ is the wave vector of j-th segment, and x_3 represents the position of the j-th segment in the unit cell, i.e., $x_3 \in \left[h^{(1)} + h^{(2)} + \ldots + h^{(j-1)}, h^{(1)} + h^{(2)} + \ldots + h^{(j)}\right]$ ($h^{(j)}$ represents the length of the j-th segment). Therefore, the expressions of the displacement and stress on the j-th segment of the n-th unit cell can be formulated as:

$$U_j^n(x_3) = P_j^{n+} e^{ik_j x_3} + P_j^{n-} e^{-ik_j x_3} \tag{A2}$$

$$\sigma_j^n(x_3) = C_{3333}^{(j)} ik_j (P_j^{n+} e^{ik_j x_3} - P_j^{n-} e^{-ik_j x_3}) \tag{A3}$$

According to the continuity conditions at the interface between the j-th and (j + 1)-th segments, the relations are required to satisfy:

$$u_j^n(h^{(1)} + h^{(2)} + \ldots + h^{(j)}) = u_{j+1}^n(h^{(1)} + h^{(2)} + \ldots + h^{(j)}) \tag{A4}$$

$$\sigma_j^n(h^{(1)} + h^{(2)} + \ldots + h^{(j)}) = \sigma_{j+1}^n(h^{(1)} + h^{(2)} + \ldots + h^{(j)}) \tag{A5}$$

By substituting Equations (A2) and (A3) into Equations (A4) and (A5), one can obtain that:

$$\mathbf{\Psi}_{j+1}^n = \mathbf{T}_j \mathbf{\Psi}_j^n \tag{A6}$$

where $\mathbf{\Psi}_j^n = [P_j^{n+}, P_j^{n-}]$ and $\mathbf{T}_j$ is the transfer matrix relating the j-th segment to the $(j+1)$-th segment, which can be acquired by:

$$\mathbf{T}_j = \begin{bmatrix} e^{ik_{j+1}L^{(j)}} & e^{-ik_{j+1}L^{(j)}} \\ C_{3333}^{(j+1)} ik_{j+1}e^{ik_{j+1}L^{(j)}} & -C_{3333}^{(j+1)} ik_{j+1}e^{-ik_{j+1}L^{(j)}} \end{bmatrix}^{-1} \begin{bmatrix} e^{ik_jL^{(j)}} & e^{-ik_jL^{(j)}} \\ C_{3333}^{(j)} ik_je^{ik_jL^{(j)}} & -C_{3333}^{(j)} ik_je^{-ik_jL^{(j)}} \end{bmatrix} \tag{A7}$$

here, $L^{(j)} = h^{(1)} + h^{(2)} + \ldots + h^{(j)}$. Based on Equation (A6), the relationship between the first segment of the n-th unit cell and the first segment of the $(n+1)$-th unit cell can be deduced as:

$$\mathbf{\Psi}_1^{n+1} = \mathbf{T}\mathbf{\Psi}_1^n \tag{A8}$$

where $\mathbf{T} = \mathbf{T}_m\mathbf{T}_{m-1}\cdots\mathbf{T}_1$. Additionally, the Bloch theory requires:

$$\mathbf{\Psi}_1^{n+1} = e^{iqh}\mathbf{\Psi}_1^n \tag{A9}$$

Finally, the transfer matrix $\mathbf{T}$ can be obtained by associating Equation (A8) with Equation (A9).

Appendix B. The Validation of the TMM Model

The material characteristics of phase A and B are given in Table A1, as illustrated from Ref. [88], each of which accounts for 0.5 of the entire domain. The normalized frequency is formulated by $\Omega = \omega H\sqrt{\rho_A/\mu_A}/(2\pi)$. Before performing band structure calculations, the adjustment on the equivalent modulus is illustrated in Figure A1 with respect to the varying prestress σ_0. It is clearly seen that the tensional prestress can lower the equivalent modulus while the compression prestress is able to induce opposite effects.

Table A1. The material properties of phases A and B.

Material Phase	ρ/(kg/m^3)	μ/(MPa)	β
A	1050	0.444	1
B	1260	0.888	2

Under fixed lateral equi-biaxial prestretch, the results are firstly computed by the TMM and FEM models separately in the cases of $\sigma_0 = -1$ and $\sigma_0 = 1$ and then compared with each other in Figure A2. The FEM simulation is conducted using the software COMSOL Multiphysics 6.0. As can be seen, whether for the soft PnC within tensile effects or compression effects, the band diagram acquired by the TMM agrees well with that by the FEM, which verifies the accuracy of the proposed TMM model.

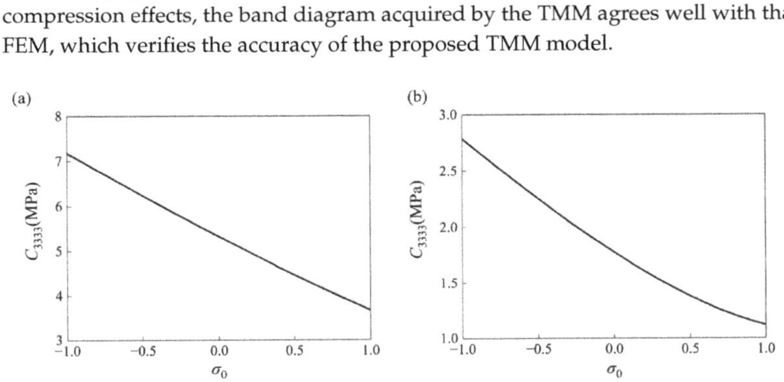

Figure A1. Variations of the simultaneous modulus C_{3333} of (**a**) material phase A and (**b**) material phase B as a function of normalized prestress σ_0 that is increased from -1 to 1.

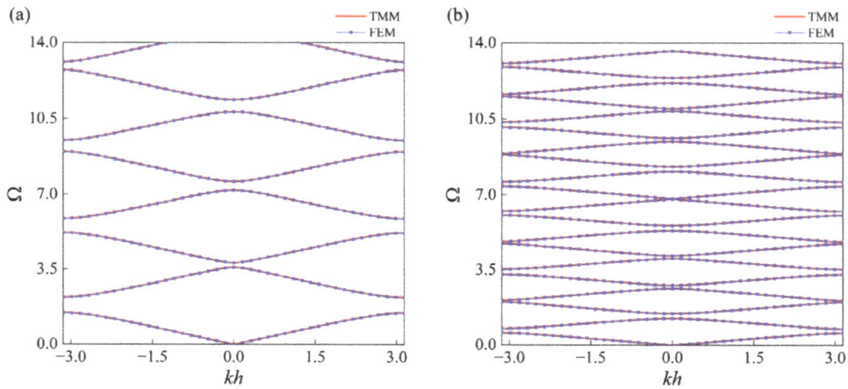

Figure A2. Comparison results of the band structures obtained by TMM and FEM for soft PnCs in the case of (**a**) $\sigma_0 = -1$ and (**b**) $\sigma_0 = 1$, where the dispersion relations from TMM to FEM are represented by the red line and the blue line with symbol, respectively.

Appendix C. Supplement Materials of DL Models

The architectures of the forward model and the inverse model are illustrated in Figure A3a,b, respectively.

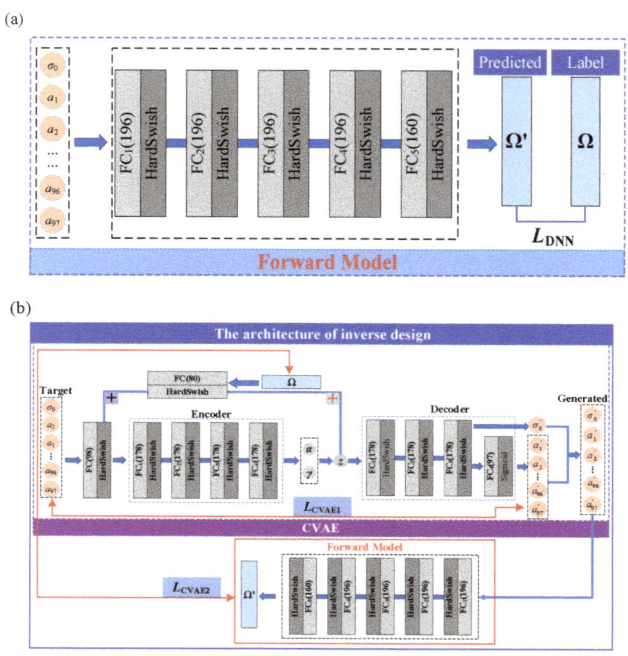

Figure A3. (**a**) The architecture of the forward model is based on the DNN, consisting of one input layer, five fully connected layers attached with activation functions (HardSwish), and one output layer.

The hidden layers contain 196, 196, 196, 196, and 160 neurons, respectively. (**b**) The architecture of the inverse model combining CVAE and a well-trained forward DNN. The design variables, including prestress and topological layout, are fed into the encoder for mapping them to the latent variables z under the behavior condition of target bounding frequencies $\boldsymbol{\Omega}$, and then z is concatenated with targets $\boldsymbol{\Omega}$, as inputs fed into the decoder. The well-trained forward model is in tandem with the CVAE; in this case, the loss function during the training of the entire inverse design contains both error terms from inverse and forward models.

Further, the accuracy of DL-based models after training is examined. The average values of MAPE of both forward DL and inverse CVAE models over the test dataset (9540 test samples) are computed and provided in Table A2. Figures A4 and A5 show the relation between the ground truth and the predictions via applying the forward model and CVAE model. It is worth noting that the predicted bounding frequencies in Figure A5 are obtained by the TMM model based on the soft PnCs generated by the CVAE. As can be seen, either the results from Table A2 or from Figure A4 to Figure A5 can demonstrate the superior capacity to generate feasible designs of soft PnCs via combining the forward DNN and the inverse CVAE.

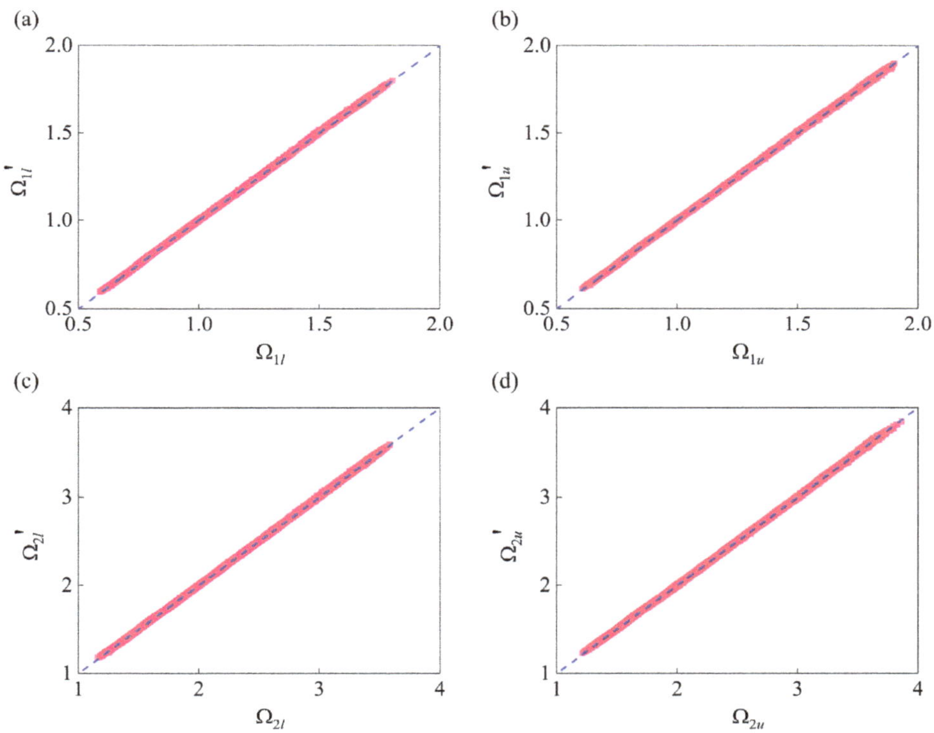

Figure A4. The predictions by the forward DNN and the ground truth of bounding frequencies over all the testing samples. The abscissa in the subfigures denotes the ground truth of bounding frequencies, and the ordinate axis denotes the prediction of bounding frequencies. In detail, (**a**,**b**) are the lower and upper boundaries of the first stop band, and (**c**,**d**) are the lower and upper boundaries of the second stop band. The blue dashed line reflects the physical meaning of "$y = x$".

Table A2. The average values of MAPE for estimating the accuracy of the forward and inverse models.

The Error Term	Forward DNN	CVAE
MAPE	0.23%	0.81%

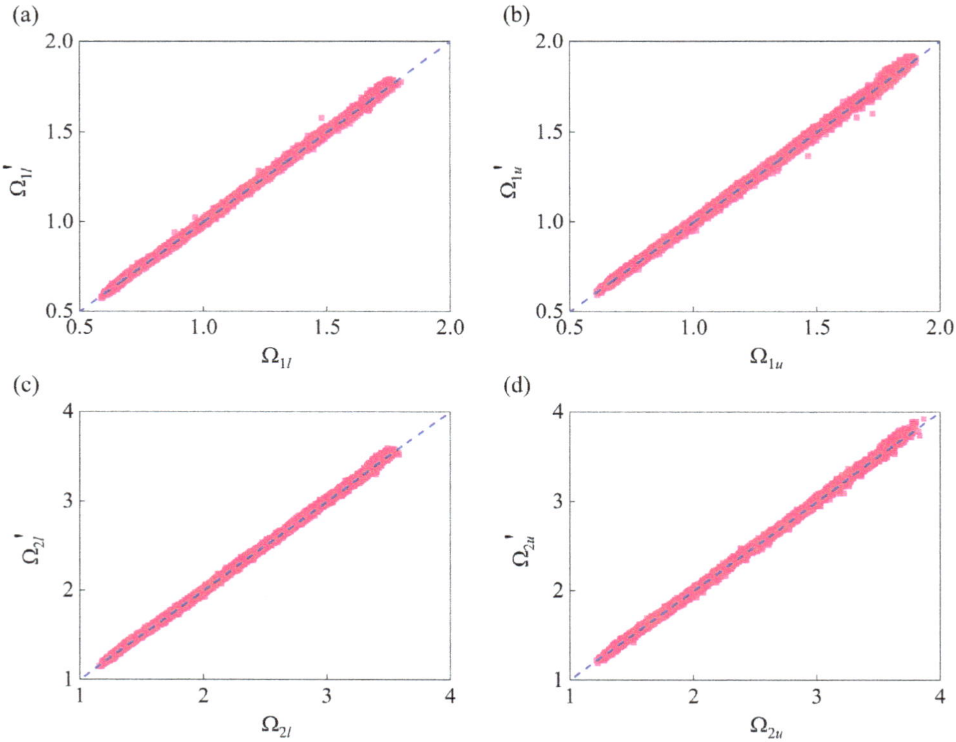

Figure A5. The predictions by the CVAE and ground truth of bounding frequencies over all the testing samples. The predictions are obtained based on the generated soft PnCs by the CVAE. The abscissa in the subfigures denotes the ground truth of bounding frequencies, and the ordinate axis denotes the prediction of bounding frequencies. In detail, (**a**,**b**) are the lower and upper boundaries of the first stop band, and (**c**,**d**) are the lower and upper boundaries of the second stop band. The blue dashed line reflects the physical meaning of "$y = x$".

References

1. Kushwaha, M.S.; Halevi, P.; Dobrzynski, L.; Djafari-Rouhani, B. Acoustic band structure of periodic elastic composites. *Phys. Rev. Lett.* **1993**, *71*, 2022–2025. [CrossRef] [PubMed]
2. Hu, G.; Tang, L.; Liang, J.; Lan, C.; Das, R. Acoustic-elastic metamaterials and phononic crystals for energy harvesting: A review. *Smart Mater. Struct.* **2021**, *30*, 085025. [CrossRef]
3. Liu, Z.; Zhang, X.; Mao, Y.; Zhu, Y.Y.; Yang, Z.; Chan, C.T.; Sheng, P. Locally Resonant Sonic Materials. *Science* **2000**, *289*, 1734–1736. [CrossRef] [PubMed]
4. Man, X.; Xia, B.; Luo, Z.; Liu, J. 3D Hilbert fractal acoustic metamaterials: Low-frequency and multi-band sound insulation. *J. Phys. D Appl. Phys.* **2019**, *52*, 195302. [CrossRef]
5. Zhao, P.; Zhang, K.; Qi, L.; Deng, Z. 3D chiral mechanical metamaterial for tailored band gap and manipulation of vibration isolation. *Mech. Syst. Sig. Process.* **2022**, *180*, 109430. [CrossRef]

6. Yang, H.; Cheng, S.; Li, X.; Yan, Q.; Wang, B.; Xin, Y.; Sun, Y.; Ding, Q.; Yan, H.; Li, Y.; et al. Innovative structural design of chiral lattices with low frequency wide multiple band gaps and vibration suppression. *Mech. Adv. Mater. Struct.* **2024**, *31*, 5823–5840. [CrossRef]
7. Xiao, P.; Miao, L.; Zheng, H.; Lei, L. Band gap characteristics of new composite multiple locally resonant phononic crystal metamaterial. *J. Phys. Condens. Mat.* **2024**, *36*, 195702. [CrossRef]
8. Deng, Z.; Zhang, B.; Zhang, K.; Peng, L.; Liu, P.; Sun, Q.; Pang, F. The coupled band gap of the double beam with attached periodic spring-mass structure. *Mech. Syst. Sig. Process.* **2024**, *208*, 111009. [CrossRef]
9. Sun, P.; Zhang, Z.; Guo, H.; Liu, N.; Wang, Y. Hierarchical square honeycomb metamaterials with low-frequency broad bandgaps and flat energy bands characteristics. *J. Appl. Phys.* **2020**, *128*, 235102. [CrossRef]
10. Xiao, L.; Iqbal, M.; Yu, X. Quasi-static band gaps in metamaterial pipes with negative stiffness resonators. *Int. J. Mech. Sci.* **2024**, *261*, 108668. [CrossRef]
11. Xu, J.; Jing, J. Low-frequency band gaps in quasi-zero stiffness locally resonant metamaterial shaft. *Int. J. Mech. Sci.* **2024**, *267*, 108992. [CrossRef]
12. Zhang, J.; Zhang, J.; Zhang, B.; An, Y.; Yang, X.; Hu, N.; Ma, L.; Peng, Y.; Wang, B. Broadband multifrequency vibration attenuation of an acoustic metamaterial beam with two-degree-of-freedom nonlinear bistable absorbers. *Mech. Syst. Sig. Process.* **2024**, *212*, 111264. [CrossRef]
13. Guo, Z.; Chen, S.; Wang, C.; Hao, C. Subwavelength bandgap phononic crystals with mixed resonant modes. *Phys. Lett. A* **2022**, *438*, 128109. [CrossRef]
14. Zhang, Y.; Zhao, Z.; Miao, X.; Wu, J.; Chen, L.; Li, B.; Liu, H.; Yang, L.; Liu, Q.; Chen, T.; et al. The mechanism of adjusting the low-frequency ultra-wide band gap within lightweight spherical superstructue. *Phys. Scr.* **2024**, *99*, 105972. [CrossRef]
15. Li, Z.; Wang, K.; Chen, T.; Cheng, L.; Xu, D.; Zhou, J. Temperature controlled quasi-zero-stiffness metamaterial beam for broad-range low-frequency band tuning. *Int. J. Mech. Sci.* **2023**, *259*, 108593. [CrossRef]
16. D'Alessandro, L.; Ardito, R.; Braghin, F.; Corigliano, A. Low frequency 3D ultra-wide vibration attenuation via elastic metamaterial. *Sci. Rep.* **2019**, *9*, 8039. [CrossRef]
17. Chen, Z.; Sun, S.; Deng, L.; Yang, J.; Zhang, S.; Du, H.; Li, W. Investigation of a new metamaterial magnetorheological elastomer isolator with tunable vibration bandgaps. *Mech. Syst. Sig. Process.* **2022**, *170*, 108806. [CrossRef]
18. Deng, J.; Yang, J.; Jiao, S.; Long, X. Band-stop characteristics of a nonlinear anti-resonant vibration isolator for low-frequency applications. *Int. J. Mech. Sci.* **2023**, *240*, 107914. [CrossRef]
19. D'Alessandro, L.; Belloni, E.; Ardito, R.; Braghin, F.; Corigliano, A. Mechanical low-frequency filter via modes separation in 3D periodic structures. *Appl. Phys. Lett.* **2017**, *111*, 231902. [CrossRef]
20. Park, S.; Jeon, W. Ultra-wide low-frequency band gap in a tapered phononic beam. *J. Sound Vib.* **2021**, *499*, 115977. [CrossRef]
21. Gu, C.; Ma, L.; Ou, Z.; Lei, F. Propagation characteristics of lamb waves in a functionally graded material plate with periodic gratings. *Mech. Adv. Mater. Struct.* **2024**, *31*, 1645–1654. [CrossRef]
22. Chen, Z.; Chong, Y.B.; Lim, K.M.; Lee, H.P. Reconfigurable 3D printed acoustic metamaterial chamber for sound insulation. *Int. J. Mech. Sci.* **2024**, *266*, 108978. [CrossRef]
23. Sun, Y.; Zhang, G.; Lee, H.P.; Zheng, H.; Luo, Z.; Li, F. Sound transmission of truss-based X-shaped inertial amplification metamaterial double panels. *Int. J. Mech. Sci.* **2024**, *283*, 109669. [CrossRef]
24. Jiang, Z.; Zhou, Y.; Zheng, S.; Liu, J.; Xia, B. Waveguides induced by replacing defects in phononic crystal. *Int. J. Mech. Sci.* **2023**, *255*, 108464. [CrossRef]
25. Lekhal, D.; Mekkakia-Maaza, N.E.; Lakhdari, A. Finite element analysis of surface elastic waveguide based on pyramidal phononic crystal. *Micro Nano Lett.* **2021**, *16*, 1–8. [CrossRef]
26. Zheng, Y.; Wang, C.; Tan, Z.; Zhang, Z.; Jiang, J.; Cheng, B.; Cao, J. Realization of a novel multimode waveguide with photonic band gap hopping based on coupled-resonator optical waveguide theory. *Results Phys.* **2023**, *45*, 106242. [CrossRef]
27. He, Z.; Zhang, G.; Chen, X.; Cong, Y.; Gu, S.; Hong, J. Elastic wave harvesting in piezoelectric-defect-introduced phononic crystal microplates. *Int. J. Mech. Sci.* **2023**, *239*, 107892. [CrossRef]
28. Gantasala, S.; Thomas, T.; Rajagopal, P. Enhanced piezoelectric energy harvesting based on sandwiched phononic crystal with embedded spheres. *Phys. Scr.* **2023**, *98*, 035029. [CrossRef]
29. Zhao, L.; Lu, Z.; Ding, H.; Chen, L. A viscoelastic metamaterial beam for integrated vibration isolation and energy harvesting. *Appl. Math. Mech.* **2024**, *45*, 1243–1260. [CrossRef]
30. Chen, Y.; Wu, B.; Su, Y.; Chen, W. Tunable Two-Way Unidirectional Acoustic Diodes: Design and Simulation. *J. Appl. Mech.* **2019**, *86*, 031010. [CrossRef]
31. Karami Mohammadi, N.; Galich, P.I.; Krushynska, A.O.; Rudykh, S. Soft Magnetoactive Laminates: Large Deformations, Transverse Elastic Waves and Band Gaps Tunability by a Magnetic Field. *J. Appl. Mech.* **2019**, *86*, 111001. [CrossRef]

32. Pierce, C.D.; Willey, C.L.; Chen, V.W.; Hardin, J.O.; Berrigan, J.D.; Juhl, A.T.; Matlack, K.H. Adaptive elastic metastructures from magneto-active elastomers. *Smart Mater. Struct.* **2020**, *29*, 065004. [CrossRef]
33. Getz, R.; Shmuel, G. Band gap tunability in deformable dielectric composite plates. *Int. J. Solids Struct.* **2017**, *128*, 11–22. [CrossRef]
34. Padmanabhan, S.; Alam, Z.; Sharma, A.K. Tunable anti-plane wave bandgaps in 2D periodic hard-magnetic soft composites. *Int. J. Mech. Sci.* **2024**, *261*, 108686. [CrossRef]
35. Montgomery, S.M.; Wu, S.; Kuang, X.; Armstrong, C.D.; Zemelka, C.; Ze, Q.; Zhang, R.; Zhao, R.; Qi, H.J. Magneto-Mechanical Metamaterials with Widely Tunable Mechanical Properties and Acoustic Bandgaps. *Adv. Funct. Mater.* **2021**, *31*, 2005319. [CrossRef]
36. Alam, Z.; Padmanabhan, S.; Sharma, A.K. Magnetically tunable longitudinal wave band gaps in hard-magnetic soft laminates. *Int. J. Mech. Sci.* **2023**, *249*, 108262. [CrossRef]
37. Dorfmann, A.; Ogden, R.W. Electroelastic waves in a finitely deformed electroactive material. *IMA J. Appl. Math.* **2010**, *75*, 603–636. [CrossRef]
38. Huang, Y.; Shen, X.D.; Zhang, C.L.; Chen, W.Q. Mechanically tunable band gaps in compressible soft phononic laminated composites with finite deformation. *Phys. Lett. A* **2014**, *378*, 2285–2289. [CrossRef]
39. Huang, Y.; Li, J.; Chen, W.; Bao, R. Tunable bandgaps in soft phononic plates with spring-mass-like resonators. *Int. J. Mech. Sci.* **2019**, *151*, 300–313. [CrossRef]
40. Wu, B.; Su, Y.; Liu, D.; Chen, W.; Zhang, C. On propagation of axisymmetric waves in pressurized functionally graded elastomeric hollow cylinders. *J. Sound Vib.* **2018**, *421*, 17–47. [CrossRef]
41. Ning, S.; Yang, F.; Luo, C.; Liu, Z.; Zhuang, Z. Low-frequency tunable locally resonant band gaps in acoustic metamaterials through large deformation. *Extrem. Mech. Lett.* **2020**, *35*, 100623. [CrossRef]
42. Wang, B.; Yang, H.G.; Xin, Y.J.; Sun, Y.T.; Cheng, S.L.; Wang, L.; Wang, S.; Zhang, Z.Z. Low frequency band gap and vibration suppression mechanism of innovative multiphase metamaterials. *Mech. Adv. Mater. Struct.* **2024**, *31*, 4839–4854. [CrossRef]
43. Jung, J.; Kim, H.; Goo, S.; Chang, K.; Wang, S. Realisation of a locally resonant metamaterial on the automobile panel structure to reduce noise radiation. *Mech. Syst. Sig. Process.* **2019**, *122*, 206–231. [CrossRef]
44. Xi, C.; Dou, L.; Mi, Y.; Zheng, H. Inertial amplification induced band gaps in corrugated-core sandwich panels. *Compos. Struct.* **2021**, *267*, 113918. [CrossRef]
45. Yang, D.; Guo, X.; Zhang, W.; Cao, D. Non-linear dynamics and bandgap control in magneto-rheological elastomers metamaterials with inertial amplification. *Thin Wall. Struct.* **2024**, *204*, 112237. [CrossRef]
46. Li, Y.; Zhang, X.; Tian, Q.; Luo, Y. Topological design of soft substrate acoustic metamaterial for mechanical tuning of sound propagation. *Compos. Struct.* **2024**, *346*, 118399. [CrossRef]
47. Xiao, P.; Miao, L.; Zheng, H.; Lei, L. Low frequency vibration reduction bandgap characteristics and engineering application of phononic-like crystal metaconcrete material. *Constr. Build. Mater.* **2024**, *411*, 134734. [CrossRef]
48. Wang, Z.W.; Chen, A.; Xu, Z.X.; Yang, J.; Liang, B.; Cheng, J.C. On-demand inverse design of acoustic metamaterials using probabilistic generation network. *Sci. China Phys. Mech.* **2023**, *66*, 224311. [CrossRef]
49. Ma, X.; Xiang, H.; Yang, X.; Xiang, J. Dual band gaps optimization for a two-dimensional phoxonic crystal. *Phys. Lett. A* **2021**, *391*, 127137. [CrossRef]
50. Yuksel, O.; Yilmaz, C. Shape optimization of phononic band gap structures incorporating inertial amplification mechanisms. *J. Sound Vib.* **2015**, *355*, 232–245. [CrossRef]
51. Jia, Z.; Bao, Y.; Luo, Y.; Wang, D.; Zhang, X.; Kang, Z. Maximizing acoustic band gap in phononic crystals via topology optimization. *Int. J. Mech. Sci.* **2024**, *270*, 109107. [CrossRef]
52. Li, Y.; Luo, Y.; Zhang, X. Topological design of phononic crystals for multiple wide band gaps. *J. Sound Vib.* **2022**, *529*, 116962. [CrossRef]
53. Wu, Q.; He, J.; Chen, W.; Li, Q.; Liu, S. Topology optimization of phononic crystal with prescribed band gaps. *Comput. Meth. Appl. Mech. Eng.* **2023**, *412*, 116071. [CrossRef]
54. He, J.; Kang, Z. Achieving directional propagation of elastic waves via topology optimization. *Ultrasonics* **2018**, *82*, 1–10. [CrossRef] [PubMed]
55. Chen, Y.; Huang, X.; Sun, G.; Yan, X.; Li, G. Maximizing spatial decay of evanescent waves in phononic crystals by topology optimization. *Comput. Struct.* **2017**, *182*, 430–447. [CrossRef]
56. Li, W.; Meng, F.; Li, Y.f.; Huang, X. Topological design of 3D phononic crystals for ultra-wide omnidirectional bandgaps. *Struct. Multidiscip. Optim.* **2019**, *60*, 2405–2415. [CrossRef]
57. Van den Boom, S.J.; Abedi, R.; van Keulen, F.; Aragón, A.M. A level set-based interface-enriched topology optimization for the design of phononic crystals with smooth boundaries. *Comput. Meth. Appl. Mech. Eng.* **2023**, *408*, 115888. [CrossRef]

58. Gao, H.; Qu, Y.; Meng, G. Topology Optimization and Wave Propagation of Three-Dimensional Phononic Crystals. *J. Vib. Acoust.* **2022**, *145*, 011002. [CrossRef]
59. Khawale, R.P.; Bhattacharyya, S.; Rai, R.; Dargush, G.F. Efficient dynamic topology optimization of 2D metamaterials based on a complementary energy formulation. *Comput. Struct.* **2024**, *299*, 107371. [CrossRef]
60. Li, W.; Hu, J.; Lu, G.; Huang, X. Topology optimization of acoustic bandgap crystals for topological insulators. *Eng. Comput.* **2024**, *40*, 2581–2594. [CrossRef]
61. Hinton, G.E.; Osindero, S.; Teh, Y.W. A Fast Learning Algorithm for Deep Belief Nets. *Neural Comput.* **2006**, *18*, 1527–1554. [CrossRef] [PubMed]
62. Krizhevsky, A.; Sutskever, I.; Hinton, G.E. ImageNet classification with deep convolutional neural networks. *Commun. ACM* **2017**, *60*, 84–90. [CrossRef]
63. Mikolov, T.; Chen, K.; Corrado, G.; Dean, J. Efficient estimation of word representations in vector space. *arXiv* **2013**, arXiv:1301.3781. [CrossRef]
64. Mikolov, T.; Sutskever, I.; Chen, K.; Corrado, G.; Dean, J. Distributed representations of words and phrases and their compositionality. *arXiv* **2013**, arXiv:1310.4546. [CrossRef]
65. Hinton, G.; Deng, L.; Yu, D.; Dahl, G.E.; Mohamed, A.-r.; Jaitly, N.; Senior, A.; Vanhoucke, V.; Nguyen, P.; Sainath, T.N. Deep neural networks for acoustic modeling in speech recognition: The shared views of four research groups. *IEEE Signal Process. Mag.* **2012**, *29*, 82–97. [CrossRef]
66. Silver, D.; Huang, A.; Maddison, C.J.; Guez, A.; Sifre, L.; Van Den Driessche, G.; Schrittwieser, J.; Antonoglou, I.; Panneershelvam, V.; Lanctot, M. Mastering the game of Go with deep neural networks and tree search. *Nature* **2016**, *529*, 484–489. [CrossRef]
67. Machado Filho, M.A.; Farmer, W.; Hsiao, C.L.; dos Santos, R.B.; Hultman, L.; Birch, J.; Ankit, K.; Gueorguiev, G.K. Density Functional Theory-Fed Phase Field Model for Semiconductor Nanostructures: The Case of Self-Induced Core–Shell InAlN Nanorods. *Cryst. Growth Des.* **2024**, *24*, 4717–4727. [CrossRef]
68. Zhang, T.; Wang, J.; Liu, Q.; Zhou, J.; Dai, J.; Han, X.; Zhou, Y.; Xu, K. Efficient spectrum prediction and inverse design for plasmonic waveguide systems based on artificial neural networks. *Photonics Res.* **2019**, *7*, 368–380. [CrossRef]
69. Liu, F.; Jiang, X.; Wang, X.; Wang, L. Machine learning-based design and optimization of curved beams for multistable structures and metamaterials. *Extrem. Mech. Lett.* **2020**, *41*, 101002. [CrossRef]
70. Chen, J.; Chen, Y.; Xu, X.; Zhou, W.; Huang, G. A physics-guided machine learning for multifunctional wave control in active metabeams. *Extrem. Mech. Lett.* **2022**, *55*, 101827. [CrossRef]
71. White, D.A.; Arrighi, W.J.; Kudo, J.; Watts, S.E. Multiscale topology optimization using neural network surrogate models. *Comput. Meth. Appl. Mech. Eng.* **2019**, *346*, 1118–1135. [CrossRef]
72. Xu, L.; Zhang, W.; Yao, W.; Youn, S.K.; Guo, X. Machine learning accelerated MMC-based topology optimization for sound quality enhancement of serialized acoustic structures. *Struct. Multidiscip. Optim.* **2024**, *67*, 85. [CrossRef]
73. Shendy, M.; Jaradat, M.A.; Alkhader, M.; Abu-Nabah, B.A.; Venkatesh, T.A. Hybrid intelligent framework for designing band gap-rich 2D metamaterials. *Int. J. Solids Struct.* **2024**, *304*, 113053. [CrossRef]
74. Kim, S.; Jwa, M.; Lee, S.; Park, S.; Kang, N. Deep learning-based inverse design for engineering systems: Multidisciplinary design optimization of automotive brakes. *Struct. Multidiscip. Optim.* **2022**, *65*, 323. [CrossRef]
75. Li, J.; Ye, H.; Zhang, X.; Wei, N. Adjustable mechanical properties design of microstructure by using generative and adversarial network with gradient penalty. *Mech. Adv. Mater. Struct.* **2024**, *31*, 1059–1070. [CrossRef]
76. Li, J.; Ye, H.; Wei, N.; Zhang, X. ResUNet involved generative adversarial network-based topology optimization for design of 2D microstructure with extreme material properties. *Math. Mech. Solids* **2024**, *29*, 1683–1698. [CrossRef]
77. Vlassis, N.N.; Sun, W. Denoising diffusion algorithm for inverse design of microstructures with fine-tuned nonlinear material properties. *Comput. Meth. Appl. Mech. Eng.* **2023**, *413*, 116126. [CrossRef]
78. Lee, D.; Youn, B.D.; Jo, S.H. Deep-learning-based framework for inverse design of a defective phononic crystal for narrowband filtering. *Int. J. Mech. Sci.* **2023**, *255*, 108474. [CrossRef]
79. Liu, D.; Tan, Y.; Khoram, E.; Yu, Z. Training deep neural networks for the inverse design of nanophotonic structures. *ACS Photonics* **2018**, *5*, 1365–1369. [CrossRef]
80. Hinton, G.E.; Salakhutdinov, R.R. Reducing the dimensionality of data with neural networks. *Science* **2006**, *313*, 504–507. [CrossRef]
81. Goodfellow, I.; Pouget-Abadie, J.; Mirza, M.; Xu, B.; Warde-Farley, D.; Ozair, S.; Courville, A.; Bengio, Y. Generative adversarial networks. *Commun. ACM* **2020**, *63*, 139–144. [CrossRef]
82. Liu, C.X.; Yu, G.L.; Zhao, G.Y. Neural networks for inverse design of phononic crystals. *AIP Adv.* **2019**, *9*, 085223. [CrossRef]
83. Li, J.; Yin, J.; Li, S.; Zhang, Z.; Liu, X. Inverse design of nano-sized FGM phononic crystals with anticipated band gaps using probabilistic generation based deep-learning network. *Eng. Struct.* **2024**, *312*, 118222. [CrossRef]

84. Chen, D.; Li, Y.; Pan, Z.; Li, X.; Xu, T.; Li, X. Low-frequency vibration bandgaps and deep learning-based intelligent design method of Y-shaped core sandwich metabeams. *Compos. Struct.* **2024**, *341*, 118214. [CrossRef]
85. Li, X.; Ning, S.; Liu, Z.; Yan, Z.; Luo, C.; Zhuang, Z. Designing phononic crystal with anticipated band gap through a deep learning based data-driven method. *Comput. Meth. Appl. Mech. Eng.* **2020**, *361*, 112737. [CrossRef]
86. Xiao, L.; Cao, Z.; Lu, H.; Cai, Y. Controllable and scalable gradient-driven optimization design for two-dimensional metamaterials based on deep learning. *Compos. Struct.* **2024**, *337*, 118072. [CrossRef]
87. Jiang, W.; Zhu, Y.; Yin, G.; Lu, H.; Xie, L.; Yin, M. Dispersion relation prediction and structure inverse design of elastic metamaterials via deep learning. *Mater. Today Phys.* **2022**, *22*, 100616. [CrossRef]
88. Sharma, A.K.; Joglekar, M.M.; Joglekar, D.M.; Alam, Z. Topology optimization of soft compressible phononic laminates for widening the mechanically tunable band gaps. *Compos. Struct.* **2022**, *289*, 115389. [CrossRef]
89. Galich, P.I.; Fang, N.X.; Boyce, M.C.; Rudykh, S. Elastic wave propagation in finitely deformed layered materials. *J. Mech. Phys. Solids* **2017**, *98*, 390–410. [CrossRef]

materials

MDPI

Article

Optimizing Carbon Dot—TiO_2 Nanohybrids for Enhanced Photocatalytic Hydrogen Evolution

Pinelopi P. Falara [1,2], Nikolaos Chatzikonstantinou [1], Adamantia Zourou [1], Polychronis Tsipas [2], Elias Sakellis [2], Eleni Alexandratou [3], Nektarios K. Nasikas [2,4], Konstantinos V. Kordatos [1,*] and Maria Antoniadou [2,5,*]

1 School of Chemical Engineering, National Technical University of Athens, 9 Iroon Polytechniou St., Zografou, 15780 Athens, Greece; pin.falara@gmail.com (P.P.F.); nick.chatziko@gmail.com (N.C.); adamantia_zourou@outlook.com.gr (A.Z.)

2 Institute of Nanoscience and Nanotechnology, National Center for Scientific Research "Demokritos", Agia Paraskevi, 15341 Athens, Greece; p.tsipas@inn.demokritos.gr (P.T.); e.sakellis@inn.demokritos.gr (E.S.); nasikas@sse.gr (N.K.N.)

3 School of Electrical and Computer Engineering, National Technical University of Athens, 9 Iroon Polytechniou St., Zografou, 15780 Athens, Greece; ealexan@central.ntua.gr

4 Department of Military Studies, Division of Mathematics and Engineering Sciences, Hellenic Army Academy, Vari, 16673 Athens, Greece

5 Department of Chemical Engineering, University of Western Macedonia, 50100 Kozani, Greece

* Correspondence: kordatos@central.ntua.gr (K.V.K.); mantoniadou@uowm.gr (M.A.)

Abstract: CDs/TiO_2 nanohybrids were synthesized and tested for photocatalytic H_2 production from aqueous media through simulated solar light-driven photocatalytic reactions. Firstly, three different types of CDs were prepared through green methods, specifically hydrothermal treatment and microwave irradiation, using citric acid and urea as precursors in varying molar ratios. After a multi-step purification procedure, impurity-free CDs were obtained. The as-synthesized CDs were thoroughly characterized using UV-Vis, FT-IR, and PL spectroscopy, along with HR-TEM. The results revealed that the size and optical and physicochemical properties of CDs can be tailored by selecting the precursors' ratio and the synthetic approach. The heterostructured CDs/TiO_2 photocatalysts were formed solvothermally and were analyzed using UV-Vis/DRS, FT-IR, and XPS techniques, which confirmed the effective incorporation of CDs and the improved properties of TiO_2. The use of sacrificial reagents is among the most common strategies for enhancing H_2 production from water through photocatalytic processes; herein, ethanol was selected as a green liquid organic hydrogen carrier. A maximum H_2 production rate of 0.906 μmol H_2/min was achieved, while the recyclability study demonstrated that the photocatalyst maintained stable performance during multiple cycles of reuse. Thus, optimizing the synthesis conditions of CDs/TiO_2 nanohybrids resulted in the creation of environmentally friendly and reusable photocatalysts.

Keywords: carbon dots; nanohybrids; photocatalysis; hydrogen production; sacrificial agent

1. Introduction

Over the past few decades, social awareness regarding environmental and energy matters has grown rapidly. Excessive use of fossil fuels as a primary energy source has led to the discharge of constantly increasing quantities of hazardous gases, such as carbon and nitrogen oxides, into the atmosphere. Meanwhile, the continuous depletion of the available fossil fuels leads to a constantly escalating energy crisis. These essential issues have directed the scientific community towards the pursuit of innovative solutions that combine reduced emissions with the development of renewable fuels [1].

Hydrogen (H_2) is considered one of the most promising energy sources, causing ever-rising interest from the scientific world. This growing attention is attributed to hydrogen's remarkable properties, including its high heating value, abundance in nature, versatility in production methods, and environmentally friendly combustion [2]. Among the various hydrogen production methods, the most common is hydrocarbon reforming, which can be performed through steam reforming, partial oxidation, or autothermal processes. However, this method is not renewable, as it relies on the use of hydrocarbons [3,4]. One of the most modern and environmentally friendly methods for hydrogen production is photocatalysis [5,6], a novel technique that utilizes solar energy to produce sustainable green hydrogen. For this purpose, the two main photocatalytic methods are water splitting [7,8] and the reforming of organic compounds [9–11]. The photocatalyst in these processes is typically an effective semiconductor exhibiting specific properties such as broad absorption spectrum, high charge separation efficiency, photostability, chemical stability, non-toxicity, low flammability, affordability, and availability [12]. Titanium dioxide (TiO_2) is the most widely recognized photocatalyst due to its exceptional properties, including suitable energy level configuration, resistance to corrosion, low harmfulness, chemical stability, and cost-effectiveness. However, despite these advantages, TiO_2 bears notable drawbacks, such as a relatively large energy gap and high electron–hole recombination rate. Thus, the material absorbs only a small fraction of the electromagnetic spectrum (wavelengths below 400 nm) and presents limited potential in practical applications [13–15].

On the other hand, carbon nanomaterials (nanotubes, fullerenes, graphene, and carbon dots) are a relatively new and rapidly advancing research area, which have applications across diverse fields such as energy [16,17], environment [18–20], and medicine [21,22]. Focusing on carbon dots (CDs), they are composed of a spherical carbon core with various functional groups (carboxyl, carbonyl, and hydroxyl groups) present on their surface [23]. CDs exhibit numerous advantageous properties, including optical characteristics such as electromagnetic radiation absorption and strong photoluminescence, high electronic conductivity, biocompatibility, high quantum efficiency, and an environmentally friendly and straightforward production process. The most commonly employed methods for synthesizing CDs are the hydrothermal and microwave irradiation methods. Both are categorized as bottom-up approaches and are distinguished by their ability to produce small-sized nanoparticles at a low cost [24]. Additionally, these methods allow the adjustment of the CDs' properties by altering factors such as temperature and precursor concentration. In recent years, the application of CDs as co-catalysts in photocatalysis, in synergy with a semiconductor, has been extensively studied in various technological fields [25–27]. CDs are widely used in photocatalytic H_2 production due to their excellent electron transfer properties, tunable band gaps and ability to facilitate charge separation, reducing electron–hole recombination. Among the most common photocatalysts combined with CDs are TiO_2, CdS, and graphitic carbon nitride (g-C_3N_4) [28]. Specifically, in the synthesis of CDs/TiO_2 nanohybrid materials, CDs—often doped with heteroatoms such as nitrogen—serve a dual purpose. Firstly, they reduce the energy gap of TiO_2, thus expanding its absorption spectrum. Secondly, they reduce the thermodynamically favored effect of recombination of charge carriers, as they have the ability of receiving excited electrons, thereby preventing their recombination with holes [29,30].

Therefore, it is proven that integrating carbon nanomaterials can improve the photocatalytic efficiency of TiO_2 by promoting charge separation and extending light absorption, providing alternatives to noble metal modifications (e.g., Pt, Au, Ag) [31]. CDs are much cheaper than noble metals like Pt, making them a more economical choice for large-scale applications. Simultaneously, carbon nanomaterials are sustainable, whereas Pt and other noble metals are scarce and expensive to extract. Also, CDs are derived from organic sources and do not pose the same environmental concerns as noble metals [32,33].

This study contributes to the advancement of photocatalytic materials, providing a comprehensive comparative analysis of three distinct types of CDs and CDs/TiO_2 hybrid materials synthesized through green and facile methods, revealing how variations in precursor ratios and synthesis techniques significantly influence their physicochemical properties, ultimately determining their effectiveness in enhancing photocatalytic performance for sustainable hydrogen production. The obtained CD materials were first synthesized via hydrothermal treatment and microwave irradiation, using citric acid and urea as precursors in two different molar ratios, and were fully characterized using ultraviolet-visible (UV-Vis) spectroscopy, Fourier transform infrared (FT-IR) spectroscopy, photoluminescence (PL) spectroscopy, and high-resolution transmission electron microscopy (HR-TEM). In addition, it is important to note that during this work, CDs devoid of impurities were obtained after the purification procedure, which was carried out in consecutive steps. The selection of both the precursors and the synthesis method significantly influences the physicochemical properties of CDs (e.g., surface chemistry, optical properties, etc.). Furthermore, the presence of abundant functional groups on the surface of CDs not only facilitates their easy modification but also enables their combination with various materials, leading to heterostructures with well-defined characteristics. Subsequently, CDs/TiO_2 nanohybrids were prepared via a solvothermal process, and their optical and structural properties were comprehensively analyzed using various techniques, including X-ray photoelectron spectroscopy (XPS). The resulting hybrid materials were employed as photocatalysts for H_2 production, utilizing an organic compound as a sacrificial agent. The findings demonstrated that H_2 evolution is significantly influenced by the nature, chemical structure, and concentration of CDs in the heterostructured photocatalysts. By optimizing the synthesis conditions of CDs/TiO_2 nanohybrids, environmentally friendly and reusable photocatalysts were developed, presenting enhanced H_2 evolution rates when ethanol, a representative member of the green liquid organic hydrogen carriers' family, was selected as the sacrificial agent.

2. Materials and Methods

2.1. Materials

All chemicals of analytical grade were utilized without further purification. Urea was obtained from Sigma-Aldrich (St. Louis, MO, USA); citric acid was purchased from Honeywell (Charlotte, NC, USA). The nanocrystalline TiO_2 powder employed in this study was commercial Degussa P25 (Evonik, Essen, Germany). Absolute ethanol (C_2H_6O, $\geq$99%) was provided by Thermo Scientific (Waltham, MA, USA) and Fischer Scientific (Hampton, NH, USA). Deionized water was used during all procedures.

2.2. Preparation of CDs and CDs/TiO_2 Nanohybrids

CDs were synthesized through two distinct methods—(i) hydrothermal treatment process and (ii) microwave irradiation—both of which are described in detail in the following section. The samples named CDs-1 and CDs-2 were prepared by a hydrothermal synthetic route using citric acid and urea in a mass ratio of 1:1 and in a molar ratio of 1:100, respectively. Briefly, the predetermined quantities of citric acid and urea were dissolved into 10.0 mL of deionized water using magnetic stirring to ensure complete dissolution. The resulting solution was then transferred to a Teflon container, which was sealed within a stainless steel autoclave. Subsequently, the autoclave was placed in an oven and heated to 200 °C for 12 h. For the synthesis of the sample labeled as CDs-2-MW, microwave irradiation was employed. More specifically, a molar ratio of citric acid to urea of 1:100 was selected, and the precursors were dissolved in 10.0 mL of deionized water. The solution was vigorously stirred for approximately 10 min and then subjected to microwave irradiation in a standard household microwave of 700 W for 4 min. The resulting product was redissolved in deionized water with the assistance of an ultrasonic bath. In order to remove by-products and non-reactants, some purification methods were employed on all CD samples. Thus, the resulting solution was centrifuged at 6000 rpm for 20 min. The precipitate was discarded; the supernatant was then filtered through standard filter paper and a 0.22 μm pore-sized PET filter. A final purification step was then carried out using dialysis membranes of 0.8–1 kDa retention capability.

For the development of CDs/TiO_2 nanohybrids, a facile solvothermal method was employed. In particular, certain amounts of the as-synthesized CDs were added to a mixture containing 400 mg of Degussa P25 dispersed in 20.0 mL of deionized water and 6.0 mL of absolute ethanol. The formed mixture was stirred at room temperature for 30 min, transferred to a Teflon-sealed autoclave and heated at 140 °C for 4 h [34]. The resulting photocatalysts were rinsed three times with deionized water, collected by centrifugation and dried at 80 °C overnight.

2.3. Characterization Methods

Ultraviolet-visible (UV-Vis) spectroscopy measurements were performed on a Varian Cary 50 UV-Vis spectrophotometer (Agilent Technologies, Santa Clara, CA, USA).

Fourier transform infrared (FT-IR) spectroscopy was carried out with a Jasco FTIR 4200 spectrometer in the range of 400–4000 cm^{-1} using KBr pellets (Jasco, Tokyo, Japan).

Fluorescence spectra of the synthesized CDs were recorded using a Perkin-Elmer (Springfield, IL, USA) LS 45 Luminescence Spectrometer. The slit width was set at 10 nm. The scan speed was adjusted to 480 nm min^{-1}.

Structural analysis of the synthesized CDs was performed using high-resolution transmission electron microscopy (HR-TEM). HR-TEM imaging was conducted on an FEI Talos F200i field-emission (scanning) transmission electron microscope (Thermo Fisher Scientific Inc., Waltham, MA, USA) operating at 200 keV, equipped with a windowless energy-dispersive spectroscopy microanalyzer (6T/100 Bruker, Hamburg, Germany). For sample preparation, a drop of CD aqueous dispersion was cast on a holey carbon film supported by a copper grid.

The optical band gap of the prepared heterostructures was measured using UV-Vis diffuse reflectance spectroscopy (UV-Vis/DRS). For this purpose, a Hitachi 3010 spectrophotometer (Hitachi, Tokyo, Japan) was used, fitted with a 60 mm diameter integrating sphere, while $BaSO_4$ served as the reference.

X-ray photoelectron spectroscopy (XPS) measurements were performed using a Mg Ka X-ray source with a photon energy of 1253.64 eV, and the spectra were collected with a PHOIBOS 100 (SPECS) hemispherical analyzer. Gaussian–Lorentzian shapes (Voigt functions) were used for the fitting of the recorded spectra after standard Shirley background subtraction.

2.4. Photocatalytic Setup for H_2 Generation and Detection

For the photocatalytic H_2 generation experiments, a cylindrical Pyrex glass reactor was used, equipped with fittings for the gas inlet–outlet. For the illumination, a 300 W Xenon lamp (Oriel, Stratford, CT, USA) was utilized, and the reactor was positioned 10 cm away. H_2 detection was conducted using an SRI 8610C gas chromatograph (SRI instruments, Torrance, CA, USA), with Argon (Ar) as the carrier gas. Calibration of the chromatograph signal was performed using a standard mixture containing 0.25% v/v H_2 in Ar. The radiation intensity at the location of the reactor was determined at 1 kW/m^2 (1 sun illumination) using an Oriel Radiant Power Meter, while samples were retrieved every 10 min with an automatic gas sampling valve to monitor H_2 concentration in the reactor output over the illumination period. For every trial, 100.0 mg of photocatalyst was suspended in 100.0 mL of a 25% v/v ethanol aqueous solution, as this concentration was found to be optimal for enhancing hydrogen production across various photocatalytic processes [35–37]. Throughout the experiments, the solution underwent constant stirring; it was first degassed with Ar flow before switching on the lamp to initiate illumination.

3. Results and Discussion

3.1. Characterization of CDs

Various techniques were applied to reveal the optical, structural and morphological properties of the as-synthesized CDs. The UV-Vis spectroscopy characterization results of the synthesized CDs are presented in Figure 1a. In the UV-Vis spectrum of the CDs-1 sample, distinct absorption peaks are observed at wavelengths of 200 nm and 338 nm. The peak at 200 nm corresponds to $\pi \rightarrow \pi^*$ electronic transitions in carbon–carbon double bonds (C=C), whereas the peak at 338 nm is attributed to $n \rightarrow \pi^*$ transitions in C=O or C=N bonds. Similarly, in the UV-Vis spectrum of the CDs-2 sample, the first peak appears at 200 nm, consistent with $\pi \rightarrow \pi^*$ transitions associated with C=C bonds. The second peak is slightly shifted to a lower wavelength, more specifically at 316 nm, indicating $n \rightarrow \pi^*$ transitions related to C=O or C=N bonds. However, there is significant differentiation in the UV-Vis spectrum of the CDs-2-MW sample, where new absorption peaks appear. Specifically, although the first peak, observed at 200 nm, remains unchanged, three new peaks emerge at 250 nm, 272 nm, and 410 nm. These peaks are attributed to $n \rightarrow \pi^*$ transitions in the electrons of the C=N, C=O, and O–H or COOH bonds, respectively. The appearance of these new peaks is possibly due to the use of microwave irradiation as a synthesis method, as they are not present in the CDs-2 dots, which share the same precursor compounds in equal amounts [38,39]. FT-IR spectroscopy was employed to identify the presence of functional groups and chemical bonds in the synthesized materials. The FT-IR spectra of CDs-1, CDs-2, and CDs-2-MW are shown in Figure 1b. More specifically, the peaks located at 1685 cm^{-1} and 1577 cm^{-1} correspond to the stretching vibrations of C=C and C=O, respectively, whereas the peak observed at 1398 cm^{-1} is assigned to the C-O bonds. Additionally, the broad peak, which is present in all samples at approximately 3150–3400 cm^{-1}, could be attributed to the stretching vibrations of both O-H and N-H bonds [18,40].

The optical properties of the synthesized CDs were analyzed using photoluminescence (PL) spectroscopy. For each measurement, the excitation wavelength was chosen based on the peaks' positions observed in the UV-Vis absorption spectra (Figure 1a), corresponding to the maximum absorption bands. The samples were appropriately diluted to prevent any limitations during instrumental measurements. Figure 1c illustrates the emission spectra of the CDs when excited at 200 nm, the wavelength corresponding to their maximum absorption. As observed from the emission spectra, the hydrothermally produced CDs-1 and CDs-2 emit in the blue region of the visible spectrum (380–500 nm). In contrast, CDs-2-MW, synthesized via microwave irradiation, exhibit a weak emission peak in the blue region; however, the sample displays a strong emission in the green region (500–570 nm), extending into the yellow region (575–585 nm). The photoluminescence of CDs-1 and CDs-2-MW is also visualized in the inset of Figure 1c. Furthermore, PL is known to be influenced by the size of the CDs. Yeh et al. [41] clarified this phenomenon, explaining that while particle size does not impact the non-bonding orbital level, significant variations are observed in the π, π^*, and σ^* orbital levels. This highlights the sensitivity of bonding (π) and anti-bonding (π^* and σ^*) orbitals to particle size. In this context, the observed redshift in the emission wavelength of CDs-2-MW, when compared to the emission of CDs-1 and CDs-2, is attributed to differences in particle size induced by the synthesis method. Based on the quantum confinement effect, it can be suggested that CDs-2-MW consist of larger nanoparticles than CDs-1 and CDs-2, as smaller particles exhibit higher-energy (shorter-wavelength) emissions.

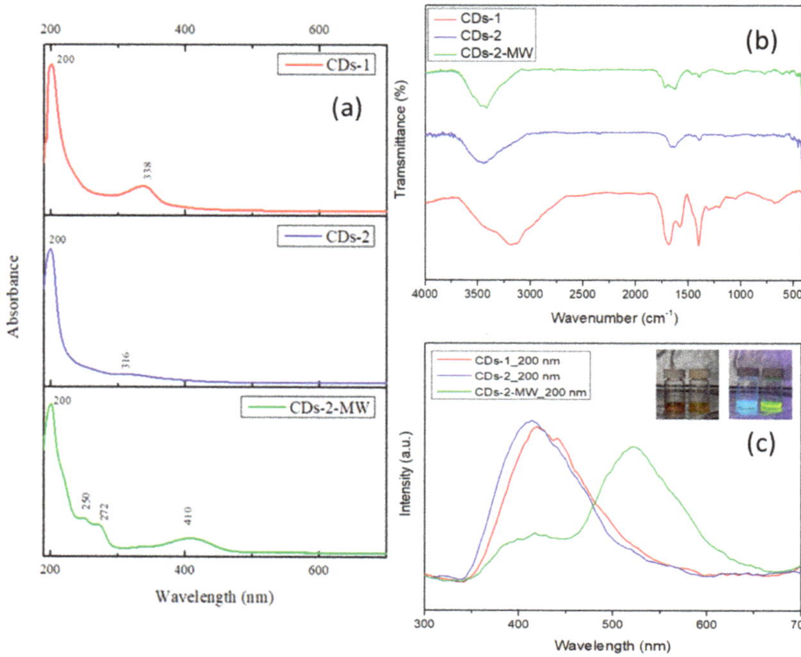

Figure 1. (**a**) UV-Vis absorption spectra of the synthesized CDs; (**b**) FT-IR spectra of the synthesized CDs; (**c**) PL emission spectra of CDs at 200 nm excitation wavelengths (inset: the photograph of CDs-1 (**left**) and CDs-2-MW (**right**) under solar and UV light).

The emission spectra of CDs-1, CDs-2, and CDs-2-MW, when excited at different wavelengths, are presented in Figure 2. CDs typically exhibit fluorescence properties that are dependent on the excitation wavelength, as commonly reported in the literature. However, there are exceptions where fluorescence emission remains unaffected by excitation wavelength. In this study, the optical behavior of CDs deviated from the usual excitation-dependent fluorescence trend, where the emission peaks gradually redshift with increasing excitation wavelength. Instead, the as-produced CDs demonstrated excitation-independent emission, suggesting that their surface states are uniform and defect-free [42,43]. Additionally, the deconvolution of one representative spectrum for each sample is provided as an inset within the corresponding graph. CDs-1 and CDs-2 exhibit comparable emission peaks, with CDs-1 showing peaks at 415.6 nm and 463.3 nm and CDs-2 displaying peaks at 402.5 nm and 448.4 nm, respectively. On the contrary, CDs-2-MW emitted at 396.8 nm, 510.6 nm, and 563.4 nm.

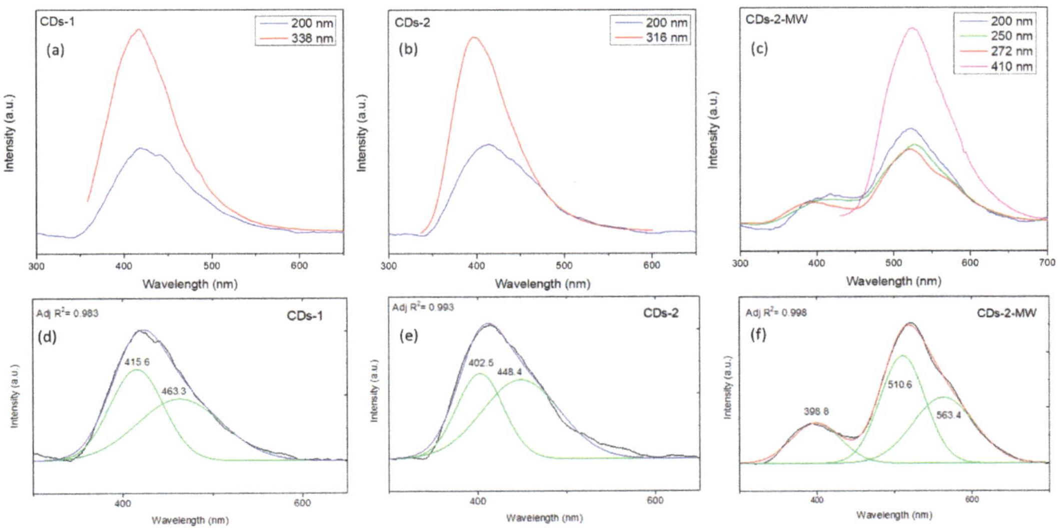

Figure 2. Emission spectra of (**a**) CDs-1; (**b**) CDs-2; and (**c**) CDs-2-MW at different excitation wavelengths, along with the deconvolution of a representative spectrum for (**d**) CDs-1; (**e**) CDs-2; and (**f**) CDs-2-MW.

In order to elucidate the morphology and structure of the prepared CDs, HR-TEM measurement was carried out. HR-TEM images of the sample CDs-1 show that the as-synthesized CDs are spherical (Figure 3a,b) with particle sizes ranging from 2.6 nm to 4.5 nm. The Gaussian distribution (Figure 3c) of a sample of 40 particles reveals the mean particle size of 3.45 ± 0.04 nm. In addition, the sample CDs-1 possesses a crystalline structure as indicated by a graphite lattice d-spacing of 0.22 nm (Figure 3d) [44].

The CDs-2-MW sample was also analyzed using HR-TEM. In Figure 4a,b, the nanoparticles appear spherical, with sizes between 5 and 8.9 nm. A Gaussian distribution (Figure 4c) based on a sample of 40 particles shows an average particle size of 6.35 ± 0.05 nm. Furthermore, the CDs-2-MW sample exhibits a crystalline structure, as evidenced by a graphite lattice d-spacing of 0.28 nm in Figure 4d [45].

The results clearly indicate that the size of the hydrothermally prepared CDs (CDs-1) is nearly half that of the microwave-synthesized CDs (CDs-2-MW). This aligns with the conclusions drawn from PL spectroscopy analysis. In summary, the hydrothermal synthesis produced smaller carbon dots, which correspond to blue light emission, whereas the microwave method yielded larger carbon dots, resulting in a redshift and green light emission.

3.2. *Characterization of CDs/TiO_2 Nanohybrids*

The CDs/TiO_2 nanohybrid photocatalysts were characterized using UV-Vis/DRS, FT-IR and XPS techniques. In order to study the optical properties of the synthesized nanohybrid materials, UV-Vis/DRS was conducted. The band gaps were calculated using the Kubelka–Munk equation and the Tauc plots (Figure 5a) [46–48]. These results indicate great potential, as they demonstrate that the incorporation of CDs enhances photon scattering, thereby improving light-harvesting efficiency. In particular, the calculated band gaps for the synthesized heterostructures are the following: pure TiO_2 exhibited a band gap of 3.21 eV, while samples CDs-1/TiO_2, CDs-2/TiO_2, and CDs-2-MW/TiO_2 presented band gaps of 3.12, 3.15, and 3.03 eV, respectively. The decrease in the band gap values of the nanohybrids when compared to pure TiO_2 can be attributed to the chemical interaction between TiO_2 and CDs, resulting in the formation of Ti–O–C bonds. Consequently, CDs play a crucial role in improving light absorption in the synthesized photocatalyst, as evidenced by previous studies [29,49].

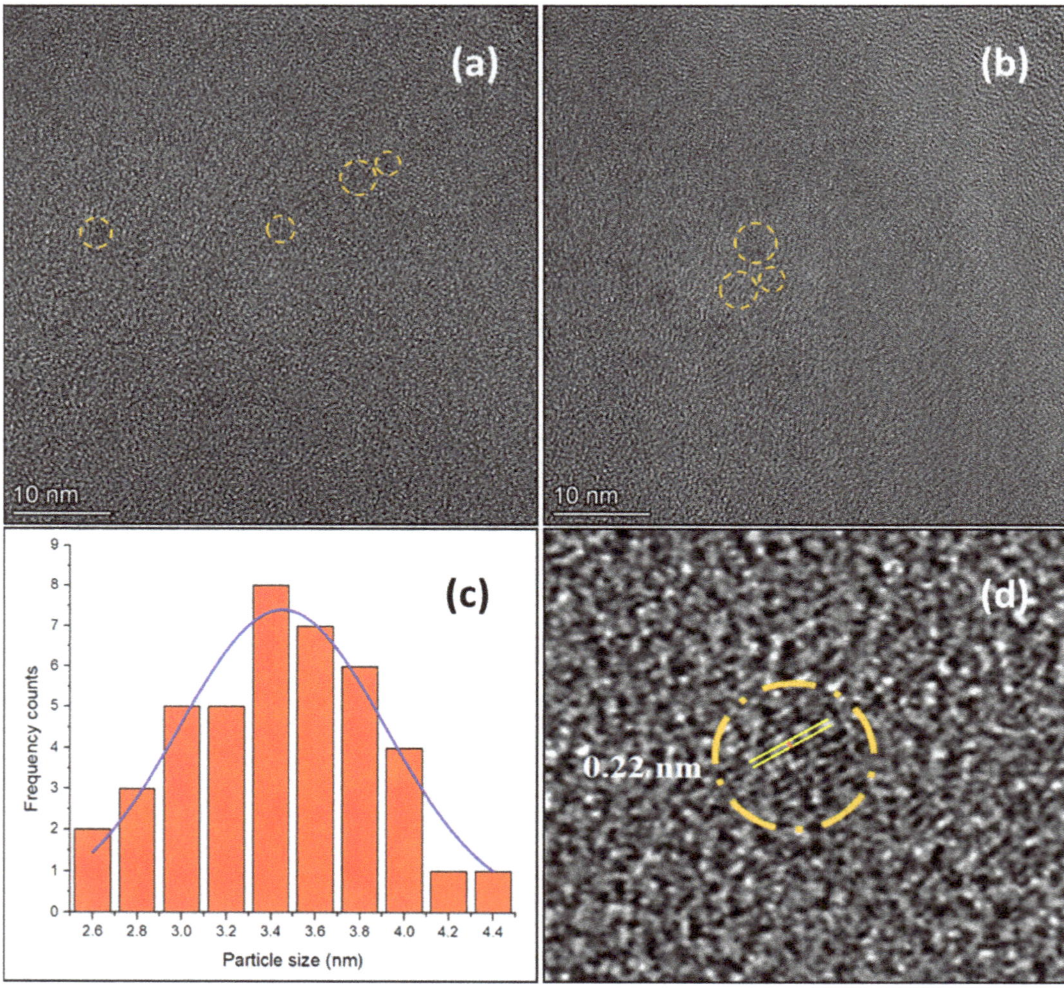

Figure 3. (**a**,**b**) HR-TEM images of CDs-1; (**c**) particle size histogram fitted with Gaussian distribution function; (**d**) graphitic core lattice.

Furthermore, FT-IR spectroscopy was conducted to identify the presence of functional groups and chemical bonds in the synthesized nanohybrids. In the FT-IR spectra of CDs/TiO_2 heterostructures, which are observed in Figure 5b, the broad peak located at ~660 cm^{-1} is characteristic of the Ti-O-Ti stretching vibrations [37]. As it is noticed, the FT-IR spectra of both TiO_2 and CDs/TiO_2 present similar results; this may be attributed to the fact that TiO_2 is in excess in the heterostructures and peaks attributed to carbon atoms bonds are likely to be obscured. Therefore, further investigation is deemed necessary in order to gain a more comprehensive understanding of the material's composition and bonding characteristics.

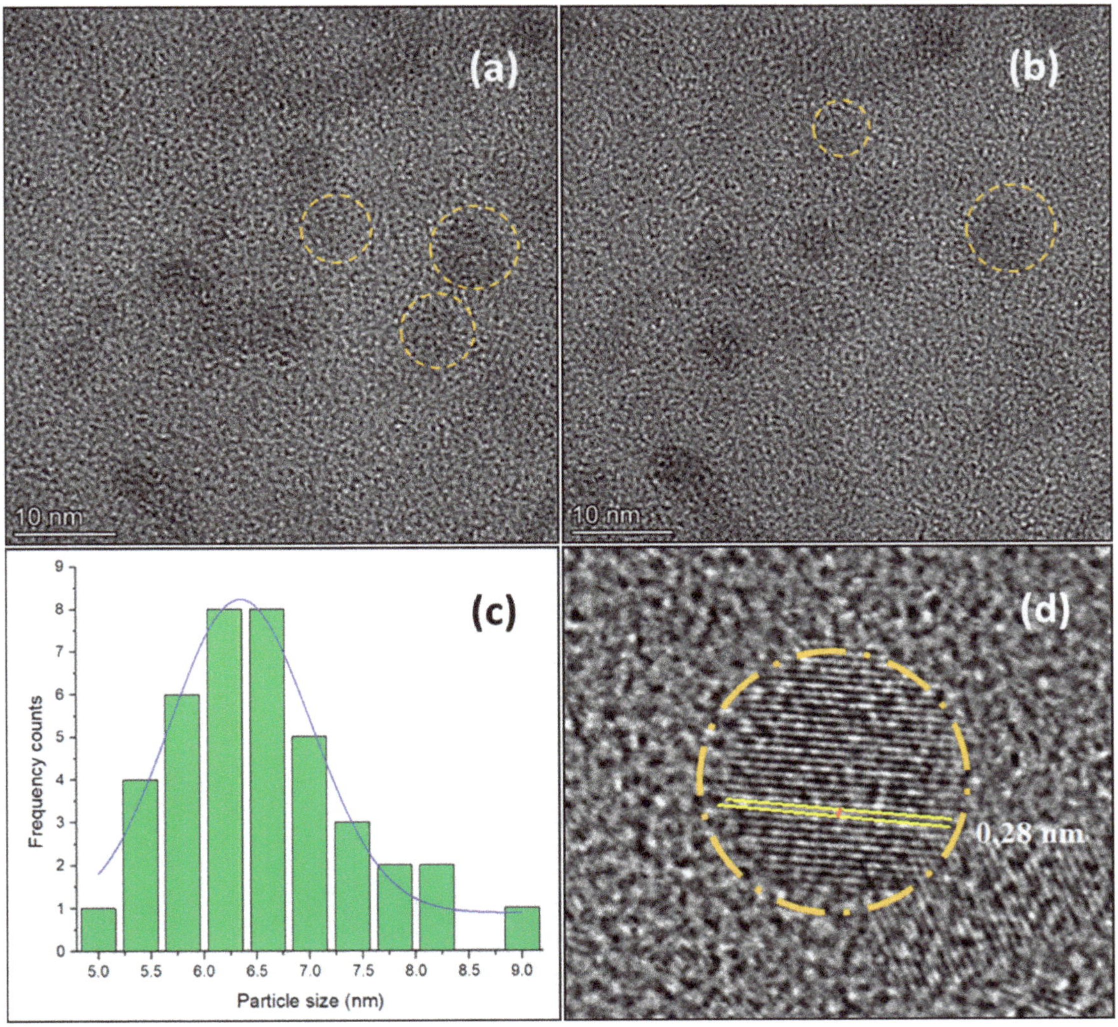

Figure 4. (**a**,**b**) HR-TEM images of CDs-2-MW; (**c**) particle size histogram fitted with Gaussian distribution function; (**d**) graphitic core lattice.

XPS measurements were performed to analyze the chemical state and the surface properties of the synthesized nanohybrids as well as to verify any incorporation of CDs into the TiO_2.

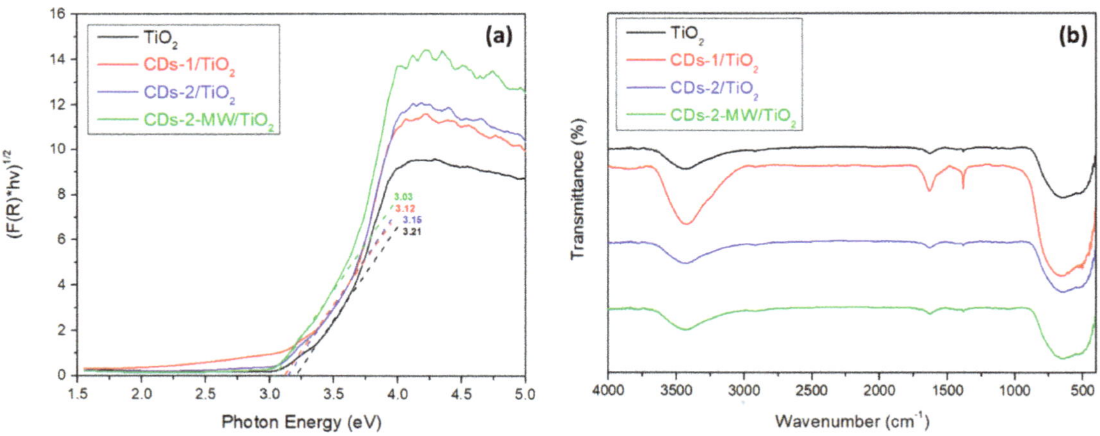

Figure 5. (**a**) Tauc plots illustrating the optical band gaps of pure TiO_2 and CDs/TiO_2 nanohybrid samples; (**b**) FT-IR spectra of TiO_2 and CDs/TiO_2 heterostructures.

As shown in Figure 6, the experimental C1s core level peak can be deconvoluted to three fitted peaks. The main peak at ~284.6 eV corresponds to C=C/C–C bonds, whereas the two relatively weak peaks observed at ~286.0 eV and ~288.5 eV correspond to C-O/C-N and HO-C=O functional groups, respectively [44,50]. Regarding the XPS spectrum of O1s (Figure 7a–c), three peaks are observed at binding energies ~529.5, ~531.6, and ~533.0 eV, which can be ascribed to Ti-O, C=O/C-O, and –OH bonds adsorbed on the surface, respectively [51–53].

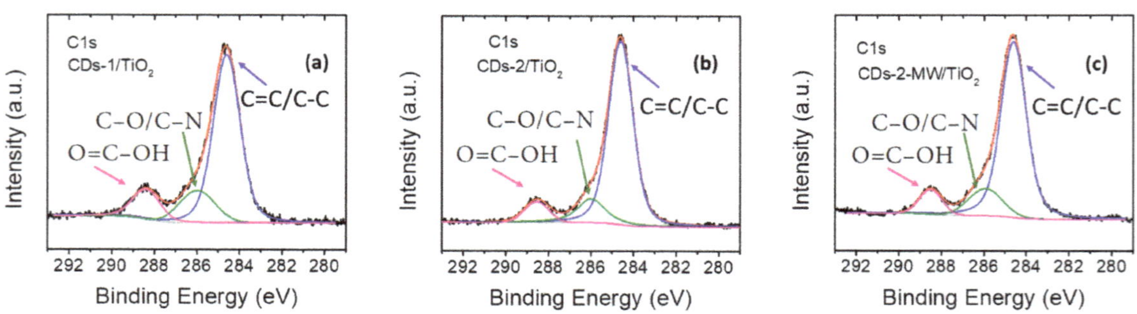

Figure 6. C1s XPS spectra of (**a**) CDs-1/TiO_2; (**b**) CDs-2/TiO_2; and (**c**) CDs-2-MW/TiO_2.

The Ti2p spectra of the three prepared heterostructures (Figure 7d–f) are similar and consist of a doublet with spin-orbit splitting of ~5.7 eV at binding energies 458.2 eV and 463.9 eV, which are attributed to Ti $2p_{3/2}$ and Ti $2p_{1/2}$, respectively. These peaks appear at higher binding energies than the corresponding Ti $2p_{3/2}$ and Ti $2p_{1/2}$ peaks of pure TiO_2. This indicates that the linkage between CDs and TiO_2 is achieved via Ti–O–C bonds [51], demonstrating the well-established concept of CD incorporation into TiO_2.

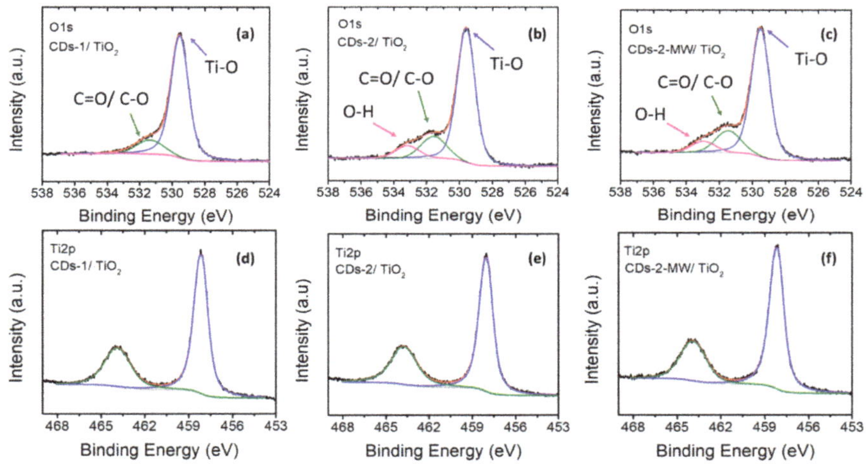

Figure 7. O1s XPS spectra (**a**) CDs-1/TiO_2; (**b**) CDs-2/TiO_2; and (**c**) CDs-2-MW/TiO_2 and Ti2p XPS spectra of (**d**) CDs-1/TiO_2; (**e**) CDs-2/TiO_2; and (**f**) CDs-2-MW/TiO_2.

3.3. Photocatalytic H_2 Production

This section presents and discusses the evaluation of the photocatalytic activity. Different volumes (50 μL, 100 μL, and 500 μL, respectively) of the prepared CDs' dispersion were added to form the corresponding CDs/TiO_2 heterostructures. The synthesized hybrid nanomaterials were assessed for their photocatalytic hydrogen production ability, using ethanol as a sacrificial agent. For each experiment, 100 mg of the respective photocatalyst was mixed with a 100 mL aqueous solution containing 25% v/v absolute ethanol. The temporal production of H_2 in the presence of CDs/TiO_2 hybrid photocatalysts is demonstrated in Figure 8a.

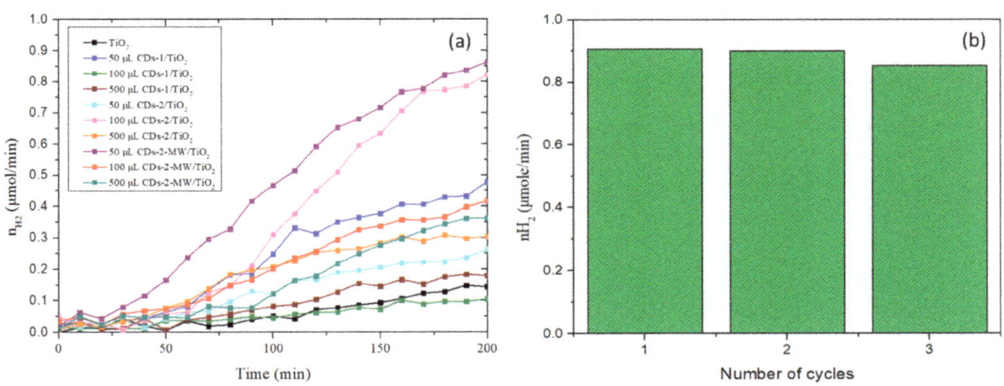

Figure 8. (**a**) Photocatalytic evolution of hydrogen via ethanol reforming in the presence of CDs/TiO_2 hybrid nanomaterials under 1-sun illumination; (**b**) recyclability tests using the sample 50 μL CDs-2-MW/TiO_2.

Regarding the hydrogen production reaction, the diagram indicates that, for most samples, hydrogen production increases over time until it reaches a plateau around 200 min. This suggests that the maximum hydrogen yield is achieved by this point, and further measurements are unlikely to show a significant increase in the rate of the produced hydrogen. The plateau characteristics were influenced by the balance between the amount of photocatalyst and fuel, along with the intensity of the incident radiation. It was observed that, with the exception of one sample (100 μL CDs-1/TiO_2), all the synthesized nanohybrids produced a greater amount of hydrogen compared to pure TiO_2. The sample that achieved the highest hydrogen production rate was 50 μL CDs-2-MW/TiO_2, followed by 100 μL CDs-2/TiO_2 and 50 μL CDs-1/TiO_2 in second and third place, respectively. More specifically, the CDs-2-MW/TiO_2 photocatalyst achieved a rate of 0.906 μmol $H_2 \cdot min^{-1}$ (543.6 $\mu mol \cdot h^{-1} \cdot g^{-1}$), compared to pure TiO_2, which reached a rate of only 0.150 μmol $H_2 \cdot min^{-1}$ (90 $\mu mol \cdot h^{-1} \cdot g^{-1}$). To ensure that even smaller quantities would not result in a higher H_2 production rate, 20 μL of CDs-2-MW was also tested for the hybrid photocatalyst. However, the rate was significantly lower, so these results are not included. It is evident that the CDs-2-MW/TiO_2 heterostructures exhibited superior photocatalytic performance, which aligns with the UV-Vis/DRS results. This hybrid displayed a smaller band gap, leading to significantly improved light-harvesting efficiency.

Additionally, it can be observed that increasing the concentration of CDs in the hybrid nanomaterial does not necessarily lead to a rise in hydrogen production or, consequently, in the photocatalytic activity. This could be because a high concentration of CDs affects light scattering, thereby reducing the amount of light absorbed by the photocatalyst. Therefore, optimizing the nanohybrid does not necessarily require increasing the content of CDs in the hybrid nanomaterial but rather finding the optimal amount that maximizes the broadening of the absorbed wavelength without compromising the photocatalyst's ability to absorb light. A further evaluation of the hydrogen production results suggests that hybrids containing CDs synthesized using citric acid and urea in a molar ratio of 1:100 (i.e., CDs-2 and CDs-2-MW) exhibited superior photocatalytic activity. This improvement is attributed to the increased nitrogen content, which is claimed to facilitate charge redistribution, thereby enhancing electron transport properties [54]. Lastly, when comparing the two methods for synthesizing CDs, the microwave-assisted approach is considered more effective for nitrogen doping due to its rapid and uniform heating [55]. This characteristic further supports the enhanced photocatalytic performance of the nanohybrid 50 μL CDs-2-MW/TiO_2. The maximum photocatalytic H_2 production rates of all prepared nanohybrids are summarized in Table 1.

Table 1. Photocatalytic H_2 evolution rates of all prepared samples.

Sample	Maximum H_2 Production Rate (μmol/min)	Maximum H_2 Production Rate (μmol/(h·g))
TiO_2	0.150	90.0
50 μL CDs-1/TiO_2	0.465	279.0
100 μL CDs-1/TiO_2	0.110	66.0
500 μL CDs-1/TiO_2	0.179	107.4
50 μL CDs-2/TiO_2	0.262	157.2
100 μL CDs-2/TiO_2	0.846	507.6
500 μL CDs-1/TiO_2	0.315	189.0
50 μL CDs-2-MW/TiO_2	0.906	543.6
100 μL CDs-2-MW/TiO_2	0.417	250.2
500 μL CDs-2-MW/TiO_2	0.378	226.8

The stability and recyclability of the CDs/TiO_2 photocatalysts were also assessed through cycling experiments, as shown in Figure 8b. Specifically, the sample 50 μL CDs-2-MW/TiO_2 was tested over three consecutive cycles in the presence of ethanol. Each cycle was performed under the same conditions outlined. After each test, the photocatalyst was thoroughly washed with deionized water. The recyclability study demonstrated that the photocatalyst maintained consistent stability during repeated use for H_2 evolution. These results indicate the potential of CDs/TiO_2 photocatalysts for practical hydrogen production applications, as their stability over multiple cycles suggests long-term durability. The synthesized heterostructures could be further investigated for integration into photocatalytic systems, such as continuous-flow reactors. Future studies focusing on extended cycling experiments and real operational conditions will be crucial to evaluate their commercial viability.

To emphasize the novelty and advancements of this study in the field of photocatalytic H_2 production, a review of the recent literature on TiO_2-based photocatalysts has been carried out. The comparison, as shown in Table 2, aims to illustrate the contribution of this research as well as to distinguish its findings from existing studies.

Table 2. Summary of recent TiO_2-based photocatalysts for H_2 evolution.

Photocatalyst	Co-Catalyst	Sacrificial Agent	Maximum H_2 Production Rate	Ref.
CDs/TiO_2	-	25% *v*/*v* ethanol	0.906 μmol/min 54.36 μmol/h or 543.6 μmol/(h·g)	This work
	-	15% *v*/*v* methanol	603.92 μmol/(h·g)	[51]
	-	25% *v*/*v* methanol	9.8 μmol/h	[56]
	- Pt	0.3 M triethanolamine	472 μmol/(h·g) 1458 μmol/(h·g)	[57]
CDs/g-C_3N_4/TiO_2	Pt	10% *v*/*v* triethanolamine	580 μmol/(h·g)	[58]
Graphene/TiO_2	Pt	25% *v*/*v* methanol	668 μmol/h	[59]
g-C_3N_4/TiO_2	-	20% *v*/*v* methanol	110 μmol/(h·g)	[60]
Black phosphorus quantum dots/TiO_2	-	20% *v*/*v* methanol	112 μmol/(h·g)	[61]
Red phosphorus/TiO_2	Pt	-	215.5 μmol/(h·g)	[62]
Au/TiO_2	-	30% *v*/*v* ethanol	360 μmol/(h·g)	[63]
Cu/TiO_2	-	10% *v*/*v* methanol	18 mmol/(h·g)	[64,65]
NiO/TiO_2	-	25% *v*/*v* methanol	337 μmol/(h·g)	[66]

4. Conclusions

Distinct types of CDs were prepared through green methods (hydrothermal treatment and microwave irradiation) using citric acid and urea as precursors in different molar ratios. Following a purification procedure carried out in consecutive steps, CDs devoid of impurities were obtained. The as-synthesized CDs were characterized thoroughly using UV-Vis, FT-IR, and PL spectroscopy, as well as HR-TEM. The results indicated that the physicochemical properties (e.g., size, surface chemistry, optoelectronic properties, etc.) of CDs strongly depend on the precursor's selection and the synthetic approach. The presence of abundant functional groups on the surface of CDs enabled their combination with TiO_2 nanoparticles, leading to CDs/TiO_2 heterostructures with well-defined characteristics prepared via a solvothermal process. The produced CDs/TiO_2 nanomaterials were thoroughly characterized using UV-Vis/DRS, FT-IR, and XPS techniques, confirming the successful incorporation of CDs and the enhanced properties of TiO_2. Subsequently, the hybrids were employed as photocatalysts for H_2 production in aqueous media utilizing ethanol as a sacrificial agent. The obtained results demonstrated that H_2 evolution is significantly influenced by the nature, chemical structure, and concentration of CDs in the heterostructured photocatalysts. In conclusion, optimizing the synthesis conditions of CDs/TiO_2 nanohybrids led to the development of environmentally friendly and reusable photocatalysts, achieving enhanced H_2 evolution rates of up to 0.906 μmol H_2/min, which may be attributed to the strong interfacial interaction between CDs and TiO_2, which promotes efficient charge separation, using ethanol as a representative green liquid organic hydrogen carrier.

Author Contributions: Conceptualization, P.P.F., K.V.K. and M.A.; methodology, N.C., P.P.F., A.Z. and P.T.; software, P.P.F., A.Z. and N.C.; validation, P.P.F., N.C., P.T., E.A. and E.S.; formal analysis, P.P.F., N.C., A.Z. and P.T.; investigation, P.P.F., N.C., A.Z. and P.T.; data curation, P.P.F., N.C., A.Z., P.T., E.S., E.A. and N.K.N.; writing—original draft preparation, P.P.F., K.V.K. and M.A.; writing—review and editing, P.P.F., K.V.K. and M.A.; visualization, P.P.F., K.V.K. and M.A.; supervision, P.P.F., K.V.K. and M.A.; project administration, P.P.F., K.V.K. and M.A. All authors have read and agreed to the published version of the manuscript.

Funding: This research received no external funding.

Institutional Review Board Statement: Not applicable.

Informed Consent Statement: Not applicable.

Data Availability Statement: The original contributions presented in this study are included in the article. Further inquiries can be directed to the corresponding authors.

Conflicts of Interest: The authors declare no conflicts of interest.

References

1. Jaiswal, K.K.; Chowdhury, C.R.; Yadav, D.; Verma, R.; Dutta, S.; Jaiswal, K.S.; Sangmesh, B.; Karuppasamy, K.S.K. Renewable and Sustainable Clean Energy Development and Impact on Social, Economic, and Environmental Health. *Energy Nexus* **2022**, *7*, 100118. [CrossRef]
2. Li, Z.; Fang, S.; Sun, H.; Chung, R.J.; Fang, X.; He, J.H. Solar Hydrogen. *Adv. Energy Mater.* **2023**, *13*, 2203019. [CrossRef]
3. Nikolaidis, P.; Poullikkas, A. A Comparative Overview of Hydrogen Production Processes. *Renew. Sustain. Energy Rev.* **2017**, *67*, 597–611. [CrossRef]
4. Nnabuife, S.G.; Darko, C.K.; Obiako, P.C.; Kuang, B.; Sun, X.; Jenkins, K. A Comparative Analysis of Different Hydrogen Production Methods and Their Environmental Impact. *Clean Technol.* **2023**, *5*, 1344–1380. [CrossRef]
5. Zhang, J.; Zhao, Y.; Qi, K.; Liu, S. yuan $CuInS_2$ Quantum-Dot-Modified g-C_3N_4 S-Scheme Heterojunction Photocatalyst for Hydrogen Production and Tetracycline Degradation. *J. Mater. Sci. Technol.* **2024**, *172*, 145–155. [CrossRef]
6. Rafique, M.; Hajra, S.; Irshad, M.; Usman, M.; Imran, M.; Assiri, M.A.; Ashraf, W.M. Hydrogen Production Using TiO_2-Based Photocatalysts: A Comprehensive Review. *ACS Omega* **2023**, *8*, 25640–25648. [CrossRef] [PubMed]
7. Zheng, D.; Xue, Y.; Wang, J.; Varbanov, P.S.; Klemeš, J.J.; Yin, C. Nanocatalysts in Photocatalytic Water Splitting for Green Hydrogen Generation: Challenges and Opportunities. *J. Clean. Prod.* **2023**, *414*, 137700. [CrossRef]
8. Gao, Y.; Xu, B.; Cherif, M.; Yu, H.; Zhang, Q.; Vidal, F.; Wang, X.; Ding, F.; Sun, Y.; Ma, D.; et al. Atomic Insights for Ag Interstitial/Substitutional Doping into $ZnIn_2S_4$ Nanoplates and Intimate Coupling with Reduced Graphene Oxide for Enhanced Photocatalytic Hydrogen Production by Water Splitting. *Appl. Catal. B Environ.* **2020**, *279*, 119403. [CrossRef]
9. Antoniadou, M.; Vaiano, V.; Sannino, D.; Lianos, P. Photocatalytic Oxidation of Ethanol Using Undoped and Ru-Doped Titania: Acetaldehyde, Hydrogen or Electricity Generation. *Chem. Eng. J.* **2013**, *224*, 144–148. [CrossRef]
10. Chung, Y.H.; Han, K.; Lin, C.Y.; O'Neill, D.; Mul, G.; Mei, B.; Yang, C.M. Photocatalytic Hydrogen Production by Photo-Reforming of Methanol with One-Pot Synthesized Pt-Containing TiO_2 Photocatalysts. *Catal. Today* **2020**, *356*, 95–100. [CrossRef]
11. Muscetta, M.; Andreozzi, R.; Clarizia, L.; Di Somma, I.; Marotta, R. Hydrogen Production through Photoreforming Processes over Cu_2O/TiO_2 Composite Materials: A Mini-Review. *Int. J. Hydrogen Energy* **2020**, *45*, 28531–28552. [CrossRef]

12. Sendão, R.M.S.; Esteves da Silva, J.C.G.; Pinto da Silva, L. Applications of Fluorescent Carbon Dots as Photocatalysts: A Review. *Catalysts* **2023**, *13*, 179. [CrossRef]
13. Racovita, A.D. Titanium Dioxide: Structure, Impact, and Toxicity. *Int. J. Environ. Res. Public Health* **2022**, *19*, 5681. [CrossRef] [PubMed]
14. Ibrahim, I.; Belessiotis, G.V.; Antoniadou, M.; Kaltzoglou, A.; Sakellis, E.; Katsaros, F.; Sygellou, L.; Arfanis, M.K.; Salama, T.M.; Falaras, P. Silver Decorated $TiO_2/g\text{-}C_3N_4$ Bifunctional Nanocomposites for Photocatalytic Elimination of Water Pollutants under UV and Artificial Solar Light. *Results Eng.* **2022**, *14*, 100470. [CrossRef]
15. Kunnath Parambil, N.S.; Raphael, S.J.; Joseph, P.; Dasan, A. Recent Advancements toward Visible-Light-Driven Titania-Based Nanocomposite Systems for Environmental Applications: An Overview. *Appl. Surf. Sci. Adv.* **2023**, *18*, 100487. [CrossRef]
16. Kothandam, G.; Singh, G.; Guan, X.; Lee, J.M.; Ramadass, K.; Joseph, S.; Benzigar, M.; Karakoti, A.; Yi, J.; Kumar, P.; et al. Recent Advances in Carbon-Based Electrodes for Energy Storage and Conversion. *Adv. Sci.* **2023**, *10*, 2301045. [CrossRef]
17. Rajeshkumar, L.; Ramesh, M.; Bhuvaneswari, V.; Balaji, D. Carbon Nano-Materials (CNMs) Derived from Biomass for Energy Storage Applications: A Review. *Carbon Lett.* **2023**, *33*, 661–690. [CrossRef]
18. Zourou, A.; Ntziouni, A.; Roman, T.; Tampaxis, C.; Steriotis, T.; Gkouzia, G.; Alff, L.; Sanchez, D.E.; Terrones, M.; Kordatos, K.V. Synthesis and Characterization of Magnetic Carbon Dots (CDs)-Based Hybrid Material as an Adsorbent for the Removal of Organic Dye from Water. *Carbon* **2024**, *230*, 119612. [CrossRef]
19. Deng, A.; Yin, Y.; Liu, Y.; Xu, Y.; He, H.; Yang, S.; Qin, Q.; Sun, D.; Li, S. Unlocking the Potential of MOF-Derived Carbon-Based Nanomaterials for Water Purification through Advanced Oxidation Processes: A Comprehensive Review on the Impact of Process Parameter Modulation. *Sep. Purif. Technol.* **2023**, *318*, 123998. [CrossRef]
20. Fan, X.; Wei, G.; Quan, X. Carbon Nanomaterial-Based Membranes for Water and Wastewater Treatment under Electrochemical Assistance. *Environ. Sci. Nano* **2022**, *10*, 11–40. [CrossRef]
21. Karagianni, A.; Tsierkezos, N.G.; Prato, M.; Terrones, M.; Kordatos, K.V. Application of Carbon-Based Quantum Dots in Photodynamic Therapy. *Carbon* **2023**, *203*, 273–310. [CrossRef]
22. Jiang, H.; Xia, C.; Lin, J.; Garalleh, H.A.; Alalawi, A.; Pugazhendhi, A. Carbon Nanomaterials: A Growing Tool for the Diagnosis and Treatment of Diabetes Mellitus. *Environ. Res.* **2023**, *221*, 115250. [CrossRef]
23. Falara, P.P.; Zourou, A.; Kordatos, K.V. Recent Advances in Carbon Dots/2-D Hybrid Materials. *Carbon* **2022**, *195*, 219–245. [CrossRef]
24. Ozyurt, D.; Al Kobaisi, M.; Hocking, R.K.; Fox, B. Properties, Synthesis, and Applications of Carbon Dots: A Review. *Carbon Trends* **2023**, *12*, 100276. [CrossRef]
25. Cringoli, M.C.; Perathoner, S.; Fornasiero, P.; Marchesan, S. Carbon Nanostructures Decorated with Titania: Morphological Control and Applications. *Appl. Sci.* **2021**, *11*, 6814. [CrossRef]
26. Zhao, H.; Yu, X.; Li, C.F.; Yu, W.; Wang, A.; Hu, Z.Y.; Larter, S.; Li, Y.; Golam Kibria, M.; Hu, J. Carbon Quantum Dots Modified TiO_2 Composites for Hydrogen Production and Selective Glucose Photoreforming. *J. Energy Chem.* **2022**, *64*, 201–208. [CrossRef]
27. Camilli, E.; Foglia, M.L.; Bravo, J.P.; Copello, G.J.; Villanueva, M.E. Comparison among OH, N and P Functionalized Carbon Quantum Dots/TiO_2 Nanocomposites for Food Industry Wastewater Remediation. *Diam. Relat. Mater.* **2024**, *145*, 111103. [CrossRef]
28. Hareesh, K. A Review on Carbon Quantum Dot/Semiconductor-Based Nanocomposites as Hydrogen Production Photocatalysts. *RSC Adv.* **2024**, *14*, 23404–23422. [CrossRef]
29. Mozdbar, A.; Nouralishahi, A.; Fatemi, S.; Talatori, F.S. The Impact of Carbon Quantum Dots (CQDs) on the Photocatalytic Activity of TiO_2 under UV and Visible Light. *J. Water Process Eng.* **2023**, *51*, 103465. [CrossRef]
30. Choi, N.; Tang, C.; Park, Y.; Du, A.; Ayoko, G.A.; Hwang, Y.; Chae, S. Visible-Light-Driven Photocatalytic Degradation of Tetracycline Using Citric Acid and Lemon Juice-Derived Carbon Quantum Dots Incorporated TiO_2 Nanocomposites. *Sep. Purif. Technol.* **2024**, *350*, 127836. [CrossRef]
31. Liu, G.; Wang, L.; Yang, H.G.; Cheng, H.M.; Lu, G.Q. Titania-Based Photocatalysts—Crystal Growth, Doping and Heterostructuring. *J. Mater. Chem.* **2010**, *20*, 831–843. [CrossRef]
32. Molochas, C.; Tsiakaras, P. Carbon-Monoxide-Tolerant Pt-Based Electrocatalysts for H_2-Pemfc Applications: Current Progress and Challenges. *Catalysts* **2021**, *11*, 1127. [CrossRef]
33. Chen, B.B.; Liu, M.L.; Huang, C.Z. Carbon Dot-Based Composites for Catalytic Applications. *Green Chem.* **2020**, *22*, 4034–4054. [CrossRef]
34. Yu, H.; Zhao, Y.; Zhou, C.; Shang, L.; Peng, Y.; Cao, Y.; Wu, L.Z.; Tung, C.H.; Zhang, T. Carbon Quantum Dots/TiO_2 Composites for Efficient Photocatalytic Hydrogen Evolution. *J. Mater. Chem. A* **2014**, *2*, 3344–3351. [CrossRef]

35. Pajares, A.; Wang, Y.; Kronenberg, M.J.; Ramírez de la Piscina, P.; Homs, N. Photocatalytic H_2 Production from Ethanol Aqueous Solution Using TiO_2 with Tungsten Carbide Nanoparticles as Co-Catalyst. *Int. J. Hydrogen Energy* **2020**, *45*, 20558–20567. [CrossRef]
36. Strataki, N.; Antoniadou, M.; Dracopoulos, V.; Lianos, P. Visible-Light Photocatalytic Hydrogen Production from Ethanol-Water Mixtures Using a Pt-CdS-TiO_2 Photocatalyst. *Catal. Today* **2010**, *151*, 53–57. [CrossRef]
37. Falara, P.P.; Antoniadou, M.; Zourou, A.; Sakellis, E.; Kordatos, K.V. Carbon Dot-Titanium Dioxide (CD/TiO_2) Nanocomposites: Reusable Photocatalyst for Sustainable H2 Production via Photoreforming of Green Organic Compounds. *Coatings* **2024**, *14*, 131. [CrossRef]
38. Vercelli, B.; Donnini, R.; Ghezzi, F.; Sansonetti, A.; Giovanella, U.; La Ferla, B. Nitrogen-Doped Carbon Quantum Dots Obtained Hydrothermally from Citric Acid and Urea: The Role of the Specific Nitrogen Centers in Their Electrochemical and Optical Responses. *Electrochim. Acta* **2021**, *387*, 138557. [CrossRef]
39. Choi, D.; Ham, S.; Jang, D.J. Visible-Light Photocatalytic Reduction of Cr(VI) via Carbon Quantum Dots-Decorated TiO_2 Nanocomposites. *J. Environ. Chem. Eng.* **2018**, *6*, 1–8. [CrossRef]
40. Falara, P.P.; Ibrahim, I.; Zourou, A.; Sygellou, L.; Sanchez, D.E.; Romanos, G.E.; Givalou, L.; Antoniadou, M.; Arfanis, M.K.; Han, C.; et al. Bi-Functional Photocatalytic Heterostructures Combining Titania Thin Films with Carbon Quantum Dots (C-QDs/TiO_2) for Effective Elimination of Water Pollutants. *Environ. Sci. Pollut. Res.* **2023**, *30*, 124976–124991. [CrossRef]
41. Yeh, T.F.; Huang, W.L.; Chung, C.J.; Chiang, I.T.; Chen, L.C.; Chang, H.Y.; Su, W.C.; Cheng, C.; Chen, S.J.; Teng, H. Elucidating Quantum Confinement in Graphene Oxide Dots Based on Excitation-Wavelength-Independent Photoluminescence. *J. Phys. Chem. Lett.* **2016**, *7*, 2087–2092. [CrossRef] [PubMed]
42. Saputra, A.M.A.; Piliang, A.F.R.; Dellyansyah; Marpongahtun; Andriayani; Goei, R.; Risky, R.R.; Gea, S. Synthesis, Properties, and Utilization of Carbon Quantum Dots as Photocatalysts on Degradation of Organic Dyes: A Mini Review. *Catal. Commun.* **2024**, *187*, 106914. [CrossRef]
43. Zhou, Y.; Zahran, E.M.; Quiroga, B.A.; Perez, J.; Mintz, K.J.; Peng, Z.; Liyanage, P.Y.; Pandey, R.R.; Chusuei, C.C.; Leblanc, R.M. Size-Dependent Photocatalytic Activity of Carbon Dots with Surface-State Determined Photoluminescence. *Appl. Catal. B Environ.* **2019**, *248*, 157–166. [CrossRef]
44. Nguyen, K.G.; Baragau, I.A.; Gromicova, R.; Nicolaev, A.; Thomson, S.A.J.; Rennie, A.; Power, N.P.; Sajjad, M.T.; Kellici, S. Investigating the Effect of N-Doping on Carbon Quantum Dots Structure, Optical Properties and Metal Ion Screening. *Sci. Rep.* **2022**, *12*, 13806. [CrossRef] [PubMed]
45. Tian, X.T.; Yin, X.B. Carbon Dots, Unconventional Preparation Strategies, and Applications Beyond Photoluminescence. *Small* **2019**, *15*, 1901803. [CrossRef] [PubMed]
46. Ibrahim, I.; Belessiotis, G.V.; Elseman, A.M.; Mohamed, M.M.; Ren, Y.; Salama, T.M.; Mohamed, M.B.I. Magnetic $TiO_2/CoFe_2O_4$ Photocatalysts for Degradation of Organic Dyes and Pharmaceuticals without Oxidants. *Nanomaterials* **2022**, *12*, 3290. [CrossRef]
47. Mishra, V.; Warshi, M.K.; Sati, A.; Kumar, A.; Mishra, V.; Kumar, R.; Sagdeo, P.R. Investigation of Temperature-Dependent Optical Properties of TiO_2 Using Diffuse Reflectance Spectroscopy. *SN Appl. Sci.* **2019**, *1*, 241. [CrossRef]
48. Elsenety, M.M.; Antoniadou, M.; Kaltzoglou, A.; Kontos, A.G.; Philippopoulos, A.I.; Mitsopoulou, C.A.; Falaras, P. Synthesis, Characterization of $((CH_3)_3S)_2SnI_{6-N}Cl_n$ and $((CH_3)_3S)_2SnI_{6-N}Br_n$ (N = 1, 2) Perovskites and Use in Dye-Sensitized Solar Cells. *Mater. Chem. Phys.* **2020**, *239*, 122310. [CrossRef]
49. Martins, N.C.T.; Ângelo, J.; Girão, A.V.; Trindade, T.; Andrade, L.; Mendes, A. N-Doped Carbon Quantum Dots/TiO_2 Composite with Improved Photocatalytic Activity. *Appl. Catal. B Environ.* **2016**, *193*, 67–74. [CrossRef]
50. Dager, A.; Uchida, T.; Maekawa, T.; Tachibana, M. Synthesis and Characterization of Mono-Disperse Carbon Quantum Dots from Fennel Seeds: Photoluminescence Analysis Using Machine Learning. *Sci. Rep.* **2019**, *9*, 14004. [CrossRef]
51. Huang, X.; Sun, L.; Liu, X.; Ge, M.; Zhao, B.; Bai, Y.; Wang, Y.; Han, S.; Li, Y.; Han, Y.; et al. Increase and Enrichment of Active Electrons by Carbon Dots Induced to Improve TiO_2 Photocatalytic Hydrogen Production Activity. *Appl. Surf. Sci.* **2023**, *630*, 157494. [CrossRef]
52. Zeng, Y.; Chen, D.; Chen, T.; Cai, M.; Zhang, Q.; Xie, Z.; Li, R.; Xiao, Z.; Liu, G.; Lv, W. Study on Heterogeneous Photocatalytic Ozonation Degradation of Ciprofloxacin by TiO_2/Carbon Dots: Kinetic, Mechanism and Pathway Investigation. *Chemosphere* **2019**, *227*, 198–206. [CrossRef] [PubMed]
53. Tong, S.; Zhou, J.; Ding, L.; Zhou, C.; Liu, Y.; Li, S.; Meng, J.; Zhu, S.; Chatterjee, S.; Liang, F. Preparation of Carbon Quantum Dots/TiO_2 Composite and Application for Enhanced Photodegradation of Rhodamine B. *Colloids Surfaces A Physicochem. Eng. Asp.* **2022**, *648*, 129342. [CrossRef]
54. Rani, U.A.; Ng, L.Y.; Ng, C.Y.; Mahmoudi, E.; Ng, Y.S.; Mohammad, A.W. Sustainable Production of Nitrogen-Doped Carbon Quantum Dots for Photocatalytic Degradation of Methylene Blue and Malachite Green. *J. Water Process Eng.* **2021**, *40*, 101816. [CrossRef]

55. Xu, C.; Xiao, X.; Cai, C.; Cheng, Q.; Zhu, L.; Zhang, J.; Wei, B.; Wang, H. Insight into the Differences in Carbon Dots Prepared from Fish Scales Using Conventional Hydrothermal and Microwave Methods. *Environ. Sci. Pollut. Res.* **2023**, *30*, 54616–54627. [CrossRef] [PubMed]
56. Shi, R.; Li, Z.; Yu, H.; Shang, L.; Zhou, C.; Waterhouse, G.I.N.; Wu, L.Z.; Zhang, T. Effect of Nitrogen Doping Level on the Performance of N-Doped Carbon Quantum Dot/TiO_2 Composites for Photocatalytic Hydrogen Evolution. *ChemSusChem* **2017**, *10*, 4650–4656. [CrossRef]
57. Sargin, I.; Yanalak, G.; Arslan, G.; Patir, I.H. Green Synthesized Carbon Quantum Dots as TiO_2 Sensitizers for Photocatalytic Hydrogen Evolution. *Int. J. Hydrogen Energy* **2019**, *44*, 21781–21789. [CrossRef]
58. Hu, Z.; Shi, D.; Wang, G.; Gao, T.; Wang, J.; Lu, L.; Li, J. Carbon Dots Incorporated in Hierarchical Macro/Mesoporous g-C_3N_4/TiO_2 as an All-Solid-State Z-Scheme Heterojunction for Enhancement of Photocatalytic H_2 Evolution under Visible Light. *Appl. Surf. Sci.* **2022**, *601*, 154167. [CrossRef]
59. Cheng, P.; Yang, Z.; Wang, H.; Cheng, W.; Chen, M.; Shangguan, W.; Ding, G. TiO_2-Graphene Nanocomposites for Photocatalytic Hydrogen Production from Splitting Water. *Int. J. Hydrogen Energy* **2012**, *37*, 2224–2230. [CrossRef]
60. Girish, Y.R.; Udayabhanu; Alnaggar, G.; Hezam, A.; Nayan, M.B.; Nagaraju, G.; Byrappa, K. Facile and Rapid Synthesis of Solar-Driven TiO_2/g-C_3N_4 Heterostructure Photocatalysts for Enhanced Photocatalytic Activity. *J. Sci. Adv. Mater. Devices* **2022**, *7*, 100419. [CrossRef]
61. Guan, R.; Wang, L.; Wang, D.; Li, K.; Tan, H.; Chen, Y.; Cheng, X.; Zhao, Z.; Shang, Q.; Sun, Z. Boosting Photocatalytic Hydrogen Production via Enhanced Exciton Dissociation in Black Phosphorus Quantum Dots/TiO_2 Heterojunction. *Chem. Eng. J.* **2022**, *435*, 135138. [CrossRef]
62. Huang, G.; Ye, W.; Lv, C.; Butenko, D.S.; Yang, C.; Zhang, G.; Lu, P.; Xu, Y.; Zhang, S.; Wang, H.; et al. Hierarchical Red Phosphorus Incorporated TiO_2 Hollow Sphere Heterojunctions toward Superior Photocatalytic Hydrogen Production. *J. Mater. Sci. Technol.* **2022**, *108*, 18–25. [CrossRef]
63. Kunthakudee, N.; Puangpetch, T.; Ramakul, P.; Serivalsatit, K.; Hunsom, M. Light-Assisted Synthesis of Au/TiO_2 Nanoparticles for H_2 Production by Photocatalytic Water Splitting. *Int. J. Hydrogen Energy* **2022**, *47*, 23570–23582. [CrossRef]
64. Valero, J.M.; Obregón, S.; Colón, G. Active Site Considerations on the Photocatalytic H_2 Evolution Performance of Cu-Doped TiO_2 Obtained by Different Doping Methods. *ACS Catal.* **2014**, *4*, 3320–3329. [CrossRef]
65. Obregón, S.; Muñoz-Batista, M.J.; Fernández-García, M.; Kubacka, A.; Colón, G. Cu-TiO_2 Systems for the Photocatalytic H_2 Production: Influence of Structural and Surface Support Features. *Appl. Catal. B Environ.* **2015**, *179*, 468–478. [CrossRef]
66. Li, L.; Cheng, B.; Wang, Y.; Yu, J. Enhanced Photocatalytic H2-Production Activity of Bicomponent NiO/TiO_2 Composite Nanofibers. *J. Colloid Interface Sci.* **2015**, *449*, 115–121. [CrossRef] [PubMed]

materials

MDPI

Review

Dynamical Projective Operatorial Approach (DPOA): Theory and Applications to Pump–Probe Setups and Semiconductors

Amir Eskandari-asl [1] and Adolfo Avella [1,2,3,*]

[1] Dipartimento di Fisica "E.R. Caianiello", Università degli Studi di Salerno, I-84084 Fisciano, SA, Italy; aeskandariasl@unisa.it
[2] CNR-SPIN c/o Università degli Studi di Salerno, I-84084 Fisciano, SA, Italy
[3] CNISM c/o Università degli Studi di Salerno, I-84084 Fisciano, SA, Italy
* Correspondence: a.avella@unisa.it; Tel.: +39-089-96-9131

Abstract: This manuscript reviews our recently developed theory, the dynamical projective operatorial approach (DPOA), for studying pump–probe setups in ultra-fast regimes. After reviewing the general formulation of the DPOA, we focus on its lattice version and provide a formalism that is particularly suitable for several pumped semiconductors. Within the DPOA, we also compute the TR-ARPES signal through out-of-equilibrium Green's functions and establish an out-of-equilibrium counterpart of the fluctuation–dissipation theorem. Moreover, we generalize the linear response theory to pumped systems and address, within the DPOA, the differential transient optical properties, providing an overall robust theoretical and computational framework for studying pump–probe setups. Considering a minimal model for a semiconductor, we illustrate the capabilities of the DPOA and discuss several features emerging in this case study that are relevant to real materials.

Keywords: pump–probe setups; semiconductors; dynamical projective operatorial approach (DPOA)

1. Introduction

In the recent few decades, by the so-called pump–probe setups being developed, it became possible to unravel out-of-equilibrium particle dynamics in femtosecond and even sub-femtosecond time regimes [1–8]. The importance of these studies is two-fold. On the application side, gaining control over systems in such short time scales paves the way towards ultra-fast electronics. On the fundamental physics side, the real-time dynamics of pump-induced charge excitations determines the underlying physical mechanisms [9–13].

The pump pulse is usually in the range of energy gaps between the valence band (VB) and conduction band (CB), which would be in the IR regime for typical semiconductors. Pumping the system excites electrons and changes their energy distributions, which is a crucial subject of investigation. In general, there are three approaches to reading out the response of a system to an ultra-short, intense pump pulse. One approach is to study the high-harmonic generation induced by the pump pulse [14–23], which is not discussed in this manuscript. Another important approach is to study the time-resolved angle-resolved photoemission spectroscopy (TR-ARPES) signal [13,24–31], which is an out-of-equilibrium version of ARPES [32–38]. It uses a probe pulse to eject electrons from the sample and analyzes their energy and momentum distribution. These distributions contain information about out-of-equilibrium bands and their occupations and vary depending on the time delay between pump and probe pulses.

The third and last approach is centered around studying the transient optical response of the pumped system to a very low-intensity probe pulse [9,10,20,39–54]. The most convenient probe pulse has a high frequency, which is used to excite electrons from the core bands to the VBs and CBs near the Fermi surface. The pump pulse excites electrons and holes in these CBs and VBs, while the probe pulse reveals the details of these excitations through exciting core-band electrons to them, which are readable from the optical response of the system to the probe pulse.

The aforementioned experimental advances demand the development of appropriate theoretical tools to both simulate and analyze the experimental results. The most standard approach for this aim is the time-dependent density functional theory (TD-DFT) [47,54–61]. Despite several advantages of TD-DFT, it unfortunately has two significant disadvantages. First, the computational cost is very high, as it demands considerable resources, and the calculations usually take very long to compute [62]. Second, it is not very easy to gain deep insights into the individual roles of different mechanisms and understand how they interplay with each other [62].

Alongside the *ab initio* approaches, there are model-Hamiltonian approaches that can be used for the pump–probe setups [47,63–65], with parameters that are usually obtained from equilibrium DFT calculations [66]. These approaches straightforwardly solve the second problem of TD-DFT, as one can turn on and off different terms of the governing Hamiltonian to see how the final results are affected. However, it is generally challenging to handle the numerous complications of *real* setups within model-Hamiltonian approaches, and this may result in oversimplification [62].

To resolve some of the issues mentioned above, we have devised a novel model-Hamiltonian method, the dynamical projective operatorial approach (DPOA) [52,67,68]. The DPOA proves to be able to analyze complicated *real* setups with several bands and unravel the photo injection mechanisms in ultra-fast regimes very affordably [52,53]. The DPOA uses projection matrices to obtain the time evolution of composite operators [69–74]; hence, in principle, it is capable of obtaining any out-of-equilibrium multi-time multi-particle quantities of interest. The DPOA is not restricted to any specific type of materials. Moreover, for instance, it can also handle strongly correlated systems [75]. However, this manuscript primarily focuses on DPOA's applications to pumped semiconductors [52,76–80].

In this paper, we review our previous results [67,68], providing a clear understanding of how the DPOA should be applied to real systems. Using a simple model, we present the features that can be expected in pumped semiconductors, either via the measurement of the TR-ARPES signal or the transient optical properties. The rest of this manuscript is organized as follows: In Section 2, we present the theoretical framework of the DPOA. Section 2.1 introduces the general DPOA theory. In Section 2.2, the formalism is adapted to lattice systems. In Section 2.3, the formulation for computing the TR-ARPES signal is given. In Section 2.4, a generalized linear response theory for computing transient optical properties is reported. In Section 3, we consider a simplified modelization of a semiconductor and, through the application of theDPOA, we study its TR-ARPES signal and transient optical properties. Finally, in Section 4, we summarize and draft some conclusions.

2. Theory

2.1. Dynamical Projective Operatorial Approach (DPOA): General Theory

Consider a general time-independent Hamiltonian $\mathcal{H}$ which describes a system in equilibrium. It is always possible to find some sets of *composite* operators $\mathcal{C}_\alpha^\dagger = \left(\mathcal{C}_{\alpha,1}^\dagger, \ldots, \mathcal{C}_{\alpha,a}^\dagger, \ldots\right)$, which close their hierarchy of the equations of motion (EOMs), where α identifies the set's number [81–96]. The total number of (disjoint) sets is determined by the number of (independent—non-correlated) degrees of freedom characterizing the system [69–74].

In the presence of a time-dependent external perturbation (e.g., an electromagnetic pump pulse), the Hamiltonian becomes time-dependent: $\mathcal{H} \to \mathcal{H}(t)$. Such a perturbation usually affects only the the single-particle term of the Hamiltonian [66] and the *closure* of the hierarchy of the EOMs of the $\mathcal{C}_\alpha$ is preserved, as follows:

$$\mathrm{i}\hbar\partial_t \mathcal{C}_\alpha(t) = [\mathcal{C}_\alpha(t), \mathcal{H}(t)] = \Xi_\alpha(t) \cdot \mathcal{C}_\alpha(t), \tag{1}$$

where $\cdot$ is the matrix product in the operatorial space, and $\Xi_\alpha(t)$ and $\mathcal{C}_\alpha(t)$ are the time-dependent energy matrix and eigenoperators within the set α in the Heisenberg picture, respectively. The dynamical projective operatorial approach (DPOA) exploits that [67]

$$\mathcal{C}_\alpha(t) = P_\alpha(t, t_{\text{ini}}) \cdot \mathcal{C}_\alpha(t_{\text{ini}}) \quad \forall t \geq t_{\text{ini}}, \tag{2}$$

where $P_\alpha(t, t_{\text{ini}})$ is the dynamical projection matrix. In the following, we choose t_{ini} to be any time before the application of the pump pulse (e.g., $t_{\text{ini}} \to -\infty$), so that $\mathcal{C}_\alpha(t_{\text{ini}})$ stands for the operatorial basis describing the equilibrium system. Moreover, for the sake of simplicity, we select $P_\alpha(t, t_{\text{ini}}) \to P_\alpha(t)$. The EOM for $P_\alpha(t)$ reads as [67]

$$\mathrm{i}\hbar\partial_t P_\alpha(t) = \Xi_\alpha(t) \cdot P_\alpha(t), \tag{3}$$

which should be solved with the initial condition $P_\alpha(t_{\text{ini}}) = \mathbf{1}$.

For a general pumped system, we have $\Xi_\alpha(t) = \Xi_\alpha^{\text{eq}} + \Xi_\alpha^{\text{pu}}(t)$, in which Ξ_α^{eq} describes the system at equilibrium. At equilibrium, the solution of Equation (3) is $P_\alpha^{\text{eq}}(t) = \mathrm{e}^{-\frac{\mathrm{i}}{\hbar}(t-t_{\text{ini}})\Xi_\alpha^{\text{eq}}}$. It is worth noting that by rewriting $P_\alpha(t) = P_\alpha^{\text{eq}}(t) \cdot P_\alpha^{\text{int}}(t) = \mathrm{e}^{-\frac{\mathrm{i}}{\hbar}(t-t_{\text{ini}})\Xi_\alpha^{\text{eq}}} \cdot P_\alpha^{\text{int}}(t)$, we obtain the following reduced equation of motion:

$$\mathrm{i}\hbar\partial_t P_\alpha^{\text{int}}(t) = \Xi_\alpha^{\text{pu,int}}(t) \cdot P_\alpha^{\text{int}}(t), \tag{4}$$

where $\Xi_\alpha^{\text{pu,int}}(t) = \mathrm{e}^{\frac{\mathrm{i}}{\hbar}(t-t_{\text{ini}})\Xi_\alpha^{\text{eq}}} \cdot \Xi_\alpha^{\text{pu}}(t) \cdot \mathrm{e}^{-\frac{\mathrm{i}}{\hbar}(t-t_{\text{ini}})\Xi_\alpha^{\text{eq}}}$. It is straightforward to show that, from Equation (4), one can obtain the following equivalent integro-differential equation:

$$P_\alpha^{\text{int}}(t) = \mathbf{1} - \frac{\mathrm{i}}{\hbar}\int_{t_{\text{ini}}}^{t} dt' \Xi_\alpha^{\text{pu,int}}(t') \cdot P_\alpha^{\text{int}}(t'). \tag{5}$$

2.2. Pumped Lattice Systems out of Equilibrium

The DPOA can be applied to any kind of system: atoms, molecules, organic structures, lattice systems, etc. In this work, we focus on pumped lattice systems. Starting from the results of the many available DFT codes for the target equilibrium system and *wannierizing* them (for example, by exploiting the Wannier90 code [97]), one obtains a tight-binding quadratic Hamiltonian, which allows us to consider the effects of applying external EM fields (see, for instance, [52]). One should keep in mind that such an approach neglects the *dynamical* Coloumb interaction (i.e., its differential contribution to the out-of-equilibrium dynamics), even though it considers the *static* Coloumb interaction (through the exchange and correlation integrals within the equilibrium DFT calculation), which is fundamental for correctly describing the bands of the target system, for example, by opening and determining the value of the band gaps.

We consider an electromagnetic pump pulse with vector potential $\boldsymbol{A}(t)$ and electric field $\boldsymbol{E}(t) = -\partial_t \boldsymbol{A}(t)$ applied to our system after some time t_{ini}. In the dipole gauge, the related time-dependent Hamiltonian is given by

$$\mathcal{H}(t) = \sum_{\mathbf{k},\nu,\nu'} \tilde{c}^{\dagger}_{\mathbf{k},\nu}(t)\tilde{\Xi}_{\mathbf{k},\nu,\nu'}(t)\tilde{c}_{\mathbf{k},\nu'}(t), \tag{6}$$

where $\tilde{c}_{\mathbf{k},\nu}(t)$ is the electronic annihilation operator, $\mathbf{k}$ is the lattice momentum in the first Brillouin zone (FBZ), and ν, which also includes the spin, refers to the set of quantum numbers characterizing a localized state (e.g., a maximally localized Wannier state) and [66,67]

$$\tilde{\Xi}_{\mathbf{k},\nu,\nu'}(t) = \tilde{T}_{\mathbf{k}+\frac{e}{\hbar}A(t),\nu,\nu'} + eE(t) \cdot \tilde{\boldsymbol{D}}_{\mathbf{k}+\frac{e}{\hbar}A(t),\nu,\nu'}. \tag{7}$$

$\tilde{T}_{\mathbf{k},\nu,\nu'}$ and $\tilde{\boldsymbol{D}}_{\mathbf{k},\nu,\nu'}$ are the hopping and dipole matrix elements, respectively. We have adopted a notation in which any operator (in either first or second quantization) written in the basis of the localized states carries the over-script $\sim$. The value of the electronic charge is indicated by $e > 0$. Moreover, $\cdot$, in this and similar cases, stands for the dot product between two vectors in the Cartesian space. Equation (7) can be considered as the generalization of the Peierls substitution [98,99] to multi-band systems [66].

It is convenient to work within a basis in which the equilibrium Hamiltonian, $\tilde{\Xi}_{\mathbf{k},\nu,\nu'}(t \leq t_{\text{ini}}) = \tilde{T}_{\mathbf{k},\nu,\nu'}$, is diagonal. The transformation to such a basis, which we call the band basis, can be performed through a unitary matrix $\Omega_{\mathbf{k},\nu,n}$ satisfying

$$\sum_{\nu,\nu'} \Omega^{\dagger}_{\mathbf{k},n,\nu}\tilde{T}_{\mathbf{k},\nu,\nu'}\Omega_{\mathbf{k},\nu',n'} = \delta_{n,n'}\varepsilon_{\mathbf{k},n}, \tag{8}$$

where $\delta_{n,n'}$ stands for the Kronecker delta, n is the index of the energy band and $\varepsilon_{\mathbf{k},n}$ is the band energy. Using $\Omega_{\mathbf{k},\nu,n}$, it is straightforward to transform any matrix from the localized basis to the band basis:

$$M_{\mathbf{k},n,n'} = \sum_{\nu,\nu'} \Omega^{\dagger}_{\mathbf{k},n,\nu}\tilde{M}_{\mathbf{k},\nu,\nu'}\Omega_{\mathbf{k},\nu',n'}, \tag{9}$$

where M can be any of T, $\boldsymbol{D}$, Ξ, etc. In order to clearly distinguish the operators in the two bases, we denote the operators in the band basis without the over-script$\sim$used for the operators in the localized basis. The annihilation operator of an electron in the band n and with momentum $\mathbf{k}$ is as follows:

$$c_{\mathbf{k},n}(t) = \sum_{\nu} \Omega^{\dagger}_{\mathbf{k},n,\nu}\tilde{c}_{\mathbf{k},\nu}(t). \tag{10}$$

Out of equilibrium, we have $c_{\mathbf{k}}(t) = P_{\mathbf{k}}(t) \cdot c_{\mathbf{k}}(t_{\text{ini}})$, where $P_{\mathbf{k}}(t_{\text{ini}}) = \mathbf{1}$ and $i\hbar\partial_t P_{\mathbf{k}}(t) = \Xi_{\mathbf{k}}(t) \cdot P_{\mathbf{k}}(t)$. Moreover, the out-of-equilibrium number of electrons in band n with momentum $\mathbf{k}$, $N_{\mathbf{k},n}(t) = \left\langle c^{\dagger}_{\mathbf{k},n}(t)c_{\mathbf{k},n}(t)\right\rangle$, can be calculated as

$$N_{\mathbf{k},n}(t) = \sum_{n'} P_{\mathbf{k},n,n'}(t)f_{\mathbf{k},n'}P^{\dagger}_{\mathbf{k},n',n}(t), \tag{11}$$

where $f_{\mathbf{k},n}$ is the Fermi distribution function.

In *real* materials, one usually deals with several bands and a dense $\mathbf{k}$ grid: computing the out-of-equilibrium Hamiltonian terms (matrices with band indexes per each momentum value in the grid and at each instant of time) is extremely time-consuming. A very efficient way to noticeably reduce the computational cost is to expand any matrix $M_{\mathbf{k}+\frac{e}{\hbar}A(t)}$ in powers of the vector potential [67]:

$$M_{\mathbf{k}+\frac{e}{\hbar}A(t)}(t) = \sum_{m=0}^{\infty} \frac{1}{m!} \Omega_{\mathbf{k}}^{\dagger} \cdot \left[\partial_{k_A}^{(m)} \tilde{M}_{\mathbf{k}} \right] \cdot \Omega_{\mathbf{k}} \left(\frac{e}{\hbar} A(t) \right)^m. \tag{12}$$

The vector potential is expressed as $\boldsymbol{A}(t) = A(t)\hat{A}$, where $\hat{A}$ is the polarization of the pulse and $A(t)$ is its amplitude. $\partial_{k_A}^{(m)}$ is the m-th partial derivative in the direction of $\hat{A}$. Judiciously truncating the *Peierls expansion*, Equation (12), one needs to compute only once the properly chosen number of derivatives per each momentum value and use them at all times. The m-th partial derivatives can be computed as follows:

$$\partial_{k_A}^{(m)} \tilde{M}_{\mathbf{k}} = \sum_{\mathbf{i}} (-\mathrm{i}\hat{A} \cdot \mathbf{R}_{\mathbf{i}})^m \mathrm{e}^{-\mathrm{i}\mathbf{k}\cdot\mathbf{R}_{\mathbf{i}}} \tilde{M}_{\mathbf{R}_{\mathbf{i}}}, \tag{13}$$

where $\tilde{M}_{\mathbf{R}_{\mathbf{i}}}$ is the matrix in the direct space, as computed via the wannierization procedure. For the hopping matrix, $M = T$, we call the term $\Omega_{\mathbf{k}}^{\dagger} \cdot \left[\partial_{k_A}^{(m)} \tilde{T}_{\mathbf{k}} \right] \cdot \Omega_{\mathbf{k}}$with $m = 1$ ($m = 2$) the velocity (inverse-mass) term. However, even though the derivative is not applied to the transformation matrix, $\Omega_{\mathbf{k}}$, the resulting quantity is not the velocity (inverse mass) relative to the energy bands.

2.3. Green's Functions and TR-ARPES Signal

Green's function (GF) formalism provides a very versatile machinery to compute several relevant properties of a system. The retarded, G^R, and the lesser, $G^<$, GFs, are defined as follows:

$$G_{\mathbf{k},n,n'}^{R}(t,t') = -\mathrm{i}\theta(t-t') \left\langle \left\{ c_{\mathbf{k},n}(t), c_{\mathbf{k},n'}^{\dagger}(t') \right\} \right\rangle, \tag{14}$$

$$G_{\mathbf{k},n,n'}^{<}(t,t') = \mathrm{i} \left\langle c_{\mathbf{k},n'}^{\dagger}(t') c_{\mathbf{k},n}(t) \right\rangle. \tag{15}$$

Using the DPOA, one can obtain non-equilibrium GFs in terms of the dynamical projection matrices P as

$$G_{\mathbf{k},n,n'}^{R}(t,t') = -\mathrm{i}\theta(t-t') \sum_{m} P_{\mathbf{k},n,m}(t) P_{\mathbf{k},n',m}^{\star}(t'), \tag{16}$$

$$G_{\mathbf{k},n,n'}^{<}(t,t') = \mathrm{i} \sum_{m,m'} (\delta_{m,m'} - \rho_{\mathbf{k},m,m'}(t_{\mathrm{ini}})) P_{\mathbf{k},n,m}(t) P_{\mathbf{k},n',m'}^{\star}(t'), \tag{17}$$

where $\rho_{\mathbf{k},m,m'}(t_{\mathrm{ini}})$ is the single-particle density matrix at equilibrium. In the band basis in which the equilibrium Hamiltonian is diagonal, $\delta_{n,n'} - \rho_{\mathbf{k},n,n'}(t_{\mathrm{ini}}) = \delta_{n,n'} f_{\mathbf{k},n}$.

For determining the out-of-equilibrium particle distribution over the energy of a pumped system, one can analyze the TR-ARPES signal [100–103], which plays almost the same role out of equilibrium of the spectral function in equilibrium. In this case, in addition to the pump pulse, one needs to consider a probe pulse too. The probe pulse ejects the electrons out of the system so that their energy and momentum can be measured: this is known as the TR-ARPES signal. This, for a probe pulse centered at t_{pr}, is proportional to (see [67] and the references therein)

$$I_{\mathbf{k}}^{<}(\omega, t_{\mathrm{pr}}) = \frac{\tau_{\mathrm{pr}}}{\sqrt{8\pi \ln 2}} \int_{-\infty}^{+\infty} dt_1 \int_{-\infty}^{+\infty} dt_2 S_{\mathrm{pr}}(t_1 - t_{\mathrm{pr}}) \times \\ \times S_{\mathrm{pr}}(t_2 - t_{\mathrm{pr}}) \Im \left[e^{\mathrm{i}\omega(t_1 - t_2)} \mathrm{Tr}\left[G_{\mathbf{k}}^{<}(t_1, t_2) \right] \right], \tag{18}$$

where $S_{\rm pr}(t-t_{\rm pr}) = \frac{2\sqrt{\ln 2}}{\sqrt{\pi}\tau_{\rm pr}} e^{-4\ln 2 (t-t_{\rm pr})^2/\tau_{\rm pr}^2}$ is the probe–pulse envelope. The normalization factor is chosen in such a way that $I^<(\omega, t_{\rm pr})$ is normalized to the total number of particles at momentum **k**,

$$\int_{-\infty}^{+\infty} d\omega I^<_{\mathbf{k}}(\omega, t_{\rm pr}) = \sum_n N_{\mathbf{k},n}. \tag{19}$$

Although $I^<_{\mathbf{k}}(\omega, t_{\rm pr})$ provides information about the occupied states, the out-of-equilibrium TR-ARPES *bands* are computed using a retarded signal defined as [67]

$$I^R_{\mathbf{k}}(\omega, t_{\rm pr}) = -\frac{\tau_{\rm pr}}{\sqrt{2\pi \ln 2}} \int_{-\infty}^{+\infty} dt_1 \int_{-\infty}^{+\infty} dt_2 S_{\rm pr}(t_1 - t_{\rm pr}) \times \\ \times S_{\rm pr}(t_2 - t_{\rm pr}) \Im\left[e^{i\omega(t_1-t_2)} \operatorname{Tr}\left[G^R_{\mathbf{k}}(t_1, t_2)\right]\right]. \tag{20}$$

In the band basis, we have [67]

$$I^<_{\mathbf{k}}(\omega, t_{\rm pr}) = \sum_{n,n'} L_{\mathbf{k},n;n'}(\omega, t_{\rm pr}) f_{\mathbf{k},n'}, \tag{21}$$

$$I^R_{\mathbf{k}}(\omega, t_{\rm pr}) = \sum_{n,n'} L_{\mathbf{k},n;n'}(\omega, t_{\rm pr}), \tag{22}$$

where

$$L_{\mathbf{k},n;n'}(\omega, t_{\rm pr}) = \frac{\tau_{\rm pr}}{2\sqrt{2\pi \ln 2}} \left| \int_{-\infty}^{+\infty} dt_1 S_{\rm pr}(t_1 - t_{\rm pr}) e^{i\omega t_1} P_{\mathbf{k},nn'}(t_1) \right|^2, \tag{23}$$

which shows that the TR-ARPES signal is non-negative. It is worth noting that Equations (21) and (22) act as the out-of-equilibrium counterpart of the fluctuation–dissipation theorem.

2.4. Out-of-Equilibrium Optical Properties

In studying the out-of-equilibrium optical properties of a pumped system, we need to consider an overall electromagnetic pulse composed of both a pump and a probe pulse. The system's optical response (reflection and/or absorption) is generally studied as a function of the time delay between the pump and probe pulses. Analyzing such a variation, one obtains information about the ultra-fast pump-induced effects on the system [9,10,20,39–43,47,49,50,52–54].

Since the probe pulse is very weak by definition, one can compute the transient optical properties of the system through a generalized linear response theory (GLRT) [68] that fully takes into account that the system is under the effect of the pump pulse. Deriving such a GLRT requires the following [68]: (i) to obtain the light-matter Hamiltonian for a general realistic lattice system and a general electromagnetic pulse; (ii) to derive the expression of the electric current for such a pumped system; (iii) to determine the expression of the optical conductivity of a pumped system in this framework; (iv) to obtain the expressions of the transient reflectivity and absorption; (v) to devise a numerical framework that effectively permits us to compute these quantities for *real* materials (many conduction and valence bands involved and very dense momentum grids).

Within this framework, exploiting the DPOA, the out-of-equilibrium time-dependent optical conductivity of a pumped system probed at time $t_{\rm pr}$ (i.e., the center of the probe pulse is at time $t_{\rm pr}$), $\sigma(t, t_{\rm pr})$, reads as follows: [68]

$$\sigma(t, t_{\rm pr}) = \sigma_1(t, t_{\rm pr}) + \sigma_2(t, t_{\rm pr}), \tag{24}$$

where

$$\sigma_1(t,t_{\rm pr}) = \frac{{\rm i}e}{\hbar\mathcal{V}}\theta(t-t_{\rm pr})\times$$
$$\times\sum_{\boldsymbol{k}}\sum_{n_1n_2n_3n_4}\sum_{n_1'n_2'} \boldsymbol{J}_{\boldsymbol{k},n_1n_2}(t)\int_{t_{\rm ini}}^{t}dt'\Big(-\frac{e}{\hbar}\boldsymbol{V}_{\boldsymbol{k},n_3n_4}(t')\theta(t'-t_{\rm pr})+e\boldsymbol{D}_{\boldsymbol{k},n_3n_4}(t')\delta(t'-t_{\rm pr})\Big)$$
$$\times P^{\star}_{\boldsymbol{k}n_1n_1'}(t)P_{\boldsymbol{k}n_2n_2'}(t)P^{\star}_{\boldsymbol{k}n_3n_2'}(t')P_{\boldsymbol{k}n_4n_1'}(t')\Big(f_{\boldsymbol{k}n_1'}-f_{\boldsymbol{k}n_2'}\Big). \quad (25)$$

which can be rewritten in a computationally more efficient way as follows:

$$\sigma_1(t,t_{\rm pr}) = \frac{{\rm i}e}{\hbar v_{\rm uc}N_{\rm grid}}\theta(t-t_{\rm pr})\sum_{\boldsymbol{k}\in{\rm grid}}{\rm Tr}\{Z_{\boldsymbol{k}}(t)\cdot[Y_{\boldsymbol{k}}(t)-Y_{\boldsymbol{k}}(t_{\rm pr})+X_{\boldsymbol{k}}(t_{\rm pr})]\}, \quad (26)$$

where $N_{\rm grid}$ is the number of points on the grid which samples the FBZ, and $v_{\rm uc}$ is the volume of unit cell. Thus, the full volume of the system is $\mathcal{V} = v_{\rm uc}N_{\rm grid}$. The matrices $Z_{\boldsymbol{k}}(t)$, $Y_{\boldsymbol{k}}(t)$ and $X_{\boldsymbol{k}}(t)$ read as follows:

$$Z_{\boldsymbol{k},n_1'n_2'}(t) = \sum_{n_1n_2}\boldsymbol{J}_{\boldsymbol{k},n_1n_2}(t)P^{\star}_{\boldsymbol{k},n_1n_1'}(t)P_{\boldsymbol{k},n_2n_2'}(t)\Big(f_{\boldsymbol{k}n_1'}-f_{\boldsymbol{k}n_2'}\Big), \quad (27)$$

$$Y_{\boldsymbol{k},n_2'n_1'}(t) = -\frac{e}{\hbar}\sum_{n_3n_4}\int_{t_{\rm ini}}^{t}dt'\boldsymbol{V}_{\boldsymbol{k},n_3n_4}(t')P^{\star}_{\boldsymbol{k},n_3n_2'}(t')P_{\boldsymbol{k},n_4n_1'}(t'), \quad (28)$$

$$X_{\boldsymbol{k},n_2'n_1'}(t) = e\sum_{n_3n_4}\boldsymbol{D}_{\boldsymbol{k},n_3n_4}(t)P^{\star}_{\boldsymbol{k},n_3n_2'}(t)P_{\boldsymbol{k},n_4n_1'}(t), \quad (29)$$

and

$$\boldsymbol{J}_{\boldsymbol{k},n_1n_2}(t) = \frac{1}{\hbar}\boldsymbol{\eta}_{\boldsymbol{k},n_1n_2}(t) - \frac{\rm i}{\hbar}[\boldsymbol{D}_{\boldsymbol{k}}(t),\boldsymbol{T}_{\boldsymbol{k}}(t)]_{n_1n_2}, \quad (30)$$

$$\boldsymbol{\eta}_{\boldsymbol{k},n_1n_2}(t) = \sum_{\nu_1\nu_2}\Omega^{*}_{\boldsymbol{k}\nu_1n_1}\Big(\boldsymbol{\nabla}_{\boldsymbol{k}}\tilde{T}_{\boldsymbol{k}+\frac{e}{\hbar}\boldsymbol{A}_{\rm pu}(t),\nu_1\nu_2}\Big)\Omega_{\boldsymbol{k}\nu_2n_2}, \quad (31)$$

$$\boldsymbol{V}_{\boldsymbol{k},n_1n_2}(t) = \boldsymbol{\eta}_{\boldsymbol{k},n_1n_2}(t) + e\boldsymbol{\Lambda}_{\boldsymbol{k},n_1n_2}(t)\cdot\boldsymbol{E}_{\rm pu}(t), \quad (32)$$

$$\boldsymbol{\Lambda}_{\boldsymbol{k},n_1n_2}(t) = \sum_{\nu_1\nu_2}\Omega^{*}_{\boldsymbol{k}\nu_1n_1}\Big(\boldsymbol{\nabla}_{\boldsymbol{k}}\tilde{\boldsymbol{D}}_{\boldsymbol{k}+\frac{e}{\hbar}\boldsymbol{A}_{\rm pu}(t),\nu_1\nu_2}\Big)\Omega_{\boldsymbol{k}\nu_2n_2}, \quad (33)$$

where the notation $[\Phi,\Psi]_{n_1n_2} = \sum_{n'}(\Phi_{n_1n'}\Psi_{n'n_2}-\Psi_{n_1n'}\Phi_{n'n_2})$ is used for the commutator.

Moreover [68],

$$\sigma_2(t,t_{\rm pr}) = \frac{e}{\mathcal{V}}\theta(t-t_{\rm pr})\sum_{n_1n_2\boldsymbol{k}}\frac{\delta\boldsymbol{J}_{\boldsymbol{k},n_1n_2}(t)}{\delta\boldsymbol{A}(t)}N_{\boldsymbol{k}n_1n_2}(t), \quad (34)$$

where

$$N_{\boldsymbol{k}n_1n_2}(t) = \sum_{n'}P^{\star}_{\boldsymbol{k}n_1n'}(t)P_{\boldsymbol{k}n_2n'}(t)f_{\boldsymbol{k}n'}, \quad (35)$$

and

$$\frac{\delta\boldsymbol{J}_{\boldsymbol{k},n_1n_2}(t)}{\delta\boldsymbol{A}(t)} = \frac{e}{\hbar^2}\xi_{\boldsymbol{k},n_1n_2}(t) - \frac{{\rm i}e}{\hbar^2}[\boldsymbol{\Lambda}_{\boldsymbol{k}}(t),T_{\boldsymbol{k}}(t)]_{n_1n_2} - \frac{{\rm i}e}{\hbar^2}[\boldsymbol{D}_{\boldsymbol{k}}(t),\boldsymbol{\eta}_{\boldsymbol{k}}(t)]_{n_1n_2}, \quad (36)$$

in which

$$\xi_{\boldsymbol{k},n_1n_2}(t) = \sum_{\nu_1\nu_2}\Omega^{*}_{\boldsymbol{k}\nu_1n_1}\Big[\nabla^2_{\boldsymbol{k}}\tilde{T}_{\boldsymbol{k}+\frac{e}{\hbar}\boldsymbol{A}_{\rm pu}(t),\nu_1\nu_2}\Big]\Omega_{\boldsymbol{k}\nu_2n_2}. \quad (37)$$

Having the optical conductivity $\sigma(t, t_{\text{pr}})$, one performs the Fourier transformation with respect to $t - t_{\text{pr}}$ to obtain it as a function of the probe–pulse frequency, ω, as

$$\sigma(\omega, t_{\text{pr}}) = \int_{-\infty}^{+\infty} e^{i(\omega + i0^+)(t - t_{\text{pr}})} \sigma(t, t_{\text{pr}}) dt, \tag{38}$$

where 0^+ is a damping factor. In real systems, this damping factor is finite. In numerical simulations, one needs to choose its value while considering the specific system under study and the details of the simulation (see ref. [68] for a detailed discussion on this issue). However, to study pump-induced dynamics, the system should have dampings small enough such that the damping time ($\sim 2\pi/0^+$) is much larger than the time scales of those pump-induced dynamics.

A noticeable simplification in the calculations is achieved if one restricts the upper limit of the integral given in Equation (38) to some $t_{\text{fin}} > t_{\text{pr}}$ at which the pump pulse is negligible and writes

$$\sigma(\omega, t_{\text{pr}}) = \int_{-\infty}^{t_{\text{fin}}} e^{i(\omega + i0^+)(t - t_{\text{pr}})} \sigma(t, t_{\text{pr}}) dt + \sigma^{\text{a.p.}}(\omega, t_{\text{pr}}), \tag{39}$$

where $\sigma^{\text{a.p.}}(\omega, t_{\text{pr}}) = \sigma_1^{\text{a.p.}}(\omega, t_{\text{pr}}) + \sigma_2^{\text{a.p.}}(\omega, t_{\text{pr}})$ in which [68]

$$\sigma_1^{\text{a.p.}}(\omega, t_{\text{pr}}) = -\frac{ie}{\hbar \mathcal{V}} \sum_{\boldsymbol{k}} \{ W_{\boldsymbol{k}}(\omega, t_{\text{fin}}, t_{\text{pr}}) + \text{Tr}[Q_{\boldsymbol{k}}(\omega, t_{\text{fin}}, t_{\text{pr}}) \cdot S_{\boldsymbol{k}}(t_{\text{fin}}, t_{\text{pr}})] \}, \tag{40}$$

where

$$Q_{\boldsymbol{k},n'n}(\omega, t_{\text{fin}}, t_{\text{pr}}) = \sum_{n_1 n_2} P^*_{\boldsymbol{k},n_2 n'}(t_{\text{fin}}) \boldsymbol{J}_{\boldsymbol{k},n_2 n_1} P_{\boldsymbol{k},n_1 n}(t_{\text{fin}}) \times \\ \times \frac{i e^{(i\omega - 0^+)(t_{\text{fin}} - t_{\text{pr}})}}{(\omega - \omega_{\boldsymbol{k},n_1 n_2} + i0^+)} (f_{\boldsymbol{k}n'} - f_{\boldsymbol{k}n}), \tag{41}$$

$$S_{\boldsymbol{k}}(t_{\text{fin}}, t_{\text{pr}}) = -Y_{\boldsymbol{k}}(t_{\text{fin}}) + Y_{\boldsymbol{k}}(t_{\text{pr}}) - X_{\boldsymbol{k}}(t_{\text{pr}}), \tag{42}$$

$$W_{\boldsymbol{k}}(\omega, t_{\text{fin}}, t_{\text{pr}}) = -\frac{e}{\hbar} \sum_{n_1 n_2} \frac{e^{(i\omega - 0^+)(t_{\text{fin}} - t_{\text{pr}})} \boldsymbol{J}_{\boldsymbol{k},n_2 n_1}}{\omega - \omega_{\boldsymbol{k},n_1 n_2} + i0^+} \times \\ \times \sum_n \left(\frac{\eta_{\boldsymbol{k},n_1 n}}{\omega - \omega_{\boldsymbol{k},n n_2} + i0^+} N_{\boldsymbol{k},n_2 n}(t_{\text{fin}}) - \frac{\eta_{\boldsymbol{k},n n_2}}{\omega - \omega_{\boldsymbol{k},n_1 n} + i0^+} N_{\boldsymbol{k},n n_1}(t_{\text{fin}}) \right), \tag{43}$$

and [68]

$$\sigma_2^{\text{a.p.}}(\omega, t_{\text{pr}}) = \frac{e}{\mathcal{V}} \sum_{\boldsymbol{k}} \sum_{n_1 n_2} \frac{i N_{\boldsymbol{k},n_1 n_2}(t_{\text{fin}}) e^{(i\omega - 0^+)(t_{\text{fin}} - t_{\text{pr}})} \frac{\delta J_{\boldsymbol{k}.n_1 n_2}}{\delta A}}{\omega_{\boldsymbol{k},n_1 n_2} + \omega + i0^+}. \tag{44}$$

We consider a case in which both the pump and probe pulses are linearly polarized and both of them have the same polarization. Therefore, the optical conductivity reduces to $\sigma(\omega, t_{\text{pr}}) = \hat{A} \cdot \boldsymbol{\sigma}(\omega, t_{\text{pr}}) \cdot \hat{A}$. Correspondingly, the dielectric function is

$$\epsilon(\omega, t_{\text{pr}}) = 1 + \frac{i}{\omega \epsilon_0} \sigma(\omega, t_{\text{pr}}), \tag{45}$$

where ϵ_0 is the vacuum dielectric constant.

If the probe pulse frequencies are much larger than the pump pulse ones, the transient reflectivity of an s-polarized probe pulse of central frequency ω and incident angle θ is given by

$$R_\theta(\omega, t_{\rm pr}) = \left| \frac{\cos\theta - \sqrt{\epsilon(\omega, t_{\rm pr}) - \sin^2\theta}}{\cos\theta + \sqrt{\epsilon(\omega, t_{\rm pr}) - \sin^2\theta}} \right|^2 . \tag{46}$$

Following the experimental measurements, we also compute the relative differential transient reflectivity, $\delta_r R_\theta(\omega, t_{\rm pr})$, defined as [52]

$$\delta_r R_\theta(\omega, t_{\rm pr}) = \frac{R_\theta(\omega, t_{\rm pr}) - R_\theta^{\rm eq}(\omega)}{R_\theta^{\rm eq}(\omega)}, \tag{47}$$

where $R_\theta^{\rm eq}(\omega)$ is the equilibrium reflectivity. To gain more information about the features appearing in $\delta_r R_\theta(\omega, t_{\rm pr})$, one can perform a Fourier transformation with respect to the time $t_{\rm pr}$:

$$\delta_r R_\theta(\omega, \omega') = \int_{-\infty}^{+\infty} e^{i\omega t_{\rm pr}} \delta_r R_\theta(\omega, t_{\rm pr}) dt_{\rm pr}. \tag{48}$$

Up until now, we have been considering reflectivity. However, it is also possible to study the transient absorption coefficient of the pumped system, which is given by

$$\alpha(\omega, t_{\rm pr}) = \frac{\omega}{n_{\rm refr}(\omega, t_{\rm pr}) c} \Im[\epsilon(\omega, t_{\rm pr})], \tag{49}$$

where c is the speed of light in a vacuum and $n_{\rm refr}(\omega, t_{\rm pr}) = \Re\left[\sqrt{\epsilon(\omega, t_{\rm pr})}\right]$ is the real out-of-equilibrium refractive index. In our calculations, we compute the transient differential absorption coefficient as

$$\delta\alpha(\omega, t_{\rm pr}) = \alpha(\omega, t_{\rm pr}) - \alpha^{\rm eq}(\omega), \tag{50}$$

where $\alpha^{\rm eq}(\omega)$ stands for the absorption coefficient at equilibrium.

3. The Case Study of a Three-Band Semiconductor

In both cases, the TR-ARPES signal and transient optical properties, to analyze the phenomenology that emerges in a pump–probe setup, it is helpful to consider a simple model describing a semiconductor. In this way, without becoming lost in the complications of any particular *real* material, one can construct a dictionary of phenomena coming from different mechanisms. Dealing with *real* experimental setups, such a dictionary can be extremely useful in analyzing the results.

Our semiconductor model consists of a cubic lattice with a lattice constant a and three bands: a valence band (VB), a conduction band (CB) , and a *core* band. The dispersions of VB and CB can be obtained by the following hopping matrix elements: $\tilde{T}_{0,1,1} = -1.65\Delta$, $\tilde{T}_{0,2,2} = 1.35\Delta$, $\tilde{T}_{\mathbf{a},1,1} = 0.2\Delta$, $\tilde{T}_{\mathbf{a},2,2} = -0.15\Delta$, and $\tilde{T}_{\mathbf{a},1,2} = \tilde{T}_{\mathbf{a},2,1} = -0.1\Delta$. Here, $\tilde{T}_{\boldsymbol{R},\nu,\nu'}$ is the hopping matrix element between localized Wannier states with indices ν and ν' centered at a relative distance $\boldsymbol{R}$, and $\mathbf{a} \in \{a(\pm1,0,0), a(0,\pm1,0), a(0,0,\pm1)\}$. The unit of energy is Δ, which determines the band gap. For example, $\Delta = 0.5\,\mathrm{eV}$ results in a band gap of $0.75\,\mathrm{eV}$ at Γ. By diagonalizing the matrix $\tilde{T}$, one obtains the dispersions of VB and CBs. On the other hand, to simplify the model and focus on the effects of the pump pulse and not on the specific features of the *core* band and of its coupling to VB and CB, we consider a flat *core* band at energy $\varepsilon_{\rm core} = -50\Delta$ and no hopping between the *core* band and the other two bands. The *core* band is coupled to the VB and CB via a local momentum-independent dipole matrix element $\boldsymbol{D} = \mathrm{i}0.05a\hat{\mathbf{j}}$. For the case of the TR-ARPES signal, we consider the *main* cubic path in the FBZ, while for studying the transient optical properties, we sample the $\mathbf{k}$ space by a cubic $32 \times 32 \times 32$ grid, which includes Γ. The damping factor is chosen as $0^+ = 0.05\Delta/\hbar$.

The vector potential of the pump pulse is $\boldsymbol{A}_{\mathrm{pu}}(t) = A_{\mathrm{pu}}(t)\hat{\mathbf{j}}$ where $A_{\mathrm{pu}}(t)$ is

$$A_{\mathrm{pu}}(t) = A_0 e^{-(4\ln 2)t^2/\tau_{\mathrm{pu}}^2} \cos(\omega_{\mathrm{pu}} t). \tag{51}$$

Accordingly, we assume that the center of the pump pulse is taken as the origin of the time axis. Hence, the center of the probe pulse, t_{pr}, coincides with the time delay between the pump and probe pulses. The pump amplitude is $A_0 = 0.4\pi\hbar/ae$, the FWHM of the pump pulse is $\tau_{\mathrm{pu}} = 7\hbar/\Delta$, and the pump frequency is $\omega_{\mathrm{pu}} = 2.33\Delta/\hbar$.

3.1. TR-ARPES Signals

In Figure 1, we show the maps of the retarded and lesser TR-ARPES signals, when the center and the FWHM of the probe pulse coincides with those of the pump pulse. The local maxima of the TR-ARPES signal can be considered as the TR-ARPES *bands*. In the proximity of the equilibrium VB and CB (shown by the red lines), we have the main TR-ARPES *bands* that are just slightly shifted. Meanwhile, in equilibrium, we only have two bands in the energy range of Figure 1, where, under the application of the pump pulse, new side bands emerge. Some of these sidebands are$\sim n\hbar\omega_{\mathrm{pu}}$ (n being a non-zero integer), apart from the main VB or CB, and we call them photon sidebands (PSBs), or more precisely, n-photon PSBs. We also witness the emergence of some other kinds of sidebands in the proximity of the main bands, which we called envelope-Peierls sidebands (EPSBs) [67]. The retarded signal features both VB and CB and all their sidebands, regardless of their occupation. On the other hand, the lesser signal, which is directly measurable in experiments, just shows the occupied portion of the band structure. Hence, it features only the VB and its sidebands.

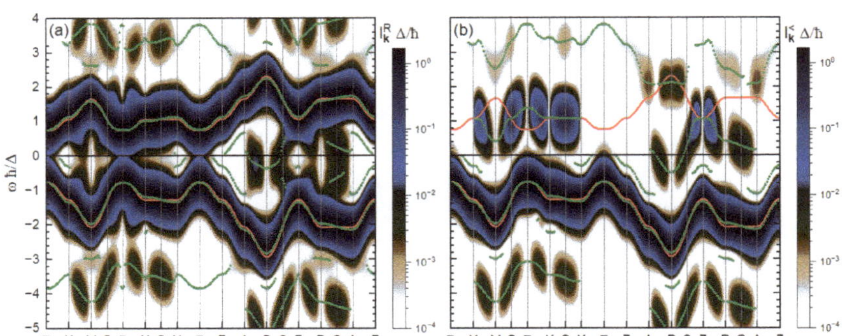

Figure 1. (**a**) The retarded, $I^R_{\mathbf{k}}$, and (**b**) the lesser, $I^<_{\mathbf{k}}$, TR-ARPES signals along the *main* path, for the case where the centers and the FWHM of the pump and probe pulses coincide with each other (i.e., $t_{\mathrm{pr}} = 0$ and $\tau_{\mathrm{pr}} = \tau_{\mathrm{pu}}$). The equilibrium band energies are shown with solid red lines, and the local maxima of the signal at each $\mathbf{k}$ point are shown with green dots. The latter indicate the out-of-equilibrium TR-ARPES bands.

The origin of the one-photon PSBs is the velocity term in the *Peierls expansion*. One can simply show that this velocity is proportional to $\sin(ak_y)$ in our model, which vanishes on the planes **Γ-X-A-Z** and **Y-M-B-D**, preventing the appearance of one-photon PSBs at these **k**-points. The inverse-mass term is proportional to $\cos(ak_y)$. Therefore, at **S**, **C**, middle points of the lines **X-M**, **A-B**, **Z-D** and **Γ-Y**, it vanishes identically. The vanishing of inverse-mass terms results in no shift of the TR-ARPES bands with respect to the equilibrium bands, but it results in the disappearance of the EPSBs. This clarifies how the symmetries of the system can affect the emerging features in the TR-ARPES signals, namely the band shifts and the pump-induced sidebands.

3.2. Transient Optical Properties

In Figure 2, the transient optical properties of the system under the application of the pump pulse are shown. The low-frequency pump pulse excites the electrons from the VB to the CB, while a high-frequency probe pulse, which is in the frequency range of the energy gaps between the *core* band and VB/CB, allows us to detect the effects of the pump pulse in transient optical properties.

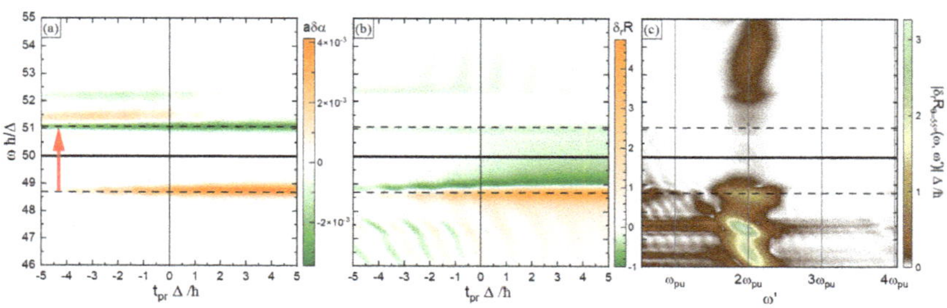

Figure 2. (**a**) Differential absorption coefficient (Equation (50)), as a function of t_{pr} and ω. The absorption coefficient is made to be dimensionless by being multiplied by the lattice constant a. The red arrow shows the electronic excitation via the one-photon resonance with the pump pulse. (**b**) Relative differential reflectivity for $\theta = 55^\circ$ (Equation (47)) as a function of t_{pr} and ω. (**c**) The frequency content of relative differential reflectivity, $|\delta_r R_{\theta=55^\circ}(\omega, \omega')|$ (see Equation (48)). The low-frequency region, $\omega' < 0.2\Delta/\hbar$, is not shown.

Figure 2a reports the differential absorption coefficient, as shown in Equation (50). The one-photon resonances between the VB and CB mainly occur at the indicated dashed lines, where the absorption coefficient shows its most significant changes. This can be understood as follows: The pump pulse strongly excites electrons from the VB to the CB at resonant $\boldsymbol{k}$ points, and it creates holes in the former and electrons in the latter. The holes in the VB make it possible for the high-frequency probe pulse to excite electrons from the *core* band to the VB, which results in an increase in the absorption coefficient. On the other hand, because of the Pauli exclusion principle, the electrons in the CB leave less vacancies for the probe-induced excitations from the *core* band. Consequently, we have a reduction in the absorption coefficient at the corresponding probe frequencies.

Figure 2b reports the relative differential reflectivity at the angle $\theta = 55^\circ$ (see Equation (47)). At the absorption edges, where the absorption coefficient undergoes the most significant changes (indicated with dashed lines), $\delta_r R$ approaches zero. Therefore, we obtain very narrow white strips. This can be very helpful in finding the connection between the experimental results for $\delta_r R$ and the band structure of the system.

Figure 2c reports $|\delta_r R_{\theta=55^\circ}(\omega, \omega')|$, which is the frequency content of $\delta_r R_{\theta=55^\circ}$ (see Equation (48)). The oscillations of the relative differential reflectivity (in t_{pr}, with respect to which the Fourier transformation is performed) are in the proximity of twice the pump frequency, $\omega' = 2\omega_{\mathrm{pu}}$, and there is no odd component. This shows that summing over all $\mathbf{k}$ points in the FBZ cancels out the odd frequencies that originate from the velocities.

4. Summary and Perspectives

In this manuscript, we reviewed our recently developed model-Hamiltonian approach, the dynamical projective operatorial approach (DPOA) [52,53,67,68]. The main ingredients of the DPOA and its applications to pumped semiconductors are reported and explained in detail. After providing the general theory of the DPOA, we clarified how it can exploit the equilibrium *ab initio* results for investigating pumped semiconductors through the study of the TR-ARPES signal and of the transient optical properties.

For the case of the TR-ARPES signal, we presented the methodology to compute the standard experimentally measurable signal, the lesser signal. Moreover, we computed another (theoretical) TR-ARPES signal, the retarded one, which can be used to study the out-of-equilibrium bands of pumped systems. In the case of transient optical properties, the results of our generalized linear response theory are presented and used to analyze a pumped system.

The DPOA can be applied to *real* materials (with [52] being an example). However, to illustrate different kinds of phenomena that can emerge in such pump–probe setups, we presented a simple yet rich model for a semiconductor. For the TR-ARPES signal, we discussed the emergence of different kinds of sidebands, the roles of system symmetries, and the relevance of the resonances. For the case of transient optical properties, we established the relations between the system band structure, the pump-induced excitations, the transient optical response of the system to the probe pulse, and the frequency content of absorption and reflectivity.

The DPOA is a versatile approach that can be applied to highly complicated systems with an affordable computational cost.

Funding: The authors acknowledge the support provided by MUR under Project PNRR MUR Missione 4 (SPOKE 2) TOPQIN "TOPological Qubit In driveN and reconfigurable heterostructures".

Institutional Review Board Statement: Not applicable.

Informed Consent Statement: Not applicable.

Data Availability Statement: The raw data supporting the conclusions of this article will be made available by the authors on request.

Acknowledgments: The authors thank Claudio Giannetti, Matteo Lucchini, Stefano Pittalis, and Carlo Andrea Rozzi for their insightful discussions.

Conflicts of Interest: The authors declare no conflicts of interest. The funders had no role in the design of the study; in the collection, analyses, or interpretation of data; in the writing of the manuscript; or in the decision to publish the results.

References

1. Brabec, T.; Krausz, F. Intense few-cycle laser fields: Frontiers of nonlinear optics. *Rev. Mod. Phys.* **2000**, *72*, 545. [CrossRef]
2. Krausz, F.; Ivanov, M. Attosecond physics. *Rev. Mod. Phys.* **2009**, *81*, 163. [CrossRef]
3. Schnyder, A.P.; Manske, D.; Avella, A. Resonant generation of coherent phonons in a superconductor by ultrafast optical pump pulses. *Phys. Rev. B* **2011**, *84*, 214513. [CrossRef]
4. Krausz, F.; Stockman, M.I. Attosecond metrology: From electron capture to future signal processing. *Nat. Photonics* **2014**, *8*, 205–213. [CrossRef]
5. Aoki, H.; Tsuji, N.; Eckstein, M.; Kollar, M.; Oka, T.; Werner, P. Nonequilibrium dynamical mean-field theory and its applications. *Rev. Mod. Phys.* **2014**, *86*, 779–837. [CrossRef]
6. Calegari, F.; Sansone, G.; Stagira, S.; Vozzi, C.; Nisoli, M. Advances in attosecond science. *J. Phys. B At. Mol. Opt. Phys.* **2016**, *49*, 062001. [CrossRef]

7. Gandolfi, M.; Celardo, G.L.; Borgonovi, F.; Ferrini, G.; Avella, A.; Banfi, F.; Giannetti, C. Emergent ultrafast phenomena in correlated oxides and heterostructures. *Phys. Scr.* **2017**, *92*, 034004. [CrossRef]
8. Borrego-Varillas, R.; Lucchini, M.; Nisoli, M. Attosecond spectroscopy for the investigation of ultrafast dynamics in atomic, molecular and solid-state physics. *Rep. Prog. Phys.* **2022**, *85*, 066401. [CrossRef] [PubMed]
9. Zürch, M.; Chang, H.T.; Borja, L.J.; Kraus, P.M.; Cushing, S.K.; Gandman, A.; Kaplan, C.J.; Oh, M.H.; Prell, J.S.; Prendergast, D.; et al. Direct and simultaneous observation of ultrafast electron and hole dynamics in germanium. *Nat. Commun.* **2017**, *8*, 15734. [CrossRef] [PubMed]
10. Zürch, M.; Chang, H.T.; Kraus, P.M.; Cushing, S.K.; Borja, L.J.; Gandman, A.; Kaplan, C.J.; Oh, M.H.; Prell, J.S.; Prendergast, D.; et al. Ultrafast carrier thermalization and trapping in silicon-germanium alloy probed by extreme ultraviolet transient absorption spectroscopy. *Struct. Dyn.* **2017**, *4*, 044029. [CrossRef] [PubMed]
11. Kaplan, C.J.; Kraus, P.M.; Ross, A.D.; Zürch, M.; Cushing, S.K.; Jager, M.F.; Chang, H.T.; Gullikson, E.M.; Neumark, D.M.; Leone, S.R. Femtosecond tracking of carrier relaxation in germanium with extreme ultraviolet transient reflectivity. *Phys. Rev. B* **2018**, *97*, 205202. [CrossRef]
12. Perfetti, L.; Loukakos, P.A.; Lisowski, M.; Bovensiepen, U.; Wolf, M.; Berger, H.; Biermann, S.; Georges, A. Femtosecond dynamics of electronic states in the Mott insulator 1T-TaS2 by time resolved photoelectron spectroscopy. *New J. Phys.* **2008**, *10*, 053019. [CrossRef]
13. Schmitt, F.; Kirchmann, P.S.; Bovensiepen, U.; Moore, R.G.; Rettig, L.; Krenz, M.; Chu, J.H.; Ru, N.; Perfetti, L.; Lu, D.; et al. Transient electronic structure and melting of a charge density wave in TbTe3. *Science* **2008**, *321*, 1649–1652. [CrossRef]
14. von der Linde, D.; Engers, T.; Jenke, G.; Agostini, P.; Grillon, G.; Nibbering, E.; Mysyrowicz, A.; Antonetti, A. Generation of high-order harmonics from solid surfaces by intense femtosecond laser pulses. *Phys. Rev. A* **1995**, *52*, R25. [CrossRef]
15. Norreys, P.A.; Zepf, M.; Moustaizis, S.; Fews, A.P.; Zhang, J.; Lee, P.; Bakarezos, M.; Danson, C.N.; Dyson, A.; Gibbon, P.; et al. Efficient Extreme UV Harmonics Generated from Picosecond Laser Pulse Interactions with Solid Targets. *Phys. Rev. Lett.* **1996**, *76*, 1832–1835. [CrossRef] [PubMed]
16. Chin, A.H.; Calderón, O.G.; Kono, J. Extreme midinfrared nonlinear optics in semiconductors. *Phys. Rev. Lett.* **2001**, *86*, 3292. [CrossRef] [PubMed]
17. Ghimire, S.; DiChiara, A.D.; Sistrunk, E.; Agostini, P.; DiMauro, L.F.; Reis, D.A. Observation of high-order harmonic generation in a bulk crystal. *Nat. Phys.* **2011**, *7*, 138–141. [CrossRef]
18. Luu, T.T.; Garg, M.; Kruchinin, S.Y.; Moulet, A.; Hassan, M.T.; Goulielmakis, E. Extreme ultraviolet high-harmonic spectroscopy of solids. *Nature* **2015**, *521*, 498–502. [CrossRef]
19. Han, S.; Kim, H.; Kim, Y.W.; Kim, Y.J.; Kim, S.; Park, I.Y.; Kim, S.W. High-harmonic generation by field enhanced femtosecond pulses in metal-sapphire nanostructure. *Nat. Commun.* **2016**, *7*, 13105. [CrossRef]
20. Borja, L.J.; Zürch, M.; Pemmaraju, C.D.; Schultze, M.; Ramasesha, K.; Gandman, A.; Prell, J.S.; Prendergast, D.; Neumark, D.M.; Leone, S.R. Extreme ultraviolet transient absorption of solids from femtosecond to attosecond timescales. *J. Opt. Soc. Am. B* **2016**, *33*, C57–C64. [CrossRef]
21. You, Y.S.; Reis, D.A.; Ghimire, S. Anisotropic high-harmonic generation in bulk crystals. *Nat. Phys.* **2017**, *13*, 345–349. [CrossRef]
22. Liu, H.; Li, Y.; You, Y.S.; Ghimire, S.; Heinz, T.F.; Reis, D.A. High-harmonic generation from an atomically thin semiconductor. *Nat. Phys.* **2017**, *13*, 262–265. [CrossRef]
23. Jiang, S.; Gholam-Mirzaei, S.; Crites, E.; Beetar, J.E.; Singh, M.; Lu, R.; Chini, M.; Lin, C. Crystal symmetry and polarization of high-order harmonics in ZnO. *J. Phys. B At. Mol. Opt. Phys.* **2019**, *52*, 225601. [CrossRef]
24. Rohwer, T.; Hellmann, S.; Wiesenmayer, M.; Sohrt, C.; Stange, A.; Slomski, B.; Carr, A.; Liu, Y.; Avila, L.M.; Kalläne, M.; et al. Collapse of long-range charge order tracked by time-resolved photoemission at high momenta. *Nature* **2011**, *471*, 490–493. [CrossRef]
25. Smallwood, C.L.; Hinton, J.P.; Jozwiak, C.; Zhang, W.; Koralek, J.D.; Eisaki, H.; Lee, D.H.; Orenstein, J.; Lanzara, A. Tracking Cooper pairs in a cuprate superconductor by ultrafast angle-resolved photoemission. *Science* **2012**, *336*, 1137–1139. [CrossRef]
26. Hellmann, S.; Rohwer, T.; Kalläne, M.; Hanff, K.; Sohrt, C.; Stange, A.; Carr, A.; Murnane, M.; Kapteyn, H.; Kipp, L.; et al. Time-domain classification of charge-density-wave insulators. *Nat. Commun.* **2012**, *3*, 1069. [CrossRef]
27. Papalazarou, E.; Faure, J.; Mauchain, J.; Marsi, M.; Taleb-Ibrahimi, A.; Reshetnyak, I.; van Roekeghem, A.; Timrov, I.; Vast, N.; Arnaud, B.; et al. Coherent Phonon Coupling to Individual Bloch States in Photoexcited Bismuth. *Phys. Rev. Lett.* **2012**, *108*, 256808. [CrossRef]
28. Wang, Y.; Steinberg, H.; Jarillo-Herrero, P.; Gedik, N. Observation of Floquet-Bloch states on the surface of a topological insulator. *Science* **2013**, *342*, 453–457. [CrossRef]
29. Johannsen, J.C.; Ulstrup, S.; Cilento, F.; Crepaldi, A.; Zacchigna, M.; Cacho, C.; Turcu, I.C.E.; Springate, E.; Fromm, F.; Raidel, C.; et al. Direct View of Hot Carrier Dynamics in Graphene. *Phys. Rev. Lett.* **2013**, *111*, 027403. [CrossRef]

30. Rameau, J.; Freutel, S.; Kemper, A.; Sentef, M.A.; Freericks, J.; Avigo, I.; Ligges, M.; Rettig, L.; Yoshida, Y.; Eisaki, H.; et al. Energy dissipation from a correlated system driven out of equilibrium. *Nat. Commun.* **2016**, *7*, 13761. [CrossRef]
31. Reimann, J.; Schlauderer, S.; Schmid, C.; Langer, F.; Baierl, S.; Kokh, K.; Tereshchenko, O.; Kimura, A.; Lange, C.; Güdde, J.; et al. Subcycle observation of lightwave-driven Dirac currents in a topological surface band. *Nature* **2018**, *562*, 396–400. [CrossRef]
32. Smith, N.; Traum, M. Angular-resolved ultraviolet photoemission spectroscopy and its application to the layer compounds Ta Se 2 and Ta S 2. *Phys. Rev. B* **1975**, *11*, 2087. [CrossRef]
33. Himpsel, F.J.; Eastman, D.E. Experimental energy dispersions for valence and conduction bands of palladium. *Phys. Rev. B* **1978**, *18*, 5236–5239. [CrossRef]
34. Kampf, A.P.; Schrieffer, J.R. Spectral function and photoemission spectra in antiferromagnetically correlated metals. *Phys. Rev. B* **1990**, *42*, 7967–7974. [CrossRef] [PubMed]
35. Smith, K.E.; Kevan, S.D. The electronic structure of solids studied using angle resolved photoemission spectroscopy. *Prog. Solid State Chem.* **1991**, *21*, 49–131. [CrossRef]
36. Damascelli, A.; Hussain, Z.; Shen, Z.X. Angle-resolved photoemission studies of the cuprate superconductors. *Rev. Mod. Phys.* **2003**, *75*, 473. [CrossRef]
37. Hüfner, S. *Photoelectron Spectroscopy: Principles and Applications*; Springer Science & Business Media: Berlin/Heidelberg, Germany, 2013.
38. Sobota, J.A.; He, Y.; Shen, Z.X. Angle-resolved photoemission studies of quantum materials. *Rev. Mod. Phys.* **2021**, *93*, 025006. [CrossRef]
39. Schultze, M.; Bothschafter, E.M.; Sommer, A.; Holzner, S.; Schweinberger, W.; Fiess, M.; Hofstetter, M.; Kienberger, R.; Apalkov, V.; Yakovlev, V.S.; et al. Controlling dielectrics with the electric field of light. *Nature* **2013**, *493*, 75–78. [CrossRef]
40. Stojchevska, L.; Vaskivskyi, I.; Mertelj, T.; Kusar, P.; Svetin, D.; Brazovskii, S.; Mihailovic, D. Ultrafast switching to a stable hidden quantum state in an electronic crystal. *Science* **2014**, *344*, 177–180. [CrossRef]
41. Schultze, M.; Ramasesha, K.; Pemmaraju, C.; Sato, S.; Whitmore, D.; Gandman, A.; Prell, J.S.; Borja, L.; Prendergast, D.; Yabana, K.; et al. Attosecond band-gap dynamics in silicon. *Science* **2014**, *346*, 1348–1352. [CrossRef]
42. Lucchini, M.; Sato, S.A.; Ludwig, A.; Herrmann, J.; Volkov, M.; Kasmi, L.; Shinohara, Y.; Yabana, K.; Gallmann, L.; Keller, U. Attosecond dynamical Franz-Keldysh effect in polycrystalline diamond. *Science* **2016**, *353*, 916–919. [CrossRef]
43. Mashiko, H.; Oguri, K.; Yamaguchi, T.; Suda, A.; Gotoh, H. Petahertz optical drive with wide-bandgap semiconductor. *Nat. Phys.* **2016**, *12*, 741–745. [CrossRef]
44. Avella, A.; Buonavolontà, C.; Guarino, A.; Valentino, M.; Leo, A.; Grimaldi, G.; de Lisio, C.; Nigro, A.; Pepe, G. Disorder-sensitive pump-probe measurements on $Nd_{1.83}Ce_{0.17}CuO_{4\pm\delta}$ films. *Phys. Rev. B* **2016**, *94*, 115426. [CrossRef]
45. Nembrini, N.; Peli, S.; Banfi, F.; Ferrini, G.; Singh, Y.; Gegenwart, P.; Comin, R.; Foyevtsova, K.; Damascelli, A.; Avella, A.; et al. Tracking local magnetic dynamics via high-energy charge excitations in a relativistic Mott insulator. *Phys. Rev. B* **2016**, *94*, 201119. [CrossRef]
46. Novelli, F.; Giovannetti, G.; Avella, A.; Cilento, F.; Patthey, L.; Radovic, M.; Capone, M.; Parmigiani, F.; Fausti, D. Localized vibrations in superconducting $YBa_2Cu_3O_7$ revealed by ultrafast optical coherent spectroscopy. *Phys. Rev. B* **2017**, *95*, 174524. [CrossRef]
47. Schlaepfer, F.; Lucchini, M.; Sato, S.A.; Volkov, M.; Kasmi, L.; Hartmann, N.; Rubio, A.; Gallmann, L.; Keller, U. Attosecond optical-field-enhanced carrier injection into the GaAs conduction band. *Nat. Phys.* **2018**, *14*, 560–564. [CrossRef]
48. Giusti, F.; Marciniak, A.; Randi, F.; Sparapassi, G.; Boschini, F.; Eisaki, H.; Greven, M.; Damascelli, A.; Avella, A.; Fausti, D. Signatures of Enhanced Superconducting Phase Coherence in Optimally Doped $Bi_2Sr_2Y_{0.08}Ca_{0.92}Cu_2O_{8+\delta}$ Driven by Midinfrared Pulse Excitations. *Phys. Rev. Lett.* **2019**, *122*, 067002. [CrossRef]
49. Kaplan, C.J.; Kraus, P.M.; Gullikson, E.M.; Borja, L.J.; Cushing, S.K.; Zürch, M.; Chang, H.T.; Neumark, D.M.; Leone, S.R. Retrieval of the complex-valued refractive index of germanium near the M4,5 absorption edge. *J. Opt. Soc. Am. B* **2019**, *36*, 1716. [CrossRef]
50. Geneaux, R.; Marroux, H.J.; Guggenmos, A.; Neumark, D.M.; Leone, S.R. Transient absorption spectroscopy using high harmonic generation: A review of ultrafast X-ray dynamics in molecules and solids. *Philos. Trans. R. Soc. A* **2019**, *377*, 20170463. [CrossRef]
51. Giusti, F.; Montanaro, A.; Marciniak, A.; Randi, F.; Boschini, F.; Glerean, F.; Jarc, G.; Eisaki, H.; Greven, M.; Damascelli, A.; et al. Anisotropic time-domain electronic response in cuprates driven by midinfrared pulses. *Phys. Rev. B* **2021**, *104*, 125121. [CrossRef]
52. Inzani, G.; Adamska, L.; Eskandari-asl, A.; Palo, N.D.; Dolso, G.L.; Moio, B.; D'Onofrio, L.J.; Lamperti, A.; Molle, A.; Borrego-Varillas, R.; et al. Field-driven attosecond charge dynamics in germanium. *Nat. Photonics* **2023**, *17*, 1059. [CrossRef]
53. Inzani, G.; Eskandari-asl, A.; Adamska, L.; Moio, B.; Dolso, G.L.; Palo, N.D.; D'Onofrio, L.J.; Lamperti, A.; Molle, A.; Rozzi, C.A.; et al. Photoinduced charge carrier dynamics in germanium. *Il Nuovo Cimento C* **2023**, *46*, 110.
54. Neufeld, O.; Tancogne-Dejean, N.; De Giovannini, U.; Hübener, H.; Rubio, A. Attosecond magnetization dynamics in non-magnetic materials driven by intense femtosecond lasers. *NPJ Comput. Mater.* **2023**, *9*, 39. [CrossRef]

55. De Giovannini, U.; Brunetto, G.; Castro, A.; Walkenhorst, J.; Rubio, A. Inside Cover: Simulating Pump–Probe Photoelectron and Absorption Spectroscopy on the Attosecond Timescale with Time-Dependent Density Functional Theory (ChemPhysChem 7/2013). *ChemPhysChem* **2013**, *14*, 1298–1298. [CrossRef]
56. De Giovannini, U.; Hubener, H.; Rubio, A. Monitoring electron-photon dressing in WSe2. *Nano Lett.* **2016**, *16*, 7993–7998. [CrossRef] [PubMed]
57. Wopperer, P.; De Giovannini, U.; Rubio, A. Efficient and accurate modeling of electron photoemission in nanostructures with TDDFT. *Eur. Phys. J. B* **2017**, *90*, 51. [CrossRef]
58. De Giovannini, U.; Hübener, H.; Rubio, A. A first-principles time-dependent density functional theory framework for spin and time-resolved angular-resolved photoelectron spectroscopy in periodic systems. *J. Chem. Theory Comput.* **2017**, *13*, 265–273. [CrossRef]
59. Pemmaraju, C.D.; Vila, F.D.; Kas, J.J.; Sato, S.A.; Rehr, J.J.; Yabana, K.; Prendergast, D. Velocity-gauge real-time TDDFT within a numerical atomic orbital basis set. *Comput. Phys. Commun.* **2018**, *226*, 30–38. [CrossRef]
60. Tancogne-Dejean, N.; Rubio, A. Atomic-like high-harmonic generation from two-dimensional materials. *Sci. Adv.* **2018**, *4*, eaao5207. [CrossRef]
61. De Giovannini, U.; Castro, A. Real-time and real-space time-dependent density-functional theory approach to attosecond dynamics. *Attosecond Mol. Dyn.* **2018**, *13*, 424–461.
62. Armstrong, G.S.; Khokhlova, M.A.; Labeye, M.; Maxwell, A.S.; Pisanty, E.; Ruberti, M. Dialogue on analytical and ab initio methods in attoscience. *Eur. Phys. J. D* **2021**, *75*, 209. [CrossRef] [PubMed]
63. Sato, S.A.; Lucchini, M.; Volkov, M.; Schlaepfer, F.; Gallmann, L.; Keller, U.; Rubio, A. Role of intraband transitions in photocarrier generation. *Phys. Rev. B* **2018**, *98*, 035202. [CrossRef]
64. Broers, L.; Mathey, L. Detecting light-induced Floquet band gaps of graphene via trARPES. *Phys. Rev. Res.* **2022**, *4*, 013057. [CrossRef]
65. D'Onofrio, L.J.; Eskandari-asl, A.; Avella, A. Time evolution of energies and populations in germanium perturbed by a near-infrared pulse on the atto-second scale. *J. Magn. Magn. Mater.* **2022**, *546*, 168785. [CrossRef]
66. Schüler, M.; Marks, J.A.; Murakami, Y.; Jia, C.; Devereaux, T.P. Gauge invariance of light-matter interactions in first-principle tight-binding models. *Phys. Rev. B* **2021**, *103*, 155409. [CrossRef]
67. Eskandari-asl, A.; Avella, A. Time-resolved ARPES signal in pumped semiconductors within the dynamical projective operatorial approach. *Phys. Rev. B* **2024**, *110*, 094309. [CrossRef]
68. Eskandari-asl, A.; Avella, A. Generalized linear response theory for pumped systems and its application to transient optical properties. *Phys. Rev. A* **2024**, *110*, 043520. [CrossRef]
69. Mancini, F.; Avella, A. The Hubbard model within the equations of motion approach. *Adv. Phys.* **2004**, *53*, 537–768. [CrossRef]
70. Avella, A.; Mancini, F. Underdoped cuprate phenomenology in the two-dimensional Hubbard model within the composite operator method. *Phys. Rev. B* **2007**, *75*, 134518. [CrossRef]
71. Avella, A.; Mancini, F. The Composite Operator Method (COM). In *Strongly Correlated Systems: Theoretical Methods*; Avella, A., Mancini, F., Eds.; Springer: Berlin/Heidelberg, Germany, 2012; pp. 103–141. [CrossRef]
72. Avella, A. The Hubbard model beyond the two-pole approximation: A composite operator method study. *Eur. Phys. J. B* **2014**, *87*, 45. [CrossRef]
73. Avella, A. Composite operator method analysis of the underdoped cuprates puzzle. *Adv. Condens. Matter Phys.* **2014**, *2014*, 515698. [CrossRef]
74. Di Ciolo, A.; Avella, A. The composite operator method route to the 2D Hubbard model and the cuprates. *Condens. Matter. Phys.* **2018**, *21*, 33701. [CrossRef]
75. Eskandari-asl, A.; Avella, A. Out-of-equilibrium Hubbard dimer within DPOA. In Proceedings of the Advances in Ultrafast Condensed Phase Physics IV, Strasbourg, France, 7–11 April 2024; Volume 12992, pp. 65–68.
76. Inzani, G.; Adamska, L.; D'Onofrio, L.J.; Moio, B.; Dolso, G.L.; Di Palo, N.; Eskandari-asl, A.; Lamperti, A.; Molle, A.; Rozzi, C.A.; et al. Attosecond Inter-and Intra-Band Charge Carrier Dynamics in Germanium. In Proceedings of the International Conference on Ultrafast Phenomena, Montreal, QC, Canada, 18–22 July 2022; Optica Publishing Group: Washington, DC, USA, 2022; pp. W4A-32.
77. Inzani, G.; Adamska, L.; Eskandari-asl, A.; Di Palo, N.; Dolso, G.L.; Moio, B.; D'Onofrio, L.J.; Lamperti, A.; Molle, A.; Borrego-Varillas, R.; et al. Light-Driven Attosecond Photoinjection in Germanium. In Proceedings of the The European Conference on Lasers and Electro-Optics, Munich, Germany, 26–30 June 2023; Optica Publishing Group: Washington, DC, USA, 2023; p. cg_2_2.
78. Di Palo, N.; Inzani, G.; Adamska, L.; Eskandari-asl, A.; Dolso, G.; Moio, B.; D'Onofrio, L.; Lamperti, A.; Molle, A.; Borrego-Varillas, R.; et al. Attosecond photoinjection dynamics in germanium. In Proceedings of the Advances in Ultrafast Condensed Phase Physics IV, Strasbourg, France, 7–11 April 2024; p. PC129920O.

79. Eskandari-asl, A.; Avella, A. Dynamical projective operatorial approach (DPOA) for pump-probe setups in the ultrafast regime. In Proceedings of the Advances in Ultrafast Condensed Phase Physics IV, Strasbourg, France, 7–11 April 2024; Volume 12992, pp. 69–73.
80. Inzani, G.; Adamska, L.; Eskandari-asl, A.; Di Palo, N.; Dolso, G.L.; Moio, B.; D'Onofrio, L.J.; Lamperti, A.; Molle, A.; Borrego-Varillas, R.; et al. Photoexcitation in germanium probed by attosecond transient reflectivity spectroscopy. In Proceedings of the Ultrafast Phenomena and Nanophotonics XXVIII, San Francisco, CA, USA, 29 January–1 February 2023; Volume 12884, pp. 31–36.
81. Avella, A.; Mancini, F.; Villani, D.; Matsumoto, H. The superconducting gap in the two-dimensional Hubbard model. *Phys. C Supercond.* **1997**, *282*, 1757–1758. [CrossRef]
82. Avella, A.; Mancini, F.; Villani, D. The overdoped regime in La2- xSrxCuO4. *Solid State Commun.* **1998**, *108*, 723–725. [CrossRef]
83. Sanchez, M.; Avella, A.; Mancini, F. The half-filled Hubbard chain in the Composite Operator Method: A comparison with Bethe Ansatz. *Europhys. Lett.* **1998**, *44*, 328. [CrossRef]
84. Avella, A.; Mancini, F.; Villani, D. Incommensurate spin fluctuations in the two-dimensional t-t′-U model. *Phys. Lett. A* **1998**, *240*, 235–240. [CrossRef]
85. Avella, A.; Mancini, F.; Villani, D.; Siurakshina, L.; Yushankhai, V.Y. The Hubbard model in the two-pole approximation. *Int. J. Mod. Phys. B* **1998**, *12*, 81–97. [CrossRef]
86. Villani, D.; Lange, E.; Avella, A.; Kotliar, G. Two-Scale Analysis of the SU(N) Kondo Model. *Phys. Rev. Lett.* **2000**, *85*, 804–807. [CrossRef]
87. Avella, A.; Mancini, F.; Villani, D.; Matsumoto, H. The two-dimensional t-t′-U model as a minimal model for cuprate materials. *Eur. Phys. J. B-Condens. Matter Complex Syst.* **2001**, *20*, 303–311. [CrossRef]
88. Avella, A.; Mancini, F.; Münzner, R. Antiferromagnetic phase in the Hubbard model by means of the composite operator method. *Phys. Rev. B* **2001**, *63*, 245117. [CrossRef]
89. Avella, A.; Mancini, F.; Turkowski, V. Bosonic sector of the two-dimensional Hubbard model studied within a two-pole approximation. *Phys. Rev. B* **2003**, *67*, 115123. [CrossRef]
90. Avella, A.; Mancini, F.; Saikawa, T. The 2-site Hubbard and t-J models. *Eur. Phys. J. B-Condens. Matter Complex Syst.* **2003**, *36*, 445–473. [CrossRef]
91. Mancini, F.; Avella, A. Equation of motion method for composite field operators. *Eur. Phys. J. B-Condens. Matter Complex Syst.* **2003**, *36*, 37–56. [CrossRef]
92. Avella, A.; Mancini, F. The Hubbard model with intersite interaction within the Composite Operator Method. *Eur. Phys. J. B-Condens. Matter Complex Syst.* **2004**, *41*, 149–162. [CrossRef]
93. Avella, A.; Krivenko, S.; Mancini, F.; Plakida, N. Self-energy corrections to the electronic spectrum of the Hubbard model. *J. Magn. Magn. Mater.* **2004**, *272*, 456–457. [CrossRef]
94. Odashima, S.; Avella, A.; Mancini, F. High-order correlation effects in the two-dimensional Hubbard model. *Phys. Rev. B* **2005**, *72*, 205121. [CrossRef]
95. Avella, A.; Mancini, F. Exact solution of the one-dimensional spin-3/2 Ising model in magnetic field. *Eur. Phys. J. B-Condens. Matter Complex Syst.* **2006**, *50*, 527–539. [CrossRef]
96. Avella, A.; Mancini, F.; Mancini, F.P.; Plekhanov, E. Emery vs. Hubbard model for cuprate superconductors: A composite operator method study. *Eur. Phys. J. B* **2013**, *86*, 265. [CrossRef]
97. Mostofi, A.A.; Yates, J.R.; Pizzi, G.; Lee, Y.S.; Souza, I.; Vanderbilt, D.; Marzari, N. An updated version of wannier90: A tool for obtaining maximally-localised Wannier functions. *Comput. Phys. Commun.* **2014**, *185*, 2309. [CrossRef]
98. Peierls, R. Zur theorie des diamagnetismus von leitungselektronen. *Z. Phys.* **1933**, *80*, 763–791. [CrossRef]
99. Ismail-Beigi, S.; Chang, E.K.; Louie, S.G. Coupling of nonlocal potentials to electromagnetic fields. *Phys. Rev. Lett.* **2001**, *87*, 087402. [CrossRef] [PubMed]
100. Freericks, J.K.; Krishnamurthy, H.R.; Pruschke, T. Theoretical Description of Time-Resolved Photoemission Spectroscopy: Application to Pump-Probe Experiments. *Phys. Rev. Lett.* **2009**, *102*, 136401. [CrossRef] [PubMed]
101. Sentef, M.; Kemper, A.F.; Moritz, B.; Freericks, J.K.; Shen, Z.X.; Devereaux, T.P. Examining electron-boson coupling using time-resolved spectroscopy. *Phys. Rev. X* **2013**, *3*, 041033. [CrossRef]
102. Freericks, J.; Krishnamurthy, H.; Sentef, M.; Devereaux, T. Gauge invariance in the theoretical description of time-resolved angle-resolved pump/probe photoemission spectroscopy. *Phys. Scr.* **2015**, *2015*, 014012. [CrossRef]
103. Schüler, M.; Sentef, M.A. Theory of subcycle time-resolved photoemission: Application to terahertz photodressing in graphene. *J. Electron Spectrosc. Relat. Phenom.* **2021**, *253*, 147121. [CrossRef]

materials

Article

Straight Disclinations in Fractional Nonlocal Medium

Tamara Kyrylych * and Yuriy Povstenko

Department of Mathematics and Computer Science, Faculty of Science and Technology, Jan Dlugosz University in Czestochowa, al. Armii Krajowej 13/15, 42-200 Czestochowa, Poland; j.povstenko@ujd.edu.pl
* Correspondence: t.kyrylych@ujd.edu.pl

Abstract: The constitutive equation for a nonlocal stress tensor is represented as an integral with the suitable kernel function. In this paper, the nonlocality kernel is chosen as the Green function of the Cauchy problem for the fractional diffusion equation with the Caputo derivative with respect to the nonlocality parameter. The solutions of nonlocal elasticity problems for the straight wedge and twist disclinations in an infinite medium are obtained in the framework of this new nonlocal theory of elasticity. The Laplace integral transform with respect to the nonlocality parameter is used. It is necessary to emphasize that the transition from the nonlocal to local stress tensor is described by the limiting value of the nonlocality parameter $\tau \to 0$. The obtained stress fields do not contain nonphysical singularities at the disclination lines.

Keywords: wedge disclination; twist disclination; nonlocal elasticity; nonlocality kernel; fractional differential equation; Caputo derivative; Mittag–Leffler function

1. Introduction

Real solids contain a large number of different types of defects. In 1907, Volterra [1] considered elasticity problems for a long homogeneous isotropic hollow cylinder. The studied objects were called distortions by Volterra. Later on, defects of the translation type were named dislocations [2], whereas the defects of the rotational type were named disclinations [3]. Translation is described by the Burgers vector $\mathbf{b}$, while rotation is associated with the Frank vector $\mathbf{\Omega}$. In 1934, Taylor [4], Orowan [5], and Polanyi [6] showed that the plastic deformation in crystalline solids can be explained in terms of the theory of Volterra dislocations. This insight was critical in developing the modern science of solid mechanics. The literature on dislocations is quite voluminous; we refer to the well-known books [7–13], where additional references can be found. For a long time, disclinations were in the shadow of dislocations, but this situation was changed after the appearance of a series of papers by deWit [14,15] (see, for example, [16–23], and the extensive bibliography in [24–28]). There are strict relations between dislocations and disclinations [15,29,30]. Dislocations can end at disclinations. Disclinations can be modelled by groups of dislocations with some density. In this case, the stress fields of disclinations can be obtained as the corresponding integrals of the stress fields caused by dislocation arrays. However, identifying which description is more appropriate in a specific case is an issue [26].

The problem of modeling stress and strain fields around dislocations and disclinations has attracted much attention from researchers. Effective methods for classical elasticity cannot describe the situation in the immediate vicinity of an imperfection, as nonphysical singularities appear. Therefore, regular attempts have been made to improve the classical elastic models of crystal defects, for instance, combining the elastic and discrete approaches for a better description of the highly distorted region near a defect. In the Frenkel–Kontorova model [31], the dislocation is considered as a set of particles coupled by nearest-nieghbour interaction and moving in a periodic potential. The modern state of studies based on the Frenkel–Kontorova model is presented in [32,33]. Another model that takes into account the discrete structure of the crystal and describes the core of the dislocation is the Peierls–Nabarro model [34,35]. The assumption of this model is that the dislocation is characterized by the elastic energy due to a finite density of dislocations and the misfit energy, which results from the nonlinear atomic interaction in the glide plane. The semi-discrete approach, according to which the crystal with dislocation is divided into two parts (a discrete lattice and an elastic continuum), has been discussed extensively in [36–39].

Great advances in improving classical elasticity solutions for dislocations and disclinations have been made by the application of different generalized theories. Straight dislocations [40–45] and disclinations [46] were studied in different theories of nonlocal elasticity, as well as in the framework of the comparable strain gradient elasticity [47–52]. The interested reader is also referred to the approach developed in the papers [53–55].

In the present paper, the solutions of nonlocal elasticity problems for the straight wedge and twist disclinations in an infinite solid are obtained in the framework of the new nonlocal theory of elasticity, in which the nonlocality kernel is chosen as the Green function of the Cauchy problem for the fractional diffusion operator with the Caputo derivative with respect to the nonlocality parameter. In the literature, other kernels were also considered as the Green functions for some differential operators [56–59].

To solve these problems, the components of the Laplacian of the symmetric second-order tensor in cylindrical coordinates are needed. Such components were obtained in the paper [46]. A coupled systems of equations are split, introducing new unknown functions. The Laplace transform with respect to the nonlocality parameter is used. The immediate application of the Hankel transform is impossible due to the logarithmic term appearing in the classical elasticity solution. For this reason, a special analysis of the modified Bessel functions of the peculiar argument has been carried out.

The paper is organized as follows. The basic equations of the fractional nonlocal elasticity are presented in Section 2. The solution for the straight wedge disclination is obtained in Section 3. Section 4 is dedicated to the solution for the straight twist disclination. Finally, concluding remarks are presented in Section 5.

2. Fractional Nonlocal Elasticity

According to the theory of nonlocal elasticity, the stress tensor at a reference point of a solid depends not only on the strains at this point but also on strains at all other points of the solid. For example, for a linear isotropic nonlocal elastic body

$$\mathbf{t}(\mathbf{x}) = \int_V K(|\mathbf{x}-\mathbf{x}'|)\,\sigma(\mathbf{x}')\,dv(\mathbf{x}'), \tag{1}$$

$$\sigma(\mathbf{x}') = 2\mu\,\mathbf{e}(\mathbf{x}') + \lambda\,\mathrm{tr}\,\mathbf{e}(\mathbf{x}')\mathbf{I}, \tag{2}$$

where $\sigma(\mathbf{x}')$ and $\mathbf{t}(\mathbf{x})$ are the classical and nonlocal stress tensors, $\mathbf{x}'$ and $\mathbf{x}$ are the running and reference points, $\mathbf{e}$ is the linear strain tensor, λ and μ denote the Lamé constants, and $\mathbf{I}$ stands for the unit tensor.

The properties of the nonlocal kernel $K(\mathbf{x})$ were investigated by Eringen [43,44]. In particular, Eringen considered the nonlocal kernel as the Green function of the Cauchy problem for the diffusion operator:

$$\frac{\partial K(\mathbf{x},\tau)}{\partial \tau} = \Delta\, K(\mathbf{x},\tau), \tag{3}$$

$$\tau = 0: \quad K(\mathbf{x},\tau) = \delta(\mathbf{x}), \tag{4}$$

which results in the kernel

$$K(\mathbf{x},\tau) = \frac{1}{(2\sqrt{\pi\tau})^n} \exp\left(-\frac{|\mathbf{x}|^2}{4\tau}\right), \tag{5}$$

with $n = 1, 2, 3$ correlated with a number of spatial variables.

This approach results in the Cauchy problem for the nonlocal stress tensor

$$\frac{\partial \mathbf{t}(\mathbf{x},\tau)}{\partial \tau} = \Delta\, \mathbf{t}(\mathbf{x},\tau), \tag{6}$$

$$\tau = 0: \quad \mathbf{t}(\mathbf{x},\tau) = \boldsymbol{\sigma}(\mathbf{x}). \tag{7}$$

Fractional calculus has many applications in different areas of science and was also used in the formulation of nonlocal elasticity approaches (see, for example, [60–65]). Fractional nonlocal elasticity theory with the nonlocality kernel being the Green function of the Cauchy problem for fractional diffusion equation with the Caputo derivative with respect to nonlocality parameter $\partial^\alpha/\partial\tau^\alpha$ and fractional Laplace operator (Riesz operator) $(-\Delta)^{\beta/2}$ was proposed in [66]. In the present paper, we restrict our attention to the case $\beta = 2$. In the proposed theory [66], the nonlocal modulus satisfies the equation

$$\frac{\partial^\alpha K(\mathbf{x},\tau)}{\partial \tau^\alpha} = a\,\Delta\, K(\mathbf{x},\tau), \quad 0 < \alpha \le 1, \tag{8}$$

$$\tau = 0: \quad K(\mathbf{x},\tau) = \delta(\mathbf{x}). \tag{9}$$

The Caputo fractional derivative [67,68] is defined as

$$\frac{d^\alpha f(\tau)}{d\tau^\alpha} = \frac{1}{\Gamma(m-\alpha)} \int_0^\tau (\tau - u)^{m-\alpha-1} \frac{d^m f(u)}{du^m}\, du, \qquad m - 1 < \alpha < m. \tag{10}$$

Here, $\Gamma(\alpha)$ is the gamma function. The Caputo derivative (10) has the Laplace transform rule:

$$\mathcal{L}\left\{\frac{\mathrm{d}^\alpha f(\tau)}{\mathrm{d}\tau^\alpha}\right\} = s^\alpha f^*(s) - \sum_{k=0}^{m-1} f^{(k)}(0^+) s^{\alpha-1-k}, \qquad m - 1 < \alpha < m, \tag{11}$$

where the asterisk denotes the Laplace transform, with s being the transform variable.

Instead of the solution (5), the nonlocal kernel $K(\mathbf{x},\tau)$ for a different number n of spatial variables is expressed in terms of the Mittag–Leffler function, Mainardi function, and Wright function (see [66]).

The Cauchy problem for the nonlocality kernel $K(\mathbf{x},\tau)$ (8) and (9) results in the Cauchy problem for the nonlocal stress tensor

$$\frac{\partial^\alpha \mathbf{t}(\mathbf{x},\tau)}{\partial \tau^\alpha} = a\,\Delta\, \mathbf{t}(\mathbf{x},\tau), \quad 0 < \alpha \le 1, \tag{12}$$

$$\tau = 0: \quad \mathbf{t}(\mathbf{x},\tau) = \boldsymbol{\sigma}(\mathbf{x}). \tag{13}$$

It is necessary to accentuate that in Equations (3), (6), (8) and (12), the letter τ does not denote time; τ is the nonlocality parameter. The conditions (7) and (13) do not correspond to the initial time value, but describe the transition from the nonlocal to local stress tensor.

Eringen in his publications [43,44] showed how the nonlocality parameter τ can be connected with atomic lattice theory. The parameter τ is associated with a characteristic length ratio l/L, where l is an internal characteristic length (for example, the lattice parameter) and L is an external characteristic length (wavelength, solid size, etc.). Eringen [43,44] and Kunin [69,70] provided a detailed analysis of relations between nonlocal elasticity theory and Debye quasi-continuum and the Born-Kármán model of atomic lattice dynamics and atomic dispersion curves. Similar considerations can be applied to the proposed theory. We also note that Eringen [42,44] considered Equations (3) and (6) without the coefficient a; we have written the coefficient a in Equations (8) and (12) to have the possibility of studying different limiting cases of the solutions.

3. Straight Wedge Disclination

Assume the z-axis as the disclination line and consider the straight wedge disclination with the Frank vector $\mathbf{\Omega} = (0, 0, \Omega_3)$. A drawing of the wedge disclination is presented in Figure 1.

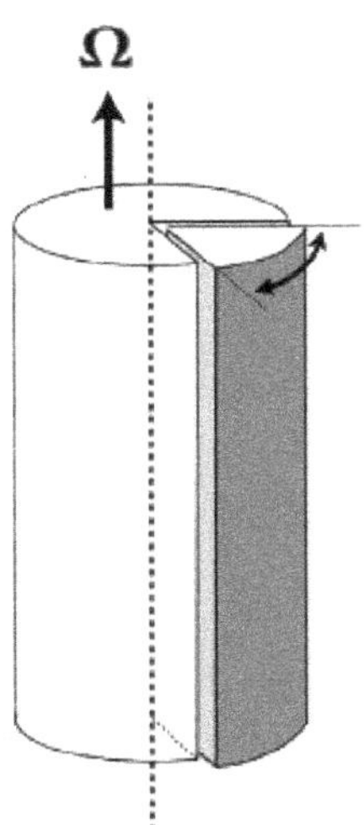

Figure 1. Schematic representation of a straight wedge disclination with the Frank vector Ω_3.

Under the conditions of the plane strain, the classical elasticity stress tensor for the straight wedge disclination has the following components [15] in cylindrical coordinates:

$$\sigma_{rr} = A\left[\ln(r/l) + \frac{\nu}{1-2\nu}\right], \tag{14}$$

$$\sigma_{\theta\theta} = A\left[\ln(r/l) + \frac{1-\nu}{1-2\nu}\right], \tag{15}$$

$$\sigma_{zz} = A\left[2\nu\ln(r/l) + \frac{\nu}{1-2\nu}\right], \tag{16}$$

$$\sigma_{r\theta} = 0, \qquad \sigma_{rz} = 0, \qquad \sigma_{\theta z} = 0, \tag{17}$$

where

$$A = \frac{\mu\Omega_3}{2\pi(1-\nu)}, \tag{18}$$

where ν is the Poisson ratio.

It should be noted that deWit [15] wrote the corresponding equations with the term $\ln r$. We have slightly changed Equations (14)–(16) in comparison with the original equations of deWit, introducing the nondimensional quantity r/l under the logarithm symbol. The specific expression for the characteristic length l will be discussed below.

From Equations (12) and (13), using the components of the Laplacian of the stress tensor presented in Appendix A and the components of the local stress tensor (14)–(17), we obtain

$$\frac{\partial^\alpha t_{rr}}{\partial \tau^\alpha} = a\left[\frac{\partial^2 t_{rr}}{\partial r^2} + \frac{1}{r}\frac{\partial t_{rr}}{\partial r} - \frac{2}{r^2}\left(t_{rr} - t_{\theta\theta}\right)\right], \tag{19}$$

$$\tau = 0: \quad t_{rr} = A\left[\ln(r/l) + \frac{\nu}{1-2\nu}\right]; \tag{20}$$

$$\frac{\partial^\alpha t_{\theta\theta}}{\partial \tau^\alpha} = a\left[\frac{\partial^2 t_{\theta\theta}}{\partial r^2} + \frac{1}{r}\frac{\partial t_{\theta\theta}}{\partial r} + \frac{2}{r^2}\left(t_{rr} - t_{\theta\theta}\right)\right], \tag{21}$$

$$\tau = 0: \quad t_{\theta\theta} = A\left[\ln(r/l) + \frac{1-\nu}{1-2\nu}\right]; \tag{22}$$

$$\frac{\partial^\alpha t_{zz}}{\partial \tau^\alpha} = a\left[\frac{\partial^2 t_{zz}}{\partial r^2} + \frac{1}{r}\frac{\partial t_{zz}}{\partial r}\right], \tag{23}$$

$$\tau = 0: \quad t_{zz} = A\left[2\nu\ln(r/l) + \frac{\nu}{1-2\nu}\right]. \tag{24}$$

Now, we introduce two auxiliary functions

$$f(r,\tau) = t_{rr}(r,\tau) + t_{\theta\theta}(r,\tau), \tag{25}$$

$$g(r,\tau) = t_{rr}(r,\tau) - t_{\theta\theta}(r,\tau) \tag{26}$$

which allow us to split Equations (19)–(22):

$$\frac{\partial^\alpha f}{\partial \tau^\alpha} = a\left(\frac{\partial^2 f}{\partial r^2} + \frac{1}{r}\frac{\partial f}{\partial r}\right), \tag{27}$$

$$\tau = 0: \quad f = A\left[2\ln(r/l) + \frac{1}{1-2\nu}\right], \tag{28}$$

$$\frac{\partial^\alpha g}{\partial \tau^\alpha} = a\left(\frac{\partial^2 g}{\partial r^2} + \frac{1}{r}\frac{\partial g}{\partial r} - \frac{4}{r^2}g\right), \tag{29}$$

$$\tau = 0: \quad g = -A. \tag{30}$$

In this case,

$$t_{rr}(r,\tau) = \frac{1}{2}[f(r,\tau) + g(r,\tau)], \tag{31}$$

$$t_{\theta\theta}(r,\tau) = \frac{1}{2}[f(r,\tau) - g(r,\tau)], \tag{32}$$

$$t_{zz}(r,\tau) = \nu f(r,\tau). \tag{33}$$

The Laplace transform with respect to the nonlocality parameter τ gives

$$\frac{\partial^2 f^*}{\partial r^2} + \frac{1}{r}\frac{\partial f^*}{\partial r} - \frac{s^\alpha}{a} f^* = -\frac{s^{\alpha-1}}{a} A\left[2\ln(r/l) + \frac{1}{1-2\nu}\right], \tag{34}$$

$$\frac{\partial^2 g^*}{\partial r^2} + \frac{1}{r}\frac{\partial g^*}{\partial r} - \frac{4}{r^2} g^* - \frac{s^\alpha}{a} g^* = \frac{s^{\alpha-1}}{a} A. \tag{35}$$

The solutions of equations of such a type are expressed in terms of the modified Bessel functions of the zeroth and second order

$$I_0\left(\sqrt{\frac{s^\alpha}{a}}\, r\right), \quad K_0\left(\sqrt{\frac{s^\alpha}{a}}\, r\right), \quad I_2\left(\sqrt{\frac{s^\alpha}{a}}\, r\right), \quad K_2\left(\sqrt{\frac{s^\alpha}{a}}\, r\right). \tag{36}$$

The modified Bessel functions $I_0(r)$ and $I_2(r)$ have exponential divergence at infinity, and the modified Bessel functions $K_0(r)$ and $K_2(r)$ have the following expansion at the origin [71]:

$$K_0(r) \sim -\ln r, \qquad K_2(r) \sim \frac{2}{r^2} - \frac{1}{2}. \tag{37}$$

The regularity requirement of the solution at the origin allows us to interprete the constant l as $l = \sqrt{a/s^\alpha}$ and obtain the solution of Equations (34) and (35) as:

$$f^*(r,s) = \frac{2A}{s} K_0\left(\sqrt{\frac{s^\alpha}{a}}\, r\right) + \frac{A}{s}\left[2\ln\left(\sqrt{\frac{s^\alpha}{a}}\, r\right) + \frac{1}{1-2\nu}\right], \tag{38}$$

$$g^*(r,s) = -\frac{2A}{s} K_2\left(\sqrt{\frac{s^\alpha}{a}}\, r\right) - \frac{A}{s} + \frac{4aA}{s^{1+\alpha} r^2}. \tag{39}$$

There are no explicit formulae for the inverse Laplace transforms

$$\mathcal{L}^{-1}\left\{\frac{1}{s} K_0\left(\sqrt{\frac{s^\alpha}{a}}\, r\right)\right\} \quad \text{and} \quad \mathcal{L}^{-1}\left\{\frac{1}{s} K_2\left(\sqrt{\frac{s^\alpha}{a}}\, r\right)\right\} \tag{40}$$

for arbitrary α. For this reason, we introduce the functions

$$f^*_{(0)}(r,s) = \frac{2A}{s} K_0\left(\sqrt{\frac{s^\alpha}{a}}\, r\right) \quad \text{and} \quad g^*_{(2)}(r,s) = -\frac{2A}{s} K_2\left(\sqrt{\frac{s^\alpha}{a}}\, r\right) \tag{41}$$

and treat them in the following way.

Using Equation (A8) from Appendix B, we evaluate the Hankel transform of the zeroth order of the function $f^*_{(0)}(r,s)$,

$$\widehat{f^*_{(0)}}(\xi,s) = \frac{2aA}{s(s^\alpha + a\xi^2)} = \frac{2A}{\xi^2}\left(\frac{1}{s} - \frac{s^{\alpha-1}}{s^\alpha + a\xi^2}\right), \tag{42}$$

where the hat denotes the Hankel transform with respect to the radial coordinate r, and ξ is the transform variable.

The inverse Laplace transforms [72]

$$\mathcal{L}^{-1}\left\{\frac{\ln s}{s}\right\} = -\ln\tau - \gamma, \tag{43}$$

where $\gamma = 0.57721566\ldots$ is the Euler constant, and [67,68]

$$\mathcal{L}^{-1}\left\{\frac{s^{\alpha-1}}{s^{\alpha}+a\xi^{2}}\right\}=E_{\alpha}\left(-a\xi^{2}\tau^{\alpha}\right), \tag{44}$$

where $E_\alpha(z)$ is the Mittag–Leffler function

$$E_{\alpha}(z)=\sum_{k=0}^{\infty}\frac{z^{k}}{\Gamma(\alpha k+1)},\quad \alpha>0,\ z\in C, \tag{45}$$

and a subsequent inverse Hankel transform allows us to obtain the expression for the function $f(r,\tau)$,

$$\begin{aligned} f(r,\tau) &= A\left[2\ln\left(\frac{r}{\sqrt{a}\tau^{\alpha/2}}\right)+\frac{1}{1-2\nu}-\alpha\gamma\right] \\ &+ 2A\int_0^{\infty}\frac{1-E_{\alpha}\left(-a\xi^{2}\tau^{\alpha}\right)}{\xi}J_0(r\xi)\,\mathrm{d}\xi. \end{aligned} \tag{46}$$

Similar treatment is carried out for the function $g^*_{(2)}(r,s)$. Applying the Hankel transform of the second order to $g^*_{(2)}(r,s)$ and taking into account integral (A9) from Appendix B, after some mathematical transformations, we arrive at the expression

$$\begin{aligned} g(r,\tau) &= -A-2A\int_0^{\infty}\frac{1-E_{\alpha}\left(-a\xi^{2}\tau^{\alpha}\right)}{\xi}J_0(r\xi)\mathrm{d}\xi \\ &+ \frac{4A}{r}\int_0^{\infty}\frac{1-E_{\alpha}\left(-a\xi^{2}\tau^{\alpha}\right)}{\xi^{2}}J_1(r\xi)\mathrm{d}\xi \end{aligned} \tag{47}$$

or, using the recurrence equation for the function $J_2(r)$ in terms of $J_0(r)$ and $J_1(r)$,

$$g(r,\tau)=-2A\int_0^{\infty}\frac{E_{\alpha}\left(-a\xi^{2}\tau^{\alpha}\right)}{\xi}J_2(r\xi)\,\mathrm{d}\xi. \tag{48}$$

Equation (48) can also be obtained applying the Hankel transform of the second order immediately to Equation (35), taking into account integral (A10) from Appendix B. This confirms the interpretation of the characteristic length as $l=\sqrt{a/s^{\alpha}}$ and the approach based on introducing the functions $f^*_{(0)}(r,s)$ and $g^*_{(2)}(r,s)$ by Equation (41). It should be emphasized that application of the Hankel transform immediately to Equation (38) is impossible due to the logarithmic term in the right-hand side of this equation.

Finally, the components of the nonlocal stress tensor take the form

$$\begin{aligned} t_{rr}(r,\tau) &= \frac{A}{2}\left[2\ln\left(\frac{r}{\sqrt{a}\tau^{\alpha/2}}\right)+\frac{1}{1-2\nu}-\alpha\gamma\right] \\ &+ A\int_0^{\infty}\frac{1-E_{\alpha}\left(-a\xi^{2}\tau^{\alpha}\right)}{\xi}J_0(r\xi)\mathrm{d}\xi \\ &- A\int_0^{\infty}\frac{E_{\alpha}\left(-a\xi^{2}\tau^{\alpha}\right)}{\xi}J_2(r\xi)\mathrm{d}\xi, \end{aligned} \tag{49}$$

$$\begin{aligned} t_{\theta\theta}(r,\tau) &= \frac{A}{2}\left[2\ln\left(\frac{r}{\sqrt{a}\tau^{\alpha/2}}\right)+\frac{1}{1-2\nu}-\alpha\gamma\right] \\ &+ A\int_0^{\infty}\frac{1-E_{\alpha}\left(-a\xi^{2}\tau^{\alpha}\right)}{\xi}J_0(r\xi)\mathrm{d}\xi \\ &+ A\int_0^{\infty}\frac{E_{\alpha}\left(-a\xi^{2}\tau^{\alpha}\right)}{\xi}J_2(r\xi)\mathrm{d}\xi, \end{aligned} \tag{50}$$

or, accounting for the recurrence equation for the Bessel function,

$$\begin{aligned} t_{rr}(r,\tau) &= A\left[\ln\left(\frac{r}{\sqrt{a}\tau^{\alpha/2}}\right)+\frac{\nu}{1-2\nu}-\frac{1}{2}\alpha\gamma\right] \\ &+ \frac{2A}{r}\int_0^\infty \frac{1-E_\alpha(-a\xi^2\tau^\alpha)}{\xi^2}J_1(r\xi)\mathrm{d}\xi, \end{aligned} \tag{51}$$

$$\begin{aligned} t_{\theta\theta}(r,\tau) &= A\left[\ln\left(\frac{r}{\sqrt{a}\tau^{\alpha/2}}\right)+\frac{1-\nu}{1-2\nu}-\frac{1}{2}\alpha\gamma\right] \\ &+ 2A\int_0^\infty \frac{1-E_\alpha(-a\xi^2\tau^\alpha)}{\xi}J_0(r\xi)\mathrm{d}\xi \\ &- \frac{2A}{r}\int_0^\infty \frac{1-E_\alpha(-a\xi^2\tau^\alpha)}{\xi^2}J_1(r\xi)\mathrm{d}\xi. \end{aligned} \tag{52}$$

Consider particular cases of the obtained solution. If $\alpha = 1$, then $E_1(-a\xi^2\tau) = \exp(-a\xi^2\tau)$, and using integrals (A11)–(A13) from Appendix B, we arrive at

$$t_{rr}(r,\tau) = A\left[\ln\left(\frac{r}{\sqrt{a\tau}}\right)+\frac{\nu}{1-2\nu}-\frac{1}{2}\gamma\right]+\frac{2Aa\tau}{r^2}\left[1-\mathbf{E}_2\left(\frac{r^2}{4a\tau}\right)\right], \tag{53}$$

$$\begin{aligned} t_{\theta\theta}(r,\tau) &= A\left[\ln\left(\frac{r}{\sqrt{a\tau}}\right)+\frac{1-\nu}{1-2\nu}-\frac{1}{2}\gamma\right] \\ &- \frac{2Aa\tau}{r^2}\left[1-\exp\left(-\frac{r^2}{4a\tau}\right)\right]+\frac{A}{2}\mathbf{E}_1\left(\frac{r^2}{4a\tau}\right). \end{aligned} \tag{54}$$

Another particular case corresponds to $\alpha \to 0$ with $E_0(-a\xi^2) = 1/(1+a\xi^2)$, which results in

$$t_{rr}(r,\tau) = A\left[\ln\left(\frac{r}{\sqrt{a}}\right)+\frac{\nu}{1-2\nu}+\frac{2a}{r^2}-\frac{2\sqrt{a}}{r}K_1\left(\frac{r}{\sqrt{a}}\right)\right], \tag{55}$$

$$t_{\theta\theta}(r,\tau) = A\left[\ln\left(\frac{r}{\sqrt{a}}\right)+\frac{1-\nu}{1-2\nu}-\frac{2a}{r^2}+K_0\left(\frac{r}{\sqrt{a}}\right)+K_2\left(\frac{r}{\sqrt{a}}\right)\right]. \tag{56}$$

To obtain the solution (55), (56), we use integrals (A15) and (A16) from Appendix B and the recurrence equations for the modified Bessel functions $K_n(r)$. Both particular cases $\alpha = 1$ and $\alpha \to 0$ were earlier studied in [46].

Figure 2 shows the dependence of nondimensional stress on nondimensional distance (similarity variable) with

$$\bar{t}_{rr} = \frac{t_{rr}}{A}, \qquad \bar{r} = \frac{r}{\sqrt{a}\tau^{\alpha/2}}. \tag{57}$$

In numerical simulations, we assume $\nu = 0.25$. It follows from Figure 2 that at the disclination line $\bar{t}_{rr}(0) \approx 1.1$. Hence, from Equations (18) and (57), for example, taking $\Omega_3 = \pi/30$, we obtain $t_{rr}(0) \approx 0.024\,\mu$. This approximate estimate is quite acceptable.

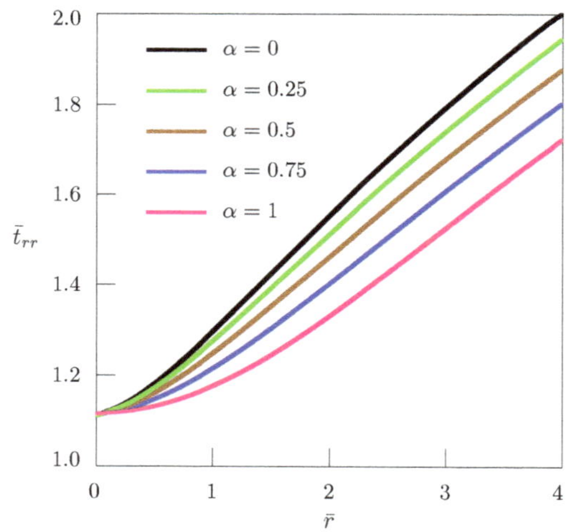

Figure 2. Dependence of stress $\bar{t}_{rr}$ on distance for straight wedge disclination.

4. Straight Twist Disclination

A drawing of the straight twist disclination with Frank vector $\mathbf{\Omega} = (0, \Omega_2, 0)$ is shown in Figure 3.

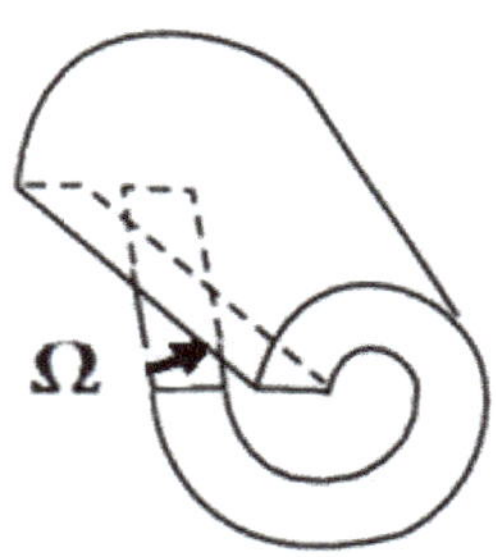

Figure 3. Schematic representation of a straight twist disclination with the Frank vector Ω_2.

The classical elasticity stress tensor for such a twist disclination has the following components [15] in cylindrical coordinates (r, θ, z):

$$\sigma_{rr} = -B\,\frac{z}{r}\,\sin\theta, \tag{58}$$

$$\sigma_{\theta\theta} = -B\,\frac{z}{r}\,\sin\theta, \tag{59}$$

$$\sigma_{zz} = -2\nu\,B\,\frac{z}{r}\,\sin\theta \tag{60}$$

$$\sigma_{r\theta} = B\,\frac{z}{r}\,\cos\theta, \tag{61}$$

$$\sigma_{rz} = -B(1-2\nu)\ln(r/l)\sin\theta, \tag{62}$$

$$\sigma_{\theta z} = -B[(1-2\nu)\ln(r/l)+1]\cos\theta, \tag{63}$$

where

$$B = \frac{\mu\,\Omega_2}{2\pi(1-\nu)}. \tag{64}$$

We assume that the components of the nonlocal stress tensor are represented as

$$t_{rr}(r,\theta,z,\tau) = T_{rr}(r,z,\tau)\,\sin\theta, \tag{65}$$

$$t_{\theta\theta}(r,\theta,z,\tau) = T_{\theta\theta}(r,z,\tau)\,\sin\theta, \tag{66}$$

$$t_{zz}(r,\theta,z,\tau) = T_{zz}(r,z,\tau)\,\sin\theta, \tag{67}$$

$$t_{r\theta}(r,\theta,z,\tau) = T_{r\theta}(r,z,\tau)\,\cos\theta, \tag{68}$$

$$t_{rz}(r,\theta,\tau) = T_{rz}(r,\tau)\,\sin\theta, \tag{69}$$

$$t_{\theta z}(r,\theta,\tau) = T_{\theta z}(r,\tau)\,\cos\theta. \tag{70}$$

Using the components of the Laplacian of the stress tensor (see Appendix A), from Equations (12), (13), (58)–(63) and (65)–(70), we obtain a coupled system of equations for determining the unknown coefficients:

$$\frac{\partial^\alpha T_{rr}}{\partial\tau^\alpha} = a\left[\frac{\partial^2 T_{rr}}{\partial r^2} + \frac{1}{r}\frac{\partial T_{rr}}{\partial r} - \frac{1}{r^2}\,T_{rr} + \frac{4}{r^2}\,T_{r\theta} - \frac{2}{r^2}\,(T_{rr} - T_{\theta\theta})\right], \tag{71}$$

$$\tau = 0: \quad T_{rr} = -B\,\frac{z}{r}; \tag{72}$$

$$\frac{\partial^\alpha T_{\theta\theta}}{\partial\tau^\alpha} = a\left[\frac{\partial^2 T_{\theta\theta}}{\partial r^2} + \frac{1}{r}\frac{\partial T_{\theta\theta}}{\partial r} - \frac{1}{r^2}\,T_{\theta\theta} - \frac{4}{r^2}\,T_{r\theta} + \frac{2}{r^2}\,(T_{rr} - T_{\theta\theta})\right], \tag{73}$$

$$\tau = 0: \quad T_{\theta\theta} = -B\,\frac{z}{r}; \tag{74}$$

$$\frac{\partial^\alpha T_{zz}}{\partial\tau^\alpha} = a\left[\frac{\partial^2 T_{zz}}{\partial r^2} + \frac{1}{r}\frac{\partial T_{zz}}{\partial r} - \frac{1}{r^2}\,T_{zz}\right], \tag{75}$$

$$\tau = 0: \quad T_{zz} = -2\nu\,B\,\frac{z}{r}; \tag{76}$$

$$\frac{\partial^\alpha T_{r\theta}}{\partial\tau^\alpha} = a\left[\frac{\partial^2 T_{r\theta}}{\partial r^2} + \frac{1}{r}\frac{\partial T_{r\theta}}{\partial r} - \frac{5}{r^2}\,T_{r\theta} + \frac{2}{r^2}\,(T_{rr} - T_{\theta\theta})\right], \tag{77}$$

$$\tau = 0: \quad T_{r\theta} = B\,\frac{z}{r}; \tag{78}$$

$$\frac{\partial^\alpha T_{rz}}{\partial\tau^\alpha} = a\left[\frac{\partial^2 T_{rz}}{\partial r^2} + \frac{1}{r}\frac{\partial T_{rz}}{\partial r} - \frac{2}{r^2}\,(T_{rz} - T_{\theta z})\right], \tag{79}$$

$$\tau = 0: \quad T_{rz} = -B(1-2\nu)\ln(r/l); \tag{80}$$

$$\frac{\partial^\alpha T_{\theta z}}{\partial\tau^\alpha} = a\left[\frac{\partial^2 T_{\theta z}}{\partial r^2} + \frac{1}{r}\frac{\partial T_{\theta z}}{\partial r} + \frac{2}{r^2}\,(T_{rz} - T_{\theta z})\right], \tag{81}$$

$$\tau = 0: \quad T_{\theta z} = -B[(1-2\nu)\ln(r/l) + 1]. \tag{82}$$

First, we study Equations (71)–(78). To split a system of coupled equations, we introduce the new unknown functions

$$f = T_{rr} + T_{\theta\theta}, \tag{83}$$

$$g = T_{rr} - T_{\theta\theta} + 2T_{r\theta}, \tag{84}$$

$$h = T_{rr} - T_{\theta\theta} - 2T_{r\theta}, \tag{85}$$

and obtain

$$\frac{\partial^{\alpha} f}{\partial \tau^{\alpha}} = a\left[\frac{\partial^2 f}{\partial r^2} + \frac{1}{r}\frac{\partial f}{\partial r} - \frac{1}{r^2} f\right], \tag{86}$$

$$\tau = 0: \quad f = -2B\,\frac{z}{r}; \tag{87}$$

$$\frac{\partial^{\alpha} g}{\partial \tau^{\alpha}} = a\left[\frac{\partial^2 g}{\partial r^2} + \frac{1}{r}\frac{\partial g}{\partial r} - \frac{1}{r^2} g\right], \tag{88}$$

$$\tau = 0: \quad g = 2B\,\frac{z}{r}; \tag{89}$$

$$\frac{\partial^{\alpha} h}{\partial \tau^{\alpha}} = a\left[\frac{\partial^2 h}{\partial r^2} + \frac{1}{r}\frac{\partial h}{\partial r} - \frac{9}{r^2} h\right], \tag{90}$$

$$\tau = 0: \quad h = -2B\,\frac{z}{r}. \tag{91}$$

The Laplace transform with respect to the nonlocality parameter τ and the Hankel transform with respect to the radial coordinate r (of the first order for the functions f and g and of the third order for the function h) give

$$f = -2Bz \int_0^{\infty} E_{\alpha}\left(-a\xi^2\tau^{\alpha}\right) J_1(r\xi)\,\mathrm{d}\xi, \tag{92}$$

$$g = 2Bz \int_0^{\infty} E_{\alpha}\left(-a\xi^2\tau^{\alpha}\right) J_1(r\xi)\,\mathrm{d}\xi, \tag{93}$$

$$h = -2Bz \int_0^{\infty} E_{\alpha}\left(-a\xi^2\tau^{\alpha}\right) J_3(r\xi)\,\mathrm{d}\xi. \tag{94}$$

The components of the nonlocal stress tensor have the following form:

$$t_{rr} = -\frac{Bz}{2}\int_0^{\infty} E_{\alpha}\left(-a\xi^2\tau^{\alpha}\right)[J_1(r\xi) + J_3(r\xi)]\mathrm{d}\xi\,\sin\theta, \tag{95}$$

$$t_{\theta\theta} = -\frac{Bz}{2}\int_0^{\infty} E_{\alpha}\left(-a\xi^2\tau^{\alpha}\right)[3J_1(r\xi) - J_3(r\xi)]\mathrm{d}\xi\,\sin\theta, \tag{96}$$

$$t_{zz} = -2B\,\nu\,z\int_0^{\infty} E_{\alpha}\left(-a\xi^2\tau^{\alpha}\right) J_1(r\xi)\,\mathrm{d}\xi\,\sin\theta, \tag{97}$$

$$t_{r\theta} = Bz\int_0^{\infty} E_{\alpha}\left(-a\xi^2\tau^{\alpha}\right)[J_1(r\xi) + J_3(r\xi)]\mathrm{d}\xi\,\cos\theta. \tag{98}$$

For $\alpha = 1$, using integrals (A18) and (A19) from Appendix B, we obtain

$$t_{rr} = -\frac{Bz}{r}\left\{1 - \frac{4a\tau}{r^2}\left[1 - \exp\left(-\frac{r^2}{4a\tau}\right)\right]\right\}\sin\theta, \tag{99}$$

$$t_{\theta\theta} = -\frac{Bz}{r}\left\{-1 + \left(2 + \frac{4a\tau}{r^2}\right)\left[1 - \exp\left(-\frac{r^2}{4a\tau}\right)\right]\right\}\sin\theta, \tag{100}$$

$$t_{zz} = -\frac{2B\,\nu\,z}{r}\left[1 - \exp\left(-\frac{r^2}{4a\tau}\right)\right]\sin\theta, \tag{101}$$

$$t_{r\theta} = \frac{Bz}{r}\left\{1 - \frac{4a\tau}{r^2}\left[1 - \exp\left(-\frac{r^2}{4a\tau}\right)\right]\right\}\cos\theta. \tag{102}$$

For $\alpha \to 0$, taking into account integrals (A15)–(A17) from Appendix B and the recurrence equations for the modified Bessel functions $K_n(r)$, we obtain

$$t_{rr} = -\frac{Bz}{r}\left[1 - \frac{4a}{r^2} + 2K_2\left(\frac{r}{\sqrt{a}}\right)\right]\sin\theta, \tag{103}$$

$$t_{\theta\theta} = -\frac{Bz}{r}\left[1 + \frac{4a}{r^2} - 2K_2\left(\frac{r}{\sqrt{a}}\right) - \frac{2r}{\sqrt{a}}K_1\left(\frac{r}{\sqrt{a}}\right)\right]\sin\theta, \tag{104}$$

$$t_{zz} = -\frac{2B\nu z}{r}\left[1 - \frac{r}{\sqrt{a}}K_1\left(\frac{r}{\sqrt{a}}\right)\right]\sin\theta, \tag{105}$$

$$t_{r\theta} = \frac{Bz}{r}\left[1 - \frac{4a}{r^2} + 2K_2\left(\frac{r}{\sqrt{a}}\right)\right]\cos\theta. \tag{106}$$

The solutions for the particular cases $\alpha = 1$ and $\alpha \to 0$ were considered in [46].

Figures 4–6 present the dependence of nondimensional stresses on distance. Nondimensional quantities are introduced as

$$\bar{t}_{ij} = \frac{t_{ij}}{B}, \qquad \bar{r} = \frac{r}{\sqrt{a}\tau^{\alpha/2}}, \qquad \bar{z} = \frac{z}{\sqrt{a}\tau^{\alpha/2}}. \tag{107}$$

In our computation, we have taken $\nu = 0.25$, $\theta = \pi/2$, $\bar{z} = 1$.

The classical elasticity solution (red curves in Figures 4–6) has nonphysical singularity at the disclination line. The solutions obtained in the framework of the fractional nonlocal elasticity are free from such singularities. Depending on the order of the fractional derivative, the approximate estimate of maximum values of stresses, based on the numerical simulation and using Equations (64) and (107) for $\Omega_2 = \pi/30$, gives quite acceptable values: $t_{rr} \approx 0.0044\,\mu - 0.0058\,\mu$, $t_{\theta\theta} \approx 0.0098\,\mu - 0.0121\,\mu$, $t_{zz} \approx 0.0035\,\mu - 0.0044\,\mu$.

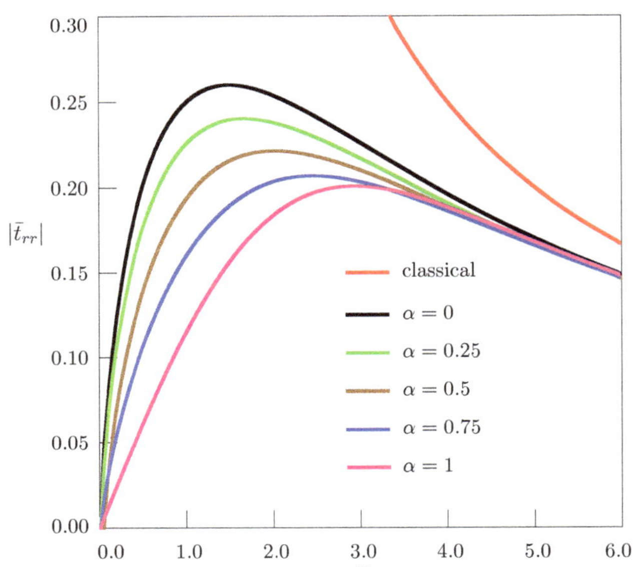

Figure 4. Dependence of stress $\bar{t}_{rr}$ on distance for straight twist disclination.

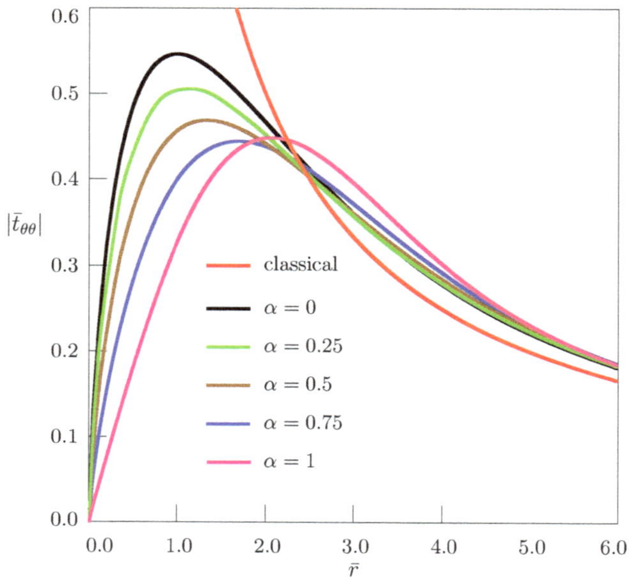

Figure 5. Dependence of stress $\bar{t}_{\theta\theta}$ on distance for straight twist disclination.

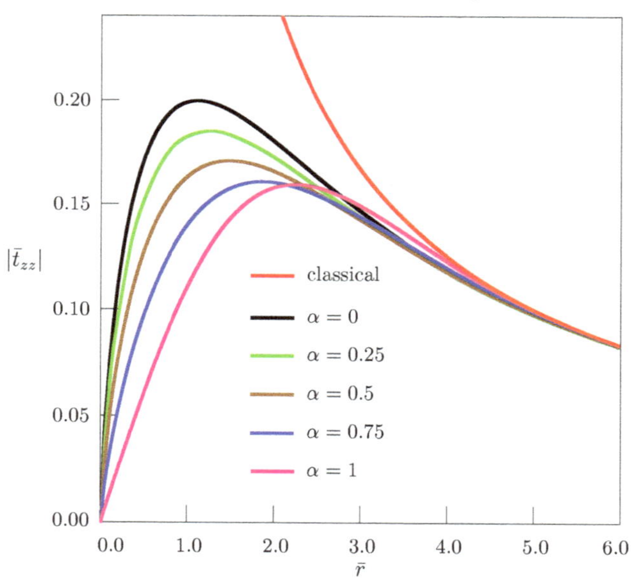

Figure 6. Dependence of stress $\bar{t}_{zz}$ on distance for straight twist disclination.

Next, we deal with Equations (79)–(82) and introduce two new unknown functions

$$p = T_{rz} + T_{\theta z}, \tag{108}$$

$$q = T_{rz} - T_{\theta z}, \tag{109}$$

satisfying the split system of equations

$$\frac{\partial^\alpha p}{\partial \tau^\alpha} = a\left(\frac{\partial^2 p}{\partial r^2} + \frac{1}{r}\frac{\partial p}{\partial r}\right), \tag{110}$$

$$\tau = 0: \quad p = -B[2(1-2\nu)\ln(r/l) + 1], \tag{111}$$

$$\frac{\partial^\alpha q}{\partial \tau^\alpha} = a\left(\frac{\partial^2 q}{\partial r^2} + \frac{1}{r}\frac{\partial q}{\partial r} - \frac{4}{r^2}q\right), \tag{112}$$

$$\tau = 0: \quad q = B. \tag{113}$$

A system of equations of such a type was considered in the case of the wedge disclination (see the corresponding Equations (27)–(30)). Repeating the examination carried out in Section 2, we can immediately write the sought components of the nonlocal stress tensor (compare Equations (49), (50), (114) and (115)):

$$\begin{aligned} t_{rz}(r,\tau) &= \left\{ -\frac{1}{2}B(1-2\nu)\left[2\ln\left(\frac{r}{\sqrt{a}\tau^{\alpha/2}}\right) + \frac{1}{1-2\nu} - \alpha\gamma\right] \right. \\ &- B(1-2\nu)\int_0^\infty \frac{1-E_\alpha\left(-a\xi^2\tau^\alpha\right)}{\xi} J_0(r\xi)\mathrm{d}\xi \\ &+ \left. B\int_0^\infty \frac{E_\alpha\left(-a\xi^2\tau^\alpha\right)}{\xi} J_2(r\xi)\mathrm{d}\xi \right\} \sin\theta, \end{aligned} \tag{114}$$

$$\begin{aligned} t_{\theta z}(r,\tau) &= \left\{ -\frac{1}{2}B(1-2\nu)\left[2\ln\left(\frac{r}{\sqrt{a}\tau^{\alpha/2}}\right) + \frac{1}{1-2\nu} - \alpha\gamma\right] \right. \\ &- B(1-2\nu)\int_0^\infty \frac{1-E_\alpha\left(-a\xi^2\tau^\alpha\right)}{\xi} J_0(r\xi)\mathrm{d}\xi \\ &- \left. B\int_0^\infty \frac{E_\alpha\left(-a\xi^2\tau^\alpha\right)}{\xi} J_2(r\xi)\mathrm{d}\xi \right\} \cos\theta. \end{aligned} \tag{115}$$

Similarly to Equations (53)–(56), the solutions for the limiting cases $\alpha = 1$ and $\alpha \to 0$ can also be written substituting the components t_{rr} and $t_{\theta\theta}$ by the components t_{rz} and $t_{\theta z}$, respectively.

5. Concluding Remarks

We have studied the nonlocal stresses caused by straight wedge and twist disclinations in an infinite medium in the framework of the new theory of fractional nonlocal elasticity. In this new theory, the constitutive equation for the nonlocal stress tensor is formulated as an integral of the local stress tensor with the nonlocality kernel (the weight function) being the Green function of the Cauchy problem for the fractional diffusion operator with the Caputo fractional derivative of the order $0 < \alpha < 1$ with respect to the nonlocality parameter. The existence of different nonlocality kernels allows one to choose the most suitable one for the specific problem. The particular cases of the solutions for $\alpha \to 1$ and $\alpha \to 0$ coincide with those known in the literature. The obtained stress fields do not contain the nonphysical singularities at the disclination line that appear in the classical local solutions. The approximate estimates of maximum values of nonlocal stress are quite acceptable from a physical point of view and are comparable with similar estimates for dislocations carried out by Eringen [40–44] (see also [73–75]). The stress problems for straight disclinations are more complicated than the corresponding problems for straight dislocations due to the logarithmic terms in the local elasticity solutions. The nonlocal theory eliminates stress singularities at the origin, but the logarithmic divergence remains at infinity. Such a divergence at infinity disappears in a finite domain, for example, for loading free boundary conditions (see, for example, [21,46]). The proposed fractional nonlocal theory of elasticity can be useful for better matching the theory of elasticity and atomic lattice theory. The importance of such matching was emphasized in many publications.

In the description of defects in solids, theories that use couple stress are of great importance. In future studies, we will analyze this aspect of nonlocal elasticity theory. The first steps in this direction have been made in the publications [76,77].

Author Contributions: Conceptualization, Y.P. and T.K.; methodology, Y.P. and T.K.; validation, T.K.; formal analysis, T.K.; investigation, Y.P. and T.K.; software, T.K.; supervision, Y.P.; writing—original draft preparation, Y.P. and T.K.; writing—review and editing, Y.P. All authors have read and agreed to the published version of the manuscript.

Funding: This research received no external funding.

Institutional Review Board Statement: Not applicable.

Informed Consent Statement: Not applicable.

Data Availability Statement: The original contributions presented in this study are included in the article. Further inquiries can be directed to the corresponding author.

Acknowledgments: T.K. thanks the Technical University of Kosice, Slovakia, for their hospitality during her internship under Grant No. DEC-2023/07/X/ST1/00354 from the National Science Centre, Poland.

Conflicts of Interest: The authors declare no conflicts of interest.

Nomenclature

The following abbreviations are used in this manuscript:

A	proportionality coefficient for wedge disclination stresses
a	coefficient in the diffusion equation for the nonlocal kernel
B	proportionality coefficient for twist disclination stresses
E_α	Mittag–Leffler function
$\mathbf{E}_\alpha$	exponential integral
$\mathbf{e}$	linear strain tensor
f	auxiliary function
g	auxiliary function
h	auxiliary function
$\mathbf{I}$	unit tensor
I_n	modified Bessel function
J_n	Bessel function
K	nonlocality kernel
K_n	modified Bessel function
l	characteristic length
p	auxiliary function
q	auxiliary function
r	cylindrical coordinate
s	Laplace transform variable
T_{ij}	coefficients of expressions for components of nonlocal stress tensor t_{ij}

$\mathbf{t}$	nonlocal stress tensor
$\mathbf{x}$	reference point
$\mathbf{x}'$	running point
z	cylindrical coordinate
α	order of fractional derivative with respect to nonlocality parameter
Γ	gamma function
γ	Euler constant
Δ	Laplace operator
δ	Dirac's delta function
θ	cylindrical coordinate
λ	Lamé constant
μ	Lamé constant
ν	Poisson ratio
ξ	Hankel transform variable
σ	local stress tensor
τ	nonlocality parameter
$\mathbf{\Omega}$	Frank vector

Appendix A

The components of the Laplacian of the symmetric second-order tensor in cylindrical coordinates (r, θ, z) were obtained in [46] (see also [44,78,79]):

$$(\Delta t)_{rr} = \Delta(t_{rr}) - \frac{4}{r^2}\frac{\partial t_{r\theta}}{\partial \theta} - \frac{2}{r^2}(t_{rr} - t_{\theta\theta}), \tag{A1}$$

$$(\Delta t)_{\theta\theta} = \Delta(t_{\theta\theta}) + \frac{4}{r^2}\frac{\partial t_{r\theta}}{\partial \theta} + \frac{2}{r^2}(t_{rr} - t_{\theta\theta}), \tag{A2}$$

$$(\Delta t)_{r\theta} = \Delta(t_{r\theta}) - \frac{4}{r^2}t_{r\theta} + \frac{2}{r^2}\frac{\partial}{\partial \theta}(t_{rr} - t_{\theta\theta}), \tag{A3}$$

$$(\Delta t)_{rz} = \Delta(t_{rz}) - \frac{1}{r^2}t_{rz} - \frac{2}{r^2}\frac{\partial t_{\theta z}}{\partial \theta}, \tag{A4}$$

$$(\Delta t)_{\theta z} = \Delta(t_{\theta z}) - \frac{1}{r^2}t_{\theta z} + \frac{2}{r^2}\frac{\partial t_{rz}}{\partial \theta}, \tag{A5}$$

$$(\Delta t)_{zz} = \Delta(t_{zz}), \tag{A6}$$

where

$$\Delta f = \frac{\partial^2 f}{\partial r^2} + \frac{1}{r}\frac{\partial f}{\partial r} + \frac{1}{r^2}\frac{\partial^2 f}{\partial \theta^2} + \frac{\partial^2 f}{\partial z^2}. \tag{A7}$$

Appendix B

In this Appendix, we present the integrals used in the paper.

Integrals (A8)–(A10) are taken from [80]:

$$\int_0^\infty K_0(ar)J_0(br)r\,\mathrm{d}r = \frac{1}{a^2+b^2}, \tag{A8}$$

$$\int_0^\infty K_2(ar)J_2(br)r\,\mathrm{d}r = \frac{b^2}{a^2(a^2+b^2)}, \tag{A9}$$

$$\int_0^\infty J_2(bx)x\,\mathrm{d}x = \frac{2}{b^2}. \tag{A10}$$

Integrals (A11) and (A12) are evaluated by the authors:

$$\int_0^\infty \frac{1}{x}\left(1-\mathrm{e}^{-ax^2}\right)J_0(bx)\mathrm{d}x = \frac{1}{2}\,\mathbf{E}_1\left(\frac{b^2}{4a}\right), \tag{A11}$$

$$\int_0^\infty \frac{1}{x^2}\left(1-\mathrm{e}^{-ax^2}\right)J_1(bx)\mathrm{d}x = \frac{a}{b}\left[1-\mathbf{E}_2\left(\frac{b^2}{4a}\right)\right]. \tag{A12}$$

Here, $\mathbf{E}_n(z)$ is the exponential integral [71]

$$\mathbf{E}_n(z) = \int_1^\infty \frac{\mathrm{e}^{-zt}}{t^n}\,\mathrm{d}t \tag{A13}$$

with the following recurrence equation:

$$\mathbf{E}_{n+1}(z) = \frac{1}{n}\left[\mathrm{e}^{-z} - z\,\mathbf{E}_n(z)\right]. \tag{A14}$$

For the exponential integral $\mathbf{E}_n(z)$, we have used the notation $\mathbf{E}_n(z)$ to prevent confusion with the Mittag–Leffler function $E_\alpha(z)$.

$$\int_0^\infty \frac{x}{x^2+a^2}\,J_0(bx)\mathrm{d}x = K_0(ab), \tag{A15}$$

$$\int_0^\infty \frac{1}{x^2+a^2}\,J_1(bx)\mathrm{d}x = \frac{1}{a}\left[\frac{1}{ab} - K_1(ab)\right], \tag{A16}$$

$$\int_0^\infty \frac{1}{x^2+a^2}\,J_3(bx)\mathrm{d}x = \frac{1}{a}\left[\frac{1}{ab} - \frac{8}{a^3b^3} + K_3(ab)\right], \tag{A17}$$

$$\int_0^\infty \exp\left(-ax^2\right)J_1(bx)\mathrm{d}x = \frac{1}{b}\left[1-\exp\left(-\frac{b^2}{4a}\right)\right], \tag{A18}$$

$$\int_0^\infty \exp\left(-ax^2\right)J_3(bx)\mathrm{d}x = \frac{1}{b}\left[1-\frac{8a}{b^2}+\left(1+\frac{8a}{b^2}\right)\exp\left(-\frac{b^2}{4a}\right)\right]. \tag{A19}$$

$K_n(z)$ is the modified Bessel function of the second kind.

References

1. Volterra, V. Sur l'équilibre des corps élastiques multiplement connexes. *Ann. Sci. Éc. Norm. Supér.* **1907**, *24*, 401–517. [CrossRef]
2. Love, A.E.H. *Treatise on the Mathematical Theory of Elasticity*, 3rd ed.; Cambridge University Press: Cambridge, UK, 1920.
3. Frank, F.C. I. Liquid crystals. On the theory of liquid crystals. *Disc. Faraday Soc.* **1958**, *25*, 19–28. [CrossRef]
4. Taylor, G.L. The mechanism of plastic deformation of crystals. I–II. *Proc. Roy. Soc. A* **1934**, *145*, 362–388.
5. Orowan, E. Zur Kristallplastizität. III. Über die Mechanismus des Gleitvorganges. *Z. Phys.* **1934**, *89*, 634–659. [CrossRef]
6. Polanyi, M. Über eine Art Gitterstörung, die einem Kristall plastisch machen könnte. *Z. Phys.* **1934**, *89*, 660–664. [CrossRef]
7. Friedel, J. *Dislocations*; Pergamon Press: New York, NY, USA, 1964.
8. Nabarro, F.R.N. *Theory of Crystal Dislocations*; Clarendon Press: Oxford, UK, 1967.
9. Teodosiu, C. *Elastic Models of Crystal Defects*; Springer: Berlin, Germany, 1982.
10. Weertman, J.; Weertman, J.R. *Elementary Dislocation Theory*; Oxford University Press: Oxford, UK, 1992.
11. Cai, W.; Nix, W.D. *Imperfections in Crystalline Solids*; Cambridge University Press: Cambridge, UK, 2016.
12. Anderson, P.M.; Hirth, J.P.; Lothe, J. *Theory of Dislocations*, 3rd ed.; Cambridge University Press: Cambridge, UK, 2017.
13. Saka, H. *Classical Theory of Crystal Dislocations*; World Scientific: Singapore, 2017.
14. deWit, R. Linear theory of static disclinations. In *Fundamental Aspects of Dislocation Theory*; National Bureau of Standards Special Publ. 317; U.S. Government Publishing Office: Washington, DC, USA, 1970; Volume 1, pp. 651–680.

15. deWit, R. Theory of disclinations: IV. Straight disclinations. *J. Res. Nat. Bureau Stand.—A. Phys. Chem.* **1973**, *77*, 607–658. [CrossRef]
16. Likhachev, V.A.; Khairov, R.Y. *Introduction to the Theory of Disclinations*; Leningrad University Press: Leningrad, Russia, 1975. (In Russian)
17. Romanov, A.E.; Vladimirov, V.I. Disclinations in solids. *Phys. Stat. Sol.* **1983**, *78*, 11–34. [CrossRef]
18. Vladimirov, V.I.; Romanov, A.E. *Disclinations in Crystals*; Nauka: Leningrad, Russia, 1986. (In Russian)
19. Romanov, A.E. Mechanics and physics of disclinations in solids. *Eur. J. Mech. A/Solids* **2003**, *22*, 727–741. [CrossRef]
20. Romanov, A.E.; Kolesnikova, A.L. Application of disclination concept to solid structures. *Prog. Mater. Sci.* **2009**, *54*, 740–769. [CrossRef]
21. Kolesnikova, A.L.; Gutkin, M.Y.; Proskura, A.V.; Morozov, N.F.; Romanov, A.E. Elastic fields of straight wedge disclinations axially piercing bodies with spherical free surfaces. *Int. J. Solids Struct.* **2016**, *99*, 82–96. [CrossRef]
22. Fressengeas, C.; Sun, X. On the theory of dislocation and generalized disclination fields and its application to straight and stepped symmetrical tilt boundaries. *J. Mech. Phys. Solids* **2020**, *143*, 104092. [CrossRef]
23. Bamney, D.; Reyes, R.; Capolungo, L.; Spearot, D.E. Disclination-dislocation based model for grain boundary stress field evolution due to slip transmission history and influence on subsequent dislocation transmission. *J. Mech. Phys. Solids* **2022**, *165*, 104920. [CrossRef]
24. Zhang, C.; Acharya, A. On the relevance of generalized disclinations in defect mechanics. *J. Mech. Phys. Solids* **2018**, *119*, 188–223. [CrossRef]
25. Fressengeas, C.; Taupin, V. Revisiting the application of field dislocation and disclination mechanics to grain boundaries. *Metals* **2020**, *10*, 1517. [CrossRef]
26. Hirth, J.P.; Hirth, G.; Wang, J. Disclinations and disconnections in minerals and metals. *Proc. Nat. Acad. Sci. USA* **2020**, *117*, 196–204. [CrossRef]
27. Geier, M.; Fulga, I.C.; Lau, A. Bulk-boundary-defect correspondence at disclinations in rotation-symmetric topological insulators and superconductors. *SciPost Phys.* **2021**, *10*, 92. [CrossRef]
28. Demouchy, S.; Thieme, M.; Barou, F.; Beausir, B.; Taupin, V.; Cordier, P. Dislocation and disclination densities in experimentally deformed polycrystalline olivine. *Eur. J. Mineral.* **2023**, *35*, 219–242. [CrossRef]
29. deWit, R. Relation between dislocations and disclinations. *J. Appl. Phys.* **1971**, *42*, 3304–3308. [CrossRef]
30. Marcinkowski, M.J. Dislocations and disclinations in a new light. *Arch. Mech.* **1990**, *42*, 279–289.
31. Kontorova, T.A.; Frenkel, J.I. Theory of plastic deformation and twinning. *J. Exp. Theor. Phys.* **1938**, *8*, 89–95. (In Russian)
32. Braun, O.M.; Kivshar, Y.S. Nonlinear dynamics of the Frenkel–Kontorova model. *Phys. Rep.* **1998**, *306*, 1–108. [CrossRef]
33. Braun, O.M.; Kivshar, Y.S. *The Frenkel–Kontorova Model: Concepts, Methods and Applications*; Springer: Berlin, Germany, 2004.
34. Peierls, R.E. The size of a dislocation. *Proc. Phys. Soc. Lond.* **1940**, *52*, 34–37. [CrossRef]
35. Nabarro, F.R.N. Dislocations in a simple cubic lattice. *Proc. Phys. Soc. Lond.* **1947**, *59*, 256–272. [CrossRef]
36. Bullough, R.; Tewary, V.K. Lattice theory of dislocations. In *Dislocations in Solids, Vol. 2. Dislocations in Crystals*; Nabarro, F.R.N., Ed.; North-Holland: Amsterdam, The Netherlands, 1979; pp. 1–66.
37. Duesbery, M.S. Modeling of the dislocation core. In *Dislocations in Solids, Vol. 8. Basic Problems and Applications*; Nabarro, F.R.N., Ed.; North-Holland: Amsterdam, The Netherlands, 1989; pp. 67–173.
38. Raabe, D. *Computational Materials Science. The Simulation of Materials Microstructures and Properties*; Wiley: Weinheim, Germany, 1998.
39. Tewary, V.K. Lattice-statics model for edge dislocations in crystal. *Phil. Mag. A* **2000**, *80*, 1445–1452. [CrossRef]
40. Eringen, A.C. Edge dislocation in nonlocal elasticity. *Int. J. Eng. Sci.* **1977**, *15*, 177–183. [CrossRef]
41. Eringen, A.C. Screw dislocation in nonlocal elasticity. *J. Phys. D Appl. Phys.* **1977**, *10*, 671–678. [CrossRef]
42. Eringen, A.C. On differential equations of nonlocal elasticity and solutions of screw dislocation and surface waves. *J. Appl. Phys.* **1983**, *54*, 4703–4710. [CrossRef]
43. Eringen, A.C. Vistas on nonlocal continuum physics. *Int. J. Eng. Sci.* **1992**, *30*, 1551–1565. [CrossRef]
44. Eringen, A.C. *Nonlocal Continuum Field Theories*; Springer: New York, NY, USA, 2002.
45. Lazar, M.; Agiasofitou, E. Screw dislocation in nonlocal anisotropic elasticity. *Int. J. Eng. Sci.* **2011**, *49*, 1404–1414. [CrossRef]
46. Povstenko, Y. Straight disclinations in nonlocal elasticity. *Int. J. Eng. Sci.* **1995**, *33*, 575–582. [CrossRef]
47. Gutkin, M.Y.; Aifantis, E.C. Screw dislocation in gradient elasticity. *Scr. Mater.* **1996**, *35*, 1353–1358. [CrossRef]
48. Gutkin, M.Y.; Aifantis, E.C. Edge dislocation in gradient elasticity. *Scr. Mater.* **1997**, *36*, 129–135. [CrossRef]
49. Lazar, M. The fundamentals of non-singular dislocations in the theory of gradient elasticity: Dislocation loops and straight dislocations. *Int. J. Solids Struct.* **2013**, *50*, 352–362. [CrossRef]
50. Lazar, M. Non-singular dislocation continuum theories: Strain gradient elasticity versus Peierls-Nabarro model. *Phil. Mag.* **2017**, *97*, 3246–3275. [CrossRef]

51. Tsagrakis, I.; Yasnikov, I.S.; Aifantis, E.C. Gradient elasticity for disclinated micro crystals. *Mech. Res. Commun.* **2018**, *93*, 159–162. [CrossRef]
52. Lazar, M.; Po, G. Non-singular straight dislocations in anisotropic crystals. *J. Mater. Sci. Mater. Theory* **2024**, *8*, 5. [CrossRef]
53. Wu, M.S. A revisit of the elastic fields of straight disclinations with new solutions for a rigid core. *Acta Mech.* **2019**, *230*, 2505–2520. [CrossRef]
54. Wu, M.S. A wedge disclination in a nonlinear elastic cylinder. *Math. Mech. Solids* **2019**, *24*, 2030–2046. [CrossRef]
55. Wu, M.S. Elastic fields of a wedge disclination in a functionally graded cylinder. *Mech. Mater.* **2021**, *157*, 103835. [CrossRef]
56. Lazar, M.; Maugin, G.A.; Aifantis, E.C. On a theory of nonlocal elasticity of bi-Helmholtz type and some applications. *Int. J. Solids Struct.* **2006**, *43*, 1404–1421. [CrossRef]
57. Romano, G.; Barretta, R. Nonlocal elasticity in nanobeams: The stress-driven integral model. *Int. J. Eng. Sci.* **2017**, *115*, 14–27. [CrossRef]
58. Romano, G.; Barretta, R. Stress-driven versus strain-driven nonlocal integral model for elastic nano-beams. *Compos. Part B* **2017**, *114*, 184–188. [CrossRef]
59. Apuzzo, A.; Barretta, R.; Faghidian, S.A.; Luciano, R.; Marotti de Sciarra, F. Free vibrations of elastic beams by modified nonlocal strain gradient theory. *Int. J. Eng. Sci.* **2018**, *133*, 99–108. [CrossRef]
60. Di Paola, M.; Zingales, M. Long-range cohesive interactions of non-local continuum faced by fractional calculus. *Int. J. Solids Struct.* **2008**, *45*, 5642–5659. [CrossRef]
61. Challamel, N.; Zorica, D.; Atanacković, T.M.; Spasić, D.T. On the fractional generalization of Eringen's nonlocal elasticity for wave propagation. *Comptes Rendus Mécanique* **2013**, *341*, 298–303. [CrossRef]
62. Carpinteri, A.; Cornetti, P.; Sapora, A. Nonlocal elasticity: An approach based on fractional calculus. *Meccanica* **2014**, *49*, 2551–2569. [CrossRef]
63. Tarasov, V.E. Lattice model of fractional gradient and integral elasticity: Long-range interaction of Grünwald–Letnikov–Riesz type. *Mech. Mater.* **2014**, *70*, 106–114. [CrossRef]
64. Tarasov, V.E. Fractional nonlocal continuum mechanics and microstructural models. In *Handbook of Nonlocal Continuum Mechanics for Materials and Structures*; Voyiadjis, G.Z., Ed.; Springer: Cham, Switzerland, 2019; pp. 839–849.
65. Povstenko, Y. Generalized theory of diffusive stresses associated with the time-fractional diffusion equation and nonlocal constitutive equations for the stress tensor. *Comput. Math. Appl.* **2019**, *78*, 1819–1825. [CrossRef]
66. Povstenko, Y. Fractional nonlocal elasticity and solutions for straight screw and edge dislocations. *Phys. Mesomech.* **2020**, *23*, 547–555. [CrossRef]
67. Podlubny, I. *Fractional Differential Equations*; Academic Press: San Diego, CA, USA, 1999.
68. Kilbas, A.A.; Srivastava, H.M.; Trujillo, J.J. *Theory and Applications of Fractional Differential Equations*; Elsevier: Amsterdam, The Netherlands, 2006.
69. Kunin, I.A. *Elastic Media with Microstructure I. One-Dimensional Models*; Springer: Berlin, Germany, 1982.
70. Kunin, I.A. *Elastic Media with Microstructure II. Three-Dimensional Models*; Springer: Berlin, Germany, 1983.
71. Abramowitz, M.; Stegun, I.A. (Eds.) *Handbook of Mathematical Functions with Formulas, Graphs and Mathematical Tables*; Dover: New York, NY, USA, 1972.
72. Doetsch, G. *Anleitung zum Praktischer Gebrauch der Laplace-Transformation und der Z-Transformation*; Springer: München, Germany, 1967.
73. Lawn, B.R.; Wilshaw, T.R. *Fracture of Brittle Solids*; Cambridge University Press: Cambridge, UK, 1975.
74. Kelly, A. *Strong Solids*, 2nd ed.; Clarendon Press: Oxford, UK, 1973.
75. Kelly, A.; Knowles, K.M. *Crystallography and Crystal Defects*, 2nd ed.; John Wiley & Sons: Chichester, UK, 2012.
76. Povstenko, Y. The mathematical theory of defects in a Cosserat continuum. *J. Math. Sci.* **1992**, *62*, 2524–2530. [CrossRef]
77. Povstenko, Y. Stress functions for continua with couple stresses. *J. Elast.* **1994**, *36*, 99–116. [CrossRef]
78. Ghavanloo, E.; Rafii-Tabar, H.; Fazelzadeh, S.A. *Computational Continuum Mechanics of Nanoscopic Structures: Nonlocal Elasticity Approaches*; Springer: Cham, Switzerland, 2019.
79. Povstenko, Y. *Fractional Thermoelasticity*, 2nd ed.; Springer: Cham, Switzerland, 2024.
80. Prudnikov, A.P.; Brychkov, Y.A.; Marichev, O.I. *Integrals and Series, Vol. 2: Special Functions*; Gordon & Breach Science Publishers: Amsterdam, The Netherlands, 1983.

materials

MDPI

Article

Study of ZrO_2 Gate Dielectric with Thin SiO_2 Interfacial Layer in 4H-SiC Trench MOS Capacitors

Qimin Huang [1,2,3], Yunduo Guo [1,2,3], Anfeng Wang [1,2,3], Zhaopeng Bai [1,2,3], Lin Gu [1,2,3], Zhenyu Wang [1,2,3], Chengxi Ding [1,2,3], Yi Shen [1,2,3], Hongping Ma [1,2,3,*] and Qingchun Zhang [1,2,3]

1 Institute of Wide Bandgap Semiconductors and Future Lighting, Academy for Engineering & Technology, Fudan University, Shanghai 200433, China
2 Shanghai Research Center for Silicon Carbide Power Devices Engineering & Technology, Fudan University, Shanghai 200433, China
3 Institute of Wide Bandgap Semiconductor Materials and Devices, Research Institute of Fudan University in Ningbo, Ningbo 315327, China
* Correspondence: hpma@fudan.edu.cn

Abstract: The transition of SiC MOSFET structure from planar to trench-based architectures requires the optimization of gate dielectric layers to improve device performance. This study utilizes a range of characterization techniques to explore the interfacial properties of ZrO_2 and SiO_2/ZrO_2 gate dielectric films, grown via atomic layer deposition (ALD) in SiC epitaxial trench structures to assess their performance and suitability for device applications. Scanning electron microscopy (SEM) and atomic force microscopy (AFM) measurements showed the deposition of smooth film morphologies with roughness below 1 nm for both ZrO_2 and SiO_2/ZrO_2 gate dielectrics, while SE measurements revealed comparable physical thicknesses of 40.73 nm for ZrO_2 and 41.55 nm for SiO_2/ZrO_2. X-ray photoelectron spectroscopy (XPS) shows that in SiO_2/ZrO_2 thin films, the binding energies of Zr $3d_{5/2}$ and Zr $3d_{3/2}$ peaks shift upward compared to pure ZrO_2. Electrical characterization showed an enhancement of E_{BR} (3.76 to 5.78 $MV{\cdot}cm^{-1}$) and a decrease of I_{ON_EBR} (1.94 to 2.09×10^{-3} $A{\cdot}cm^{-2}$) for the SiO_2/ZrO_2 stacks. Conduction mechanism analysis identified suppressed Schottky emission in the stacked film. This indicates that the incorporation of a thin SiO_2 layer effectively mitigates the small bandgap offset, enhances the breakdown electric field, reduces leakage current, and improves device performance.

Keywords: SiO_2/ZrO_2; ZrO_2; trench structure; ALD; gate dielectric; MOS capacitor

1. Introduction

The current SiC MOSFET market is experiencing a new phase of growth, fueled by the rapid expansion of the renewable energy sector [1]. In this context, high-performance SiC MOSFETs are progressively replacing traditional silicon-based power devices [2]. Among materials, SiC has attracted increasing interest in power electronics and renewable energy conversion systems owing to its low on-resistance, high voltage resistance, high-frequency switching capability, and excellent high-temperature reliability [3]. The evolution of SiC MOSFETs from traditional planar to trench-based architectures has led to notable performance improvements, such as higher current density, reduced on-resistance, faster switching speeds, and enhanced blocking voltages [4,5]. However, the introduction of trench structures has also brought technical challenges to the gate dielectric layer [6,7]. Advanced methods such as electron holographic tomography can now be used to study the internal device potential [8]. Despite the high critical breakdown electric field of SiC, the electric field concentrations at the trench corners may subject the oxide layer to excessive stress, severely impacting device reliability and stability [8–10]. Therefore, optimizing the design of the gate oxide layer and addressing interface engineering challenges in trench structures have become critical research and development priorities in the field of SiC devices [11–13].

Various alternative gate dielectric materials have so far been explored to address these challenges [14]. Wirths et al. demonstrated that integrating high-k dielectrics in place of traditional SiO_2 in planar SiC MOSFETs effectively mitigates interface trap density, stabilizes threshold voltage behavior, and enhances operational reliability, thereby advancing device performance [15–17]. Among high-k materials, Al_2O_3, including certain ALON compounds, has been thoroughly investigated [18–21]. However, zirconia (ZrO_2) has emerged as a promising candidate due to its high dielectric constant and wide bandgap (5.8 eV) [18]. In this view, Wang et al. investigated the impact of inserting a thin SiO_2 interlayer between ZrO_2 and SiC using first-principles calculations, predicting that such a structure could increase band offset, reduce leakage current, and preserve the dielectric advantages of ZrO_2 [22]. However, the traditional thermal oxidation process of SiC/SiO_2 interfaces often results in the low density of interface states (D_{it}), reducing channel mobility [23]. This problem can be solved by optimizing interface engineering to lower interface trap density as a critical focus.

In this study, high-performance gate oxide layers with exceptional uniformity were synthesized using atomic layer deposition (ALD) for the comparative assessment of interfacial characteristics in ZrO_2-based and SiO_2/ZrO_2 trench capacitors. Compared to previous studies, this work provides a more complete experimental assessment of SiO_2/ZrO_2 gate dielectric stacks in trench MOS structures. The trench depth, width, and the growth of high-k thin films were studied by scanning electron microscopy (SEM). Comprehensive material validation was performed through multi-technique analysis, including surface morphology examination with atomic force microscopy (AFM), optoelectronic property evaluation by spectroscopic ellipsometry (SE), and chemical composition analysis via X-ray photoelectron spectroscopy (XPS). SE analysis provided critical data on layer thickness and optical bandgap, revealing that the integration of SiO_2 interfacial layers systematically modulated the energy band structure. Key electrical parameters, such as the dielectric breakdown field intensity (E_{BR}) and conduction current at the breakdown (I_{ON_EBR}), were quantified through current–voltage (I-V) profiling. Additionally, capacitance–voltage (C-V) hysteresis analysis allowed for precise determination of D_{it} and effective oxide charge (N_{eff}), while further investigation of leakage current mechanisms was conducted through I-V curve modeling, providing a detailed understanding of the dominant transport pathways.

Compared to ZrO_2 alone, the fabricated SiO_2/ZrO_2 stacked dielectric exhibited a higher E_{BR} and lower I_{ON_EBR}, underscoring the feasibility and potential of SiO_2/ZrO_2 stacked dielectrics as a promising solution for gate oxide applications in trench-structured SiC MOSFETs.

2. Experimental Section

The sample preparation process of the films and a scheme of the final product are depicted in Figure 1. The experiment commenced with a 4H-SiC epitaxial wafer (n-doped, (0001) orientation with 4° off-axis) as the substrate. A 2.5 μm-thick SiO_2 masking layer was initially grown via chemical vapor deposition (CVD, SINO-Plasma 8000, amteglobal, Wuxi, China). Next, photolithography (ABM/6/350/NUV/DCCD/M, ABM Inc., San Jose, CA, USA) was employed to pattern the mask, defining trench arrays with 3 μm width.

Following SiO_2 layer patterning through dry etching, the residual photoresist was stripped via solvent treatment, enabling subsequent reactive ion etching of the SiC substrate to define trench architectures. The finalized structures exhibited precise dimensional control with trench widths of 3 μm, vertical depths of 1.5 μm, and inter-trench spacing of 5 μm. To ensure mask integrity, buffered oxide etch (BOE) was selectively applied to remove residual oxides from unpatterned regions.

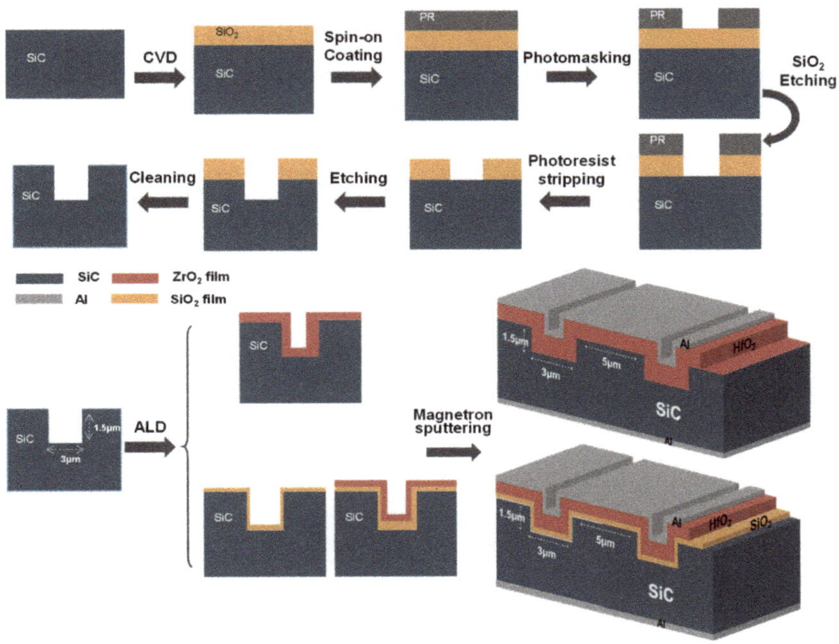

Figure 1. Fabrication process flow and final trench MOS device structure.

After the formation of clean trench structures, the gate dielectric layer was deposited using an atomic layer deposition (ALD, Exploiter E200SP, SuperALD, Shenzhen, China) system at 270 °C. For the single-layer ZrO_2 film, 400 deposition cycles were sufficient. For the SiO_2/ZrO_2 stacked dielectrics, sequential deposition of 10 nm SiO_2 (precursor: plasma) and 30 nm ZrO_2 (precursor: $CpZr(NMe_2)_3$) resulted in a composite dielectric layer with a total thickness of approximately 40 nm after 400 cycles, demonstrating excellent thickness uniformity. Finally, 200 nm aluminum electrodes were deposited on both device surfaces via magnetron sputtering, with standardized 200×200 μm^2 test contacts formed. A schematic representation of the obtained ZrO_2 thin film and SiO_2/ZrO_2 stacked film samples is also shown in Figure 1.

Following etching and ALD, the cleaved samples were fixed onto the sample stage using conductive adhesive. To improve conductivity, a thin layer of gold was applied to the sample surface using sputtering for 30 s. The cross sections were analyzed using a scanning electron microscope (SEM, Sigma 300, ZEISS, Oberkochen, Germany) to evaluate the trench etching quality and the uniformity of the film deposition within the trench.

Additional analysis of the ALD-deposited films was performed by determining the film thickness through spectroscopic ellipsometry (SE, Woollam RC2, J.A. Woollam, Lincoln, NE, USA). The SiC substrate was characterized using B-spline curves and the Cauchy dispersion model to determine the film thickness and optical parameters, enabling detailed characterization of the gate dielectric layer. The surface morphology was analyzed using atomic force microscopy (AFM, BRUKER BRK0003, Bruker Corporation, Karlsruhe, Germany) over a 5 μm × 5 μm area, with raw data processed by NanoScope analysis (3.0) (Billerica, MA, USA) software to determine surface roughness. X-ray photoelectron spectroscopy (XPS, ESCALAB 250Xi, Thermo Scientific, Waltham, MA, USA) examined the chemical states of the films, with all XPS spectra calibrated to the C 1s peak at a binding energy of 284.8 eV.

The electrical characterization of thin-film devices was systematically performed utilizing a TS2000-HP probe system (MPI Corporation, Hsinchu, Taiwan) coupled with a Keithley 4200A-S parametric analyzer (Tektronix Inc., Beaverton, OR, USA). Current–voltage characteristics were recorded through progressive elevation of the gate bias until dielectric failure occurred, enabling determination of both E_{BR} and I_{ON_EBR} in metal–oxide–semiconductor structures. Capacitance–voltage profiling was implemented through gate voltage scanning under alternating current excitation (30 mV amplitude) across multiple frequency settings. Subsequent analysis of capacitance–voltage relationships permitted quantitative evaluation of D_{it} and N_{eff} in the fabricated capacitors. All measurements were performed at room temperature.

3. Results

In this study, ZrO_2 and SiO_2/ZrO_2 gate dielectric films were grown by atomic layer deposition, and their characteristics were studied by various analytical techniques. The trench structures observed by SEM in Figure 2a revealed the deposition films with trench widths of 3 μm, vertical depths of 1.5 μm, and a trench pitch of 5 μm. The SEM image in Figure 2b shows a gate oxide layer that was grown using ALD. The high-k layer demonstrated excellent uniformity and consistency across the entire trench surface, even at the trench corners. Thus, the ALD process can be used to deposit films with precisely controlled film thickness and uniformity, outperforming the traditional dry oxidation method.

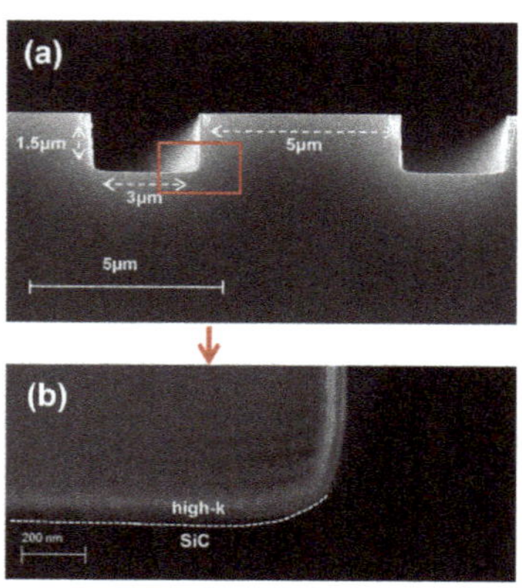

Figure 2. SEM images of (**a**) the trench structure and (**b**) the magnified view of the corner region marked by the red box in (**a**) after the ALD process.

The surface morphologies and roughnesses of ZrO_2 thin films, SiO_2/ZrO_2 stacked films, and the interfacial SiO_2 layer analyzed by AFM are compared in Figure 3. Both types of films exhibited relatively smooth surfaces, with a roughness value of 0.401 nm for ZrO_2 thin films and 0.756 nm for SiO_2/ZrO_2 stacked films. Further analysis of the surface profiles revealed roughness within a sub-1 nm range, indicating the deposition of films with a smooth surface finish. By comparison, the surface roughness of the SiO_2/ZrO_2 stacked films was slightly higher than that of single-layer ZrO_2 films. Such an increase in roughness can be attributed to the physical and chemical differences between SiO_2 and ZrO_2, which might induce interfacial stress or defects during the deposition process. Additionally, the multilayer nature of the stacked structure may have contributed to cumulative roughness since each additional layer introduced a minor increment to the overall surface roughness

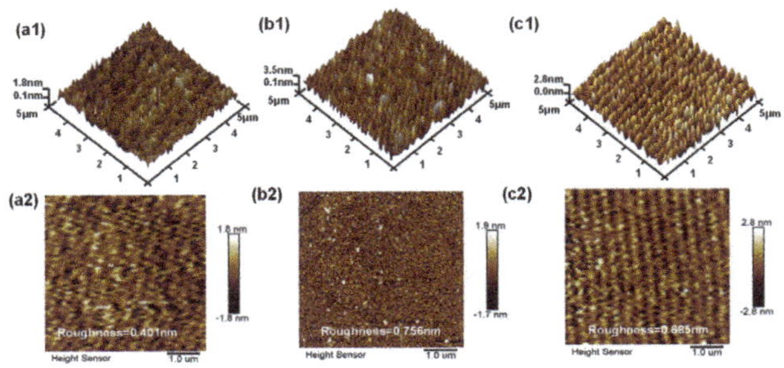

Figure 3. 3D and 2D AFM images within a 5 × 5 μm^2 area are shown for: (**a1**,**a2**) ZrO_2, (**b1**,**b2**) SiO_2/ZrO_2, (**c1**,**c2**) interlayer SiO_2.

Spectroscopic ellipsometry (SE) was employed to systematically analyze multiple regions of the deposited thin films to ensure measurement reliability and determine their thicknesses. The ZrO_2 single-layer film exhibited an average thickness of 40.73 nm, while the SiO_2 and ZrO_2 layers in the SiO_2/ZrO_2 bilayer structure measured 10.73 nm and 30.82 nm, respectively. Figure 4a,b present the experimental Ψ and Δ parameters, while the corresponding refractive index (n) and extinction coefficient (k) are shown in Figure 4c.

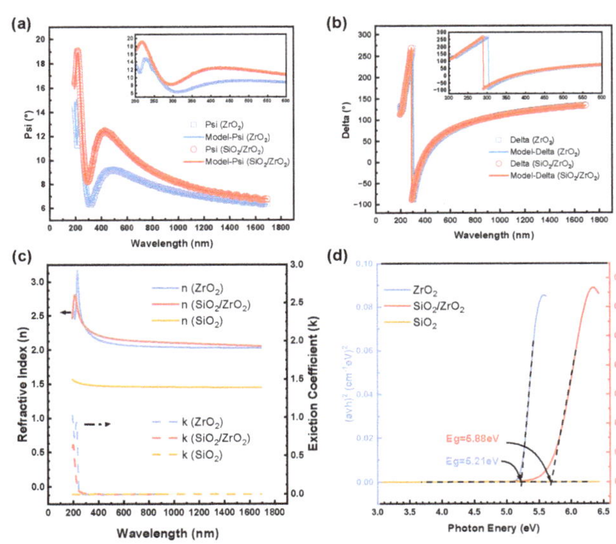

Figure 4. The SE analysis of ZrO_2 and SiO_2/ZrO_2 film heterostructures: (**a**) Ψ and (**b**) Δ components of complex reflectance. (**c**) Refractive index (n) and extinction coefficient (k) profiles. (**d**) Tauc plot ($(\alpha h\nu)^2$ vs. photon energy) for bandgap determination.

The optical bandgap can be determined using the Tauc method, which is based on the relationship $(\alpha h\nu)^2 \propto (h\nu - E_g)$, where α is the absorption coefficient, ν is photon frequency, and E_g is the optical bandgap [24]. This method relies on the fact that, for semiconductors, most electrons in the valence and conduction bands are located near the bandgap. When the energy of incident photons approaches the bandgap, electrons can absorb the photon energy. In this method, the absorption coefficient (α) is calculated using the relationship $\alpha = 4\pi k/\lambda$, where λ is the wavelength of the incident light. The results are displayed in Figure 4. The optical bandgap is extracted by plotting $(\alpha h\nu)^2$ versus the photon energy $(h\nu)$ and then linearly extrapolating the results to zero [25].

Using this approach, the optical bandgaps of the ZrO_2 and SiO_2/ZrO_2 composite films were determined to be 5.21 eV and 5.88 eV, respectively [26]. Despite the relatively thin SiO_2 layer, its inherent high bandgap (approximately 9 eV) contributed to bandgap tuning in the composite film, thereby enhancing the device's breakdown voltage and reducing leakage current. Therefore, the presence of the SiO_2 buffer layer in the stack structure can significantly enhance the overall device performance by adjusting the bandgap [27,28].

XPS analysis was employed to probe the elemental composition and chemical bonding states of the dielectric surfaces. Figure 5a displays the full-range XPS spectra for both ZrO_2 and SiO_2/ZrO_2 films. The O 1s core-level XPS spectra and peak deconvolutions for both thin films are presented in Figure 5b. For the ZrO_2 thin film, the O 1s peak can be deconvoluted into two components at 529.6 eV and 530.7 eV. By comparison, the O 1s peak of SiO_2/ZrO_2 thin film can be deconvoluted into two components at 530.3 eV and 531.4 eV.

As shown in Figure 5c, he Zr 3d core-level XPS spectra exhibit well-resolved spin–orbit doublets (Zr $3d_{5/2}$ and Zr $3d_{3/2}$) for both ZrO_2 and SiO_2/ZrO_2 thin films, with a characteristic splitting energy of 2.4 eV. For the pristine ZrO_2 film, the binding energies of Zr $3d_{5/2}$ and Zr $3d_{3/2}$ were determined to be 181.8 eV and 184.2 eV, respectively [29]. Upon SiO_2 interlayer incorporation, these values showed a systematic positive shift to 182.4 eV and 184.8 eV. The experimentally observed area ratio between Zr $3d_{5/2}$ and Zr $3d_{3/2}$ peaks was calculated to be 59:41 for ZrO_2 and 58:42 for SiO_2/ZrO_2, closely approximating the theoretical 3:2 ratio dictated by spin–orbit coupling effects [30]. This ratio is consistent with the expected energy level splitting determined by the spin–orbit coupling effect. Spin–orbit coupling, which is the interaction between the spin angular momentum and the orbital angular momentum of an electron, causes energy level splitting. In this case, the spin–orbit coupling of the Zr 3d orbitals results in a consistent area ratio of the Zr $3d_{5/2}$ and Zr $3d_{3/2}$ peaks, reflecting the similar electronic states of Zr atoms in both samples.

The O 1s core-level and Si 2p XPS spectra of the SiO_2 interlayer, along with their peak deconvolutions, are presented in Figure 5d,e. The O 1s peak can be deconvoluted into two components at 532.7 eV and 533.9 eV [31]. The stronger covalent bonding in SiO_2 may have resulted in a higher electron cloud density around oxygen atoms, leading to higher binding energy of the O 1s peak when compared to that in ZrO_2.

Overall, the XPS results demonstrate that in the SiO_2/ZrO_2 thin film, the binding energies of both the Zr $3d_{5/2}$ and Zr $3d_{3/2}$ peaks are shifted upward compared to those in pure ZrO_2. This indicates that charge redistribution or polarization at the interface has altered the electronic environment of the Zr atoms, leading to an increase in their binding energies.

As illustrated in Figure 6a,b, the O 1s energy loss spectra (ELS) for both ZrO_2 and SiO_2/ZrO_2 were analyzed by examining the energy differences between the O 1s peak positions. This approach enables the determination of the material's band gap, as the onset of energy loss corresponds to the excitation of electrons from the valence band to the conduction band. The energy separation between the O 1s peak and the threshold energy loss provides an estimate of the band gap. Using this method, the band gaps were found to be 5.6 eV for ZrO_2 and 5.7 eV for SiO_2/ZrO_2 [32]. The discrepancy between E_g values derived from ELS and Tauc methods can be attributed to their distinct probing mechanisms. ELS emphasizes localized electronic transitions near the surface, while Tauc analysis reflects bulk optical absorption. These results align with the optical bandgap values obtained from previous fitting analysis.

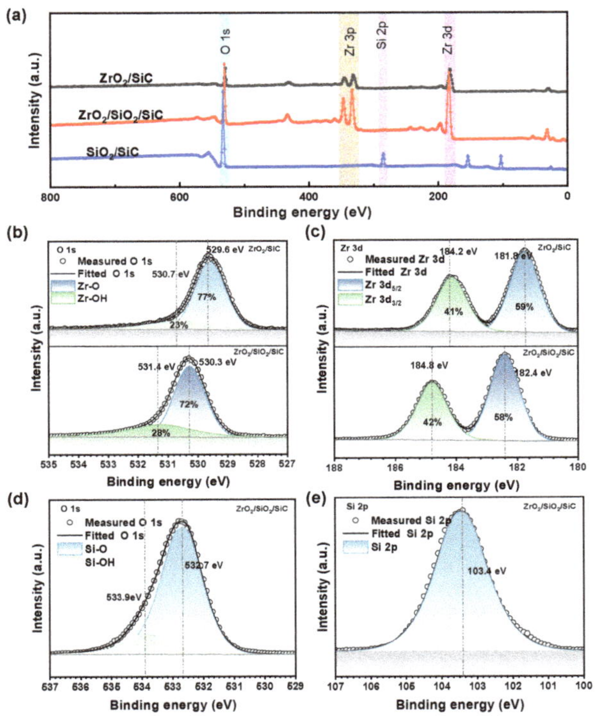

Figure 5. (**a**) XPS spectra of ZrO_2 and SiO_2/ZrO_2, (**b**) O 1s core level and (**c**) Zr 3d core level spectra of ZrO_2 and SiO_2/ZrO_2, (**d**) O 1s core level and (**e**) Si 2p core level spectra of the intermediate SiO_2 layer.

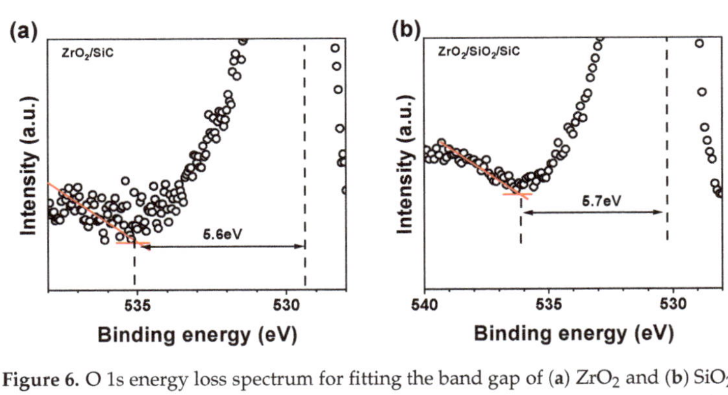

Figure 6. O 1s energy loss spectrum for fitting the band gap of (**a**) ZrO_2 and (**b**) SiO_2/ZrO_2.

The comparative current density–voltage (J-V) characteristics of ZrO_2-based MOS capacitors and their SiO_2/ZrO_2 stacked counterparts are illustrated in Figure 7. Current density values were normalized to the square electrode area using the relationship $J = \frac{I}{A}$, where A donates the electrode area and I represents the measured current [33]. The J-V curves of both devices exhibited plateaus at lower voltages, followed by a gradual increase as the electric field intensified. However, no distinct breakdown phenomenon was observed for the single-layer ZrO_2 capacitor, even at higher electric fields. Interestingly, the conduction current stabilized at sustained levels, displaying typical features of a metastable "soft breakdown" state [34]. The analysis revealed that the single-layer ZrO_2 devices failed at a dielectric breakdown voltage of 19.1 V, corresponding to a critical field intensity of 3.76 MV·cm^{-1}, which is notably lower than the theoretical breakdown field of bulk ZrO_2. By contrast, the SiO_2/ZrO_2 stacked heterostructure exhibited significantly improved dielectric strength, with a breakdown voltage of 24.9 V and a corresponding field strength of 5.78 MV·cm^{-1}. The improvement in breakdown performance stems from two synergistic mechanisms: on one hand, the wide bandgap of SiO_2 (9 eV, higher than ZrO_2's 5.8 eV) enables the formation of a more favorable staggered band alignment at the ZrO_2/SiC interface; on the other hand, the effective reduction in oxygen vacancy density substantially suppresses trap-assisted conduction pathways mediated by defect states, thereby collectively enhancing the dielectric properties. Therefore, the addition of the SiO_2 layer not only reduced the I_{ON_EBR} but also enhanced the E_{BR}, highlighting the critical role of the stacked structure in optimizing dielectric performance.

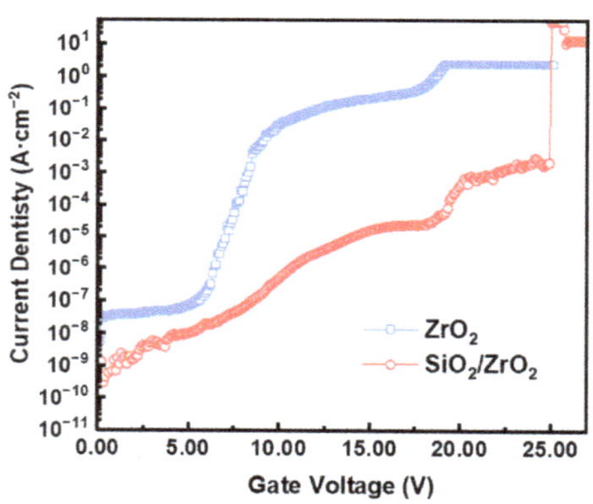

Figure 7. J-V curves of MOS capacitors.

A methodical analysis of charge transport phenomena under low-leakage conditions was performed via rigorous numerical simulations across diverse gate electric field regimes. Linear regions of experimental current–voltage profiles were systematically examined to quantify key transport parameters. These parameters were subsequently implemented within computational models aligned with established theoretical frameworks, enabling precise reconstruction of individual conduction pathways.

Current leakage phenomena in MOS capacitor architectures originate from five principal mechanisms: (i) direct quantum tunneling across ultrathin dielectrics, (ii) trap-assisted tunneling (TAT) [35], (iii) Schottky emission [36], (iv) field-enhanced Poole–Frenkel (P-F) conduction [37], and (v) Fowler–Nordheim (F-N) tunneling [38]. Given that direct tunneling occurs primarily in thin oxide films, this study focused on a detailed characterization of TAT, Schottky emission, P-F emission, and F-N tunneling processes. The governing equations for each transport mechanism, along with their associated physical constants and fitting parameters, are summarized in Table 1.

Table 1. Conduction plots and analytical model equations of different gate leakage current mechanisms.

Conduction Mechanisms	Conduction Plot	Analytical Model Equations	
TAT	$\ln(JE)$ vs. $\frac{1}{E}$	$J_{TAT} = \frac{2qc_T N_T \phi_T \exp\left(-\frac{D\Phi_T^{\frac{3}{2}}}{E_{ox}}\right)}{3E_{ox}}$	[35]
Schottky emission	$\ln(J)$ vs. $\sqrt{E}$	$J_{SE} = A^* T^2 \exp\left(\frac{-q\left(\Phi_B - \sqrt{\frac{qE_{ox}}{4\Pi\varepsilon_{ox}}}\right)}{kT}\right)$	[36]
P-F emission	$\ln\left(\frac{J}{E}\right)$ vs. $\sqrt{E}$	$J_{PF} = qN_c\mu E_{ox}\exp\left(\frac{-q\left(\phi_t - \sqrt{\frac{qE_{ox}}{\pi\varepsilon_0 k_r}}\right)}{kT}\right)$	[37]
F-N tunneling	$\ln\left(\frac{J}{E^2}\right)$ vs. $\frac{1}{E}$	$J_{FN} = AE_{ox}{}^2\exp\left(-\frac{B}{E_{ox}}\right)$	[38]

The TAT effect occurs when electrons tunnel into trap states within the dielectric film, followed by a transition into the conduction band of the dielectric. As shown in Figure 8a, both ZrO_2 and SiO_2/ZrO_2 dielectric exhibited TAT behavior, indicating the formation of traps within the samples. These traps facilitated the capture and release of electrons at low voltages, creating well-defined TAT conduction pathways [39]. At low-to-moderate bias conditions, TAT is the dominant conduction mechanism, as the applied electric field is insufficient to trigger significant thermal excitation of carriers.

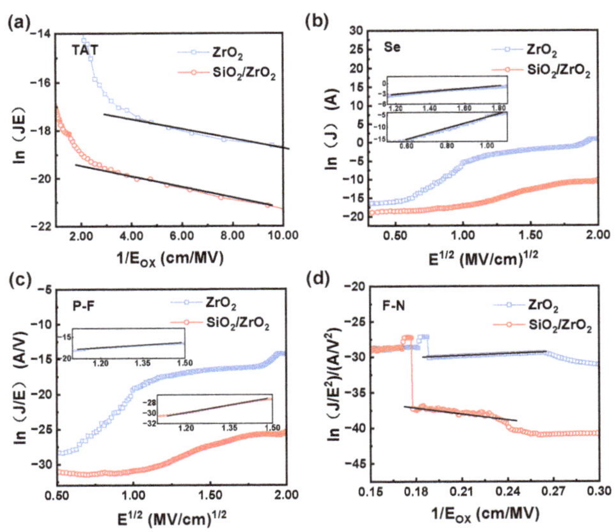

Figure 8. Electric fields of ZrO_2 and SiO_2/ZrO_2 dielectric MOS capacitors for (**a**) TAT, (**b**) Schottky emission, (**c**) F-N emission, and (**d**) F-N tunneling model fitting plots.

However, as the applied voltage increases, TAT transitions into P-F emission, where charge carriers are thermally excited from bulk dielectric trap states to the conduction band. Under applied electric fields, charge carriers are thermally activated from bulk dielectric trap states to the conduction band, leading to P-F emission [40]. As depicted in Figure 8c, this emission mechanism was primarily observed in both samples under intermediate to strong field conditions, with its dominance increasing proportionally to the magnitude of the applied bias.

Schottky emission refers to the thermally activated transfer of electrons from the semiconductor to the conduction band of the insulator, similar to thermionic emission observed in metal–semiconductor junctions [41]. As shown in Figure 8b, the lower barrier height of ZrO_2 facilitated Schottky emission, while the higher barrier provided by the SiO_2 interlayer effectively suppressed this mechanism, reducing leakage current in the SiO_2/ZrO_2 stack. Additionally, tunneling effects at moderate electric field strengths caused a sharp increase in current, which contributed to the reduced the dielectric breakdown field intensity (E_{BR}) and enhanced leakage current [42]. This observation is consistent with previous findings in high-k dielectric stacks, where SiO_2 interlayers have been demonstrated to effectively block electron injection and suppress Schottky emission [43].

Under elevated gate bias conditions, electrons traversed the triangular energy barrier via quantum tunneling mechanisms, initiating F-N tunneling phenomena under intense electric fields. This behavior was consistently observed in both dielectric configurations, with the characteristic tunneling profiles comprehensively illustrated in Figure 8d. This transition from P-F to F-N conduction is dictated by the electric field strength and the barrier shape at the dielectric interface.

In summary, the leakage current behavior in ZrO_2 thin films was predominantly governed by multiple mechanisms: TAT, Schottky emission, P-F emission, and F-N tunneling. The SiO_2/ZrO_2 stacked structure demonstrated leakage characteristics primarily dominated by TAT, P-F emission, and F-N tunneling, with SiO_2 interlayer serving as a key modulator of charge transport pathways. Notably, Schottky emission was identified as a major contributor to leakage current variability in ZrO_2 films, highlighting its distinct role compared to the bilayer configuration. This difference underscores the efficacy of the SiO_2 interlayer in suppressing specific conduction mechanisms while maintaining overall dielectric integrity.

The interfacial properties of MOS capacitors were investigated by measuring the C-V curves of ZrO_2 and SiO_2/ZrO_2 trench MOS capacitors at frequencies of 1 kHz, 10 kHz, 100 kHz, and 1 MHz. As shown in Figure 9a,b, minimal curve scatterings were observed in both samples, indicating the high quality of both gate dielectrics.

To further characterize the interfacial properties, normalized C-V hysteresis measurements were performed at a frequency of 1 MHz. As shown in Figure 9c, under an AC signal of 30 mV, the voltage was swept from negative bias to positive bias and back to negative bias. This cyclic testing protocol effectively captured the dynamic charge trapping and de-trapping behaviors at the dielectric–semiconductor interfaces, with the width of the closed hysteresis loop serving as a key indicator of interfacial quality and its potential degradation or improvement. For the SiO_2/ZrO_2 stacked structure, the weighted average method was employed to calculate its effective permittivity ($\varepsilon_{ox} = \frac{d_{SiO_2}}{d_{total}} \cdot \varepsilon_{SiO_2} + \frac{d_{ZrO_2}}{d_{total}} \cdot \varepsilon_{ZrO_2}$). The effective permittivity of the SiO_2/ZrO_2 stack was calculated to be approximately 16 [42]. The flat-band voltage (V_{FB}) values for ZrO_2 and SiO_2/ZrO_2 MOS capacitors were determined as 4.3 V and 3.4 V, respectively [22]. The reduction in flat-band voltage shift highlighted the significant impact of the introduced SiO_2 layer into the stack.

During high-frequency C-V testing, boundary traps exhibit a delayed response to small AC signal variations due to their slower charge capture and release dynamics. However, these traps can gradually charge or discharge as the DC gate voltage is swept. As a result, when the gate voltage is swept from negative to positive or vice versa, the charge state of these interface traps affects the C-V curve, leading to hysteresis between the high-frequency C-V measurements. This hysteresis occurs because the interface traps are unable to fully charge or discharge at high frequencies, leading to a mismatch between charge accumulation/release and the voltage sweep, thus causing hysteresis [44]. Since the unstacked ZrO_2 dielectric displayed more significant hysteresis than the SiO_2/ZrO_2 stacked dielectric, the lower hysteresis in the stacked structure suggests a reduction in slow trap density, which aligns with the quantitatively lower the low density of interface states (D_{it}) values obtained through the high–low frequency method.

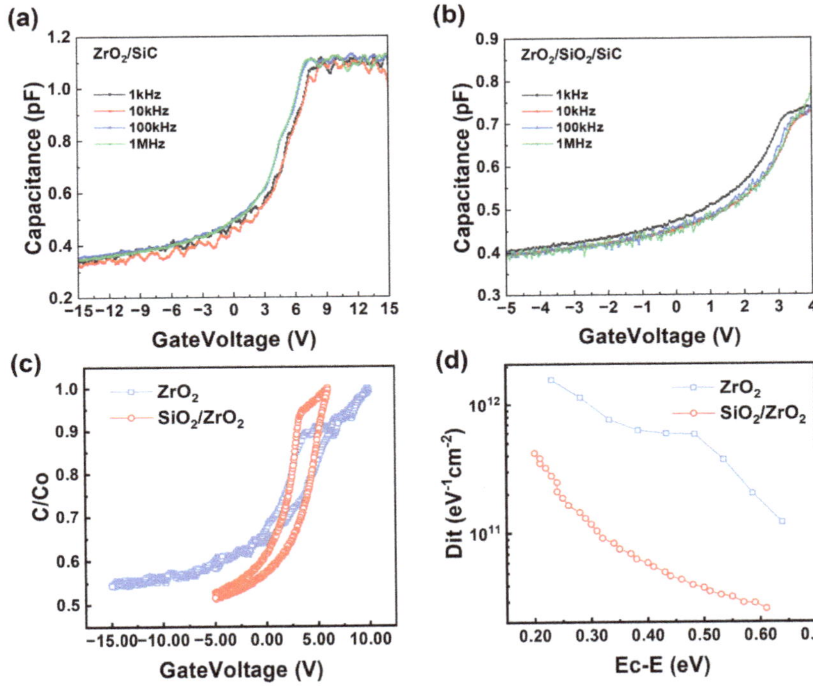

Figure 9. C-V curves of (**a**) ZrO_2 MOS and (**b**) SiO_2/ZrO_2 MOS at 1 kHz, 10 kHz, 100 kHz, and 1 MHz, (**c**) normalized C-V hysteresis curves of MOS capacitors with ZrO_2 and SiO_2/ZrO_2 dielectrics, and (**d**) calculated D_{it} plots.

As shown in Figure 9d, the D_{it} values calculated at E_C-0.2 eV by the high–low frequency method were estimated to be 1.58×10^{12} $eV^{-1} \cdot cm^{-2}$ for ZrO_2 and 4.19×10^{11} $eV^{-1} \cdot cm^{-2}$ for the SiO_2/ZrO_2 stack [45]. Thus, both gate dielectric samples contained certain defects, aligning with the TAT mechanism observed under moderate electric fields. The lower D_{it} value of the SiO_2/ZrO_2 can be attributed to the passivation effect of the SiO_2 interlayer [46]. Material characterization via XPS and SE previously demonstrated that SiO_2 forms a stable, uniform interface layer with reduced oxygen vacancy density, which correlates directly with the measured improvement in D_{it}. On the other hand, the SiO_2 interlayer serves as a buffer, alleviating the lattice mismatch between ZrO_2 and the SiC substrate. In the stacked structure, the buffering effect of the SiO_2 layer isolated ZrO_2 from direct contact with SiC, mitigating the impact of oxygen vacancies on D_{it}.

Furthermore, the degradation of channel mobility is primarily influenced by acoustic phonon scattering, surface roughness scattering, and interface defect scattering, with interface defect scattering being the dominant factor. This suggests that the suppression of interface defects via the SiO_2 interlayer can mitigate mobility degradation, thereby improving channel conduction efficiency and device reliability. The increase in D_{it} may also be related to defects introduced during the etching process, where interface quality could be improved through annealing. Nevertheless, the D_{it} values of both ZrO_2 and SiO_2/ZrO_2 samples were overall relatively low, indicating good electrical performance.

The N_{eff} within the oxide layer can be quantitatively derived from the flat-band voltage (V_{FB}) [23]. The N_{eff} of ZrO_2 and SiO_2/ZrO_2 dielectric MOS capacitors were calculated as -3.63×10^{12} cm^{-2} and -1.71×10^{12} cm^{-2}. As summarized by the electrical and interfacial properties of the MOS capacitors in Table 2, the negative N_{eff} values contributed to the forward shift of the flat-band voltage V_{FB}. The fixed charges or interface traps within the dielectric layer, particularly in the presence of defective oxide layers, can capture or release electrons, thereby influencing the V_{FB}. The negative fixed charges in the oxide layer cause the accumulation of electrons at the semiconductor–oxide interface when the gate voltage is low, resulting in a forward shift of the V_{FB} [47]. For n-type semiconductors, this shift causes an increased accumulation of electrons at the interface. The introduction of a SiO_2 interlayer into the SiO_2/ZrO_2 stacked structure reduced V_{FB} shifts and lowered the N_{eff}, thereby improving the interface quality of the devices. This improvement can primarily be attributed to the SiO_2 interlayer's ability to mitigate the effects of interface traps and other charge-related defects within the dielectric stack, resulting in enhanced electrical performance and reliability.

Table 2. Electrostatic and interfacial performance metrics of MOS capacitor structures by ZrO_2 and SiO_2/ZrO_2 dielectric.

	E_{BR} (MV·cm^{-1})	V_{FB} (V)	E_{BR} (MV·cm^{-1})	I_{ON_EBR} (A·cm^{-2})	D_{it} (eV^{-1}·cm^{-2}) E_C-0.2 eV	N_{eff} (cm^{-2})
ZrO_2	3.76	4.3	3.76	1.94	1.58×10^{12}	-3.63×10^{12}
SiO_2/ZrO_2	5.78	3.4	5.78	2.09×10^{-3}	4.19×10^{11}	-1.71×10^{12}

4. Conclusions

In summary, this study successfully fabricated MOS capacitors featuring ALD-grown ZrO_2 and SiO_2/ZrO_2 gate dielectric layers on 4H-SiC trench structures. The AFM measurements showed roughnesses of both films less than 1 nm, indicating relatively smooth surfaces. The band gap energy of the SiO_2/ZrO_2 film is higher than that of the ZrO_2 film, as evidenced by both the O 1s energy loss spectrum fitting and the $(\alpha h\nu)^2$ versus photon energy analysis. The electrostatic evaluations highlighted distinct variations in dielectric behavior. The ZrO_2 single-layer configuration displayed an E_{BR} of 3.76 MV·cm^{-1}, accompanied by I_{ON_EBR} reaching 1.94 A·cm^{-2}. Mechanistic investigations identified TAT, Schottky emission, PF emission, and F-N tunneling as the principal conduction pathways governing leakage in the monolithic ZrO_2 system. By comparison, the SiO_2/ZrO_2 bilayer architecture achieved an enhanced E_{BR} of 5.78 MV·cm^{-1} with I_{ON_EBR} reduced to 2.09×10^{-3} A·cm^{-2}, reflecting a 54% improvement in E_{BR} in leakage current compared to the ZrO_2 monolayer. Within the stacked configuration, leakage mechanisms were predominantly mediated by TAT, PF emission, and F-N tunneling, with Schottky emission effectively suppressed through the interfacial SiO_2 layer. The C-V hysteresis tests indicated a larger hysteresis in ZrO_2, consistent with calculated D_{it}. As a result, the higher interface trap density in ZrO_2 thin films had a more pronounced impact on the electrical performance. The incorporation of a thin SiO2 interlayer effectively addressed the small bandgap offset, enhancing the bandgap from 5.21 eV to 5.88 eV and reducing I_{ON_EBR}. However, the etching process induces interfacial defects in the trench structures, which could negatively impact device performance. Post-treatment techniques showed promise in mitigating these interface issues, with future investigations potentially focusing on methods such as annealing to further enhance interface quality. Additionally, research into other high-k materials, such as Y_2O_3, Ta_2O_5, and La_2O_3, could be considered. This comparative study provides a comprehensive evaluation of the electrical behavior of ZrO_2 and SiO_2/ZrO_2 stacked dielectric configurations in SiC trench capacitors, offering valuable insights for optimizing future SiC trench-gate MOSFET designs.

Author Contributions: Q.H.: methodology, data curation, formal analysis, writing—original draft. Y.G.: formal analysis, investigation, validation. A.W.: investigation, validation. Z.B.: investigation, validation. L.G.: investigation, writing—review and editing. Z.W.: investigation. C.D.: validation, writing—review and editing. Y.S.: validation, writing—review and editing. H.M.: conceptualization, software, supervision, funding acquisition, project administration, writing—review and editing. Q.Z.: formal analysis, investigation, validation, supervision. All authors have read and agreed to the published version of the manuscript.

Funding: This work is supported by the National Key R&D Program of China (2023YFB4606300), National Natural Science Foundation of China (No. 62474049), the Science and Technology Innovation Plan of Shanghai Science and Technology Commission (Nos. 21DZ1100800, 23ZR1405300, 20501110700, and 20501110702), and the Joint Research Plan of sci-tech innovation community in Yangtze River Delta (No.2023CSJG0600).

Institutional Review Board Statement: Not applicable.

Informed Consent Statement: Not applicable.

Data Availability Statement: The original contributions presented in this study are included in the article. Further inquiries can be directed to the corresponding author.

Conflicts of Interest: The authors declare no conflict of interest.

References

1. Kimoto, T. Material Science and Device Physics in SiC Technology for High-Voltage Power Devices. *Jpn. J. Appl. Phys.* **2015**, *54*, 040103. [CrossRef]
2. Town, G.; Taghizadeh, S.; Deilami, S. Review of Fast Charging for Electrified Transport: Demand, Technology, Systems, and Planning. *Energies* **2022**, *15*, 1276. [CrossRef]
3. She, X.; Huang, A.Q.; Lucia, O.; Ozpineci, B. Review of Silicon Carbide Power Devices and Their Applications. *IEEE Trans. Ind. Electron.* **2017**, *64*, 8193–8205. [CrossRef]
4. Baliga, B.J. *Fundamentals of Power Semiconductor Devices*; Springer: Cham, Switzerland, 2019.
5. Zhang, Y.; Dong, D.; Li, Q.; Zhang, R.; Udrea, F.; Wang, H. Wide-Bandgap Semiconductors and Power Electronics as Pathways to Carbon Neutrality. *Nat. Rev. Electr. Eng.* **2025**, *2*, 155–172. [CrossRef]
6. Jayant Baliga, B. Silicon Carbide Power Devices: Progress and Future Outlook. *IEEE J. Emerg. Sel. Top. Power Electron.* **2023**, *11*, 2400–2411. [CrossRef]
7. Baliga, B.J. *Wide Bandgap Semiconductor Power Devices*; Woodhead Publishing: Cambridge, UK, 2018.
8. Wolf, D.; Lubk, A.; Prete, P.; Lovergine, N.; Lichte, H. 3D Mapping of Nanoscale Electric Potentials in Semiconductor Structures Using Electron-Holographic Tomography. *J. Phys. D Appl. Phys.* **2016**, *49*, 364004. [CrossRef]
9. Yu, L.C.; Dunne, G.T.; Matocha, K.S.; Cheung, K.P.; Suehle, J.S.; Sheng, K. Reliability Issues of SiC MOSFETs: A Technology for High-Temperature Environments. *IEEE Trans. Device Mater. Reliab.* **2010**, *10*, 418–426. [CrossRef]
10. Wei, J.; Wei, Z.; Fu, H.; Cao, J.; Wu, T.; Sun, J.; Zhu, X.; Li, S.; Zhang, L.; Liu, S.; et al. Review on the Reliability Mechanisms of SiC Power MOSFETs: A Comparison between Planar-Gate and Trench-Gate Structures. *IEEE Trans. Power Electron.* **2023**, *38*, 8990–9005. [CrossRef]
11. Qian, J.; Shi, L.; Jin, M.; Bhattacharya, M.; Shimbori, A.; Yu, H.; Houshmand, S.; White, M.H.; Agarwal, A.K. An Investigation of Body Diode Reliability in Commercial 1.2 KV SiC Power MOSFETs with Planar and Trench Structures. *Micromachines* **2024**, *15*, 177. [CrossRef]

12. Takaya, H.; Misumi, T.; Fujiwara, H.; Ito, T. 4H-SiC Trench MOSFET with Low On-Resistance at High Temperature. In Proceedings of the 2020 32nd International Symposium on Power Semiconductor Devices and ICs (ISPSD), Vienna, Austria, 13–18 September 2020; pp. 118–121. [CrossRef]
13. Huang, H.-L.; Hsuesh, L.-T.; Tu, Y.-C.; Tsui, B.-Y. A 4H-SiC Trench MOS Capacitor Structure for Sidewall Oxide Characteristics Measurement. In Proceedings of the 2024 IEEE 36th International Conference on Microelectronic Test Structures (ICMTS), Edinburgh, UK, 15–18 April 2024; pp. 1–5. [CrossRef]
14. Siddiqui, A.; Khosa, R.Y.; Usman, M. High-*k* Dielectrics for 4H-Silicon Carbide: Present Status and Future Perspectives. *J. Mater. Chem. C* **2021**, *9*, 5055–5081. [CrossRef]
15. Race, S.; Kumar, P.; Natzke, P.; Kovacevic-Badstuebner, I.; Bathen, M.E.; Grossner, U.; Romano, G.; Arango, Y.; Bolat, S.; Wirths, S.; et al. Gate Impedance Analysis of SiC Power MOSFETs with SiO_2 and High-k Dielectric. In Proceedings of the 2023 35th International Symposium on Power Semiconductor Devices and ICs (ISPSD), Hong Kong, China, 28 May–1 June 2023. [CrossRef]
16. Wirths, S.; Mihaila, A.; Romano, G.; Schneider, N.; Ceccarelli, E.; Alfieri, G.; Arango, Y.; Knoll, L. Study of 1.2kV High-k SiC Power MOSFETS under Harsh Repetitive Switching Conditions. In Proceedings of the 2021 33rd International Symposium on Power Semiconductor Devices and ICs (ISPSD), Nagoya, Japan, 30 May–3 June 2021; pp. 107–110. [CrossRef]
17. Wirths, S.; Arango, Y.C.; Prasmusinto, A.; Alfieri, G.; Bianda, E.; Mihaila, A.; Kranz, L.; Bellini, M.; Knoll, L. Vertical 1.2kV SiC Power MOSFETs with High-k/Metal Gate Stack. In Proceedings of the 2019 31st International Symposium on Power Semiconductor Devices and ICs (ISPSD), Shanghai, China, 19–23 May 2019. [CrossRef]
18. Idris, M.I.; Wright, N.G.; Horsfall, A.B. Effect of Post Oxide Annealing on the Electrical and Interface 4H-SiC/Al_2O_3 MOS Capacitors. *Mater. Sci. Forum* **2018**, *924*, 486–489. [CrossRef]
19. Khosa, R.Y.; Thorsteinsson, E.B.; Winters, M.; Rorsman, N.; Karhu, R.; Hassan, J.; Sveinbjörnsson, E.Ö. Electrical Characterization of Amorphous Al_2O_3 Dielectric Films on N-Type 4H-SiC. *AIP Adv.* **2018**, *8*, 025304. [CrossRef]
20. Hosoi, T.; Kashiwagi, Y.; Hosaka, S.; Yamamoto, K.; Aketa, M.; Nakamura, T.; Kimoto, T.; Shimura, T.; Watanabe, H. Comprehensive and Systematic Design of Metal/High-K Gate Stack for High-Performance and Highly Reliable SiC Power MOSFET. *Jpn. J. Appl. Phys.* **2019**, *59*, 021001. [CrossRef]
21. Ye, G.; Wang, H.; Ji, R. Band Alignment between 4H-SiC and Atomic-Layer-Deposited ZrO_2Determined by X-ray Photoelectron Spectroscopy. *Appl. Phys. Express* **2015**, *8*, 091302. [CrossRef]
22. Wang, Z.; Zhang, Z.; Shao, C.; Robertson, J.; Liu, S.; Guo, Y. Tuning the High-κ Oxide (HfO_2, ZrO_2)/4H-SiC Interface Properties with a SiO_2 Interlayer for Power Device Applications. *Appl. Surf. Sci.* **2020**, *527*, 146843. [CrossRef]
23. Guo, Z.; Wu, J.; Tian, R.; Wang, F.; Xu, P.; Yang, X.; Fan, Z.; Yang, F.; He, Z. Extraction of the Trench Sidewall Capacitances in an N-Type 4H-SiC Trench Metal– Oxide–Semiconductor Structure. *IEEE Trans. Electron. Devices* **2021**, *68*, 2879–2885. [CrossRef]
24. Fowler, R.H.; Nordheim, L. Electron Emission in Intense Electric Fields. *Proc. R. Soc. Lond. A* **1928**, *119*, 173–181. [CrossRef]
25. Nguyen, T.T.; Jung, D.H.; Lee, J.J.; Lee, H. Optical Study of $Ga_{2-X}Sn_xO_3$ ($0 \leq X \leq 0.7$) Thin Films Using Spectroscopic Ellipsometry and Cathodoluminescence. *Thin Solid Film.* **2023**, *764*, 139611. [CrossRef]
26. Zhang, Q.X. The Electrical Performances and Leakage Current Conduction Mechanism of $Al_2O_3/ZrO_2/SiO_2/ZrO_2/Al_2O_3$ MIM Capacitors. *Integr. Ferroelectr.* **2021**, *217*, 233–239. [CrossRef]
27. Haider, M.; Shaim, A.; Elsayed-Ali, H.E. Boron and Barium Incorporation at the 4H-SiC/SiO_2 Interface Using a Laser Multi-Charged Ion Source. *J. Mater. Sci. Mater. Electron.* **2021**, *32*, 16079–16087. [CrossRef]
28. Behera, M.K.; Yarbrough, K.A.; Bahoura, M. Study of Dielectric Properties in Stacked High-k Dielectric Metal Oxide Semiconductor Capacitor (MOSCAP) Devices. *J. Mater. Sci. Mater. Electron.* **2024**, *35*, 1316. [CrossRef]
29. Shishodia, G.; Gupta, S.; Pahwa, N.; Shishodia, P.K. ZrO_2 Nanoparticles Synthesized by the Sol–Gel Method: Dependence of Size on PH and Annealing Temperature. *J. Electron. Mater.* **2024**, *53*, 5159–5168. [CrossRef]
30. Jung, E.S.; Yoo, J.; Choi, T.M.; Lee, H.R.; Lee, C.Y.; Kim, D.H.; Pyo, S.G. Performance Variation with Pristine and Doped High-K Materials via Atomic Layer Deposition. *Int. J. Appl. Ceram. Technol.* **2024**, *15001*. [CrossRef]
31. Ye, G.; Wang, H.; Arulkumaran, S.; Ng, G.I.; Li, Y.; Liu, Z.H.; Ang, K.S. Band Alignment between GaN and ZrO_2 Formed by Atomic Layer Deposition. *Appl. Phys. Lett.* **2014**, *105*, 022106. [CrossRef]
32. Cañas, J.; Alba, G.; Leinen, D.; Lloret, F.; Gutierrez, M.; Eon, D.; Pernot, J.; Gheeraert, E.; Araujo, D. Diamond/γ-Alumina Band Offset Determination by XPS. *Appl. Surf. Sci.* **2020**, *535*, 146301. [CrossRef]
33. Lee, T.-E.; Toprasertpong, K.; Takenaka, M.; Takagi, S. Re-Examination of Effects of ALD High-k Materials on Defect Reduction in SiGe Metal–Oxide–Semiconductor Interfaces. *AIP Adv.* **2021**, *11*, 085021. [CrossRef]
34. Martínez-Puente, M.A.; Horley, P.; Aguirre-Tostado, F.S.; López-Medina, J.; Borbón-Núñez, H.A.; Tiznado, H.; Susarrey-Arce, A.; Martínez-Guerra, E. ALD and PEALD Deposition of HfO2 and Its Effects on the Nature of Oxygen Vacancies. *Mater. Sci. Eng. B* **2022**, *285*, 115964. [CrossRef]

35. Esakki, P.; Kailath, B.J. Extraction and Analysis of Gate Leakage Current Mechanism in Silicon Carbide (SiC) MIS Capacitors. *IETE J. Res.* **2019**, *68*, 592–601. [CrossRef]
36. Reddy, V.R.; Prasad, C.V. Surface Chemical States, Electrical and Carrier Transport Properties of Au/ZrO_2/N-GaN MIS Junction with a High-k ZrO_2 as an Insulating Layer. *Mater. Sci. Eng. B* **2018**, *231*, 74–80. [CrossRef]
37. Chaudhuri, S.K.; Nag, R.; Mandal, K.C. A Novel Ni/Y_2O_3/4H-SiC Heteroepitaxial Metal–Oxide–Semiconductor (MOS) Betavoltaic Cell. *J. Mater. Sci. Mater. Electron.* **2023**, *34*, 543. [CrossRef]
38. Król, K.; Kwietniewski, N.; Gierałtowska, S.; Wachnicki, Ł.; Sochacki, M. Electronic Properties of Stacked ZrO_2 Films Fabricated by Atomic Layer Deposition on 4H-SiC. *Acta Phys. Pol. A* **2017**, *132*, 329–331. [CrossRef]
39. Cheong, K.Y.; Bahng, W.; Kim, N.-K. Analysis of Charge Conduction Mechanisms in Nitrided SiO_2 Film on 4H SiC. *Phys. Lett. A* **2008**, *372*, 529–532. [CrossRef]
40. Liu, J.; Liao, M.; Imura, M.; Tanaka, A.; Iwai, H.; Koide, Y. Low On-Resistance Diamond Field Effect Transistor with High-k ZrO_2 as Dielectric. *Sci. Rep.* **2014**, *4*, 6395. [CrossRef]
41. Sasikumar, K.; Bharathikannan, R.; Raja, M. Effect of Annealing Temperature on Structural and Electrical Properties of Al/ZrO_2/P-Si MIS Schottky Diodes. *Silicon* **2018**, *11*, 137–143. [CrossRef]
42. Galizia, B.; Fiorenza, P.; Schilirò, E.; Pecz, B.; Foragassy, Z.; Greco, G.; Saggio, M.; Cascino, S.; Lo Nigro, R.; Roccaforte, F. Towards Aluminum Oxide/Aluminum Nitride Insulating Stacks on 4H–SiC by Atomic Layer Deposition. *Mater. Sci. Semicond. Process.* **2024**, *174*, 108244. [CrossRef]
43. Lin, L.M.; Lai, P.T. Improved High-Field Reliability for a SiC Metal–Oxide–Semiconductor Device by the Incorporation of Nitrogen into Its HfTiO Gate Dielectric. *J. Appl. Phys.* **2007**, *102*, 054515. [CrossRef]
44. Jia, Y.; Lv, H.; Tang, X.; Han, C.; Song, Q.; Zhang, Y.; Zhang, Y.; Dimitrijev, S.; Han, J.; Haasmann, D. Influence of Various NO Annealing Conditions on N-Type and P-Type 4H-SiC MOS Capacitors. *J. Mater. Sci. Mater. Electron.* **2019**, *30*, 10302–10310. [CrossRef]
45. Seo, Y.; Lee, S.; An, I.; Song, C.; Jeong, H. Conduction Mechanism of Leakage Current due to the Traps in ZrO_2 Thin Film. *Semicond. Sci. Technol.* **2009**, *24*, 115016. [CrossRef]
46. Kurniawan, T.; Wong, Y.H.; Yew Cheong, K.; Moon, J.H.; Bahng, W.; Abdul Razak, K.; Lockman, Z.; Joon Kim, H.; Kim, N.-K. Effects of Post-Oxidation Annealing Temperature on ZrO_2 Thin Film Deposited on 4H-SiC Substrate. *Mater. Sci. Semicond. Process.* **2011**, *14*, 13–17. [CrossRef]
47. Noborio, M.; Suda, J.; Kimoto, T. 4H-SiC MIS Capacitors and MISFETs with Deposited SiN_x/SiO_2 Stack-Gate Structures. *IEEE Trans. Electron Devices* **2008**, *55*, 2054–2060. [CrossRef]

materials

Article

Determination of Quantum Yield in Scattering Media Using Monte Carlo Photoluminescence Cascade Simulation and Integrating Sphere Measurements

Philip Gelbing *, Joachim Jelken, Florian Foschum and Alwin Kienle

Institut für Lasertechnologien in der Medizin und Meßtechnik, Universität Ulm, Helmholtzstr. 12, 89081 Ulm, Germany
* Correspondence: philip.gelbing@ilm-ulm.de

Abstract: Accurate determination of the quantum yield (Φ_f) in scattering media is essential for numerous scientific and industrial applications, but it remains challenging due to re-absorption and scattering-induced biases. In this study, we present a GPU-accelerated Monte Carlo simulation framework that solves the full fluorescence radiative transfer equation (FRTE), incorporating spectrally dependent absorption, scattering, and fluorescence cascade processes. The model accounts for re-emission shifts, energy scaling due to the Stokes shift and implements a digital optical twin of the experimental setup, including the precise description of the applied integrating sphere. Using Rhodamine 6G in both ethanol and PDMS matrices, we demonstrate the accuracy of the method by comparing simulated reflectance and transmission spectra with independent experimental measurements. Φ_f and emission distributions are optimized using a Levenberg–Marquardt algorithm. The obtained quantum yields agree well with literature values for Rhodamine 6G. This approach eliminates the need for empirical correction factors, enabling the reliable determination of actual, undistorted emission spectra and the Φ_f in complex scattering media.

Keywords: quantum yield; fluorescence; luminescence; photoluminescence; monte carlo simulation; scattering media; radiative transfer equation; integrating sphere; rhodamine 6g

1. Introduction

Photoluminescence materials play a central role in numerous scientific and industrial applications, including imaging, diagnostics, and DNA sequencing. In the lighting industry, they are essential for the development of white light LEDs, in which they serve as wavelength converters. This technology is characterized by high energy efficiency and longevity, whereby the quantum yield (Φ_f) is a decisive parameter for the performance of these materials and a direct measure of luminescence efficiency [1,2]. This paper primarily discusses fluorescence, which is the rapid luminescent emission of light that occurs when fluorescent dyes, such as Rhodamine 6G, absorb photons. However, our methodological framework is applicable to various types of photoluminescence. The exact determination of the quantum yield is a complex challenge, especially in scattering media. It is influenced by various factors, including the chemical structure of the fluorophore, the ambient temperature, interactions with the solvent, and quenching processes such as internal transformations or non-radiative relaxations. Two main optical methodological approaches are used in the literature to measure the quantum yield—relative and absolute measurements [3–5]. In the context of relative measurement, the emission of a sample characterized by an unknown quantum yield is compared with that of a reference sample, which possesses a known quantum yield, under conditions that are identical to those of the reference sample. This method is widely employed, yet it possesses significant limitations, outlined as follows: The choice of reference materials with well characterized optical properties is limited, and matching the optical properties of the sample and the reference, especially with respect to absorption and scattering, is difficult to achieve in practice. In strongly scattering samples, re-absorption of scattered photons can significantly complicate measurements, leading to an overestimation of absorption and an underestimation of emission. Notably, re-absorption effects are not limited to scattering media. They also appear in non-scattering or weakly scattering media, particularly when high concentrations or strongly absorbing fluorophores are involved. This systematic bias can lead to an underestimation of the quantum yield. Consequently, optically thin solutions are typically employed to mitigate this issue [6–10]. In contrast to the relative method, absolute measurements do not necessitate a reference sample and can thus be utilized for a broader range of materials. Integrating spheres are commonly used in this context. Without the use of an integrating sphere, a spectrometer measurement would merely detect a fraction of the emitted light, as this depends on the scattering, the refractive index, the geometry of the sample, and other factors, ensuring more accurate results [3,11–13]. The quantum yield is then determined by comparing the total number of photons emitted to the total number of photons absorbed.

In a multitude of technical and biological systems, the processes of scattering and multiple scattering result in complex light propagation, particularly in highly scattering media such as luminescence phosphor layers or biological tissues. The interplay between absorption, fluorescence, and scattering is particularly pronounced in these contexts. The inability to dilute samples to reduce scattering often presents a significant challenge in determining the fluorescence quantum yield and the true emission spectrum in these and other scattering media.

As outlined in [14], a primary issue is that the fraction of emitted photons captured by the detector is highly dependent on the scattering and absorption properties of the samples. Accurately determining the exact proportion of photons absorbed, particularly those absorbed by the fluorophore rather than by other non-fluorescent components, is essential for measuring the correct Φ_f. Additional complications result from inner filter effects and re-absorption processes as follows: The excitation light can undergo significant attenuation within the sample, resulting in spatially non-uniform fluorescence emission. Such spatial gradients become particularly pronounced in slab geometry, where fluorescence intensity is typically stronger on the side facing the excitation source. Furthermore, emitted photons may be re-absorbed before escaping the medium, particularly in systems with small Stokes shifts, where the absorption and emission spectra overlap significantly. These effects, along with the wavelength dependence of the medium's optical properties, profoundly alter the light propagation within the sample. This leads to distortions in both the spectral shape and the absolute intensity of the measured fluorescence and absorption signals. Consequently, if these factors are not appropriately accounted for or corrected, the experimentally determined quantum yield may substantially deviate from the intrinsic Φ_f of the fluorophore.

A range of methodologies for addressing these effects can be found in the literature. In [15], an analytical approach is described to correct re-absorption effects in solid-state samples. Samples with varying thicknesses are prepared to visualize how re-absorption impacts measurements. While re-absorption effects are minimal in thin layers, they become significant in thicker layers. To ensure the accuracy of the quantum yield determination, the absorption and emission spectra of both thin and thick samples are meticulously recorded. Crucially, the approach assumes availability of an undistorted emission spectrum, typically obtained from very thin films or highly diluted solutions where re-absorption is negligible. The probability of re-absorption, denoted by a, is calculated using the overlap integral between absorption and emission spectra. This calculation integrates the normalized emission spectrum with the absorption coefficient based on Beer–Lambert law principles. The corrected quantum yield is subsequently calculated using the following formula:

$$\Phi_f = \frac{\Phi_{f,\mathrm{obs}}}{1 - a + a\,\Phi_{f,\mathrm{obs}}}. \tag{1}$$

A similar but extended approach is adopted in [16], where Φ_f of scattering media is determined using an integrating sphere under both direct and indirect illumination. The emission spectra that result from this measurement can be used as reliable indicators of undistorted emission. The same overlap-integral formalism and correction formula (Equation (1)) are then applied. This approach has been found to significantly reduce the effects of scattering and to provide more precise quantum yield measurements, especially when dealing with highly scattering media. Another approach to correcting spectral distortions by re-absorption in scattering media is described in [17]. The authors also create samples with different layer thicknesses to compare the undistorted (true) emission spectrum with the spectrum distorted by re-absorption.

While the approaches mentioned make important contributions to the determination of quantum yield in scattering media, they have specific drawbacks. They rely on *post hoc* analytical corrections based on indirect estimations of re-absorption probability rather than modeling the full propagation of light within the medium. Additionally, the requirement to prepare and analyze multiple samples with varying layer thicknesses imposes considerable experimental effort. This process is not only time consuming and resource intensive, but also introduces the potential for systematic errors, particularly in the case of inhomogeneous, very thin, or polydisperse samples.

In order to correctly determine the quantum yield under simultaneous scattering and absorption of excitation and fluorescence photons and avoid the inclusion of correction factors and sample-by-sample empirical fixes, it may be advisable to employ a physically-based model that takes into account the entire light propagation process, including absorption, scattering, and fluorescence. A promising method for describing these effects is the Fluorescent Radiative Transfer Equation (FRTE), an extension of the classical radiative transfer equation that explicitly incorporates the spectral properties of fluorescence, including wavelength shift and re-absorption [18]. Recent studies have demonstrated that the FRTE facilitates more precise modeling of light propagation in fluorescent scattering media, particularly by accounting for multiple scattering and spectrally varying absorption effects [19–22]. In this study, the FRTE is therefore employed as a foundation for the quantitative analysis of quantum yield in scattering media. In order to accurately model light propagation in fluorescence media, the utilization of commercial ray-tracing software often proves to be inadequate [21]. This is because such models typically operate in two separate steps: First, light propagation is simulated for the excitation wavelengths. Then, based on the absorbed energy and quantum yield, the emission distribution is calculated. This simplification neglects the re-absorption effects, which are especially relevant in highly fluorescent and scattering media. As a consequence, the resulting simulations can significantly deviate from experimental observations. An alternative approach was proposed in [20], which, to the best of our knowledge, constitutes the first and so far only Monte Carlo solution for broadband fluorescence systems that accounts for the full fluorescence cascade, including absorption and scattering. The model, implemented in Matlab, was validated against published transmitted-flux data. However, the computational power of this approach is limited, and it is not known whether it scales well for complex geometries and material compositions. Our current approach employs a GPU-accelerated Monte Carlo method that considers the entirety of the fluorescence cascade as well, thereby enabling the precise calculation of light propagation in complex media comprising various fluorescent, absorbing, and scattering components. Unlike previous methods, our approach involves the experimental determination and integration of specific optical properties of scattering materials, thereby ensuring a realistic simulation over a broad wavelength spectrum. A key advantage of the proposed method is the utilization of GPU parallelization, which substantially improves computational efficiency and enables simulations of full fluorescence light propagation in complex geometries over extensive excitation spectra. Another essential aspect is the direct consideration of the experimental setup in the model, particularly the integrating sphere detector, ensuring close correspondence between simulation and measurement. This integration allows precise determination of optical and fluorescent properties, such as quantum yield and wavelength-dependent emission probabilities, directly from a single measurement. Prior experimental methods frequently encountered systematic uncertainties due to inadequate modeling of complex scattering phenomena and neglected spectral overlaps between absorption and emission, issues our current methodology specifically addresses.

2. Theory

2.1. Radiative Transfer Equation

In order to understand the employed model that has been utilized, we apply the RTE for unpolarized light. This equation is widely accepted as the standard formula for describing the propagation of light in scattering media, where interference effects can be neglected. For the steady-state case at a fixed wavelength, the equation is as follows:

$$\hat{s} \cdot \nabla L(\vec{r}, \hat{s}) \;+\; \mu_t L(\vec{r},\, \hat{s}) \;=\; \mu_s \int L(\vec{r},\, \hat{s}')p(\hat{s},\, \hat{s}')d\hat{s}' \;+\; \varepsilon(\vec{r},\, \hat{s})\,. \tag{2}$$

In this context, $L(\vec{r}, \hat{s})$ is defined as *radiance* at the position $\boldsymbol{r}$ in the direction of the unit vector $\hat{s}$ (per unit area and per unit solid angle). The *extinction coefficient* is defined as

$$\mu_t = \mu_a + \mu_s\,, \tag{3}$$

including the *absorption coefficient* μ_a and the *scattering coefficient* μ_s. The *phase function* $p(\hat{s}, \hat{s}')$ indicates how light is redistributed from the direction $\hat{s}'$ into the direction $\hat{s}$. The term $\varepsilon(\boldsymbol{r}, \hat{s})$ denotes additional radiation sources [23]. When fluorescence is considered, it is imperative to extend the RTE to account for the spectral dependence of the optical properties. This entails consideration of the entire wavelength spectrum rather than a single wavelength only. The extended *FRTE* is expressed as follows:

$$\begin{aligned}
&\hat{s} \cdot \nabla L(\lambda, \boldsymbol{r}, \hat{s}) + \mu_t(\lambda) L(\lambda, \boldsymbol{r}, \hat{s}) = \\
&\mu_s(\lambda) \int_{4\pi} L(\lambda, \boldsymbol{r}, \hat{s}')\, p(\lambda, \hat{s}, \hat{s}')\, d\Omega' + \varepsilon(\lambda, \boldsymbol{r}, \hat{s}) \\
&+ \frac{1}{4\pi} \Phi_f \int \mu_{a,f}(\lambda') \left[\int_{4\pi} L(\lambda', \boldsymbol{r}, \hat{s}')\, d\Omega' \right] P_f(\lambda)\, d\lambda'\,.
\end{aligned} \tag{4}$$

The additional term represents the secondary light source generated by fluorescence within the media. The absorption of the fluorophore at wavelength λ' is quantified by the spectral absorption profile $\mu_{a,f}(\lambda')$. The radiation at emission wavelength λ is assumed to be isotropically emitted. The quantum yield Φ_f is a material-dependent constant, as it describes the ratio of emitted to absorbed photons:

$$\Phi_f = \frac{N_{\text{em}}}{N_{\text{abs}}} \,. \tag{5}$$

The energy transfer processes within fluorescent molecules can be effectively visualized using the Jablonski diagram [1,24]. Following photon absorption, a molecule is excited into a higher vibronic state of an electronic level (e.g., S_1 or S_2). According to the Franck–Condon principle [25,26], these transitions occur vertically in the Jablonski diagram due to the disparity between rapid electronic excitation and slower nuclear motion. As a result, molecular geometry remains unchanged during excitation. Subsequently, the molecule undergoes ultrafast internal conversion, typically within femto- to picoseconds. This process dissipates excess vibrational energy non-radiatively to the environment, leading to relaxation of the molecule to the lowest vibronic level of S_1. By definition, this non-radiative relaxation occurs without photon emission. Fluorescence emission then takes place through the radiative transition from the lowest excited singlet state (S_1) to the ground electronic state (S_0), accompanied by the emission of a photon. This behavior is encapsulated by the *Kasha rule* [27], which states that fluorescence predominantly arises from the lowest excited singlet state (S_1), regardless of the initially excited electronic state (e.g., S_2 or higher). As fluorescence emission consistently occurs from the same initial state (S_1) and is predominantly determined by molecular structure, the quantum yield Φ_f is effectively independent of the excitation wavelength λ'. In the absence of additional decay pathways, such as non-radiative quenching or energy transfer mechanisms, the fluorescence quantum yield remains constant for a given material, irrespective of the specific excitation wavelength [1]. The spectral distribution of emitted photons is characterized by a normalized probability density function (PDF), denoted as $\widetilde{P}_f(\lambda)$, satisfying the following normalization condition:

$$\int_0^\infty \widetilde{P}_f(\lambda)\, d\lambda = 1 \,. \tag{6}$$

Due to refraction, scattering, and re-absorption effects within the medium, this probability distribution does not directly correspond to the experimentally measured emission spectrum. Instead, it represents the intrinsic probability distribution for the emission of a single photon from an individual fluorescence event.

2.2. Monte Carlo Implementation

The FRTE (4) is solved using a GPU-accelerated Monte Carlo simulation. The GPU kernel is implemented in OpenCL, while host-side control, data preprocessing, and post-processing are handled via Python (version 3.13) and ISO C++ 14-Standard. All simulations were performed on a workstation equipped with an AMD Ryzen 9 7900X CPU (12 cores, 4.7 GHz; Advanced Micro Devices, Santa Clara, CA, USA) and a single NVIDIA GeForce RTX 4070 Ti GPU (NVIDIA Corporation, Santa Clara, CA, USA), using CUDA version 12.3. For each of the N simulated photons, pseudo-random numbers (PRNs) determine the initial position, propagation angle, and wavelength. The starting wavelengths are sampled from the experimentally measured spectrum, incorporating the hyperchromator calibration and spectrometer instrumental response (see Section 2.4). The optical properties (μ_s, μ_a, phase function, and refractive index) are assigned for both non-fluorescent and fluorescent components at each wavelength. For each material, these are determined prior to simulation using the method developed by us and described in [28]. The optical properties of each material are represented by the weighted contributions of its constituent components, which may include fluorescent and non-fluorescent fractions. Each component retains a distinct set of wavelength-dependent optical properties. At each step in the simulation, the probability of absorption versus scattering is calculated. The decision process for each photon is summarized in Algorithm 1, which calls for either the absorption (Algorithm 2) or the scattering (Algorithm 3) subroutine, depending on the outcome.

Algorithm 1 Photon interaction: absorption vs. scattering

Require: Photon with wavelength λ and weight w_λ
Require: Absorption: $\mu_a^{\text{base}}(\lambda)$, $\mu_a^{\text{fluo}}[i](\lambda)$
Require: Scattering coefficient $\mu_s^{\text{tot}}(\lambda)$
Require: Random number $\xi \sim \mathcal{U}[0,1)$
Compute total absorption: $\mu_a^{\text{tot}} = \mu_a^{\text{base}} + \sum_i \mu_a^{\text{fluo}}[i]$
Compute absorption probability: $P_{\text{abs}} = \mu_a^{\text{tot}} / (\mu_a^{\text{tot}} + \mu_s^{\text{tot}})$
Draw random number ξ
if $\xi < P_{\text{abs}}$ **then**
    Call Algorithm 2: Absorption
else
    Call Algorithm 3: Wavelength-Dependent Scattering
end if

Algorithm 2 Absorption with fluorescence and isotropic re-emission

Require: Quantum yield $\Phi_f[i]$, emission $CDF[i](\lambda)$
Require: Random numbers $\xi_0, \xi_1, \xi_2 \sim \mathcal{U}[0,1)$
Draw $\xi_0 \in [0,1)$
if $\xi_0 < \mu_a^{\text{base}} / \mu_a^{\text{tot}}$ **then**
    Photon is absorbed in non-fluorescent base material (no re-emission)
else
    for each fluorescent component i **do**
        Accumulate: $\mu_a^{\text{cum}} \mathrel{+}= \mu_a^{\text{fluo}}[i]$
        if $\xi_0 < \mu_a^{\text{cum}} / \mu_a^{\text{tot}}$ **then**
            Draw $\xi_1 \in [0,1)$
            if $\xi_1 > \Phi_f[i]$ **then**
                Photon is absorbed in component [i] (no re-emission)
            else
                Photon is re-emitted
                Update: $\lambda \leftarrow \lambda'$, $w_\lambda \leftarrow w_{\lambda'} \cdot \frac{\lambda'}{\lambda}$
                Emit isotropically:
                Draw $\xi_1, \xi_2 \in [0,1)$
                Compute: $\cos\theta = 1 - 2\xi_1, \quad \phi = 2\pi\xi_2$
                Set direction: $\vec{v}_{\text{photon}} \leftarrow (\sin\theta\cos\phi, \sin\theta\sin\phi, \cos\theta)$
            end if
            break
        end if
    end for
end if

Photons propagate through the medium until absorption or interaction with a boundary occurs [23]. Absorption is treated distinctly for fluorescent and non-fluorescent components. The total absorption coefficient is defined as the sum of all component absorptions, and the component in which absorption occurs is probabilistically determined according to the relative contributions of the components. A PRN is used to select the absorbing component (Algorithm 2). If the photon is absorbed in a non-fluorescent region, it is terminated.

If absorbed in a fluorescent component, re-emission is determined probabilistically using Φ_f. If another PRN exceeds the quantum yield, the photon is terminated; in all other cases, it is re-emitted with a new wavelength. To model the intrinsic emission probability $\widetilde{P}_f(\lambda)$ introduced in Equation (6), we employ a normalized Gaussian mixture model (GMM) with three components. Each component represents a Gaussian distribution defined by its mean μ_i, standard deviation σ_i, and weight w_i, as follows:

$$\widetilde{P}_f(\lambda) \approx \sum_{i=1}^{3} w_i \cdot \mathcal{N}(\lambda \mid \mu_i, \sigma_i^2), \quad \text{with} \sum_{i=1}^{3} w_i = 1. \tag{7}$$

To enforce normalization, only two of the weights are treated as free parameters, with the third defined as $w_3 = 1 - w_1 - w_2$. The model is therefore parameterized by eight independent variables as follows: three means μ_i, three standard deviations σ_i, and two weights. The cumulative distribution function (CDF) associated with $\widetilde{P}_f(\lambda)$ is used to generate random emission wavelengths during the Monte Carlo sampling process. However, the true probability distribution of re-emitted wavelengths is not directly accessible. Due to multiple cycles of re-absorption and re-emission, the effective emission spectrum differs from the initially measured fluorescence spectrum. As a pragmatic starting point, the GMM parameters are initialized by fitting the model to the experimentally measured fluorescence spectrum, which serves as a proxy for the intrinsic emission probability distribution. These initial parameters are subsequently refined by comparing the results from a complete Monte Carlo simulation, accounting for the full complexity of light propagation in the sample, with experimental measurements. Initially, we explored other models, such as cubic spline models, due to their ability to flexibly represent complex emission spectra, which may be necessary for other luminescent materials. However, the increased number of fitting parameters and their interdependencies reduced numerical robustness and computational efficiency during optimization.

The assignment of a new wavelength upon re-emission is achieved by sampling a PRN from the CDF. Each time a photon is re-emitted at a different wavelength, its energy is modified accordingly. This energy change can be handled in two ways as follows: either by retrospectively computing the ratio of final to initial wavelengths during postprocessing, or by continuously propagating a photon weight variable throughout the simulation. The latter method, which is adopted here, enables the use of a broad excitation spectrum and allows direct normalization to the total number of incident photons. To account for the shift in energy, the photon weight is updated according to

$$w_\lambda = w_{\lambda'} \left(\frac{\lambda'}{\lambda} \right), \tag{8}$$

where $w_{\lambda'}$ and λ' denote the photon's weight and wavelength prior to emission, and w_λ and λ after emission. If neither absorption or emission occurs, the photon undergoes scattering.

In materials such as Rhodamine 6G, where the fluorophore itself exhibits negligible scattering, the new propagation direction is determined by the phase function of the non-fluorescent base material. In heterogeneous media composed of multiple components with distinct scattering coefficients, the respective phase function is selected according to their relative contribution to the total scattering. Our system exhibits strong forward scattering. The Henyey–Greenstein phase function was selected due to its computational efficiency and analytical tractability when sampling scattering angles. It supports wavelength-dependent anisotropy $g(\lambda)$ and is therefore compatible with spectrally resolved light propagation in fluorescent materials. Alternative phase functions may be implemented for strongly anisotropic or size-dependent scattering, but these were not required for the present system. If a photon exits the media, it may reach the virtual integrating sphere detector either in forward (transmission) or backward (reflectance) directory, depending on its trajectory and angular distribution. In the present study, we assume a slab geometry for all samples, where the forward and backward directions correspond to transmission and reflectance, respectively. By modeling the individual components of a sample separately in the Monte Carlo simulation, a more detailed representation of photon propagation becomes possible. This allows for the isolated study of fluorescent and non-fluorescent contributions to the measured signal. In theory, this approach enables the estimation of the quantum yield of complex material combinations. Materials such as Rhodamine contribute minimally to scattering, making the scattering coefficient of added scatterers the dominant factor. In contrast, other luminescent materials may exhibit strong intrinsic scattering. This explicit modeling allows complex cascades of light propagation to be simulated, for example in the context of optimizing phosphor-converted white LEDs.

Algorithm 3 Wavelength-dependent scattering

Require: Scattering coefficients $\mu_s^j(\lambda)$ for each material component j
Require: Phase function (e.g. Henyey–Greenstein) $p_j(\theta; g_j(\lambda))$ for each j
Require: Random number $\xi \sim \mathcal{U}[0,1)$

Compute total scattering coefficient: $\mu_s^{\text{tot}} = \sum_j \mu_s^j(\lambda)$
Draw $\xi \in [0,1)$
Initialize cumulative scattering: $\mu_s^{\text{cum}} \leftarrow 0$
for each scattering component j **do**
    Update cumulative scattering: $\mu_s^{\text{cum}} += \mu_s^j(\lambda)$
    if $\xi < \mu_s^{\text{cum}} / \mu_s^{\text{tot}}$ **then**
        Sample scattering angle θ from phase function $p_j(\theta; g_j(\lambda))$
        break
    end if
end for

2.3. Integrating Sphere Model

To accurately compare the measurement with the Monte Carlo simulation, the specific characteristics of the experimental environment must be taken into account. Photons emerging from a sample slab, either in transmission or reflectance, can enter the integrating sphere depending on their exit angle, position, and the sphere's geometry. The measurement setup, as delineated in [29], utilizes a 150 mm diameter integrating sphere. The integrating sphere is coupled to a spectrometer (HR6 XR, OceanOptics, Orlando, FL, USA) with a 2048-pixel CCD array. Equipped with a 300 lines/mm grating and a 100 µm entrance slit, it has an optical resolution of 3.7 nm over a 200–1100 nm spectral range. The detector exhibits a dynamic range of approximately 12,000:1. To calibrate the system response, including the integrating sphere, the coupling optics and spectrometer, a broadband reference source with known output spectrum was measured using a power meter. This enabled the conversion of measured counts to radiant power. Detector linearity was independently verified in the relevant intensity range and found to be sufficient. In order to minimize the influence of the sample on the internal mixing process, the sample is mounted externally to the sphere. Excitation for reflectance measurements is achieved through an opposing port, thereby ensuring clear separation between transmitted and reflected light, thus reducing systematic errors. A schematic overview of the setup is shown in Figure 1, highlighting the external sample placement and separation of the transmission and reflection beams. The inner surface of the sphere is coated with barium sulfate, and the setup contains four ports, as follows: a reflection port, a normalization port, a sample port, and a detector port. The use of a baffle is avoided, as this has been shown to disrupt the internal mixing process within the sphere. However, this configuration allows for the direct illumination of the detector field of view (FOV) by photons reflected by the sample, emitted or re-emitted due to fluorescence. A proportion of the emitted light may also escape through one of the open ports, contributing no signal, while the remaining photons interact with the inner sphere wall. The reflectance of the photon is determined by the coating's properties, and the photons undergo multiple diffuse reflections within the sphere. Consequently, the simulation must account for the spatial and angular distribution of energy that reaches various components of the sphere, depending on the wavelength and optical properties of the sample. The final fraction of light detected by the spectrometer is computed using the following analytically derived model, which is extended to incorporate spectral dependence, based on [29]

$$\Phi_D(\lambda) = \Phi_{D0}(\lambda) + \frac{A_{\text{det}}}{A_{\text{sph}}} \sum_{n=1}^{N} \rho_n(\lambda)\Phi_n(\lambda) \left(1 - \sum_{l=1}^{N} \rho_l(\lambda) \frac{A_l}{A_{\text{sph}}}\right)^{-1}. \tag{9}$$

In this context, the symbol $\Phi_{D0}(\lambda)$ denotes the fraction of energy at wavelength λ that directly reaches the detector FOV without internal scattering inside the sphere. The second term accounts for the geometric series of reflected contributions from all the inner sphere surfaces n, including the ports and the sample. Each surface n is characterized by its area A_n and wavelength-dependent reflectance $\rho_n(\lambda)$, and it is irradiated by radiant power $\Phi_n(\lambda)$. The value of $\Phi_n(\lambda)$ can be determined by taking into account the angular distribution of the light entering the integrating sphere. For light scattered and emitted by the sample, this angular distribution is obtained from the Monte Carlo simulation described in Section 2.2. The open ports are modeled with $\rho = 0$. To determine the spectral reflectance of the sample, an auxiliary Monte Carlo simulation was conducted with the same optical properties as in the initial simulation. Due to the geometry of the inner sphere and the diffuse mixing process within it, the sample is reached by photons with a quasi-isotropic angular distribution, which can be well approximated by a cosine-distributed source [29].

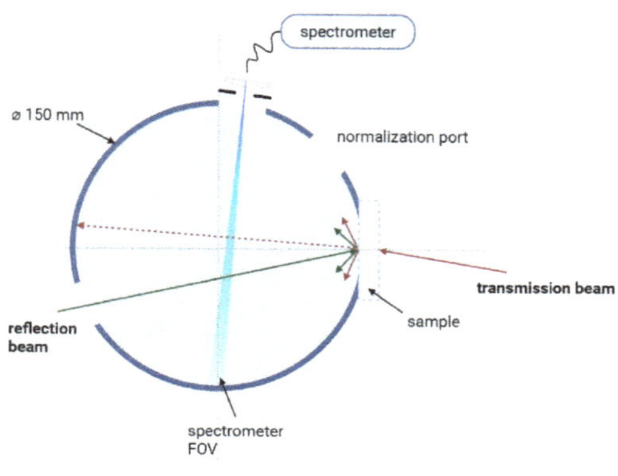

Figure 1. Schematic of the integrating sphere setup used in the measurement. The sample is mounted externally, and photons are either reflected back into the sphere or transmitted through the sample.

2.4. Source Model

To correctly determine the excitation spectrum $\Phi_s(\lambda)$ incident on the sample during measurement, it is important to recognize that the spectrum that is recorded by the spectrometer, $\Phi_D(\lambda)$, does not directly represent the true source spectrum prior to illumination. Several correction steps must therefore be applied to account for the specific properties of the measurement setup. First, the reflectance of the integrating sphere wall as a function of wavelength, $\rho_{\text{wall}}(\lambda)$, was experimentally determined from calibration measurements at each wavelength. Assuming an open-port configuration with no direct excitation of the detector field of view (i.e., $\Phi_{D0}(\lambda) = 0$), the effective source intensity in transmission geometry was obtained by inverting the integrating sphere model from Equation (9) to

$$\Phi_s(\lambda) = \frac{\Phi_D(\lambda) \cdot A_{\text{sph}} \cdot \left(1 - \rho_{\text{wall}}(\lambda) \cdot \frac{A_{\text{wall}}}{A_{\text{sph}}}\right)}{A_{\text{det}} \cdot \rho_{\text{wall}}(\lambda)}. \tag{10}$$

In the case of reflectance measurements, where the excitation light is redirected onto the sample by means of a reflective mirror, the mirror's reflectance, $\rho_{\text{mirror}}(\lambda)$, is included. The effective excitation spectrum $\Phi_s^{\text{rem}}(\lambda)$ at the sample position is then given by

$$\Phi_s^{\text{rem}}(\lambda) = \frac{\Phi_D(\lambda) \cdot \frac{A_{\text{sph}}}{A_{\text{det}}} \cdot \left(1 - \left[\rho_{\text{wall}}(\lambda) \cdot \frac{A_{\text{wall}}}{A_{\text{sph}}} + \rho_{\text{mirror}}(\lambda) \cdot \frac{A_{\text{mirror}}}{A_{\text{sph}}}\right]\right)}{\rho_{\text{mirror}}(\lambda)}. \tag{11}$$

All measured spectra were normalized to a common integration time to account for wavelength-dependent acquisition times used to optimize the signal-to-noise ratio. The finite spectral resolution of the detection system broadens the recorded spectrum relative to the physical illumination spectrum. The spectrometer's instrument function was independently measured using a reference light source with a negligible intrinsic linewidth. This instrument function was then used to numerically deconvolve the measured spectrum. The reconstructed excitation function, $\Phi_s(\lambda)$, is used in the Monte Carlo simulation to sample the initial wavelengths at the start of each photon trajectory, as well as the total intensity of the source. This ensures consistency with the spectral resolved excitation conditions of the experimental setup.

2.5. Sample Preparation and Reference Dye

As a reference dye, we used *Rhodamine 6G (CAS 989-38-8)* supplied by AAT Bioquest, Inc. For the experimental determination of Φ_f, we dissolved Rhodamine 6G in ethanol at different concentrations. Three solutions were prepared as follows: $c_1 = 3.8\,\mu\text{M}$, $c_2 = 7.9\,\mu\text{M}$, and $c_3 = 15.4\,\mu\text{M}$, corresponding to a simple scaling where $c_3 \approx 2c_2 \approx 4c_1$. Since the dye itself exhibits negligible scattering, we added 1% polystyrene scatterers with a 2.4 µm diameter to the solution to introduce controlled scattering conditions. The solutions were poured into custom-made cuvettes consisting of two 1 mm-thick BK7 glass plates, which were separated by a 4 mm spacer. After filling, the cuvettes were sealed with Parafilm to prevent evaporation. In the simulation, the refractive indices of both the BK7 glass and the surrounding air were taken into account. In addition to the liquid-phase samples, we also investigated Rhodamine 6G embedded in a solid matrix. For this, a Rhodamine-ethanol solution was mixed into a *polydimethylsiloxane (PDMS)* matrix (Elastosil M 4641 A/B, Wacker Chemie AG, Munich, Germany), together with *zirconium doxide* (*ZrO_2*) scatterers (diameter: 800 nm; US Research Nanomaterials Inc., Houston, TX, USA), resulting in a reduced scattering coefficient $\mu_s' \approx 12\,\text{mm}^{-1}$. Two dye concentrations were prepared as follows: $c_a = 4.7\,\mu\text{M}$ and $c_b = 2.35\,\mu\text{M}$. In the ethanol-based solutions, polystyrene particles were preferred over ZrO_2, due to their slower sedimentation kinetics, which ensures a more homogeneous distribution throughout the measurement duration, including optical property characterization. After determining the optical properties of the samples [28], the fluorescence quantum yield Φ_f and the parameters of the emission PDF, modeled as a GMM, were optimized. This was achieved by fitting Monte Carlo simulations to the experimentally recorded reflectance and transmission spectra using a Levenberg–Marquardt algorithm [30]. The optimization minimized the root-mean-square error (RMSE) between simulated and measured intensity of the sample.

3. Results

3.1. Quantum Yield in Ethanol

Figure 2 and Table 1 show the fitted spectra and quantum yields for the three Rhodamine 6G concentrations. The statistical uncertainty is derived from the square root of the diagonal elements of the covariance matrix obtained in the Levenberg–Marquardt optimization. The excitation peak, visible in both transmission and reflectance spectra in Figure 2, decreases with increasing dye concentration. This is attributed to stronger absorption of the excitation light and thus enhanced fluorescence emission. The spectral shape and position of this peak are determined by the optical properties of the sample. Additional contributing factors include the integrating sphere model and the calibrated source spectrum. Notably, these are independent of the fitted Φ_f or the emission PDF. Therefore, deviations between the simulation and measurement in the relative heights of the excitation peaks can be indicative of errors in the optical properties: for instance, an overestimated scattering coefficient would decrease the transmission peak while increasing the reflectance signal. This effect is analogous to that of an incorrect sample thickness. According to the light propagation model [29], an error in the sample thickness translates roughly linearly into an error in the retrieved optical parameters (e.g., a 5% thickness error propagates almost linearly to $\approx$5% in μ_s and μ_a). This, in turn, affects the predicted fluorescence and excitation characteristics. Hence, the excitation peak serves as an internal consistency check, unaffected by the outcome of the optimization. In contrast, the magnitude of the fluorescence signal is highly sensitive to the value of Φ_f. An overestimated Φ_f would lead to an excess emission in both transmission and reflectance spectra, while an underestimated Φ_f to lack of emission. All values are reported with two decimal places.

Table 1. Fitted quantum yields Φ_f for Rhodamine 6G in ethanol at three concentrations. Statistical errors are derived from the Levenberg–Marquardt fit.

Concentration	Φ_f	Statistical Error	95% CI
c_1	0.91	$\pm$0.006	[0.90, 0.92]
c_2	0.94	$\pm$0.005	[0.93, 0.95]
c_3	0.93	$\pm$0.003	[0.92, 0.94]

Published quantum yields for Rhodamine 6G in ethanol vary depending on method and concentration, but typically fall within the range of 0.89–0.95 [3,5,31–34]. It should be emphasized that these reference values are based on low-scattering systems or pure solutions. Our method, in contrast, allows for the retrieval of quantum yield even in the presence of significant scattering. This was made possible by explicitly modeling the scattering contribution of the added microspheres and simulating complex light propagation in layered or heterogeneous samples. Würth et al. [33] reported values of 0.92 ± 0.05 using an absolute method based on a calibrated integrating sphere. Their values are concentration-dependent and corrected for re-absorption. They also fall within the range of our results. Overall, the retrieved values of $\Phi_f = 0.91$, 0.94, and 0.93 are within their respective error ranges, demonstrating agreement with the literature. Nevertheless, differences between individual measurements warrant further analysis. Empirical evidence indicates that the quantum yield for Rhodamine 6G is concentration-dependent [31,35,36]. Lower dye concentrations might also present higher uncertainties due to weaker emission signals and sensitivity to calibration errors. Larger relative errors can be observed in conditions with lower absorption (μ_a), such as in sample c_1, and scattering (μ_s) coefficients [29]. Variations between assumed model parameters and actual physical sample conditions, or errors in detector sensitivity calibration across wavelengths, can impact accuracy, especially for small optical thicknesses. Quantum efficiency can also exhibit temperature dependence even for small changes, typically decreasing with increasing temperatures [37]. Temperature fluctuations during measurements or differences from the literature conditions may therefore contribute additional variance to the quantum yields.

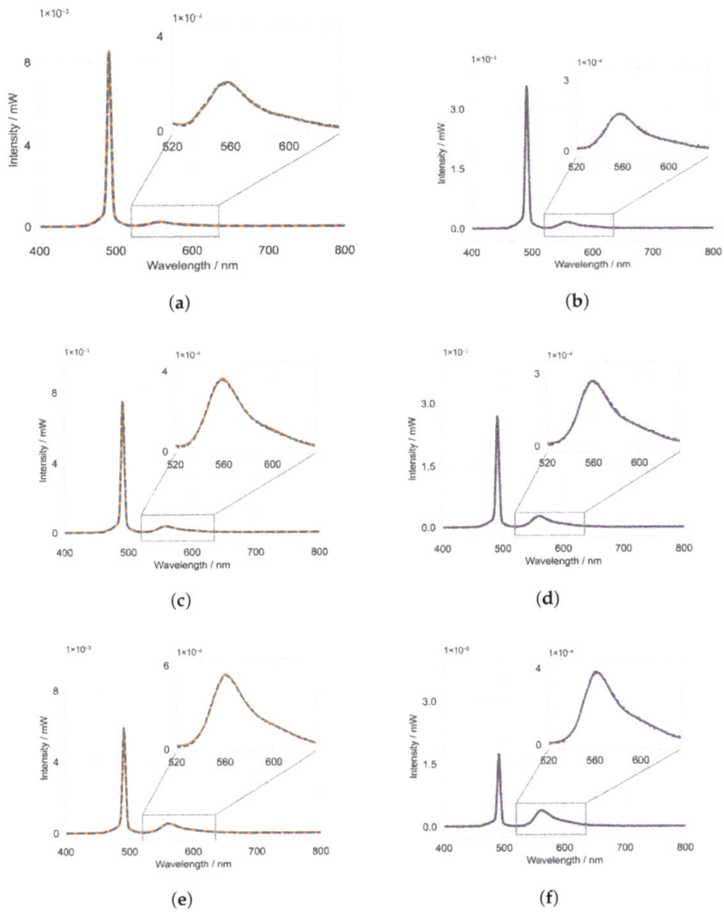

Figure 2. Reflectance (**left**) and Transmission (**right**) spectra for three concentrations c_1, c_2, and c_3. Subfigures are arranged by concentration from top to bottom and labeled as (**a**) c_1 reflectance, (**b**) c_1 transmission, (**c**) c_2 reflectance, (**d**) c_2 transmission, (**e**) c_3 reflectance, and (**f**) c_3 transmission. The gray dashed line represents the experimental measurement, the orange solid line corresponds to the simulated reflectance, and the purple solid line corresponds to the simulated transmission.

3.2. *Effect of Re-Absorption on the Quantum Yield*

Conventional absolute integrating sphere protocols estimate Φ_f by dividing the detected emission by the total number of *all primary* absorption events of the sample. This approach is valid only for dilute, non-scattering samples in which the absorption and emission spectra have negligible overlap. However, in practical scenarios, additional mechanisms can bias the determination of Φ_f.

One such mechanism is the re-absorption of fluorescence photons when emission and absorption bands overlap. Photons initially emitted via fluorescence can be re-absorbed and re-emitted multiple times before escaping the sample. Each absorption-emission cycle typically red shifts the spectrum, increasing the probability of loss through non-radiative decay pathways, thereby reducing the detected emission. The probability of these cascading absorption–emission events increases with optical thickness and is further amplified by multiple scattering. As a result, the measured fluorescence underestimates the true amount of photons generated by the initial excitation events.

This effect is captured by our simulations, which distinguish between photons absorbed at the excitation wavelength and subsequent fluorescent photons, while tracking the history of absorption–emission cycles. If we quantify all fluorescent photons exiting the sample via either reflectance or transmission and divide by the number of primary absorption events, we obtain an uncorrected quantum yield of $\Phi_f^{\mathrm{uncorr}} = 0.87$ for concentration c_3. This contrasts with the intrinsic quantum yield obtained from our simulation, $\Phi_f^{\mathrm{true}} = 0.93$.

Figure 3a illustrates this discrepancy. The blue dashed curve represents the initial GMM fit to the measured reflectance spectrum of the c_3 sample. The solid yellow curve shows the final fitted emission probability distribution used in the Monte Carlo simulations to assign photon wavelengths after each emission event. A distinct ~5 nm blue shift and an increased probability at shorter wavelengths is observed compared to the measured spectrum. This reflects the cumulative effect of re-absorption events, which favors the survival of photons having a shorter wavelength. Figure 3b further decomposes this phenomenon by grouping detected reflectance photons according to the number of absorption-emission cycles they underwent. Each additional cycle shifts the spectrum incrementally toward longer wavelengths, indicating energy loss through sequential re-emission.

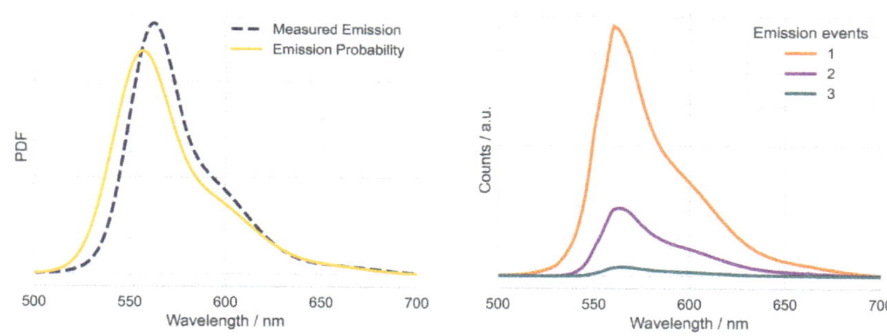

(**a**) Measured and intrinsic emission spectra.

(**b**) Fluorescence photon counts by wavelength conversion events number.

Figure 3. (**a**) Normalized emission spectra: the measured spectrum (recorded using an integrating sphere setup) is compared to the intrinsic emission PDF used in the simulation. Both are normalized to unit area for comparability. (**b**) Distribution of simulated fluorescence photon counts in reflectance, grouped by the number of emission (wavelength conversion) events each photon underwent. Data shown for sample c_3.

3.3. Other Distortion Effects

Scattering not only increases the probability of re-absorption but also introduces additional systematic distortions [14]. In strongly scattering media, the excitation light penetrates only a short distance into the slab geometry, localizing fluorescence generation near the illuminated surface. This spatial inhomogeneity is accounted for in our Monte Carlo simulation. In contrast, conventional integrating sphere setups, where the sample is placed inside the sphere, introduce additional mixing [16,33]. Photons strike the sample from all angles multiple times, averaging out directional effects and artificially increasing re-excitation probability. To avoid this mixing artifact and distinguish between the direction of reflectance and transmission, we positioned the sample outside the sphere during our measurements.

Scattering can also redirect photons away from fluorophores and increase the probability of their absorption by non-fluorescent components, such as the cuvette glass, solvent, or scatterers. If these effects are not distinguished, the number of true excitation events is overestimated. For example, neglecting to differentiate these absorption sites results in an even lower quantum yield of $\Phi_f^{gross} = 0.85$ for sample c_3. Our simulation framework resolves this ambiguity by modeling scattering and absorption for each material component individually, enabling accurate retrieval of Φ_f even in complex, highly scattering systems.

3.4. Quantum Yield in PDMS

In addition to the ethanol-based samples, Rhodamine 6G was also embedded in a PDMS matrix to evaluate fluorescence efficiency in a solid scattering media. The procedure is identical to that in the ethanol samples, but the optical properties of the PDMS matrix were used in the simulation. The fitted quantum yields for two concentrations c_a and c_b are presented in Table 2. Compared to the ethanol solutions, a decrease in Φ_f is observed.

Table 2. Fitted quantum yields Φ_f for Rhodamine 6G embedded in a PDMS matrix. Statistical errors are derived from the Levenberg–Marquardt fit.

Concentration	Φ_f	Statistical Error	95% CI
c_a	0.75	±0.003	[0.74, 0.75]
c_b	0.78	±0.003	[0.78, 0.79]

The reduction in quantum yield is consistent with known solvent effects on Rhodamine dyes. Ethanol, being a polar and hydrogen-bonding solvent, stabilizes the excited state and reduces non-radiative decay [38,39]. In contrast, PDMS is non-polar and aprotic, lacking such stabilizing interactions [40]. These differences may reduce excited-state stabilization and increase the probability of non-radiative pathways, such as internal conversion [41]. Additionally, rigid matrices like PDMS have been reported to influence molecular relaxation dynamics, which could further affect Φ_f [40]. While these effects are plausible contributors, they are not directly measured in this study and should be interpreted with caution. These effects collectively reduce the radiative efficiency and therefore the quantum yield of the dye.

4. Conclusions

In this work, we presented a GPU-accelerated Monte Carlo framework for the quantitative determination of the fluorescence quantum yield in complex, highly scattering media. The spectral modeling of the fluorescence cascade was based on experimentally determined optical properties. This allowed us to correctly account for re-absorption-induced signal distortions depending on the light propagation. Explicit consideration of the experimentally applied integrating sphere geometry created a digital optical twin with which the simulation and sample could be compared spectrally resolved. This enabled a precise reconstruction of the measured emission spectrum and determination of the fluorescence quantum yield. In contrast to conventional methods for determining quantum yields in scattering media, our approach solved the complete radiative transfer equation with fluorescence term. This approach obviated the necessity for empirical correction factors. The agreement between the quantum yields of Rhodamine 6G in ethanol is obtained and the existing literature values served to confirm the reliability of the model, even in the presence of additional scatterers. The applicability of the model was further demonstrated for solid-state systems, where the embedding matrix exerted a significant influence on the radiation efficiency. Our methodology can be applied to a wide range of luminescent materials beyond Rhodamine 6G. However, achieving accurate results depends on appropriately modeling the emission probability distribution. Further research may encompass the exploration of light propagation in multilayer or anisotropic geometries, the optimization of phosphors in LEDs, and the development of models for time-resolved fluorescence processes.

Author Contributions: Conceptualization, P.G.; methodology, P.G. and F.F.; software, P.G. and F.F.; validation, P.G., F.F. and J.J.; formal analysis, P.G., F.F. and J.J.; investigation, P.G., F.F. and J.J.; resources, A.K.; data curation, P.G. and J.J.; writing—original draft preparation, P.G.; writing—review and editing, P.G., J.J., F.F. and A.K.; visualization, P.G.; supervision, A.K. and F.F.; project administration, A.K. and F.F.; funding acquisition, A.K. All authors have read and agreed to the published version of the manuscript.

Funding: This research was funded by Invest BW (VDI/VDE Innovation + Technik GmbH), within the project AdelPhos. This research is part of the Zentrum zur Bestimmung der optischen Eigenschaften (ZOE, Center of the Determination of Optical Properties of Scattering Media) at the ILM Ulm.

Data Availability Statement: The data that support the findings of this study are available from the corresponding author, P.G., upon reasonable request due to privacy.

Acknowledgments: We thank Karsten Pink for his assistance with the measurements. We also acknowledge David Hevisov and Dominik Reitzle for the fruitful discussions and helpful suggestions throughout the project.

Conflicts of Interest: The authors declare no conflicts of interest.

References

1. Lakowicz, J.R. *Principles of Fluorescence Spectroscopy*; Springer Science & Business Media: New York, NY, USA, 2007.
2. Rurack, K. Fluorescence Quantum Yields: Methods of Determination and Standards. In *Standardization and Quality Assurance in Fluorescence Measurements I: Techniques*; Resch-Genger, U., Ed.; Springer: Berlin/Heidelberg, Germany, 2008; pp. 101–145. [CrossRef]
3. Würth, C.; Grabolle, M.; Pauli, J.; Spieles, M.; Resch-Genger, U. Relative and absolute determination of fluorescence quantum yields of transparent samples. *Nat. Protoc.* **2013**, *8*, 1535–1550. [CrossRef]
4. Martini, M.; Montagna, M.; Ou, M.; Tillement, O.; Roux, S.; Perriat, P. How to measure quantum yields in scattering media: Application to the quantum yield measurement of fluorescein molecules encapsulated in sub-100 nm silica particles. *J. Appl. Phys.* **2009**, *106*, 094304. [CrossRef]
5. Fischer, M.; Georges, J. Fluorescence quantum yield of rhodamine 6G in ethanol as a function of concentration using thermal lens spectrometry. *Chem. Phys. Lett.* **1996**, *260*, 115–118. [CrossRef]
6. Levitus, M. Tutorial: Measurement of fluorescence spectra and determination of relative fluorescence quantum yields of transparent samples. *Methods Appl. Fluoresc.* **2020**, *8*, 033001. [CrossRef]

7. Brouwer, A.M. Standards for photoluminescence quantum yield measurements in solution (IUPAC Technical Report). *Pure Appl. Chem.* **2011**, *83*, 2213–2228. [CrossRef]
8. Resch-Genger, U.; Hoffmann, K.; Nietfeld, W.; Engel, A.; Neukammer, J.; Nitschke, R.; Ebert, B.; Macdonald, R. How to Improve Quality Assurance in Fluorometry: Fluorescence-Inherent Sources of Error and Suited Fluorescence Standards. *J. Fluoresc.* **2005**, *15*, 337–362. [CrossRef]
9. Parker, C.A.; Rees, W.T. Correction of fluorescence spectra and measurement of fluorescence quantum efficiency. *Analyst* **1960**, *85*, 587–600. [CrossRef]
10. Ware, W.R.; Rothman, W. Relative fluorescence quantum yields using an integrating sphere. The quantum yield of 9,10-diphenylanthracene in cyclohexane. *Chem. Phys. Lett.* **1976**, *39*, 449–453. [CrossRef]
11. Gaigalas, A.K.; Wang, L. Measurement of the Fluorescence Quantum Yield Using a Spectrometer with an Integrating Sphere Detector. *J. Res. Natl. Inst. Stand. Technol.* **2008**, *113*, 17–28. [CrossRef]
12. Porrès, L.; Holland, A.; Pålsson, L.O.; Monkman, A.P.; Kemp, C.; Beeby, A. Absolute Measurements of Photoluminescence Quantum Yields of Solutions Using an Integrating Sphere. *J. Fluoresc.* **2006**, *16*, 267–273. [CrossRef]
13. Leyre, S.; Coutino-Gonzalez, E.; Joos, J.J.; Ryckaert, J.; Meuret, Y.; Poelman, D.; Smet, P.F.; Durinck, G.; Hofkens, J.; Deconinck, G.; et al. Absolute determination of photoluminescence quantum efficiency using an integrating sphere setup. *Rev. Sci. Instrum.* **2014**, *85*, 123115. [CrossRef]
14. Lagorio, M.G. Determination of Fluorescence Quantum Yields in Scattering Media. *Methods Appl. Fluoresc.* **2020**, *8*, 043001. [CrossRef]
15. Ahn, T.S.; Al-Kaysi, R.; Mueller, A.; Wentz, K.; Bardeen, C. Self-absorption correction for photoluminescence quantum yields from integrating sphere measurements. *Rev. Sci. Instrum.* **2007**, *78*, 086105. [CrossRef]
16. Würth, C.; Resch-Genger, U. Determination of Photoluminescence Quantum Yields of Scattering Media with an Integrating Sphere: Direct and Indirect Illumination. *Appl. Spectrosc.* **2015**, *69*, 749–759. [CrossRef]
17. Lagorio, M.G.; San Román, E. How Does Light Scattering Affect Luminescence? Fluorescence Spectra and Quantum Yields in the Solid Phase. *J. Chem. Educ.* **2002**, *79*, 1362. [CrossRef]
18. Liemert, A.; Reitzle, D.; Kienle, A. Analytical solutions of the radiative transport equation for turbid and fluorescent layered media. *Sci. Rep.* **2017**, *7*, 3819. [CrossRef]
19. Ma, Y.; Wang, M.; Sun, J.; Hu, R.; Luo, X. Phosphor modeling based on fluorescent radiative transfer equation. *Opt. Express* **2018**, *26*, 16442–16455. [CrossRef]
20. Yalçın, R.A.; Ertürk, H. Monte Carlo method solution of the broadband fluorescent radiative transfer equation considering fluorescent cascade. *Appl. Opt.* **2021**, *60*, 1068–1077. [CrossRef]
21. Leyre, S.; Ryckaert, J.; Acuna, P.; Audenaert, J.; Meuret, Y.; Hofkens, J.; Durinck, G.; Deconinck, G.; Hanselaer, P. Taking the spectral overlap between excitation and emission spectra of fluorescent materials into account with Monte Carlo simulations. In *Optical Modelling and Design III*; SPIE: Bellingham, WA, USA, 2014; Volume 9131, p. 913103. [CrossRef]
22. Sandilands, L.J. Modeling fluorescence reemission in the one-dimensional radiative transfer problem using the P3 approximation. *J. Opt. Soc. Am. A* **2021**, *38*, 1631. [CrossRef]
23. Martelli, F.; Binzoni, T.; Liemert, A.; Del Bianco, S.; Kienle, A. *Light Propagation Through Biological Tissue and Other Diffusive Media: Theory, Solutions, and Validations*, 2nd ed.; SPIE: Bellingham, WA, USA, 2022. [CrossRef]
24. Jabłoński, A. Über den Mechanismus der Photolumineszenz von Farbstoffphosphoren. *Z. Für Phys.* **1935**, *94*, 38–46. [CrossRef]
25. Franck, J.; Dymond, E.G. Elementary processes of photochemical reactions. *Trans. Faraday Soc.* **1926**, *21*, 536–542. [CrossRef]
26. Condon, E.U. Nuclear Motions Associated with Electron Transitions in Diatomic Molecules. *Phys. Rev.* **1928**, *32*, 858–872. [CrossRef]
27. Kasha, M. Characterization of electronic transitions in complex molecules. *Discuss. Faraday Soc.* **1950**, *9*, 14–19. [CrossRef]
28. Jelken, J.; Brall, T.; Gelbing, P.; Foschum, F.; Kienle, A. Characterization of the Optical Properties of Photoluminescent Turbid Media Using an Integrating Sphere and Monte Carlo Simulations. *Materials* **2024**, *17*, 6072. [CrossRef]
29. Foschum, F.; Bergmann, F.; Kienle, A. Precise determination of the optical properties of turbid media using an optimized integrating sphere and advanced Monte Carlo simulations. Part 1: Theory. *Appl. Opt.* **2020**, *59*, 3203–3215. [CrossRef]
30. Marquardt, D.W. An Algorithm for Least-Squares Estimation of Nonlinear Parameters. *J. Soc. Ind. Appl. Math.* **1963**, *11*, 431–441. [CrossRef]
31. Penzkofer, A.; Leupacher, W. Fluorescence behaviour of highly concentrated rhodamine 6G solutions. *J. Lumin.* **1987**, *37*, 61–72. [CrossRef]
32. Rohwer, L.S.; Martin, J.E. Measuring the absolute quantum efficiency of luminescent materials. *J. Lumin.* **2005**, *115*, 77–90. [CrossRef]

33. Würth, C.; González, M.G.; Niessner, R.; Panne, U.; Haisch, C.; Genger, U.R. Determination of the absolute fluorescence quantum yield of rhodamine 6G with optical and photoacoustic methods—Providing the basis for fluorescence quantum yield standards. *Talanta* **2012**, *90*, 30–37. [CrossRef]
34. Magde, D.; Wong, R.; Seybold, P.G. Fluorescence Quantum Yields and Their Relation to Lifetimes of Rhodamine 6G and Fluorescein in Nine Solvents: Improved Absolute Standards for Quantum Yields. *Photochem. Photobiol.* **2002**, *75*, 327–334. [CrossRef]
35. Zehentbauer, F.M.; Moretto, C.; Stephen, R.; Thevar, T.; Gilchrist, J.R.; Pokrajac, D.; Richard, K.L.; Kiefer, J. Fluorescence spectroscopy of Rhodamine 6G: Concentration and solvent effects. *Spectrochim. Acta. Part A Mol. Biomol. Spectrosc.* **2014**, *121*, 147–151. [CrossRef]
36. Kumar, B.R.; Basheer, N.S.; Kurian, A.; George, S.D. Study of concentration-dependent quantum yield of Rhodamine 6G by gold nanoparticles using thermal-lens technique. *Appl. Phys. B* **2014**, *115*, 335–342. [CrossRef]
37. Kubin, R.F.; Fletcher, A.N. Fluorescence quantum yields of some rhodamine dyes. *J. Lumin.* **1982**, *27*, 455–462. [CrossRef]
38. Zhang, X.F.; Zhang, Y.; Liu, L. Fluorescence lifetimes and quantum yields of ten rhodamine derivatives: Structural effect on emission mechanism in different solvents. *J. Lumin.* **2014**, *145*, 448–453. [CrossRef]
39. Zakerhamidi, M.S.; Moghadam, M.; Ghanadzadeh, A.; Hosseini, S. Anisotropic and isotropic solvent effects on the dipole moment and photophysical properties of rhodamine dyes. *J. Lumin.* **2012**, *132*, 931–937. [CrossRef]
40. Amemori, S.; Kikuchi, K.; Mizuno, M. Poly(dimethylsiloxane) and oligo(dimethylsiloxane) solvent effects on aromatic donor–acceptor interactions. *Chem. Commun.* **2021**, *57*, 1141–1144. [CrossRef]
41. López Arbeloa, F.; López Arbeloa, T.; López Arbeloa, I.; Costela, A.; García-Moreno, I.; Figuera, J.; Amat-Guerri, F.; Sastre, R. Relations between photophysical and lasing properties of rhodamines in solid polymeric matrices. *Appl. Phys. B* **1997**, *64*, 651–657. [CrossRef]

materials

MDPI

Article

Elastic Anisotropy in BCC Ti-X Alloys (X = V, Nb, Ta) Determined from First Principles

Cyprian Sobczak [1,2], Piotr Kwasniak [2], Pawel Strak [1,*], Marek Muzyk [2] and Stanislaw Krukowski [1,*]

1 Institute of High Pressure Physics, Polish Academy of Sciences, Sokolowska 29/37, 01-142 Warsaw, Poland; csobczak@unipress.waw.pl
2 Multidisciplinary Research Center, Cardinal Stefan Wyszynski University in Warsaw, Dewajtis 5, 01-815 Warsaw, Poland; m.muzyk@uksw.edu.pl (M.M.)
* Correspondence: strak@unipress.waw.pl (P.S.); stach@unipress.waw.pl (S.K.)

Abstract: Elastic isotropy is a phenomenon in which a material responds uniformly to stress, regardless of its direction. In the case of cubic crystals, which possess distinct crystallographic directions, this represents a remarkable manifestation of quantum mechanics in macroscopic objects. Such behavior of a crystal cannot be explained within the framework of classical physics. The phenomenon is closely related to the balancing of internal forces resulting from Coulomb interactions, Pauli repulsion, and the overlap in the bands when stress is applied to the crystal. On the macroscopic level, this corresponds to the relationship between elastic constants given by $2\,C_{44}/(C_{11} - C_{12}) = 1$. The subject of the present work is to demonstrate the influence of the number of valence electrons per atom in binary titanium alloys with vanadium, niobium, and tantalum on the shape of the anisotropy curve. The result of the work is the identification of a new Ti-53Nb alloy exhibiting elastic isotropy, and the demonstration that this phenomenon cannot occur for TiTa alloys, in the range of mechanical stability of these alloys. This study includes a summary of the main trends exhibited by the elastic constants, Young's modulus, and bulk modulus of the discussed Ti-based alloys, based on ab initio methods. Additionally, the work addresses the well-known difficulty in determining the elastic constants of vanadium and niobium, along with a proposed solution that offers significant improvement in reproducing experimental results compared to the conventional use of the PBE (Perdew–Burke–Ernzerhof) functional.

Keywords: titanium alloys; ab initio modelling; elastic anisotropy

1. Introduction

Elastic isotropy is a property of a material manifested by an identical response to applied stress, regardless of the crystallographic direction. Technically, this means that both the Young's modulus *E* and the shear modulus *G* are the same for all crystallographic directions [1,2]. Such a unique property of materials is a kind of anomaly in crystalline structures—it occurs, among others, in tungsten and in the Ti-72V alloy [3].

Crystals with elastically isotropic properties distribute stress equally to neighboring grains during force application, which makes the material more resistant to fatigue. This enables increased durability, which is particularly important in medical (implants), structural, and aerospace applications [4,5]. Elastic isotropy gives extraordinary strength to polycrystalline metal material against cracking under thermal and mechanical stress. Despite the obvious advantages, elastic isotropy in metals remains a relatively poorly studied feature with the exception of the Tian et al. work [6]. It cannot be explained solely by electrostatic mechanisms, and in addition, it has deep quantum foundations.

The mechanical properties of a material are influenced by the following components: W_e—electrostatic interactions between ions and electrons and W_r—repulsions resulting from the overlap of electronic bands originating from individual ions [7,8]. The final and probably dominant contribution comes from subtle electronic effects related to the value of the Fermi energy and the shape of the Fermi surface. This component, denoted as W_f, has a significant influence on the shear modulus, and thus on the elastic constant C_{44} (in cubic structures), and therefore clearly depends on the number of electrons in the *d*-shell. A strong correlation has been observed between the *e/a* ratio (the number of electrons per atom) in transition metal alloys and the Zener anisotropy factor *A* [3,9]. Zener initiated considerations on the anisotropy of cubic crystals by introducing classification criteria for materials. Based on the mechanical stability condition ($C' > 0$, where $C' = (C_{11} - C_{12})/2$), he observed that for $A = C_{44}/C'$, the material exhibits elastic isotropy. Zener [9] emphasized the important role of the relationship between the response of the second and nearest neighbors in the BCC structure to changes in interatomic distance, highlighting its significance in the phenomenon of elastic isotropy (he suggested that second neighbors respond more strongly to distance changes). It is currently believed that elastic isotropy in crystals occurs when the contributions W_e, W_r, and W_f mutually compensate one another. The relation of the C′ elastic modulus to stability of BCC transition metals was the subject of the study of Fisher et al. [10].

Proceeding from extremal principles of elasticity, Ranganathan et al. [11] introduced a new universal anisotropy index. Furthermore, they established special relationships between the proposed anisotropy index and the existing anisotropy measures for special cases. A new elastic anisotropy diagram was constructed for over 100 different crystals (from cubic through triclinic), demonstrating that the proposed anisotropy measure is applicable to all types of elastic single crystals, and thus fills an important void in the existing literature. Another example of definition of an anisotropic factor was proposed by Kube [12].

The importance of the Fermi surface shape for the physical properties of metals was already considered earlier in the context of Van Hove singularities [13] and the associated Lifshitz topological transitions [14]. This phenomenon is, in essence, the deformation of the Fermi surface, resulting in its constriction. The appearance of the so-called "neck" results in a sudden change in physical properties, such as electrical resistivity or a drop in the values of elastic constants. Changes in these physical quantities can, to some extent, be controlled by influencing the Fermi level via pressure, magnetic field, or the number of electrons in the conduction band [15–19]. It is assumed that the unusual shape of the Fermi surface is the source for the change of the characteristic distribution of the *A*(*e/a*) function (where e/a is weakly bound electrons from the outer shells per atom), which takes a parabolic form, with a visible minimum of the anisotropy factor [3,20,21].

Although the condition for elastic isotropy is quite clear, it presents many challenges in the design of such alloys. Ab initio methods are very well suited to reproducing certain general trends exhibited by materials depending on the chemical composition. However, when predicting specific material properties, difficulties arise due to the influence of phonons or the temperature dependence of elastic constants. In comparisons with experimental data, contamination by unwanted elements is inevitable, which also affects the relationship between theory and measurement. In theoretical considerations of elastic isotropy in transition metals, a particularly important issue turns out to be the precise determination of the elastic constant C_{44}. In some elements, such as Mo, Ta, or W, this problem does not manifest significantly, or it can be controlled by an appropriate set of parameters that allows good agreement with experimental results. In the case of metals such as V and Nb, however, this difficulty becomes severe, leading to a reduction of the anisotropy factor by almost half. The problem of drastically underestimated C_{44} in V and Nb has been known for many years; researchers attribute its cause to the overly localized metallic bonding within the PBE framework [22–24].

In this article, we provide an ab initio modelling of titanium containing alloys with certain metals. We relate the anisotropy factor with respect to the number of valence electrons per atom (e/a) and also elastic constants, bulk modulus, and Young modulus. The main finding is that we found a new alloy, i.e., Ti-53Nb, which is isotropic. We hope that the theoretical work presented in this paper will inspire experimentalists to obtain this new alloy, which then could lead to practical applications.

2. Methodology

2.1. Computational Setup

This study was conducted within the framework of density functional theory (DFT), implemented in the VASP computational package [25–28], which utilizes the Projector Augmented-Wave method (PAW) [29]. The applied functional approximation was based on the generalized gradient approximation (GGA), using the PBE functional [30]. In the case of corrections for vanadium and niobium (the RP-Revised PBE from Hammer et al. [31]) and ML (PW86R exchange [32] + PBE correlation [30]) functionals were used, which will be discussed in more detail when addressing those specific cases.

For titanium, vanadium, and tantalum, *pv*-type PBE pseudopotentials were applied, allowing for semi-core *p*-electrons, with a plane-wave cutoff energy of 500 eV. Niobium required the use of *sv*-type pseudopotentials (PBE), which include deeper *s*-electrons, and a plane-wave cutoff energy of 600 eV. Atomic models consisting of 128 atoms ($4 \times 4 \times 4$ supercell, with periodic boundary conditions) were generated using the ATAT program via the SQS (special quasi-random structure) method [33]. The structure is created based on a statistical Monte Carlo method and the measurement of correlations between neighboring atoms, aiming to achieve maximum disorder [33]. The *k*-point sampling, in the case of alloys with niobium and tantalum, required a $4 \times 4 \times 4$ mesh, whereas for alloys with vanadium (RP), a $3 \times 3 \times 3$ mesh was used with a cutoff energy of 600 eV. While testing the nonstandard ML functional for the Ti–Nb alloy variant, it also required adjusting the plane-wave cutoff to 550 eV due to high RAM (random access memory) usage. The ground state of each structure was determined based on both volume and ionic relaxation, using the Hellmann–Feynman force convergence criterion. The relaxation was stopped when the difference in the forces acting on all atoms were less than 0.02 eV/Å. The SCF (self-consistent field) loop was stopped when energies of the two consecutive iterations were less than 2.5×10^{-6} eV. For alloys containing vanadium and niobium, volume and ionic relaxation were performed separately due to the stability of the systems within the considered composition range, which yielded results consistent with single-step relaxation. In the case of alloys containing tantalum, at very low titanium concentrations, the cells tended to tilt slightly, but without significant deviation from the BCC structure. Ti-Ta alloys required simultaneous volume and ionic relaxation to obtain a state free of internal stresses.

To calibrate the parameters for Ti, a brief test was performed on a single HCP (Hexagonal Close-Packed) unit cell using the pv pseudopotential, a $16 \times 16 \times 10$ k-point mesh with 500 eV plane-wave cutoff, and the remaining settings as specified above.

2.2. Determination of Elastic Constants

By determining the ground state of a crystal structure, we obtain information about its total internal energy (excluding phonon contributions). If the considered structure is then subjected to a small, permanent deformation and subsequently relaxed again, we can determine the energy required to induce such a deformation. In other words, we can calculate the difference between the energy of the ground state and the deformed state and relate it to the well-known Taylor series expansion of Hooke's law [34–36]:

$$E(V,\varepsilon) = E_0 + V_0\left(\sigma_{ij}\varepsilon_{ij} + \frac{1}{2}C_{ijkl}\varepsilon_{ij}\varepsilon^{kl}\right) \tag{1}$$

where σ_{ij} is the stress tensor, ε_{ij} is the strain tensor, E_0 is the ground state energy (without deformation), and V_o is the volume of the undeformed structure. C_{ijkl} is the stiffness tensor, which, in the case of a BCC structure, reduces to only three independent constants, expressed in Voigt notation as C_{11}, C_{12}, and C_{44}. The middle term in Equation (1) is neglected due to its large contribution from the first derivative with respect to strain. In this method, attention is focused on the quadratic term, which contains the factor representing the relationship with the elastic constants. Therefore, by properly selecting the deformation, the values of the elastic constants can be obtained directly from the energy. Deformation modeling is carried out by multiplying the edges of the cubic unit cell by appropriate deformation matrices (Equations (2)–(4)). The matrix generated according to this scheme forms the basis for the structure, already including the desired deformation. (Equations (2)–(4) constitute not only one valid set of deformation matrices) [35–38].

$$D_I = \begin{bmatrix} 1+\varepsilon & 0 & 0 \\ 0 & 1-\varepsilon & 0 \\ 0 & 0 & 1/(1-\varepsilon^2) \end{bmatrix} \tag{2}$$

$$D_{II} = \begin{bmatrix} 1+\varepsilon & 0 & 0 \\ 0 & 1+\varepsilon & 0 \\ 0 & 0 & 1+\varepsilon \end{bmatrix} \tag{3}$$

$$D_{III} = \begin{bmatrix} 1 & \varepsilon/2 & \varepsilon/2 \\ \varepsilon/2 & 1 & \varepsilon/2 \\ \varepsilon/2 & \varepsilon/2 & 1 \end{bmatrix} \tag{4}$$

The strain values e adopted for all considered alloys are −0.015, −0.01, −0.005, 0.005, 0.01, and 0.015. For such prepared deformations, the energy expansions presented in Equations (5)–(7) are applied, neglecting higher-order terms.

$$E_I(V,\varepsilon) = E_0 + V_0\left[(C_{11} - C_{12})\varepsilon^2 + O(\varepsilon^4)\right] \tag{5}$$

$$E_{II}(V,\varepsilon) = E_0 + V_0\varepsilon(\sigma_1 + \sigma_2 + \sigma_3) + V_0\left[\frac{3}{2}(C_{11} + 2C_{12})\varepsilon^2 + O\left(\varepsilon^4\right)\right] \tag{6}$$

$$E_{III}(V,\varepsilon) = E_0 + V_0\left[\frac{3}{2}C_{44}\varepsilon^2 + O\left(\varepsilon^4\right)\right] \tag{7}$$

Based on the matrices and Equations (2)–(7), three sets of energy versus deformation functions are constructed, each exhibiting a parabolic distribution. The slope coefficients are represented by the elastic constants, multiplied by appropriate factors depending on the applied deformations. After determining the constants, each structure is rotated, and the procedure is repeated using different model orientation. Such approach minimizes an influence of possible atomic segregation on the evaluation of mechanical properties. Based on the thus-determined elastic constants, the bulk modulus B and Young's modulus E are calculated using the well-known Equations (8) and (9) [35].

$$B = \frac{1}{3}(\overline{C}_{11} + 2\overline{C}_{12}) \tag{8}$$

$$E = \frac{1}{2}\left[\frac{9}{1/B + 15/(\overline{C}_{11} - \overline{C}_{12} + 3\overline{C}_{44})} + \frac{5}{5/9B + 4/3(\overline{C}_{11} - \overline{C}_{12}) + 1/\overline{C}_{44}}\right] \tag{9}$$

where the symbols C denote the average values of the elastic constants obtained from each of the three spatial orientations of the model.

3. Results and Discussion

The previously mentioned issues related to performing calculations using the PBE functional for pure V and Nb cannot be resolved by appropriately adjusting the parameters. As a measure of last resort, the possibility of forcing the use of other functionals in the VASP computational package functionals incompatible with the PBE pseudopotentials was employed. The results of this workaround turned out to be surprisingly close to the experimental data. Due to the fact that such a method of performing calculations may raise concerns, the anisotropy curves for the TiV and TiNb alloys were also determined using the conventional method with PBE. Table 1 presents a summary of the results for the elastic constants, elastic moduli, and the anisotropy factor for the pure elements.

Table 1. The comparison of the results obtained in this work using the PBE functional, as well as RP for vanadium and ML for niobium, with experimental data and ab initio results reported by other researchers [39].

Metal	Source	Functional	B [GPa]	G [GPa]	E [GPa]	C_{11} [GPa]	C_{12} [GPa]	C_{44} [GPa]	A
V	This work	PBE	192	50	137	258	139	18	0.3
V	this work	RP	174	54	147	268	127	43	0.61
V	Ab initio [39]	PBE	183	41	114	270	140	25	0.38
V	Exp. [39]	-	156	48	130	229	119	43	0.78
Nb	This work	PBE	169	31	87	243	132	15	0.26
Nb	This work	ML	163	35	99	234	128	24	0.45
Nb	Ab initio [39]	PBE	172	32	91	247	134	16	0.28
Nb	Exp. [39]	-	171	40	111	246	134	29	0.52
Ta	This work	PBE	195	61	166	259	163	70	1.46
Ta	ab initio [39]	PBE	190	63	170	257	156	71	1.41
Ta	Exp. [39]	-	192	70	187	261	157	82	1.58
W	This work	PBE	300	127	335	474	214	126	0.97
W	Ab initio [39]	PBE	304	156	400	527	192	149	0.89
W	Exp. [39]	-	314	163	418	533	205	163	0.99

Titanium in its pure form does not adopt the BCC structure below 882 °C. Consequently, attempts to model the elastic constants of BCC Ti using standard DFT (which assumes 0 K) are unsuccessful. Such attempts yield results $E < 0$ and $A < 0$, yet we can find these values in the literature [40,41]. An additional complication is the lack of experimental elastic-constant data for BCC Ti. For this reason, we selected to test parameters on a single HCP unit cell—the stable structure of Ti below 882 °C—and to compare the resulting elastic constants with experimental data. The results summarized in Table 2 indicate a close similarity between the findings of this study, Chihi et al. ab inito modelling [42] and the experimental data for HCP Ti [43].

Table 2. Summary of the ab initio results from this study, those reported in [42], and the experimental values obtained for pure HCP Ti at 4 K [43].

Metal	Source	Functional	C_{11} [GPa]	C_{12} [GPa]	C_{33} [GPa]	C_{13} [GPa]	C_{44} [GPa]
Ti	This work	PBE	176	84	201	74	46
Ti	Ab initio [42]	PBE	180	53	169	65	42
Ti	Exp. [43]	-	176	94	191	68	51

3.1. TiV Alloys

As shown in Figure 1, the plot presenting the results for TiV alloys can be reasonably well approximated by a second-degree polynomial. The points obtained using PBE (red curve) show very little scatter around the curve; a slightly larger deviation is observed for the curve calculated with the RP functional [38] (orange curve), which, on the other hand, reproduces the properties of pure vanadium much better. In the case of the RP method, the anisotropy factor for pure vanadium is 0.61, whereas for PBE it is 0.3. In reference [18], using molecular dynamics methods, researchers achieved significantly better results than those from PBE, yet still far from the experimental values, in which the anisotropy factor is 0.78. The brown points represent the results of experimental measurements for TiV alloys. The blue line is an approximation of experimental results, but it was not derived from any specific alloy; instead, it refers to various transition metal alloys, including the elastically isotropic Ti-72V alloy. The PBE line captures the slope of the distribution well; however, despite the interesting fact that the PBE line almost intersects the RP line exactly at the point corresponding to elastic isotropy for *e/a* = 4.45, this is not the *e/a* value at which the phenomenon is actually observed in experiments. What should be emphasized is that the authors of [41] obtained an accurate result for an elastically isotropic alloy using ab initio methods; however, they employed the virtual crystal approximation (VCA) to generate structures, rather than SQS as in this work (A = 1, for Ti-73V). Despite the precise agreement between the value predicted in [41] and the experimental results for the Ti–72V alloy, pure vanadium still lies beyond the method's reach (according to [41], the anisotropy factor for vanadium is 0.37). Both methods (RP and PBE), however, confirm the existence of a point with elastically isotropic properties. Both curves determined in the present work allow for observation of the parabolic character of the distribution, resulting from the minimum of A occurring at e/a = 4.91.

3.2. TiNb Alloys

In the case of TiNb alloys, once again, a much better approximation for pure niobium is provided by the ML functional (yellow curve in Figure 2). It allows for obtaining an anisotropy factor of 0.45, while the result from PBE is A = 0.26 (green line), with the experimentally determined value being 0.52. For TiNb alloys, no experimental results are found in the literature; therefore, the results were compared with theoretical studies [45,46] (olive line). The results turned out to be very similar for lower *e/a* values, particularly in the region of expected elastic isotropy. According to reference [45], the value A = 1 is obtained near *e/a* = 4.51; for PBE in that study, it is *e/a* = 4.43, and using ML it is *e/a* = 4.45.

All predictions confirm the existence of elastic isotropy. Analysis of the anisotropy curve for Ti–Nb, as well as for Ti–V, was also undertaken in [47], but the Ti–Nb alloys considered there lie in the e/a range of 4.2–4.4, whereas the present work examines the anisotropy behavior in the range e/a = 4.5–5.0. As emphasized in Refs. [20,23,24], properly determining the elastic constants for pure niobium is an extremely problematic task. For the ML functional calculations are time-consuming and numerically unstable; in several cases, they even required removal of some points on the strain parabola due to large deviations. Nevertheless, the good approximation of results on this functional for niobium-based alloys is confirmed. For TiNb alloys, as in the case of TiV, the molecular dynamics methods were applied [46]. The results obtained in the present work are not consistent with those from Refs. [48,49]. However, they are close to the results reported in Ref. [40]. A reliable modeling of the behavior of titanium–niobium-based alloys can serve as a valuable direction in the development of these alloys in the context of shape memory alloys [48,49]. Figure 2 presents a comparison of lines obtained using PBE (green line) and ML (yellow line), as well as the line obtained in Ref. [45], that was also determined using ab initio methods. The illustration shows, as in the case of vanadium, the lines obtained for transition metals and the experimental point that were determined only for pure niobium. Results in Figure 2 prove that the minimum of the anisotropy curve for TiNb alloys is obtained when using the ML functional at e/a = 4.91. In the case of PBE, the regression curve also indicates the presence of a minimum, although it is not highlighted by a specific point.

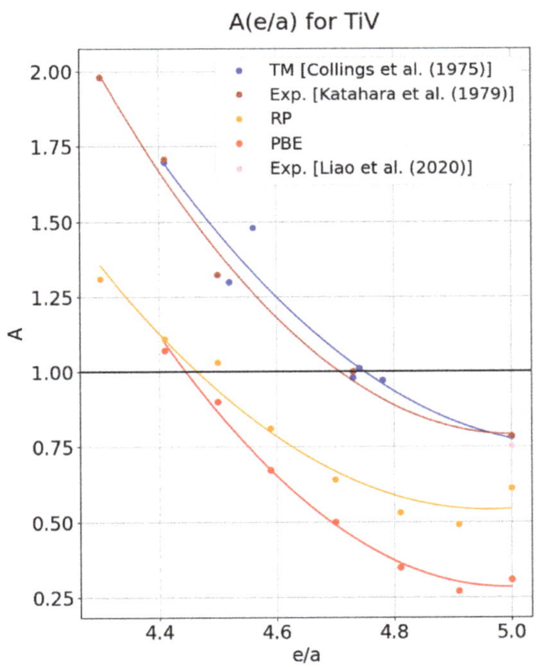

Figure 1. Relation of the anisotropy factor with respect to the number of valence electrons per atom (e/a), determined experimentally for transition metals (blue line) [3], and experimental results for vanadium (pink point) [41], as well as for TiV alloys determined experimentally (brown line) [44], using the RP functional (orange line), and using PBE (red line).

Similar analyses of elastic isotropy were reported in [51], but over a composition (concentration) range that precludes a meaningful comparison (%at.: 10–45).

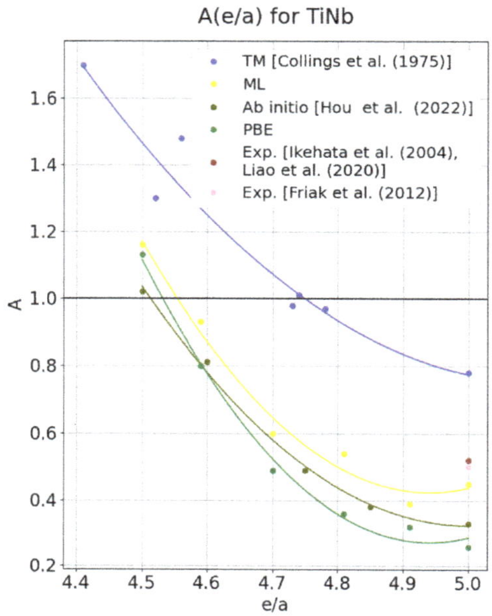

Figure 2. Relation of the anisotropy factor with respect to the number of valence electrons per atom (*e/a*) determined experimentally for transition metals (blue curve) [3], and for TiNb alloys determined using the ML functional (yellow line), PBE (green curve), and ab initio by another author (olive line) [45]. The brown and pink point represents the experimentally determined value for pure Nb [39,41,50].

3.3. *TiTa Alloys*

The anisotropy curve for Ti–Ta alloys exhibits a very high scatter of data points around the regression curve. This issue is not limited to the curve obtained in the present work using PBE; it is also visible in other authors' results (Refs. [45,52,53]). The curves shown in Figure 3 display a fairly similar character, decreasing gently with increasing e/a. The high degree of similarity among the anisotropy curves reported by other authors influenced the decision in this work to discard the point corresponding to e/a = 4.5, where the curve took the value A = 1.53. Including this point drastically affected the course of the curve, strongly flattening it. A likely reason for such a strong deviation from the trend line for the remaining points is the effect of atomic arrangement within the cell. In mixtures of elements such as Ti and V, the atomic distribution in the cell has a less drastic impact on the alloy's mechanical properties. In the case of Ti and Ta, differences in atomic size may prove significant. Supporting this argument is the observation that for low Ti contents in Ti–Ta, all curves lie close to each other, which changes markedly as e/a decreases, i.e., as the Ti concentration increases. The size of the cell considered also appears to have a significant effect. In Refs. [45,53], the calculations were carried out for 16-atom cells, which, under periodic boundary conditions, inevitably impose a strong ordering of the lattice. The study in Ref. [52] used 54 atoms, whereas the present work uses 128.

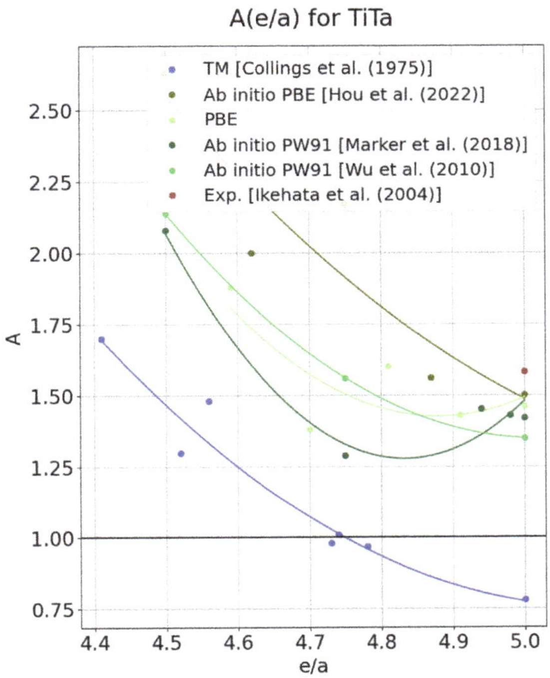

Figure 3. The anisotropy factor in function of the number of valence electrons per atom (*e/a*), determined experimentally for transition metals (blue points) [3] and for TiTa alloys calculated using the PBE functional (dark green [52], light green line [53]), and by ab initio calculations in Ref. [45] (olive line). The brown point represents the experimentally determined value for pure Ta [39].

Despite the use of the PW91 functional in Refs. [52,53], the anisotropy curves do not show particularly strong similarity. We therefore conclude that the size difference between Ti and Ta unexpectedly increases the dependence on the type of atomic mixing in the supercell model. There is, however, no doubt that within the e/a range considered here, no elastically isotropic alloy is observed; moreover, judging from the character of the curves, no binary Ti–Ta alloy exhibits elastic isotropy. Although two of the four curves appear to possess a minimum in the anisotropy, the analysis of the elastic constants in Figure 3 does not indicate anomalies that could be associated with a van Hove effect.

3.4. Comparison of Anisotropy Curves for Ti-X Alloys (X = V, Nb, Ta, Mo)

Figure 4 presents a comparison of all lines considered, in addition to the violet line obtained in Ref. [25], representing TiMo alloys investigated using the same method. It is plain to see that the presented lines differ among themselves, particularly in the case of TiNb alloys, which appear to intersect the lines for TiV and TiMo. The TiTa line exhibits a completely different character than the other ones, not reaching the value $A = 1$ at any point. It should therefore be assumed that the trends determined using PBE indicate more complex relationships for the anisotropy factor than those merely based on the number of electrons per atom. Otherwise, one would expect the possibility that these lines would be by a simple translation by a constant value.

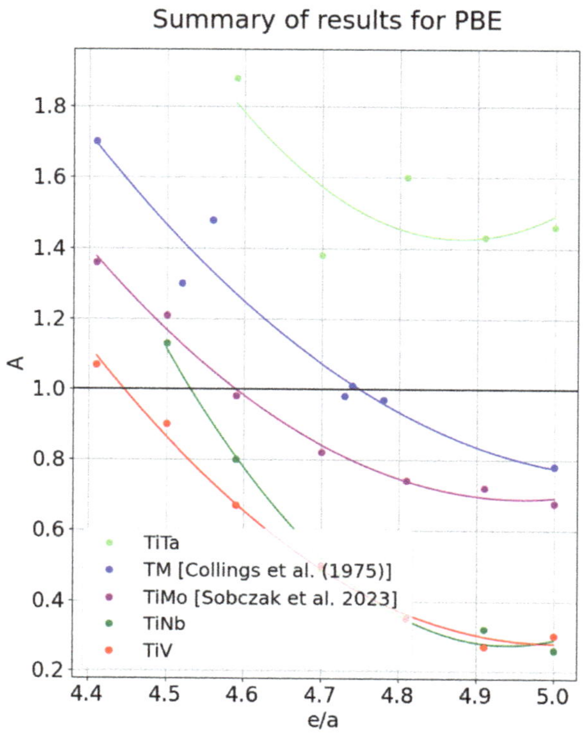

Figure 4. Comparison of anisotropy determined in this work for TiV alloys (red line); TiNb (green line); TiTa (light green line), and the results for TiMo alloy (violet line) obtained in Ref. [25]. The blue line represents the approximation for transition metals [3].

Based on the presented results, the conjecture in Ref. [3]—that the dependence of A on e/a has a decisive influence on the distribution of points along the curve—can therefore be ruled out. This, in turn, shows that the attempt to describe all transition metal alloys by approximating them with a single anisotropy curve, as done in Ref. [3], is insufficient.

3.5. General Trends in Ti-X Alloys (X = V, Nb, Ta, Mo)

The trends exhibited by titanium alloys with the considered transition metals as a function of the *e/a* ratio are presented in Figure 5. An increase of almost all elastic constants with the *e/a* ratio is noticeable.

An exception is provided by the Ti–V and Ti–Nb alloys, for which a decrease in the C44 constant with increasing e/a is evident. In the Ti–V alloy, a minimum of C44 is observed near $e/a \approx 4.91$, after which it increases slightly at $e/a = 5$. In view of the side effects noted in the introduction on the van Hove effect—such as unexpected changes in mechanical properties—one may surmise that at $e/a = 4.91$ when the Fermi surface assumes an unusual shape. Although the difference in the behavior of C44 is visible, it is not dramatic.

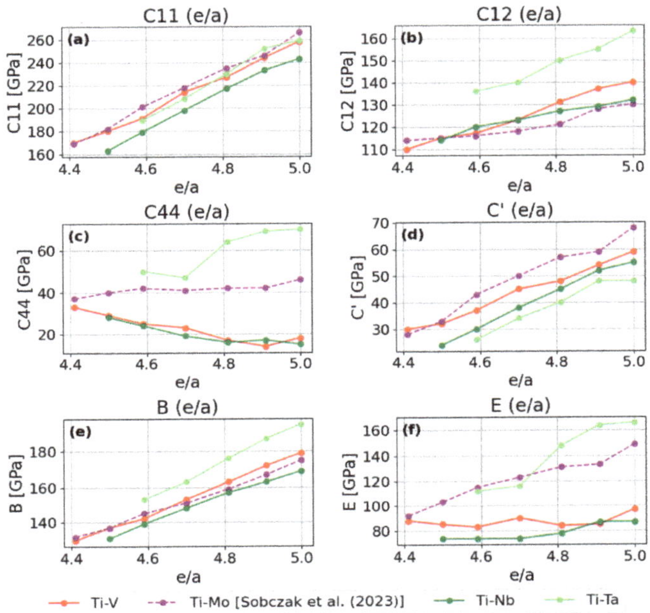

Figure 5. A comparison of trends observed in Ti-X alloys (X = V, Nb, Ta, Mo [25]): (**a**) C_{11} elastic constant, (**b**) C_{12} elastic constant, (**c**) C_{44} elastic constant, (**d**) $C' = (C_{11} - C_{12})/2$, (**e**) bulk modulus B, and (**f**) Young's modulus E, determined using PBE.

For Ti–Nb, the C44(e/a) trend appears to reach its lowest point at pure niobium, without pronounced deviations. Based on the set of elastic constants, one cannot unambiguously identify a point that would suggest the presence of a van Hove effect known to occur in niobium under pressure. In the case of the Ti–Mo alloy, the C44(e/a) curve does not suggest any anomalous effects. The slope of the curve does change, but only very slightly. For Ti–Ta, although there is a point that differs from the others in the C44(e/a) profile, it appears quite early and therefore cannot be associated with the minimum in the anisotropy curve. Moreover, as noted in the section on Ti–Ta alloys, at high titanium contents (i.e., lower e/a) these alloys are more sensitive to the atomic mixing configuration, which manifests under shear deformation of the cell. For this reason, based on the elastic constants alone, one cannot conclude that an anomaly is present in Ti–Ta.

To establish whether a deformation of the Fermi surface is present, density-of-states (DOS) curves must be determined.

The bulk modulus B for all of the alloys considered shows the same increasing trend, with comparable linear slopes. The Young's modulus E likewise increases for the Ti–Ta and Ti–Mo alloys. For Ti–V alloys, E falls in the range 83–97 [GPa], whereas for Ti–Nb alloys, it lies in the range 74–87 [GPa].

Table 3 compiles binary Ti alloys predicted to be elastically isotropic, determined theoretically and compared with experimental results where possible. The immediately apparent discrepancy—namely, the experiment indicating elastic isotropy for Ti–72V versus the markedly different prediction for Ti–45V—can be attributed to the fact that even approximating the anisotropy factor for pure vanadium carries substantial uncertainty. In the section devoted to Ti–V alloys, we explained the considerable effort made to approximate as accurately as possible the mechanical properties determined for V, which ultimately did not yield the desired outcome. The method employed in the present work also produces a correct result for W, for which A = 0.97, very close to the A = 1 value for which tungsten is well known. For the Ti–53Nb and Ti–30Mo [25] alloys, the theoretical predictions cannot be directly confronted with measurements. Nevertheless, the Ti–53Nb alloy is particularly promising and awaits experimental confirmation of its elastic isotropy.

Table 3. The known isotropically elastic alloys, and the results determined by PBE.

Alloy	e/a Exp.	% at. Exp.	e/a PBE	% at. PBE	A Exp.	A PBE
TiV	4.72 [3]	Ti-72V	4.44	Ti-45V	1 [3]	1
TiNb	-	-	4.53	Ti-53Nb	-	1
TiMo	-	-	4.59 [25]	Ti-30Mo	-	0.98 [25]
W	6 [39]	W	6	W	1	0.97

4. Summary and Conclusions

One benefit of the present work was to highlight the relationship between the anisotropy factor and the ratio of weakly bound valence electrons per atom (e/a) in binary titanium alloys with transition metals. This relationship is evident—though it differs from alloy to alloy—which provides an important guideline for further studies on Ti–transition-metal alloys. Moreover, as a byproduct, we attempted to detect an anomaly in the C44 constant as a function of e/a, which could indicate a van Hove effect associated with a stress- or electron-density-induced deformation of the Fermi surface. Such an effect should manifest as an atypical drop in C44 in the alloy under study. This type of observation was made only for vanadium, where a slight decrease of C44 was recorded at e/a = 4.91. Further development of the discussion of anomalies related to the Fermi surface shape will require presenting density-of-states (DOS) curves, and a definitive resolution will demand appropriate experiments that map the Fermi surface.

The main goal of the work was to determine—or ascertain the existence of—compositions with elastically isotropic properties (A(e/a) = 1) in Ti–Nb and Ti–Ta alloys. We also performed tests intended to reproduce, theoretically, known isotropic cases such as Ti–72V and W. While with the present method, the agreement between prediction and experiment was confirmed for tungsten, the theoretical confirmation of the Ti–72V composition in Ti–V was not successful. The composition indicated by the theoretical approach was Ti–45V, which we attribute to the difficulty of correctly determining the anisotropy factor for pure vanadium. It is well known that PBE does not accurately capture all bonding effects in this element.

Using the method presented here, we established that an elastically isotropic Ti–Ta alloy does not exist. We also identified an elastically isotropic Ti–53Nb alloy. Because of the presence of unusual electronic effects in pure niobium (though not as pronounced as in vanadium), other functionals were also tested. The ML functional, which gives a good approximation of the anisotropy factor for pure Nb, when applied to Ti–Nb yielded nearly the same result, indicating an isotropic composition at e/a = 4.55. Obtaining nearly identical compositions with two different functionals supports the reliability of this prediction. To determine unambiguously whether Ti–53Nb is indeed elastically isotropic, experimental verification will be required.

A further research perspective in this field involves determining analogous curves for hafnium- and zirconium-based alloys with early transition metals, thereby providing a complete picture of the character of anisotropy curves. Advancing knowledge of this unusual phenomenon may, in the future, lead to the development of structural materials with greater fatigue resistance, enabling industry—and even the construction sector—to save substantial costs otherwise spent on renovation and maintenance.

Author Contributions: Conceptualization, P.K. and P.S.; methodology, C.S. and P.K.; software, C.S.; validation, C.S., P.K. and P.S.; formal analysis, C.S.; investigation, C.S.; resources, C.S.; data curation, C.S. and P.S.; writing—original draft preparation, C.S., P.S. and S.K.; writing—review and editing, C.S., P.S. and S.K.; visualization, C.S.; supervision, M.M. and S.K.; project administration, M.M. All authors have read and agreed to the published version of the manuscript.

Funding: This research received no external funding.

Data Availability Statement: The original contributions presented in this study are included in the article. Further inquiries can be directed to the corresponding authors.

Acknowledgments: This research was carried out with the support of the Interdisciplinary Centre for Mathematical and Computational Modelling at the University of Warsaw (ICM UW) under grant no. GB99-2112 and also of the Multidisciplinary Research Centre, Cardinal Stefan Wyszynski University in Warsaw, Dewajtis 5, 01-815 Warsaw, Poland for the use of the (CATO) supercomputer.

Conflicts of Interest: The authors have no conflicts to disclose.

References

1. Meyers, M.; Chawla, K. *Mechanical Behavior of Materials*; Cambridge University Press: Cambridge, UK, 2009.
2. Zener, C. Contributions to the theory of beta-phase alloys. *Phys. Rev.* **1947**, *71*, 846–851. [CrossRef]
3. Collings, E.; Gegel, H. *Physic of Solid Solution Strengthening*; Plenum Press: New York, NY, USA, 1975.
4. Kwasniak, P.; Wróbel, J.; Garbacz, H. Origin of low Young modulus of multicom-ponent, biomedical Ti alloys-Seeking optimal elastic properties through a first principles investigation. *J. Mech. Behav. Biomed. Mater.* **2018**, *88*, 352–361. [CrossRef]
5. Giroud, T.; Villechaise, P.; Naït-Ali, A.; Mellier, D.; Hémery, S. Anisotropy in tensile properties of high strength metastable β titanum alloys. *Mater. Des.* **2024**, *247*, 113401. [CrossRef]
6. Tian, F.; Varga, L.K.; Chen, N.; Shen, J.; Vitos, L. Ab initio design of elastically isotropic TiZrNbMoVx high-entropy alloys. *J. Alloys Compd.* **2014**, *599*, 19–25. [CrossRef]
7. Fuchs, K. The elastic constants and specific heats of the alkali metals. *Proc. R. Soc. London Ser. A-Math. Phys. Sci.* **1936**, *157*, 444–450. [CrossRef]
8. Fuchs, K.; Fowler, R.H. A quantum mechanical investigation of the cohesive forces of metallic copper. *Proc. R. Soc. London Ser. A-Math. Phys. Sci.* **1935**, *151*, 585–602. [CrossRef]
9. Zener, C. *Elasticity and Anelasticity*; University of Chicago Press: Chicago, IL, USA, 1948.
10. Fisher, E.; Dever, D. Relation of the C′ elastic modulus to stability of B.C.C. transition metals. *Acta Met.* **1970**, *18*, 265–269. [CrossRef]
11. Ranganathan, S.I.; Ostoja-Starzewski, M. Universal Elastic Anisotropy Index. *Phys. Rev. Lett.* **2008**, *101*, 055504. [CrossRef]
12. Kube, C.M. Elastic anisotropy of crystals. *AIP Adv.* **2016**, *6*, 095209. [CrossRef]
13. Van Hove, L. The occurrence of singularities in the elastic frequency distribution of crystal. *Phys. Rev. B* **1953**, *89*, 1189. [CrossRef]
14. Lifshitz, M. Anomalies of electron characteristic of a metal in the high pressure region. *Sov. Phys. JETP* **1960**, *11*, 1130–1135.
15. Reiter, F.; Marmodoro, A.; Mardare, A.I.; Mardare, C.C.; Hassel, A.W.; Ernst, A.; Hoffmann, M. Revisiting electronic topological transitions in the silver-palladium (Ag_cPd_{1-c}) solid solution: An experimental and theoretical investigation. *Materials* **2024**, *17*, 2743. [CrossRef]
16. Sun, W.; Luo, W.; Feng, Q.; Ahuja, R. Anisotropic distortion and Lifshitz transition in α-Hf under pressure. *Phys. Rev. B* **2017**, *95*, 115130. [CrossRef]
17. Koudela, D.; Richter, M.; Möbius, A.; Koepernik, K.; Eschrig, H. Lifshitz transiyions and elastic properties of osmium under pressure. *Phys. Rev. B* **2006**, *74*, 214103. [CrossRef]
18. Weck, P.F.; Townsend, J.P.; Cochrane, K.R.; Crockett, S.D.; Moore, N.W. Shock compression of niobium from first-principles. *J. Appl. Phys.* **2019**, *125*, 245905. [CrossRef]
19. Wang, Y.X.; Geng, H.Y.; Wu, Q.; Chen, X.R.; Sun, Y. First-Principles investigation of elastic anomalies in niobium at high pressure and temperature. *J. Appl. Phys.* **2017**, *122*, 235903. [CrossRef]

20. Yan, G.-Q.; Cheng, X.-L.; Zhang, H. Kohn anomaly and van Hove singularity in IV^B and V^B group transition metals nitrides and carbides. *J. Phys. Condens. Matter* **2019**, *32*, 075401. [CrossRef] [PubMed]
21. Delaire, O. The Phonon Entropy of Transition Metals and Alloys: Effects of Impurities and of a Martensitic Phase Transition. Ph.D. Thesis, California Institute of Technology, Pasadena, CA, USA, 2006.
22. Skripnyak, N.; Ponomareva, A.; Belov, M.; Abrikosov, I. Ab initio calculations of elastic properties of alloys with mechanical instability Application to BCC Ti-V alloys. *Mater. Des.* **2018**, *140*, 357–365. [CrossRef]
23. Wang, Y.X.; Geng, H.Y.; Wu, Q.; Chen, X.R. Orbital localization error of density functional theory in shear properties of vanadium and niobium. *J. Chem. Phys.* **2020**, *152*, 024118. [CrossRef]
24. Wang, Y.; Wu, H.; Liu, Y.; Wang, H.; Chen, X.; Geng, H. Recent progress in phase stability and elastic anomalies of group VB transition metals. *Crystals* **2022**, *12*, 1762. [CrossRef]
25. Sobczak, C.; Kwasniak, P.; Muzyk, M.; Strak, P.; Prima, F. Introduction to elastically isotropic β- Ti alloys. *Comput. Mater. Sci.* **2023**, *230*, 112496. [CrossRef]
26. Kresse, G.; Hafner, J. Ab initio molecular dynamics for liquid metals. *Phys. Rev. B* **1993**, *47*, 558–561. [CrossRef]
27. Kresse, G.; Furthmüller, J. Efficient iterative schemes for ab initio total-energy calculations using a plane-wave basis set. *J. Phys. Rev. B* **1996**, *54*, 11169–11186. [CrossRef]
28. Vienna Ab Initio Simulation Package. Available online: https://www.vasp.at (accessed on 1 September 2025).
29. Blöchl, P.E. Projector augmented-wave method. *Phys. Rev. B* **1994**, *50*, 17953–17979. [CrossRef]
30. Perdew, J.P.; Burke, K.; Ernzerhof, M. Generalized Gradient Approximation Made Simple. *Phys. Rev. Lett.* **1996**, *77*, 3865–3868. [CrossRef] [PubMed]
31. Hammer, B.; Hansen, L.B.; Nørskov, J.K. Improved adsorption energetics within density-functional theory using revised Perdew-Burke-Ernzerhof functionals. *Phys. Rev. B* **1999**, *59*, 7413. [CrossRef]
32. Lee, K.; Murray, É.D.; Kong, L.; Lundqvist, B.I.; Langreth, D.C. Higher-accuracy van der Waals density functional. *Phys. Rev. B* **2010**, *82*, 081101. [CrossRef]
33. Zunger, A.; Wei, S.-H.; Ferreira, L.G.; Bernard, J.E. Special quasirandom structures. *Phys. Rev. Lett.* **1990**, *65*, 353–356. [CrossRef]
34. Mehl, M.J. Pressure dependence of the elastic moduli in aluminum-rich Al-Li compounds. *Phys. Rev. B* **1993**, *47*, 2493–2500. [CrossRef] [PubMed]
35. Jamal, M.; Asadabadi, S.J.; Ahmad, I.; Aliabad, H.A.R. Elastic constants of cubic crystals, Comput. *Comput. Mater. Sci.* **2014**, *95*, 592–599. [CrossRef]
36. Mishra, P.; Kumar, P.; Neelakantan, L.; Adlakha, I. First-principles prediction of electrochemical polarization and mechanical behavior in Mg based intermetallics. *omput. Mater. Sci.* **2022**, *214*, 111667. [CrossRef]
37. Kumar, P.; Adlakha, I. Effect of Interstitial Hydrogen on Elastic Behavior of Metals: An Ab-Initio Study. *J. Eng. Mater. Technol.* **2023**, *145*, 011003. [CrossRef]
38. Zhou, W.; Souissi, M.; Abe, T.; Sahara, R.; Sit, P.H.-L.; Tsuchiya, K. Evaluating the phase stability of binary titanum alloys Ti-X(X=Mo, Nb, Al and Zr) using first-principles calculations and Debye model. *Calphad* **2020**, *71*, 102207. [CrossRef]
39. Ikehata, H.; Nagasako, N.; Furuta, T.; Fukumoto, A.; Miwa, K.; Saito, T. First-principles calculations for development of low elastic modulus Ti alloys. *Phys. Rev. B* **2004**, *70*, 174113. [CrossRef]
40. Marker, C.; Shang, S.L.; Zhao, J.C.; Liu, Z.K. Elastic knowledge base of bcc Ti alloys from first-principles calculations and CALPHAD-based modeling. *CMS* **2017**, *140*, 121–139. [CrossRef]
41. Liao, M.; Liu, Y.; Cui, P.; Qu, N.; Zhou, F.; Yang, D.; Han, T.; Lai, Z.; Zhu, J. Modeling of alloying effect on elastic properties in BCC Nb-Ti-V-Zr solid solution: From unary to quaternary. *CMS* **2020**, *172*, 109289. [CrossRef]
42. Chihi, T.; Fatmi, M.; Ghebouli, B.; Gebouli, M.A. Ab initio study of the parent (BCC) and martensitic (HCP) phases of nonferrous Ti, Zr, and HF metals. *Chin. J. Phys.* **2016**, *54*, 127–134. [CrossRef]
43. Joost, W.J.; Ankem, S.; Kuklja, M.M. Amodified embedded atom method potential for the titanum-oxygen system. *Model. Simul. Mater. Sci. Eng.* **2015**, *23*, 015006. [CrossRef]
44. Katahara, K.W.; Manghnani, M.H.; Fisher, E.S. Pressure derivatives of the elastic moduli of BCC Ti-V-Cr, Nb-Mo and Ta-W alloys. *J. Phys. F Met. Phys.* **1979**, *9*, 773. [CrossRef]
45. Hou, S.; Li, J.; Wang, Y.; Yang, D.; Wan, Z. Study on stability and elastic properties of β-TiX (X=Nb, Ta) alloys from first-principles calculation. *Sec. Biomater.* **2022**, *9*, 932007. [CrossRef]
46. Wang, H.; Huang, B.; Hu, W.; Huang, J. Studying plastic deformation mechanism in β-Ti-Nb, alloys by molecular dynamic simulations. *Metals* **2024**, *14*, 318. [CrossRef]
47. Wang, P.; Todai, M.; Nakano, T. Beta titanum single crystal with bone-like elastic modulus and large crystallographic elastic anisotropy. *J. Alloys Compd.* **2019**, *782*, 667–671. [CrossRef]

48. Golasiński, K.M.; Maj, M.; Urbański, L.; Staszczak, M.; Gradys, A.; Pieczyska, E.A. Experimental study of thermomechanical behaviour of Gum Metal during cyclic tensile loadings: The quantitative contribution of IRT and DIC. *Quant. Infrared Thermogr. J.* **2022**, *21*, 259–276. [CrossRef]
49. Golasiński, K.M.; Maj, M.; Tasaki, W.; Pieczyska, E.A.; Kim, H.Y. Full-Field Deformation Study of Ti–25Nb, Ti–25Nb–0.3O and Ti–25Nb–0.7O Shape Memory Alloys During Tension Using Digital Image Correlation. *Met. Mater. Trans. A* **2024**, *55*, 2509–2518. [CrossRef]
50. Friák, M.; Counts, W.A.; Ma, D.; Sander, B.; Holec, D.; Raabe, D.; Neugebauer, J. Theory-Guided materials desing of multi-Phase TiNb alloys with Bone-matching elastic properties. *Materials* **2012**, *5*, 1853–1872. [CrossRef]
51. Chakraborty, T.; Rogal, J. Trends in elastic properties of Ti–Ta alloys from first-principles calculations. *J. Phys. Condens. Matter* **2021**, *33*, 035701. [CrossRef]
52. Marker, C.; Shang, S.-L.; Zhao, J.-C.; Liu, Z.-K. Effects of alloying elements on the elastic properties od bcc Ti-X alloys, from first-principles calculation. *Comput. Mater. Sci.* **2018**, *142*, 215–226. [CrossRef]
53. Wu, C.; Xin, Y.; Wang, X.; Lin, J. Effects of Ta content on the phase stability and elastic properties of β Ti-Ta alloys from first principles calculation. *Solid State Sci.* **2010**, *12*, 2120–2124. [CrossRef]

materials

MDPI

Review

Electrocaloric Effect on Lead-Free Ferroelectrics: Challenges in Identifying Trends and Evaluating Predictive Models

Magdalena Krupska-Klimczak [1,2,*], Michał Frontczak [1], Zdobysław Świerczyński [1], Serhii Semenov [1], Dariusz Kajewski [3] and Irena Jankowska-Sumara [2]

[1] Institute of Security and Computer Science, University of National Education Commission, Podchorążych 2, 30-084 Kraków, Poland; michal.frontczak1@uken.krakow.pl (M.F.); serhii.semenov@uken.krakow.pl (S.S.)

[2] Faculty of Exact and Natural Sciences, University of National Education Commission, Podchorążych 2, 30-084 Kraków, Poland; irena.jankowska-sumara@uken.krakow.pl

[3] Institute of Physics, University of Silesia in Katowice, ul. 75 Pułku Piechoty 1, 41-500 Chorzów, Poland; dariusz.kajewski@us.edu.pl

* Correspondence: magdalena.krupska-klimczak@uken.krakow.pl

Abstract: The electrocaloric effect (ECE) has become one of the most intensively studied topics in ferroelectrics, with dozens of new papers that report experimental results and provide increasingly extensive data compilations every year. However, the heterogeneity of the literature, arising from differences in compositions, dopants, preparation routes, measurement protocols, and analysis methods, makes direct comparison between studies highly problematic. In this work, we focus on barium titanate ($BaTiO_3$) as a representative lead-free ferroelectric system. $BaTiO_3$ was chosen because, within this class of materials, it offers by far the largest body of reported ECE results, obtained under a wide range of experimental conditions, thus allowing for the most comprehensive characterization. Using this example, we explore whether meaningful patterns related to the influence of chemical substitution on the magnitude and temperature dependence of the ECE can be discerned. In addition, we critically examine why certain comparisons reported in the literature may be misleading or inherently unreliable. Finally, we discuss predictive approaches, including those employing artificial intelligence algorithms, and evaluate their applicability and limitations in modeling the electrocaloric response.

Keywords: electrocaloric effect; lead-free ceramics; machine learning

1. Introduction

The electrocaloric effect (ECE) has attracted significant interest over recent decades due to its potential for enabling efficient, environmentally friendly solid-state cooling technologies [1]. This phenomenon, which involves reversible temperature changes in certain materials upon application or removal of an electric field, is seen by many researchers as a promising alternative to traditional refrigeration methods that rely on greenhouse gases [2]. However, despite the enthusiasm surrounding ECE, the field remains controversial. On one hand, the vast potential of ECE is widely acknowledged, with numerous reports highlighting large temperature changes in specific materials under certain conditions [3]. On the other hand, no practical electrocaloric (EC) material or device has yet been developed that consistently delivers the performance required for widespread technological adoption [4]. This contradiction underscores the complexities inherent in understanding and harnessing the ECE.

Initial research efforts into ECE largely focused on individual experimental measurements conducted on select materials, often under narrowly defined conditions [5,6]. These studies provided valuable, detailed insights into the fundamental properties and behaviors of candidate EC materials, contributing to a growing but fragmented knowledge base. However, isolated data points alone are insufficient to comprehend the underlying mechanisms governing ECE fully or to guide the rational design of improved materials. As research in this field expanded, numerous experimental results accumulated in the literature, spanning different material compositions, processing methods, and measurement techniques [1,7,8]. Such a wealth of data should, in principle, facilitate the identification of consistent trends and empirical relationships critical for advancing ECE technology.

Yet, the reality is more complicated. The literature on EC materials is characterized by considerable variability and a lack of systematic organization. Different synthesis and experimental protocols, diverse material systems, and inconsistent reporting standards contribute to a highly heterogeneous dataset. This absence of uniformity makes it challenging to extract clear correlations or to construct predictive models based purely on traditional data analysis. In particular, studies on barium titanate $BaTiO_3$-based ceramics, which are among the most extensively researched EC materials, exemplify this issue. Results for $BaTiO_3$ ceramics display significant scatter, with reported values for key EC parameters differing widely depending on subtle variations in material composition, processing, and measurement methodology [9–11].

This highlights a pressing need for systematic data acquisition, standardization, and curation tools to aggregate and harmonize the available data, enabling more reliable analysis and trend discovery.

The intrinsic complexity of the ECE itself further compounds the challenges in data handling. Predicting the magnitude and behavior of ECE in a given material system requires understanding a confluence of factors, including crystal structure, phase transitions, defect chemistry, and electric field interactions, all of which can vary in subtle and nonlinear ways. As a result, conventional theoretical and empirical approaches often fall short in delivering accurate, generalizable predictions. In this context, modern data-driven techniques such as machine learning (ML) have emerged as promising alternatives. Machine learning algorithms can uncover hidden patterns within complex datasets and predict EC performance based on a range of material descriptors. Several recent publications have demonstrated the feasibility of applying ML models to ECE data [12–15], offering new pathways for accelerated materials discovery and optimization.

Nonetheless, despite the promise of machine learning, significant limitations remain. The scarcity of large, high-quality, consistent datasets restricts robust models' training and validation [16]. Moreover, many ML approaches currently function as "black boxes," providing limited physical interpretability of their predictions, which hinders their acceptance and utility within the materials science community [17]. These drawbacks point to the necessity of integrating domain knowledge with advanced data analytics and ensuring rigorous data collection protocols. Furthermore, practical deployment of ML-based predictive tools requires careful consideration of model uncertainty, transferability, and scalability [18,19].

In this context, the present work is structured to provide a comprehensive and critical overview of the current state of research on the ECE. We begin by summarizing the principal experimental methodologies employed for measuring ECE, highlighting their strengths and inherent limitations. This discussion serves as a foundation for evaluating the reliability and comparability of published results. Building upon this, we present a focused analysis of literature data concerning $BaTiO_3$-based ceramics, encompassing both undoped and doped compositions [20–76]. By systematically compiling and organizing these data, we aim to identify possible trends and correlations that may shed light on the factors governing EC performance. At the same time, we critically assess whether such an effort is justified in view of the significant variability and lack of standardization in the literature. Finally, we review the predictive approaches available in the current body of research, with particular emphasis on machine learning–driven strategies. By evaluating both their achievements and shortcomings, we aim to delineate the prospects and limitations of prediction tools as a step toward more rational and reliable design of EC materials.

2. Materials and Methods

2.1. Measurement Techniques of the Electrocaloric Effect

The ECE can be determined by two main methods: direct measurements, in which the actual change in sample temperature or heat transfer is recorded upon application of an electric field, and indirect methods, in which the effect is calculated from other measured quantities using the laws of thermodynamics [77–79]. Due to experimental difficulties, i.e., the lack of commercial measurement systems for direct characterization of ECE and the need to meet stringent conditions (e.g., adiabatic), in practice, most experimental data come from indirect methods as will be shown later in this article.

2.1.1. Formalism Based on Maxwell's Relations

The indirect method based on Maxwell's relations determines the entropy change in the system from measurements of polarization as a function of applied electric field and temperature (Figure 1a) and subsequently calculates the ECE using thermodynamic relations [77] (Figure 1b). This method relies on the Maxwell relation for thermodynamic systems under an electric field, which links the entropy change (ΔS) to the change in electric field intensity (ΔE) and the polarization change (ΔP) to the temperature change (ΔT) of the material:

$$\left(\frac{\partial S}{\partial E}\right)_T = \left(\frac{\partial P}{\partial T}\right)_E$$

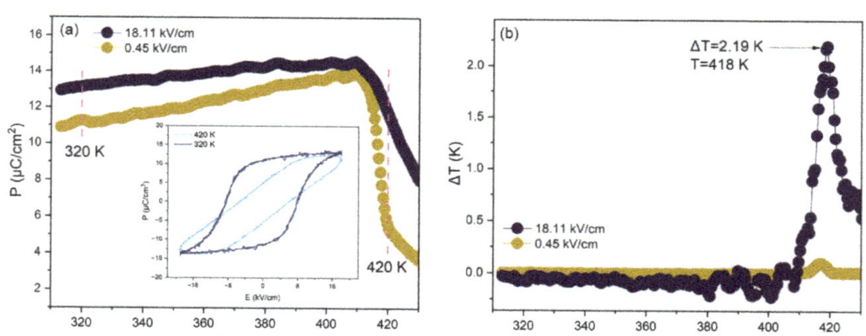

Figure 1. Example of the indirect method for determining the electrocaloric effect: (**a**) temperature dependence of polarization with inset hysteresis loops, and (**b**) the corresponding ΔT calculated using Maxwell's relations.

As a result of integration of this dependence, the isothermal change in entropy ΔS caused by the change in field from E_1 to E_2 can be determined:

$$\Delta S(E_1 \rightarrow E_2, T) = \int_{E_1}^{E_2} \left(\frac{\partial P}{\partial T}\right)_E dE$$

In turn, the expected adiabatic temperature change ΔT_{ad} related to the ECE (assuming no heat exchange with the environment) is:

$$\Delta T_{ad} = -\frac{T}{\rho C}\int_{E_1}^{E_2}\left(\frac{\partial P}{\partial T}\right)_E dE$$

The above formula results from the fact that ΔS corresponds to the entropy change in an isothermal process, while in adiabatic conditions, this entropy change results in an opposite temperature change (negative sign, because an increase in entropy causes a decrease in ΔT under adiabatic conditions). In the above equations, C denotes the specific heat (at constant pressure, or possibly at a constant field E) of the material, and ρ is the sample density. It is usually assumed that C does not depend strongly on the electric field and its values measured at $E = 0$ can be used (unless a sharp phase transition occurs). However, the change in heat capacity with temperature $C = C(T)$ within the measurement range is typically taken into account.

In practice, the implementation of the Maxwell formalism requires measuring the polarization hysteresis loop $P(E)$ as a function of temperature. Typically, a series of $P(E)$ measurements is made at a series of constant temperatures, uniformly covering the range of interest (e.g., around the phase transition temperature of the material, where the effect is strongest). Hysteresis loop measurements are performed under quasi-static conditions, with slow field changes, often on the order of 10^{-1}–10^{0} Hz, so that the sample remains close to thermodynamic equilibrium and the Maxwell relation is applicable (irreversible effects related to energy dissipation are neglected) [80,81]. The sample must be equipped with electrodes and placed in a temperature-stabilized measurement chamber (e.g., a thermostatic oven or a cryostat) with accurate temperature control and measurement. In addition to the hysteresis loop, the specific heat of the material must be determined (for the calculation of ΔT_{ad}). Usually, separate measurements (e.g., DSC) or literature data for the tested substance are used.

The Maxwell approach is very popular due to its experimental accessibility. A standard hysteresis loop measurement setup is sufficient to determine the effect, without the need for direct detection of thermal signals. Most published ECE results were obtained using this method, integrating $P(T,E)$ data. This method allows for the determination of both the isothermal entropy change ΔS and the corresponding adiabatic temperature change ΔT_{ad} over a wide range of temperatures and fields. The limitations are essentially the temperature range of the measuring equipment and the maximum field strength that can be applied to the sample (limited by the electrical strength of the material). However, it should be emphasized that the computational nature of the Maxwell method poses certain limitations and potential sources of error. First of all, Maxwell's relation is strictly valid only for reversible and quasi-static processes. If significant hysteresis, dielectric losses, or frequency dependence are present in the polarization, the resulting calculated ΔT_{ad} may deviate from reality. In particular, in relaxor-type materials, where local inhomogeneities occur (so-called polar nanoregions) and a strong dependence on the field application rate, Maxwell's method may lead to significant discrepancies with direct measurements. For example, it has been found that in such cases the indirectly calculated temperature change can even be of the opposite sign (so-called apparent negative electrocaloric effect), while accurate direct measurements show only a positive effect [82]. The uncertainty of Maxwell's method results from the cumulation of the uncertainty of the polarization and temperature measurements, as well as from the procedure of differentiation and integration of the data—the discrete and often limited dataset $P(T)$ requires approximation (e.g., function fitting) and numerical differentiation, which may introduce errors. Furthermore, the accuracy of the estimate of ΔT_{ad} depends on the precision of the knowledge of the specific heat *C*; any uncertainty or neglect (e.g., neglecting changes in *C*(*T*)) linearly affects the calculated ΔT_{ad}. In summary, the Maxwell formalism is a powerful analytical tool for ECE evaluation, simple to implement and widely used, but it requires careful data acquisition and may fail in situations deviating from the thermodynamic equilibrium of the sample.

The ECE can also be evaluated by applying the indirect Maxwell relation to pyroelectric measurements [83,84]. In this approach, the key quantity is the temperature derivative of the polarization at constant electric field, which corresponds to the pyroelectric coefficient. Experimentally, this coefficient is obtained by monitoring the pyroelectric current during a controlled temperature sweep under a fixed applied field. If the sample has surface area A and the heating or cooling rate is $\beta = \frac{dT}{dt}$, the current I_p is related to the pyroelectric coefficient through $p(T,E) = \frac{I_p(T,E)}{(A\beta)}$. Measuring this current as a function of temperature and field thus yields the desired $\left(\frac{\partial P}{\partial T}\right)_E$.

Once the pyroelectric coefficient has been determined, it can be introduced into the Maxwell relation linking entropy and polarization. For an adiabatic change in the electric field from zero to *E*, the EC temperature change is calculated according to:

$$\Delta T(T,E) = -\frac{T}{\rho c}\int_0^E p(T,E')dE',$$

where ρ and c are the density and specific heat of the material, respectively. In practice, the integral over field is evaluated numerically using the set of pyroelectric data recorded at different bias fields, and the temperature dependence of the heat capacity is either measured independently or taken from literature values.

It must be emphasized, however, that the accuracy of this approach is affected by the complexity of the pyroelectric response itself. The measured pyroelectric coefficient generally includes both a primary contribution, which reflects the intrinsic change in spontaneous polarization with temperature at constant strain, and a secondary contribution, which originates from thermally induced strains coupled through the piezoelectric effect. Because it is experimentally challenging to separate these two terms, the pyroelectric data used in the Maxwell relation may not represent the purely intrinsic polarization change. As a consequence, the EC temperature change derived from such measurements can be overestimated or underestimated depending on the magnitude and sign of the secondary contribution. This limitation should be kept in mind when interpreting EC results obtained by the pyroelectric indirect method.

2.1.2. Differential Scanning Calorimetry (DSC)

Differential scanning calorimetry (DSC) is a direct thermal measurement method, but it can be considered an indirect technique for determining the ECE, as it provides information on the entropy change based on the measured heat flow [85]. DSC involves the precise measurement of the heat flux difference between a sample and a reference material during a controlled temperature program. In the context of ECE, modified commercial DSC devices are used, allowing for the application of a high voltage to the sample during the measurement. This allows for direct measurement of the heat exchanged by the EC material when the electric field changes, and then the corresponding entropy change ΔS and (if C is known) the adiabatic temperature change ΔT_{ad} can be determined [77,79]. In other words, DSC records the thermal effect accompanying the polarization or depolarization of the sample. If the measurement is performed under isothermal conditions (ambient/sample temperature maintained constant), the released or absorbed heat Q related to the ECE corresponds to the entropy change $\Delta S = Q/T$. In turn, if the sample changes temperature (e.g., in almost adiabatic conditions inside the DSC), the recorded temperature curve can be used to calculate ΔT_{ad} directly. Most often, however, the isothermal or quasi-isothermal mode is used: the sample is kept at a constant temperature and at some point, the electric field is turned on (or off), observing in the DSC thermogram, respectively, an exothermic peak (heat release upon material polarization) or an endothermic peak (heat absorption upon depolarization). The equivalence of the areas of both peaks (of opposite sign) proves the reversibility of the effect and the absence of irreversible losses [77,80]. On this basis, the value of the exchanged heat can be calculated, and hence ΔS. Alternatively, for materials exhibiting phase transitions, constant-field scanning measurements are performed: the sample is heated or cooled at a given rate under various constant field strengths E (so-called isofield measurements), and DSC curves are recorded. A set of such curves reveals the shift in the phase transition temperature under the influence of the field (e.g., the shift T_C) and the change in the enthalpy of the transition. Using the Clausius–Clapeyron equation or direct integration of heat flux differences, the entropy and temperature change associated with the EC effect can be determined.

The advantage of DSC is its high sensitivity and precision in detecting enthalpy changes. Modern differential calorimeters can detect thermal effects on the order of microjoules, enabling the observation of even very small ECE [77,78]. Furthermore, DSC records the full thermodynamic response of the sample, including both reversible effects such as ECE and potential side processes (e.g., unwanted Joule's losses), allowing them to be distinguished and corrected. The literature emphasizes that DSC is a more accurate alternative to simpler heat measurements (e.g., using thermocouples). Ensuring stable conditions and a controlled temperature profile in DSC minimizes noise and random errors; typical measurement repeatability is very good, and calorimetric signals at the level of a few milliwatts are easily distinguished from the background. Furthermore, DSC allows for continuous study of the EC effect as a function of temperature (by recording the signal during temperature scanning), which provides rich information, for example, about the temperature range in which the effect occurs or the complexity of phase transitions.

The limitations and challenges of the DSC method include the need to modify the apparatus. Special measurement cells with high voltage feed-throughs are required, which present technical difficulties and can restrict the available field range (typically to several dozen kV/cm) [78]. The sample must be small (usually several dozen mg or less) and properly prepared: it is placed in a measuring crucible, often as a thin plate or layer with integrated electrodes. Avoiding measurement artifacts caused by Joule's heating is essential. If the tested material has finite electrical conductivity or substantial dielectric losses, then the current passing through the sample during field application generates additional heat, which overlaps with the EC signal. In a well-conducted experiment, leakage currents should be kept below 10^{-8} A so that resistive heat in the sample remains negligible. This is practically verified by observing the DSC signal: after switching off the field, it should return to the baseline, indicating no further sample heating. Otherwise, it is necessary to improve sample insulation or adjust for losses. Another limitation is the relatively slow rate at which the electric field can be changed in DSC. Although the field usually changes abruptly, the measurement itself has limited temporal resolution (on the order of seconds). Therefore, this technique records the equilibrium or slowly varying thermal response; it does not capture the ultrafast dynamics of the effect (for such purposes, thermal imaging with an infrared camera with high sampling rate is employed). The uncertainty of DSC results mainly stems from errors in heat flux calibration and background subtraction. The baseline must be determined with high precision (for example, by performing a separate measurement with the sample absent an electric field) and the field signal subtracted to extract the pure ECE. Small variations in the sample's heat capacity or minor thermal drifts can lead to an uncertainty in the calculated ΔS of several percent. Nonetheless, a correctly performed DSC measurement offers very reliable quantitative ECE results, often serving as a benchmark for verifying indirect Maxwell-based methods.

In summary, DSC enables the direct determination of the ECE by measuring heat transfer. Typical experiments are performed at temperatures within the range of interest for applications (e.g., near room temperature for cooling materials) using an electric field of several to several tens of kV/cm. Experimental conditions are selected to ensure near reversibility of the process (low leakage currents, repeatable on/off cycles) and good thermal signal resolution (slower temperature scanning improves enthalpy's resolution at the expense of measurement time).

2.1.3. Heat Flux Measurement Methods (Direct Calorimetry)

The third group of methods consists of techniques involving the direct measurement of the heat flux released or absorbed by the sample during polarization changes. Unlike DSC, these methods often use simpler calorimetric systems designed specifically for EC measurements. An example is quasi-adiabatic calorimetry with thermocouples. In this configuration, the EC material is thermally isolated (placed in a container with very good thermal insulation), and one or more thermocouples are placed near the sample to measure the temperature difference between the sample and the environment [8]. When an external electric field is applied (or is disconnected), the dipoles in the material become ordered (or deordered), resulting in the release or absorption of a certain amount of heat. Under ideal adiabatic conditions, this would cause a change in the sample temperature ΔT. However, in practice, some heat flows to the environment—and it is this heat flux that is detected by thermoelectric sensors. The signal from a thermocouple (or a system of thermocouples connected in series to increase sensitivity, so-called thermopile) allows for the determination of a small temperature difference ΔT between the sample and the housing, proportional to the heat flow. By measuring this signal as a function of time, the total exchanged energy Q can be determined by integrating the signal (e.g., thermoelectric voltage) over the duration of the effect. Then, similarly to the DSC method, the entropy change $\Delta S = Q/T$ (assuming a known mean process temperature) and the adiabatic temperature change

$$\Delta T_{ad} = -\frac{T\Delta S}{C}$$

(taking into account the sample's heat capacity) are calculated. Alternatively, if the measuring system is sensitive enough to detect only the sample's ΔT (with minimized heat loss), this result directly reflects the ECE.

Heat flux measurement methods are characterized by simpler equipment than DSC. Often, the construction of such a calorimeter involves shielding the sample with an insulating material and installing temperature sensors, which can be achieved relatively easily even in laboratory conditions. This allows for the study of samples of various sizes, including larger ones than in DSC (limited by the crucible size and the required low mass to avoid signal attenuation). Furthermore, these systems allow for faster electrical field changes—it is possible to study the pulsed action of the electric field and the dynamic thermal response of the material [86,87]. They are often used to test prototype cooling elements, where the efficiency of heat transfer during cyclic polarization of the material is important. In the literature, such methods are also referred to as quasi-direct methods because they measure heat (or temperature) directly but usually require some interpretation and intermediate calculations to obtain ΔT_{ad}.

The disadvantages and limitations of this group of methods include lower accuracy and resolution compared to professional DSCs. Signals from thermocouples (in the microvolt range) require amplification and careful calibration. The achieved sensitivity is limited. The minimum detectable ΔT is usually several milliKelvins, while the best DSCs can indirectly detect fractions of a milliKelvin corresponding to the thermal effect. Furthermore, it is difficult to achieve perfectly adiabatic conditions: there are always small heat losses to the environment, which may result in a slightly underestimated measured signal. In practice, calorimetric corrections are introduced. For example, a calibration procedure involves generating a known amount of heat within the system (e.g., by a Joule current pulse across a built-in resistor) and measuring the sensor response. Based on this, the effective time constant of the system and the proportionality of the signal are determined. It has been reported that corrections of 20–30% are necessary in typical experiments to compensate for thermal losses and imperfect insulation [77,80]. Despite these difficulties, once the system is calibrated, reproducible quantitative results can be obtained [86]. It should also be noted that thermocouple methods integrate the signal across the entire sample, making it more difficult to distinguish local effects or material inhomogeneities, unlike, for example, thermal imaging cameras, which depict the distribution of ΔT across the sample surface.

Typical experimental conditions for heat flux measurement methods involve maintaining the sample at a constant ambient temperature (often room temperature or another temperature of interest) and suddenly applying or removing an electric field with a defined amplitude. A short-duration thermal signal is recorded (the thermocouple voltage peak) lasting from fractions of a second to several seconds, depending on the rate of heat exchange with the environment. The sample is often placed in a vacuum or an inert gas atmosphere to eliminate convection and moisture (which could affect the signal). The electric field used is often similar to that used in other methods (on the order of several hundred kV/m); however, ensuring good electrical insulation within the calorimeter is crucial to avoid current leakage through the thermocouples or the chamber walls. These measurements are typically performed at a single temperature, possibly repeated at different sample temperatures to determine the $\Delta T_{\mathrm{ad}}(T)$ curve, which, however, requires time-consuming stabilization of the sample at the next temperature level.

In terms of uncertainty, the dominant bias is the systematic error associated with the non-ideal adiabatic nature of the system. If some heat escapes to the environment before it is measured, the resulting ΔS will be underestimated. Therefore, the aforementioned calibration is crucial. Additionally, there are random errors in thermocouple voltage measurement (electronic noise, amplifier zero drift, etc.). The total uncertainty of the obtained ΔT_{ad} is usually greater than in the DSC method, estimated at several percent. Nevertheless, heat flux measurement methods are a valuable complement to ECE studies, especially at the stage of prototype cooling devices, where obtaining a direct thermal signal from the material is crucial.

In summary, the indirect method based on Maxwell's equations provides quick and relatively simple estimates but is prone to systematic errors and is predictive in nature. DSC offers high sensitivity and provides rich information about phase transitions, but its results do not directly reflect the actual adiabatic ΔT. The heat flux method, on the other hand, allows for the observation of the effect under dynamic conditions, but at the cost of lower sensitivity and the need for precise thermal calibration. In practice, the best results are obtained by combining these methods: theoretical predictions based on Maxwell's relations allow for the expected range of the effect, DSC enables analysis of the thermodynamic mechanisms of the phenomenon, and heat flux measurements bring the experiment closer to the actual application conditions. The combined use of these techniques allows for both quantitative and qualitative characterization of the ECE in a given material.

2.2. Literature Data Collection

To provide a consistent foundation for trend analysis, we systematically collected EC data for $BaTiO_3$-based ceramics from the available literature and presented it in the Appendix A. The search used major scientific databases such as Web of Science and Scopus, with keywords including "electrocaloric effect", "$BaTiO_3$ ceramics", and "doped $BaTiO_3$".

The dataset was deliberately restricted to $BaTiO_3$-based ceramics, including both undoped and doped compositions, while excluding single crystals, thin films, and other forms. This choice was motivated by several factors. $BaTiO_3$ is among the most extensively studied ferroelectric materials, providing a sufficiently large body of high-quality experimental data to enable meaningful statistical analysis. Furthermore, its relatively low Curie temperature (T_C), compared to other ferroelectrics such as PZT, allows for EC measurements to be conducted under more accessible laboratory conditions. The ceramic form was specifically selected due to its technological relevance and the well-established processing methods, which reduce variability associated with microstructural differences that are more pronounced in single crystals or thin films. Focusing on ceramics thus ensures a consistent material platform for trend analysis and provides a reliable foundation for developing predictive models of the ECE.

For each selected study, quantitative EC data were extracted directly from tables provided by the authors or, when tables were unavailable, approximated from published graphs. In the latter case, the maximum ΔT reported was identified, and its corresponding temperature and electric field were recorded. Whenever possible, the measurement methodology (direct or indirect) was also noted, as this distinction is essential for evaluating the comparability and reliability of reported results. This dual approach allowed us to include well-tabulated datasets and works where results were only graphically represented, thereby maximizing the coverage of the literature.

For each entry, information was organized in a standardized tabular format comprising measurement temperature T, EC temperature change ΔT, applied electric field E, EC strength $\Delta T/E$, measurement technique (direct or indirect), and literature reference. An example of this organization is presented in Table 1. This structured approach enabled subsequent visualization and analysis of potential trends across different studies.

Table 1. An example of the data organization. Symbols: ΔT—electrocaloric temperature change, E—applied electric field, $\Delta T/E$—electrocaloric strength, I—indirect.

Composition	ΔT [K]	E [kV/cm]	T [K]	Δ [K·cm/kV]	Method	Ref.
$Ba_{0.94}Sm_{0.04}TiO_3$	0.920	30	350	0.0307	I	[22]
$Ba_{0.94}Dy_{0.04}TiO_3$	1.040	30	412	0.0347	I	
$Ba_{0.94}La_{0.04}TiO_3$	0.680	30	319	0.0227	I	
$Ba_{0.94}Ce_{0.04}TiO_3$	0.720	30	325	0.0240	I	
$Ba_{0.94}Nd_{0.04}TiO_3$	0.790	30	337	0.0263	I	
$Ba_{0.94}Eu_{0.04}TiO_3$	0.890	30	359	0.0297	I	
$Ba_{0.94}Gd_{0.04}TiO_3$	0.860	30	367	0.0287	I	

It should be emphasized, however, that the extracted dataset remains affected by the intrinsic heterogeneity of the literature. Variability in measurement protocols, reporting standards, and data presentation introduces unavoidable uncertainties, particularly in cases where values were digitized from figures rather than obtained directly from numerical tables. As such, the resulting compilation should be regarded as a representative but non-exhaustive dataset, suitable for identifying general tendencies rather than providing definitive quantitative benchmarks. This aspect will be discussed in more detail in Section 4.

The dopants identified across the surveyed studies span a broad range of the periodic table (Figure 2), reflecting the diverse strategies employed to tailor the EC response of $BaTiO_3$-based ceramics. Substitutions at the A-site, such as Ca, Sr, and La, are commonly used to adjust the lattice constant and shift the Curie temperature (T_C), optimizing the working temperature window [88–90].

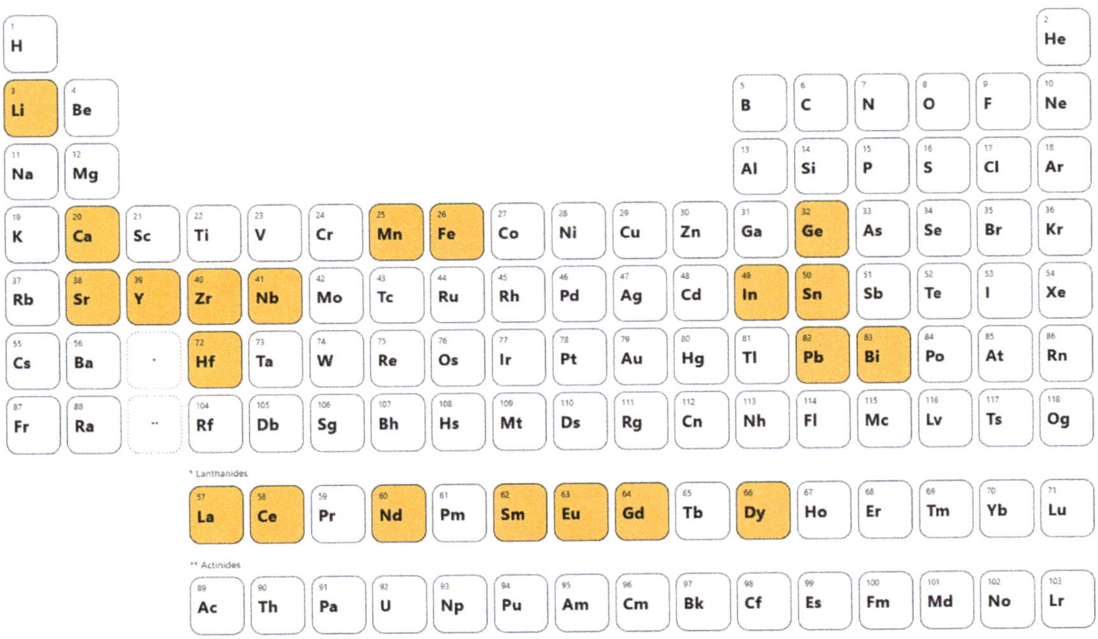

Figure 2. Elements reported in the literature as dopants in $BaTiO_3$, highlighted on the periodic table.

At the B-site, elements including Zr [91,92], Hf [93], Nb [94], Mn [95], Fe [96], Ge [97], and In [98] play a crucial role in modifying polarization behavior, broadening phase transitions, or influencing electrical conductivity through donor–acceptor effects. Heavy p-block elements such as Bi [99] and Pb [34] introduce strong polarizability due to their lone-pair electrons, often enhancing ferroelectric distortions. Meanwhile, rare-earth dopants (La, Ce, Nd, Sm, Eu, Gd, Dy) are frequently employed to fine-tune Tc, suppress leakage currents, and stabilize the microstructure [100]. The wide chemical diversity of dopants highlights the absence of a single universal strategy; instead, different substitutions are chosen depending on whether the primary aim is to shift transition temperatures, maximize ΔT, improve dielectric stability, or reduce losses under high electric fields. It should be emphasized, however, that at this stage the overview serves only to illustrate the compositional landscape of the studies included in this review. At the same time, the impact of specific dopants on the magnitude of ΔT will be analyzed and discussed in detail in Section 3.1.

The analysis of the collected dataset indicates a clear predominance of a single measurement approach: among the 58 studies considered, 37 employed the indirect method, 16 the direct method, and only 5 utilized both techniques concurrently (Figure 3). This disparity may be attributed to practical considerations, as the indirect method is generally easier to implement and provides rapid estimates of the ECE. However, it should be noted that indirect measurements represent approximations, as the effect is calculated rather than directly observed. In contrast, direct measurements yield the actual EC response but necessitate specialized equipment and more rigorous experimental conditions, limiting their widespread adoption.

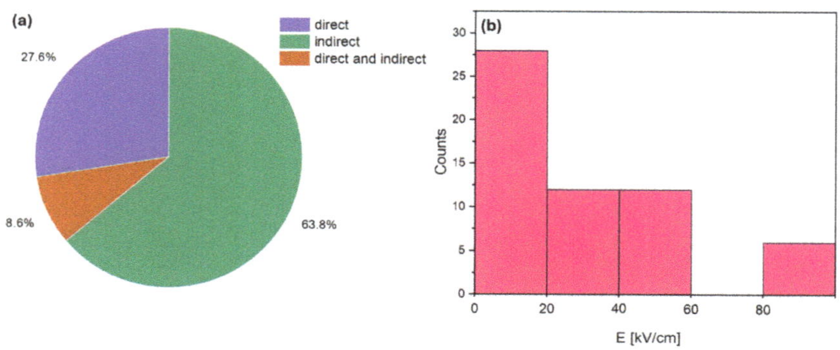

Figure 3. (**a**) Distribution of EC measurements according to the applied method (indirect, direct, or combined). (**b**) Histogram of electric field strengths employed in the reported studies.

Importantly, the concurrent application of both methods on the same samples would allow for a more robust evaluation of the discrepancies between calculated and measured values. Such data are critical for developing reliable machine-learning-based predictive models, as they enable calibration against actual material behavior and reduce systematic uncertainties inherent to either method individually. Therefore, while the assembled dataset facilitates identification of general trends in $BaTiO_3$-based ceramics, its heterogeneity and the limited use of dual-method measurements constitute a significant constraint on constructing accurate and generalizable predictive tools.

Regarding the magnitude of the applied electric field, expressed in kV/cm, a majority of the studies investigated relatively low fields: in 28 of the 58 studies, the field was below 20 kV/cm, in 12 studies it ranged from 21 to 40 kV/cm, in another 12 studies from 41 to 60 kV/cm, and only 6 studies applied fields above 80 kV/cm, with no studies reported in the intermediate range of 61–80 kV/cm. Studies applying very high fields, exceeding 100 kV/cm, remain scarce due to the significant experimental challenges they pose. At such field strengths, the risk of dielectric breakdown, sample degradation, and irreversible changes in microstructure increases, limiting the reproducibility and reliability of the measured EC response. Maintaining uniform fields across ceramic samples also requires careful electrode design and precise control of sample geometry. From an application perspective, the predominance of lower fields reflects technologically accessible and safe operating conditions, as high fields are generally impractical for solid-state cooling devices due to safety concerns, increased energy consumption, and reduced long-term reliability. Nevertheless, data obtained at very high fields are valuable for fundamental studies and calibrating predictive models, as they provide insights into the ultimate EC performance limits of $BaTiO_3$-based ceramics.

3. Results

3.1. Literature Overview of Electrocaloric Effect in $BaTiO_3$-Based Materials

The comparative plots presented in Figure 4 provide an overview of the ECE in $BaTiO_3$-based materials, grouped according to the type of dopant. Panels (a)–(c) separate the results into A-site dopants, B-site dopants, and combined A- and B-site co-doping, while panel (d) compiles all available data into a single diagram. The vertical axis represents the EC temperature change ΔT, while the horizontal axis corresponds to the measurement temperature T. The magnitude of the applied electric field E is indicated by the color scale, with darker shades representing stronger fields. In this comparative analysis, the dataset is limited to results obtained under electric fields below 50 kV/cm in order to ensure a consistent basis for comparison and to exclude extreme field conditions that may not be representative of practical device operation.

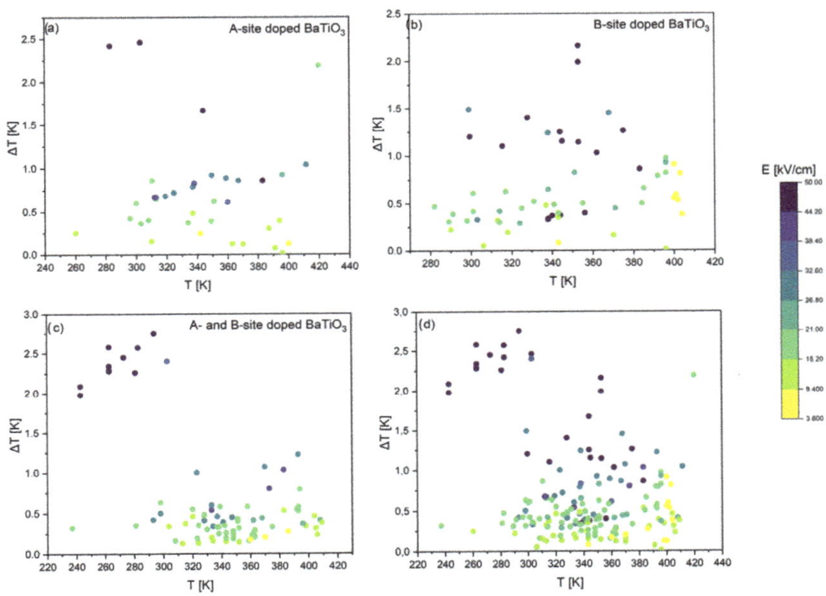

Figure 4. Comparative electrocaloric temperature change (ΔT) as a function of measurement temperature for $BaTiO_3$ ceramics: (**a**) A-site doped, (**b**) B-site doped, (**c**) co-doped at A- and B-sites, and (**d**) all compositions combined [20–26,28–31,33,34,36–48,50–68,70–76]. Color scale indicates the applied electric field strength; only data obtained below 50 kV/cm are included.

Several general trends emerge from these charts. Firstly, the most significant ΔT values, often exceeding 2 K, are consistently associated with darker colors, confirming that strong EC responses are typically observed only under applied fields of about 50 kV/cm [42], although some compositions also display relatively high $\Delta T/E$ ratios under more moderate fields (note that the Table A1 of the Appendix A includes additional compositions showing $\Delta T > 2$ K; however, these values are obtained under very high electric fields, exceeding 100 kV/cm [27,32,69]). Secondly, the majority of responses cluster in the range of 300–380 K, close to the Curie temperature of $BaTiO_3$, reinforcing that the ECE is maximized in the vicinity of the ferroelectric–paraelectric transition.

A-site substitutions usually yield moderate ΔT values of 0.5–1 K, with only a few outliers exceeding 2 K [34,42], and their responses are relatively broadly distributed in temperature. In contrast, B-site doping produces a broader spread of results, ranging from weak (<0.5 K) (e.g., [28,41,55–57,59,60,62,63]) up to strong (>2 K) (e.g., [40]) responses, indicating that B-site substitutions exert a stronger influence on polarization dynamics and phase stability. The highest responses, in some cases above 2.5 K [42,47], are recorded for co-doped compositions, underlining the synergistic benefits of simultaneously tuning both A- and B-sites.

Based on all collected data, the comparative analysis was subsequently divided into four focused groups, each discussed separately in the following sections. Since Ca, Sr, and Zr are among the most frequently used dopants in the literature, their datasets are presented and analyzed in greater detail (Figures 5–7). It was also decided to discuss doping with rare-earth (Figure 8) elements, as they introduce local lattice distortions and polarization fluctuations through their unique ionic size and 4f electronic structure, enabling enhanced tunability and stabilization of the EC response. In cases where individual studies report complete compositional series, only the most representative results are included in the figures to ensure clarity and balanced comparison. In these detailed plots, the visualization is slightly different from that presented in Figure 4: the bubble size represents the applied electric field E, while the color scale corresponds to the EC strength ($\Delta T/E$).

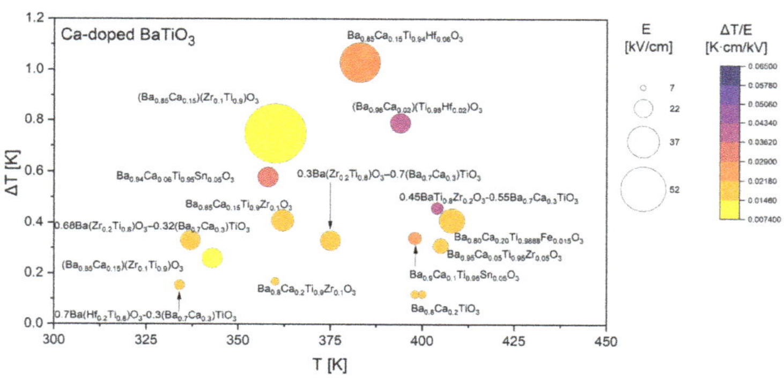

Figure 5. Electrocaloric response of Ca-doped $BaTiO_3$: ΔT versus T [26,28,36,37,43,47–52,65–67,73]. Bubble size represents the applied electric field, while color indicates electrocaloric strength ($\Delta T/E$). Only data obtained below 60 kV/cm are shown.

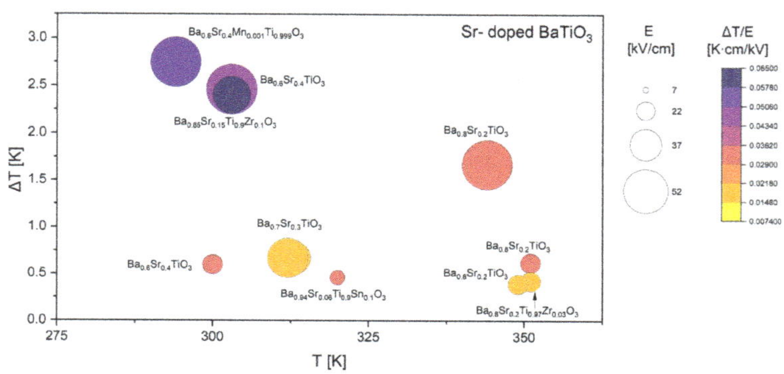

Figure 6. Electrocaloric response of Sr-doped $BaTiO_3$: ΔT versus T [41,42,46,53,54,68,70,75]. Bubble size represents the applied electric field, while color indicates electrocaloric strength ($\Delta T/E$). Only data obtained below 60 kV/cm are shown.

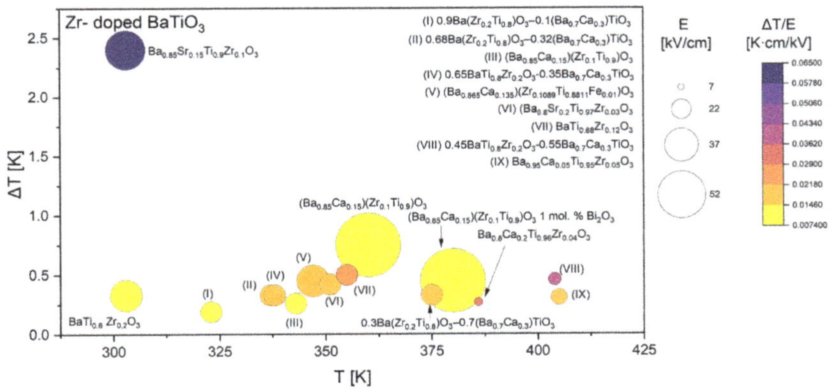

Figure 7. Electrocaloric response of Zr-doped $BaTiO_3$: ΔT versus T [45,46,49–52,59,60,65,67,73,74]. Bubble size represents the applied electric field, while color indicates electrocaloric strength ($\Delta T/E$). Only data obtained below 60 kV/cm are shown.

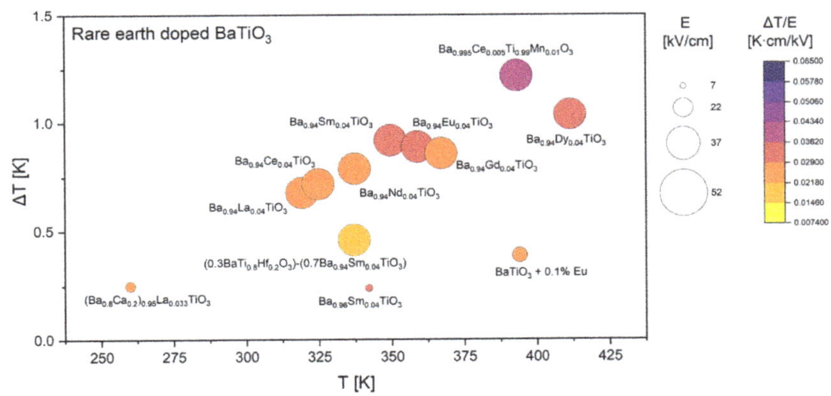

Figure 8. Electrocaloric response of rare earth-doped $BaTiO_3$: ΔT versus T [21,22,24,29,36,38]. Bubble size represents the applied electric field, while color indicates electrocaloric strength ($\Delta T/E$). Only data obtained below 60 kV/cm are shown.

The EC performance of Ca-doped $BaTiO_3$ presented in Figure 5 is strongly dependent on both the level of substitution and the presence of additional co-dopants such as Zr [47,49–51,67,73], Sn [26,66], Hf [28,37,48], and Fe [43]. Ca-modified compounds display a wide range of ΔT values, from below 0.1 K to nearly 1.0 K. This variability reflects the complex role of Ca^{2+} ions, which are smaller than Ba^{2+} and thus induce stronger lattice distortions while simultaneously shifting the ferroelectric transition temperature. The strongest responses are observed for multi-phase or co-doped systems rather than for simple Ca substitution. For example, $Ba_{0.85}Ca_{0.15}Ti_{0.94}Hf_{0.06}O_3$ [37] and $(Ba_{0.85}Ca_{0.15})(Zr_{0.1}Ti_{0.9})O_3$ [50] compositions reach ΔT values close to 1 K, highlighting the beneficial effect of simultaneously tuning the A- and B-sites. Similarly, $Ba_{0.94}Ca_{0.06}Ti_{0.95}Sn_{0.05}O_3$ [66] shows an enhanced effect compared to simple Ca doping, again pointing to the role of co-substitution in broadening phase transitions and optimizing the entropy change. The large bubble sizes associated with these compositions indicate that such improvements are achieved under relatively high fields, but the corresponding $\Delta T/E$ values confirm that the intrinsic response is also significant.

A clear trend can be seen in the temperature dependence of the effect. The strongest EC responses cluster around 340–380 K, close to the Curie temperature of $BaTiO_3$, which confirms that the mechanism is dominated by entropy changes near the ferroelectric–paraelectric phase boundary. At lower or higher temperatures, the effect is consistently weaker, as demonstrated by compositions such as $Ba_{0.8}Ca_{0.2}TiO_3$ [36,65] or $0.7Ba(Hf_{0.2}Ti_{0.8})O_3$–$0.3(Ba_{0.7}Ca_{0.3})TiO_3$ [48], both of which exhibit only minor ΔT values. This behavior again emphasizes that compositional tuning of the transition temperature is essential for maximizing EC strength.

Overall, the results demonstrate that Ca-doping in $BaTiO_3$ provides significant compositional flexibility but also introduces sensitivity to stoichiometry and co-dopant choice. While simple Ca substitution yields only modest EC responses, carefully engineered multi-doped systems such as $Ba_{0.85}Ca_{0.15}Ti_{0.94}Hf_{0.06}O_3$ [37] or $Ba_{0.94}Ca_{0.06}Ti_{0.95}Sn_{0.05}O_3$ [66] show promising behavior, with ΔT values approaching 1 K near the ferroelectric transition. These findings highlight that the design of Ca-doped $BaTiO_3$ electrocalorics should focus on balancing the destabilizing effect of Ca substitution with co-dopants that tune the transition temperature and enhance the entropy change.

Strontium substitution in $BaTiO_3$ presented in Figure 6 leads to a markedly different EC behavior compared to Ca-doped systems. The most striking feature of the Sr-doped compositions is the exceptionally high ΔT values, in some cases exceeding 2.5 K, which are among the strongest responses reported for $BaTiO_3$-based materials. In particular, $Ba_{0.6}Sr_{0.4}TiO_3$ and $Ba_{0.6}Sr_{0.4}Mn_{0.001}Ti_{0.999}O_3$ [42] exhibit large ECE in the temperature range around 290–300 K, with $\Delta T/E$ that are significantly higher than those observed for Ca-doped counterparts. The incorporation of Mn in small amounts further enhances the response, suggesting that subtle modifications of defect chemistry and domain wall pinning can play a synergistic role in maximizing the entropy change.

Compared to Ca substitution, where excessive doping suppresses ferroelectric order, Sr doping up to 40% appears to maintain strong polarization dynamics while shifting the ferroelectric transition temperature to lower values. This shift results in EC maxima appearing near room temperature, a particularly attractive feature for applications in solid-state cooling. The data also show that the ECE in Sr-doped $BaTiO_3$ is less reliant on high electric fields compared to other dopant systems. Large ΔT values are obtained even under moderate field strengths, as indicated by the relatively high $\Delta T/E$ values in the purple region of the color scale. Note that if it were otherwise, the bubbles would be colored in the middle of the scale. It suggests that Sr doping effectively enhances the intrinsic EC coefficient by tuning the dielectric susceptibility near the transition. The results demonstrate that Sr substitution is highly effective in producing strong EC responses, with ΔT values exceeding 2 K in compositions with high Sr content at relatively low fields (below 50 kV/cm). The combination of large intrinsic $\Delta T/E$, operation near room temperature, and reduced dependence on extreme electric fields makes Sr-doped $BaTiO_3$ one of the most promising lead-free EC materials among the dopant families considered. The additional enhancement observed for Mn-modified Sr-doped $BaTiO_3$ [42] further highlights the importance of co-doping strategies to fine-tune the interplay between lattice distortions, defect chemistry, and polarization dynamics.

Zirconium-modified $BaTiO_3$ compositions presented in Figure 7 display a distinct EC behavior compared to Ca- or Sr-doped systems. In these materials, the amount of Zr incorporated into the lattice exerts a decisive influence on phase stability and the position of the ferroelectric transition. At low substitution levels ($x < 0.1$) [46,65,67], the EC temperature change remains comparable to that of undoped $BaTiO_3$, typically in the range of 0.5–0.8 K [29,32,89]. With increasing Zr content, however, a systematic reduction in the maximum ΔT is observed, often falling below 0.3 K [50,52]. This decline can be attributed to the broadening of the ferroelectric–paraelectric transition and the partial suppression of long-range ferroelectric order, both of which diminish the entropy change associated with polarization switching. Despite the overall reduction in ΔT, compositions such as $BaZr_{0.12}Ti_{0.88}O_3$ [60] or $0.45BaTi_{0.8}Zr_{0.2}O_3$–$0.55Ba_{0.7}Ca_{0.3}TiO_3$ [73] exhibit relatively favorable $\Delta T/E$ values. That indicates that Zr substitution can enhance dielectric susceptibility and the intrinsic EC coefficient under certain conditions, even though the absolute temperature changes remain modest compared to those achieved in Sr-doped systems, where values exceeding 2 K are reported. On the other hand, the $BaTi_{1-x}Zr_xO_3$ (BZT) solutions above the composition x = 0.2 exhibit relaxor behavior [32]. It is well known that the magnitude of ΔT in relaxors is highly competitive with that in normal ferroelectrics and the operational temperature window can be quite wide [32,74,101,102].

The temperature dependence of the effect further supports this interpretation: the EC maxima shift to lower temperatures with increasing Zr concentration, reflecting the downward movement of the Curie temperature, yet the overall magnitude of the response decreases simultaneously. The most favorable performance is typically observed around 320–350 K, where the modified transition still retains sufficient entropy change to drive a measurable effect. Taken together, these findings suggest that Zr doping in $BaTiO_3$ is best regarded as a structural stabilizer rather than a direct enhancer of the EC response. While selected compositions can yield useful $\Delta T/E$ ratios [60,73], the maximum ΔT values are consistently lower than in Ca- or Sr-modified systems. As a result, Zr-doped $BaTiO_3$ appears less promising for practical EC cooling applications, although it may play an important role in multiphase or co-doped systems where phase-boundary effects can be exploited.

The discussion concludes with an overview of rare-earth substituted $BaTiO_3$ compositions. Looking at Figure 8 it can be seen that across the investigated compositions, the magnitude of the EC temperature change is typically on the order of 1 K, regardless of whether light rare-earth elements such as La [22,36], Ce [22,24], Nd [22], or Sm [21,22,38], or heavier elements such as Eu [22,29], Gd [22], and Dy [22] are employed. This observation suggests that the ECE is not strongly dependent on the atomic number of the rare-earth dopant itself, but rather on how the substitution perturbs the local lattice environment and the ferroelectric transition temperature.

A notable exception is the co-doped $Ba_{0.995}Ce_{0.005}Ti_{0.9}Mn_{0.01}O_3$ [24] composition, which exhibits the highest $\Delta T/E$ in this dataset. The cooperative effect of Ce and Mn highlights the potential of co-doping strategies to optimize both phase stability and domain dynamics to maximize the caloric response. The role of dopant concentration is equally crucial. The strongest effects are consistently observed near a rare-earth substitution level of approximately four percent. Even small deviations from this composition, as demonstrated by the comparison between $Ba_{0.96}Sm_{0.04}TiO_3$ [22] and $Ba_{0.94}Sm_{0.04}TiO_3$ [38], can drastically reduce the ECE despite the application of comparable electric fields. This sensitivity indicates that the EC response is highly dependent on maintaining the delicate balance of lattice distortion and polar instability induced by the dopant ions.

Temperature dependence further reinforces the connection between the ECE and ferroelectric phase transitions. The majority of strong responses are concentrated in the range of 330–380 K, corresponding to the vicinity of the Curie temperature of $BaTiO_3$. At lower temperatures, particularly in the case of Ca- and La-modified compositions [36] measured between 250 and 300 K, the magnitude of ΔT is significantly reduced.

Taken together, these results demonstrate that rare-earth doping in $BaTiO_3$ provides a robust pathway to achieving ECEs on the order of 1 K, with the strongest enhancements arising from optimized dopant concentrations near the ferroelectric transition temperature. The pronounced response observed for the Ce–Mn co-doped system [24] suggests that rationally designed multi-element doping strategies may represent a promising route for further improvement of lead-free EC materials.

Although non-negligible, the electric field's influence appears secondary to compositional and phase-related factors. Several compositions exhibit relatively strong ECEs under moderate applied fields, while others show weak responses even when subjected to higher fields. That underscores that tailoring the chemical composition is more effective than merely increasing the external stimulus.

The lack of standardized reporting practices in the ECE literature, particularly with regard to applied fields, sample microstructure, and measurement protocols, further limits comparability. We therefore strongly support the development of community-wide guidelines for reporting EC data, which would significantly facilitate both experimental cross-validation and machine-learning-based predictive modeling described in the next section.

3.2. *Electrocaloric Effect Predictions from Machine Learning Models in the Literature*

In recent years, the application of machine learning (ML) has emerged as a powerful strategy for predicting the ECE and accelerating the discovery of new ferroelectric materials. A number of recent studies [12–15] have established methodological frameworks that combine chemically informed descriptors, advanced regression models, and rigorous validation protocols. These approaches not only provide accurate predictive tools but also enable insights into the underlying physical mechanisms that govern polarization and entropy changes. Below, we present a detailed discussion of four representative contributions.

The predictive modeling of EC materials relies heavily on ML techniques capable of capturing complex, nonlinear relationships between chemical composition, structural features, and functional properties. Among the most commonly employed methods are decision trees and their ensemble extensions [12,13,15]. Decision trees are non-parametric models that recursively partition a dataset into subsets that are as homogeneous as possible with respect to the target property, such as adiabatic temperature change (ΔT) or isothermal entropy change (ΔS). Their transparent structure allows for the identification of key descriptors influencing the EC response.

While individual decision trees are easy to interpret [103], they may suffer from overfitting and limited predictive accuracy. To overcome these limitations, ensemble learning methods such as boosting are often applied [104]. Boosting algorithms iteratively combine multiple weak learners to form a stronger model. In Adaptive Boosting (AdaBoost), for instance, misclassified observations are given higher weight in successive iterations, improving the model's ability to handle difficult-to-predict samples [105]. Gradient boosting, particularly its high-performance implementation XGBoost, optimizes an objective function using gradient descent in function space, offering both high accuracy and scalability [106]. These methods are especially effective in materials science, where datasets may include hundreds of descriptors derived from elemental properties, crystal structures, and processing conditions.

In the context of EC materials, ML workflows generally begin with the generation of chemically and physically informed descriptors, followed by feature selection procedures to remove redundant or irrelevant variables. This ensures robust and interpretable models that not only predict ΔT and ΔS with high accuracy but also highlight the material features most strongly correlated with EC performance [12]. The integration of decision-tree-based models and boosting algorithms has thus become a standard approach for accelerating the discovery and optimization of novel ferroelectric materials [13].

Gong et al. [12] developed one of the earliest large-scale ML frameworks for predicting ΔT in EC ceramics. Their workflow began with the systematic collection of data from literature, yielding 97 ceramic compositions and over 4000 datapoints. Each material was encoded using Magpie descriptors, which generate chemically informed features from elemental properties. To avoid redundancy, descriptors were filtered through variance analysis, Pearson correlation thresholds, and backward elimination.

The predictive model was built using eXtreme Gradient Boosting (XGBoost) regression, with hyperparameters optimized via an exhaustive grid search covering thousands of parameter combinations. The authors also implemented t-distributed stochastic neighbor embedding (t-SNE) and k-means clustering to guide the train–test split, ensuring that structurally distinct materials were fairly represented in both sets. Model development was carried out using Scikit-learn, TensorFlow, and PyTorch.

Performance was assessed by comparing predicted and experimental ΔT values. The results ($R^2 \approx 0.77$, RMSE ≈ 0.38 K) showed close agreement, as illustrated in the parity plot Figure 9). Importantly, the model generalized beyond the training set, predicting unseen compositions with consistent accuracy.

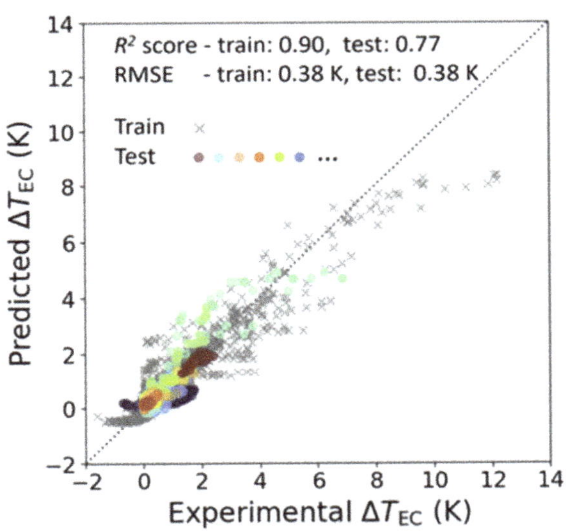

Figure 9. Parity plot of the predicted EC temperature change from one of the high-performing XGBoost models versus the experimental values. The gray crosses denote the training materials, the colored dots denote the test materials (different colors are used to distinguish different material compositions). Reprinted from Ref. [12].

To gain physical insights, the authors analyzed feature importance using impurity-based metrics and correlation heatmaps. This revealed that electronegativity differences and ionic charge states were the dominant descriptors, directly linking the ML model to established ferroelectric theory, in which bond polarizability and charge asymmetry govern dipole reorientation. The feature importance plot (Figure 10) highlights these descriptors as the key drivers of ΔT.

Finally, the model was extended to unexplored compositions, predicting ΔT values above 2 K under 100 kV cm^{-1} for several lead-free candidates at room temperature.

This result suggests practical applicability for environmentally friendly cooling devices (Figure 11).

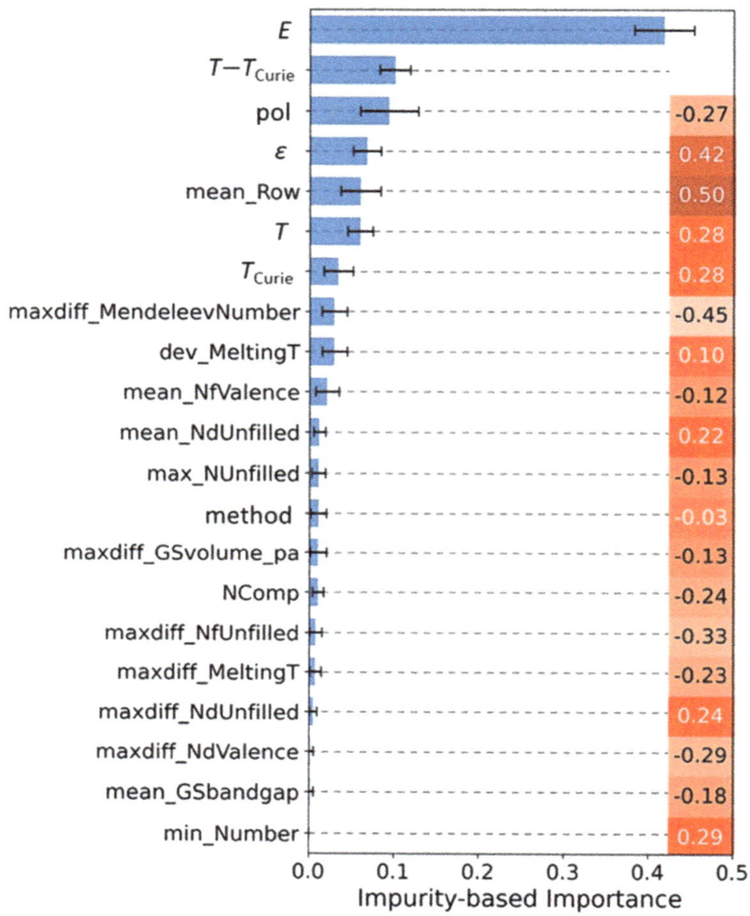

Figure 10. Impurity-based feature importance analysis. The error bar comes from the standard deviation of 100 random seeds. The color map shows the Pearson correlation coefficient of each feature with the predicted ΔT_{EC} in the model shown in Figure 9 after ruling out the influence of E and T-T_{Curie} by setting E = 100 kV/cm and T = T_{Curie}. Reprinted from Ref. [12].

For illustration, we show here a comparison of the EC response for $Ba_{0.92}Ca_{0.08}Ti_{0.92}Zr_{0.08}O_3$ with literature-reported experimental ECE data (ΔT = 0.38 K, E = 15 kV/cm, T = 409 K) [67], as well as for $Ba_{0.91}Ca_{0.09}Ti_{0.86}Zr_{0.14}O_3$ with ML-predicted values (ΔT = 2.6 K, E = 100 kV/cm, T = 318 K) compiled by Gong et al. It should be emphasized that the properties of $Ba_{0.91}Ca_{0.09}Ti_{0.86}Zr_{0.14}O_3$ have been reported in the literature [107], but no experimental EC values are available for it. Upon checking the Gong et al. database, we did not find evidence that the experimental values for $Ba_{0.92}Ca_{0.08}Ti_{0.92}Zr_{0.08}O_3$ were included in the ML training dataset. Although slight differences in composition and measurement conditions should be considered, the comparison suggests that the ML predictions are reasonably representative and provide a meaningful benchmark against experimental trends. The observed difference in the temperature at which ΔT is measured can be attributed primarily to compositional effects, while larger applied electric effects increase the magnitude of ΔT. However, we do not aim to evaluate the complete accuracy of these values here.

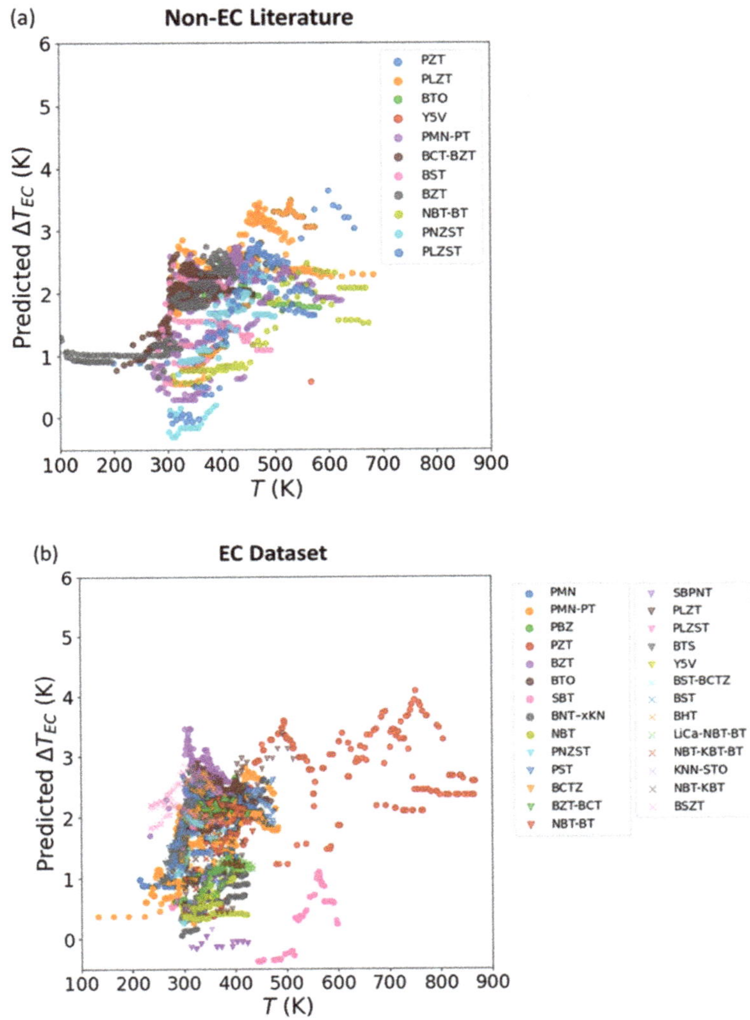

Figure 11. Predicted ΔT_{EC} at 100 kV/cm and varying temperature for ferroelectric ceramics. (**a**) predictions on the non-EC literature, (**b**) Predictions on the EC dataset. Reprinted from Ref. [12].

Su et al. [13] focused specifically on $BaTiO_3$-based ceramics, constructing a data-driven framework that distinguished between indirect and direct EC measurements. Two independent regression models were trained: (i) an SVR with Gaussian kernels and (ii) a random forest regression, supplemented by a support vector classification (SVC) model for predicting structural phase transitions.

Descriptors included ionic radii, electronegativity, tolerance factor, Mendeleev numbers, and dopant encoding. To reduce redundancy, pairwise correlation filtering was applied. Training was conducted in RStudio using the e1071 package, and ensembles were built via bootstrap resampling. Hyperparameters were tuned by grid search with cross-validation, ensuring robustness across multiple dataset partitions.

Model performance is shown in Figure 12. Both indirect and direct SVR ensembles reproduced experimental ΔT trends in unseen test sets, confirming that the models captured the nonlinear relationship between composition, structure, and EC response.

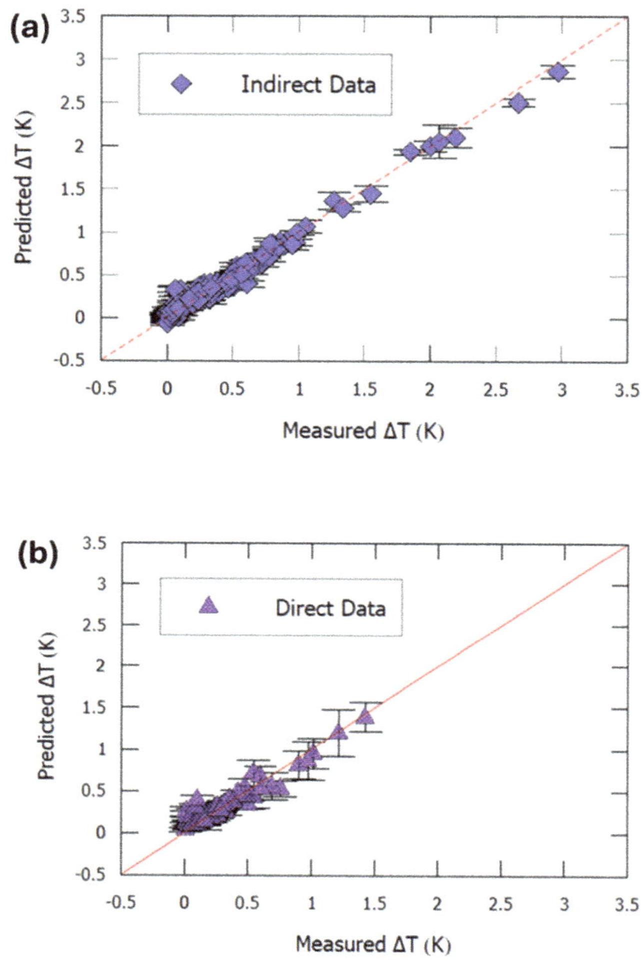

Figure 12. Performance plots of an ensemble of SVR models for (**a**) indirect and (**b**) direct data. The experimental ΔT is the x-axis and predictions from the trained models are on the y-axis. The error bars are the standard deviation of the predicted ΔT. All data points shown here correspond to those present in the independent test set, which the trained models have not seen before. (**a**) Indirect ensemble of SVR models with 20 bootstrap replications with the input descriptors: rTol_Fac, r_B, B_EN, D_EN, and T_K. (**b**) Direct ensemble of SVR models with 40 bootstrap replications with the input descriptors: rTol_Fac, r_B, B_EN, E and T_K. Reprinted with permission from Ref. [13]. Copyright 2021 American Chemical Society.

Beyond accuracy, Su et al. [13] also emphasized accessibility. They deployed their trained models in an interactive Shiny web application, allowing researchers to query predictions for new compositions in real time. From a physical perspective, the models highlighted that the tolerance factor and electronegativity differences were central to describing the EC response, consistent with long-standing concepts in perovskite ferroelectrics.

Nevertheless, the authors noted that dataset size and diversity remain limiting factors. They recommended systematic data sharing and automated literature extraction as future priorities to accelerate the adoption of informatics in EC research.

Bayir and Mensur [14] advanced the field by integrating XGBoost regression with Bayesian optimization for high-dimensional exploration of BCZT ceramics ($Ba_xCa_{1-x}Zr_yTi_{1-y}O_3$). Their dataset comprised 2188 entries, including compositional ratios, processing parameters, and experimental conditions. Features combined Magpie descriptors (average atomic weight, electronegativity, atomic volume) with Curie temperature, explicitly encoding the role of phase transitions.

To ensure reliability, the dataset was split using group-aware cross-validation and ShuffleSplit, preventing data leakage from single publications. Bayesian optimization, implemented with the Optuna framework and the Tree-structured Parzen Estimator (TPE) algorithm, efficiently tuned hyperparameters while minimizing mean absolute error (MAE).

The resulting model achieved $R^2 \approx 0.99$ and MAE ≈ 0.02 °C, as shown in Figure 13, indicating nearly perfect predictive accuracy within the dataset. Feature importance analysis identified electric field strength, measurement temperature, and proximity to the Curie point as the most significant drivers of ΔT_ec, echoing fundamental physical principles of entropy maximization near ferroelectric–paraelectric transitions.

Finally, Bayesian optimization was applied to identify optimal processing conditions, yielding a BCZT composition ($Ba_{0.85}Ca_{0.15}Zr_{0.4}Ti_{0.6}$) calcined at 1090 °C (3 h) and sintered at 1300 °C (2 h), predicted to achieve ΔT_ec ≈ 1.03 °C at 24 °C under 40 kV cm^{-1}. This showcases how ML not only reproduces physical trends but also directly informs materials design.

Last, Yuan et al. [15] proposed a surrogate modeling framework, leveraging polarization data as a proxy for EC properties. Instead of directly modeling ΔT, they constructed models predicting Curie temperature (Tc) and polarization, then linked these to entropy change (ΔS) through a physics-informed surrogate.

Their dataset consisted of 195 $BaTiO_3$-based compounds, characterized by five descriptors: NCT, D, tolerance factor, atomic volume, and electronegativity parameter (ENP). A support vector regression (SVR) with radial kernel was trained to predict Tc, using 1000 bootstrap resamples with 10-fold CV to ensure robustness. The model reproduced Tc with high accuracy, as shown in Figure 14.

Building on this, a Gaussian process regression (kriging-based, exponential kernel) was employed to model polarization behavior. Training began with 230 known compositions, after which the model was extended to 250,000 virtual compounds, enabling rapid high-throughput screening. Predictions aligned well with experimental polarization values (Figure 14b), validating the descriptors' ability to capture dipole alignment mechanisms.

By coupling polarization predictions to a Landau-based thermodynamic model, the authors established a linear link between polarization and ΔS in moderate regimes, demonstrating that polarization can serve as a reliable surrogate for the ECE. This approach bridges the gap between data-driven methods and physical models, providing both predictive power and interpretability.

Across these four studies, a consistent methodological pattern emerges:

- Feature engineering: Magpie descriptors, ionic radii, tolerance factors, and polarization parameters;

- Dimensionality reduction and filtering: variance analysis, correlation thresholds, backward elimination;
- Models: ensemble regression (XGBoost, random forest, SVR), surrogate modeling (Gaussian process), and classification (SVC);
- Validation: bootstrap resampling, group-aware cross-validation, ShuffleSplit, and exhaustive grid/Bayesian optimization;
- Software frameworks: Scikit-learn, TensorFlow, PyTorch, RStudio (e1071), Optuna.

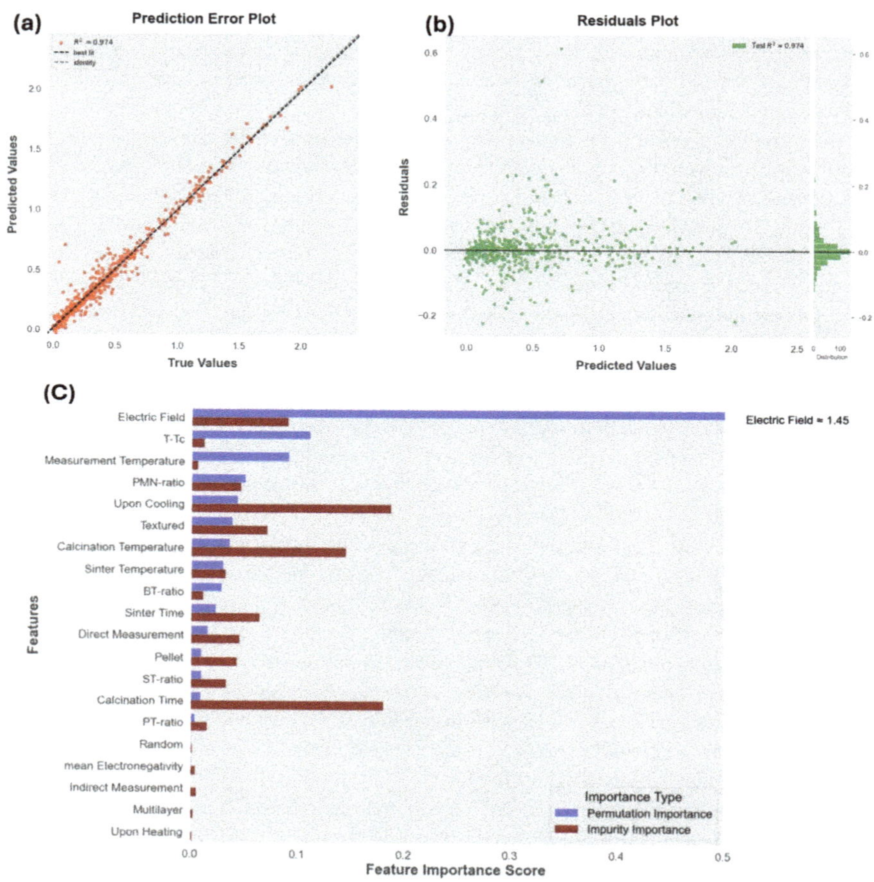

Figure 13. (**a**) Evaluation of XGBoost model performance for ΔT_{EC} prediction. Predicted versus true ΔT_{EC} values, (**b**) prediction residuals (difference between predicted and true values), (**c**) Feature importance analysis for ΔT_{EC} prediction using impurity-based ranking, SHAP values, and Pearson correlation. Reprinted from Ref. [14].

A structured overview of datasets, features, algorithms, validation strategies, and key results is provided in Table 2, which enables direct comparison across all four approaches.

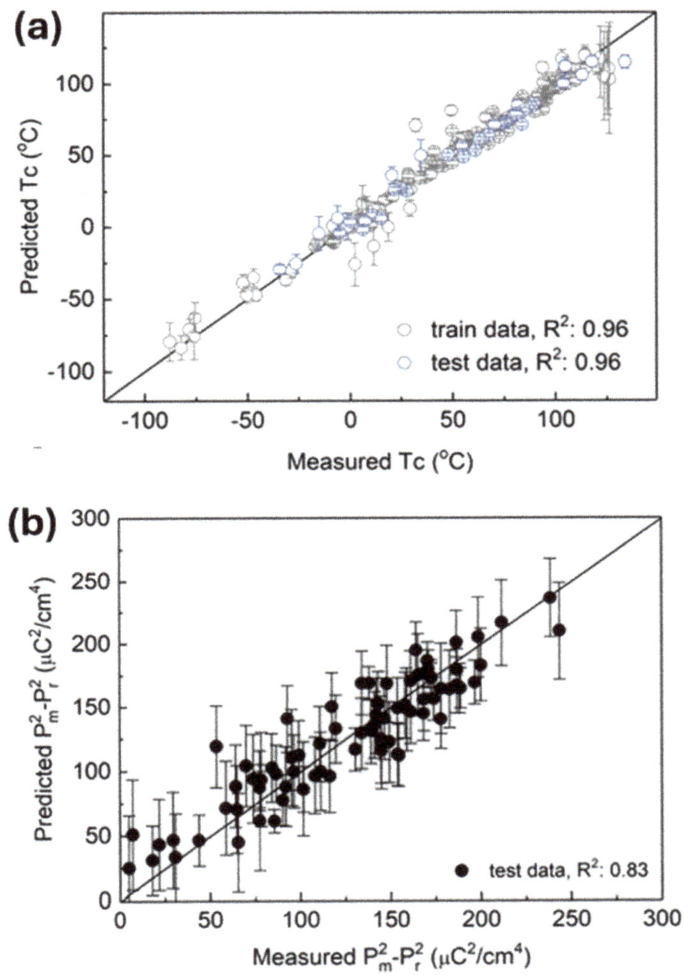

Figure 14. (**a**) The ML model for *Tc*, showing the performance on the training data (gray points) as well as on the unseen test data (blue poins). (**b**) The ML model for polarization, comparing the model prediction to measurement. Data points in the test data are shown to validate the generalizability of ML model. Reprinted with permission from Ref. [15]. Copyright 2022 Elsevier.

Table 2. Overview of machine learning approaches applied to electrocaloric materials in the literature.

Study	Dataset Size and Type	Features	ML Methods	Validation Strategy	Strengths	Limitations
Gong et al. [12]	~97 compositions, >4000 datapoints (literature)	Magpie descriptors, dielectric constant, Curie temperature, experimental conditions	XGBoost regression	t-SNE + k-means clustering, extensive grid search	$R^2 \approx 0.77$, RMSE ~0.38 K; captured physical trends (EN differences, ionic charge)	Dependent on heterogeneous literature data

Table 2. *Cont.*

Study	Dataset Size and Type	Features	ML Methods	Validation Strategy	Strengths	Limitations
Su et al. [13]	$BaTiO_3$-based ceramics, indirect and direct data separated	Ionic radii, electronegativity, tolerance factor, dopant encoding	SVR (Gaussian kernels), Random Forest, ensemble models	Bootstrap resampling, grid search + cross-validation	Robust predictions for both direct and indirect datasets; interactive web app	Limited by dataset size and diversity
Bayir and Mensur [14]	2188 entries (BCZT ceramics, literature)	Magpie descriptors, Curie temperature, processing parameters	XGBoost + Bayesian optimization (Optuna TPE)	ShuffleSplit + group-aware cross-validation	$R^2 \approx 0.99$, MAE ≈ 0.02 °C; identified optimal BCZT composition	High accuracy but dataset-specific
Yuan et al. [15]	~195 compositions ($BaTiO_3$-based)	Compositional descriptors (NCT, tolerance factor, atomic volume, EN, etc.)	SVR (radial kernel), Gaussian process surrogate models	Bootstrap (1000 samples) + 10-fold CV	Predicted Tc and polarization; surrogate model links polarization $\rightarrow \Delta S$	Requires indirect inference, not direct ΔT data

4. Discussion

The analysis of the literature dataset confirms that $BaTiO_3$-based ceramics remain the most widely studied family of lead-free EC materials, yet they also exemplify the intrinsic variability of reported results. Despite numerous experimental reports, values of the EC temperature change ΔT differ substantially between studies, even for nominally similar compositions. This scatter arises not only from intrinsic compositional effects but also from extrinsic factors such as grain size, porosity, electrode configuration, and above all, the measurement protocol employed. In particular, indirect results obtained via Maxwell's relations may diverge significantly from direct calorimetric values. Such discrepancies highlight the necessity of cautious interpretation when comparing literature values.

Nevertheless, some robust qualitative patterns can be identified. Among A-site dopants, Sr substitution is particularly effective in enhancing the EC response, often shifting the peak ΔT toward room temperature and delivering values exceeding 2 K under moderate fields [42]. Ca-doping alone results in modest effects, but co-doping at both the A- and B-site (e.g., Ca–Hf or Ca–Sn systems [37,66]) enhances the response significantly. At the B-site, Zr substitution tends to suppress the maximum ΔT by broadening the ferroelectric transition and inducing relaxor behavior, although certain compositions exhibit favorable $\Delta T/E$ ratios [60,73]. It should be emphasized that although discrepancies between indirect and direct measurements are frequently observed, particularly in relaxor-type compositions, for certain solid solutions such as BZT–BCT [51], good agreement has been reported (0.318 K and 0.33 K for indirect and direct measurements, respectively, in the $0.68Ba(Zr_{0.2}Ti_{0.8})O_3$–$0.32(Ba_{0.7}Ca_{0.3})TiO_3$). However, the inconsistency in other cases underscores that the reliability of indirect methods must be evaluated on a case-by-case basis, ideally by cross-checking against direct calorimetric techniques such as DSC.

Rare-earth substitutions typically produce responses of about 1 K, with optimized concentrations around 4% giving the best performance, and co-doping (e.g., Ce–Mn) further improving the effect [24]. These results demonstrate that while the absolute magnitude of ΔT remains variable, directional trends associated with specific dopant families are consistently observed across multiple independent studies.

The most critical weakness of the current body of work is the lack of full comparability between reported results. Literature data originate from different laboratories, using diverse synthesis routes, sintering conditions, and analysis procedures. This diversity introduces large systematic uncertainties and prevents truly quantitative benchmarking of EC performance across studies. Digitization of graphs, varying reporting standards, and incomplete descriptions of experimental conditions further exacerbate these issues. Consequently, while the identified trends are qualitatively meaningful, they cannot be considered precise or universal design rules.

Given these challenges, the establishment of standardized measurement and reporting practices is essential. Systematic documentation of applied electric fields, temperature ranges, grain size, density, and more would greatly enhance comparability across studies. Moreover, open repositories of raw polarization and calorimetry data, annotated with metadata, would provide a foundation for building benchmark datasets. Such standardized datasets are particularly critical for compositions like $BaTiO_3$, $Ba(Zr,Ti)O_3$–$Ba(Ca,Ti)O_3$, and $(Ba,Sr)TiO_3$, which could serve as community reference systems. Adopting these practices would significantly reduce uncertainty, facilitate meaningful cross-validation between direct and indirect results, and accelerate the development of predictive models.

In this context, machine learning (ML) has emerged as a promising but still early-stage tool for predicting EC behavior and accelerating materials discovery. The models developed so far successfully capture qualitative composition–property correlations across large, heterogeneous datasets, providing valuable guidance on dopant effects and substitutional trends. From a physical perspective, ML consistently identifies descriptors directly linked to ferroelectric mechanisms, such as electronegativity differences, ionic charge states, tolerance factor, and proximity to the Curie temperature [108,109]. These variables are central to dipole reorientation and entropy change, and their recurrence across different models reinforces the physical validity of the predictions. Notably, surrogate approaches linking polarization to entropy change have provided a bridge between data-driven methods and Landau-type thermodynamic theory [110].

From an informatics perspective, ensemble methods such as XGBoost and kernel-based regressors like SVR demonstrate strong predictive performance even on relatively small and heterogeneous datasets, particularly when combined with rigorous validation strategies such as bootstrap resampling, ShuffleSplit, or Bayesian optimization [12–15]. However, the predictive accuracy of ML is fundamentally limited by the scatter and inconsistency of the training data. Reported ΔT values vary not only with composition but also with field strength and measurement method, and ML models inevitably inherit this variability. As a result, current ML outcomes should be regarded as qualitative indicators rather than quantitative forecasts for specific synthesis routes. Moreover, most frameworks operate as black boxes, offering limited interpretability. Addressing this limitation requires the incorporation of physics-informed descriptors (already included in the models), such as tolerance factors or ionic radius variance, which indirectly capture structural distortions that drive ferroelectric transitions. Equally important is the inclusion of systematic uncertainty quantification through ensemble methods, Gaussian process regression, or other probabilistic approaches, so that predictions are accompanied by confidence intervals rather than single-point estimates.

It is important to situate ML within the broader computational landscape. Physics-based methods, including effective Hamiltonian models [111], first-principles (ab initio) simulations [112], and second-principles models (i.e., simplified, parametrized approaches derived from first-principles calculations) [113], offer complementary strengths. These methods capture intrinsic polarization dynamics, entropy changes, and phase-transition behavior with high physical interpretability, although they often involve significant computational cost and limited scalability. In contrast, ML approaches are data-driven, enabling rapid screening of large chemical spaces and excelling at uncovering empirical trends across diverse experimental datasets. To illustrate these complementary roles, Table 3 provides a comparative overview of their input requirements, strengths, limitations, and potential applications.

Table 3. Overview of physics-based methods and machine learning approaches for predicting the electrocaloric effect. MC—Monte Carlo, MD—Molecular Dynamics.

Method	Mechanism Captured	Typical Simulation Scale	Strengths	Limitations	Ref.
Effective Hamiltonian	Polar soft modes, coupling to strain; can include disorder if parameterized	10^3–10^5 unit cells, MC/MD	Mechanistic insight into ΔT, efficient for phase transitions, established for BTO/PTO, BZT	Requires careful DFT parametrization; difficult to include extrinsic defects; relaxors need large cells	[111]
Second principles	Anharmonic couplings, elasticity, mesoscale domains	10^5–10^7 atoms, longer MD	Captures domain/nanoregion effects; scalable to realistic microstructures	Parametrization heavy, limited transferability; still idealized vs. experiments	[113]
First principles MD	Full electronic structure, phonons, vibrational entropy	$\leq 10^3$ atoms, short times	Highest fidelity per atom; captures electronic contributions	Too small/short to capture domains/disorder; very costly	[112]
Machine learning	Correlations between composition/process descriptors and ΔT	data-dependent/no single typical scale	Fast screening; finds statistical trends	Accuracy limited by data quality; limited mechanistics insights	[12–15]

Looking forward, the reliable prediction and rational design of next-generation lead-free EC materials will require an integrated approach that combines standardized experimental protocols, uncertainty-aware ML models, and physics-informed mechanistic simulations. A next-generation ML framework could merge physics-informed surrogate models with advanced regressors, hybrid architectures, and transfer learning strategies, complemented by automated data augmentation (e.g., from high-throughput DFT or literature mining). Such a convergence of approaches would improve scalability, interpretability, and generalizability beyond $BaTiO_3$-based systems, providing a robust platform to guide the discovery of new high-performance EC materials and bringing the field closer to practical solid-state cooling applications.

5. Conclusions

This work provided a comprehensive overview of the EC behavior of $BaTiO_3$-based lead-free ceramics, focusing on the role of chemical substitution and the potential of emerging predictive approaches. The analysis of literature data confirmed several robust qualitative tendencies: the strongest EC responses occur close to the Curie temperature, typically within 300–380 K; Sr substitution produces particularly high ΔT values, often above 2 K under moderate fields, making it one of the most promising A-site strategies; Ca substitution alone yields modest responses, but co-doping at both the A- and B-sites (e.g., Ca-Hf or Ca-Sn) leads to clear improvements; Zr substitution broadens phase transitions and stabilizes the structure, often reducing the maximum ΔT while in some cases enhancing $\Delta T/E$; and rare-earth dopants generally provide moderate responses around 1 K, with optimized concentrations near 4% and further enhancements in co-doped systems such as Ce-Mn.

At the same time, the compilation of results highlights that reported ΔT values for nominally similar compositions often differ significantly due to differences in sample preparation, microstructure, and measurement protocols. Such discrepancies underline the importance of careful interpretation and, wherever possible, cross-validation of results obtained with different experimental approaches.

Machine learning (ML) is emerging as a complementary tool for analyzing these heterogeneous data. Current models successfully capture qualitative composition- property correlations and consistently identify descriptors closely linked to ferroelectric mechanisms, providing useful guidance on substitution strategies. However, their predictive accuracy is limited by the variability and incompleteness of the available datasets, and ML outcomes should therefore be considered qualitative rather than quantitative. Incorporating physics-informed descriptors, uncertainty quantification, and improved standardization of reported experimental data will be key to enhancing the reliability of ML predictions. ML frameworks could also benefit from integration with insights obtained from first- and second-principles approaches, thereby combining rapid trend identification with mechanistic understanding.

In conclusion, meaningful insights into the EC behavior of $BaTiO_3$-based ceramics can already be drawn from the literature, but further progress toward reliable prediction and practical optimization will depend on coordinated improvements in experimental standardization, data availability, and the integration of complementary predictive methods. Such convergence provides the most promising pathway for the rational design of next-generation lead-free EC materials, bringing environmentally friendly solid-state cooling technologies closer to application.

Author Contributions: Conceptualization, M.K.-K. and I.J.-S.; methodology, M.K.-K., D.K. and I.J.-S.; software, M.F. and Z.Ś.; validation, S.S.; formal analysis, M.K.-K. and D.K.; investigation, M.K.-K., D.K. and I.J.-S.; resources, I.J.-S.; data curation, M.F., Z.Ś. and S.S.; writing—original draft preparation, M.K.-K. and D.K.; writing—review and editing, M.K.-K., D.K., S.S. and I.J.-S.; visualization, M.K.-K.; supervision, I.J.-S.; project administration, M.K.-K.; funding acquisition, M.K.-K. All authors have read and agreed to the published version of the manuscript.

Funding: This research received no external funding.

Data Availability Statement: No new data were created or analyzed in this study. Data sharing is not applicable to this article.

Conflicts of Interest: The authors declare no conflicts of interest.

Appendix A

To provide a systematic overview of the reported EC properties of $BaTiO_3$-based ceramics, the results collected from the literature are summarized in Table A1. The individual formulations have been reproduced exactly as presented in the original article, which employs varying notational conventions and may therefore appear somewhat inconsistent. The dataset includes both undoped and doped compositions, with key parameters described in Section 2.2. It should be emphasized that many of the values were obtained by digitizing published graphs rather than from tabulated data. As a result, small deviations relative to the original sources may occur, although every effort was made to preserve accuracy and reliability.

Table A1. $BaTiO_3$ data from the literature. The individual formulations have been reproduced exactly as presented in the original article, which employs varying notational conventions. M denotes method, I—indirect, and D—direct.

Composition	ΔT [K]	E [kV/cm]	T [K]	$\Delta T/E$ [K·cm/kV]	M	Ref.
$BaTi_{0.975}Y_{0.025}O_3$	1.260	50.0	375	0.0252	I	[20]
$BaTi_{0.9625}Y_{0.0375}O_3$	1.030	50.0	362	0.0206	I	
$BaTi_{0.95}Y_{0.05}O_3$	1.150	50.0	345	0.0230	I	
$BaTi_{0.9375}Y_{0.0625}O_3$	0.370	50.0	340	0.0074	I	
$(0.6BaTi_{0.8}Hf_{0.2}O_3)$–$(0.4Ba_{0.94}Sm_{0.04}TiO_3)$	0.340	30.0	334	0.0113	I	[21]
$(0.5BaTi_{0.8}Hf_{0.2}O_3)$–$(0.5Ba_{0.94}Sm_{0.04}TiO_3)$	0.390	30.0	341	0.0130	I	
$(0.3BaTi_{0.8}Hf_{0.2}O_3)$–$(0.7Ba_{0.94}Sm_{0.04}TiO_3)$	0.460	30.0	337	0.0153	I	
$Ba_{0.94}Sm_{0.04}TiO_3$	0.920	30.0	350	0.0307	I	[22]
$Ba_{0.94}Dy_{0.04}TiO_3$	1.040	30.0	412	0.0347	I	
$Ba_{0.94}La_{0.04}TiO_3$	0.680	30.0	319	0.0227	I	
$Ba_{0.94}Ce_{0.04}TiO_3$	0.720	30.0	325	0.0240	I	
$Ba_{0.94}Nd_{0.04}TiO_3$	0.790	30.0	337	0.0263	I	
$Ba_{0.94}Eu_{0.04}TiO_3$	0.890	30.0	359	0.0297	I	
$Ba_{0.94}Gd_{0.04}TiO_3$	0.860	30.0	367	0.0287	I	
$BaTiO_3$ + 4% Li	0.545	20.0	392	0.0273	I	[23]
$BaTiO_3$ + 6% Li	0.575	20.0	393	0.0288	I	
$BaTiO_3$ + 8% Li	0.525	20.0	393	0.0263	I	
$Ba_{0.995}Ce_{0.005}Ti_{0.99}Mn_{0.01}O_3$	1.220	30.0	393	0.0407	I	[24]
$Ba_{0.985}Ce_{0.015}Ti_{0.99}Mn_{0.01}O_3$	1.070	30.0	370	0.0357	I	
$Ba_{0.97}Ce_{0.03}Ti_{0.99}Mn_{0.01}O_3$	0.410	30.0	328	0.0137	I	
$Ba_{0.96}Ce_{0.04}Ti_{0.99}Mn_{0.01}O_3$	0.500	30.0	298	0.0167	I	
$Ba_{0.955}Ce_{0.045}Ti_{0.99}Mn_{0.01}O_3$	0.420	30.0	293	0.0140	I	
$BaTiO_3$	0.010	10.0	396	0.0010	I	[25]
$BaTi_{0.98}Sn_{0.02}O_3$	0.160	9.8	370	0.0163	I	
$BaTi_{0.95}Sn_{0.05}O_3$	0.079	3.6	343	0.0218	I	
$BaTi_{0.9}Sn_{0.1}O_3$	0.050	10.1	306	0.0049	I	
$Ba_{0.98}Ca_{0.02}Ti_{0.95}Sn_{0.05}O_3$	0.200	7.9	370	0.0253	I	[26]
$Ba_{0.95}Ca_{0.05}Ti_{0.95}Sn_{0.05}O_3$	0.530	10.8	383	0.0490	I	
$Ba_{0.9}Ca_{0.1}Ti_{0.95}Sn_{0.05}O_3$	0.340	12.7	398	0.0268	I	
$0.8BaTi_{0.82}Zr_{0.18}O_3$–$0.2BaTi_{0.89}Sn_{0.11}O_3$	3.500	100.0		0.0350	D	[27]
$BaTi_{0.82}Hf_{0.18}O_3$	0.330	20.0	n/i	0.0165	I	[28]
$BaTi_{0.82}Hf_{0.18}O_3$	0.370	20.0	n/i	0.0185	D	
$(Ba_{0.98}Ca_{0.02})(Ti_{0.98}Hf_{0.02})O_3$	0.790	20.0	394	0.0395	I	
$(Ba_{0.87}Ca_{0.13})(Ti_{0.87}Hf_{0.13})O_3$	0.470	20.0	328	0.0235	I	
$(Ba_{0.87}Ca_{0.13})(Ti_{0.87}Hf_{0.13})O_3$	0.480	20.0	328	0.0240	D	
$(Ba_{0.86}Ca_{0.14})(Ti_{0.86}Hf_{0.14})O_3$	0.350	20.0	324	0.0175	I	
$(Ba_{0.82}Ca_{0.18})(Ti_{0.82}Hf_{0.18})O_3$	0.160	20.0	308	0.0080	I	
$BaTiO_3$ + 0.1% Eu	0.390	14.0	394	0.0279	I	[29]
$BaTiO_3$ + 1% Eu	0.300	10.0	387	0.0300	I	
$BaTiO_3$ + 2% Eu	0.120	10.0	363	0.0120	I	
$BaTiO_3$ + 3% Eu	0.070	10.0	391	0.0070	I	
$BaTiO_3$	0.970	20.0	396	0.0485	I	[30]
$BaTi_{0.998}Nb_{0.002}O_3$	0.820	20.0	396	0.0410	I	
$BaTi_{0.996}Nb_{0.004}O_3$	0.790	20.0	392	0.0395	I	
$BaTi_{0.994}Nb_{0.006}O_3$	0.660	20.0	385	0.0330	I	
$BaTi_{0.992}Nb_{0.008}O_3$	0.500	20.0	384	0.0250	I	
$BaTi_{0.99}Nb_{0.01}O_3$	0.440	20.0	n/i	0.0220	I	

Table A1. *Cont.*

Composition	ΔT [K]	E [kV/cm]	T [K]	$\Delta T/E$ [K·cm/kV]	M	Ref.
$BaTi_{0.94}Sn_{0.06}O_3$-Y	1.100	50.0	316	0.0220	I	[31]
$BaTi_{0.92}Sn_{0.08}O_3$-Y	1.250	50.0	344	0.0250	I	
$BaTi_{0.89}Sn_{0.11}O_3$-Y	1.400	50.0	328	0.0280	I	
$BaTi_{0.87}Sn_{0.13}O_3$-Y	1.200	50.0	300	0.0240	I	
$Ba(Zr_{0.15}Ti_{0.85})O_3$	4.200	150.0	342	0.0280	D	[32]
$Ba(Zr_{0.2}Ti_{0.8})O_3$	4.500	145.0	312	0.0310	D	
$(Ba_{0.86}Ca_{0.14})_{0.995}La_{0.005}Ti_{0.92}Sn_{0.08}O_3$	0.850	40.0		0.0213	D	[33]
$(Ba_{0.86}Ca_{0.14})_{0.99}La_{0.01}Ti_{0.92}Sn_{0.08}O_3$	1.160	60.0		0.0193	D	
$(Ba_{0.86}Ca_{0.14})_{0.985}La_{0.015}Ti_{0.92}Sn_{0.08}O_3$	1.450	100.0		0.0145	D	
$(Ba_{0.86}Ca_{0.14})_{0.98}La_{0.02}Ti_{0.92}Sn_{0.08}O_3$	1.600	140.0		0.0114	D	
$(Ba_{0.86}Ca_{0.14})_{0.975}La_{0.025}Ti_{0.92}Sn_{0.08}O_3$	1.620	160.0		0.0101	D	
$Ba_{0.96}Pb_{0.04}TiO_3$	2.190	18.0	420	0.1217	I	[34]
$Ba_{0.99}La_{0.01}TiO_3$	0.600	60.0	420	0.0100	I	[35]
$Ba_{0.98}La_{0.02}TiO_3$	0.650	60.0	370	0.0108	I	
$Ba_{0.97}La_{0.03}TiO_3$	0.750	60.0	327	0.0125	I	
$Ba_{0.96}La_{0.04}TiO_3$	1.810	60.0	303	0.0302	I	
$(Ba_{0.8}Ca_{0.2})TiO_3$	0.120	8.0	400	0.0151	I	[36]
$(Ba_{0.8}Ca_{0.2})_{0.98}La_{0.013}TiO_3$	0.115	9.9	370	0.0116	I	
$(Ba_{0.8}Ca_{0.2})_{0.97}La_{0.02}TiO_3$	0.150	9.9	310	0.0152	I	
$(Ba_{0.8}Ca_{0.2})_{0.95}La_{0.033}TiO_3$	0.250	9.9	260	0.0253	I	
$Ba_{0.85}Ca_{0.15}Ti_{0.94}Hf_{0.06}O_3$	1.030	40.0	383	0.0258	D	[37]
$Ba_{0.85}Ca_{0.15}Ti_{0.9}Hf_{0.1}O_3$	0.800	40.0	373	0.0200	D	
$Ba_{0.85}Ca_{0.15}Ti_{0.85}Hf_{0.15}O_3$	0.540	40.0	333	0.0135	D	
$Ba_{0.96}Sm_{0.04}TiO_3$	0.240	7.0	342	0.0343	D	[38]
$Ba_{0.7}Sr_{0.3}TiO_3$ + 1mol% Br^+ + 0.3 mol% Mn^{2+}	0.650	20.0	313	0.0325	I	[39]
$Ba_{0.7}Sr_{0.3}TiO_3$ + 1mol% Br^+ + 0.5 mol% Mn^{2+}	0.860	20.0	311	0.0430	I	
$Ba_{0.7}Sr_{0.3}TiO_3$ + 1mol% Br^+ + 1.5 mol% Mn^{2+}	0.400	20.0	308	0.0200	I	
$BaTi_{0.9}Hf_{0.1}O_3$	1.990	50.0	353	0.0398	D	[40]
$0.2BaTi_{0.93}Sn_{0.07}O_3$–$0.8BaTi_{0.9}Hf_{0.1}O_3$	2.160	50.0	353	0.0432	D	
$BaTi_{0.93}Sn_{0.07}O_3$	1.140	50.0	353	0.0228	D	
$BaTi_{0.9}Sn_{0.1}O_3$	0.480	15.0	337	0.0320	I	[41]
$Ba_{0.94}Sr_{0.06}Ti_{0.9}Sn_{0.1}O_3$	0.460	15.0	320	0.0307	I	
$Ba_{0.93}Sr_{0.07}Ti_{0.9}Sn_{0.1}O_3$	0.370	15.0	315	0.0247	I	
$Ba_{0.92}Sr_{0.08}Ti_{0.9}Sn_{0.1}O_3$	0.340	15.0	304	0.0227	I	
$Ba_{0.6}Sr_{0.4}TiO_3$	2.460	50.0	303	0.0492	D	[42]
$Ba_{0.6}Sr_{0.4}Mn_{0.001}Ti_{0.999}O_3$	2.750	50.0	294	0.0550	D	
$Ba_{0.6}Sr_{0.4}Mn_{0.002}Ti_{0.998}O_3$	2.570	50.0	283	0.0514	D	
$Ba_{0.6}Sr_{0.4}Mn_{0.003}Ti_{0.997}O_3$	2.450	50.0	273	0.0490	D	
$Ba_{0.6}Sr_{0.4}Mn_{0.004}Ti_{0.996}O_3$	2.290	50.0	263	0.0458	D	
$Ba_{0.6}Sr_{0.4}Mn_{0.005}Ti_{0.995}O_3$	2.090	50.0	243	0.0418	D	
$Ba_{0.65}Sr_{0.35}TiO_3$	2.420	50.0	283	0.0484	D	
$Ba_{0.65}Sr_{0.35}Mn_{0.001}Ti_{0.999}O_3$	2.580	50.0	263	0.0516	D	
$Ba_{0.65}Sr_{0.35}Mn_{0.002}Ti_{0.998}O_3$	2.340	50.0	263	0.0468	D	
$Ba_{0.65}Sr_{0.35}Mn_{0.003}Ti_{0.997}O_3$	2.280	50.0	263	0.0456	D	
$Ba_{0.65}Sr_{0.35}Mn_{0.004}Ti_{0.996}O_3$	2.260	50.0	281	0.0452	D	
$Ba_{0.65}Sr_{0.35}Mn_{0.005}Ti_{0.995}O_3$	1.980	50.0	243	0.0396	D	
$Ba_{0.80}Ca_{0.20}Ti_{0.9925}Fe_{0.01}O_3$	0.370	25.0	396	0.0148	I	[43]
$Ba_{0.80}Ca_{0.20}Ti_{0.9888}Fe_{0.015}O_3$	0.410	25.0	408	0.0164	I	
$Ba_{0.80}Ca_{0.20}Ti_{0.985}Fe_{0.02}O_3$	0.440	25.0	368	0.0176	I	

Table A1. *Cont.*

Composition	ΔT [K]	E [kV/cm]	T [K]	$\Delta T/E$ [K·cm/kV]	M	Ref.
$Ba(Ge_{0.02}Ti_{0.98})O_3$	0.550	8.0	400	0.0692	I	[44]
$Ba(Ge_{0.03}Ti_{0.97})O_3$	0.590	8.0	401	0.0742	I	
$Ba(Ge_{0.05}Ti_{0.95})O_3$	0.530	8.0	402	0.0667	I	
$Ba(Ge_{0.06}Ti_{0.94})O_3$	0.810	8.0	403	0.1019	I	
$Ba(Ge_{0.09}Ti_{0.91})O_3$	0.380	8.0	404	0.0478	I	
$Ba(Ge_{0.06}Ti_{0.94})O_3$	0.904	8.0	400	0.1130	D	
$(Ba_{0.865}Ca_{0.135})(Zr_{0.1089}Ti_{0.8811}Fe_{0.01})O_3$	0.450	30.0	347	0.0150	I	[45]
$Ba_{0.8}Sr_{0.2}TiO_3$	0.620	20.0	351	0.0310	I	[46]
$Ba_{0.8}Sr_{0.2}Ti_{0.97}Zr_{0.03}O_3$	0.420	20.0	351	0.0210	I	
$Ba_{0.8}Sr_{0.2}Ti_{0.95}Zr_{0.05}O_3$	0.430	20.0	351	0.0215	I	
$Ba_{0.8}Sr_{0.2}Ti_{0.93}Zr_{0.07}O_3$	0.400	20.0	351	0.0200	I	
$Ba_{0.8}Sr_{0.2}Ti_{0.9}Zr_{0.1}O_3$	0.350	20.0	342	0.0175	I	
$Ba_{0.85}Ca_{0.15}Ti_{0.9}Zr_{0.1}O_3$	0.410	22.0	362	0.0186	I	[47]
$Ba_{0.85}Ca_{0.1}Sr_{0.05}Ti_{0.9}Zr_{0.1}O_3$	0.600	30.0	333	0.0200	I	
$Ba_{0.85}Ca_{0.05}Sr_{0.1}Ti_{0.9}Zr_{0.1}O_3$	1.000	30.0	323	0.0333	I	
$Ba_{0.85}Sr_{0.15}Ti_{0.9}Zr_{0.1}O_3$	2.400	37.0	303	0.0649	I	
$0.8Ba(Hf_{0.2}Ti_{0.8})O_3$-$0.2(Ba_{0.7}Ca_{0.3})TiO_3$	0.130	10.0	324	0.0130	I	[48]
$0.7Ba(Hf_{0.2}Ti_{0.8})O_3$-$0.3(Ba_{0.7}Ca_{0.3})TiO_3$	0.155	10.0	334	0.0155	I	
$0.6Ba(Hf_{0.2}Ti_{0.8})O_3$-$0.4(Ba_{0.7}Ca_{0.3})TiO_3$	0.140	10.0	352	0.0140	I	
$0.5Ba(Hf_{0.2}Ti_{0.8})O_3$-$0.5(Ba_{0.7}Ca_{0.3})TiO_3$	0.240	10.0	362	0.0240	D	
$0.8Ba(Hf_{0.2}Ti_{0.8})O_3$-$0.2(Ba_{0.7}Ca_{0.3})TiO_3$	0.125	10.0	313	0.0125	D	
$0.7Ba(Hf_{0.2}Ti_{0.8})O_3$-$0.3(Ba_{0.7}Ca_{0.3})TiO_3$	0.200	10.0	333	0.0200	D	
$0.6Ba(Hf_{0.2}Ti_{0.8})O_3$-$0.4(Ba_{0.7}Ca_{0.3})TiO_3$	0.150	10.0	343	0.0150	D	
$0.5Ba(Hf_{0.2}Ti_{0.8})O_3$-$0.5(Ba_{0.7}Ca_{0.3})TiO_3$	0.270	10.0	363	0.0270		
$(Ba_{0.85}Ca_{0.15})(Zr_{0.1}Ti_{0.9})O_3$	0.750	60.0	360	0.0125	D	[49]
$(Ba_{0.85}Ca_{0.15})(Zr_{0.1}Ti_{0.9})O_3$ 1 mol. % Bi_2O_3	0.450	60.0	380	0.0075	D	
$(Ba_{0.85}Ca_{0.15})(Zr_{0.1}Ti_{0.9})O_3$	0.260	20.0	343	0.0130	I	[50]
$(Ba_{0.85}Ca_{0.15})(Zr_{0.1}Ti_{0.9})O_3$	0.464	60.0	373	0.0077	D	
$0.68Ba(Zr_{0.2}Ti_{0.8})O_3$–$0.32(Ba_{0.7}Ca_{0.3})TiO_3$	0.318	20.0	352	0.0159	I	[51]
$0.65Ba(Zr_{0.2}Ti_{0.8})O_3$–$0.35(Ba_{0.7}Ca_{0.3})TiO_3$	0.188	20.0	348	0.0094	I	
$0.63Ba(Zr_{0.2}Ti_{0.8})O_3$–$0.37(Ba_{0.7}Ca_{0.3})TiO_3$	0.155	20.0	359	0.0078	I	
$0.6Ba(Zr_{0.2}Ti_{0.8})O_3$–$0.4(Ba_{0.7}Ca_{0.3})TiO_3$	0.225	20.0	361	0.0113	I	
$0.55Ba(Zr_{0.2}Ti_{0.8})O_3$–$0.45(Ba_{0.7}Ca_{0.3})TiO_3$	0.205	20.0	363	0.0103	I	
0.68Ba(Zr0.2Ti0.8)O3–0.32(Ba0.7Ca0.3)TiO3	0.330	20.0	337	0.0165	D	
$0.65Ba(Zr_{0.2}Ti_{0.8})O_3$–$0.35(Ba_{0.7}Ca_{0.3})TiO_3$	0.250	20.0	338	0.0125	D	
$0.63Ba(Zr_{0.2}Ti_{0.8})O_3$–$0.37(Ba_{0.7}Ca_{0.3})TiO_3$	0.195	20.0	343	0.0098	D	
$0.6Ba(Zr_{0.2}Ti_{0.8})O_3$–$0.4(Ba_{0.7}Ca_{0.3})TiO_3$	0.230	20.0	348	0.0115	D	
$0.55Ba(Zr_{0.2}Ti_{0.8})O_3$–$0.45(Ba_{0.7}Ca_{0.3})TiO_3$	0.255	20.0	362	0.0128	D	
$0.9Ba(Zr_{0.2}Ti_{0.8})O_3$–$0.1(Ba_{0.7}Ca_{0.3})TiO_3$	0.190	20.0	323	0.0095	I	[52]
$0.7Ba(Zr_{0.2}Ti_{0.8})O_3$–$0.3(Ba_{0.7}Ca_{0.3})TiO_3$	0.300	20.0	330	0.0150	I	
$0.65Ba(Zr_{0.2}Ti_{0.8})O_3$–$0.35(Ba_{0.7}Ca_{0.3})TiO_3$	0.280	20.0	348	0.0140	I	
0.5 $0.5Ba(Zr_{0.2}Ti_{0.8})O_3$–$0.5(Ba_{0.7}Ca_{0.3})TiO_3$	0.290	20.0	363	0.0145	I	
$0.3Ba(Zr_{0.2}Ti_{0.8})O_3$–$0.7(Ba_{0.7}Ca_{0.3})TiO_3$	0.330	20.0	375	0.0165	I	
$Ba_{0.6}Sr_{0.4}TiO_3$	0.600	20.0	300	0.0300	D	[53]
$Ba_{0.7}Sr_{0.3}TiO_3$	0.670	33.0	313	0.0203	I	[54]
$Ba_{0.8}Sr_{0.2}TiO_3$	0.830	33.0	338	0.0252	I	
$Ba_{0.9}Sr_{0.1}TiO_3$	0.610	33.0	360	0.0185	I	
$BaTi_{0.99}In_{0.01}O_{2.995}$	0.220	10.0	290	0.0220	I	[55]
$BaTi_{0.97}In_{0.03}O_{2.985}$	0.290	26.0	324	0.0112	I	
$BaTi_{0.95}In_{0.05}O_{2.975}$	0.420	26.0	314	0.0162	I	

Table A1. *Cont.*

Composition	ΔT [K]	E [kV/cm]	T [K]	$\Delta T/E$ [K·cm/kV]	M	Ref.
$BaTi_{0.944}Y_{0.056}O_{2.972}$	0.400	45.0	356	0.0089	I	[56]
$BaTi_{0.942}Y_{0.058}O_{2.971}$	0.375	45.0	344	0.0083	I	
$BaTi_{0.94}Y_{0.06}O_{2.97}$	0.340	45.0	338	0.0076	I	
$BaTi_{0.938}Y_{0.062}O_{2.969}$	0.340	45.0	338	0.0076	I	
$BaTi_{0.936}Y_{0.064}O_{2.968}$	0.330	45.0	338	0.0073	I	
$BaTi_{0.97}Hf_{0.03}O_3$	0.190	10.0	318	0.0190	D	[57]
$BaTi_{0.89}Hf_{0.11}O_3$	0.350	10.0	343	0.0350	D	
$BaTi_{0.83}Hf_{0.17}O_3$	0.320	10.0	313	0.0320	D	
$BaTiO_3$	0.925	24.0	396	0.0385	I	[58]
$BaTi_{0.99}Ce_{0.1}O_3$	0.450	24.0	371	0.0188	I	
$BaTi_{0.88}Ce_{0.12}O_3$	0.825	24.0	351	0.0344	I	
$BaTi_{0.85}Ce_{0.15}O_3$	0.650	24.0	338	0.0271	I	
$BaTi_{0.8}\ Zr_{0.2}O_3$	0.330	30.0	303	0.0110	I	[59]
$BaTi_{0.88}Zr_{0.12}O_3$	0.500	20.0	355	0.0250	D	[60]
$BaTi_{0.8}Zr_{0.2}O_3$	0.300	20.0	314	0.0150	D	
0.05 $BaTi_{0.95}Sn_{0.05}O_3$	1.450	30.0	368	0.0483	D	[61]
$0.1BaTi_{0.9}Sn_{0.1}O_3$	1.240	30.0	338	0.0413	D	
$BaTi_{0.85}Sn_{0.15}O_3$	1.490	30.0	299	0.0497	D	
$BaTi_{0.9}Sn_{0.1}O_3$	0.400	20.0	343	0.0200	I	[62]
$BaTi_{0.88}Sn_{0.12}O_3$	0.450	20.0	325	0.0225	I	
$BaTi_{0.85}Sn_{0.15}O_3$	0.420	20.0	301	0.0210	I	
$BaTi_{0.82}Sn_{0.18}O_3$	0.318	20.0	298	0.0159	I	
$BaTi_{0.92}Sn_{0.08}O_3$	0.490	20.0	341	0.0245	D	[63]
$BaTi_{0.89}Sn_{0.11}O_3$	0.630	20.0	317	0.0315	D	
$BaTi_{0.86}Sn_{0.14}O_3$	0.390	20.0	291	0.0195	D	
$BaTi_{0.85}Sn_{0.15}O_3$	0.310	20.0	289	0.0155	D	
$BaTi_{0.92}Sn_{0.08}O_3$	0.520	20.0	331	0.0260	I	[64]
$BaTi_{0.895}Sn_{0.105}O_3$	0.610	20.0	301	0.0305	I	
$BaTi_{0.86}Sn_{0.14}O_3$	0.470	20.0	282	0.0235	I	
$Ba_{0.8}Ca_{0.2}TiO_3$	0.120	8.0	398	0.0151	I	[65]
$Ba_{0.8}Ca_{0.2}Ti_{0.96}Zr_{0.04}O_3$	0.270	8.0	386	0.0340	I	
$Ba_{0.8}Ca_{0.2}Ti_{0.9}Zr_{0.1}O_3$	0.170	8.0	360	0.0214	I	
$Ba_{0.94}Ca_{0.06}Ti_{0.95}Sn_{0.05}O_3$	0.580	20.0	358	0.0290	I	[66]
$Ba_{0.94}Ca_{0.06}Ti_{0.925}Sn_{0.075}O_3$	0.580	20.0	339	0.0290	I	
$Ba_{0.94}Ca_{0.06}Ti_{0.9}Sn_{0.1}O_3$	0.550	20.0	320	0.0275	I	
$Ba_{0.94}Ca_{0.06}Ti_{0.875}Sn_{0.125}O_3$	0.630	20.0	298	0.0315	I	
$Ba_{0.94}Ca_{0.06}Ti_{0.85}Sn_{0.15}O_3$	0.350	20.0	281	0.0175	I	
$Ba_{0.94}Ca_{0.06}Ti_{0.8}Sn_{0.2}O_3$	0.320	20.0	237	0.0160	I	
$Ba_{0.95}Ca_{0.05}Ti_{0.95}Zr_{0.05}O_3$	0.310	15.0	405	0.0207	I	[67]
$Ba_{0.92}Ca_{0.08}Ti_{0.92}Zr_{0.08}O_3$	0.380	15.0	409	0.0253	I	
$Ba_{0.9}Ca_{0.1}Ti_{0.92}Zr_{0.08}O_3$	0.230	15.0	406	0.0153	I	
$Ba_{0.8}Sr_{0.2}TiO_3$	0.390	20.0	349	0.0195	I	[68]
$Ba_{0.75}Sr_{0.25}TiO_3$	0.370	20.0	334	0.0185	I	
$Ba_{0.7}Sr_{0.3}TiO_3$	0.360	20.0	303	0.0180	I	
$Ba_{0.65}Sr_{0.35}TiO_3$	0.420	20.0	296	0.0210	I	
$Ba_{0.8}Zr_{0.2}TiO_3$	2.780	161.0	350	0.0173	I	[69]
$Ba_{0.7}Sr_{0.3}TiO_3$ con	0.670	40.0	312	0.0168	I	[70]
$Ba_{0.7}Sr_{0.3}TiO_3$ SPS	1.850	90.0	312	0.0206	I	
$Ba_{0.7}Sr_{0.3}Ti_{0.997}Mn_{0.003}O_3$ SPS	2.530	120.0	303	0.0211	I	
$Bi_{0.5}Na_{0.5}TiO_3$–$0.06BaTiO_3$	0.860	50.0	383	0.0172	D	[71]
$Ba_{0.95}Ca_{0.05}Ti_{0.94}Sn_{0.06}O_3$	1.600	180.0	300	0.0089	I	[72]
$0.45BaTi_{0.8}Zr_{0.2}O_3$-$0.55Ba_{0.7}Ca_{0.3}TiO_3$	0.460	12.0	404	0.0383	I	[73]

Table A1. *Cont.*

Composition	ΔT [K]	E [kV/cm]	T [K]	$\Delta T/E$ [K·cm/kV]	M	Ref.
$0.65BaTi_{0.8}Zr_{0.2}O_3$-$0.35Ba_{0.7}Ca_{0.3}TiO_3$	0.330	20.0	338	0.0165	D	[74]
$Ba_{0.8}Sr_{0.2}TiO_3$	1.670	50.0	344	0.0334	I	[75]
$Ba_{0.65}Sr_{0.35}TiO_3$ Con	0.830	90.0	303	0.0092	I	[76]
$Ba_{0.65}Sr_{0.35}TiO_3$ SPS	2.100	90.0	303	0.0233	I	
$Ba_{0.65}Sr_{0.35}Ti_{0.997}Mn_{0.003}O_3$ SPS	3.080	130.0	293	0.0237	I	

References

1. Greco, A.; Masselli, C. Electrocaloric Cooling: A Review of the Thermodynamic Cycles, Materials, Models, and Devices. *Magnetochemistry* **2020**, *6*, 67. [CrossRef]
2. Singh, S.K.; Rakshit, D.; Kumar, K.R.; Agarwal, A. Recent advancements and sustainable solutions in adsorption-based cooling systems integrated with renewable energy sources and industrial waste heat: A review. *Clean. Eng. Technol.* **2024**, *23*, 100827. [CrossRef]
3. Guzmán-Verri, G.G.; Littlewood, P.B. Why is the electrocaloric effect so small in ferroelectrics? *APL Mater.* **2016**, *4*, 064106. [CrossRef]
4. Klinar, K.; Law, J.Y.; Franco, V.; Moya, X.; Kitanovski, A. Perspectives and Energy Applications of Magnetocaloric, Pyromagnetic, Electrocaloric, and Pyroelectric Materials. *Adv. Energy Mater.* **2024**, *14*, 2401739. [CrossRef]
5. Bai, Y.; Ding, K.; Zheng, G.P.; Shi, S.Q.; Qiao, L.J.; Guo, D. The Electrocaloric Effect in $BaTiO_3$ Thick Film Multilayer Structure at High Electric Field. *Key Eng. Mater.* **2012**, *512*, 1304–1307. [CrossRef]
6. Mischenko, A.S.; Zhang, Q.; Scott, J.F.; Whatmore, R.W.; Mathur, N.D. Giant Electrocaloric Effect in Thin-Film $PbZr_{0.95}Ti_{0.05}O_3$. *Science* **2006**, *311*, 1270–1271. [CrossRef]
7. Sun, Y.; Shirsath, S.E.; Zhang, S.; Wang, D. A reflection on recent efforts in optimization of cooling capacity of electrocaloric thin films. *APL Mater.* **2023**, *11*, 090602. [CrossRef]
8. Shi, J.; Han, D.; Li, Z.; Yang, L.; Lu, S.-G.; Zhong, Z.; Chen, J.; Zhang, Q.; Qian, X. Electrocaloric Cooling Materials and Devices for Zero-Global-Warming-Potential, High-Efficiency Refrigeration. *Joule* **2019**, *3*, 1200–1225. [CrossRef]
9. Bai, Y.; Han, X.; Zheng, X.-C.; Qiao, L. Both High Reliability and Giant Electrocaloric Strength in $BaTiO_3$ Ceramics. *Sci. Rep.* **2013**, *3*, 2895. [CrossRef]
10. Patel, S.; Kumar, M. Influence of grain size on the electrocaloric and pyroelectric properties in non-reducible $BaTiO_3$ ceramics. *AIP Adv.* **2020**, *10*, 085302. [CrossRef]
11. Han, X.; Bai, Y.; Qiao, L.J. The Electrocaloric Effect of $BaTiO_3$ Ceramics Using Hydrothermal Synthesized Nano-Sized Starting Powders. *Adv. Mater. Res.* **2013**, *624*, 138–141. [CrossRef]
12. Gong, J.; Chu, S.; Mehta, R.K.; McGaughey, A.J.H. XGBoost model for electrocaloric temperature change prediction in ceramics. *npj Comput. Mater.* **2022**, *8*, 140. [CrossRef]
13. Su, M.; Grimes, R.; Garg, S.; Xue, D.; Balachandran, P.V. Machine-Learning-Enabled Prediction of Adiabatic Temperature Change in Lead-Free $BaTiO_3$-Based Electrocaloric Ceramics. *ACS Appl. Mater. Interfaces* **2021**, *13*, 53475–53484. [CrossRef]
14. Bayir, M.C.; Mensur, E. Electrocaloric property optimization of PMN-PT ceramics using predictive modeling. *J. Mater. Sci. Mater. Electron.* **2025**, *36*, 1097. [CrossRef]
15. Yuan, R.; Liu, Z.; Xu, Y.; Yin, R.; He, J.; Bai, Y.; Zhou, Y.; Li, J.; Xue, D.; Lookman, T. Optimizing Electrocaloric Effect in Barium Titanate-Based Room Temperature Ferroelectrics: Combining Landau Theory, Machine Learning and Synthesis. *Acta Mater.* **2022**, *235*, 118054. [CrossRef]
16. Xu, P.; Ji, X.; Li, M.; Lu, W. Small data machine learning in materials science. *npj Comput. Mater.* **2023**, *9*, 42. [CrossRef]
17. Oviedo, F.; Ferres, J.L.; Buonassisi, T.; Butler, K.T. Interpretable and Explainable Machine Learning for Materials Science and Chemistry. *Accounts Mater. Res.* **2022**, *3*, 597–607. [CrossRef]
18. Chang, R.; Wang, Y.-X.; Ertekin, E. Towards overcoming data scarcity in materials science: Unifying models and datasets with a mixture of experts framework. *npj Comput. Mater.* **2022**, *8*, 242. [CrossRef]
19. Kailkhura, B.; Gallagher, B.; Kim, S.; Hiszpanski, A.; Han, T.Y.-J. Reliable and explainable machine-learning methods for accelerated material discovery. *npj Comput. Mater.* **2019**, *5*, 108. [CrossRef]
20. Zhao, Y.; Wang, Q.; Lv, J.H.; Zhao, X.; Wang, C.M. Dielectric properties and electrocaloric effect of yttri-um-modified $BaTiO_3$ ceramics. *Ceram. Int.* **2021**, *47*, 18610–18618. [CrossRef]
21. Zhang, B.; Lou, X.; Zheng, K.; Xie, X.; Shi, P.; Guo, M.; Zhu, X.; Gao, Y.; Liu, Q.; Kang, R. Enhanced electrocaloric effect in the Sm and Hf co-doped $BaTiO_3$ ceramics. *Ceram. Int.* **2021**, *47*, 1101–1108. [CrossRef]

22. Han, F.; Bai, Y.; Qiao, L.-J.; Guo, D. A systematic modification of the large electrocaloric effect within a broad temperature range in rare-earth doped $BaTiO_3$ ceramics. *J. Mater. Chem. C* **2016**, *4*, 1842–1849. [CrossRef]
23. Guvenc, C.M.; Adem, U. Influence of aging on electrocaloric effect in Li^+ doped $BaTiO_3$ ceramics. *J. Alloys Compd.* **2019**, *791*, 674–680. [CrossRef]
24. Liu, S.; Xie, Q.; Zhang, L.; Zhao, Y.; Wang, X.; Mao, P.; Wang, J.; Lou, X. Tunable electrocaloric and energy storage behavior in the Ce, Mn hybrid doped $BaTiO_3$ ceramics. *J. Eur. Ceram. Soc.* **2018**, *38*, 4664–4669. [CrossRef]
25. Upadhyay, S.K.; Reddy, V.R.; Bag, P.; Rawat, R.; Gupta, S.M.; Gupta, A. Electro-caloric effect in lead-free Sn doped $BaTiO_3$ ceramics at room temperature and low applied fields. *Appl. Phys. Lett.* **2014**, *105*, 112907. [CrossRef]
26. Upadhyay, S.K.; Fatima, I.; Reddy, V.R. Study of electro-caloric effect in Ca and Sn co-doped $BaTiO_3$ ceramics. *Mater. Res. Express* **2017**, *4*, 046303. [CrossRef]
27. Hou, Y.; Yang, L.; Qian, X.; Zhang, T.; Zhang, Q.M. Electrocaloric response near room temperature in Zr- and Sn-doped $BaTiO_3$ systems. *Philos. Trans. R. Soc. A Math. Phys. Eng. Sci.* **2016**, *374*, 20160055. [CrossRef]
28. Zhao, L.; Ke, X.; Zhou, Z.; Liao, X.; Li, J.; Wang, Y.; Wu, M.; Li, T.; Bai, Y.; Ren, X. Large electrocaloric effect over a wide temperature range in $BaTiO_3$-modified lead-free ceramics. *J. Mater. Chem. C* **2019**, *7*, 1353–1358. [CrossRef]
29. Gwizd, P.; Sitko, D.; Jankowska-Sumara, I.; Krupska-Klimczak, M. The electrocaloric effect in $BaTiO_3$:Eu ceramics determined by an indirect method. *Phase Transit.* **2021**, *94*, 192–198. [CrossRef]
30. Bai, Y.; Han, X.; Qiao, L.-J. Effect of donor doping in B sites on the electrocaloric effect of $BaTi_{1-x}Nb_xO_3$ ceramics. *RSC Adv.* **2015**, *5*, 71873–71877. [CrossRef]
31. Qiang, H.; Fu, S.; Xu, Z. Electrocaloric Effect in Sn- and Y-Codoped $BaTiO_3$ Ceramics. *Phys. Status Solidi (A)* **2021**, *218*, 2100090. [CrossRef]
32. Qian, X.; Ye, H.; Zhang, Y.; Gu, H.; Li, X.; Randall, C.A.; Zhang, Q.M. Giant Electrocaloric Response over a Broad Temperature Range in Modified $BaTiO_3$ Ceramics. *Adv. Funct. Mater.* **2014**, *24*, 1300–1305. [CrossRef]
33. Meng, Y.; Tang, S.; Wang, Z.; Niu, X.; Zhang, H.; Wang, D.; Bai, Y.; Peng, B.; Lu, S.-G.; Ke, Q.; et al. Significantly enhanced electrocaloric effect by composition modulation in lead-free $BaTiO_3$-based ceramics. *J. Mater.* **2024**, *11*, 100903. [CrossRef]
34. Krupska-Klimczak, M.; Ziewiec, K.; Jankowska-Sumara, I. The Influence of Measurement Conditions on the Electrocaloric Effect in Ferroelectric Ceramics of Modified Barium Titanate. *Materials* **2024**, *17*, 3329. [CrossRef]
35. Lv, Z.; Wei, J.; Yang, T.; Sun, Z.; Xu, Z. Manipulation of Curie temperature and ferroelectric polarization for large electrocaloric strength in BaTiO3-based ceramics. *Ceram. Int.* **2020**, *46*, 14978–14984. [CrossRef]
36. Asbani, B.; Gagou, Y.; Trček, M.; Dellis, J.-L.; Amjoud, M.; Lahmar, A.; Mezzane, D.; Kutnjak, Z.; El Marssi, M. Dielectric permittivity enhancement and large electrocaloric effect in the lead free $(Ba_{0.8}Ca_{0.2})_{1-x}La_{2x/3}TiO_3$ ferroelectric ceramics. *J. Alloys Compd.* **2018**, *730*, 501–508. [CrossRef]
37. Zhao, C.; Huang, Y.; Wu, X.; Gao, M.; Lin, T.; Lin, C. Evolution behavior of electrocaloric effect in $BaTiO_3$-based ceramics with different phase structures and phase transitions. *J. Am. Ceram. Soc.* **2024**, *107*, 195–204. [CrossRef]
38. Achkar, M.; Fasquelle, D.; Duponchel, B.; Poupin, C.; Sahraoui, A.H.; Longuemart, S. Electrocaloric characterization of samarium doped barium titanate ceramics synthesized by sol-gel process. *Ceram. Int.* **2024**, *50*, 13794–13801. [CrossRef]
39. Xu, Z.; Wang, J.; Chen, Y. Improved electrocaloric effect of $Ba_{0.7}Sr_{0.3}TiO_3$ ceramics doped with B and Mn. *J. Sol-Gel Sci. Technol.* **2023**, *107*, 483–489. [CrossRef]
40. Niu, X.; Jian, X.; Gong, W.; Liang, W.; Gong, X.; Zhang, G.; Jiang, S.; Yu, K.; Zhao, X.; Yao, Y.; et al. Field-driven merging of polarizations and enhanced electrocaloric effect in $BaTiO_3$-based lead-free ceramics. *J. Adv. Ceram.* **2022**, *11*, 1777–1788. [CrossRef]
41. Qi, S.; Zhang, G.; Duan, L.; Zeng, T.; Cao, J. Electrocaloric effect in Pb-free Sr-doped $BaTi_{0.9}Sn_{0.1}O_3$ ceramics. *Mater. Res. Bull.* **2017**, *91*, 31–35. [CrossRef]
42. Niu, X.; Jian, X.; Chen, X.; Li, H.; Liang, W.; Yao, Y.; Tao, T.; Liang, B.; Lu, S.-G. Enhanced electrocaloric effect at room temperature in Mn^{2+} doped lead-free (BaSr)TiO_3 ceramics via a direct measurement. *J. Adv. Ceram.* **2021**, *10*, 482–492. [CrossRef]
43. Redhu, P.; Hooda, A.; Sharma, P.; Dahiya, S.; Punia, R.; Tandon, R.P. Study of energy storage and electrocaloric behavior of lead-free Fe-doped BCT ceramics. *Ferroelectrics* **2020**, *569*, 136–147. [CrossRef]
44. Asbani, B.; Gagou, Y.; Ben Moumen, S.; Dellis, J.-L.; Lahmar, A.; Amjoud, M.; Mezzane, D.; El Marssi, M.; Rozic, B.; Kutnjak, Z. Large Electrocaloric Responsivity and Energy Storage Response in the Lead-Free $Ba(Ge_xTi_{1-x})O_3$ Ceramics. *Materials* **2022**, *15*, 5227. [CrossRef] [PubMed]
45. Patel, S.; Chauhan, A.; Vaish, R. Multiple caloric effects in $(Ba_{0.865}Ca_{0.135}Zr_{0.1089}Ti_{0.8811}Fe_{0.01})O_3$ ferroelectric ceramic. *Appl. Phys. Lett.* **2015**, *107*, 042902. [CrossRef]
46. Şanlı, K.; Adem, U. Electrocaloric properties of $Ba_{0.8}Sr_{0.2}Ti_{1-x}Zr_xO_3$ ($0 \leq x \leq 0.1$) system: The balance between the nature of the phase transition and phase coexistence. *Ceram. Int.* **2020**, *46*, 2213–2219. [CrossRef]

47. Patel, S.; Chauhan, A.; Vaish, R. Large room temperature electrocaloric strength in bulk ferroelectric ceramics: An optimum solution. *Phase Transit.* **2016**, *89*, 1019–1028. [CrossRef]
48. Wu, P.; Lou, X.; Li, J.; Li, T.; Gao, H.; Wu, M.; Wang, S.; Wang, X.; Bian, J.; Hao, X. Direct and indirect measurement of electrocaloric effect in lead-free (100-x)Ba($Hf_{0.2}Ti_{0.8}$)O_3-x($Ba_{0.7}Ca_{0.3}$)TiO_3 ceramics near multi-phase boundary. *J. Alloys Compd.* **2017**, *725*, 275–282. [CrossRef]
49. Weyland, F.; Hayati, R.; Novak, N. Tuning of electrocaloric performance in ($Ba_{0.85}Ca_{0.15}$)($Zr_{0.1}Ti_{0.9}$)O_3 by induced relaxor-like behavior. *Ceram. Int.* **2019**, *45*, 11408–11412. [CrossRef]
50. Lu, S.; Chen, G.; Zhang, Y.; Zhao, Z.; Li, F.; Lv, Z.; Ma, Z.; Wang, D.; Lu, C.; Li, S. Electrocaloric effect in lead-free 0.5Ba($Zr_{0.2}Ti_{0.8}$)O_3-0.5(Ba0.7$Ca_{0.3}$)TiO_3 ceramic measured by direct and indirect methods. *Ceram. Int.* **2018**, *44*, 21950–21955. [CrossRef]
51. Sanlialp, M.; Shvartsman, V.V.; Acosta, M.; Lupascu, D.C.; Alford, N. Electrocaloric Effect in Ba(Zr,Ti)O_3–(Ba,Ca)TiO_3 Ceramics Measured Directly. *J. Am. Ceram. Soc.* **2016**, *99*, 4022–4030. [CrossRef]
52. Bai, Y.; Han, X.; Qiao, L. Optimized electrocaloric refrigeration capacity in lead-free $(1-x)BaZr_{0.2}Ti_{0.8}O_3$-$xBa_{0.7}Ca_{0.3}TiO_3$ ceramics. *Appl. Phys. Lett.* **2013**, *102*, 252904. [CrossRef]
53. Zhang, H.F.; Chen, X.; Gao, J.; Lam, K.H.; Liang, S.D.; Fei, L.F.; Mak, C.L.; Chen, J.F. The Fabrication and Electrocaloric Effect of Bimodal-Grain Structure ($Ba_{0.60}Sr_{0.40}$)TiO_3 Using the Induced Abnormal Grain Growth Method. *IOP Conf. Ser. Mater. Sci. Eng.* **2019**, *678*, 012138. [CrossRef]
54. Srikanth, K.S.; Patel, S.; Vaish, R. Electrocaloric behavior and temperature dependent scaling of dynamic hysteresis of BaxSr1-xTiO_3 (x = 0.7, 0.8 and 0.9) bulk ceramics. *J. Aust. Ceram. Soc.* **2018**, *54*, 439–450. [CrossRef]
55. Zhao, Y.; Liu, X.; Wu, S.; Chen, X. Effect of phase transition on electrocaloric effect in Indium substituted $BaTiO_3$ ceramics. *J. Alloys Compd.* **2020**, *822*, 153632. [CrossRef]
56. Zhao, Y.; Liu, X.; Wu, J.; Wu, S.; Chen, X. Electrocaloric effect in relaxor ferroelectric Ba($Ti_{1-x}Y_x$)$O_{3-x/2}$ ceramics over a broad temperature range. *J. Alloys Compd.* **2017**, *729*, 57–63. [CrossRef]
57. Li, J.; Zhang, D.; Qin, S.; Li, T.; Wu, M.; Wang, D.; Bai, Y.; Lou, X. Large room-temperature electrocaloric effect in lead-free BaHf TiO_3 ceramics under low electric field. *Acta Mater.* **2016**, *115*, 58–67. [CrossRef]
58. Srikanth, K.; Vaish, R. Enhanced electrocaloric, pyroelectric and energy storage performance of $BaCe_xTi_{1-x}O_3$ ceramics. *J. Eur. Ceram. Soc.* **2017**, *37*, 3927–3933. [CrossRef]
59. Maiwa, H. Temperature dependences of the electromechanical and electrocaloric properties of Ba(Zr,Ti)O_3 and (Ba,Sr)TiO_3 ceramics. *Jpn. J. Appl. Phys.* **2017**, *56*, 10PC05. [CrossRef]
60. Sanlialp, M.; Shvartsman, V.V.; Faye, R.; Karabasov, M.O.; Molin, C.; Gebhardt, S.; Defay, E.; Lupascu, D.C. Quasi-adiabatic calorimeter for direct electrocaloric measurements. *Rev. Sci. Instruments* **2018**, *89*, 034903. [CrossRef] [PubMed]
61. Li, J.; Li, D.-D.; Lu, B.; Yao, Y.-B.; Tao, T.; Liang, B.; Lu, W.-H.; Luo, G.-L.; Lu, S.-G. Large electrocaloric effect obtained in Ba(Sn_xTi_{1-x})O_3 lead-free ceramics using direct and indirect measurements. *J. Adv. Dielectr.* **2018**, *8*, 1850038. [CrossRef]
62. Zhang, X.; Wu, L.; Gao, S.; Liu, J.Q.; Xu, B.; Xia, Y.D.; Yin, J.; Liu, Z.G. Large electrocaloric effect in Ba($Ti_{1-x}Sn_x$)O_3 ceramics over a broad temperature region. *AIP Adv.* **2015**, *5*, 047134. [CrossRef]
63. Sanlialp, M.; Luo, Z.; Shvartsman, V.V.; Wei, X.; Liu, Y.; Dkhil, B.; Lupascu, D.C. Direct measurement of electrocaloric effect in lead-free Ba(Sn_xTi_{1-x})O_3 ceramics. *Appl. Phys. Lett.* **2017**, *111*, 173903. [CrossRef]
64. Luo, Z.; Zhang, D.; Liu, Y.; Zhou, D.; Yao, Y.; Liu, C.; Dkhil, B.; Ren, X.; Lou, X. Enhanced electrocaloric effect in lead-free $BaTi_{1-x}Sn_xO_3$ ceramics near room temperature. *Appl. Phys. Lett.* **2014**, *105*, 102904. [CrossRef]
65. Asbani, B.; Dellis, J.-L.; Lahmar, A.; Courty, M.; Amjoud, M.; Gagou, Y.; Djellab, K.; Mezzane, D.; Kutnjak, Z.; El Marssi, M. Lead-free $Ba_{0.8}Ca_{0.2}$(ZrxTi_{1-x})O_3 ceramics with large electrocaloric effect. *Appl. Phys. Lett.* **2015**, *106*, 042902. [CrossRef]
66. Wang, X.; Tian, F.; Zhao, C.; Wu, J.; Liu, Y.; Dkhil, B.; Zhang, M.; Gao, Z.; Lou, X. Giant electrocaloric effect in lead-free $Ba_{0.94}Ca_{0.06}Ti_{1-x}Sn_xO_3$ ceramics with tunable Curie temperature. *Appl. Phys. Lett.* **2015**, *107*, 252905. [CrossRef]
67. Singh, G.; Tiwari, V.S.; Gupta, P.K. Electro-caloric effect in ($Ba_{1-x}Ca_x$)($Zr_{0.05}Ti_{0.95}$)O_3: A lead-free ferroelectric material. *Appl. Phys. Lett.* **2013**, *103*, 202903. [CrossRef]
68. Bai, Y.; Han, X.; Ding, K.; Qiao, L. Combined effects of diffuse phase transition and microstructure on the electrocaloric effect in $Ba_{1-x}Sr_xTiO_3$ ceramics. *Appl. Phys. Lett.* **2013**, *103*, 162902. [CrossRef]
69. Zhang, Y. Thermal hysteresis and electrocaloric effect in $Ba_{1-x}Zr_xTiO$. *J. Phys. Chem. Solids* **2018**, *115*, 326–331. [CrossRef]
70. Liu, X.Q.; Chen, T.T.; Fu, M.S.; Wu, Y.J.; Chen, X.M. Electrocaloric effects in spark plasma sintered $Ba_{0.7}Sr_{0.3}TiO_3$-based ceramics: Effects of domain sizes and phase constitution. *Ceram. Int.* **2014**, *40*, 11269–11276. [CrossRef]
71. Li, F.; Chen, G.; Liu, X.; Zhai, J.; Shen, B.; Li, S.; Li, P.; Yang, K.; Zeng, H.; Yan, H. Type–I pseudo–first–order phase transition induced electrocaloric effect in lead–free $Bi_{0.5}Na_{0.5}TiO_3$–0.06$BaTiO_3$ ceramics. *Appl. Phys. Lett.* **2017**, *110*, 182904. [CrossRef]
72. Du, H.; Chang, Y.; Li, C.; Hu, Q.; Pang, J.; Sun, Y.; Weyland, F.; Novak, N.; Jin, L. Ultrahigh room temperature electrocaloric response in lead-free bulk ceramics via tape casting. *J. Mater. Chem. C* **2019**, *7*, 6860–6866. [CrossRef]

73. Singh, G.; Bhaumik, I.; Ganesamoorthy, S.; Bhatt, R.; Karnal, A.K.; Tiwari, V.S.; Gupta, P.K. Electro-caloric effect in $0.45BaZr_{0.2}Ti_{0.8}O_3$-$0.55Ba_{0.7}Ca_{0.3}TiO_3$ single crystal. *Appl. Phys. Lett.* **2013**, *102*, 082902. [CrossRef]
74. Sanlialp, M.; Shvartsman, V.V.; Acosta, M.; Dkhil, B.; Lupascu, D.C. Strong electrocaloric effect in lead-free $0.65Ba(Zr_{0.2}Ti_{0.8})O_3$-$0.35(Ba_{0.7}Ca_{0.3})TiO_3$ ceramics obtained by direct measurements. *Appl. Phys. Lett.* **2015**, *106*, 062901. [CrossRef]
75. Bai, Y.; Han, F.; Xie, S.; Li, J.; Qin, S.; Li, J.; Qiao, L.; Guo, D. Thickness dependence of electrocaloric effect in high-temperature sintered $Ba_{0.8}Sr_{0.2}TiO_3$ ceramics. *J. Alloys Compd.* **2018**, *736*, 57–61. [CrossRef]
76. Liu, X.Q.; Chen, T.T.; Wu, Y.J.; Chen, X.M.; Valant, M. Enhanced Electrocaloric Effects in Spark Plasma-Sintered $Ba_{0.65}Sr_{0.35}TiO_3$-Based Ceramics at Room Temperature. *J. Am. Ceram. Soc.* **2013**, *96*, 1021–1023. [CrossRef]
77. Kutnjak, Z.; Rožič, B.; Pirc, R. Electrocaloric Effect: Theory, Measurements, and Applications. In *Wiley Encyclopedia of Electrical and Electronics Engineering*; John Wiley & Sons, Inc.: Hoboken, NJ, USA, 2015; pp. 1–19. [CrossRef]
78. Valant, M. Electrocaloric materials for future solid-state refrigeration technologies. *Prog. Mater. Sci.* **2012**, *57*, 980–1009. [CrossRef]
79. Liu, Y.; Scott, J.F.; Dkhil, B. Direct and indirect measurements on electrocaloric effect: Recent developments and perspectives. *Appl. Phys. Rev.* **2016**, *3*, 031102. [CrossRef]
80. Pirc, R.; Kutnjak, Z.; Blinc, R.; Zhang, Q.M. Electrocaloric effect in relaxor ferroelectrics. *J. Appl. Phys.* **2011**, *110*, 074113. [CrossRef]
81. Cheng, X.; Weyland, F.; Novak, N.; Li, Y. Indirect Electrocaloric Evaluation: Influence of Polarization Hysteresis Measurement Frequency. *Phys. Status Solidi (A)* **2019**, *216*, 1900684. [CrossRef]
82. Bai, Y.; Zheng, G.; Shi, S. Direct measurement of giant electrocaloric effect in $BaTiO_3$ multilayer thick film structure beyond theoretical prediction. *Appl. Phys. Lett.* **2010**, *96*, 192902. [CrossRef]
83. Kaddoussi, H.; Lahmar, A.; Gagou, Y.; Asbani, B.; Dellis, J.-L.; Cordoyiannis, G.; Allouche, B.; Khemakhem, H.; Kutnjak, Z.; El Marssi, M. Indirect and direct electrocaloric measurements of $(Ba_{1-x}Ca_x)(Zr_{0.1}Ti_{0.9})O_3$ ceramics (x = 0.05, x = 0.20). *J. Alloys Compd.* **2016**, *667*, 198–203. [CrossRef]
84. Asbani, B.; Dellis, J.L.; Gagou, Y.; Kaddoussi, H.; Lahmar, A.; Amjoud, M.; El Marssi, M. Electrocaloric effect in $Ba_{0.2}Ca_{0.8}Ti_{0.95}Ge_{0.05}O_3$ determined by a new pyroelectric method. *Europhys. Lett.* **2015**, *111*, 57008. [CrossRef]
85. Sebald, G.; Guyomar, D.; Agbossou, A. On thermoelectric and pyroelectric energy harvesting. *Smart Mater. Struct.* **2009**, *18*, 125006. [CrossRef]
86. Moya Raposo, X.E. Effect of inactive volume on thermocouple measurements of electrocaloric temperature change in multilayer capacitors of $0.9Pb(Mg_{1/3}Nb_{2/3})O_3$–$0.1PbTiO_3$. *J. Phys. D Appl. Phys.* **2017**, *50*, 42. [CrossRef]
87. Liu, Y.; Dkhil, B.; Defay, E. Spatially resolved imaging of electrocaloric effect and the resultant heat flux in multilayer capacitors. *ACS Energy Lett.* **2016**, *1*, 521–528. [CrossRef]
88. Borkar, S.N.; Aggarwal, P.; Deshpande, V. Effect of calcium substitution on structural, dielectric, ferroelectric, piezoelectric, and energy storage properties of $BaTiO_3$. *Curr. Appl. Phys.* **2022**, *39*, 205–213. [CrossRef]
89. Ali, S.; Ahmad, S.; Farooq, J.; Uddin, S.; Chen, B.; Ullah, I.; Alam, M.M. The effect of A-site strontium substitution on the energy storage properties of ferroelectric $BaZr_{0.2}Ti_{0.8}O_3$ system. *Ceram. Int.* **2025**, *51*, 14821–14833. [CrossRef]
90. Younas, U.; Atif, M.; Anjum, A.; Nadeem, M.; Ali, T.; Shaheen, R.; Khalid, W.; Ali, Z. Fabrication of La^{3+} doped $Ba_{1-x}La_xTiO_3$ ceramics with improved dielectric and ferroelectric properties using a composite-hydroxide-mediated method. *RSC Adv.* **2023**, *13*, 5293–5306. [CrossRef]
91. Xu, Q.; Li, Z. Dielectric and ferroelectric behaviour of Zr-doped $BaTiO_3$ perovskites. *Process. Appl. Ceram.* **2020**, *14*, 188–194. [CrossRef]
92. Tang, Q.; Shi, Z.; Xia, S.; Bie, X.; Yang, Y.; Bian, D.; Xu, D.; Fan, R. Enhanced dielectric properties of Sr^{2+} and Zr^{4+} doped $BaTiO_3$ colossal permittivity metamaterials. *EPJ Appl. Metamaterials* **2024**, *11*, 13. [CrossRef]
93. Anwar, S.; Sagdeo, P.; Lalla, N. Ferroelectric relaxor behavior in hafnium doped barium-titanate ceramic. *Solid State Commun.* **2006**, *138*, 331–336. [CrossRef]
94. Brzozowski, E.; Castro, M. Influence of Nb^{5+} and Sb^{3+} dopants on the defect profile, PTCR effect and GBBL characteristics of $BaTiO_3$ ceramics. *J. Eur. Ceram. Soc.* **2004**, *24*, 2499–2507. [CrossRef]
95. Si, F.; Tang, B.; Fang, Z.; Zhang, S. Nb-Doped $0.8BaTiO_3$-$0.2Bi(Mg_{0.5}Ti_{0.5})O_3$ Ceramics with Stable Dielectric Properties at High Temperature. *Crystals* **2017**, *7*, 168. [CrossRef]
96. Islam, A.; Momin, A.; Nesa, M. Effect of Fe doping on the structural, optical and electronic properties of $BaTiO_3$: DFT based calculation. *Chin. J. Phys.* **2019**, *60*, 731–738. [CrossRef]
97. Horchidan, N.; Lukacs, V.A.; Asandulesa, M.; Stoian, G.; Ciomaga, C.E.; Mitoseriu, L. Exploring grain size and composition effects on the functional properties of $BaGe_xTi_{1-x}O_3$ ceramics. *J. Eur. Ceram. Soc.* **2025**, *45*, 116997. [CrossRef]
98. Sun, J.; Wang, S.; Tong, L.; Li, Q.; Yu, Y.; Li, Y.; Huang, S.; Guo, Y.; Wang, C. Enhanced dielectric properties in (In, Nb) co-doped $BaTiO_3$ ceramics. *Mater. Lett.* **2017**, *200*, 51–54. [CrossRef]

99. Sareecha, N.; Shah, W.; Mirza, M.; Maqsood, A.; Awan, M. Electrical investigations of Bi-doped $BaTiO_3$ ceramics as a function of temperature. *Phys. B Condens. Matter* **2018**, *530*, 283–289. [CrossRef]
100. Ismail, F.A.; Osman, R.A.M.; Idris, M.S. Review on dielectric properties of rare earth doped barium titanate. In Proceedings of the 2nd International Conference on Functional Materials and Metallurgy (ICOFM 2016), Penang, Malaysia, 28 May 2016; p. 090005.
101. Lu, S.G.; Rožič, B.; Zhang, Q.M.; Kutnjak, Z.; Pirc, R.; Lin, M.; Li, X.; Gorny, L. Comparison of directly and indirectly measured electrocaloric effect in relaxor ferroelectric polymers. *Appl. Phys. Lett.* **2010**, *97*, 202901. [CrossRef]
102. Rožič, B.; Kosec, M.; Uršič, H.; Holc, J.; Malič, B.; Zhang, Q.M.; Blinc, R.; Pirc, R.; Kutnjak, Z. Influence of the critical point on the electrocaloric response of relaxor ferroelectrics. *J. Appl. Phys.* **2011**, *110*, 064118. [CrossRef]
103. Quinlan, J.R. Induction of decision trees. *Mach. Learn.* **1986**, *1*, 81–106. [CrossRef]
104. Mohammed, A.; Kora, R. A comprehensive review on ensemble deep learning: Opportunities and challenges. *J. King Saud Univ. Comput. Inf. Sci.* **2023**, *35*, 757–774. [CrossRef]
105. Mayr, A.; Binder, H.; Gefeller, O.; Schmid, M. The evolution of boosting algorithms. *Methods Inf. Med.* **2014**, *53*, 419–427. [PubMed]
106. Chen, T.; Guestrin, C. Xgboost: A scalable tree boosting system. In Proceedings of the 22nd ACM Sigkdd International Conference on Knowledge Discovery and Data Mining, San Francisco, CA, USA, 13–17 August 2016; pp. 785–794.
107. Liu, W.; Ren, X. Large Piezoelectric Effect in Pb-Free Ceramics. *Phys. Rev. Lett.* **2009**, *103*, 257602. [CrossRef] [PubMed]
108. He, J.; Li, J.; Liu, C.; Wang, C.; Zhang, Y.; Wen, C.; Xue, D.; Cao, J.; Su, Y.; Qiao, L.; et al. Machine learning identified materials descriptors for ferroelectricity. *Acta Mater.* **2021**, *209*, 116815. [CrossRef]
109. He, J.; Wang, C.; Li, J.; Liu, C.; Xue, D.; Cao, J.; Su, Y.; Qiao, L.; Lookman, T.; Bai, Y. Machine learning assisted prediction of dielectric temperature spectrum of ferroelectrics. *J. Adv. Ceram.* **2023**, *12*, 1793–1804. [CrossRef]
110. Sun, Z.; Gao, R.; Wang, P.; Liu, X.; Bai, Y.; Luo, J.; Yang, H.; Hu, W. Accelerating the prediction of remanent polarization in multicomponent ferroelectrics by using variational autoencoder-based data augmentation. *J. Mater. Chem. C* **2025**. [CrossRef]
111. Beckman, S.P.; Wan, L.F.; Barr, J.A.; Nishimatsu, T. Effective Hamiltonian methods for predicting the electrocaloric behavior of $BaTiO_3$. *Mater. Lett.* **2012**, *89*, 254–257. [CrossRef]
112. Marathe, M.; Grünebohm, A.; Nishimatsu, T.; Entel, P.; Ederer, C. First-principles-based calculation of the electrocaloric effect in $BaTiO_3$: A comparison of direct and indirect methods. *Phys. Rev. B* **2016**, *93*, 054110. [CrossRef]
113. Wojdeł, J.C.; Hermet, P.; Ljungberg, M.P.; Ghosez, P.; Íñiguez, J. First-principles model potentials for lattice-dynamical studies: General methodology and example of application to ferroic perovskite oxides. *J. Physics Condens. Matter* **2013**, *25*, 305401. [CrossRef]

materials

MDPI

Communication

Observation of Excitonic Doublet Structure, Biexcitons and Their Temperature Dependence in High-Quality β-InSe Single Crystals

Tran Thi Thu Huong [1], Long V. Le [1,*], Nguyen Thu Loan [1], Man Hoai Nam [1], Tien-Thanh Nguyen [1], Thi Thuong Huyen Tran [1], Ung Thi Dieu Thuy [1], Thi Huong Nguyen [2] and Tae Jung Kim [3,*]

[1] Institute of Materials Science, Vietnam Academy of Science and Technology, Hanoi 100000, Vietnam; huongttt@ims.vast.ac.vn (T.T.T.H.); loannt@ims.vast.ac.vn (N.T.L.); nammh@ims.vast.ac.vn (M.H.N.); ntthanh@ims.vast.ac.vn (T.-T.N.); huyenttt@ims.vast.ac.vn (T.T.H.T.); dieuthuy@ims.vast.vn (U.T.D.T.)

[2] Department of Physics, Nha Trang University, Nha Trang 650000, Vietnam

[3] Department of Physics, Kyung Hee University, Seoul 02447, Republic of Korea

* Correspondence: longlv@ims.vast.ac.vn (L.V.L.); tjkim@khu.ac.kr (T.J.K.)

Abstract: We present a systematic study of the fundamental optical properties of indium selenide (InSe) single crystals over a temperature range of 17 K to 300 K. The high structural quality of the β-polytype crystals was confirmed through X-ray diffraction, Raman spectroscopy, and high-resolution transmission electron microscopy, demonstrating excellent crystallinity and a nearly stoichiometric In:Se ratio. The temperature-dependent absorption and photoluminescence (PL) spectra are characterized by a prominent free exciton (FX) resonance. At 17 K, the photoluminescence spectrum exhibits a distinct fine-structure splitting of the Wannier–Mott exciton, yielding a triplet state at 1.333 eV and a singlet state at 1.336 eV. Additionally, a biexciton (XX) is localized at an energy of 1.322 eV as confirmed by the nonlinear dependence of intensity on excitation power density. At low temperatures, the absorption spectrum exhibits the free exciton ground state ($n = 1$) at 1.338 eV together with the first excited state ($n = 2$) at 1.350 eV. We systematically tracked and analyzed the temperature evolution of these quasiparticle energies. These findings enhance our understanding of the intrinsic many-body interactions in high-quality InSe, providing essential parameters for advancing its applications in innovative optoelectronic and quantum light-emitting devices.

Keywords: InSe single crystal; Mott-Wannier excitons; biexcitons

1. Introduction

Layered group-III monochalcogenides, including GaSe, GaS, and InSe, have attracted considerable attention due to their unique electronic and optical properties, which stem from their van der Waals-bonded structures and anisotropic band dispersions. Among these materials, the ε- and β-phases of indium selenide are particularly noteworthy, exhibiting a direct band gap in the near-infrared region (~1.35 eV at low temperature and ~1.25 eV at room temperature) [1–5], high electron mobility [6,7], and enhanced environmental stability compared to other layered semiconductors [8,9]. These characteristics make InSe a promising platform for optoelectronic devices [10–12], photovoltaics [13–15], and quantum light sources [16,17].

The optical response of InSe is strongly influenced by excitonic effects, owing to its relatively large exciton binding energy (14–20 meV) [18,19]. At low temperatures, photoluminescence (PL) spectra reveal distinct excitonic features, with previous studies reporting peaks attributed to free excitons, donor–acceptor recombinations, and impurity-related transitions [20–23]. However, despite extensive research on excitonic complexes in two-dimensional transition metal dichalcogenides (TMDs) [24–26] and colloidal quantum wells [27–29], the finer aspects of exciton physics in bulk InSe including doublet splitting and biexciton formation remain inadequately explored.

Excitonic doublet structures, often referred to as fine-structure splitting (FSS), can arise from electron–hole exchange interactions [30], crystal-field asymmetry [31], or interlayer coupling [32]. Such splitting has been extensively documented in semiconductor quantum dots, where FSS values generally range from a few μeV to several meV. Control over this splitting is crucial for applications in entangled photon emission [33]. In layered GaSe, a related III–VI compound, the singlet–triplet splitting of Wannier–Mott excitons has been reported with energy differences of several meV [34,35]. These findings suggest that excitonic doublet structures may also exist in InSe, although experimental confirmation is limited.

The earliest investigation of the optical properties of InSe single crystals was conducted by Damon and Redington in 1954 [36], wherein the optical absorption was examined in mechanically cleaved layers with thicknesses ranging from 0.5 to 80 μm. In 1968, Andriyashik et al. [37] reported the absorption and reflection of InSe single crystals in the energy region of 1 to 5 eV and in the temperature range of 90 to 293 K. A decade later, Camssel et al. [1] studied the temperature dependence of the excitonic absorption edge from 1.6 to 300 K using a 7-μm-thick InSe sample and observed up to the second excited excitonic state ($n = 3$). In 2003 and 2004, Ates et al. [38,39] investigated the temperature-dependent influence of an external electric field on the absorption edge of InSe and InSe:Ho single crystals, and reported that the application of the field induces a shift of the absorption edge toward lower energies in both samples. Specifically, a shift of about 21 meV at 10 K and 11 meV at room temperature was observed for a 5 kV/cm electric field in the pure InSe sample, while a shift of about 23 meV at 10 K and 17 meV at 240 K was noted for a 7.5 kV/cm electric field in the InSe:Ho sample.

In 1982, Abha and Warrier [21] performed the first photoluminescence (PL) study of InSe single crystals over a temperature range of 12 to 300 K, using an Argon ion laser with a 4880 Å line and 500 mW power output. Three peaks at 1.311, 1.272, and 1.227 eV at 12 K were detected, corresponding to impurity-to-band transitions, donor-acceptor type transitions, and transitions within an impurity vacancy complex. Two years later, Gnatenko et al. [19] examined the luminescence spectrum of InSe single crystals doped with chromium (Cr) at concentration of 0.001% across various temperatures (4.5 to 50 K). Their results showed the free exciton emissions characterized by a doublet structure (E_1^A = 1.3371 eV and E_2^A = 1.3365 eV), attributed to valence band splitting due to interlayer interactions. Additionally, two bound exciton states were observed at 1.3343 eV and 1.3234 eV, identified as phonon satellites of the free exciton with optical phonon energies at 17 and 110 cm^{-1}, respectively. Recently, Ertap et al. [40] reported PL studies of undoped and boron-doped InSe crystals in the 10–100 K temperature range. At 12 K, PL spectra exhibited four emission bands: 1.337, 1.304, 1.283, and 1.236 eV for pure InSe; 1.335, 1.305, 1.284, and 1.242 eV for 0.1% B-doped InSe; and 1.330, 1.306, 1.284, and 1.244 eV for 0.5% B-doped InSe. These bands are associated with radiative recombination of direct free excitons ($n = 1$), impurity-band transitions, donor–acceptor recombinations, and defect-related transitions involving impurity atoms, vacancy complexes, and structural imperfections.

In this study, we present an in-depth investigation of the optical properties of high-quality β-InSe single crystals from 17 to 300 K. X-ray diffraction, Raman spectroscopy, and high-resolution TEM confirm excellent crystallinity and a near-stoichiometric composition. Temperature-dependent absorption and photoluminescence (PL) spectra exhibit pronounced excitonic signatures, dominated by a strong free-exciton resonance. At 17 K, the PL resolves the Wannier—Mott exciton fine structure into a triplet at 1.333 eV and a singlet at 1.336 eV; a biexciton line appears at 1.322 eV and is verified by its nonlinear power scaling. The absorption spectrum further resolves the free-exciton ground state (n = 1, 1.338 eV) and first excited state (n = 2, 1.350 eV). Tracking these features across temperature provides new insight into excitonic doublet splitting, biexciton formation, and the thermal stability of quasiparticles in β-InSe.

2. Materials and Methods

2.1. Sample Growth and Preparation

β-InSe single crystals were grown using a temperature gradient method. A stoichiometric mixture of high-purity indium (In, 99.999%) and selenium (Se, 99.99%) powders in a 1:1 molar ratio was sealed in a cleaned, double-walled quartz ampoule under vacuum (~10^{-4} Torr). The ampoule was placed vertically inside an electric furnace and underwent a multi-stage heating process. It was first heated to 723 K at a rate of 20 K·h^{-1} to initiate the reaction. The temperature was then raised to 1023 K at 10 K·h^{-1} and held for 16 h to complete the reaction. The ampoule was subsequently cooled to 723 K at 1 K·h^{-1}, followed by a faster cooling to room temperature at 20 K·h^{-1}. The cylindrical part of the grown ingot has a diameter of approximately 15 mm and a length of about 20 mm.

For absorption measurements, an InSe slab with a thickness of approximately 50 μm was prepared by mechanical exfoliation using Scotch tape, as illustrated in Figure 1c. In addition, InSe flakes were produced via ultrasonic exfoliation in ethanol and subsequently employed for X-ray diffraction (XRD) and transmission electron microscopy (TEM) analyses.

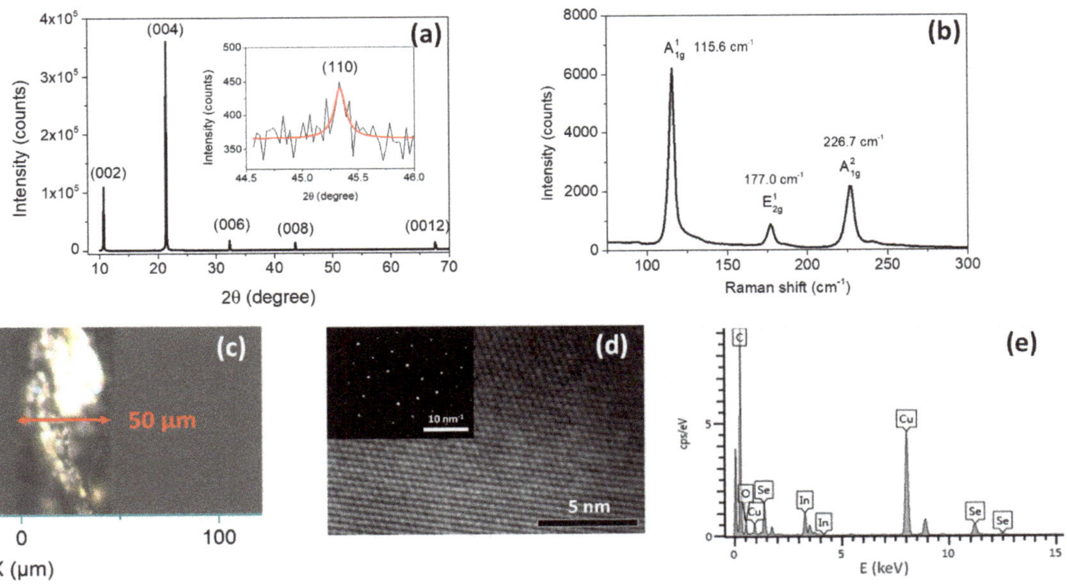

Figure 1. (**a**) XRD pattern of InSe flakes deposited on a glass substrate. (**b**) Raman spectrum of the InSe single crystal characterized on the c-plane. (**c**) Optical image showing the thickness of the InSe thin slab used for the transmission measurements. (**d**) HR-TEM image of the sample captured in the c-plane, with the SAED pattern shown in the inset. (**e**) EDS of the sample.

2.2. Characterization

X-ray diffraction patterns were recorded on a Bruker D8 Advance diffractometer (Bruker, Billerica, MA, USA) with Cu Kα radiation (λ = 1.5406 Å) over a 2θ range of 10–70° at a scan rate of 2.4°/min. Raman spectra were acquired using an XploRA PLUS spectrometer (Horiba, Kyoto, Japan) with a 532 nm excitation source, a 1% neutral-density filter, and a 2400 grooves/mm grating. The laser was focused with a 100× objective lens (NA = 0.8), yielding a spot size of ~1 μm, and backscattered signals were collected under ambient conditions.

Transmission electron microscopy (TEM) was performed on a JEM-2010 microscope (JEOL, Kyoto, Japan) operating at 200 kV. High-resolution TEM (HRTEM) and selected-area electron diffraction (SAED) were employed to examine the structural and crystallographic properties of the InSe flakes.

Temperature-dependent transmission and photoluminescence (PL) measurements were carried out using a closed-cycle cryostat (APD Cryogenics, Allentown, PA, USA) evacuated to ~10^{-6} mbar (10^{-4} Pa) with a turbomolecular pump. Transmission spectra were obtained with a 50 W halogen lamp through a 2 mm pinhole on the cold finger. PL excitation was provided by a 355 nm laser diode (Teem Photonics, Grenoble, France) at 45° incidence, with an excitation power density of 22.54 W/cm^2. Emission spectra were collected using an iHR550 spectrometer (Horiba, Kyoto, Japan) equipped with a 150 grooves/mm grating.

3. Results and Discussion

The phase purity and structural characteristics of the InSe flakes were first examined by XRD, as shown in Figure 1a. The diffraction peaks at 10.69°, 21.42°, 32.32°, 43.54°, and 67.56° are indexed to the (002), (004), (006), (008), and (0012) planes, characteristic of the hexagonal phase of InSe. These reflections are consistent with earlier reports and correspond well to the standard JCPDS card No. 87-34-1431 [41,42]. In addition, the appearance of the (110) reflection at 45.34° is attributed to the random orientation of exfoliated InSe flakes. The calculated lattice parameters based on the hexagonal crystal system are a = 4.00 Å and c = 16.62 Å. Raman spectroscopy provided further verification of the crystal structure. As shown in Figure 1b, the Raman spectrum exhibits three characteristic modes: A^1_{1g}, E^1_{2g}, and A^2_{1g} at 115.6, 177.0 and 226.7 cm^{-1}, respectively, which are in excellent agreement with previous studies [43,44]. These vibrational fingerprints unambiguously confirm the β-polytype of InSe. High-resolution TEM analysis in Figure 1d revealed well-resolved lattice fringes of the c-plane, while the corresponding SAED pattern displayed six-fold rotational symmetry, further corroborating the hexagonal lattice. The stoichiometric composition is determined by the energy dispersive X-ray spectrum (EDS), as shown in Figure 1e, with an In:Se ratio of approximately 1.08:1. The C, O, and Cu signals are attributable to the TEM support grid rather than the InSe sample [45].

The temperature-dependent absorption of InSe single crystals in the range of 17 K to 300 K is shown in Figure 2a. The absorption coefficient α was obtained from transmission spectra using the relation $\alpha = 1/d\log(I_0/I)$, where d is the sample thickness as shown in Figure 1c, I_0 the transmitted intensity without the sample, and I the transmitted intensity with the sample. A clear blue shift and sharpening of the excitonic features and band edge are observed with decreasing temperature. At low temperatures, the excited excitonic state (n = 2) is resolved. The evolution of PL spectra, excited by a 355 nm laser with a power density of 22.54 W.cm^{-2}, is presented in Figure 2b. The PL intensity increases markedly as the temperature decreases, with the dominant emission peak assigned to the recombination of free excitons (FX), specifically the n = 1 state associated with the direct band gap, in agreement with previous reports [18,19,21,23,40]. Additional peaks appear at lower energies relative to the FX emission at low temperatures, which may originate from biexcitons (XX), as discussed later. The high-energy FX feature corresponds to the ground states of Wannier-Mott excitons formed by electrons at the conduction band minimum and holes at the valence band maximum at the Γ-point [46–48]. A peak at about 1.27 eV is attributed to donor-acceptor pair recombination, while an even lower-energy feature is likely associated with deep-level defect states [21].

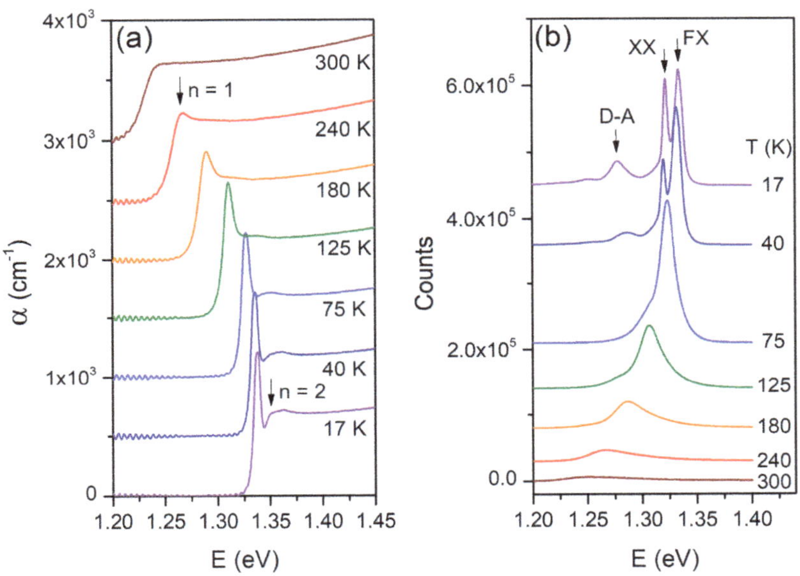

Figure 2. Temperature-dependent (**a**) absorption and (**b**) photoluminescence spectra of the InSe thin slab measured between 17 and 300 K. For clarity, representative spectra at selected temperatures are displayed with vertical offsets.

To gain further insight into the low-temperature optical response of InSe single crystals, Figure 3a,b shows the PL and absorption spectra measured at 17 and 300 K, respectively. At 17 K, the free exciton (FX) ground state exhibits a fine-structure splitting into a triplet (1.333 eV) and a singlet (1.336 eV), corresponding to an energy separation of 3 meV. This value is consistent with the earlier report of the γ-InSe given by Paraskevopoulos et al. [49], but significantly larger than the 0.6 meV splitting observed at 4.5 K by Gnatenko et al. [50]. As pointed out by Costa et al. [51], the β- and γ-InSe polytypes exhibit essentially identical electronic properties. Due to the weak Van der Waals nature of the interlayer interactions, the arrangement of the layers does not significantly influence the electronic structure, resulting in similar photoluminescence (PL) properties. The discrepancy may arise from differences in polytype, excitation source, or crystal quality. The origin of this doublet structure has remained unclear until now. In Ref. [50], this feature is considered as a doublet structure due to valence band splitting by interlayer interaction. In contrast, Ref. [49] assignes these features as a bound exciton and an excitonic peak. The clear appearance of the excited state (n = 2) is observed in the absorption spectrum at 17 K and a peak in higher energy side is band-to-band absorption. At 300 K, the PL spectrum is broad and asymmetric, featuring an extended high-energy tail accurately modeled by an empirical asymmetric double-sigmoid function [52,53].

The temperature-dependent shifts in the exciton and biexciton energies can be quantitatively described by the following relation:

$$E(T) = E_0 + AT + B\frac{1}{e^{\hbar\omega_{ph}/k_BT} - 1}, \quad (1)$$

where the linear term represents the contribution from lattice thermal expansion, while the second term accounts for the electron–phonon interaction effects on the electronic energy [54].

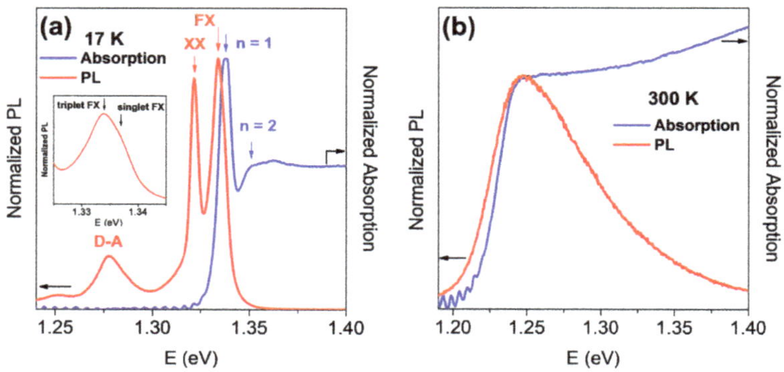

Figure 3. Photoluminescence and absorption spectra of a InSe thin slab obtained at two temperatures: (**a**) 17 and (**b**) 300 K. The inset in (**a**) indicates the singlet and triplet states of PL spectrum at 17 K.

Figure 4a,b shows the temperature dependence of the ground-state exciton and biexciton energies in absorption and PL, respectively; symbols denote experimental data and solid lines are fits. The corresponding best-fit parameters to Equation (1) are listed in Table 1. The fitted phonon energies $\hbar\omega_{ph}$ of the ground exciton states for both absorption and PL is 8.6 ± 3.4 and 10.0 ± 1.2 meV, respectively, comparable to the A_{1g}^1 optical phonon at 14.3 meV, as shown in Figure 1b. It is worth to emphasize that fitting values of data below 200 K bring the closer results to optical mode A_{1g}^1 because position of peaks is more identical. The phonon energy associated with the biexciton is in close agreement with the E_{2g}^1 optical mode (21.5 meV). The redshift of the biexciton occurs more rapidly than that of the exciton, and this phenomenon has also been observed in cubic ZnS single crystals [55] and in the layered structure of Sb_2Se_3 [56].

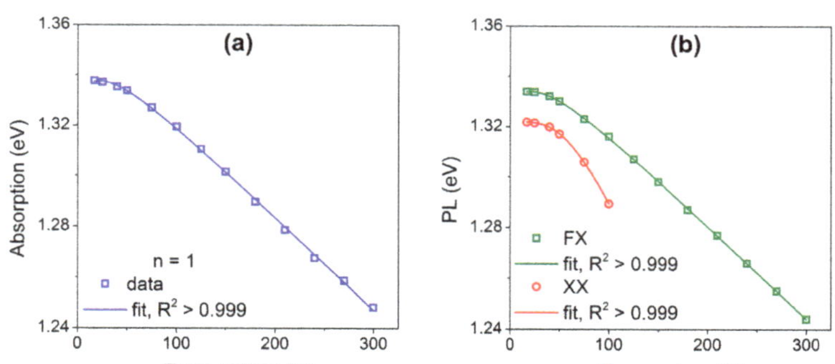

Figure 4. Temperature dependence of the ground-state exciton (squares) and biexciton (circles) energies in InSe single crystal. Experimental data are shown as points; solid lines are fits for (**a**) absorption and (**b**) photoluminescence.

Figure 5 presents the PL spectra of an InSe single crystal recorded at 17 K under varying excitation power densities. The spectra illustrate the evolution of radiative recombination pathways as the excitation power increases from 0.225 W/cm^2 to 22.54 W/cm^2. At lower excitation levels, the emission is primarily dominated by a broad peak centered around 1.275 eV, which can be attributed to donor–acceptor pair (D–A) recombination. As the power density increases, this peak rises rapidly and subsequently saturates, reflecting the characteristic behavior associated with D–A recombination. A similar saturation effect is observed for the feature at approximately 1.25 eV, which is associated with deep-level defect states.

Table 1. Parameters from the best fits to Equation (1) describing the temperature-dependent exciton and biexciton energies in absorption and PL.

	Exciton States	E_0 (eV)	A (meVK)	B (meVK)	$\hbar\omega_{ph}$ (meV)
Absorption	$n = 1$	1.336 ± 0.005	0.1 ± 0.2	−47 ± 7	8.6 ± 3.4
PL	FX ($n = 1$)	1.333 ± 0.001	0.04 ± 0.06	−47 ± 2	10.0 ± 1.2
	XX	1.322 ± 0.001	−0.03 ± 0.01	−194 ± 13	17.4 ± 0.7

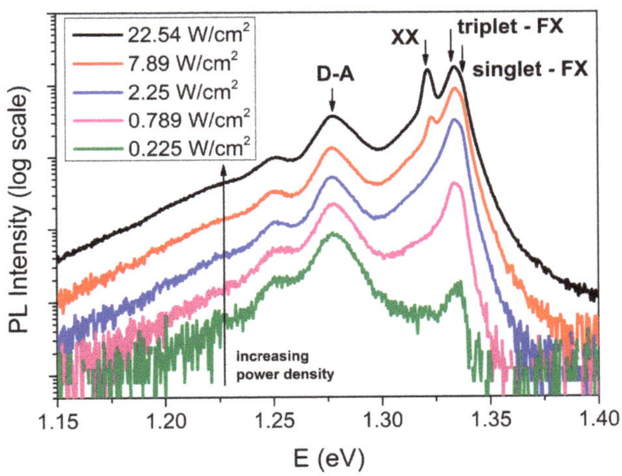

Figure 5. Log-scale luminescence spectra of an InSe single crystal at 17 K, measured under 355 nm excitation as a function of excitation power density.

At elevated excitation densities, additional high-energy features emerge, corresponding to excitonic transitions. The peaks designated as "singlet-FX" and "triplet-FX" exhibit an almost linear increase in intensity with excitation power, thereby confirming their identification as free exciton recombination. At sufficiently high power densities, a new peak appears at 1.322 eV (labeled "XX"), attributed to biexciton emission. The formation of biexcitons necessitates a substantial exciton population, making this feature prominent only under strong excitation conditions. The detection of biexciton emission at 17 K aligns with previous findings reported at 5 K [46].

4. Conclusions

In conclusion, we investigated the optical properties of high-quality β-InSe single crystals over a temperature range of 17–300 K. Structural analyses using XRD, Raman spectroscopy, and HRTEM confirmed excellent crystallinity and near-stoichiometric composition. Temperature-dependent absorption and photoluminescence spectra revealed strong excitonic features dominated by free exciton resonances. At 17 K, the PL spectra clearly showed a fine-structure splitting of the Wannier–Mott exciton into a triplet state at 1.333 eV and a singlet state at 1.336 eV, corresponding to a separation of approximately 3 meV. Additionally, a biexciton emission at 1.322 eV was identified and verified through its nonlinear dependence on excitation power density. The temperature evolution of the exciton and biexciton peaks exhibited a redshift behavior consistent with lattice expansion and electron–phonon interactions. These findings provide direct evidence of excitonic doublet splitting and biexciton formation in bulk β-InSe, underscoring its potential for future optoelectronic and quantum photonic applications.

Author Contributions: Conceptualization, writing—original draft, funding acquisition, T.T.T.H. and L.V.L.; investigation, sample characterization, formal analysis, N.T.L., M.H.N., T.-T.N., T.T.H.T., U.T.D.T. and T.H.N.; review and editing, funding acquisition, T.J.K. All authors have read and agreed to the published version of the manuscript.

Funding: This research was funded by the Institute of Materials Science—Vietnam Academy of Science and Technology (Grant No. CSCL04.11/23-24). Additionally, this research was funded by the National Research Foundation of Korea (NRF) grant funded by the Korean Government (MSIT) (RS-2020-NR049536).

Data Availability Statement: The raw data supporting the conclusions of this article will be made available by the authors on request.

Acknowledgments: T.T.T.H. and L.V.L. acknowledges Objective Lab for Agriculture-Bio-Medicine and Energy of VAST/IMS for using the Lab's equipment.

Conflicts of Interest: The authors declare no conflicts of interest.

References

1. Camassel, J.; Merle, P.; Mathieu, H.; Chevy, A. Excitonic Absorption Edge of Indium Selenide. *Phys. Rev. B* **1978**, *17*, 4718–4725. [CrossRef]
2. Wu, M.; Xie, Q.; Wu, Y.; Zheng, J.; Wang, W.; He, L.; Wu, X.; Lv, B. Crystal Structure and Optical Performance in Bulk γ-InSe Single Crystals. *AIP Adv.* **2019**, *9*, 25013. [CrossRef]
3. Boukhvalov, D.W.; Gürbulak, B.; Duman, S.; Wang, L.; Politano, A.; Caputi, L.S.; Chiarello, G.; Cupolillo, A. The Advent of Indium Selenide: Synthesis, Electronic Properties, Ambient Stability and Applications. *Nanomaterials* **2017**, *7*, 372. [CrossRef]
4. Sang, D.K.; Wang, H.; Qiu, M.; Cao, R.; Guo, Z.; Zhao, J.; Li, Y.; Xiao, Q.; Fan, D.; Zhang, H. Two Dimensional β-InSe with Layer-Dependent Properties: Band Alignment, Work Function and Optical Properties. *Nanomaterials* **2019**, *9*, 82. [CrossRef]
5. Isik, M.; Gasanly, N.M. Temperature-Tuned Band Gap Characteristics of InSe Layered Semiconductor Single Crystals. *Mater. Sci. Semicond. Process.* **2020**, *107*, 104862. [CrossRef]
6. Li, Z.; Liu, J.; Ou, H.; Hu, Y.; Zhu, J.; Huang, J.; Liu, H.; Tu, Y.; Qi, D.; Hao, Q.; et al. Enhancement of Carrier Mobility in Multilayer InSe Transistors by van Der Waals Integration. *Nanomaterials* **2024**, *14*, 382. [CrossRef]
7. Bandurin, D.A.; Tyurnina, A.V.; Yu, G.L.; Mishchenko, A.; Zólyomi, V.; Morozov, S.V.; Kumar, R.K.; Gorbachev, R.V.; Kudrynskyi, Z.R.; Pezzini, S.; et al. High Electron Mobility, Quantum Hall Effect and Anomalous Optical Response in Atomically Thin InSe. *Nat. Nanotechnol.* **2017**, *12*, 223–227. [CrossRef]
8. Island, J.O.; Steele, G.A.; van der Zant, H.S.J.; Castellanos-Gomez, A. Environmental Instability of Few-Layer Black Phosphorus. *2D Mater.* **2015**, *2*, 11002. [CrossRef]
9. Kong, D.; Cha, J.J.; Lai, K.; Peng, H.; Analytis, J.G.; Meister, S.; Chen, Y.; Zhang, H.-J.; Fisher, I.R.; Shen, Z.-X.; et al. Rapid Surface Oxidation as a Source of Surface Degradation Factor for Bi2Se3. *ACS Nano* **2011**, *5*, 4698–4703. [CrossRef] [PubMed]
10. Xu, Y.; Zhou, J.; Tian, D.; Fu, Z.; Huang, Y.; Feng, W. Photoelectrochemical-Type Photodetectors Based on Ball Milling InSe for Underwater Optoelectronic Devices. *Nanomaterials* **2025**, *15*, 3. [CrossRef] [PubMed]
11. Tamalampudi, S.R.; Villegas, J.E.; Dushaq, G.; Sankar, R.; Paredes, B.; Rasras, M. High-Speed Waveguide-Integrated InSe Photodetector on SiN Photonics for Near-Infrared Applications. *Adv. Photonics Res.* **2023**, *4*, 2300162. [CrossRef]
12. Tamalampudi, S.R.; Dushaq, G.; Rasras, M.S. An Ultra-High-Speed Vertically Illuminated Self-Driven Lateral Asymmetric InSe Photodetector. *Nanoscale* **2025**, *17*, 14206–14214. [CrossRef]

13. Huang, X.; Cao, Q.; Wan, M.; Song, H.-Z. Electronic and Optical Properties of BP, InSe Monolayer and BP/InSe Heterojunction with Promising Photoelectronic Performance. *Materials* **2022**, *15*, 6214. [CrossRef]
14. Ma, Z.; Li, R.; Xiong, R.; Zhang, Y.; Xu, C.; Wen, C.; Sa, B. InSe/Te van Der Waals Heterostructure as a High-Efficiency Solar Cell from Computational Screening. *Materials* **2021**, *14*, 3768. [CrossRef]
15. Al Garni, S.E.; Darwish, A.A.A. Photovoltaic Performance of TCVA-InSe Hybrid Solar Cells Based on Nanostructure Films. *Sol. Energy Mater. Sol. Cells* **2017**, *160*, 335–339. [CrossRef]
16. Zhao, H.; Hus, S.M.; Chen, J.; Yan, X.; Lawrie, B.J.; Jesse, S.; Li, A.-P.; Liang, L.; Htoon, H. Telecom-Wavelength Single-Photon Emitters in Multilayer InSe. *ACS Nano* **2025**, *19*, 6911–6917. [CrossRef]
17. Blundo, E.; Tuzi, F.; Cuccu, M.; Re Fiorentin, M.; Pettinari, G.; Patra, A.; Cianci, S.; Kudrynskyi, Z.R.; Felici, M.; Taniguchi, T.; et al. Giant Light Emission Enhancement in Strain-Engineered InSe/MS2 (M = Mo or W) van Der Waals Heterostructures. *Nano Lett.* **2025**, *25*, 3375–3382. [CrossRef]
18. Shubina, T.V.; Desrat, W.; Moret, M.; Tiberj, A.; Briot, O.; Davydov, V.Y.; Platonov, A.V.; Semina, M.A.; Gil, B. InSe as a Case between 3D and 2D Layered Crystals for Excitons. *Nat. Commun.* **2019**, *10*, 3479. [CrossRef] [PubMed]
19. Gnatenko, Y.P.; Zhirko, Y.I. Exciton Absorption and Luminescence of Indium Selenide Crystals. *Phys. Status Solidi* **1987**, *142*, 595–604. [CrossRef]
20. Balkanski, M.; Julien, C.; Jouanne, M. Electron and Phonon Aspects in a Lithium Intercalated InSe Cathode. *J. Power Sources* **1987**, *20*, 213–219. [CrossRef]
21. Abha; Warrier, A.V.R. Photoluminescence Studies on the Layer Semiconductor InSe. *J. Appl. Phys.* **1982**, *53*, 5169–5171. [CrossRef]
22. Bacıoğlu, A.; Ertap, H.; Karabulut, M.; Mamedov, G.M. Sub-Bandgap Analysis of Boron Doped InSe Single Crystals by Constant Photocurrent Method. *Opt. Mater.* **2014**, *37*, 70–73. [CrossRef]
23. Abay, B.; Efeoğlu, H.; Yoğurtçu, Y.K. Low-Temperature Photoluminescence of n-InSe Layer Semiconductor Crystals. *Mater. Res. Bull.* **1998**, *33*, 1401–1410. [CrossRef]
24. Wang, G.; Chernikov, A.; Glazov, M.M.; Heinz, T.F.; Marie, X.; Amand, T.; Urbaszek, B. Colloquium: Excitons in Atomically Thin Transition Metal Dichalcogenides. *Rev. Mod. Phys.* **2018**, *90*, 21001. [CrossRef]
25. Stevens, C.E.; Paul, J.; Cox, T.; Sahoo, P.K.; Gutiérrez, H.R.; Turkowski, V.; Semenov, D.; McGill, S.A.; Kapetanakis, M.D.; Perakis, I.E.; et al. Biexcitons in Monolayer Transition Metal Dichalcogenides Tuned by Magnetic Fields. *Nat. Commun.* **2018**, *9*, 3720. [CrossRef]
26. Mueller, T.; Malic, E. Exciton Physics and Device Application of Two-Dimensional Transition Metal Dichalcogenide Semiconductors. *NPJ 2D Mater. Appl.* **2018**, *2*, 29. [CrossRef]
27. Grim, J.Q.; Christodoulou, S.; Di Stasio, F.; Krahne, R.; Cingolani, R.; Manna, L.; Moreels, I. Continuous-Wave Biexciton Lasing at Room Temperature Using Solution-Processed Quantum Wells. *Nat. Nanotechnol.* **2014**, *9*, 891–895. [CrossRef]
28. Emara, M.M.; Burya, S.J.; Van Patten, P.G. Enhancement of Biexciton Generation via Resonance Energy Transfer in Nanoclusters of Colloidal CdTe Quantum Dots for Optical Gain Applications. *Mater. Chem. Phys.* **2022**, *277*, 125540. [CrossRef]
29. Macias-Pinilla, D.F.; Planelles, J.; Climente, J.I. Biexcitons in CdSe Nanoplatelets: Geometry, Binding Energy and Radiative Rate. *Nanoscale* **2022**, *14*, 8493–8500. [CrossRef]
30. de Carvalho, L.C.; Schleife, A.; Fuchs, F.; Bechstedt, F. Valence-Band Splittings in Cubic and Hexagonal AlN, GaN, and InN. *Appl. Phys. Lett.* **2010**, *97*, 232101. [CrossRef]
31. Shikanai, A.; Azuhata, T.; Sota, T.; Chichibu, S.; Kuramata, A.; Horino, K.; Nakamura, S. Biaxial Strain Dependence of Exciton Resonance Energies in Wurtzite GaN. *J. Appl. Phys.* **1997**, *81*, 417–424. [CrossRef]
32. Wang, T.; Miao, S.; Li, Z.; Meng, Y.; Lu, Z.; Lian, Z.; Blei, M.; Taniguchi, T.; Watanabe, K.; Tongay, S.; et al. Giant Valley-Zeeman Splitting from Spin-Singlet and Spin-Triplet Interlayer Excitons in WSe2/MoSe2 Heterostructure. *Nano Lett.* **2020**, *20*, 694–700. [CrossRef] [PubMed]
33. Akopian, N.; Lindner, N.H.; Poem, E.; Berlatzky, Y.; Avron, J.; Gershoni, D.; Gerardot, B.D.; Petroff, P.M. Entangled Photon Pairs from Semiconductor Quantum Dots. *Phys. Rev. Lett.* **2006**, *96*, 130501. [CrossRef]
34. Le, L.V.; Huong, T.T.; Nguyen, T.-T.; Nguyen, X.A.; Nguyen, T.H.; Cho, S.; Kim, Y.D.; Kim, T.J. The Wannier-Mott Exciton, Bound Exciton, and Optical Phonon Replicas of Single-Crystal GaSe. *Crystals* **2024**, *14*, 539. [CrossRef]
35. Dey, P.; Paul, J.; Moody, G.; Stevens, C.E.; Glikin, N.; Kovalyuk, Z.D.; Kudrynskyi, Z.R.; Romero, A.H.; Cantarero, A.; Hilton, D.J.; et al. Biexciton Formation and Exciton Coherent Coupling in Layered GaSe. *J. Chem. Phys.* **2015**, *142*, 212422. [CrossRef]
36. Damon, R.W.; Redington, R.W. Electrical and Optical Properties of Indium Selenide. *Phys. Rev.* **1954**, *96*, 1498–1500. [CrossRef]
37. Andriyashik, M.V.; Sakhnovskii, M.Y.; Timofeev, V.B.; Yakimova, A.S. Optical Transitions in the Spectra of the Fundamental Absorption and Reflection of InSe Single Crystals. *Phys. Status Solidi* **1968**, *28*, 277–285. [CrossRef]

38. Ateş, A.; Gürbulak, B.; Yıldırım, M.; Doğan, S. Electric Field Influence on Absorption Measurement in InSe Single Crystal. *Phys. E Low-Dimens. Syst. Nanostructures* **2003**, *16*, 274–279. [CrossRef]
39. Ateş, A.; Gürbulak, B.; Yildirim, M.; Tüzemen, S. Absorption Measurements in InSe:Ho Single Crystal under an Electric Field. *Czechoslov. J. Phys.* **2004**, *54*, 377–385. [CrossRef]
40. Ertap, H.; Bacıoğlu, A.; Karabulut, M. Photoluminescence Properties of Boron Doped InSe Single Crystals. *J. Lumin.* **2015**, *167*, 227–232. [CrossRef]
41. Ertap, H.; Karabulut, M. Structural and Electrical Properties of Boron Doped InSe Single Crystals. *Mater. Res. Express* **2019**, *6*, 35901. [CrossRef]
42. Taurino, A.; Signore, M.A. Concurrent Growth of InSe Wires and In2O3 Tulip-like Structures in the Au-Catalytic Vapour-Liquid-Solid Process. *Mater. Res. Express* **2015**, *2*, 65001. [CrossRef]
43. Carlone, C.; Jandl, S. Second Order Raman Spectrum and Phase Transition in InSe. *Solid State Commun.* **1979**, *29*, 31–33. [CrossRef]
44. Borodin, B.R.; Eliseyev, I.A.; Galimov, A.I.; Kotova, L.V.; Durnev, M.V.; Shubina, T.V.; Yagovkina, M.A.; Rakhlin, M.V. Indirect-to-Direct Band-Gap Transition in Few-Layer β-InSe as Probed by Photoluminescence Spectroscopy. *Phys. Rev. Mater.* **2024**, *8*, 14001. [CrossRef]
45. Baracchini, E.; Bianco, C.; Crosera, M.; Filon, F.L.; Belluso, E.; Capella, S.; Maina, G.; Adami, G. Nano-and Submicron Particles Emission during Gas Tungsten Arc Welding (GTAW) of Steel: Differences between Automatic and Manual Process. *Aerosol Air Qual. Res.* **2018**, *18*, 579–589. [CrossRef]
46. Godzaev, M.O.; Sernelius, B.E. Electron-Hole Liquid in Layered InSe: Comparison of Two-and Three-Dimensional Excitonic States. *Phys. Rev. B* **1986**, *33*, 8568–8581. [CrossRef]
47. Segura, A.; Bouvier, J.; Andrés, M.V.; Manjón, F.J.; Muñoz, V. Strong Optical Nonlinearities in Gallium and Indium Selenides Related to Inter-Valence-Band Transitions Induced by Light Pulses. *Phys. Rev. B* **1997**, *56*, 4075–4084. [CrossRef]
48. Le, L.V.; Kim, T.J.; Nguyen, X.A.; Choi, J.; Kim, Y.D. Spectroscopic Ellipsometry Investigation of Temperature-Dependent Dielectric Functions and Critical Points of β-InSe. *Phys. Status Solidi* **2025**, *262*, 2400591. [CrossRef]
49. Paraskevopoulos, K.M.; Julien, C.; Balkanski, M. Fine Structure in the Free Exciton Region of the Emission Spectrum of γ-InSe. *Phys. Status Solidi* **1987**, *143*, 741–748. [CrossRef]
50. Gnatenko, Y.P.; Skubenko, P.A.; Zhirko, Y.I.; Fialkovskaya, O.V. Emission of Free and Bound Excitons in GaSe and InSe Crystals in Direct and Indirect Transition Region. *J. Lumin.* **1984**, *31–32*, 472–475. [CrossRef]
51. Gomes da Costa, P.; Dandrea, R.G.; Wallis, R.F.; Balkanski, M. First-Principles Study of the Electronic Structure of γ-InSe and β-InSe. *Phys. Rev. B* **1993**, *48*, 14135–14141. [CrossRef]
52. Ran, W.; Zhang, Z.; Ma, X.; Sun, G.; Yan, T. A Novel Optical Temperature Sensor Based on Boltzmann Function in $BiZn_2PO_6$ Phosphor. *J. Lumin.* **2023**, *255*, 119562. [CrossRef]
53. Krustok, J.; Collan, H.; Yakushev, M.; Hjelt, K. The Role of Spatial Potential Fluctuations in the Shape of the PL Bands of Multinary Semiconductor Compounds. *Phys. Scr.* **1999**, *1999*, 179. [CrossRef]
54. Antonioli, G.; Bianchi, D.; Emiliani, U.; Podini, P.; Franzosi, P. Optical Properties and Electron-Phonon Interaction in GaSe. *Nuovo Cim. B* **1979**, *54*, 211–227. [CrossRef]
55. Liem, N.Q.; Quang, V.X.; Thanh, D.X.; Lee, J.I.; Kim, D. Temperature Dependence of Biexciton Luminescence in Cubic ZnS Single Crystals. *Solid State Commun.* **2001**, *117*, 255–259. [CrossRef]
56. Krustok, J.; Kondrotas, R.; Nedzinskas, R.; Timmo, K.; Kaupmees, R.; Mikli, V.; Grossberg, M. Identification of Excitons and Biexcitons in Sb_2Se_3 under High Photoluminescence Excitation Density. *Adv. Opt. Mater.* **2021**, *9*, 2100107. [CrossRef]

materials

MDPI

Article

Flux-Dependent Superconducting Diode Effect in an Aharonov–Bohm Interferometer

Yu-Mei Gao [1], Hao-Yuan Yang [2], Feng Chi [1,*], Zi-Chuan Yi [1] and Li-Ming Liu [1]

1 School of Electronic and Information Engineering, University of Electronic Science and Technology of China, Zhongshan Institute, Zhongshan 528400, China; yumeigao@zsc.edu.cn (Y.-M.G.); yizichuan@zsc.edu.cn (Z.-C.Y.); liulmxps@zsc.edu.cn (L.-M.L.)

2 South China Academy of Advanced Optoelectronics, South China Normal University, Guangzhou 510006, China; 2024024413@m.scnu.edu.cn

* Correspondence: chifeng@semi.ac.cn

Abstract: We theoretically investigate the supercurrent and superconducting diode effect (SDE) in an Aharonov–Bohm (AB) interferometer sandwiched between two aluminium-based superconducting leads. The interferometer features a quantum dot (QD), which is created in an indium arsenide (InAs) semiconductor nanowire by local electrostatic gating, inserted in one of its arms and a magnetic flux threading through the ring structure. The magnetic flux breaks the system time-reversal symmetry by modulating the quantum phase difference between electronic transport through the QD path and the direct arm, which enhances constructive interference in one direction and destructive interference in the other. This leads to a discrepancy between the magnitudes of the forward and reverse critical supercurrents and is the core mechanism that induces the SDE. We demonstrate that the critical supercurrents exhibit Fano line shapes arising from the interference between discrete Andreev bound states in the QD and continuous states in the direct arm. It is found that when electron transport is dominated by the QD-containing path as compared to the direct arm path of the interferometer, the diode efficiency reaches a maximum, with values as high as 80%. In contrast, when the direct arm path dominates transport, the diode efficiency becomes weak. This attenuation is attributed to the participation of higher-order quantum interference processes, which disrupt the nonreciprocal supercurrent balance. Importantly, the proposed AB interferometer system has a relatively simple structure, and the realization of the SDE within it is feasible using current nano-fabrication technologies.

Keywords: supercurrent; superconducting diode effect; quantum dot; Aharonov–Bohm interferometer; magnetic flux

1. Introduction

The superconducting diode effect (SDE), which refers to the difference in magnitude of critical supercurrents flowing in opposite directions, has been recently proposed and demonstrated in diverse superconducting systems [1–7]. In the conventional semiconductor p-n junction, imbalance of chemical potential distribution breaks system inversion symmetry and leads to nonreciprocal electric currents [8]. A semiconductor diode with resistance depending on the current flow direction is a kind of key component in electronic circuits. But the unavoidable heat generation by charge currents under external bias voltage in semiconductor diodes is a bottleneck problem impeding further integration of the circuit. On the contrary, the supercurrent is driven by the superconducting phase difference in Josephsoun junctions in the absence of bias voltage or resistance and thus there is no Joule heating [1,2,4,9]. The superconductor diode, then, should be a low-energy-consumption component in superconducting devices for quantum computing and communication [4,10–12]. From the basic research aspect, the SDE is a powerful means to study diverse topics in condensed matter physics, for example the connections between symmetry [13], spin–orbit coupling [14–17], vortex dynamics [18], and unconventional superconductors [19]. Until now, the SDE has been demonstrated in various noncentrosymmetric superlattices [5], Josephson junctions [20], nano-fabricated devices [21], and thin films under a magnetic field [22].

In general, the SDE originates from breaking of either inversion or time-reversal symmetry, or both [2–4,10]. The inversion symmetry can be broken by the asymmetry of a stacked heterostructure [5], an artificial device's interface [6] or edge [23], and a crystal lattice [7]. The time-reversal asymmetry usually is achievable with the help of external magnetic fields [9] or mechanism relating to the nature of the magnetic layers [24], valley polarization [25], interface magnetism [6,26], etc. In multi-terminal or multi-path superconducting systems, the magnetic field generates a magnetic flux, which will lead to the time-reversal symmetry being broken and induce a nonreciprocal supercurrent through the superconducting quantum interference effect [27–31]. For example, a four-terminal double-loop superconducting quantum interference device (SQUID) was recently realized in an indium arsenide/aluminum two-dimensional heterostructure [32]. A remarkable SDE characterized by a diode efficiency as large as 34% was observed with the help of magnetic fluxes and gate voltages. Subsequently, a simplified two-terminal double-loop SQUID inserted between Josephson junctions was studied [33,34]. A nonreciprocal critical supercurrent emerges due to the lack of both inversion and time-reversal symmetries by threading two magnetic fluxes individually through the two loops.

In the last two decades, interferometers coupled to superconductor leads with semiconductor quantum dots (QDs) inserted in their arms have received extensive attention [27,35–43]. Since the size of QDs is in the nanometer scale in two of the three directions, the confined electron exhibits a discrete electronic state and fully adjustable quantum levels [44,45]. These unique properties make semiconductor QDs have extremely broad applications in single-electron devices, memory devices, and various optical devices. The confined electron in hybridized superconductor/QD devices can couple to Cooper pairs in the superconductors and form Andreev bound states (ABSs) in the superconducting gap, becoming the carrier of the supercurrent [27,35–43,46–49]. Recently, an AB interferometer with two QDs on each of its arms was sandwiched between two superconductors in experiments [40–42]. The amplitude, transport directions, and period with respect to the phase difference of the supercurrent in these devices can be efficiently varied by the combined actions of the ABSs, magnetic flux, and the spin–orbit interactions (SOIs) of Rashba type on the QDs. Besides the phase difference between the superconductor leads, the phase arisen from the SOI may also serve as a driving force for the supercurrent and results in the SDE [43]. A significant SDE in an AB interferometer with a QD inserted in one of its two arms was also proposed theoretically very recently [16]. The working mechanism relies on the Zeeman splitting from a vertically applied magnetic field and the SOI-induced phase. With the help of gate voltages that adjust the QD's energy levels, a large value of diode efficiency reaching up to 70% can be achieved.

Here we propose a system for the SDE in Figure 1a. It consists of an AB interferometer sandwiched between two superconductor leads. The supercurrent driven by the two superconductors' phase difference becomes nonreciprocal due to a magnetic flux penetrating through the ring and by properly changing the QD's energy level. As compared to previous similar setups [16,43], both the configuration and working mechanism of the present one are quite simple, as it requires a single-level QD and magnetic flux only. The AB effect predicts that the phase of the wavefunction of electrons flowing in a metal ring is affected by an externally applied magnetic flux that passes through the ring, even if the electrons do not directly experience the magnetic field *per se* [50]. This theoretical prediction of Aharonov and Bohm is based on the fact that the magnetic vector potential, rather than the magnetic field, is the underlying fundamental physical entity from a quantum mechanical point of view. Now, this prediction has been clarified experimentally. While originally the AB effect referred to zero magnetic field experienced by the electrons, later it was clarified that this was not a necessity [28,30–33]. Our results indicate that when the transport processes are dominated by the the QD path as compared to the direct arm path, the maximum value of the diode efficiency can be enhanced up to 80%. But if the direct arm path dominates, the diode efficiency value becomes small.

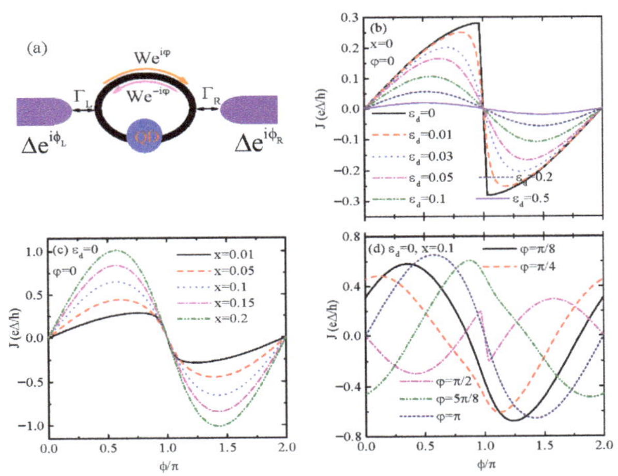

Figure 1. (**a**) Schematic plot of the device with an Aharonov–Bohm (AB) interferometer, in which a quantum dot (QD) is situated in one arm and a magnetic flux φ penetrates through the ring. The two superconductor leads with an energy gap Δ and phases $\phi_{L/R}$ couple to the QD with strengths of $\Gamma_{L/R}$ and to each other through a tunnel barrier with strength $We^{i\varphi}$. It is noted that due to the applied magnetic flux, which breaks the system time-reversal symmetry, a phase factor φ is added in W and then the forward tunnel amplitude $We^{i\varphi}$ is different from the backward tunnel one $We^{-i\varphi}$. (**b**–**d**) Show the Josephson current J as a function of the phase difference ϕ for the indicated parameters. In the QD–ring structure ($x \neq 0$), the magnitudes of the forward and backward Josephson currents become different from each other due to the existence of magnetic flux φ and induce a diode effect as in (**d**).

Beyond the fundamental exploration of quantum interference and nonreciprocal supercurrent dynamics, the AB interferometer-based superconducting diode proposed in this work exhibits substantial potential for practical applications in advanced superconducting electronics [10]. First, in the field of quantum computing, where low energy consumption and minimal decoherence are critical, this diode's zero-Joule-heating supercurrent transport and high tunability via QD energy levels or magnetic flux make it an ideal candidate for on-chip signal rectification and quantum bit control [10,51]. It can efficiently filter out reverse supercurrent noise that would otherwise disrupt qubit coherence, while its high maximum diode efficiency ensures minimal energy loss, which is a key advantage over existing SDE devices with lower efficiency [10,11,51]. Second, in superconducting microwave circuits for quantum communication [31], the diode's simple structure, compatible with current nano-fabrication, enables monolithic integration with other superconducting components, facilitating the development of compact, high-performance signal processing modules such as superconducting rectifiers or isolators [12]. Finally, in low-temperature electronic systems such as cryogenic sensors, the diode's ability to operate without resistive heating reduces the load on cryocoolers, extending system lifetime and lowering operational costs [10,11]. These applications not only leverage the unique strengths of the proposed system but also address critical bottlenecks in current superconducting technology, paving the way for more efficient and integrated quantum and low-temperature electronic devices.

2. Model and Method

The present device, with two superconductor leads that couple to a QD and simultaneously to each other through a tunnel barrier serving as the direct arm, is shown in Figure 1a. Here we consider a single-level QD, and the system Hamiltonian is modeled by [16,21,35–37],

$$\begin{aligned} H &= \sum_{\sigma} \varepsilon_d d_{\sigma}^{\dagger} d_{\sigma} + \sum_{\alpha,k,\sigma} \varepsilon_{\alpha,k\sigma} C_{\alpha,k\sigma}^{\dagger} C_{\alpha,k\sigma} + \sum_{\alpha,k} (\Delta_{\alpha} e^{i\phi_{\alpha}} C_{\alpha,k\uparrow}^{\dagger} C_{\alpha,-k\downarrow} + \text{H.c.}) \\ &+ \sum_{\alpha,k,\sigma} (V_{\alpha} C_{\alpha,k\sigma}^{\dagger} d_{\sigma} + \text{H.c.}) + \sum_{k,k',\sigma} (W^{-i\varphi} C_{L,k\sigma}^{\dagger} C_{R,k'\sigma} + \text{H.c.}), \end{aligned} \tag{1}$$

in which $d_{\sigma}^{\dagger}$ (d_{σ}) creates (annihilates) an electron in the QD with energy level ε_d and spin direction $\sigma = \uparrow, \downarrow$. The operator $C_{\alpha,k\sigma}^{\dagger}(C_{\alpha,k\sigma})$ denotes creation (annihilation) of electrons in the superconductor lead $\alpha(\alpha = L, R)$ with an energy of $\varepsilon_{\alpha,k\sigma}$, energy gap Δ_{α}, and superconducting phase ϕ_{α}. Here we study the supercurrent arisen from the phase difference $\phi_L - \phi_R$ in the absence of bias voltage. In addition, chemical potentials of the left and right leads are set to be $\mu_L = \mu_R = 0$. The quantity V_{α} in the forth term of Hamiltonian (1) describes the hopping amplitude between QD and lead-α. The last term in Hamiltonian (1) is for the direct coupling between the left and right superconductor leadss, with a hopping amplitude of W. Due to the applied vertical magnetic field with strength B, a phase factor φ is added in W with $\varphi = \int \vec{A} \cdot d\vec{r} / \varphi_0$, in which $\vec{A} = (-By, 0, 0)$ and $\varphi_0 = \hbar/e$ are individually the vector potential and flux quanta [35–37].

The Josephson supercurrent J through the QD-AB ring is obtained from the time evolution of the particle number operator of electrons $N_L = \sum_{k,\sigma} C_{L,K\sigma}^{\dagger} C_{L,k\sigma}$ in the left superconductor lead [35–37,47,48]:

$$\begin{aligned} J &= -e < \frac{dN_L}{dt} > \\ &= \frac{4e}{\hbar} \int \frac{d\varepsilon}{2\pi} \text{Re}[V_L G_{dL,11}^{<}(\varepsilon) + W e^{-i\varphi} G_{RL,11}^{<}(\varepsilon)], \end{aligned} \tag{2}$$

in which Green's functions $\mathbf{G}_{dL}^{<}(\varepsilon)$ and $\mathbf{G}_{RL}^{<}(\varepsilon)$ are individually the Fourier transformation of $\mathbf{G}_{dL}^{<}(t,t)$ and $\mathbf{G}_{RL}^{<}(t,t)$. In the Nambu representation, they are given by [21,37],

$$\mathbf{G}_{dL}^{<}(\varepsilon) = i \sum_k \begin{bmatrix} < C_{L,k\uparrow}^{\dagger}(t) d_{\uparrow}(t) > & < C_{L,k\uparrow}^{\dagger}(t) d_{\downarrow}^{\dagger}(t) > \\ < C_{L,-k\downarrow}(t) d_{\uparrow}(t) > & < C_{L,-k\downarrow}(t) d_{\downarrow}^{\dagger}(t) > \end{bmatrix}, \tag{3}$$

and

$$\mathbf{G}_{dL}^{<}(\varepsilon) = i \sum_{k,k'} \begin{bmatrix} < C_{L,k\uparrow}^{\dagger}(t) C_{R,k'\uparrow}(t) > & < C_{L,k\uparrow}^{\dagger}(t) C_{R,-k'\downarrow}^{\dagger}(t) > \\ < C_{L,-k\downarrow}(t) C_{R,k'\uparrow}(t) > & < C_{L,-k\downarrow}(t) C_{R,-k'\downarrow}^{\dagger}(t) > \end{bmatrix}. \tag{4}$$

Notice that the Josephson current arises from the superconductors' phase difference without bias voltage, and then the device is in equilibrium and Green's function obeys the fluctuation–dissipation theorem [21,37] $\mathbf{G}^{<}(\varepsilon) = -f(\varepsilon)[\mathbf{G}^{r}(\varepsilon) - \mathbf{G}^{a}(\varepsilon)]$, where $f(\varepsilon) = 1/[1 + e^{\varepsilon/k_B T}]$ is the equilibrium Fermi distribution function and k_B and T are individually the Boltzmann constant and system temperature T. $\mathbf{G}^{r/a}(\varepsilon)$ is the retarded/advanced Green function and $\mathbf{G}^{a}(\varepsilon) = [\mathbf{G}^{r}(\varepsilon)]^{\dagger}$. We next calculate the retarded Green function $\mathbf{G}^{r}(\varepsilon)$ with the help of the Dyson equation technique [21,37]

$$\mathbf{G}^{r}(\varepsilon) = \mathbf{g}^{r}(\varepsilon) + \mathbf{g}^{r}(\varepsilon)\mathbf{\Sigma}^{r}(\varepsilon)\mathbf{G}^{r}(\varepsilon), \tag{5}$$

in which $\mathbf{G}^{r}(\varepsilon)$ is defined in a space composed of three regions: the left and right superconductor leads and the QD,

$$\mathbf{G}^{r}(\varepsilon) = \begin{bmatrix} \mathbf{G}^{r}_{LL}(\varepsilon) & \mathbf{G}^{r}_{LR}(\varepsilon) & \mathbf{G}^{r}_{LD}(\varepsilon) \\ \mathbf{G}^{r}_{RL}(\varepsilon) & \mathbf{G}^{r}_{RR}(\varepsilon) & \mathbf{G}^{r}_{RD}(\varepsilon) \\ \mathbf{G}^{r}_{DL}(\varepsilon) & \mathbf{G}^{r}_{DR}(\varepsilon) & \mathbf{G}^{r}_{DD}(\varepsilon) \end{bmatrix}. \tag{6}$$

The self-energy $\mathbf{\Sigma}^{r}$ in Equation (5) is defined similarly as

$$\mathbf{\Sigma}^{r} = \begin{bmatrix} 0 & \mathbf{\Sigma}^{r}_{LR} & \mathbf{\Sigma}^{r}_{L} \\ \mathbf{\Sigma}^{r\dagger}_{LR} & 0 & \mathbf{\Sigma}^{r}_{R} \\ \mathbf{\Sigma}^{r\dagger}_{L} & \mathbf{\Sigma}^{r\dagger}_{R} & 0 \end{bmatrix}, \tag{7}$$

in which the diagonal 2×2 matrices $\mathbf{\Sigma}^{r}_{L/R}$ and $\mathbf{\Sigma}^{r}_{LR}$ represent coupling between the left/right superconductor with the QD and with each other, respectively. They are given individually by $\mathbf{\Sigma}^{r}_{L/R} = \mathrm{diag}(V_{L/R}, -V_{L/R})$ and $\mathbf{\Sigma}^{r}_{LR} = \mathrm{diag}(We^{i\varphi}, -We^{-i\varphi})$. $\mathbf{g}^{r}(\varepsilon)$ in Equation (5) stands for the free retarded Green function of the system without coupling between the leads and the QD with $\mathbf{g}^{r}(\varepsilon) = \mathrm{diag}[\mathbf{g}^{r}_{LL}(\varepsilon), \mathbf{g}^{r}_{RR}(\varepsilon), \mathbf{g}^{r}_{DD}(\varepsilon)]$. Green's function of the isolated leads is written as [16,21]

$$\mathbf{g}^{r}_{\alpha\alpha} = -i\pi\rho_\alpha(\varepsilon)\gamma_\alpha(\varepsilon)\begin{bmatrix} 1 & -\dfrac{\Delta_\alpha}{\varepsilon}e^{-i\phi_\alpha} \\ -\dfrac{\Delta_\alpha}{\varepsilon}e^{i\phi_\alpha} & 1 \end{bmatrix}, \tag{8}$$

where $\rho_\alpha(\varepsilon)$ is the normal density of states of lead-α. The quantity $\gamma_\alpha(\varepsilon)$ is given by [21,47]

$$\gamma_\alpha(\varepsilon) = \frac{|\varepsilon|\vartheta(|\varepsilon| - \Delta_\alpha)}{\sqrt{\varepsilon^2 - \Delta_\alpha^2}} + \frac{\varepsilon\vartheta(\Delta_\alpha - |\varepsilon|)}{i\sqrt{\Delta_\alpha^2 - \varepsilon^2}}, \tag{9}$$

in which $\vartheta(x) = 1$ when $x > 0$ and $\vartheta(x) = 0$ otherwise. For an isolated QD, Green's function is solved as $\mathbf{g}^{r}_{DD}(\varepsilon) = \mathrm{diag}[1/(\varepsilon - \varepsilon_d + i\delta), 1/(\varepsilon + \varepsilon_d + i\delta)]$ [16,21].

To study the SDE effect, we introduce the diode efficiency given by [21]

$$\eta = \frac{J_{c+} - |J_{c-}|}{(J_{c+} + |J_{c-}|)/2}, \tag{10}$$

in which J_{c+} (J_{c-}) is the positive (negative) critical current, which is obtained by choosing the maximum (minimum) Josephson current in a 2π period of the superconductor phase [16,21,46,51]. As a key parameter distinguishing superconductors from ordinary conductors, the experimentally measurable critical current determines the stability of superconducting devices and systems during operation [51].

3. Numerical Results

During numerical calculations, we focus on the zero-temperature condition ($T = 0$) and assume identical supreconducting gaps of the two leads ($\Delta_L = \Delta_R = \Delta$), in which $\Delta \equiv 1$ is set to be the energy unit [16,21,37]. It is further assumed that the central region composed of the QD and the direct arm is connected to the leads with the same strengths described by the line-width function $\Gamma = 2\pi\rho_\alpha V_\alpha^2$ and $x = (\pi\rho_\alpha W)^2$, respectively. Figure 1b–d show the Josephson current J varying with respect to the superconductors' phase difference $\phi = \phi_L/2 = -\phi_R/2$. When the direct arm is disabled ($x = 0$), as shown in Figure 1b, the present system reduces to a QD sandwiched between two superconductors (S-QD-S), and the $J - \phi$ curve obeys the relation of $J(\phi) = -J(\phi + \pi)$, which is a signature of symmetric ABS contributions. If the QD's energy level ε_d aligns with the Fermi energy of the leads $E_F = 0$, the $J - \phi$ curve shows an approximately triangular line shape and jumps from positive to negative at $\phi = \pi$. This is because the ABSs degenerate at the Fermi energy [46] and eliminate nonreciprocity (no SDE). When the QD's level is shifted away from the Fermi energy ($\varepsilon_d \neq 0$), the $J - \phi$ is usually a sinuous line shape. Now the supercurrent is suppressed by increasing ε_d. The reason is that now the ABSs depart from the Fermi energy and break down the resonance, leading to the suppression of the current [37,46–48]. As shown in Figure 1c, when the two superconductor leads are directly coupled ($x \neq 0$) for $\varepsilon_d = 0$, the amplitude of J is obviously enhanced due to the additional electron transport channel. Now the $J - \phi$ curve becomes a sinuous shape with retained $J(\phi) = -J(\phi + \pi)$ property without magnetic flux as the time-reversal symmetry is preserved [37,46]. In the presence of magnetic flux $\varphi \neq 0$, Figure 1d shows that the Josephson current is strongly varied, and now the current–phase curve does not preserve the sinusoidal-like shape [37]. The current can become non-zero at $\phi = n\pi$ with integer $n = 0, \pm1, \pm2, \cdots\cdots$. Moreover, the maxima of the forward and reverse currents can be different, or in other words, the magnitudes of the positive and negative critical currents are not equal ($J_{c+} \neq |J_{c-}|$). The nonreciprocity of the supercurrent arises because φ modulates the phase difference between electrons traversing the QD and direct arm, creating constructive interference in one direction and destructive in the other. Figure 1 confirms that the magnetic flux φ is mandatory for breaking symmetry and inducing the SDE.

To fully understand the nonreciprocal Josephson current, we plot the current-carrying density of states (CCDOS) $j(\varepsilon) = \mathrm{Re}[V_L G^<_{dL,11}(\varepsilon) + We^{-i\varphi} G^<_{RL,11}(\varepsilon)]$ [21,48], which quantifies how electronic states contribute to the supercurrent and directly links ABS behavior to nonreciprocal current for the cases with and without the magnetic flux in Figures 2a and 2b, respectively. As is well known, the supercurrent J has two parts; one is the continuous part due to electrons with energy ε outside the energy gap of the superconductors, whereas the other is a discrete one from electrons with energy within the gap [46–48]. The continuous part can be calculated directly by the integral of the CCDOS, while the discrete part is approached by solving poles of the denominator of the CCDOS, which are the ABSs. Generally, the supercurrent amplitude arisen from the discrete spectrum will be stronger than that due to the continuous spectrum. Moreover, the ABSs are paired with energy of opposite signs. Under the condition of zero temperature, only those with an energy of $[-\Delta, 0]$ relate to the current. Affected by direct connection of the two leads, instead of one bound state in a S-QD-S junction in the region of $[-\Delta, 0]$, there are two bound states arisen from hybridization of the QD E^-_{QD} and the arm E^-_A [37]. The latter depends on the magnetic flux through $E^-_A = -\Delta\sqrt{1 - \frac{4x}{(1+x)^2}\sin^2(\varphi/2)}$. At $\varphi = 0$, as shown in Figure 2a, the CCDOS exhibits a symmetric δ-like peak due to the ABS satisfying the relation of $E^-(\phi/2) = -E^-(3\phi/2)$. This symmetry ensures that the positive and negative cancel out $|J_{c+}| = |J_{c-}|$, as the ABS contribution to current is identical in both directions. In the presence of magnetic flux ($\varphi = \pi/4$, for example) that disrupts ABS symmetry, Figure 2b shows that the bound state near Δ is shifted from a positive value to a negative one, whereas that near $E_F = 0$ remains almost unchanged for $\phi = \pi/2$, leading to an obviously reduced positive current. While for $\phi = 3\pi/2$, the original bound state near Δ is suppressed to zero, that near E_F survives and the negative current will be slightly reduced as compared to the case of the positive current, leading to a nonreciprocal current and the SDE.

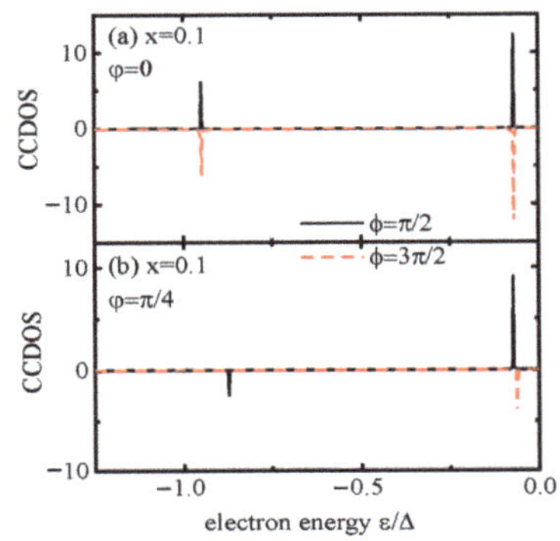

Figure 2. Current-carrying density of states (CCDOS) as a function of energy ε for $\varphi = 0$ in (**a**) and $\varphi = \pi/2$ in (**b**), respectively. Other parameters are line-width function $\Gamma = 0.1$ and dot level $\varepsilon_d = 0$.

We then study the diode efficiency ζ and critical currents $J_{c+}(J_{c-})$ through the system in Figure 3. For the case of magnetic flux phase $\varphi = n\pi$ ($n = 0, \pm1, \pm2, \cdots\cdots$), the current J is antisymmetric with respect to $\varphi = n\pi$ and $J_{c+} = |J_{c-}|$, leading to zero diode efficiency ($\zeta = 0$). Apart from $\varphi = n\pi$, the Josephson current is nonreciprocal ($J_{c+} \neq |J_{c-}|$) and induces the SDE, which is presented in Figure 3. For $\varphi = \pi/8$ or $\pi/4$, as individually shown in Figure 3a,b, the extinct SDE characterized by negative ζ occurs in dot level regimes of $\varepsilon_d < 0$. In the case of magnetic flux phase factor $\varphi = \pi/2$, ζ becomes antisymmetric due to the electron–hole symmetry given in Figure 3c. Increasing the value of $\varphi > \pi/2$, large positive diode efficiency emerges in the positive dot level regime, which is indicated in Figure 3d. It is worth noting that, due to the existence of a direct arm between the two leads, the critical current $J_{c\pm}$ as a function of the energy level of the QD ε_d exhibits asymmetric peak–dip features, which are hallmarks of Fano resonance. This arises from interference between discrete ABS states in the QD and continuous states in the direct arm [37]. In the absence of magnetic flux ($\varphi = 0$) and small x, the Fano resonant peak emerges at $\varepsilon_d = -\Gamma\sqrt{x}/(1-x)$, and the dip at approximately $\varepsilon_d = \Gamma/(2\sqrt{x})$. Correspondingly, the diode efficiency ζ versus QD level also exhibits a Fano line-shape, which holds true in the presence of magnetic flux. This tunability is critical for experimenters; i.e., by adjusting ε_d to align with Fano peaks, one can optimize ζ without modifying the device's physical structure.

Figure 4 presents a counter plot of the diode efficiency varying with the QD's level and the flux φ. Several characteristics are worth noting: (1) The value of the diode efficiency can reach as large as ±0.8 for small $x = 0.025$ in a relatively wide regime of ε_d and φ (Figure 4a) and then decreases with increasing x, as in Figure 4b–d. In previous theoretical work, the diode efficiency could reach as high as 70% in the presence of a magnetic field and spin–orbit interaction in a similar AB-QD ring structure [16]. When the QD is coupled to a magnetic impurity, the diode efficiency is about 40%, which is the typical value found in experiments [30–32]. The broad high-efficiency region in Figure 4a ($x = 0.025$) is ideal for experiments, as it tolerates small variations in ε_d and φ, which are common in fabrication, without significant efficiency loss. (2) For a small x value, ζ changes in a relative small regime of φ because the direct path is almost closed, as shown in Figure 4a. With increasing x, the oscillation of $\zeta - \varphi$ is strong, as given in Figure 4b–d. The oscillation shape obviously departs from the $\sin\varphi$ or $\cos\varphi$ shape [37]. This is because larger x strengthens the direct arm, introducing higher-order tunneling processes that disrupt ABS symmetry and reduce nonreciprocity. (3) The diode efficiency obeys the relation of $\zeta(\varepsilon_d, \varphi) = \zeta(-\varepsilon_d, \varphi + \pi)$ because of the electron–hole symmetry and $J_c(\varepsilon_d, \varphi) = J_c(-\varepsilon_d, \varphi + \pi)$. ζ is a periodic function of magnetic flux φ of period π when $\varepsilon_d = 0$ and period 2π for the case of $\varepsilon_d \neq 0$. This result is the same as that of the critical current, which was found in Ref. [37]. This periodicity shift is a unique signature of the AB interferometer: φ modulates the quantum phase of the direct arm, and ε_d breaks electron–hole symmetry to alter the period.

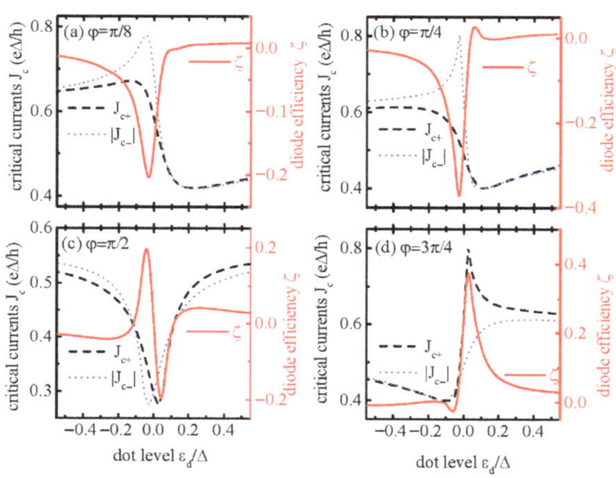

Figure 3. Positive critical Josephson current J_{c+}, absolute value of the negative Josephson current $|J_{c-}|$, and the diode efficiency ζ as functions of the dot level ε_d for $\varphi = \pi/8$ in (**a**), $\varphi = \pi/4$ in (**b**), $\varphi = \pi/2$ in (**c**), and $\varphi = 3\pi/4$ in (**d**). Other parameters are $x = \Gamma = 0.1$.

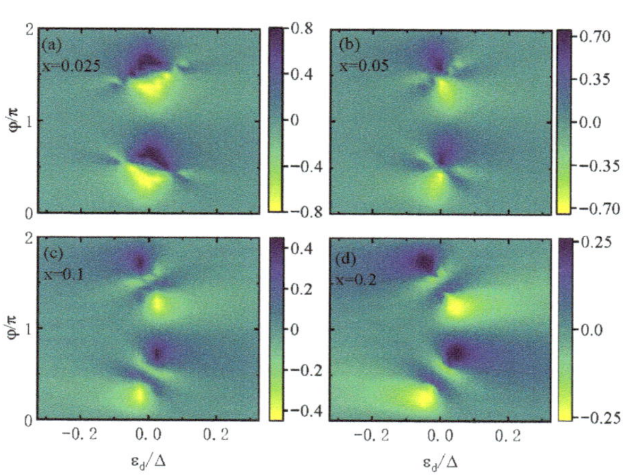

Figure 4. Counter plot of the diode efficiency ζ varying with respect to the dot level ε_d and magnetic flux φ for direct coupling strength $x = 0.025$ in (**a**), $x = 0.05$ in (**b**), $x = 0.1$ in (**c**), and $x = 0.2$ in (**d**). The line-width function is fixed at $\Gamma = 0.1$.

We then examine the impacts of direct coupling strength between the leads x on the diode efficiency for different QD levels ε_d in Figure 5a and line-width functions Γ in Figure 5b. Since the diode efficiency satisfies the relation of $\zeta(\varepsilon_d, \varphi) = \zeta(-\varepsilon_d, \varphi + \pi)$, we only present the cases of negative QD levels $\varepsilon_d < 0$ and $\varphi = \pi/4$. As shown in Figure 5a, ζ is negative for the chosen parameters. For all $\varepsilon_d < 0$, the absolute value of ζ first increases with x up to about 0.05, reaches a maximum, and then decreases as x exceeds $\Gamma = 0.1$. This non-monotonic trend reflects the "path dominance transition"; i.e., when $x < \Gamma$ the QD path dominates the transport processes as compared with the direct arm path, and ζ is large because increasing x adds a weak direct arm that enhances quantum interference with the QD path and strengthens ABS asymmetry. When $x > \Gamma$, where the direct arm path dominates the transport, the absolute value of the diode efficiency is relatively small and depends nonlinearly on the value of x, as higher-order tunneling processes such as multiple electron scattering between leads reduce ABS contributions and weaken ζ. These results are confirmed by Figure 5b, where the diode efficiency shows similar behaviors to those in Figure 5a. For larger Γ (e.g., $\Gamma = 0.3$), the maximum ζ shifts to larger $x \approx 0.15$, as a stronger QD–superconductor coupling delays the transition to direct path dominance. This confirms that Γ can be tuned to extend the high-efficiency regime to larger x, which is useful for balancing efficiency with current-carrying capacity.

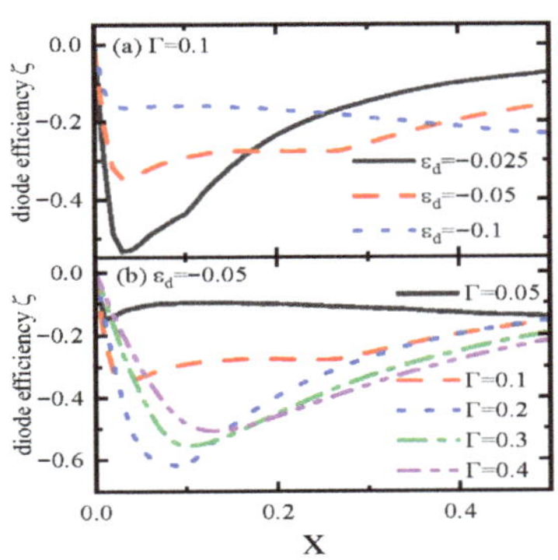

Figure 5. Diode efficiency ζ versus x for different values of ε_d in (**a**) and different Γ in (**b**). The magnetic flux is set to be $\varphi = \pi/4$.

The results of this work demonstrate that AB interferometers with a single QD are a promising platform for a high-efficiency SDE, with efficiency reaching up to 80% driven by QD-dominated transport and flux-tuned quantum interference. However, translating this performance to practical devices requires addressing several challenges. The first challenge is the stability and uniformity of magnetic flux because the SDE in this system relies on precise control of φ, including magnetic field inhomogeneity, which causes non-uniform φ across the ring, magnetic fields from nearby superconductors or qubits, or flux creep due to slow magnetic flux dissipation over time. This problem may be overcome by fabricating on-chip microcoils directly above the AB ring to generate localized, uniform magnetic fields and using flux-locked loops with SQUID sensors to stabilize φ or designing the AB ring with a symmetric, compact geometry to reduce flux inhomogeneity and edge effects [30,32,33]. The second challenge is the precise tuning of ε_d to align with the Fermi energy of superconducting leads. However, in experimental implementations, QDs are susceptible to charge noise from trapped charges in the substrate or gate dielectrics and thermal fluctuations, which cause ε_d to drift over time. This drift shifts the QD away from resonance with ABSs, reducing nonreciprocity and efficiency. This maybe resolved by employing coupled QD systems such as double QDs to create more robust resonance states that are less sensitive to local noise [41,42] or using top-gate architectures with high-k dielectrics to minimize charge trapping and improve gate capacitance, enabling finer and more stable ε_d control [45,46]. The third challenge is the temperature sensitivity of ABSs [46]. The present theoretical model assumes zero temperature, where only ABSs in the energy range $[-\Delta, 0]$ contribute to the supercurrent. At non-zero temperatures, thermal excitation generates quasi-particles that scatter ABSs, reducing their contribution to the nonreciprocal current [7,46]. Additionally, higher temperatures widen the thermal distribution of electrons, blurring the distinction between positive and negative critical currents. Potential solutions include the usage of high-critical-temperature superconductors as leads, which operate at higher temperatures while maintaining large superconducting gaps, reducing quasi-particle excitation [7].

4. Conclusions

In conclusion, this study presents a theoretical analysis of the Josephson supercurrent and the associated superconducting diode effect (SDE) in an Aharonov–Bohm (AB) interferometer, which is constructed by connecting a QD-containing arm and a direct arm to two superconducting leads. Our key results reveal that both the magnitude of the supercurrent and its oscillation period with respect to the superconducting phase difference between the two leads are strongly dependent on two critical factors: the energy level of the QD ε_d and the magnetic flux φ penetrating the AB interferometer ring. The quantum interference effect, driven by these factors, generates a nonreciprocal critical supercurrent, which is an experimentally measurable phenomenon that directly gives rise to the SDE. We find that the diode efficiency ζ adheres to specific rules: It satisfies the relation of $\zeta(\varepsilon_d, \varphi) = \zeta(-\varepsilon_d, \varphi + \pi)$ due to electron–hole symmetry and the intrinsic characteristics of the critical supercurrent. Additionally, the diode efficiency exhibits periodicity with respect to the magnetic flux: it has a period of π when the QD energy level $\varepsilon_d = 0$ and a period of 2π if $\varepsilon_d \neq 0$. A critical practical insight from this work is the achievement of high diode efficiency by optimizing three core parameters, i.e., the direct coupling strength between the superconducting leads, the energy level of the QD, and the magnetic flux through the ring. Specifically, maximizing the diode efficiency requires ensuring that electron transport is dominated by the QD-containing path rather than the direct arm path.

Author Contributions: Conceptualization, formal analysis, investigation, and writing—original draft preparation, Y.-M.G., H.-Y.Y. and F.C.; formal analysis and software, Y.-M.G., H.-Y.Y. and F.C.; conceptualization and writing—review and editing, F.C., Z.-C.Y. and L.-M.L. All authors have read and agreed to the published version of the manuscript.

Funding: This research was funded by the Education Science Planning Project of the Department of Education in Guang-dong Province (No. 2023GXJK542), the Educational Quality Project of the Department of Education in Guang-dong Province of China (No. SJD202302), the Graduate Joint Training Base (Zhongshan) Special Project in Guang-dong Province of China (No. 4YYJS04), the Project of Guangdong University Teaching Management Institute of China (No. GDZLGL2325), the China Private Universities Association project (school development, CANFZG24126), and the Educational Quality Project of Zhongshan Institute of University of Electronic Science and Technology of China (232023XYWLSZ10). Chi acknowledges support from the Guang-dong Province Education Department (Grant No. 2023KTSCX174), the Key Laboratory of Guang-dong Higher Education Institutes (Grant No. 2023KSYS011), and the Science and Technology Bureau of Zhongshan (Grant No. 2023B2035).

Institutional Review Board Statement: Not applicable.

Informed Consent Statement: Not applicable.

Data Availability Statement: The original contributions presented in this study are included in the article. Further inquiries can be directed to the corresponding author.

Conflicts of Interest: The authors declare no conflicts of interest.

References

1. Krasnov, V.M.; Oboznov, V.A.; Pedersen, N.F. Fluxon dynamics in long Josephson junctions in the presence of a temperature gradient or spatial nonuniformity. *Phys. Rev. B* **1997**, *55*, 14486. [CrossRef]
2. Hu, J.P.; Wu, C.J.; Dai, X. Proposed Design of a Josephson Diode. *Phys. Rev. Lett.* **2007**, *99*, 067004. [CrossRef]
3. Ideue, T.; Iwasa, Y. One-way supercurrent achieved in an electrically polar film. *Nature* **2020**, *584*, 349. [CrossRef]
4. Nadeem, M.; Fuhrer, M.S.; Wang, X. The superconducting diode effect. *Nat. Rev. Phys.* **2023**, *5*, 558–577. [CrossRef]
5. Ando, F.; Miyasaka, Y.; Li, T.; Ishizuka, J.; Arakawa, T.; Shiota, Y.; Moriyama, T.; Yanase, Y.; Ono, T. Observation of superconducting diode effect. *Nature* **2020**, *584*, 373–376. [CrossRef]
6. Le, T.; Pan, Z.; Xu, Z.; Liu, J.; Wang, J.; Lou, Z.; Yang, X.; Wang, Z.; Yao, Y.; Wu, C.; et al. Superconducting diode effect and interference patterns in kagome CsV_3Sb_5. *Nature* **2024**, *630*, 64–69. [CrossRef]
7. Qi, S.; Ge, J.; Ji, C.; Ai, Y.; Ma, G.; Wang, Z.; Cui, Z.; Liu, Y.; Wang, Z.; Wang, J. High-temperature field-free superconducting diode effect in high-Tc cuprates. *Nat. Commun.* **2025**, *16*, 531. [CrossRef]
8. Rudan, M. *Physics of Semiconductor Devices*, 2nd ed.; Spinger: Berlin/Heidelberg, Germany, 2018. [CrossRef]
9. Liu, F.; Itahashi, Y.M.; Aoki, S.; Dong, Y.; Wang, Z.; Ogawa, N.; Ideue, T.; Iwasa, Y. Superconducting diode effect under time-reversal symmetry. *Sci. Adv.* **2024**, *10*, eado1502. [CrossRef]
10. Ma, J.J.; Zhan, R.Y.; Lin, X. Superconducting Diode Effects: Mechanisms, Materials and Applications. *Adv. Phys. Res.* **2025**, *4*, 6. [CrossRef]
11. Kochan, D.; Strunk, C. Low-loss electronics with superconducting diodes. *Nat. Electron.* **2025**, *8*, 380–381. [CrossRef]
12. Ingla-Aynés, J.; Hou, Y.; Wang, S.; Chu, E.D.; Mukhanov, O.A.; Wei, P.; Moodera, J.S. Efficient superconducting diodes and rectifiers for quantum circuitry. *Nat. Electron.* **2025**, *8*, 411–416. [CrossRef]
13. He, J.J.; Tanaka, Y.; Nagaosa, N. A phenomenological theory of superconductor diodes. *New J. Phys.* **2022**, *24*, 053014. [CrossRef]
14. Amundsen, M.; Linder, J.; Robinson, J.W.A.; Žutić, I.; Banerjee, N. Colloquium: Spin-orbit effects in superconducting hybrid structures. *Rev. Mod. Phys.* **2024**, *96*, 021003. [CrossRef]
15. Mao, Y.; Yan, Q.; Zhuang, Y.C.; Sun, Q.F. Universal Spin Superconducting Diode Effect from Spin-Orbit Coupling. *Phys. Rev. Lett.* **2024**, *132*, 216001. [CrossRef]
16. Debnath, D.; Dutta, P. Gate-tunable Josephson diode effect in Rashba spin-orbit coupled quantum dot junctions. *Phys. Rev. B* **2024**, *109*, 174511. [CrossRef]
17. Huang, C.S.; Lin, W.N.; Lu, X.C. Anomalous angle dependence of superconducting diode effect in a mixed spin-orbit coupling system. *J. Phys. Condens. Matter* **2025**, *37*, 135807. [CrossRef] [PubMed]
18. Gutfreund, A.; Matsuki, H.; Plastovets, V.; Noah, A.; Gorzawski, L.; Fridman, N.; Yang, G.; Buzdin, A.; Millo, O.; Robinson, J.W.; et al. Direct observation of a superconducting vortex diode. *Nat. Commun.* **2023**, *14*, 1630. [CrossRef]
19. Yuan, N.F.Q.; Fu, L. Supercurrent diode effect and finite-momentum superconductors. *Proc. Natl. Acad. Sci. USA* **2022**, *119*, e2119548119. [CrossRef] [PubMed]
20. Pal, B.; Chakraborty, A.; Sivakumar, P.K.; Davydova, M.; Gopi, A.K.; Pandeya, A.K.; Krieger, J.A.; Zhang, Y.; Date, M.; Ju, S.; et al. Josephson diode effect from Cooper pair momentum in a topological semimetal. *Nat. Phys.* **2022**, *18*, 1228–1233. [CrossRef]
21. Sun, Y.F.; Mao, Y.; Sun, Q.F. Design of Josephson diode based on magnetic impurity. *Phys. Rev. B* **2023**, *108*, 214519. [CrossRef]
22. Hou, Y.; Nichele, F.; Chi, H.; Lodesani, A.; Wu, Y.; Ritter, M.F.; Haxell, D.Z.; Davydova, M.; Ilić, S.; Glezakou-Elbert, O.; et al. Ubiquitous Superconducting Diode Effect in Superconductor Thin Films. *Phys. Rev. Lett.* **2023**, *131*, 027001. [CrossRef]
23. Gupta, M.; Graziano, G.V.; Pendharkar, M.; Dong, J.T.; Dempsey, C.P.; Palmstrøm, C.; Pribiag, V.S. Gate-tunable superconducting diode effect in a three-terminal Josephson device. *Nat. Commun.* **2023**, *14*, 3078. [CrossRef]
24. Narita, H.; Ishizuka, J.; Kan, D. Magnetization Control of Zero-Field Intrinsic Superconducting Diode Effect. *Adv. Mater.* **2023**, *35*, 2304083. [CrossRef]
25. Hu, J.X.; Sun, Z.T.; Xie, Y.M.; Law, K.T. Josephson diode effect induced by valley polarization in twisted bilayer graphene. *Phys. Rev. Lett.* **2023**, *130*, 266003. [CrossRef]
26. Qiu, G.; Yang, H.Y.; Hu, L.; Zhang, H.; Chen, C.Y.; Lyu, Y.; Eckberg, C.; Deng, P.; Krylyuk, S.; Davydov, A.V.; et al. Emergent ferromagnetism with superconductivity in Fe(Te,Se) van der Waals Josephson junctions. *Nat. Commun.* **2023**, *14*, 6691. [CrossRef]

27. Souto, R.S.; Leijnse, M.; Schrade, C. Josephson Diode Effect in Supercurrent Interferometers. *Phys. Rev. Lett.* **2022**, *129*, 267702 [CrossRef]
28. Ciaccia, C.; Haller, R.; Drachmann, A.C.C.; Lindemann, T.; Manfra, M.J.; Schrade, C.; Schonenberger, C. Gate-tunable Josephson diode in proximitized InAs supercurrent interferometers. *Phys. Rev. Res.* **2023**, *5*, 033131. [CrossRef]
29. Greco, A.; Pichard, Q.; Giazotto, F. Josephson diode effect in monolithic dc-SQUIDs based on 3D Dayem nanobridges. *Appl. Phys. Lett.* **2023**, *123*, 092601. [CrossRef]
30. Paolucci, F.; De Simoni, G.; Giazotto, F. A gate-and flux-controlled supercurrent diode effect. *Appl. Phys. Lett.* **2023**, *122*, 042601. [CrossRef]
31. Cuozzo, J.J.; Pan, W.; Shabani, J.; Rossi, E. Microwave-tunable diode effect in asymmetric SQUIDs with topological Josephson junctions. *Phys. Rev. Res.* **2024**, *6*, 023011. [CrossRef]
32. Coraiola, M.; Svetogorov, A.E.; Haxell, D.Z. Flux-Tunable Josephson Diode Effect in a Hybrid Four-Terminal Josephson Junction. *ACS Nano* **2024**, *18*, 9221–9231. [CrossRef]
33. Greco, A.; Pichard, Q.; Strambini, E.; Giazotto, F. Double loop dc-SQUID as a tunable Josephson diode. *Appl. Phys. Lett.* **2023**, *125*, 072601. [CrossRef]
34. Reinhardt, S.; Ascherl, T.; Costa, A.; Berger, J.; Gronin, S.; Gardner, G.C.; Lindemann, T.; Manfra, M.J.; Fabian, J.; Kochan, D.; et al. Link between supercurrent diode and anomalous Josephson effect revealed by gate-controlled interferometry. *Nat. Commun.* **2024**, *15*, 4413. [CrossRef] [PubMed]
35. Pan, H.; Lin, T.H. Control of the supercurrent through a parallel-coupled double quantum dot system. *Phys. Rev. B* **2006**, *74*, 235312. [CrossRef]
36. Pan, H.; Lin, T.H. Tunable supercurrent in a parallel double quantum dot system. *Eur. Phys. J. B* **2007**, 57, 299–303. [CrossRef]
37. Cheng, S.G.; Xing, Y.X.; Xie, X.C.; Sun, Q.F. Supercurrent and its Fano effect in a Josephson Aharonov-Bohm ring. *Eur. Phys. J. B* **2009**, *67*, 551–557. [CrossRef]
38. Bai, L.; Jiang, L.; Zhang, Z.Z.; Duan, C.L. Tunable Josephson current in a lateral triple quatnum dot molecule. *J. Appl. Phys.* **2010**, *108*, 123714. [CrossRef]
39. Yi, G.Y.; Wang, X.Q.; Gong, W.J.; Wu, H.N.; Chen, X.H. Josephson effect in a triple-quantum-dot ring with one dot coupled to superconductors: Numerical renormalization group calculations. *Phys. Lett. A* **2016**, *380*, 1385–1391. [CrossRef]
40. Deacon, R.S.; Oiwa, A.; Sailer, J.; Baba, S.; Kanai, Y.; Shibata, K.; Hirakawa, K.; Tarucha, S. Cooper pair splitting in parallel quantum dot Josephson junctions. *Nat. Commun.* **2015**, *6*, 7446. [CrossRef]
41. DDebbarma, R.; Aspegren, M.; Bostr?m, F.V.; Lehmann, S.; Dick, K.; Thelander, C. Josephson Current via Spin and Orbital States of a Tunable Double Quantum Dot. *Phys. Rev. B* **2022**, *106*, L180507. [CrossRef]
42. Debbarma, R.; Tsintzis, A.; Aspegren, M.; Souto, R.S.; Lehmann, S.; Dick, K.; Leijnse, M.; Thelander, C. Josephson Junction $\pi - 0$ Transition Induced by Orbital Hybridization in a Double Quantum Dot. *Phys. Rev. Lett.* **2023**, *131*, 256001. [CrossRef]
43. Chi, F.; Shen, Y.; Gao, Y.; Liu, J.; Fu, Z.; Yi, Z.; Liu, L. Supercurrent and Superconducting Diode Effect in Parallel Double Quantum Dots with Rashba Spin-Orbit Interaction. *Materials* **2024**, 17, 4497. [CrossRef]
44. Zhou, W.D.; Coleman, J.J. Semiconductor quantum dots. *Curr. Opin. Solid State Mater. Sci.* **2016**, *20*, 352–360. [CrossRef]
45. García de Arquer, F.P.; Talapin, D.V.; Klimov, V.I.; Arakawa, Y.; Bayer, M.; Sargent, E.H. Semiconductor quantum dots: Technological progress and future challenges. *Science* **2021**, *373*, eaaz8541. [CrossRef]
46. Martín-Rodero, A.; Yeyati, A.L. Josephson and Andreev transport through quantum dots. *Adv. Phys.* **2011**, *60*, 899–958. [CrossRef]
47. Sun, Q.F.; Wang, J.; Lin, T.H. Control of the supercurrent in a mesoscopic four-terminal Josephson junction. *Phys. Rev. B* **2000**, *62*, 648. [CrossRef]
48. Zhu, Y.; Sun, Q.F.; Lin, T.H. Andreev bound states and the π-junction transition in a superconductor/quantum-dot/superconductor system. *J. Phys. Condens. Matter* **2001**, *13*, 8783. [CrossRef]
49. Bai, L.; Zhang, L.; Tang, F.R.; Zhang, R. Spin-related thermoelectricity in a hybrid Aharonov-Bohm interferometer with the Rashba effect. *J. Appl. Phys.* **2023**, *134*, 184304. [CrossRef]
50. Ahronov, Y.; Bohm, D. Significance of electromagnetic potentials in the quantum theory. *Phys. Rev.* **1959**, *115*, 485. [CrossRef]
51. van Dam, J.; Nazarov, Y.; Bakkers, E.P.A.M.; De Franceschi, S.; Kouwenhoven, L.P. Supercurrent reversal in quantum dots. *Nature* **2006** *442*, 667–670. [CrossRef]

materials

MDPI

Article

Comparison of Quantum Transition Characteristics of Group II–VI (ZnO), Group III–V (GaN) Compound Semiconductors, and Intrinsic (Si) Semiconductors in Response to Externally Applied Energy

Herie Park [1,2,*] and Su-Ho Lee [1,*]

[1] Department of Electrical Engineering, Dong-A University, Busan 49315, Republic of Korea
[2] Department of ICT Integrated Safe Ocean Smart Cities Engineering, Dong-A University, Busan 49315, Republic of Korea
* Correspondence: herie@dau.ac.kr (H.P.); leesuho@dau.ac.kr (S.-H.L.)

Abstract: In this paper, we study the line-shape (LS), which indicates the amount of absorbed energy, and the line-width (LW), which indicates the scattering factor, according to the vibrational direction of the externally applied energy in the electron–phonon potential interaction system of representative semiconductor bonding types, group II–VI (ZnO) and group III–V (GaN) bonded compound semiconductors and pure group IV (Si) bonded semiconductors. One of the two systems receives the externally applied energy of right-handed circular polarization vibration, and the other receives the externally applied energy of left-handed circular polarization vibration. To analyze the quantum transport, we first employ quantum transport theory (QTR) for an electron system confined within a square-well potential, where the projected Liouville equation is addressed using the balanced-average projection method. In analyzing quantum transitions, phonon emission is linked to the transition line-width (LW), whereas phonon absorption is evaluated through the transition line-shape (LS), highlighting its sensitivity to temperature and magnetic field variations. As a result of analyzing the line-width (LW), which is a quantum scattering coefficient, and the line-shape (LS), which represents the absorbed power, the absorbed power and scattering coefficient were higher for the left circularly polarized vibration under the influence of the external magnetic field. In contrast, the right polarization produced smaller values. In addition, the scattering coefficient (LW) and the absorbed power according to the bonding type of the semiconductor were the largest in Si, a group IV bonded semiconductor, followed by group III–V (GaN) and group II–VI (ZnO) bonded semiconductors.

Keywords: quantum transition characteristics; electron-phonon interaction; scattering coefficient; absorbed power

1. Introduction

As AI-related industries advance rapidly, semiconductors are increasingly required to possess diverse characteristics according to their functions. To meet these diverse demands, a deeper understanding of the quantum transition properties of semiconductor materials is essential. Therefore, this study comparatively analyzed the interactions between electron potential and phonons in ZnO, a representative group II–VI semiconductor with a typical bonding structure; GaN, a group III–V semiconductor; and Si, a pure group IV material. ZnO, with its excellent optical properties, is widely used in optoelectronics as devices and electrodes. Furthermore, its high electromechanical coupling coefficient has gained considerable attention for its applicability in piezoelectric devices. GaN, a next-generation semiconductor material, has recently attracted increasing attention as a high-speed, low-loss, and high-efficiency semiconductor stemming from its ability to surpass the physical limits of silicon and its high electron mobility, enabling high-speed switching and reduced power loss in systems [1,2]. Accordingly, so have the quantum transition properties of the two semiconductors with Si, which still serves as a representative semiconductor [3–8]. Therefore, in this study, we compared the interaction between the electron potential and phonons in accordance with the direction of the external input energy field in ZnO, GaN, and Si semiconductor materials [9–11]. The external input energy field is divided into two systems: one is an energy field applied with a right-handed circularly polarized field (RCF), while the second corresponds to an energy field applied with a left-handed circularly polarized field (LCF). This study is directed toward analyzing the line-shape (LS), which represents the degree of absorbed energy, and the line-width (LW), which reflects the scattering factor. In addition, the quantum transition characteristics were compared under two conditions. The first corresponds to a right-handed circularly polarized field, while the second involves an energy field. The analysis explores how these characteristics change with the direction of the applied field. There are numerous theories for quantum transport problems, including Boltzmann transport theory [12], Green's function approach [13,14], the force balance approach [15], Feynman's path integral formulation [16] and the projection operator techniques [17,18]. Despite the reasonableness of all these methods, research on nonlinear behavior has been restricted. A thorough view of electron transport in strong electric fields requires the inclusion of nonlinear effects. Furthermore, the method proposed by the research groups of Zwanzig and Kenkre directly employed the projection operator in the Liouville equation to propose a response function that incorporates Kubo's theory as a lowest-order approximation [9,10,19,20]. The propagator includes a nonlinear factor, but since it appears in the exponent, extending this term is difficult.

Many methods have been proposed to interpret such quantum transport, and in this study, we employed quantum transport theory (QTR) for an electron system confined in a square-well potential. The present work applies the projected Liouville equation method with the Equilibrium Average Projection Technique (EAPT), which has the advantage of excellent extensibility of quantum transport theory and of being able to obtain the quantum response function as well as the scattering factor expression at once through this extension [11,21,22]. Previous studies have been limited to unbound potential systems using the extremely weak coupling (EWC) approximation and have also suffered from the lack of intermediate states in quantum transitions. However, the advantage of using the EAPS is that it extends quantum transport theory to obtain the quantum response function and scattering factor formulas in a single step. Regarding the dependence of the line-width (LW) on the magnetic field, which represents the scattering factor, it is difficult to calculate or measure the absorbed power at various external magnetic field wavelengths, making it very difficult to analyze using other theories or experiments. However, our research team proposed that the QTR of the EAPS can directly determine the line-width (LW) of the scattering factor at various external magnetic field wavelengths through the EAPS, making it easy to verify the variation in the line-width (LW) with magnetic field [23,24].

This study mainly aimed to investigate the transition characteristics of various semiconductor materials by comparing the line-shapes (LSs) of quantum-level optical transitions, which indicate the absorbed power according to the input of right circularly polarized fields (RCFs) and left circularly polarized fields (LCFs), with the quantum-level transition line-width (LW), corresponding to the scattering factor effect, in the electron-potential–phonon interaction system of representative semiconductor materials such as ZnO, GaN, and Si [24,25].

2. Expression for Absorbed Power and Scattering Factor

When the electron field is placed in a static magnetic field $\vec{B} = B_z\hat{z}$, the single electron energy level is quantized into Landau levels. Here, we choose an electron field confined within an infinite square well potential (SQWP) extending from $z = 0$ and $z = L_z$ along a single direction, the z-direction. Also, the eigenvalues and eigenstates of the square well potential system are those in the reference [22]. Assuming that an electric field $E(t) = E_0\exp(i\omega t)$ is imposed along the z-axis, accompanied by displacement, the absorbed power appearing in the system is expressed as $P(\omega) = (E_0^2/2)\mathrm{Re}\{\sigma(\omega)\}$. Here, "Re" represents the real component, and $\sigma(\omega)$ is the photoconductivity tensor, representing the coefficient of the current expression. The absorbed power and the scattering factor represent the optical line-shape (LS) and the optical line-width (LW), respectively. The Hamiltonian describing the electron–phonon interaction is formulated as in Equation (1).

$$H_s = H_e + H_P + V = \sum_{\beta}\langle\beta|h_0|\beta\rangle a_\beta^+ a_\beta + \sum_q \hbar\omega_q b_q^+ b_q + \sum_q\sum_{\alpha,\mu} C_{\alpha,\mu}(q) a_q^+ a_\mu \left(b_q + b_{-q}^+\right) \quad (1)$$

H_e: Hamiltonian of an electron; h_0: Hamiltonian of a single-electron; H_P: Hamiltonian of a phonon; $b_1(b_2^+)$: Boson particle annihilation operator (creation operator); V: interaction Hamiltonian of electron–phonon (or impurity); $\vec{q}$: wave vector of a phonon (or impurity). The electron–phonon (or impurity) interaction Hamiltonian of the interacting system is as follows:

$$V \equiv \sum_q\sum_{\alpha,\mu} C_{\alpha,\mu}(q) a_\alpha^+ a_\mu \left(b_q + b_{-q}^+\right) \quad (2)$$

here, the electron–phonon interaction coupling matrix element $C_{\alpha,\mu}(q)$ is $C_{\alpha,\mu}(q) \equiv V_q < \alpha\left|\exp(i\vec{q}\cdot\vec{r})\right|\mu >$, and here V_q is the material's coupling coefficient and $\vec{r}$ is the electron position vector. Using the successive approximation in the reference [23], the absorbed power formula in a right-circularly polarized external magnetic field (or LS formula) is finally obtained as Equation (3). We also proposed a specific absorbed power formula that constrains the potential system in the reference [24].

$$P^R(\omega) \propto \left(\frac{e^2\omega_c^2}{\pi^2\hbar\omega}\right)\left\{\frac{\gamma_{total}^R(\omega_c)\sum_{N_\alpha}\int_{-\infty}^{\infty} dk_{z\alpha}(N_\alpha+1)(f_\alpha - f_{\alpha+1})}{(\omega-\omega_c)^2 + \left(\gamma_{total}^R(\omega_c)\right)^2}\right\} \quad (3)$$

Also, we can express the absorption power formula (or LS formula) for the energy input of the left-hand circular vibration external field as Equation (4).

$$P^{(L)}(\omega) \propto \left(\frac{e^2\omega_c^2}{\pi^2\hbar\omega}\right)\left[\frac{\gamma_{total}^{(L)}(\omega_c)\sum_{N_\alpha}\int_{-\infty}^{\infty} dk_{z\alpha}(N_\alpha+1)(f_{\alpha+1} - f_\alpha)}{(\omega-\omega_c)^2 + \left(\gamma_{total}^{(L)}(\omega_c)\right)^2}\right] \quad (4)$$

Here, the scattering factor function (or LW formula) for the energy applied to the right-hand circularly oscillating external field can be expressed as Equation (5).

$$\gamma_{total}^{(R)}(\omega) \equiv \mathrm{Re}\Xi_{kl}^{(R)}(\omega) \equiv \sum_{\mp}\sum_{N_\alpha=0}\sum_{N_\beta=0}\gamma_{\alpha,\beta}^{\mp}$$
$$= \left(\frac{\Omega}{4\pi\hbar^2 v_s}\right)\left(\frac{\pi}{L_z}(2+\delta(n_\alpha,n_\beta))\right)\left\{\frac{\sum_{\mp}\sum_{N_\alpha=0}\sum_{N_\beta=0}\int_{-\infty}^{\infty}dk_{z\alpha}\int_{-\infty}^{\infty}dq_z Y_{\alpha,\beta}^{(R)\mp}}{\sum_{N_\alpha=0}^{\infty}\int_{-\infty}^{\infty}dk_{z\alpha}(N_\alpha+1)(f_{\alpha+1}-f_\alpha)}\right\} \quad (5)$$

Also, the scattering factor function (or LW) for the energy applied to the left-hand circularly oscillating external field is given by Equation (6).

$$\gamma_{total}^{(L)}(\omega) \equiv \mathrm{Re}\Xi_{kl}^{(L)}(\omega) \equiv \sum_{\mp}\sum_{N_\alpha=0}\sum_{N_\beta=0}\gamma_{\alpha,\beta}^{(L)\mp}$$
$$= \left(\frac{\Omega}{4\pi\hbar^2 v_s}\right)\left(\frac{\pi}{L_z}(2+\delta(n_\alpha,n_\beta))\right)\left\{\frac{\sum_{\mp}\sum_{N_\alpha=0}\sum_{N_\beta=0}\int_{-\infty}^{\infty}dk_{z\alpha}\int_{-\infty}^{\infty}dq_z Y_{\alpha,\beta}^{(L)\mp}}{\sum_{N_\alpha=0}^{\infty}\int_{-\infty}^{\infty}dk_{z\alpha}(N_\alpha+1)(f_\alpha-f_{\alpha+1})}\right\} \quad (6)$$

When a circularly polarized external magnetic field of various frequencies is applied in right- or left-handed directions, the dependence of LSs and LWs on the magnetic field and temperature for tetravalent-bonded semiconductor silicon (Si) and the group III–V and group II–VI bonded ones, gallium nitride (GaN) and zinc oxide (ZnO), were numerically analyzed using Equations (3)–(6). Since the absorbed power must be calculated for various frequencies, it is difficult to obtain the LWs using other theories for the analysis. However, the Equilibrium Average Projection Scheme (EAPS) allows for the direct calculation of the LWs, which facilitates the numerical analysis. By comparing the dependence of the LWs and LSs with the magnetic field and temperature, covering transitions both between and within Landau levels, the quantum transition process in germanium can be analyzed. The material constants of ZnO, GaN, and Si semiconductors are listed in Table 1. These constants were adopted for the numerical analysis. To compare LSs and LWs according to the direction of application of two polarized oscillating external magnetic fields, the left circular oscillating external magnetic field (LCF) is abbreviated as "L" in the graph, and the right circular oscillating external magnetic field (RCF) is abbreviated as "R".

Table 1. Material constant.

Symbol	Contents	Value (ZnO)	Value (GaN)	Value (Si)
m^*	Effective mass of electron	$0.24m_0$	$0.22m_0$	$0.33m_0$
$\overline{m}$	Effective mass of hole	$1.2m_0$	$0.58m_0$	$0.58m_0$
ρ	Mass density	$4090\ \mathrm{kg/m^3}$	$4820\ \mathrm{kg/m^3}$	$2340\ \mathrm{kg/m^3}$
κ	Characteristic constant	$17.88 \times 10^{-4}\ \mathrm{eV/K}$	$8.58 \times 10^{-4}\ \mathrm{eV/K}$	$4.37 \times 10^{-4}\ \mathrm{eV/K}$
ξ	Characteristic constant	204	235	636
$\overline{K}$	Electromechanical constant	$6 \times 10^{-2}\ \mathrm{m/s}$	$2.6 \times 10^{-2}\ \mathrm{m/s}$	$2.98005 \times 10^{-2}\ \mathrm{m/s}$
$\overline{v}_s$	Speed of sound	4300.5 m/s	3045 m/s	9030 m/s
$\overline{\varepsilon}_s$	Energy gap	3.4 eV	3.4 eV	1.424 eV
L_s L_z	Length of well of z direction	20×10^{-9} m	20×10^{-9} m	20×10^{-9} m

3. The Analysis

Figure 1 compares the influence of the magnetic field on the absorption power $P^R(B)$ and $P^L(B)$ according to the line-shapes (LSs) of Si, GaN, and ZnO at temperatures of T = 50, 70, 90, 120, and 210 K, with an external magnetic field wavelength of λ = 394 μm. Figure 1 shows that the absorption power P(B) and the line-width (LW), which represents the scattering coefficient, increase with increasing temperature. These results indicate that the increase in lattice vibration due to the increase in heat increases the phonon collision effect, which can account for the resonance observed in the electron–phonon interaction system. The externally applied energy field is larger when applied by the LCF than when applied by the RCF. In addition, the absorption power $P^R(B)$ of all three materials is almost the same when applied by the LCF, but the $P^L(B)$ follows the order of Si > GaN > ZnO. This is because the difference in absorption power depends on the binding energy of the materials. These results demonstrate that the QTR of the EAPS effectively accounts for resonance phenomena, incorporating quantum transfer and scattering processes on a microscopic scale. The comparison of absorbed power P(B) in a magnetic field with a circular external field direction was first proposed by our research team. The QTR of EAPS is readily available compared to other theories, as P(B) in a circular external field is obtained by the QTR of the EAPS, which has a comparable theoretical framework.

Figure 2 shows the temperature dependences $\gamma^{(R)}(T)$ and $\gamma^{(L)}(T)$ of the scattering coefficient LWs. For all three materials (Si, GaN, and ZnO), $\gamma^{(L)}(T)$ increases with increasing temperature for external field wavelengths of λ = 220, 394, 513, 550, and 720 μm, while $\gamma^{(R)}(T)$ also increases somewhat, but the change is not large. $\gamma^{(L)}(T)$ increases in the order of Si, a tetravalent semiconductor, GaN, a group III–V bond semiconductor, and ZnO, a group II–VI bond semiconductor. This finding agrees well with the results presented in Figure 1. In addition, $\gamma^{(L)}(T)$ increases with increasing temperature for all external field wavelengths, and $\gamma^{(L)}(T)$ is larger than $\gamma^{(R)}(T)$. This result indicates that the phonon scattering effect increases with increasing temperature for all external field wavelengths and that it is affected by the direction of electron motion. This can be seen as a result of increased energy absorption and emission due to increased thermal lattice vibration as temperature increases, as illustrated in Figure 1. Furthermore, it can be seen that the binding energy of the three semiconductors also affects the scattering coefficient.

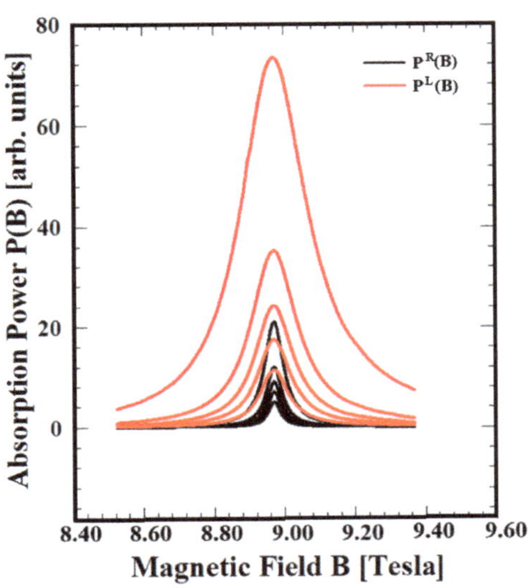

(a)

Figure 1. *Cont.*

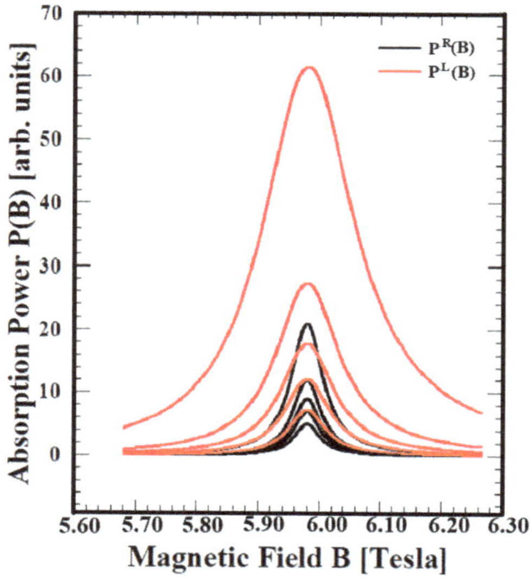

(**b**)

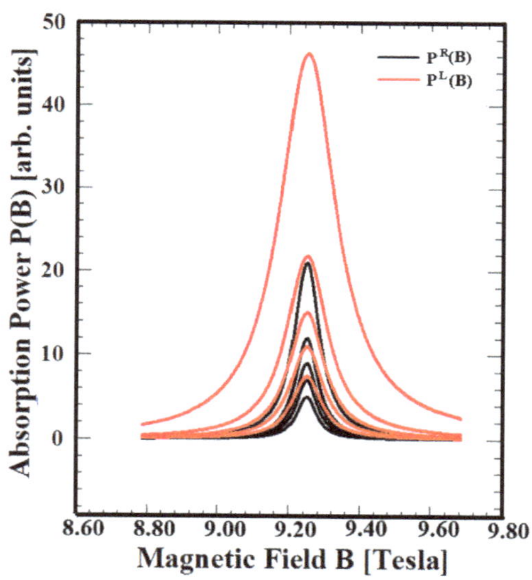

(**c**)

Figure 1. The magnetic field dependence of normalized P(B) (LSs) of Si, GaN, and ZnO. (**a**) Magnetic field dependence of normalized P(B) (LSs) of Si with λ = 394 μm at T = 50, 70, 90, 120, and 210 K (ordered from bottom upward). (**b**) Magnetic field dependence of normalized P(B) (LSs) of GaN with λ = 394 μm at T = 50, 70, 90, 120, and 210 K (ordered from bottom upward). (**c**) Magnetic field dependence of normalized P(B) (LSs) of ZnO with λ = 394 μm at T = 50, 70, 90, 120, and 210 K (ordered from bottom upward).

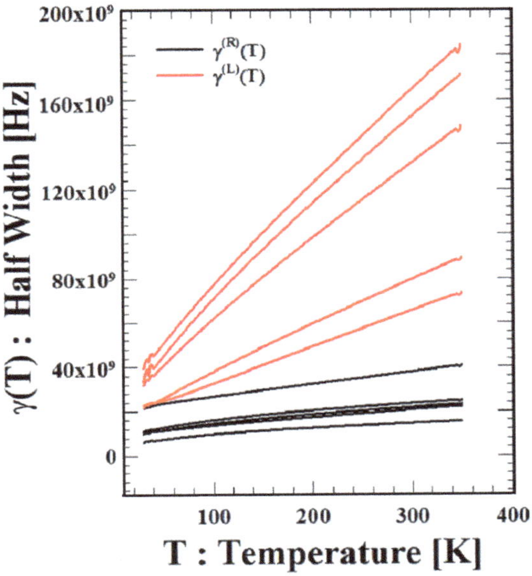

(a)

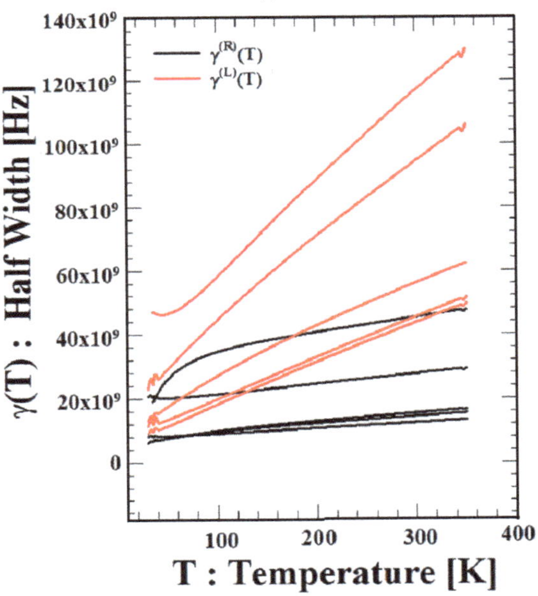

(b)

Figure 2. *Cont.*

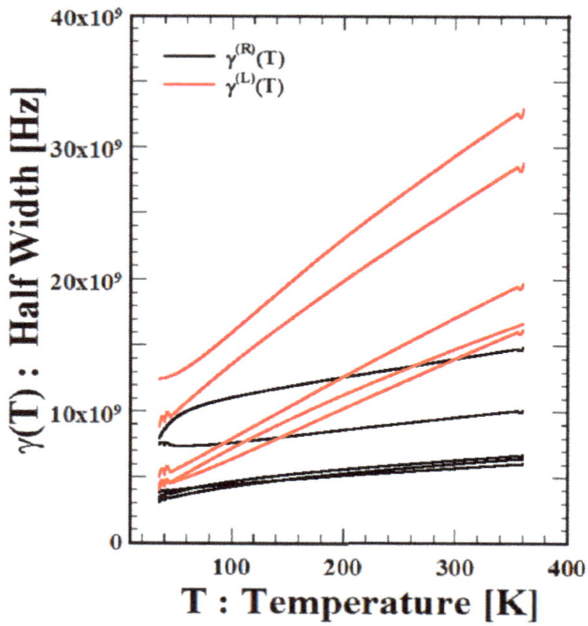

(c)

Figure 2. The temperature dependence of scattering factor (LWs), $\gamma(\mathrm{T})$, of Si, GaN, and ZnO. (**a**) Temperature dependence of QTLW of Si, $\gamma(\mathrm{T})$, with λ = 220, 394, 513, 550, and 720 μm (ordered from bottom upward). (**b**) Temperature dependence of QTLW of GaN, $\gamma(\mathrm{T})$, with λ = 220, 394, 513, 550, and 720 μm (ordered from bottom upward). (**c**) Temperature dependence of QTLW of ZnO, $\gamma(\mathrm{T})$, with λ = 220, 394, 513, 550, and 720 μm (ordered from bottom upward).

Figure 3 shows the dependence of the magnetic field on the scattering factor LWs, $\gamma^{(R)}(B)$ and $\gamma^{(L)}(B)$, of three materials (Si, GaN, and ZnO) at T = 50, 70, 90, 120, and 210 K. For all three materials, $\gamma^{(L)}(B)$ is greater than $\gamma^{(R)}(B)$ under the full set of temperature conditions. This result also gives a consistent explanation of the directional characteristics of electron motion, depending on the magnetic field direction and system conditions. All three materials increase with increasing magnetic field, but for the Si semiconductor, the value decreases in the low-field region. Furthermore, for all three materials, the scattering factor LWs increase in the high-field region, with the values increasing in the order of ZnO > GaN > Si. Analyzing the scattering factor LWs at various wavelengths as a function of the magnetic field is very helpful in characterizing the magnetic properties of materials. Analyzing the magnetic field dependence at various wavelengths through theory or experimentation is difficult. This is because it is difficult to calculate or experimentally observe absorption and emission energies at various external magnetic field wavelengths. However, the QTR of the EAPS proposed in this study is an effective tool for directly measuring long-wavelength energy at various external magnetic field wavelengths.

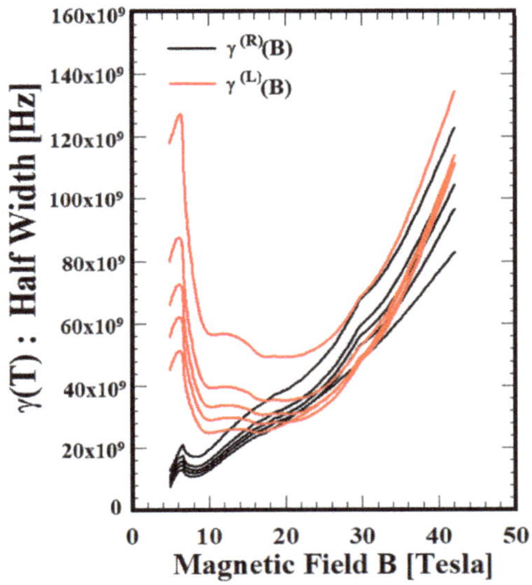

(**a**)

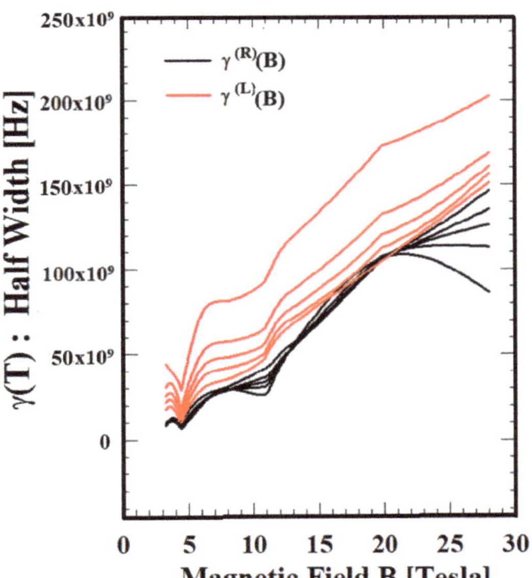

(**b**)

Figure 3. *Cont.*

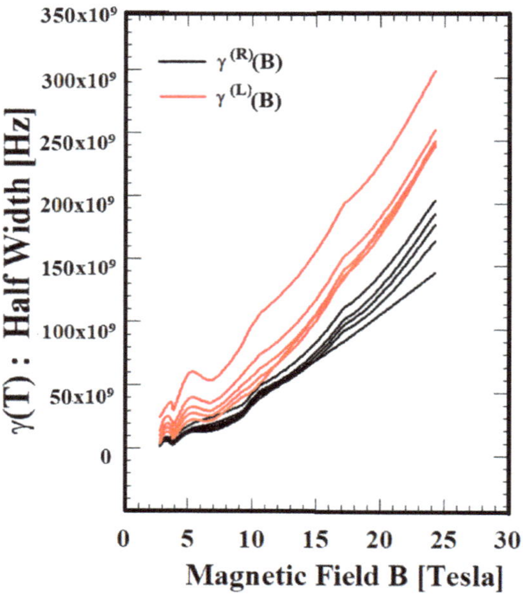

(c)

Figure 3. The magnetic field dependence of scattering factor (LWs), $\gamma(\mathrm{B})$, of Si, GaN, and ZnO. (**a**) Magnetic field dependence of LWs of Si, $\gamma(\mathrm{B})$, at T = 50, 70, 90, 120, and 210 K (ordered from bottom upward). (**b**) Magnetic field dependence of LWs of GaN, $\gamma(\mathrm{B})$, at T = 50, 70, 90, 120, and 210 K (ordered from bottom upward). (**c**) Magnetic field dependence of LWs of ZnO, $\gamma(\mathrm{B})$, at T = 50, 70, 90, 120, and 210 K (ordered from bottom upward).

For RFC, to investigate the process of the quantum transition, the total line-width (LW) is written as $\gamma_{total}^{(R)} \equiv \gamma_{em}^{(R)} + \gamma_{ab}^{(R)}$, where $\gamma_{em}^{(R)} \equiv \gamma_{intraL}^{(R)em} + \gamma_{interL}^{(R)em}$ and $\gamma_{ab}^{(R)} \equiv \gamma_{intraL}^{(R)ab} + \gamma_{interL}^{(R)ab}$ are the line-widths (LWs) of the overall phonon emission and absorption transitions, respectively. Here, $\gamma_{intraL}^{(R)em}$, $\gamma_{interL}^{(R)em}$,$\gamma_{intraL}^{(R)ab}$ and $\gamma_{interL}^{(R)ab}$ are the line-widths, and are defined, respectively, for the intra-level emission, inter-level emission, intra-level absorption, and inter-level emission-absorption transitions. For LFC, L is used instead of R in the subscript.

Figure 4 indicates the dependence of temperature on the line-widths (LWs), $\gamma_{total}^{(R)}(T)$, $\gamma_{em}^{(R)}(T)$, $\gamma_{ab}^{(R)}(\mathrm{T})$, $\gamma_{total}^{(L)}(T)$, $\gamma_{em}^{(L)}(T)$ and $\gamma_{ab}^{(L)}(T)$ of three types of semiconductors (Si, GaN, and ZnO) when the wavelength of the external applied magnetic field is λ = 513 μm. The total line-width (LW) $\gamma_{total}^{(L)}(T)$, which is equal to the sum of phonon emission and absorption, increases with increasing temperature for all three materials because both phonon emission and absorption processes increase. In addition, $\gamma_{total}^{(R)}(T)$ does not increase much because both phonon emission and absorption processes show little change with temperature. In addition, the difference between RFC and LCF is regarded as the characteristics of the electron motion direction in accordance with the applied magnetic field direction and system conditions. Although the scattering behavior of transitions involving phonon emission and absorption cannot be experimentally separated, examining the relationship between the overall scattering behavior and the contributions of the two processes can provide insight into the thermal characteristics of the scattering effects in the system. The role of the two processes might differ across various systems and cases. In this study, the order of magnitudes of the left-hand line-widths (LWs) caused by the direction of the externally applied magnetic field is $\gamma_{ab}^{(L)}(T) < \gamma_{em}^{(L)}(T) < \gamma_{total}^{(L)}(T)$, and the order of magnitudes of the right-hand line-widths (LWs) is also $\gamma_{ab}^{(R)}(T) < \gamma_{em}^{(R)}(T) < \gamma_{total}^{(R)}(T)$ for all three semiconductors in the entire temperature range. Moreover, the analysis shows the value of the phonon emission transition, $\gamma_{em}^{(L)}(T)$ or $\gamma_{em}^{(R)}(T)$ is larger than the value of the phonon absorption transition, and $\gamma_{ab}^{(L)}(T)$ or $\gamma_{ab}^{(R)}(T)$ in all temperature ranges for all three semiconductors, regardless of the orientation of the externally applied magnetic field. These results help clarify the orientation-dependent behavior of the electron in relation to the applied magnetic field orientation and system conditions.

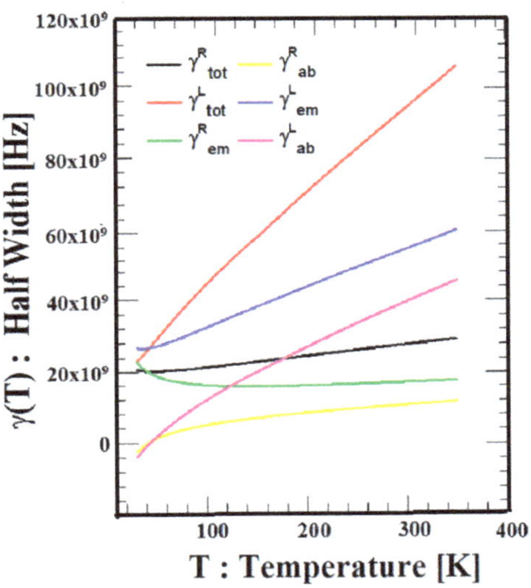

(a)

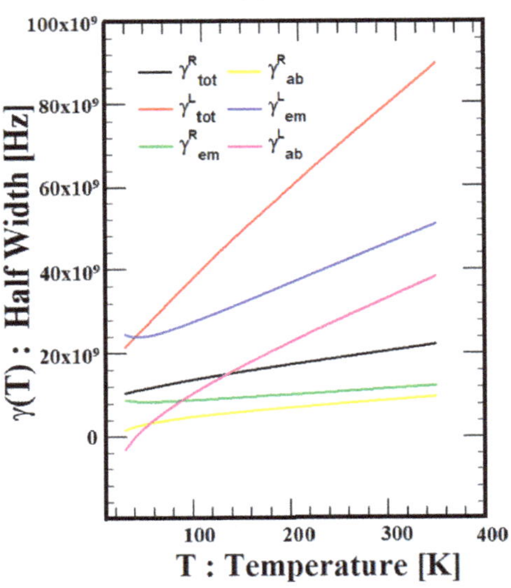

(b)

Figure 4. *Cont.*

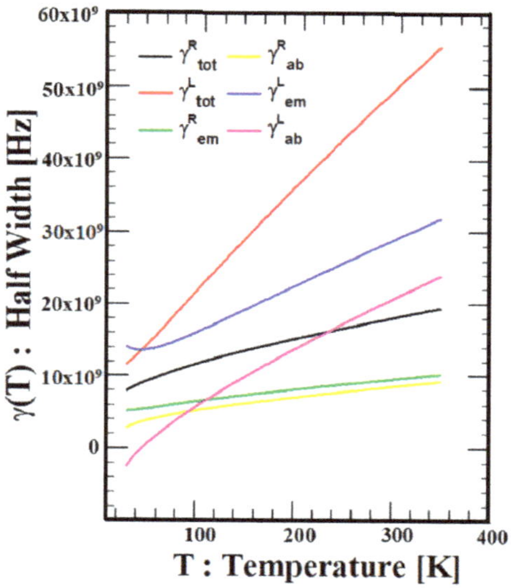

(c)

Figure 4. Temperature dependence of scattering factor (LWs), $\gamma(\mathrm{T})$, of Si, GaN, and ZnO. (**a**) Comparisons of the temperature dependence of LWs of Si, $\gamma_{total}(T)$, $\gamma_{em}(T)$, and $\gamma_{ab}(T)$ with λ = 513 μm. (**b**) Comparisons of the temperature dependence of LWs of GaN, $\gamma_{total}(T)$, $\gamma_{em}(T)$, and $\gamma_{ab}(T)$ with λ = 513 μm. (**c**) Comparisons of the temperature dependence of LWs of GaN, $\gamma_{total}(T)$, $\gamma_{em}(T)$, and $\gamma_{ab}(T)$ with λ = 513 μm.

4. Practical Implications

By examining the behavior of electron–phonon interactions within quantum wells in response to external energy, we believe it will aid in the interpretation of conduction mechanisms in various semiconductor materials. Furthermore, by analyzing energy emission and absorption at specific wavelengths, we anticipate contributing to the development of optical devices that generate new optical wavelengths and optical sensors.

5. Conclusions

This study numerically analyzed the mechanism of scattering associated with electron–phonon coupling through strain potential in three types of semiconductors, a topic of recent research. Furthermore, we examined the electron dynamics relative to the orientation of the external magnetic field. First, the results of Equations (3)–(6) proposed in this study are more rigorous than those in previous studies and are directly applicable to numerical analysis via integration in the dual wave vectors. This investigation provides a distribution function that is significantly different with other approaches. Our EAPS theory, with its fewer computational steps, allows for a much easier analysis of quantum transitions in a variety of situations than other theories. Furthermore, the EAPS theory allows for the separation of line-widths for individual quantum transitions depending on the case. It facilitates the analysis of quantum transitions, which is a key advantage of the EAPS theory. The findings provide a basis for the numerical analysis of the scattering mechanism in materials with electron strain potential–phonon interaction. The characteristics of LWs and LSs increase with temperature and magnetic field for all three types of semiconductors. Si, a tetravalent semiconductor, showed the highest absorption power and scattering factor values, followed by GaN, a group III–V bond semiconductor, and ZnO, a group

II–VI bond semiconductor. In addition, the absorption power as well as the scattering intensity increased under a leftward external magnetic field compared with the rightward orientation. In the scattering process, the scattering factor in the phonon emission transition process was larger than that in the transition process of phonon absorption, indicating that the phonon emission transition process is the main scattering process. This suggests that the EAPS theory can offer a promising means of elucidating material properties in other condensed matter systems.

Author Contributions: Conceptualization, S.-H.L. and H.P.; methodology, S.-H.L.; software, S.-H.L.; validation, S.-H.L.; formal analysis, S.-H.L.; investigation, S.-H.L.; resources, S.-H.L.; data curation, S.-H.L.; writing—original draft preparation, S.-H.L. and H.P.; writing—review and editing, S.-H.L. and H.P.; visualization, S.-H.L.; supervision, S.-H.L.; project administration, S.-H.L. and H.P.; funding acquisition, S.-H.L. All authors have read and agreed to the published version of the manuscript.

Funding: This research was supported by the Regional Innovation System & Education (RISE) program through the Institute for Regional Innovation System & Education in Busan Metropolitan City, funded by the Ministry of Education (MOE) and the Busan Metropolitan City, Republic of Korea. (2025-RISE-02-003).

Data Availability Statement: The original contributions presented in this study are included in the article. Further inquiries can be directed to the corresponding authors.

Acknowledgments: This research has been supported by Dong-A University.

Conflicts of Interest: The authors declare no conflict of interest.

References

1. Fan, B.; Zhao, X.; Zhang, J.; Sun, Y.; Yang, H.; Guo, L.J.; Zhou, S. Monolithically integrating III-Nitride quantum structure for full-spectrum white LED via bandgap engineering heteroepitaxial growth. *Laser Photonics Rev.* **2023**, *17*, 2200455. [CrossRef]
2. Song, J.; Shi, L.; Cui, S.; Meng, L.; Zhou, Q.; Jiang, J.; Jin, C.; Hu, J.; Wen, K.; Zhou, S. High-performance GaN-based green flip-chip mini-LED with lattice-compatible AlN passivation layer. *Nanomaterials* **2025**, *15*, 1048. [CrossRef]
3. Wolfe, C.M.; Stillman, G.E. *Physical Properties of Semiconductors*; Prentice-Hall: Englewood Cliffs, NJ, USA, 1989.
4. Ferry, D.K. *Semiconductors*; Macmillan: New York, NY, USA, 1991.
5. Chung, S.L. *Physics of Optoelectronic Devices*; Wiley: New York, NY, USA, 1995.
6. Arif, O.; Zannier, V.; Rossi, F.; DeMatteis, D.; Kress, K.; Zardo, I.; Sorba, L. GaAs/GaP superlattice nanowires: Growth, vibrational and optical properties. *Nanoscale* **2023**, *15*, 1145–1153. [CrossRef] [PubMed]
7. Tran, C.P.; Le, N.M.; Nguyen, D.H. Comparison of electron scattering by acoustic-phonons in two types of quantum wells with GaAs and GaN materials. *Nanoscale Adv.* **2024**, *6*, 832–845.
8. Aviv, K.; Nichalos, R.; Valeno, D.G.; Ady, A.; Javier, G.A.F.; Ido, K. Modeling quantum optical phenomena using transition currents. *Appl. Phys. Rev.* **2024**, *11*, 35. [CrossRef]
9. Jo, S.G.; Kang, N.L.; Cho, Y.J.; Choi, S.D.; Kim, S.C.; Ryu, J.Y. Modeling of the cyclotron transition theory for quasi-two-dimensional electron systems by the isolation-projection technique. *J. Korean Phys. Soc.* **1997**, *30*, 103–105.
10. Kobori, H.; Ohyama, T.; Otsuka, E. Line-width of quantum limit cyclotron resonance. I. Phonon scatterings in Ge, Si, CdS, and InSb. *J. Phys. Soc. Jpn.* **1990**, *59*, 2141–2163. [CrossRef]
11. Sug, J.Y.; Lee, S.H.; Choi, J.Y.; Sa-Gong, G.; Kim, J.J. Magnetic properties of optical quantum transition line shapes and line widths of electron–piezoelectric potential phonon interacting materials under circularly oscillating fields. *Jpn. J. Appl. Phys.* **2008**, *47*, 7757. [CrossRef]
12. Barker, J.R. Damping theory of the magnetophonon structure in the longitudinal Seebeck effect in semiconductors. *J. Phys. C Solid State Phys.* **1973**, *6*, L52. [CrossRef]
13. Ting, C.S.; Ying, S.C.; Quinn, J.J. Theory of cyclotron resonance of interacting electrons in a semiconducting surface inversion layer. *Phys. Rev. B* **1977**, *16*, 5394–5402. [CrossRef]
14. Wu, X.; Peeters, F.M.; Devreese, J.T. Theory of the cyclotron resonance spectrum of a polaron in two dimensions. *Phys. Rev. B* **1986**, *34*, 8800. [CrossRef]
15. Grigoglini, P.; Parravidini, G.P. Phonon thermal baths: A treatment in terms of reduced models. *Phys. Rev. B* **1982**, *25*, 5180. [CrossRef]

16. Kubo, R. Statistical-mechanical theory of irreversible processes. I. General theory and simple applications to magnetic and conduction problems. *J. Phys. Soc. Jpn.* **1957**, *12*, 570–586. [CrossRef]
17. Mori, H. Transport, collective motion, and Brownian motion. *Prog. Theor. Phys.* **1965**, *33*, 423–455. [CrossRef]
18. Nagano, K.; Karasudani, T.; Okamoto, H.; Mori, H. Reduced equations of motion for generalized fluxes and forces in the continued-fraction expansion. *Prog. Theor. Phys.* **1980**, *63*, 1904–1916. [CrossRef]
19. Zwanzig, R. Ensemble method in the theory of irreversibility. *J. Chem. Phys.* **1960**, *33*, 1338–1341. [CrossRef]
20. Kenkre, V.M. Integrodifferential equation for response theory. *Phys. Rev. A* **1971**, *4*, 2327. [CrossRef]
21. Lee, S.H.; Sug, J.Y. The temperature dependences of the electron-deformation potential phonon interacting system of quasi-twodimensional systems under circularly oscillating fields in GaAs and Ge. *J. Nanoelectron. Optoelectron.* **2017**, *12*, 409–412.
22. Lee, S.H.; Park, H. Magnetic and Temperature Effects on Optical Quantum Transition Line Properties in Electron-Piezoelectric Phonon Coupled Materials Under Square Well Confinement Potential. *Electronics* **2025**, *14*, 1256. [CrossRef]
23. Sug, J.Y.; Choi, S.D. Quantum transport theory based on the equilibrium density projection technique. *Phys. Rev. E* **1997**, *55*, 314–319. [CrossRef]
24. Lee, S.H.; Sug, J.Y. The temperature dependences of the electron-piezoelectric potential phonon interacting system of quasi-twodimensional systems in GaN and ZnS. *J. Nanosci. Nanotechnol.* **2018**, *18*, 7255–7258.
25. Su, C.; Zhang, C.; Zhang, B. Adjusting the structural, electronic and optical properties of CdS by the introduction of Be: A DFT study. *Mater. Today Commun.* **2022**, *31*, 103394. [CrossRef]

materials

MDPI

Communication

Vacancy and Strain Effects on the Stability and Electronic Properties of 2D-Mg Intercalated GaN

Qilin Wu [1], Shuqing Zhang [1,*], Xiaoyan Song [2] and Xinping Zhang [1]

1 Institute of Information Photonics Technology, School of Physics and Optoelectronic Engineering, Beijing University of Technology, Beijing 100124, China; qlwu2001@emails.bjut.edu.cn (Q.W.); zhangxinping@bjut.edu.cn (X.Z.)

2 College of Materials Science and Engineering, Key Laboratory of Advanced Functional Materials, Ministry of Education of China, Beijing University of Technology, Beijing 100124, China; xysong@bjut.edu.cn

* Correspondence: zhangsq@bjut.edu.cn

Abstract: The recent discovery of two-dimensional Mg (2D-Mg) intercalation in GaN has attracted increasing attention, prompting fundamental questions regarding its structural stability and electronic properties. In this work, we employ first-principles calculations to investigate the structural and electronic effects of 2D-Mg intercalation in GaN. We identify the most energetically favorable intercalation ratio of Mg, reveal the critical role of Ga vacancies in restoring semiconducting behavior, and demonstrate that compressive strain further modulates the electronic structure. In particular, the configuration with 75% Mg intercalation and nearest-neighbor Ga vacancy under compressive strain exhibits significant band gap narrowing and enhanced Mg-related acceptor activity. These findings challenge long-standing assumptions about Mg clustering and establish a mechanistic framework based on intercalation, vacancy engineering, and strain control for the design of next-generation p-type GaN devices.

Keywords: gallium nitride; 2D-Mg intercalation; Ga vacancies; strain control

1. Introduction

Gallium nitride (GaN) has emerged as a critical wide-bandgap semiconductor for advanced power electronics and optoelectronic devices, owing to its excellent intrinsic properties [1–4]. However, achieving effective p-type doping remains a major challenge that limits further performance improvements [5,6]. Mg is currently the only practical p-type dopant for GaN and has been applied in commercial optoelectronic devices, such as blue LEDs [7,8]. Nevertheless, the realization of high-performance p-type GaN remains far from ideal [9]. The deep acceptor level of Mg leads to low ionization efficiency at room temperature [10], while compensation from native defects significantly limits hole concentration and mobility [11]. These intrinsic limitations pose a critical bottleneck to the development of next-generation high-efficiency p-type GaN devices.

To address these issues, various alternative strategies have been proposed, including co-doping with additional impurities to enhance acceptor ionization [12,13], polarization-induced doping to exploit built-in electric fields in heterostructures [14], and strain engineering to modify the band structure and enhance hole transport [15–17]. While these approaches have yielded incremental improvements, they often involve complex fabrication processes or provide limited control over hole mobility and transport. Moreover, when Mg doping concentrations exceed ~10^{19} cm^{-3}, closed pyramidal inversion domains tend to form, further degrading hole conductivity due to disrupted crystal polarity and defect clustering [17–19].

Intercalation has long been recognized as an effective route to tune structural stability and electronic properties of materials [20]. Recently, a fundamentally different strategy has emerged based on the intercalation of two-dimensional Mg (2D-Mg) layer into GaN, forming so-called Mg-intercalated GaN superlattices (MiGs). The feasibility of such superlattice formation is further supported by the close lattice match between GaN and Mg, which minimizes interfacial strain and stabilizes the intercalated configuration. Experimental studies have shown that samples containing MiGs exhibit up to a six-fold enhancement in hole transport compared to non-intercalated samples, primarily attributed to strong out-of-plane compressive strain and periodic polarity inversion. However, despite these experimental advances [21], the fundamental mechanisms underlying the structural stability and electronic modulation induced by 2D-Mg intercalation remain poorly understood. Our study addresses this gap by providing a systematic theoretical investigation.

In this work, we systematically investigate the influence of 2D-Mg intercalation on the structural and electronic properties of GaN using first-principles calculations. We construct a series of intercalated models with varying Mg coverage, compare their formation energies, and identify the most stable configuration. Introducing Ga vacancies near the Mg layer to maintain charge neutrality not only restores semiconducting behavior but also generates shallow Mg acceptor states. Additionally, compressive strain along the *c*-axis further narrows the band gap and facilitates hole excitation. These results provide mechanistic insights into the role of intercalation, vacancies, and strain in modulating the electronic properties of GaN.

2. Materials and Methods

First-principles calculations are performed using projector augmented wave (PAW) method [22] as implemented in the Vienna ab initio Simulation Package (VASP) [23,24]. Structural optimization was completed until the total energy and atomic forces converged to 10^{-5} eV and 0.01 eV/Å, respectively. A plane-wave energy cutoff of 500 eV was used, and the Brillouin zone of unit cell was sampled with a Monkhorst–Pack k-point mesh [25] of $15 \times 15 \times 9$, corresponding to a maximum spacing of 0.02 Å^{-1}. The relaxed atomic structure of hexagonal GaN is shown in Figure 1a,b. In all intercalated models, a vacuum layer larger than 15 Å was included along the *c*-axis to avoid spurious interactions. Uniaxial compressive strain was applied along the [0001] direction, ranging from 0% to -10% in steps of -2%. The expressions used to calculate the formation energies are provided in the Supporting Information.

To assess the reliability of the exchange–correlation functional, we compared the Perdew–Burke–Ernzerhof (PBE) functional [26] with the Heyd–Scuseria–Ernzerhof (HSE06) hybrid functional [27] by calculating the band structure of GaN unit cell. As shown in Figure 1c,d, PBE significantly underestimates the band gap, as expected. However, the band-edge exhibit strong agreement with HSE06, and the calculated electron and hole effective masses according to the band edges are in close correspondence for both functionals; the detailed comparation is shown in Table S1 in the Supplementary Materials. Notably, the valence band dispersion obtained from PBE nearly overlaps with that of HSE06. Since this work primarily focuses on hole-related properties, and in view of the substantial computational cost of hybrid functionals, we adopted the PBE functional for all subsequent calculations.

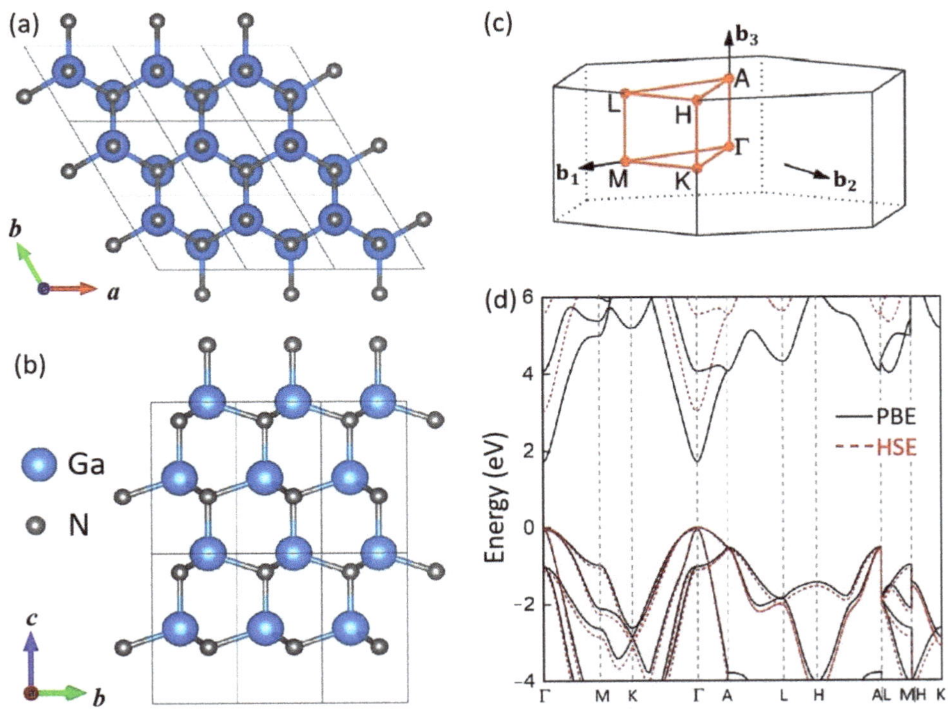

Figure 1. (**a**) Top and (**b**) side views of the atomic structure of GaN. (**c**) The first Brillouin zone for bulk GaN. (**d**) Band structure of GaN calculated by PBE and HSE functionals.

3. Results

To investigate the structural and electronic properties of 2D-Mg intercalated GaN, we first constructed models consisting of GaN slabs separated by a monolayer of Mg, as illustrated in Figure 2. Bulk GaN crystallizes in a hexagonal structure with an ABBA stacking sequence and exhibits strong spontaneous polarization along the [0001] (*c*-axis) direction. In our model, a single Mg atomic layer is inserted at the central C-site position, with GaN slabs placed on both sides in a centrosymmetric configuration. This arrangement leads to opposing polarization directions in the two GaN regions with respect to the Mg layer, resulting in an inversion-symmetric structure about the Mg plane that cannot be periodically extended along the *c*-axis. To resolve this, we constructed finite slab models with symmetric truncation at both ends. This modeling strategy is consistent with earlier theoretical work. For example, Northrup et al. [17] analyzed possible Mg incorporation configurations at inversion domain boundaries in GaN using boundary-based slab models, underscoring the necessity of non-periodic approaches to capture polarity inversion effects.

We then investigated various edge terminations by evaluating their relative stabilities. Specifically, we considered nitrogen-terminated edges with the outermost N atoms occupying the A-, B-, or C-sites, as well as a reconstructed nitrogen edge where surface N atoms form molecular N_2 dimers. These models are referred to as the N-A, N-B, N-C, and N-N_2 configurations, respectively, as shown in Figure 2. In addition, a Ga-terminated edge (Ga-edge) was also modeled for comparison.

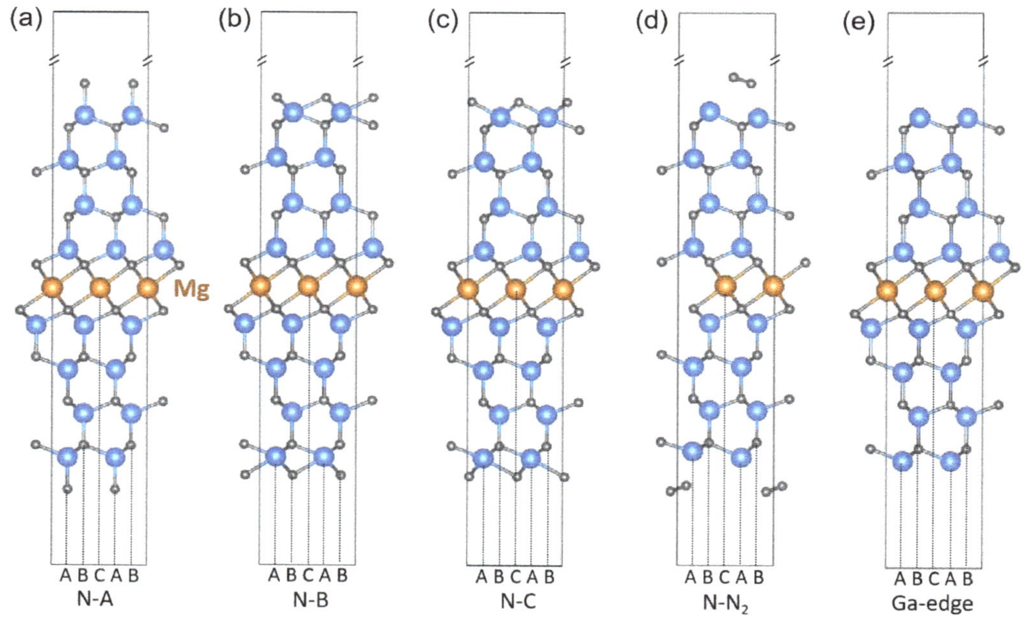

Figure 2. (**a**–**e**) Schematic illustration of five possible edge configurations for 2D-Mg intercalated GaN slab models.

The formation energies of the edge configurations in Figure 2 are summarized in Table 1. Among the nitrogen-terminated models, only the N_2-terminated edge exhibits a negative formation energy, indicating its thermodynamic favorability. The desorption energy of the N_2 dimer was calculated to be 0.11 eV using the climbing image nudged elastic band method [28] (see Figure S1 in the Supplementary Materials). This low desorption barrier falls within the typical range of physisorption energies and suggests that N_2 can readily desorb under mild thermal conditions [29]. Following N_2 desorption, the resulting structure corresponds directly to the Ga-terminated configuration. Notably, the Ga-edge structure also shows a negative formation energy, confirming its stability. Therefore, owing to the ease of N_2 desorption and the favorable energetics of the Ga edge, we adopted the Ga-terminated configuration in all subsequent calculations to ensure both thermodynamic plausibility and structural consistency.

Table 1. Formation energies of five edge configurations in Figure 2.

Edge Configuration	**N-A**	**N-B**	**N-C**	**N-N_2**	**Ga-Edge**
E_{form}/eV	6.48	5.02	5.62	−0.63	−0.51

To further understand the electronic characteristics of various edge configurations, we calculated the projected density of states (PDOS) for the five edge structures discussed above, as presented in Figure S2 of the Supplementary Materials. In all cases, the occupied Mg states are located below −3 eV, indicating that Mg in the fully 2D intercalated configuration is unlikely to serve as an effective acceptor for hole conduction. Furthermore, due to the metallic nature of the 2D-Mg layer itself, the density of states exhibits metallic characteristics across all five configurations. Experimental studies also suggest that Mg atoms are not completely intercalated at the C-sites in a continuous fashion. Instead, localized 2D-Mg sheets are typically embedded between GaN layers.

To evaluate the impact of Mg intercalation ratio on the structural stability, we constructed a series of models with nominal Mg coverage of 25%, 50%, 75%, 100%, 125%, and 150% within a 2 × 2 GaN supercell, as illustrated in Figure 3a–f. The 125% and 150% models correspond to extended configurations in which Mg atoms not only form a complete monolayer but also substitute adjacent Ga atoms, in line with experimental observations of partial Mg diffusion into Ga sites [21]. The calculated formation energies reveal that the 75% intercalation model is the most energetically favorable among all models as shown in Figure 3g. This configuration was therefore selected as the basis for subsequent analysis.

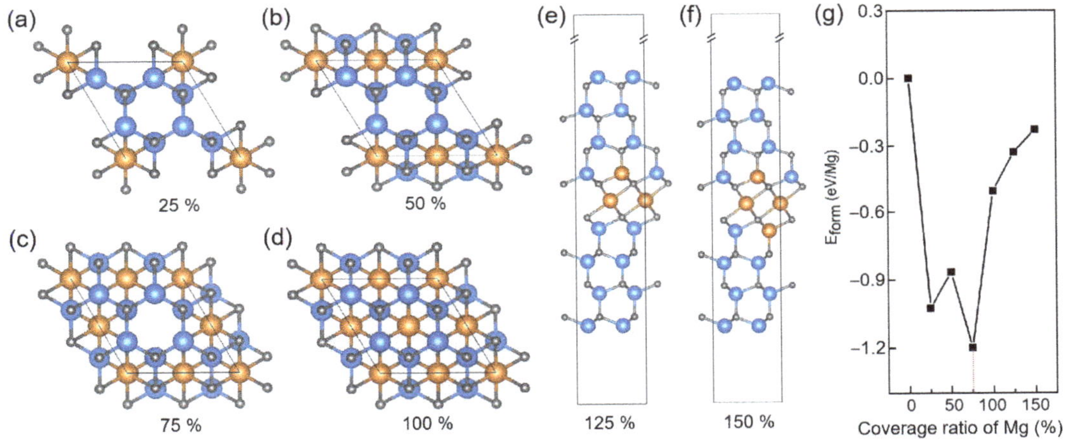

Figure 3. (**a–d**) Top views of 2 × 2 GaN supercells with Mg intercalation ratios of 25%, 50%, 75%, and 100%, respectively. (**e,f**) Side views of extended configurations with Mg coverage of 125% and 150%, where excess Mg atoms substitute neighboring Ga sites. (**g**) Formation energies of all models as a function of Mg coverage ratio.

The PDOS for these models are presented in Figure S3. The 25% and 50% intercalation structures retain semiconducting characteristics, whereas configurations with 75% Mg coverage or higher display metallic states near the Fermi level, indicating a transition from semiconducting to metallic behavior with increasing Mg concentration. In practical devices, such metallicity may provide leakage pathways and reduce efficiency.

To explore possible mechanisms for mitigating this metallicity, we next considered the role of Ga vacancies near the Mg layer. This choice is motivated both by the need to compensate for the excess electrons introduced by intercalated Mg atoms and by experimental observations showing that the Ga signal intensity adjacent to the Mg layer decreases to approximately 75% of its original value [17,21], consistent with partial Ga depletion. Accordingly, Ga vacancies were introduced into our models to examine their compensating effect and impact on electronic properties.

For the thermodynamically most stable 75% Mg coverage model (i.e., three Mg atoms contributing six excess electrons in a 2 × 2 supercell), we removed two Ga atoms to achieve charge compensation. These vacancies were introduced symmetrically, with one located above and the other below the Mg layer. Four distinct vacancy configurations were considered, denoted as V_{Ga}-1 through V_{Ga}-4, corresponding to the removal of Ga atoms at first-nearest, second-nearest, third-nearest, and outermost-layer positions relative to the Mg layer, respectively, as illustrated in Figure 4a.

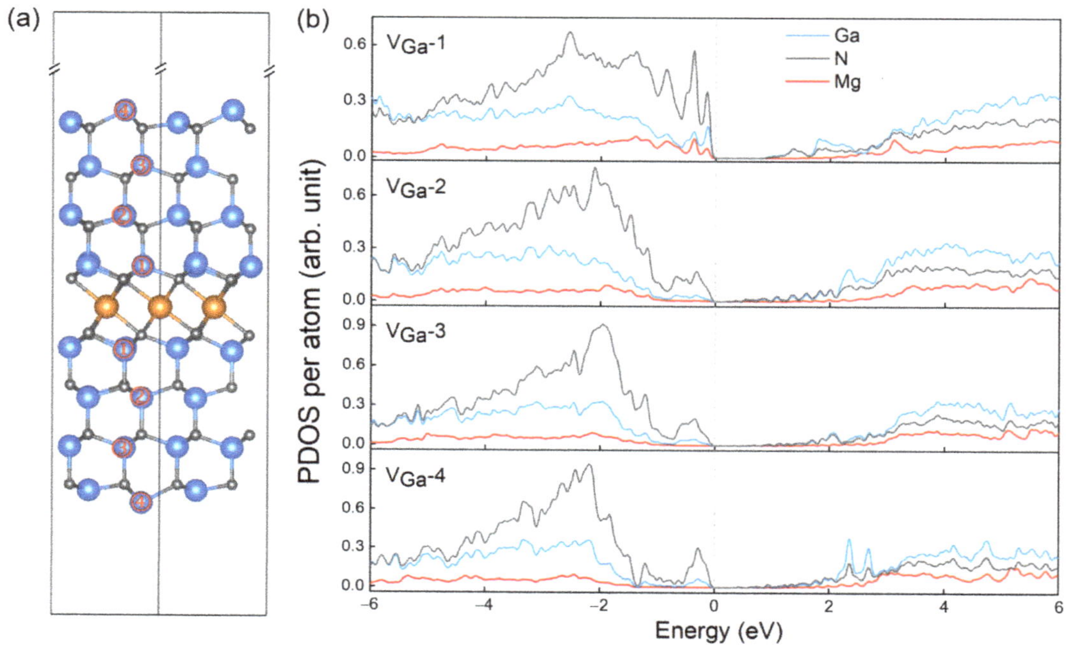

Figure 4. (**a**) Side view of the 75% Mg intercalation model, showing the positions of Ga vacancies introduced at four different sites: first-nearest (V_{Ga}-1), second-nearest (V_{Ga}-2), third-nearest (V_{Ga}-3), and outermost-layer Ga (V_{Ga}-4) relative to the Mg layer. (**b**) PDOS per atom for the four Ga-vacancy configurations.

The electronic structures of the four Ga-vacancy configurations (V_{Ga}-1 to V_{Ga}-4) under 75% Mg coverage were further analyzed by calculating their PDOS, as shown in Figure 4b. All four configurations exhibit semiconducting behavior, in sharp contrast to the metallic character of the defect-free intercalated structure (see Figure S3 for details). The calculated PBE bandgaps are 0.98, 0.32, 0.65, and 0.39 eV for V_{Ga}-1 through V_{Ga}-4, respectively. The introduction of Ga vacancies not only restores the semiconducting nature of the system but also reduces the bandgap relative to pristine GaN, thereby lowering the thermal activation energy required for hole conduction. In particular, the V_{Ga}-1 configuration, where Ga vacancies are located nearest to the Mg layer, exhibits pronounced Mg-related states near the valence band maximum, suggesting an increased likelihood of Mg atoms acting as effective acceptors. This configuration also aligns with experimental observations reporting an approximately 25% reduction in Ga signal intensity near the Mg layer [21]. According to the Bader analysis, each Mg atom loses approximately 1.6 e^- for the VGa-1 configuration, with the excess charge mainly transferred to the neighboring N atoms. This indicates that Mg sites are electronically depleted, which is consistent with the emergence of Mg-related unoccupied states near the valence band maximum observed in the PDOS.

Motivated by experimental observations indicating that 2D-Mg intercalation can induce compressive lattice strain as high as 10% along the [0001] direction of GaN [21], we systematically investigated the impact of uniaxial compression strain applied along the *c*-

axis on the bandgap evolution of the four Ga-vacancy configurations, as shown in Figure 5. For V_{Ga}-2 through V_{Ga}-4, the bandgap increases with strain, consistent with the typical behavior of conventional semiconductors such as Si, Ge, and GaN [30,31]. In contrast, the V_{Ga}-1 configuration exhibits a distinct trend: its bandgap decreases monotonically under compressive strain. This unusual response suggests enhanced transport, as the reduced bandgap lowers the thermal activation barrier for carrier excitation. Notably, at 10% compressive strain, the bandgap of V_{Ga}-1 is reduced to 0.74 eV, which is less than half the intrinsic PBE bandgap of unstrained GaN. Experimental studies on Mg-doped GaN thin films also emphasize the interplay between defect complexes and mechanical characteristics [32], supporting the relevance of strain- and defect-assisted mechanisms.

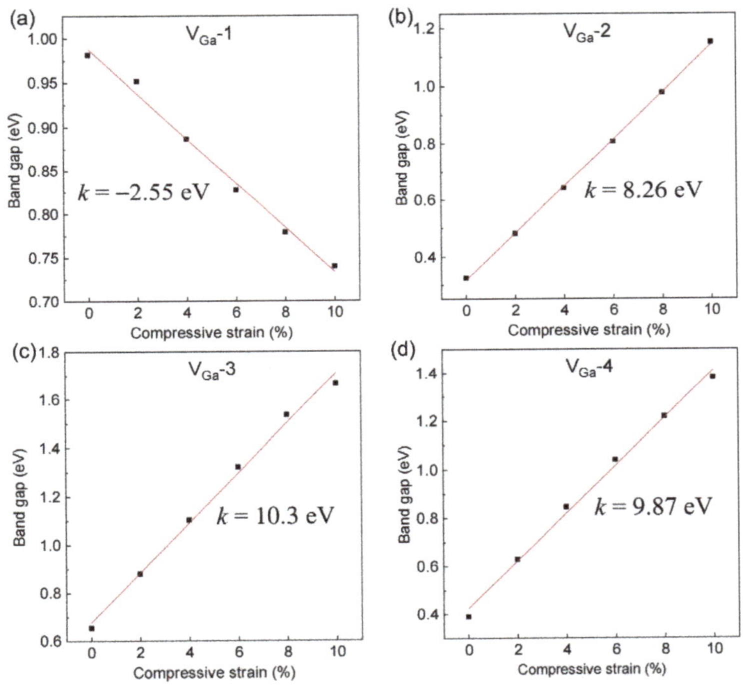

Figure 5. **(a–d)** Bandgap evolution under uniaxial compressive strain along the GaN [0001] direction for the four Ga-vacancy configurations (V_{Ga}-1 to V_{Ga}-4).

4. Conclusions

In summary, we have systematically investigated the structural and electronic properties of GaN systems with 2D-Mg intercalation using first-principles calculations. To model experimentally relevant configurations, truncated slab models with various edge terminations were constructed, and Ga-terminated edges were found to be thermodynamically favorable following N_2 desorption from nitrogen-terminated surfaces. We further explored a series of partially intercalated models with varying Mg coverage ratios, and identified that the 75% intercalation model exhibited the lowest formation energy, identifying it as the most energetically favorable structure. To maintain charge neutrality and recover semiconducting behavior, symmetric Ga vacancies were introduced near the Mg layer. Among the four vacancy configurations studied, the nearest-neighbor Ga-vacancy structure (V_{Ga}-1) not only opened a finite bandgap but also generated Mg-related acceptor states near the valence band maximum, which may enhance hole conductivity. Motivated by experimental reports of high elastic compressibility in such layered systems, we further examined the effect of uniaxial compressive strain along the *c*-axis. The VGa-1 configuration displayed an anomalous bandgap under increasing compression, potentially facilitating hole excitation.

Overall, this study provides mechanistic insight into the doping behavior of 2D-Mg intercalated GaN and its potential for improved p-type functionality, offering theoretical guidance for future experimental design of high-performance p-type GaN heterostructures.

Supplementary Materials: The following supporting information can be downloaded at: https://www.mdpi.com/article/10.3390/ma18204755/s1, Table S1: Calculated electron and hole effective masses of GaN using PBE and HSE functionals.; Figure S1: Energy profile as a function of the distance between the N_2 dimer and the GaN surface in the N–N_2; edge configuration, calculated using the climbing image nudged elastic band (CI-NEB) method; Figure S2: The projected density of states (PDOS) per atom for the five edge structures discussed in Figure 2 in the text. Figure S3: Projected density of states (PDOS) per atom for models with different Mg intercalation ratios corresponding to those in Figure 3. Figure S4. Bandgap variation of the four Ga-vacancy configurations (VGa-1 to VGa-4) under uniaxial c-axis compression and coupled biaxial strain, where perpendicular *ab*-plane relaxation is considered via the Poisson ratio (ν = 0.18) of GaN.

Author Contributions: Writing—original draft, Q.W.; Writing—review & editing, S.Z.; Supervision, X.S. and X.Z. All authors have read and agreed to the published version of the manuscript.

Funding: This research was funded by the National Natural Science Foundation of China (No. 12104025).

Data Availability Statement: The original contributions presented in this study are included in the article. Further inquiries can be directed to the corresponding author.

Conflicts of Interest: The authors declare no conflicts of interest.

References

1. Lee, H.E.; Choi, J.; Lee, S.H.; Jeong, M.; Shin, J.H.; Joe, D.J.; Kim, D.; Kim, C.W.; Park, J.H.; Lee, J.H.; et al. Monolithic flexible vertical GaN light-emitting diodes for a transparent wireless brain optical stimulator. *Adv. Mater.* **2018**, *30*, 1800649. [CrossRef]
2. Wang, L.; Xu, K.; Yu, G.; Ren, X.; Qin, X.; Zhang, L.; Chen, T.; Zhang, F.; Li, F.; Zhou, J.; et al. Enhanced luminescence efficiency in GaN-based blue laser diodes by H plasma technology. *Opt. Express* **2024**, *32*, 34492–34499. [CrossRef]
3. Zhang, L.; Xu, J.; He, J.; Cheng, Z.; Wang, W.; Liu, C.; Wang, X.; Mi, C.; Tan, W.; Zhang, Y. AlGaN/GaN heterojunction bipolar transistors with record f T/f max= 21.6/4.23 GHz. *IEEE Electron Device Lett.* **2025**, *46*, 912–915. [CrossRef]
4. Liu, X.; Wang, D.; Chen, W.; Kang, Y.; Fang, S.; Luo, Y.; Luo, D.; Yu, H.; Zhang, H.; Liang, K.; et al. Optoelectronic synapses with chemical-electric behaviors in gallium nitride semiconductors for biorealistic neuromorphic functionality. *Nat. Commun.* **2024**, *15*, 7671. [CrossRef] [PubMed]
5. Voss, L.F.; Frye, C.D.; Allen, N.A.; Harrison, S.E.; Kweon, K.; Varley, J.B.; Lordi, V.; Nikolic, R.; Anderson, T.J.; Hite, J.K. Moderate temperature Mg diffusion doping of GaN. *Meet. Abstr.* **2020**, *MA2020-02*, 1810. [CrossRef]
6. Nakamura, S.; Iwasa, N.; Masayuki Senoh, M.S.; Takashi Mukai, T.M. Hole compensation mechanism of p-type GaN films. *Jpn. J. Appl. Phys.* **1992**, *31*, 1258. [CrossRef]
7. Amano, H.; Kito, M.; Hiramatsu, K.; Akasaki, I. P-type conduction in Mg-doped GaN treated with low-energy electron beam irradiation (LEEBI). *Jpn. J. Appl. Phys.* **1989**, *28*, L2112. [CrossRef]
8. Velpula, R.T.; Jain, B.; Patel, M.; Shakiba, F.M.; Toan, N.Q.; Nguyen, H.-D.; Nguyen, H.P.T. High-efficiency InGaN blue LEDs with reduced positive sheet polarization. *Appl. Opt.* **2022**, *61*, 4967–4970. [CrossRef]
9. Narita, T.; Yoshida, H.; Tomita, K.; Kataoka, K.; Sakurai, H.; Horita, M.; Bockowski, M.; Ikarashi, N.; Suda, J.; Kachi, T.; et al. Progress on and challenges of p-type formation for GaN power devices. *J. Appl. Phys.* **2020**, *128*, 090901. [CrossRef]
10. Lyons, J.L.; Janotti, A.; de Walle, C.V. Shallow versus deep nature of Mg acceptors in nitride semiconductors. *Phys. Rev. Lett.* **2012**, *108*, 156403. [CrossRef]
11. Lyons, J.L.; de Walle, C.V. Computationally predicted energies and properties of defects in GaN. *npj Comput. Mater.* **2017**, *3*, 12. [CrossRef]

12. Kent, D.G.; Overberg, M.E.; Pearton, S.J. Co-implantation of Be+ O and Mg+ O into GaN. *J. Appl. Phys.* **2001**, *90*, 3750–3753. [CrossRef]
13. Jacobs, A.G.; Spencer, J.A.; Hite, J.K.; Hobart, K.D.; Anderson, T.J.; Feigelson, B.N. Novel Codoping Moiety to Achieve Enhanced P-Type Doping in GaN by Ion Implantation. *Phys. Status Solidi A* **2023**, *220*, 2200848. [CrossRef]
14. Simon, J.; Protasenko, V.; Lian, C.; Xing, H.; Jena, D. Polarization-induced hole doping in wide–band-gap uniaxial semiconductor heterostructures. *Science* **2010**, *327*, 60–64. [CrossRef] [PubMed]
15. Poncé, S.; Jena, D.; Giustino, F. Hole mobility of strained GaN from first principles. *Phys. Rev. B* **2019**, *100*, 085204. [CrossRef]
16. Han, B.; Sun, M.; Chang, Y.; He, S.; Zhao, Y.; Qu, C.; Qiu, W. Raman Characterization of the In-Plane Stress Tensor of Gallium Nitride. *Materials* **2023**, *16*, 2255. [CrossRef]
17. Northrup, J.E. Magnesium incorporation at (0001) inversion domain boundaries in GaN. *Appl. Phys. Lett.* **2003**, *82*, 2278–2280. [CrossRef]
18. Leroux, M.; Vennéguès, P.; Dalmasso, S.; de Mierry, P.; Lorenzini, P.; Damilano, B.; Beaumont, B.; Gibart, P.; Massies, J. Pyramidal defects in highly Mg-doped GaN: Atomic structure and influence on optoelectronic properties. *Eur. Phys. J. Appl. Phys.* **2004**, *27*, 259–262. [CrossRef]
19. Ohnishi, K.; Amano, Y.; Fujimoto, N.; Nitta, S.; Watanabe, H.; Honda, Y.; Amano, H. Electrical properties and structural defects of p-type GaN layers grown by halide vapor phase epitaxy. *J. Cryst. Growth* **2021**, *566–567*, 126173. [CrossRef]
20. Minakshi, M.; Singh, P.; Issa, T.B.; Thurgate, S.; Marco, R.D. Lithium insertion into manganese dioxide electrode in MnO_2/Zn aqueous battery: Part II. Comparison of the behavior of EMD and battery grade MnO_2 in Zn | MnO_2 | aqueous LiOH electrolyte. *J. Power Sources* **2004**, *138*, 319–322. [CrossRef]
21. Wang, J.; Cai, W.; Lu, W.; Lu, S.; Kano, E.; Agulto, V.C.; Sarkar, B.; Watanabe, H.; Ikarashi, N.; Iwamoto, T.; et al. Observation of 2D-magnesium-intercalated gallium nitride superlattices. *Nature* **2024**, *631*, 67–72. [CrossRef]
22. Blöchl, P.E. Projector augmented-wave method. *Phys. Rev. B* **1994**, *50*, 17953–17979. [CrossRef]
23. Kresse, G.; Furthmüller, J. Efficiency of ab-initio total energy calculations for metals and semiconductors using a plane-wave basis set. *Comp. Mater. Sci.* **1996**, *6*, 15–50. [CrossRef]
24. Kresse, G.; Furthmüller, J. Efficient iterative schemes for ab initio total-energy calculations using a plane-wave basis set. *Phys. Rev. B* **1996**, *54*, 11169–11186. [CrossRef] [PubMed]
25. Monkhorst, H.J.; Pack, J.D. Special points for Brillouin-zone integrations. *Phys. Rev. B* **1976**, *13*, 5188–5192. [CrossRef]
26. Perdew, J.P.; Burke, K.; Ernzerhof, M. Generalized gradient approximation made simple. *Phys. Rev. Lett.* **1996**, *77*, 3865–3868. [CrossRef]
27. Heyd, J.; Scuseria, G.E.; Ernzerhof, M. Hybrid functionals based on a screened Coulomb potential. *J. Chem. Phys.* **2003**, *118*, 8207–8215, Erratum in *J. Chem. Phys.* **2006**, *124*, 219906. [CrossRef]
28. Henkelman, G.; Uberuaga, B.P.; Jónsson, H. A climbing image nudged elastic band method for finding saddle points and minimum energy paths. *J. Chem. Phys.* **2000**, *113*, 9901–9904. [CrossRef]
29. Wang, Y.; Nieman, R.; Minton, T.K.; Guo, H. Insights into adsorption, diffusion, and reactions of atomic nitrogen on a highly oriented pyrolytic graphite surface. *J. Chem. Phys.* **2021**, *154*, 074708. [CrossRef]
30. Dreyer, C.E.; Janotti, A.; de Walle, C.V. Effects of strain on the electron effective mass in GaN and AlN. *Appl. Phys. Lett.* **2013**, *102*, 142105. [CrossRef]
31. Sukhdeo, D.S.; Nam, D.; Kang, J.-H.; Brongersma, M.L.; Saraswat, K.C. Direct bandgap germanium-on-silicon inferred from 5.7%⟨100⟩ uniaxial tensile strain. *Photon. Res.* **2014**, *2*, 8–13. [CrossRef]
32. Benzarti, Z.; Khalfallah, A.; Bougrioua, Z.; Evaristo, M.; Cavaleiro, A. Understanding the influence of physical properties on the mechanical characteristics of Mg-doped GaN thin films. *Mater. Chem. Phys.* **2023**, *307*, 128182. [CrossRef]

materials

MDPI

Article

An Enhancement in the Magnetocaloric Effect in a Composite Powder Based on Lanthanum Manganites

Fidel Ivan Reyes Patricio [1], Cristhian Antonio Taboada Moreno [1], Ana María Bolarín Miró [1], Claudia Alicia Cortés Escobedo [2], María Isabel Reyes Valderrama [1] and Félix Sánchez De Jesús [1,*]

1 Área Académica de Ciencias de la Tierra y Materiales, Universidad Autónoma del Estado de Hidalgo, Mineral de la Reforma 42184, Mexico; re295437@uaeh.edu.mx (F.I.R.P.); ta271740@uaeh.edu.mx (C.A.T.M.); abolarin@uaeh.edu.mx (A.M.B.M.); profe_5490@uaeh.edu.mx (M.I.R.V.)

2 Centro de Investigación e Innovación Tecnológica, Instituto Politécnico Nacional, Ciudad de México 02250, Mexico; ccortese@ipn.mx

* Correspondence: fsanchez@uaeh.edu.mx

Abstract: This study presents a dual-phase lanthanum manganite ceramic composite based on a mixture of equal weight ratios of $La_{0.7}Ca_{0.2}Sr_{0.1}MnO_3$ and $La_{0.7}Ca_{0.25}Sr_{0.05}MnO_3$ designed to enhance the magnetocaloric effect (MCE) of individual compounds, under a low magnetic field ($\leq$18 kOe). X-ray diffraction (XRD) analysis revealed the coexistence of two orthorhombic manganite phases corresponding to the individual compounds, with no secondary phases detected. Temperature-dependent magnetization measurements in the composite evidenced two Curie temperatures at 286.8 K and 307.6 K, reflecting the effect of Ca^{2+} and Sr^{2+} concentrations. Arrott plots and β parameters confirmed that the phase transition is of second order. Although the maximum magnetic entropy change (ΔS_M) of the composite is slightly lower than that of the individual manganite phases, its relative cooling power (RCP) reaches 188.82 $J{\cdot}kg^{-1}$, with an extended operational temperature window (OTW) of approximately 85 K, spanning from around 243 K to 328 K. This broad OTW enables efficient operation over a wider temperature range compared to similar materials, such as the individual $La_{0.7}Ca_{0.2}Sr_{0.1}MnO_3$ and $La_{0.7}Ca_{0.25}Sr_{0.05}MnO_3$ compounds, which exhibit an RCP of 55.24 and 65.12 $J{\cdot}kg^{-1}$, respectively, under a comparable magnetic field (~18 kOe). The improved magnetocaloric performance is attributed to interfacial exchange coupling and strain-mediated effects that broaden the ΔS_M response and generate a non-additive RCP. These results demonstrate that interphase coupling and microstructural tuning effectively broaden the operating temperature range for magnetic refrigeration under moderate fields, making this composite a strong candidate for practical cooling applications.

Keywords: Magnetic refrigeration; composite; magnetocaloric effect; magnetic entropy change (ΔS_M); relative cooling power (RCP); manganites; high-energy ball milling

1. Introduction

Magnetic refrigeration, exploiting the magnetocaloric effect (MCE), has attracted significant attention as a sustainable and efficient alternative to conventional gas compression cooling technologies. The MCE refers to the reversible temperature change in a material upon the application or removal of an external magnetic field under adiabatic conditions, enabling energy-efficient thermal management without environmentally noxious refrigerants [1–4]. Among magnetocaloric materials, gadolinium (Gd)-based alloys are considered the most important systems, exhibiting large magnetic entropy changes (ΔS_M) near room temperature due to their strong magnetic moment and favorable magnetic transitions. However, the practical deployment of Gd-based materials faces obstacles, including high production costs, complex synthesis routes, and long-term stability issues [3]. In contrast, ceramic magnetocaloric materials, particularly manganites of the $LaMnO_3$ family, present cost-effective and chemically stable alternatives with tunable magnetic properties achieved via substitutional doping [4–8]. The partial replacement of La^{3+} by alkaline earth metals such as Ca^{2+} or Sr^{2+} adjusts critical parameters including the Curie temperature (Tc), thereby optimizing the MCE near ambient temperatures [4]. These dopants modify the Mn^{3+}/Mn^{4+} ratio, enhancing the double-exchange (DE) interactions responsible for ferromagnetic ordering. Such modifications promote ferrimagnetic states through structural distortions involving MnO_6 octahedra and Jahn–Teller effects, which affect the magnetic entropy change and widen the operating temperature window (OTW), resulting in increased relative cooling power (RCP) [5–17]. Nevertheless, despite these advantages, manganites generally exhibit lower maximum entropy changes and relatively narrow temperature spans when compared to Gd-based materials, which constrains their overall refrigeration efficiency and limits their practical applications [6,9–11,17].

To overcome these limitations shown by doped lanthanum manganites, considerable research has been directed toward designing composite and mixed-phase manganite systems. Such strategies aim to broaden the OTW and enhance the RCP by combining phases with different Curie temperatures. Composite-based manganite has been explored to induce multiple magnetic transitions, thus expanding the temperature span of the magnetocaloric effect. In this context, Maalam et al. [8] produced $(La_{0.45}Nd_{0.25})Sr_{0.3}MnO_3/5Cu$ composites via solid-state reactions; analysis using scanning electron microscopy (SEM) revealed CuO phases located predominantly at the grain boundaries, which influenced magnetic interactions and an improvement in the magnetocaloric properties. Notably, these composites demonstrated a decrease in Tc but a more than twofold increase in the maximum ΔS_M compared to parent compounds, highlighting enhanced magnetocaloric performance. Additionally, Zhou et al. [9] explored copper doping effects in $(1-x)La_{0.67}Sr_{0.33}MnO_3/xCu$ composite coatings, finding significant impacts on the structural, electrical, and low-field magnetoresistance properties that are relevant for tuning the magnetocaloric response. Reported ΔS_M peak values in manganite composites generally range between 3 and 5 $J{\cdot}kg^{-1}{\cdot}K^{-1}$ under magnetic field changes of 1 to 5 T, with relative cooling power (RCP) values often broadened over wider working temperature intervals due to the composite nature. Focused on the use of manganites as individual compounds, Jeddi et al. [10] reported an RCP of 210 $J{\cdot}kg^{-1}{\cdot}$ and $|\Delta S_M|$ of 0.9 $J{\cdot}kg^{-1}{\cdot}K^{-1}$ at 3 T with a Tc increase from 255 to 365 K using a 40:60 weight ratio of $La_{0.6}Ca_{0.4}MnO_3$ and $La_{0.6}Sr_{0.4}MnO_3$, extending the work temperature range. Other studies such as those by Sellami et al. [11], Akça et al. [12], M'nassri et al. [13], and Yigiter et al. [14] reported ΔS_M values ranging from 1.8 to 4.1 $J{\cdot}kg^{-1}{\cdot}K^{-1}$ and an RCP up to 303 $J{\cdot}kg^{-1}{\cdot}$ under applied magnetic fields of 5 T for composites of $La_{0.7}Ca_{0.3}MnO_3$ and $La_{0.7}Ag_{0.3}MnO_3$.

Additional approaches include adjusting the proportion of rare-earth and transition metal ions and performing post-synthesis treatments such as thermal annealing to improve phase diffusion and entropy stability. For instance, Bisht et al. [15], Kılıç et al. [16], and Boubekri et al. [17] have shown a wide range of operating temperatures with stable magnetocaloric responses using optimized ceramic mixtures and simulations. Bouhani et al. [18] emphasized that maximizing ΔS_M is critical for material sensitivity, while RCP determines the cooling capacity of the material. Therefore, materials with a large RCP and a large entropy change are ideal for improving cooling efficiency.

Focused on the use of manganite for obtaining composites, Ezaami et al. [19] synthesized composites formed by a mixture of $La_{0.7}Ca_{0.25}Sr_{0.05}MnO_3$ and $La_{0.7}Ca_{0.2}Sr_{0.1}MnO_3$ compounds using the solid-state reaction method, where the Curie temperature of the composites was found to be intermediate between those of the parent compounds, exhibiting a double transition indicative of two different values. The ΔS_M values of these composites vary accordingly between those of the parents, with optimized ΔS_M maxima achieved in certain compositions [20–31]. These results show a promising way to widen the OTW to near room temperature and, at the same time, increase the RCP.

Several studies have shown that annealing or thermal treatment significantly improves the magnetocaloric properties of manganite-based composites due to the promotion of the diffusion process forming intermediate phases. For instance, in the work by El Amrani et al. [23], $Nd_{0.7}Sr_{0.3}MnO_3$:CuO composites were synthesized and heat-treated at 1473 K, leading to enhanced magnetic entropy change ($|\Delta S_M| \approx 5.14$ J·kg^{-1}·K^{-1} at 5 T) and an improved relative cooling power (RCP $\approx$ 232 J·kg^{-1}) compared to the parent manganite. Similarly, Giri et al. [24] investigated $La_{0.7}Ca_{0.3}MnO_3$-MO (MO = CuO, CoO, NiO) composites and demonstrated that the addition of these metal oxides affected the ferromagnetic transition temperature and enhanced the RCP values under applied fields, attributing part of the improvement to changes induced by thermal processing. Moreover, Yilmaz et al. [25] designed a composite consisting of $La_{0.67}Ca_{0.27}Sr_{0.06}MnO_3$ and $La_{0.7}Sr_{0.3}Mn_{0.94}Ni_{0.06}O_3$, finding that thermal treatment and appropriate phase blending yielded a broadened magnetocaloric effect due to the extended Curie temperature of the composite, with RCP values up to 317.7 J·kg^{-1} under 5 T. Similarly, Fkhar et al. [27] applied heat treatment at 1473 K for 12 h to $La_{0.45}Nd_{0.25}Sr_{0.3}MnO_3$/CuO composites, which resulted in the diffusion of both compounds, enhancing some magnetocaloric properties of the composite compared with individual compounds. These examples highlight the widespread application of post-heat treatments in manganite-based composites and emphasize the importance of controlling phase coexistence for optimizing material properties.

Therefore, this work focuses on the study of a composite powder obtained from a mixture of two lanthanum-based manganites, both doped at site A. Each individual phase exhibits a magnetocaloric effect (MCE) within a narrow temperature range, defined by its respective Curie temperature. However, for practical applications in magnetic cooling, especially at room temperature, it is essential to have a longer operating temperature range. To overcome this limitation, the present study explores a controlled 50:50 mixture of both manganite phases, with the main objective of enhancing the magnetocaloric effect by extending the effective OTW. This approach seeks to improve the RCP without the need to maximize ΔS_M. Although the maximum value of ΔS_M may remain comparable or even slightly reduced relative to that of the pure phases, the overlap of the entropy change peaks of both manganites results in a broader MCE peak, leading to a significant improvement in RCP. On the other hand, the synthesis of manganites is carried out by high-energy ball milling, a technique known to induce structural distortions and generate beneficial crystallographic defects that can enhance the magnetocaloric response. Compared to conventional wet chemical methods, this mechanochemical route offers notable advantages in terms of scalability, cost-effectiveness, and process simplicity while maintaining a high degree of reproducibility.

As aforementioned, the aim of this work is to investigate the synergistic effect of two lanthanum manganites as a composite powder on the OTW, ΔS_M, and the RCP in order to enhance their MC performance and also provide a deeper understanding of the physical mechanisms governing RCP improvement. Consequently, the results of this work contribute to the rational design of functional magnetocaloric materials for solid-state cooling technologies, emphasizing the fundamental role of synthesis techniques in adapting the properties of advanced materials. This study is focused on demonstrating that the chosen synthesis approach can enhance the magnetocaloric performance of manganite composites under practical, low-field conditions (18 kOe), making them suitable for domestic applications.

2. Materials and Methods

2.1. The Synthesis of Manganites and Preparation of the Composite

High-purity La_2O_3, SrO, CaO, and Mn_2O_3 powders, all from Sigma-Aldrich (St. Louis, MO, USA) with purities ranging from 99.8% to 99.9%, were used as precursors for the synthesis of individual manganites, $La_{0.7}Ca_{0.25}Sr_{0.05}MnO_3$ (LCSM0.05) and $La_{0.7}Ca_{0.2}Sr_{0.1}MnO_3$ (LCSM0.1), according to the following reaction:

$$0.7La_2O_3 + (0.3 - x)CaO + xSrO + Mn_2O_3 + 1/2O_2 \rightarrow 2La_{0.7}Ca_{0.3-x}Sr_xMnO_3 \quad (1)$$

For the synthesis of the individual manganites, 5 g of the stoichiometric oxide's mixtures, according to Equation (1), was introduced, together with 50 g of hardened steel balls of 12.7 mm in diameter, into a 60 mL hardened steel vial (6×10^{-5} m^3) and was subjected to a high-energy ball milling process in an air atmosphere, using a Mixer/Mill (SPEX model 8000D, Metuchen, NJ, USA), for 5 h at a ball-to-powder weight ratio of 10:1 for 5 h. The milled powders were subsequently annealed in an air atmosphere at 1473 K with a heating rate of 10 °C/min and held for 2 h and air-cooled using a tubular muffle furnace (Lindberg/Blue M model STF54459C, Riverside, MI, USA) to achieve complete crystalline phase formation, in agreement with previous studies [20]. A schematic representation of the experimental process is presented in Figure 1.

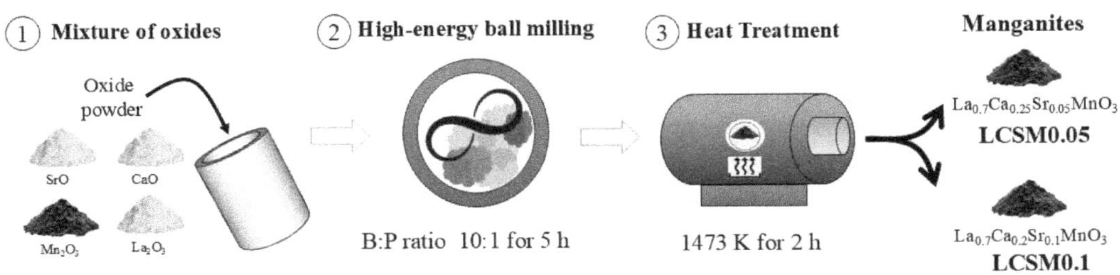

Figure 1. A schematic representation of the experimental procedure used to obtain single manganites (LCSM0.05 and LCSM0.1).

To prepare the composite, based on equal weights (50 wt. % each) of the manganites synthesized previously, LCSM0.05 and LCSM0.1, the mixture of manganites in proper proportions was attained, as shown in Figure 2. A total of 2 g of the mixture with 12 g of a hardened steel ball of 9.525 mm in diameter, using a ball-to-powder weight ratio of 60:1, was milled in the high-energy ball miller for 2 min in order to homogenize the mixture, using methanol as a process control agent, only during this step, to prevent agglomeration and ensure a uniform dispersion without modifying the stoichiometry of the pre-synthesized manganites [20]. The obtained powder was then annealed at 1473 K for 2 h as a strategy to promote particle bonding, as some studies have demonstrated the benefit of annealing in terms of the magnetocaloric properties of composites [21–27,32–41]. A flowchart of the experimental procedure is shown in Figure 2.

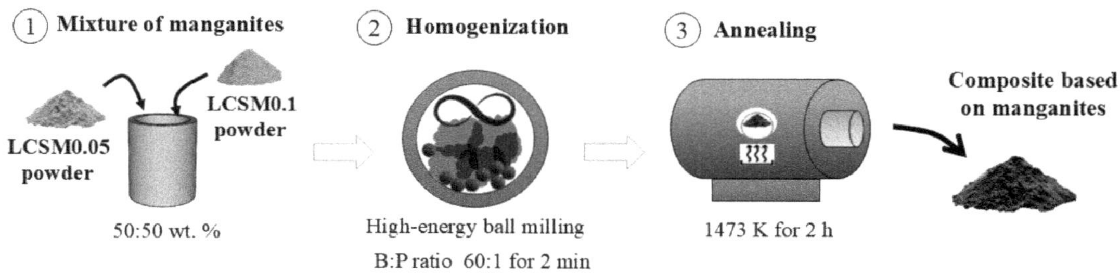

Figure 2. A schematic representation of the experimental procedure used to prepare the composite based on the synthesized manganites: LCSM0.05 and LCSM0.1.

2.2. Crystal Structure and Microstructure

The crystal structure of the synthesized manganites was analyzed using a Bruker D8 Advance X-ray diffractometer (Bruker D8 Advance, Karlsruhe, Germany) with Cu Kα radiation (λ = 1.541874 Å). Rietveld refinement was performed with Material Analysis Using Diffraction (MAUD) software Version 2.992, while structural identification was assisted by the use of the Inorganic Crystal Structure Database (ICSD). Diffraction peak profiles were modeled using a pseudo-Voigt function, combining Gaussian and Lorentzian components. The Gaussian term accounts mainly for instrumental and microstrain broadening, while the Lorentzian reflects size effects. The instrumental resolution function was calibrated with a standard SiO_2 pattern (CIF COD 1010921), and zero-shift corrections were applied, estimating a detection limit of 2.5 wt. % Lattice parameters, phase fractions, crystallite size, and microstrain were refined simultaneously in MAUD. The uncertainties of the structural parameters were derived from the covariance matrix, considering statistical propagation and parameter correlations. Additionally, crystal structure simulations were performed using Visualization for Electronic and Structural Analysis (VESTA) software Version 4.6.0. The morphology and particle size of the composite powders were examined by scanning electron microscopy (SEM, HITACHI model TM3030, Ibaraki, Japan) equipped with an energy-dispersive spectroscopy system (EDS, QUANTAX model 75, Bruker Nano GmbH, Berlin, Germany) for chemical analysis.

2.3. Magnetic and Magnetocaloric Properties

Magnetic and magnetocaloric measurements were performed using a VSM vibrating sample magnetometer (MicroSense EV7, Lowell, MA, USA); this piece of equipment has an instrumental uncertainty of 0.005 Oe in magnetic field measurement. Isothermal magnetization curves (M vs. H) were obtained in the temperature range covering the ferromagnetic-to-paramagnetic transition ($\pm$39 K around T_C, with $\Delta T = 3$ K), under applied magnetic fields from 0 to 18 kOe after zero-field cooling (ZFC). In this test, magnetization was measured in a three-step protocol: 0–1 kOe at 100 Oe intervals, 1–5 kOe at 600 Oe intervals, and 5–18 kOe at 1 kOe intervals. The Curie temperature (T_C) was determined from temperature-dependent magnetization measurements under an applied field of 10 kOe around the Curie temperature of each component. Arrott plots were created and ΔS_M and RCP were calculated from isothermal magnetic curves. Electron spin resonance (ESR) spectra were obtained in a Magnettech Miniscope MS 400 (MT MagnetTech GmbH, Berlin, Germany) spectrometer in the X-band (9.4 GHz), equipped with a JUMO dTRON304 temperature system (JUMO GmbH & Co. KG, Fulda, Germany).

The uncertainty in the $-\Delta S_M$ calculations was estimated by the error propagation of the discrete Maxwell relation, considering the instrumental error in magnetization ($\delta M = \pm 0.005$ emu), the temperature step ($\Delta T = 3$ K), the maximum applied field ($\Delta H = 1.8$ T), and the sample mass (~0.2208 g). The resulting deviation is approximately $\pm$ 0.0192 J Kg^{-1} K^{-1}. This uncertainty, which mainly arises from the temperature step and magnetization precision, was taken into account in the ΔS_M analysis and is representative of the experimental confidence range.

3. Results and Discussion

3.1. Crystal Structure

Figure 3 shows the X-ray diffraction (XRD) patterns of the individually synthesized manganites (LCSM0.05 and LCSM0.1) and the composite. The single-phase manganites show diffraction peaks corresponding to the orthorhombic perovskite structure of calcium-doped lanthanum manganite ($La_{0.65}Ca_{0.325}MnO_{2.975}$, COD 1521096, *Pbnm*). Slight shifts in the 2θ positions of certain diffraction peaks, particularly those corresponding to the (110) and (104) planes (shown in detail on the right side of Figure 3), correspond to a change in interplanar distances, depending on strontium content, as larger strontium ions (ionic radius 1.26 Å) substitute smaller calcium ions (ionic radius 1.12 Å). The XRD pattern on the composite displays the same set of diffraction peaks, corresponding to a mixture of both the orthorhombic manganites.

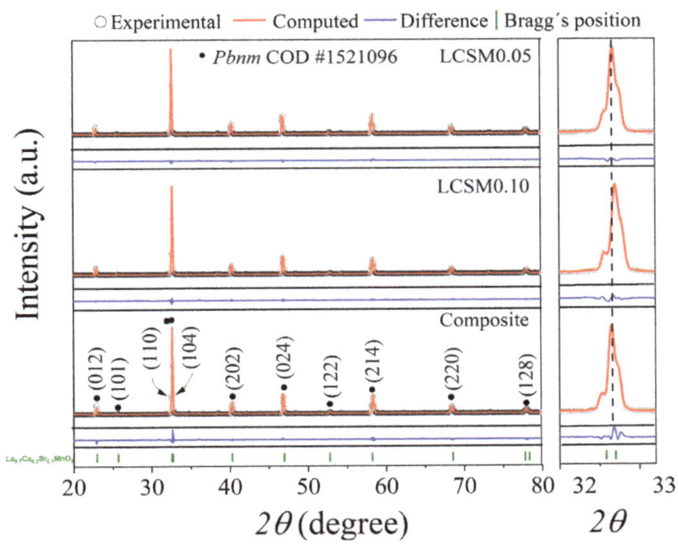

Figure 3. The X-ray diffraction (XRD) patterns of the individually synthesized manganites (LCSM0.05, LCSM0.1) and the composite.

The main diffraction peak exhibits a shift toward lower 2θ angles, indicating an expansion of the unit cell. This behavior can be attributed to the incorporation of ions with larger ionic radii, which induce lattice strain during the sintering process, while the crystal structure is preserved. The XRD patterns confirm the successful synthesis of the compounds and the obtaining of the composite based on the individual manganites.

To quantify the phases and modifications of the orthorhombic structure, Rietveld refinement was performed on the XRD patterns (Figure 3). The refinement results, summarized in Table 1, confirm that the single compounds (LCSM0.05 and LCSM0.1) crystallize in a single orthorhombic phase, while the composite consists of a mixture of both manganite phases. The comparison between the experimental data and the cell parameters of the reference XRD pattern suggests that the presence of Ca^{2+} and Sr^{2+} in the manganite crystal structure does not significantly alter the original crystal structure. Rietveld refinement determined that the weight percentage of manganites in the composite powder was 50.4 and 49.6% of LCSM0.05 and LCSM0.1, respectively.

Table 1. Phase quantification, lattice parameters, crystallite size (D), microstrain (μs), and goodness of fit obtained from Rietveld refinement of XRD patterns for LCSM 0.05, LCSM 0.1 and composite.

Sample	Phase/Space Group	Phase (wt. %)	D (nm)	Microstrain (Adim)	Lattice Parameter (Å)			Goodness of Fit	
					a	b	c	R_{wp}	χ^2
LCSM0.05	$La_{0.7}Ca_{0.25}Sr_{0.05}MnO_3$ *Pbnm*	100 ± 0.00	77.40 ± 4.26	$4.76 \times 10^{-4} \pm 3.11 \times 10^{-6}$	$5.44 \pm 6 \times 10^{-4}$	$7.76 \pm 3 \times 10^{-4}$	$5.47 \pm 6 \times 10^{-4}$	20.89	1.18
LCSM0.1	$La_{0.7}Ca_{0.2}Sr_{0.1}MnO_3$ *Pbnm*	100 ± 0.00	79.01 ± 2.26	$6.36 \times 10^{-4} \pm 3.72 \times 10^{-6}$	$5.48 \pm 1 \times 10^{-4}$	$7.69 \pm 1 \times 10^{-4}$	$5.46 \pm 9 \times 10^{-4}$	15.80	1.87
Composite	$La_{0.7}Ca_{0.25}Sr_{0.05}MnO_3$ *Pbnm*	50.40 ± 0.65	93.94 ± 5.46	$1.77 \times 10^{-4} \pm 2.46 \times 10^{-6}$	$5.48 \pm 5 \times 10^{-3}$	$7.72 \pm 5 \times 10^{-3}$	$5.47 \pm 5 \times 10^{-3}$	21.26	2.47
	$La_{0.7}Ca_{0.2}Sr_{0.1}MnO_3$ *Pbnm*	49.60 ± 1.46	100 ± 5.50	$6.09 \times 10^{-4} \pm 2.26 \times 10^{-6}$	$5.46 \pm 4 \times 10^{-3}$	$7.71 \pm 9 \times 10^{-3}$	$5.48 \pm 7 \times 10^{-3}$		

Table 1 also includes crystallite size, which is similar for the LCSM0.1 and LCSM0.05 samples, while in the case of the composite, the crystallite size of both phases is slightly larger compared to the precursor manganites, because of the heat treatment that causes recrystallization processes. Table 1 presents the refinement quality evaluated through the R_{wp} and χ^2 parameters. Although the R_{wp} values deviate from the conventional standards, likely due to experimental data quality, magnetic contributions, and background effects, the χ^2 values indicate a satisfactory agreement between observed and calculated profiles. Given that $\chi^2 = (Rwp/Rexp)^2$, this parameter provides a more reliable assessment of refinement quality, as it accounts for data statistics, background subtraction, and the consistency between experimental and simulated patterns [42].

Although the refinement yielded an adequate goodness of fit, it is plausible that minor interfacial phases formed during annealing as a result of limited Sr–Ca interdiffusion between the two manganite components. These regions, with slightly shifted stoichiometry and Mn^{3+}/Mn^{4+} ratios, could modify local exchange interactions and contribute to the broadening of the magnetic transition. Although their fraction is likely below the XRD detection limit and cannot be quantified by Rietveld refinement, their presence may be inferred from subtle features in temperature-dependent magnetization and ESR response, both sensitive to local structural and magnetic inhomogeneities.

Due to the effect of the Mn^{3+}/Mn^{4+} ratio on double-exchange magnetic interactions in manganites, an analysis of the distortion of Mn-O-Mn bond octahedra was carried out, evaluating the angles. The different oxidation state of Mn modifies the magnetic interactions and magnetocaloric properties of the composite powder, underscoring the importance of structural modifications to optimize the performance of manganite based-materials.

The bond angles between the Mn and O ions are presented in Table 2. Specifically, Mn_1-O-Mn_1 represents the angles in the b-direction, while Mn_2-O-Mn_2 corresponds to the angle in the c-direction. The data show a slight structural distortion in the composite compared to the former LCSM0.05 and LCSM0.1, a phenomenon attributed to Sr^{2+} diffusion between single compounds during annealing.

Table 2. Bond angles obtained using VESTA simulation data from XRD patterns.

Sample	Phase	Mn-O-Mn Bond Angle (Degree)	
		Mn_1-O-Mn_1	Mn_2-O-Mn_2
LCSM0.05	$La_{0.7}Ca_{0.25}Sr_{0.05}MnO_3$	160.29	161.32
LCSM0.1	$La_{0.7}Ca_{0.2}Sr_{0.1}MnO_3$	160.15	161.37
Composite	$La_{0.7}Ca_{0.25}Sr_{0.05}MnO_3$ $La_{0.7}Ca_{0.2}Sr_{0.1}MnO_3$	160.19 160.24	161.36 161.36

Manganites possess a perovskite-type structure whose distortion arises from the ionic radius mismatch between A-site dopants (La, Sr, Ca) and the host lattice and from oxygen vacancies induced by doping. These distortions, strongly influenced by the Jahn–Teller effect, modify the Mn^{3+}–Mn^{4+} ratio, affecting the Curie temperature and the magnetic interactions. While the Mn^{3+}–Mn^{4+} interaction via the double-exchange mechanism promotes ferromagnetism, the prevalence of Mn^{3+} favors antiferromagnetism; the variation in bond lengths and angles due to doping thus tunes the magnetic and magnetocaloric properties of the material [23]. In mixed-valence perovskite manganites, magnetic exchange (double-exchange and super-exchange) and therefore the Curie temperature [20] are highly sensitive to the tilt of the MnO_6 octahedron and dependent on the Mn-O-Mn geometry because the 3d–2p orbital overlap (bandwidth) depends strongly on the Mn-O-Mn angle; therefore, modest modifications in the bond angle change the effective hopping amplitude and can thus alter magnetic coupling. In the studied material, formed by a mixture of two manganites with very similar compositions, differences of a few degrees in the Mn-O-Mn bond angle are expected, as they produce important changes in magnetic properties, allowing us to modulate the magnetic properties, such as Curie temperature.

3.2. Particle Morphology and Chemical Composition

To analyze the microstructural characteristics of the composite powder, including particle morphology and size distribution, scanning electron microscopy (SEM) was employed. Representative micrographs at different magnifications are shown in Figure 4.

As can be seen in Figure 4, the composite powder shows a broad monomodal asymmetric particle size distribution (5–45 μm) with media of 21.47 μm, with a relatively wide standard deviation (3.14 μm) arising from particle agglomeration. The presence of a monomodal distribution, despite consisting of two manganites, is explained by particle agglomeration during milling and annealing, which masks any separation of individual size populations. Most particles appear to originate from the coalescence of finer grains, leading to polyhedral morphologies that suggest an incipient sintering process. This coalescence, particularly between LCSM0.05 and LCSM0.1 particles, not only promotes the mechanical consolidation of the composite but also enhances interfacial connectivity between crystallites. Such microstructural features play a decisive role in the MCE, since the density of grain boundaries and the degree of chemical homogeneity modulate the double-exchange interactions and spin alignment. Consequently, the optimized microstructure in the composite powder strengthens the cooperative magnetic response, thereby improving the magnitude and reversibility of the magnetic entropy change.

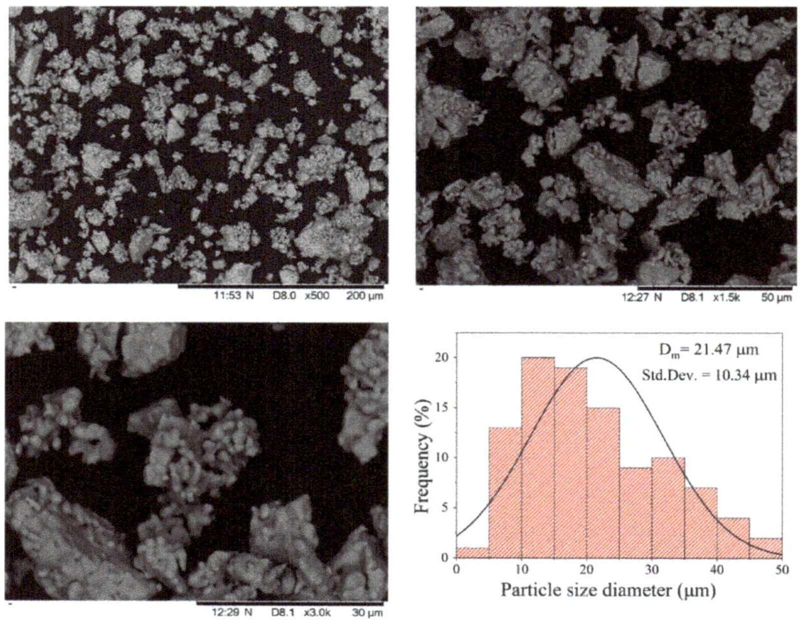

Figure 4. Representative SEM micrographs of the composite, showing particle morphology and particle size distribution.

Energy-dispersive X-ray spectroscopy (EDS) was employed to qualitatively evaluate the chemical composition and elemental distribution across the composite powder, as shown in Figure 5. The corresponding spectra display only the expected peaks of the constituent elements, without detectable impurities, and the quantified values are consistent with the compositions of the manganites forming the composite. Additionally, elemental analysis is congruent with the presence of a mixture of both manganites. While the overall distribution of elements is homogeneous, variations in strontium content are observed in certain particles, confirming that individual particles have different compositions, consistent with the coexistence of manganites with distinct stoichiometries within the composite.

3.3. Magnetic Properties

As is well known, magnetic properties are highly sensitive and critically dependent on composition and crystal structure, as previously discussed in the structural characterization. Previous results indicate that the presence of manganites with slight compositional variations in the composite powder can influence its magnetic behavior and, consequently, its magnetocaloric performance. In this study, it is anticipated that the coexistence of different manganites within the composite powder could enhance the magnetocaloric response, particularly the RCP. To test this hypothesis, magnetic hysteresis loops at room temperature (298 K) were measured for the composite and the individual compounds, and the results are shown in Figure 6. As observed in the table included as an inset in Figure 6, the pure compound LCSM0.05 exhibits paramagnetic behavior with a specific magnetization of 15.93 emu·g^{-1} at 18 kOe, whereas LCSM0.1 shows ferromagnetic ordering with a specific magnetization of 36.62 emu·g^{-1}, highlighting the impact of small amounts of Sr^{2+} on the magnetic properties of the manganite. These results are consistent with previously reported studies [2–11] and in agreement with the XRD refinement parameters listed in Tables 1 and 2, where Sr doping in LCSM0.1 induces higher structural distortion, favoring double-exchange interactions at room temperature. In Figure 6, it is also observed that the composite exhibits a slim magnetic hysteresis loop (coercive field ~11.57 Oe) with ferromagnetic ordering and a specific magnetization of 37.17 emu·g^{-1} at 18 kOe, making it an excellent soft magnetic composite powder for magnetocaloric applications, since less energy is required for cyclic magnetization and demagnetization. The magnetic response of the composite thus reflects the mass-weighted contribution of its individual constituents, confirming that its overall behavior arises from the combined properties of the two manganites [27–31].

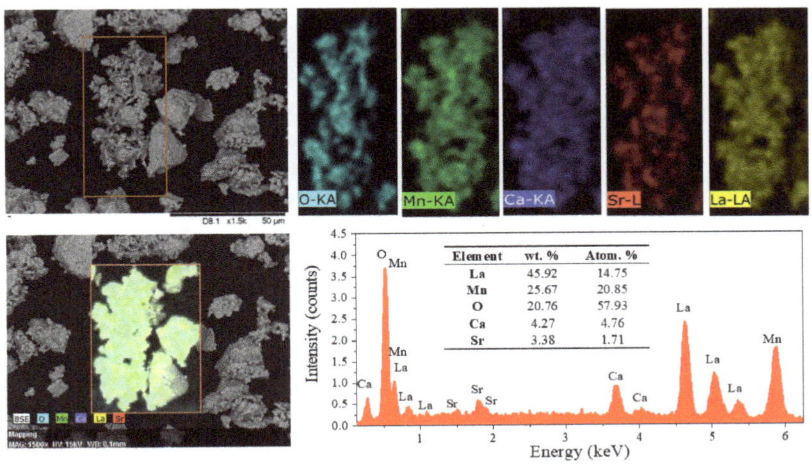

Figure 5. SEM micrographs and compositional analysis of composite. Top: Elemental maps of O, Mn, Ca, Sr, and La, evidencing homogeneous spatial distribution. Bottom left: BSE image showing particle morphology. Bottom right: EDS spectrum confirming intended stoichiometry of composite.

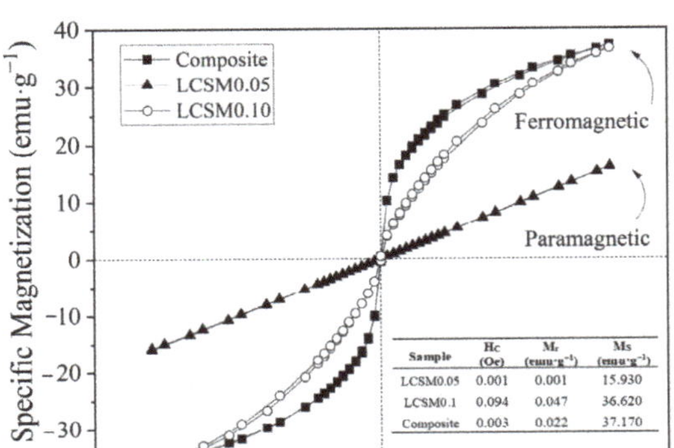

Figure 6. Magnetic hysteresis loops at room temperature (293 K) for LCSM0.05, LCSM0.1, and the composite. In the inset, the magnetic parameter, coercivity (Hc), remanent magnetization (Mr), and specific magnetization at 18 kOe (Ms) are presented.

Moreover, the temperature dependence of the specific magnetization, shown in Figure 7, was used to determine the Curie temperature (T_C) by analyzing the derivative of the magnetization curve. T_C was identified at the maximum of the first derivative (dM/dT), which corresponds to the inflection point marking the transition from the ferromagnetic to the paramagnetic state. Measurements were carried out under a constant magnetic field of 10 kOe, and the obtained values were compared with those reported in the literature and presented in Table 3. As displayed in Figure 7a, the Curie temperature was determined as 277.7 K for LCSM0.05 and 311.5 K for LCSM0.1, with both values being in agreement with those previously reported for analogous manganite systems [30]. In contrast, the composite exhibits two distinct peaks at 286.8 K and 307.6 K, reflecting the coexistence and interaction of ferromagnetic and paramagnetic phases within the material. Additionally, the curve of the composite presents a very small deviation, indicating intermediate states that arise from the phase interaction. This behavior corroborates the presence of two predominant phases in the composite while simultaneously generating intermediate responses, thereby confirming the dual-phase nature of the system. Such coexistence is particularly advantageous for applications requiring tunable magnetic properties, such as magnetic refrigeration. Furthermore, the persistence of two main transitions together with the intermediate features after thermal treatment demonstrates that the designed composite preserves the targeted multiphase structure, enabling enhanced performance through the synergistic contribution of the different phases.

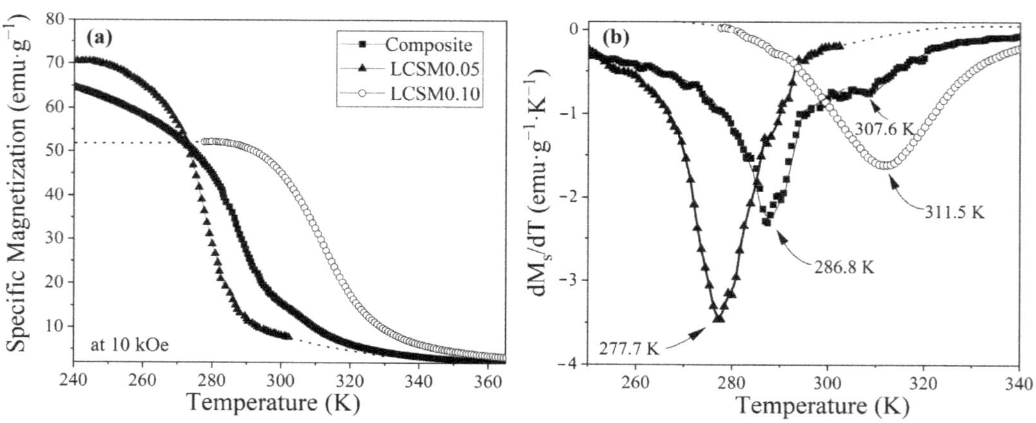

Figure 7. (**a**) Specific magnetization curves (Ms) versus temperature under applied field of 10 kOe and (**b**) dM/dT of specific magnetization curves of LCSM0.05, LCSM0.1, and composite.

The dual Curie temperatures observed in the composite sample can be quantitatively explained by the nearly equal phase fractions determined by Rietveld refinement (50.4% LCSM0.05 and 49.6% LCSM0.1). The lower T_C transition corresponds to the LCSM0.05 phase, while the higher one corresponds to LCSM0.1. The comparable weight fractions of both phases justify the presence of two well-defined peaks of similar magnitude (286.8 and 307.6 K) arising from their interactions. Thus, the overall ΔS_M response results from the additive contributions of both phases, broadening the operational temperature window between 243 and 328 K and yielding an enhanced RCP of 188.82 J·kg^{-1}, which increased by almost threefold compared with the single phases.

Table 3. Comparative magnetocaloric parameters: Curie temperature (Tc), operating temperature window (OTW), magnetic entropy change ($-\Delta S_M$), and relative cooling power (RCP) for different ceramic composites based on lanthanum manganites.

Composite	Tc (K)	$\mu_0 H$ kOe	OTW (K)	$\lvert\Delta S_M\rvert$ ($J \cdot kg^{-1} \cdot K^{-1}$)	RCP ($J \cdot kg^{-1}$)	Reference
$0.5La_{0.75}Ca_{0.25}Sr_{0.05}MnO_3/0.5La_{0.7}Ca_{0.2}Sr_{0.1}MnO_3$	286–307	18	85	5.86	189	This work
$0.5La_{0.7}Ca_{0.2}Sr_{0.1}MnO_3/0.5La_{0.7}Te_{0.3}MnO_3$	255–303	20	70	2.07	125	[7]
$0.95La_{0.45}Nd_{0.25}Sr_{0.3}MnO_3/0.05CuO$	285	15	95	2.55	75	[8]
$0.4La_{0.6}Ca_{0.4}MnO_3/0.6La_{0.6}Sr_{0.4}MnO_3$	255–350	30	250	0.92	215	[10]
$0.5La_{0.65}Ca_{0.35}MnO_3/0.5Pr_{0.5}Sr_{0.5}MnO_3$	277–298	20	100	0.68	80	[11]
$0.75La_{0.7}Ca_{0.2}Sr_{0.1}MnO_3$ $/0.25La_{0.7}Te_{0.3}MnO_3$	290	20	70	2.5	138	[12]
$0.5La_{0.7}Ca_{0.3}MnO_3/0.5La_{0.7}Ag_{0.3}MnO_3$	260–304	20	98	2.65	139	[14]
$0.75La_{0.7}Ca_{0.3}MnO_3/0.25La_{0.84}Sr_{0.16}MnO_3$	255	20	140	0.72	89	[15]
$0.5La_{0.67}Ca_{0.27}Sr_{0.06}MnO_3/0.5La_{0.7}Sr_{0.3}Mn_{0.95}Cu_{0.05}O_3$	264–360	20	75	1.82	138	[16]
$0.5La_{0.7}Ca_{0.2}Sr_{0.1}MnO_3/0.5La_{0.7}Ca_{0.15}Sr_{0.15}MnO_3$	310–336	20	90	1.26	115	[19]
$0.33Pr_{0.67}Ca_{0.33}MnO_3/0.66La_{0.67}Sr_{0.33}MnO_3$	346	20	250	0.98	77	[40]

Although the studied composite is formed by two different lanthanum manganites with a very similar composition, their Curie temperatures are different, with a difference of $\approx$33.8 K between the two components, at 277.7 K and 311.5 for LCSM0.05 and LCSM0.1, respectively, as shown in Figure 7b, which correlates with the subtle but reproducible bond angle changes and with the complementary magnetic tests presented previously.

To further corroborate our findings, we conducted electron spin resonance (ESR) analysis, a technique highly sensitive to spin dynamics and local magnetic correlations, which can reveal distinct signatures for each manganite phase and their composite. As shown in Figure 8, the ESR spectra recorded at 293 K and 378 K for the individual compounds and the composite exhibit notable differences. Both LCSM0.05 and the composite display a characteristic hump in the signal, a hallmark of the coexistence of ferromagnetic (FM) and paramagnetic (PM) regions, as commonly reported in phase-separated manganites [43]. The values of the g factor were 1.992, 1.988, and 2.179 at 293 K for LCSM0.05, LCSM0.1, and the composite, respectively. In contrast, g achieves values of 1.983, 1.998, and 2.010 for LCSM0.05, LCSM0.1, and the composite at 378 K, respectively. The value close to $g \approx 2$ observed in LCSM0.05 is typical of a paramagnetic response above its Tc, while the deviation found in LCSM0.1 reflects stronger ferromagnetic exchange near its Tc. The composite shows the highest g values at both temperatures, indicating greater structural anisotropy. In this case, some Mn-O-Mn bonds present different angles and distances, and spin–orbit coupling is uneven. As a result, not all spins reach the ferromagnetic–paramagnetic transition simultaneously but progressively under slightly different conditions, which broadens the temperature dependence of ΔS_M and the RCP simultaneously. The effective g factor was determined using the following equation [44].

$$g_{\text{eff}} = \frac{h\nu}{\mu_B H_{\text{res}}} \tag{2}$$

where h is Planck's constant, u the microwave frequency, m_B the Bohr magneton, and H_{res} the resonance magnetic field. Values of g close to 2 are typical of itinerant e_g electrons associated with Mn ions, reflecting the influence of the Mn^{3+}/Mn^{4+} ratio and their exchange interactions. The slight decrease in g at 378 K indicates a transition to predominantly paramagnetic behavior as thermal fluctuations suppress long-range FM order.

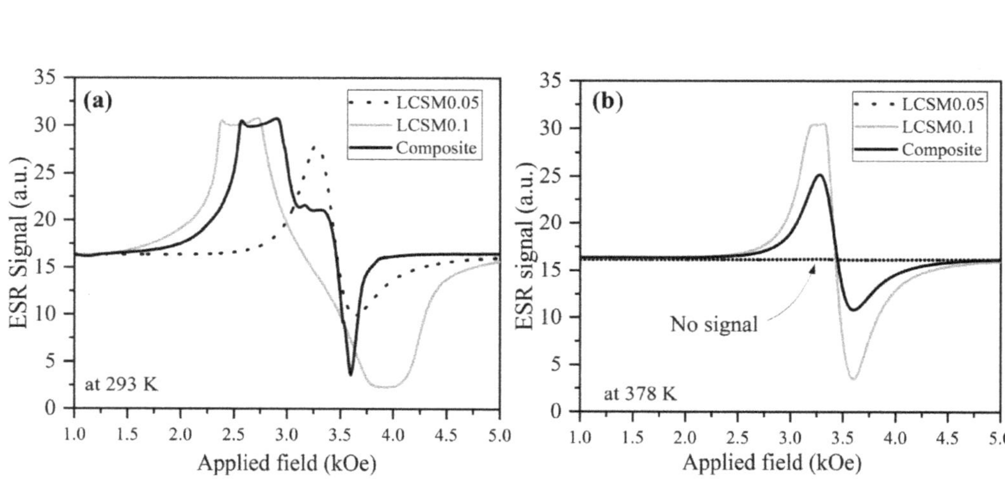

Figure 8. Electron spin resonance (ESR) spectra of LCSM0.05, LCSM0.1, and composite measured at (**a**) 293 K and (**b**) 378 K.

3.4. Magnetocaloric Properties: ΔS, OTW, and RCP

Isothermal magnetization curves M_s(H, T) were recorded at different temperature intervals, ranging from 243 K to 328 K, depending on the Curie temperature of each manganite sample, under an applied magnetic field from 0 to 18 kOe, to evaluate the magnetocaloric properties. Figure 9 shows the evolution from ferromagnetic to paramagnetic order in the three samples as the temperature increases. The widened OTW observed in the composite is attributed to the coexistence of both manganites, as previously evidenced in the magnetic hysteresis loops.

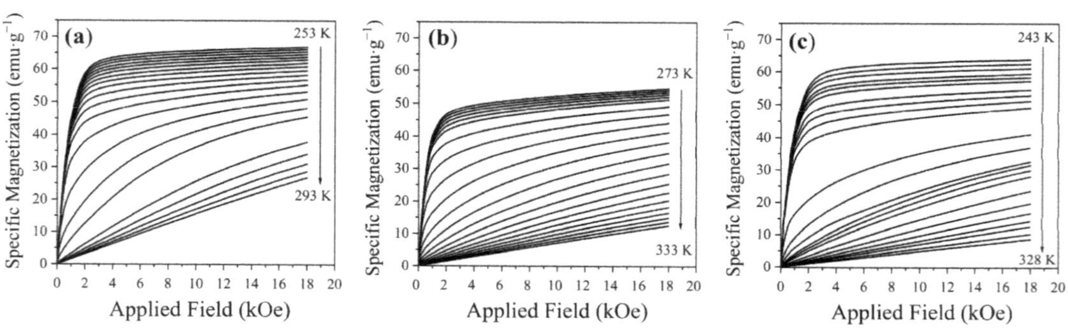

Figure 9. Isothermal magnetization curves M_s(H, T) of (**a**) LCSM0.05, (**b**) LCSM0.1, and (**c**) composite.

To determine the order of the ferromagnetic-to-paramagnetic transition, Arrott plots were constructed, representing H/M as a function of M^2, based on the isothermal magnetization data in Figure 10. This method, derived from Landau's mean-field theory, relies on the Banerjee criterion [34], according to which positive slopes of the isothermal lines correspond to a second-order transition, while negative slopes indicate a first-order one. In addition, the Landau coefficient β was analyzed using the modified Arrott–Noakes method, based on the equation of state $(H/M)^{1/\gamma} = \alpha\ \varepsilon + bM^{1/\beta}$, where $\varepsilon = (T - Tc)/Tc$. From the high-field linear region of the Arrott plots (Figure 10), the spontaneous magnetization was extracted and fitted to $M(T) \propto (-\varepsilon)^{\beta}$, yielding $\beta = 0.36 \pm 0.02$ (LCSM0.05),

0.43 ± 0.05 (LCSM0.1), and 0.37 ± 0.02–0.41 ± 0.12 (composite). The obtained critical parameter values together with the positive slopes throughout the investigated temperature range of the Arrott plots shown in Figure 10 confirm that all the samples exhibit a second-order nature of the transitions. This behavior is consistent with earlier reports on La-based manganites substituted with Ca and Sr [4,7,14,32,33], where substitution alters the Mn^{3+}/Mn^{4+} ratio and the Mn-O-Mn bond angle but maintains the dominance of the double-exchange mechanism, typically giving rise to continuous transitions. In the LCSM0.05 and LCSM0.1 manganites, the smooth decrease in magnetization with temperature reflects the gradual suppression of ferromagnetic ordering, whereas in the composite, the coexistence of both parent phases preserves the same second-order character, without evidence of discontinuities or thermal hysteresis that would suggest a first-order process. The confirmation of second-order transitions in both the single-phase manganites and the composite is technologically relevant, since such transitions are associated with reversible entropy changes, broad operating temperature windows, and minimal hysteresis losses, features that are highly desirable for efficient magnetic refrigeration.

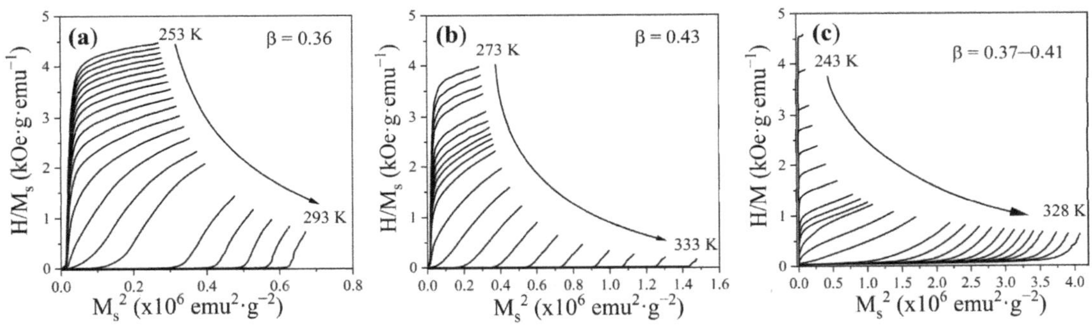

Figure 10. Arrott curves for (**a**) LCSM0.05, (**b**) LCSM0.1, and (**c**) composite.

ΔS_M, as a measurement of the magnetocaloric behavior of the materials, was calculated from the isothermal magnetization curves (Figure 9) using Maxwell's relation [26]:

$$\Delta S_M(\mathrm{T,H}) = -\mu_0 \int_0^H \left(\frac{dM}{dT}\right)_H dH \quad (3)$$

where μ_0 is the permeability of free space, M is magnetization, T is the absolute temperature, H is the applied magnetic field, and negative values denote the normal magnetocaloric effect. In practice, the integral was approximated via the trapezoidal rule and the derivative by central finite differences, yielding the following:

$$|\Delta S_M(\mathrm{T,H})| = \sum_i \frac{M_{i+1}(T_{i+1}\cdot H) - M_i(T_i\cdot H)}{T_{i+1} - T_i} \Delta \mathrm{H} \quad (4)$$

where M_i denotes the experimentally measured magnetization at temperature T_i and field H.

Plots of $-\Delta S_M$ versus temperature for the manganites LCSM0.05 and LCSM0.1 and the composite are shown in Figure 11. In each case, the maximum magnitude of ΔS_M occurs near the corresponding Curie temperature (T_C), with there being two maximum magnetic entropy changes under an applied magnetic field of 18 kOe of 5.86 and 2.34 $J\cdot kg^{-1}\cdot K^{-1}$ for the composite and 4.05 and 6.28 $J\cdot kg^{-1}\cdot K^{-1}$ for LCSM0.1 and LCSM0.05, respectively, in line with the expectations for magnetocaloric materials, since ferromagnetic order progressively decreased above T_C. Conversely, as can be observed in Figure 11, the ΔS_M versus T curve presented by the composite exhibits two well-resolved entropy maxima and a broad, nearly continuous ΔS_M response extending from ≈243 K to ≈328 K (≈85 K span) with local ΔS_M values remaining higher than 0.5 J·kg^{-1}·K^{-1} across the intermediate region under the same field, ascribed to the presence of two types of manganites with different Curie temperatures. In addition, the maximum value of ΔS_M in the composite was lower than that in its individual manganite components. This reduction is characteristic of composite systems, where interphase interactions often moderate the peaks of ΔS_M. Similar behavior is reported in composites of manganites and perovskites, where the peak shifts or diminishes due to phase mixing and distributed transitions. Maintaining stable and predictable ΔS_M over a broad temperature span is essential for reliable magnetocaloric performance [44–52].

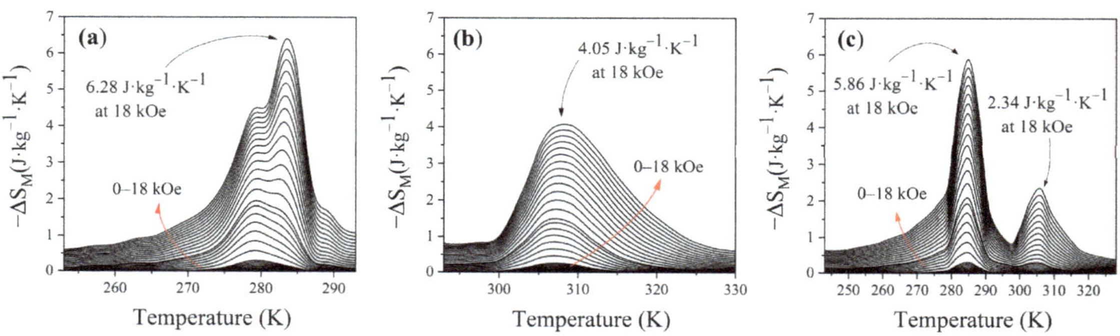

Figure 11. The temperature dependence of the entropy changes for (**a**) LCSM0.05, (**b**) LCSM0.1, and (**c**) the composite.

The relative cooling power (RCP), a commonly used figure of merit for magnetocaloric materials and a key parameter for magnetocaloric applications, was calculated as the product of the maximum magnetic entropy change (ΔS_M^{max}) and the full width at half maximum δT_{FWHM} of the $|\Delta S_M^{max}|$ peak, following the equation below [4].

$$\mathrm{RCP} = |\Delta S_M^{max}| \cdot \delta T_{\mathrm{FWHM}} \quad (5)$$

Therefore, the RCP depends not only on the peak of $-\Delta S_M$ but also on the width of the temperature interval over which significant entropy change occurs. The RCP values obtained for a magnetic applied field of 18 kOe are 55.24, 65.12, and 188.82 J·kg^{-1} for LCSM0.05, LCSM0.1, and the composite, respectively; these results confirm the increase in the RCP with the formation of the composite, as was predicted in this study. This reduction in the peak magnitude but extended operating temperature window (OTW) is attributed to the presence of multiple phases in the composite, where overlapping transitions broaden the ΔS_M profile and can increase RCP while lowering the peak of ΔS_M. As shown in Table 3, which presents the magnetocaloric parameters of several lanthanum-based manganite composites, the material developed in this work exhibits one of the highest magnetic entropy changes ($|\Delta S_M|$ = 5.86 J·kg^{-1}·K^{-1}) and a competitive relative cooling power (RCP = 188.8 J·kg^{-1}) within an extended operational range (≈85 K around room temperature), confirming its superior magnetocaloric performance and the effectiveness of the dual-phase composite approach employed in this study.

A field normalization analysis was also performed (included in the Supplementary Materials), revealing that the field dependence of the magnetic entropy change follows the relation DS α H^n. The exponent n was determined experimentally from the $|\Delta S_M|$ data, obtaining values of 1.25 and 1.99 at 243 and 328 K, respectively, with an average value of n ≈ 1.65 near the transition temperature. Using this relationship, the DS_M values were normalized to standard magnetic fields of 2 T to facilitate a direct comparison with literature data. The maximum normalized values obtained were 6.7 and 2.8 J Kg^{-1} K^{-1} at 2 T, positioning this composite material among those with superior performance compared to previously reported systems. The quantified uncertainty ensures a reliable assessment of ΔS_M behavior and supports the consistency of the obtained results.

The improved magnetocaloric performance of the studied composite is attributed to interfacial exchange and strain-mediated coupling between the two manganite phases (compounds), which modify the local Mn-O-Mn geometry and broaden the magnetic transition. This coupling results in a non-additive RCP and smoother ΔS_M, consistent with cooperative effects at the interfaces. However, further local analysis will be required to confirm this interpretation in future research.

Optimization strategies therefore must balance peak magnitude and the OTW (for example, via microstructure control, phase fraction tuning, or measurement at higher ΔH) to maximize usable refrigeration capacity for the target operating temperature range. Therefore, while composite formation led to enhancements in RCP and broadened the operating temperature window, the reduced peak values and lower intermediate temperatures pose a limitation. Optimization strategies, for example, stronger applied fields or improved composite architecture (phase proportions, microstructure), may mitigate these drawbacks and lead to materials with more uniform and elevated ΔS_M across the temperature range.

4. Conclusions

A composite powder formed by a mixture of $La_{0.7}Ca_{0.2}Sr_{0.1}MnO_3$ and $La_{0.7}Ca_{0.25}Sr_{0.05}MnO_3$ in an equal weight ratio was successfully obtained, with no evidence of secondary phases detected by X-ray diffraction and Rietveld refinement. The manganites were synthesized by the solid-state reaction route, using high-energy milling for 5 h and annealing at 1473 K for 2 h, while the composite was prepared by mixing both phases using high-energy ball milling for 2 min and applying a further heat treatment at 1473 K for 2 h. SEM and EDS analyses revealed a consolidated microstructure with good interfacial connectivity and controlled chemical heterogeneity, features that enhance the cooperative magnetic response and magnetocaloric performance of the composite. This approach confirms a simple and effective method for producing magnetocaloric composite ceramic materials. Magnetic analysis through Arrott plots shows that the composite undergoes a second-order phase transition according to Banerjee's criterion, which is favorable for magnetic refrigeration since it minimizes magnetic and thermal hysteresis. Although the maximum $|\Delta S_M|$ value of the composite is lower than that of the individual manganites, the relative cooling power reaches 188.82 J·kg^{-1}, which is considerably higher than the 55.24 and 65.12 J·kg^{-1} obtained for LCSM0.05 and LCSM0.1, respectively. The non-additive RCP, the broader and smoother ΔS_M, and EPR evidence confirm that the enhanced magnetocaloric response of the composite likely arises from interfacial exchange coupling and strain-mediated effects between the two manganite phases. This improvement points to a more efficient magnetocaloric performance of the composite, supporting its potential use in magnetic cooling applications. These results demonstrate the effectiveness of our synthesis route in enhancing magnetocaloric performance under practical, low-field conditions. Furthermore, considering practical aspects, the composite is expected to exhibit a moderate adiabatic temperature change (ΔT_{ad}) consistent with its enhanced relative cooling power, negligible magnetic hysteresis losses due to its second-order transition, and good thermal and magnetic stability over repeated cycles, which are essential attributes for reliable magnetic refrigeration applications.

Supplementary Materials: The following supporting information can be downloaded at: https://www.mdpi.com/article/10.3390/ma18214869/s1, Figure S1: SEM-EDS; Figure S2: Normalized n; Figure S3: critical parameter; Table S1: g factor; Table S2: uncertainty.

Author Contributions: F.I.R.P.: Formal analysis, methodology, investigation, writing—original draft. C.A.T.M.: Formal analysis, methodology, investigation, data curation, writing—original draft. A.M.B.M.: Conceptualization, visualization, supervision, resources, methodology, investigation, writing—review and editing. C.A.C.E.: Visualization, resources, writing—review and editing, funding acquisition. M.I.R.V.: Visualization, validation, writing—review and editing. F.S.D.J.: Conceptualization, methodology, resources, writing—review and editing, supervision, funding acquisition. All authors have read and agreed to the published version of the manuscript.

Funding: This research was funded by the Science, Humanities, Technology, and Innovation Secretary (SECIHTI) of Mexico, grant number CF-2023-I-769.

Institutional Review Board Statement: Not applicable.

Informed Consent Statement: Not applicable.

Data Availability Statement: The raw and processed data required to reproduce these findings are available for download at https://data.mendeley.com/preview/3r8g4s3cr2?a=fbe060a9-6e85-454a-9eb1-0a728613bbb2 accessed on 17 October 2025.

Acknowledgments: The authors are grateful for the financial support of the Science, Humanities, Technology, and Innovation Secretary of Mexico (SECIHTI) under grant CF-2023-I-769. F.I. Reyes-Patricio thanks SECIHTI for the scholarship granted to carry out his doctoral thesis. The authors are grateful for the technical support for doing the ESR analysis by J. Zamora under the project PAPIIT-UNAM (IA104125).

Conflicts of Interest: The authors declare no conflicts of interest.

References

1. Gong, Y.; Miao, X.; Qian, F.; Xu, F.; Caron, L. A brief review of microstructure design in transition metal-based magnetocaloric materials. *J. Phys. Condens. Matter* **2024**, *36*, 503001–503023. [CrossRef]
2. Taboada, C.A.; Sánchez, F.; Pedro, F.; Cortés, C.A.; Betancourt, J.A.; Ramírez, M.; Bolarín, A.M. Large magnetocaloric effect near to room temperature in Sr-doped $La_{0.7}Ca_{0.3}MnO_3$. *J. Magn. Magn. Mater.* **2020**, *496*, 165887–165898. [CrossRef]
3. Biswal, H.; Singh, V.; Nath, R.; Sahu, J.R. Magnetic properties and near-room-temperature large magnetocaloric effect in $(La_{1-x}Bi_x)_{0.67}Ba_{0.33}MnO_3$ (x = 0–0.3) ceramics. *Mater. Res. Bull.* **2021**, *133*, 111030–111037. [CrossRef]
4. Zhang, H.; Yan, J.; Shao, C.; Zhu, Y.; Zhu, F.; Chen, Y.; Cheng, Y.; Gu, W.; Li, H.; Liu, Z.; et al. Structural and magnetic properties of $La_{0.80-x}Eu_xNa_{0.20}MnO_3$ perovskite manganites with enhanced magnetocaloric effect by appropriate Eu-doping content. *Ceram. Int.* **2025**, *51*, 21856–21866. [CrossRef]
5. Bolarín, A.M.; Taboada, C.A.; Cortés, C.A.; Sánchez-De Jesús, F. Tuning the magnetocaloric properties of lanthanum–strontium manganite by rare-earth Nd^{3+} doping. *J. Therm. Anal. Calorim.* **2024**, *149*, 12665–12674. [CrossRef]
6. Taboada, C.A.; Bolarín, A.M.; Pedro, F.; Cortés, C.A.; Sánchez, F. Role of Gd^{3+} on the magnetocaloric properties of lanthanum-strontium manganite. *J. Magn. Magn. Mater.* **2023**, *570*, 170542–170552. [CrossRef]
7. Akça, G.; Kiliç, S.; Ekicibil, A. Magnetic and magnetocaloric properties of $0.5La_{0.7}Ca_{0.2}Sr_{0.1}MnO_3/0.5La_{0.7}Te_{0.3}MnO_3$ composite. *Cumhur. Sci. J.* **2020**, *41*, 144–151. [CrossRef]
8. Maalam, K.E.; Balli, M.; Habouti, S.; Dietze, M.; Hamedoun, M.; Hlil, E.K.; Es-Souni, M.; El Kenz, A.; Benyoussef, A.; Mounkachi, O. Composite $(La_{0.45}Nd_{0.25})Sr_{0.3}MnO_3/5CuO$ materials for magnetic refrigeration applications. *J. Magn. Magn. Mater.* **2018**, *449*, 25–32. [CrossRef]

9. Zhou, Y.; Zhu, X.; Li, S. Effects of copper doping on the structure, electrical and low-field magnetoresistance properties of (1 − x)$La_{0.67}Sr_{0.33}MnO_3$/xCu (x = 0−0.15) composite coatings. *Ceram. Int.* **2017**, *43*, 10026–10031. [CrossRef]
10. Jeddi, M.; Gharsallah, H.; Bekri, M.; Dhahri, E.; Hlil, E.K. Improvement of magnetocaloric properties around room temperature in (1−x) $La_{0.6}Ca_{0.4}MnO_3$/(x)$La_{0.6}Sr_{0.4}MnO_3$ ($0 \leq x \leq 1$) composite system. *Phase Transit.* **2020**, *93*, 311–322. [CrossRef]
11. Sellami, E.; Regaieg, Y.; Cheikhrouhou, W.; Koubaa, M.; Cheikhrouhou, A.; Njah, N. Magnetic and magnetocaloric properties of $La_{0.65}Ca_{0.35}MnO_3/La_{0.7}Ca_{0.2}Ba_{0.1}MnO_3$ and $La_{0.65}Ca_{0.35}MnO_3/Pr_{0.5}Sr_{0.5}MnO_3$ composite manganites. *J. Supercond. Nov. Magn.* **2015**, *28*, 3121–3126. [CrossRef]
12. Akça, G.; Çetin, S.K.; Ekicibil, A. Composite x$La_{0.7}Ca_{0.2}Sr_{0.1}MnO_3$/(1 − x)$La_{0.7}Te_{0.3}MnO_3$ materials: Magnetocaloric properties around room temperature. *J. Mater. Sci. Mater. Electron.* **2020**, *31*, 6796–6808. [CrossRef]
13. M'nassri, R.; Nofal, M.M.; De Rango, P.; Chniba, N. Magnetic entropy table-like shape and enhancement of refrigerant capacity in $La_{1.4}Ca_{1.6}Mn_2O_7$–$La_{1.3}Eu_{0.1}Ca_{1.6}Mn_2O_7$ composite. *RSC Adv.* **2019**, *9*, 14916–14927. [CrossRef]
14. Yigiter, N.; Pektas, M.; Kolat, V.S.; Izgi, T.; Bayri, N.; Gencer, H.; Atalay, S. Structural, magnetic, and magnetocaloric properties of (1 − x)$La_{0.7}Ca_{0.3}MnO_3$/(x)$La_{0.7}Ag_{0.3}MnO_3$ composites. *J. Mater. Sci. Mater. Electron.* **2022**, *33*, 4721–4735. [CrossRef]
15. Bisht, P.; Meenakshi; Gaur, A.; Mahato, R.N. Structural and magnetocaloric properties (0.75) $La_{0.7}Ca_{0.3}MnO_3$/(0.25) $La_{0.84}Sr_{0.16}MnO_3$ nanocomposite. *Physica B Condens. Matter* **2021**, *619*, 413215–413224. [CrossRef]
16. Kılıç, S.; Akça, G.; Aslan, M.S.; Ekicibil, A. Room-temperature magnetocaloric properties of y$La_{0.67}Ca_{0.27}Sr_{0.06}MnO_3$/1−y$La_{0.7}Sr_{0.3}$ $Mn_{0.95}Cu_{0.05}O_3$ composites. *J. Therm. Anal. Calorim.* **2024**, *149*, 4441–4451. [CrossRef]
17. Boubekri, A.; Elmaddahi, Z.; Jarmoumi, Y.; Gueddouch, K.; Farchakh, A.; Hafidi, M.E. A phenomenological approach for predicting magnetic and magnetocaloric properties in the $(La_2MnNiO_6)_x/(La_2MnCoO_6)_{1-x}$ composite. *J. Supercond. Nov. Magn.* **2024**, *37*, 1401–1410. [CrossRef]
18. Bouhani, H.; Endichi, A.; Kumar, D.; Copie, O.; Zaari, H.; David, A.; Fouchet, A.; Prellier, W.; Mounkachi, O.; Balli, M.; et al. Engineering the magnetocaloric properties of $PrVO_3$ epitaxial oxide thin films by strain effects. *Appl. Phys. Lett.* **2020**, *117*, 072402–072407. [CrossRef]
19. Ezaami, A.; Sellami, E.; Cheikhrouhou, W.; Cheikhrouhou, A. Enhanced magnetocaloric properties with a tunable Curie temperature in (1 − x)$La_{0.55}Ca_{0.35}MnO_3$/x$La_{0.7}Ca_{0.2}Sr_{0.1}MnO_3$ ($0 \leq x \leq 1$). *J. Mater. Sci. Mater. Electron.* **2017**, *28*, 16741–16746. [CrossRef]
20. Bolarín, A.M.; Taboada, C.A.; Cortés, C.A.; Rosales, O.; Torres, G.; Sánchez, F. Effect of high-energy ball milling on magnetocaloric properties of $La_{0.7}Ca_{0.2}Sr_{0.1}MnO_3$. *Appl. Phys. A* **2020**, *126*, 369–379. [CrossRef]
21. Pękała, M.; Pękała, K.; Drozd, V.; Staszkiewicz, K.; Fagnard, J.F.; Vanderbemden, P. Magnetocaloric and transport study of poly- and nanocrystalline composite manganites $La_{0.7}Ca_{0.3}MnO_3/La_{0.8}Sr_{0.2}MnO_3$. *J. Appl. Phys.* **2012**, *112*, 023906–023914. [CrossRef]
22. Chen, Z.; Han, J.; Jiang, J.; Liang, X.; Wang, H.; Wang, Y.; Wang, Z.; Pan, Z.; Sun, J.; Ma, J.; et al. Obtained large room-temperature TCR of $La_{0.67}Ca_{0.25}Sr_{0.08}MnO_3$ ceramics by optimizing sintering temperature. *Ceram. Int.* **2024**, *50*, 34680–34688. [CrossRef]
23. El Amrani, K.; Moubark, Y.; Ben Ali, M.; Toulemonde, O. Magnetocaloric effect in $Nd_{0.7}Sr_{0.3}MnO_3$:CuO composites. In Proceedings of the 24th IIR International Congress of Refrigeration, Yokohama, Japan, 16–22 August 2015.
24. Giri, S.K.; Dahal, R.H.; Casey, J.; Neupane, D.; Pathak, A.K.; Karna, S.; Mishra, S.R. Effect of metal-oxide phase on the magnetic and magnetocaloric properties of $La_{0.7}Ca_{0.3}MnO_3$ − MO (MO = CuO, CoO, NiO) composite. *Magnetochemistry* **2022**, *8*, 163–182.
25. Kılıç, S.; Akça, G.; Mehmet, S.A.; Ahmet, E. Large magnetocaloric effect in La-based manganite composites near room temperature. *J. Therm. Anal. Calorim.* **2022**, *147*, 3095–3104. [CrossRef]
26. Linh, D.C.; Viet Chinh, N.T.; Dung, N.T.; Bau, L.V.; Duc, N.H.; Manh, D.H.; Thanh, T.D. Critical behavior and room temperature magnetocaloric effect of La-doped $Pr_{0.7}Sr_{0.3}MnO_3$ compounds. *Phys. B Condens. Matter.* **2023**, *661*, 414945–414953. [CrossRef]
27. Fkhar, L.; Lamouri, R.; Mahmoud, A.; Boschini, F.; Hamedoun, M.; Ez-Zahraouy, H.; Benyoussef, A.; Hlil, E.K.; Ait Ali, M.; Mounkachi, O. Enhanced magnetic and magnetocaloric properties of $La_{0.45}Nd_{0.25}Sr_{0.3}MnO_3$/CuO composite. *J. Supercond. Nov. Magn.* **2020**, *33*, 2543–2549. [CrossRef]
28. Alonso, J.A.; Martínez-Lope, M.J.; Casais, M.T.; Fernández-Díaz, M.T. Evolution of the Jahn–Teller distortion of MnO_6 octahedra in $RMnO_3$ perovskites (R = Pr, Nd, Dy, Tb, Ho, Er, Y): A neutron diffraction study. *Inorg. Chem.* **2000**, *39*, 917–923. [CrossRef]
29. Qin, J.; Wen, Z.; Ma, B.; Wu, Z.; Lv, Y.; Yu, J.; Zhao, Y. Microstructure and magnetic properties of novel high-entropy perovskite ceramics $(Gd_{0.2}La_{0.2}Nd_{0.2}Sm_{0.2}Y_{0.2})MnO_3$. *J. Magn. Magn. Mater.* **2024**, *597*, 172010–172021. [CrossRef]
30. Jin, S.; Li, H.; Chu, K.; Yu, X.; Guan, X.; Pu, X.; Gu, X.; Liu, X. High room-temperature TCR of $La_{0.7}(K_{0.25}Sr_{0.05})$ MnO_3:xAg_2O composites obtained at optimized Ag_2O ratio. *J. Alloys Compd.* **2021**, *873*, 159762–159773. [CrossRef]
31. Krawczyk, P.A.; Jurczyszyn, M.; Pawlak, J.; Salamon, W.; Baran, P.; Kmita, A.; Gondek, Ł.; Sikora, M.; Kapusta, C.; Strączek, T.; et al. High-entropy perovskites as multifunctional metal oxide semiconductors: Synthesis and characterization of $(Gd_{0.2}Nd_{0.2}La_{0.2}Sm_{0.2}Y_{0.2})CoO_3$. *ACS Appl. Electron. Mater.* **2020**, *2*, 3211–3220. [CrossRef]

32. Qin, J.; Wen, Z.; Ma, B.; Wu, Z.; Yu, J.; Tang, L.; Lu, T.; Zhao, Y. Effect of A-site rare-earth ions on structure and magnetic properties of novel $(Ln_{0.2}Gd_{0.2}La_{0.2}Nd_{0.2}Sm_{0.2})MnO_3$ (Ln = Eu, Ho, Yb) high-entropy perovskite ceramics. *Ceram. Int.* **2024**, *50*, 26040–26048. [CrossRef]
33. Ezaami, A.; Chaaba, I.; Cheikhrouhou-Koubaa, W.; Cheikhrouhou, A.; Hlil, E.K. Enhancement of magnetocaloric properties around room temperature in $(1 - x)$ $La_{0.7}Ca_{0.25}Sr_{0.05}MnO_3/xLa_{0.7}Ca_{0.2}Sr_{0.1}MnO_3$ $(0 \leq x \leq 1)$. *J. Alloys Compd.* **2018**, *735*, 2331–2335. [CrossRef]
34. Banerjee, B.K. On a generalised approach to first and second order magnetic transitions. *Phys. Lett.* **1964**, *12*, 16–17. [CrossRef]
35. Koroleva, L.I.; Zashchirinskiĭ, D.M.; Khapaeva, T.M.; Gurskiĭ, L.I.; Kalanda, N.A.; Trukhan, V.M.; Szymczak, R.; Krzumanska, B. Effect of oxygen deficiency on the magnetic, electrical, magnetoelectric, and magnetoelastic properties of $La_{1-x}Sr_xMnO_{3-\delta}$ manganites. *Phys. Solid State* **2008**, *50*, 2298–2302. [CrossRef]
36. Halder, P.; Mukadam, M.D.; Yusuf, S.M. Direct measurement of the magnetocaloric effect in Gd based alloy. *AIP Conf. Proc.* **2020**, *2265*, 18–22. [CrossRef]
37. Bolarín, A.M.; Pedro, F.; Rosales, O.; Sánchez, F. Synergistic magnetocaloric effect in composite materials based on mixed doped lanthanum manganites. *J. Supercond. Nov. Magn.* **2024**, *37*, 469–472. [CrossRef]
38. Anwar, M.S.; Ahmed, F.; Danish, R.; Koo, B.H. Impact of Co_3O_4 phase on the magnetocaloric effect and magnetoresistance in $La_{0.7}Sr_{0.3}MnO_3/Co_3O_4$ and $La_{0.7}Ca_{0.3}MnO_3/Co_3O_4$ ceramic composites. *Ceram. Int.* **2014**, *40*, 15471–15476. [CrossRef]
39. Jeddi, M.; Gharsallah, H.; Bekri, M.; Dhahri, E.; Hlil, E.K. Structural, magnetic and magnetocaloric properties of $0.75La_{0.6}Ca_{0.4}MnO_3/0.25La_{0.6}Sr_{0.4}MnO_3$ nanocomposite manganite. *RSC Adv.* **2018**, *8*, 28649–28656. [CrossRef]
40. Das, K.; Roy Chowdhury, R.; Midda, S.; Sen, P.; Das, I. Magnetocaloric effect study of $Pr_{0.67}Ca_{0.33}MnO_3 - La_{0.67}Sr_{0.33}MnO_3$ nanocomposite. *J. Magn. Magn. Mater.* **2018**, *449*, 304–307. [CrossRef]
41. Wang, G.F.; Zhao, Z.R.; Li, H.L.; Zhang, X.F. Enhancement of refrigeration capacity and table-like magnetocaloric effect in $La_{0.8}Ca_{0.2}MnO_3/$ $La_{0.8}K_{0.2}MnO_3$ nanocrystalline composite. *Ceram. Int.* **2015**, *41*, 10061–10067. [CrossRef]
42. Toby, B.H. R factors in Rietveld analysis: How good is good enough? *Powder Diffr.* **2006**, *21*, 67–70. [CrossRef]
43. Causa, M.T.; Moreno, N.O.; Tovar, M.; Levy, P. High-temperature spin dynamics in CMR manganites: ESR and magnetization. *Phys. Rev. B* **1998**, *58*, 3233–3238. [CrossRef]
44. Autret, C.; Gervais, M.; Gervais, F.; Raimboux, N.; Simon, P. Signature of ferromagnetism, antiferromagnetism, charge ordering and phase separation by electron paramagnetic resonance study in rare earth manganites, $Ln_{1-x}A_xMnO_3$ (Ln = rare earth, A = Ca, Sr). *Solid State Sci.* **2024**, *6*, 815–824. [CrossRef]
45. Siqueira, G.O.; Belardi, R.M.; Almeida, P.H.; da Silva, C.L.; Brant, M.C.; Matencio, T.; Domingues, R.Z. Determination of the Mn^{3+}/Mn^{4+} ratio in $La_{1-x}Sr_xMnO_3 \pm d$ powders. *J. Alloys Compd.* **2012**, *521*, 50–54. [CrossRef]
46. Sakka, A.; M'nassri, R.; Nofal, M.M.; Mahjoub, S.; Cheikhrouhou-Koubaa, W.; Chniba-Boudjada, N.; Oumezzine, M.; Cheikhrouhou, A. Structure, magnetic and field dependence of magnetocaloric properties of $Pr_{0.5}RE_{0.1}Sr_{0.4}MnO_3$ (RE = Eu and Er). *J. Magn. Magn. Mater.* **2020**, *514*, 167158–167172. [CrossRef]
47. Mahjoub, S.; M'nassri, R.; Baazaoui, M.; Hlil, E.K.; Oumezzine, M. Tuning magnetic and magnetocaloric properties around room temperature via chromium substitution in $La_{0.65}Nd_{0.05}Ba_{0.3}MnO_3$ system. *J. Magn. Magn. Mater.* **2019**, *481*, 29–38. [CrossRef]
48. Griffith, L.D.; Mudryk, Y.; Slaughter, J.; Pecharsky, V.K. Material-based figure of merit for caloric materials. *J. Appl. Phys.* **2018**, *123*, 034902–034911. [CrossRef]
49. Ho, T.A.; Dang, N.T.; Phan, T.L.; Yang, D.S.; Lee, B.W.; Yu, S.C. Magnetic and magnetocaloric properties in $La_{0.7}Ca_{0.3-x}Na_xMnO_3$ exhibiting first-order and second-order magnetic phase transitions. *J. Alloys Compd.* **2016**, *676*, 305–312. [CrossRef]
50. Diana, I.P.; Vera, D.S.; Nataliya, I.C.; Vyacheslav, S.R.; Yulia, A.A.; Alexey, N.T.; Bernard, F.; Lorenzo, S.; Moulay, T.S. The structural and magnetic features of perovskite oxides $La_{1-x}Sr_xMnO_{3+\delta}$ (x = 0.05, 0.10, 0.20) depending on the strontium doping content and heat treatment. *Ceram. Int.* **2023**, *49*, 10774–10786.
51. Diana, P.; Vera, S.; Oxana, R.; Bernard, F.; Moulay, T.S. In situ high temperature XRD study of Sr-doped ceramics $La_{0.95}Sr_{0.05}MnO_{3+\delta}$. *Solid State Commun.* **2021**, *336*, 114401–114405.
52. Ferreira, M.C.; Pimentel, B.; Andrade, V.; Zverev, V.; Gimaev, R.R.; Pomorov, A.S.; Pyatakov, A.; Alekhina, Y.; Komlev, A.; Makarova, L. Understanding the Dependence of Nanoparticles Magnetothermal Properties on Their Size for Hyperthermia Applications: A Case Study for La-Sr Manganites. *Nanomaterials* **2021**, *11*, 1826. [CrossRef]

materials

MDPI

Article

Nature of the Dielectric Anomaly in $Na_{0.5}Bi_{0.5}TiO_3$–Based Ferrolectrics

Eriks Birks [1], Marija Dunce [1], Šarūnas Svirskas [2], Algimantas Kežionis [2], Juras Banys [2] and Andrei Kholkin [1,*]

1 Institute of Solid State Physics, University of Latvia, Kengaraga 8, LV-1063 Riga, Latvia; eriks.birks@cfi.lu.lv (E.B.); marija.dunce@cfi.lu.lv (M.D.)
2 Faculty of Physics, Vilnius University, Saulėtekio al. 3, LT-10257 Vilnius, Lithuania; sarunas.svirkas@ff.vu.lt (Š.S.); algimantas.kezionis@ff.vu.lt (A.K.); juras.banys@ff.vu.lt (J.B.)
* Correspondence: andrei.kholkin@cfi.lu.lv

Abstract: The temperature–frequency dependence of dielectric permittivity in $Na_{0.5}Bi_{0.5}TiO_3$ (NBT) -based compositions displays a diffused, frequency-independent maximum along with a frequency-dependent shoulder below this maximum. This behavior deviates from that of both classical ferroelectrics and conventional relaxor ferroelectrics, and its interpretation is further complicated by challenges in linking it to known structural phase transitions. This study proposes a new interpretation of the dielectric behavior of NBT-based materials through a comparative analysis of temperature–frequency permittivity data in both unpoled and poled NBT samples and $0.95Na_{0.5}Bi_{0.5}TiO_3$–$0.05CaTiO_3$ solid solution over a broad frequency range (10 Hz–100 MHz). Results reveal that the steep permittivity change between the maximum and shoulder—accompanied by pronounced thermal hysteresis—can be attributed to a phase transition between two non-ferroelectric phases. When this contribution is excluded, the dielectric response aligns with classical relaxor ferroelectric behavior. To reconcile this with other known properties of NBT, the "breathing" model is employed, offering a unified framework for understanding its relaxor-like characteristics.

Keywords: ferroelectric relaxors; phase transition; dielectric anomaly

1. Introduction

At the moment, lead-based perovskite ferroelectrics dominate this market segment, but their use is limited due to health and environmental issues [1]. NBT-based compositions are among the most promising lead-free substitutes here. Structure and phase transitions of these materials are among the aspects being in focus of the research. While the poled state of NBT at room temperature is undoubtedly ferroelectric, with R3c symmetry [2], the unpoled state does not show presence of domains or any other signs of ferroelectric state [3]. In this respect, the unpoled state and the poled state have to be considered separately. Symmetry in the unpoled state is an issue that has been debated for a long time. Seemingly, X-ray and neutron diffraction patterns satisfactorily conform to the R3c symmetry [4–6]. However, the observed complex shape of X-ray diffraction peaks [7–11], as well as simultaneous presence of superlattice reflections belonging to different symmetries [11,12] indicate a coexistence of the R3c symmetry phase with some other phase. As a result, coexistence of phases of R3c and Cc [9], R3c and Pm3-m [8], R3c and 4Pbm [13,14], as well as R3c and Pnma [10] symmetries was suggested. Unfortunately, distortions of the NBT pseudo-cubic unit cell are small, and the comparison of various coexisting phases yields similar quantitative results [10,15]. More sophisticated approaches [16–18], paying attention to differences between the local and global symmetry in the presence of specific mesoscopic structures, extend the range of possible interpretations.

In the temperature region above 300 °C, interpretation of NBT crystalline structure is more convincing: a transition to a phase with weak tetragonal P4bm symmetry takes place through the phase coexistence region at 300–320 °C (this region could be even wider) [19,20]. In all reported studies, a coexistence region with phase of R3c symmetry was found. Further, at 500–540 °C, this phase transforms into a cubic phase of Pm3-m symmetry [4].

Characterization of crystalline structures below the phase transition to the tetragonal phase is problematic. Abandoning the idea of pure R3c phase at room temperature has motivated researchers to perform more detailed studies of structure upon increasing of temperature in the range where existence of pure R3c phase was assumed earlier—up to the temperature of the phase transition to the tetragonal phase [4]. Thus, when analyzing neutron elastic scattering measurements, short-range rhombohedral (R3c) and tetragonal (P4bm) correlations were detected in the temperature range 200–300 °C [21]. While, according to TEM studies, at around 200 °C, orthorhombic sheets of Pnma symmetry appear, separating areas of R3c symmetry [16]. In such modulated mesoscopic structures, dipole moments in neighboring R3c areas should be oriented in opposite directions, creating an association with the antiferroelectric state. Upon increasing temperature, concentration of the sheets increases until, at 280 °C, a transfer to complete orthorhombic phase takes place. This phase exists in a narrow temperature range, before the phase transition to the tetragonal phase occurs at 320 °C [12]. Unfortunately, such an interpretation has not received better experimental confirmation.

The problems with identification of structure and especially the diffused changes in it dependent on temperature do not allow for a convincing connection with the anomalies observed in physical properties, first of all, dielectric permittivity. In general, the diffused peak of the temperature dependence of dielectric permittivity $\varepsilon'(T)$, which in NBT is located at a temperature around 330–340 °C, usually denoted as T_c, is almost independent of frequency. (It is important to stress that T_c here may not be a ferroelectric phase transition temperature.) However, there are some exemptions. Specifically, in NBT-$PbZrO_3$ solid solutions, a clear relaxor-characteristic dependence of $\varepsilon'(T)$ on frequency is observed in the region of T_c [22]. In addition, contribution from DC conductivity, which is dependent on the composition and increases with increasing temperature, can also create frequency-dependent contributions in the region of T_c [23].

It should be noted that the dielectric permittivity peak located at T_c is strongly asymmetric—it has a steep part in the low-temperature side and a diffused part in the high-temperature side. In ferroelectrics, a $\varepsilon'(T)$ peak usually indicates a ferroelectric phase transition. Therefore, there were attempts to apply the Curie–Weiss law, which is valid above T_c for traditional ferroelectrics, and extract its parameters also in NBT [21,24,25]. However, such an approach does not seem to be plausible in this case, as it is substantiated only for a ferroelectric–paraelectric or antiferroelectric–paraelectric phase transition. In NBT, the ferroelectric state exists only after poling and, even then, it is destroyed upon heating at the depolarization temperature T_d = 180 °C, significantly below T_c. Moreover, the Curie–Weiss law in NBT is always applied in a temperature range starting from a temperature significantly above T_c—which is not well grounded, even conceptually.

Sometimes, NBT is assumed to be in an antiferroelectric state within the temperature range between the region of dielectric permittivity shoulder and T_c [26–28]. Nevertheless, there is no evidence in experimental studies of the symmetry of the crystalline structure confirming this hypothesis [4,12,16,21], whereas the features of double polarization hysteresis loops sometimes observed in NBT [23,29] are evidence of any kind of the first-order phase transition, including a relaxor–ferroelectric one, and do not necessarily indicate the presence of an antiferroelectric phase. Therefore, there is not sufficient motivation to insist on the relation with the antiferroelectric phase, and the peak at T_c could be caused by some other reason.

In the above-considered studies, the dielectric permittivity peak at T_c in NBT-based compositions was attributed to a phase transition. In contrast, Ge et al. have found that the temperature at which tetragonal splitting appears in XRD patterns upon increasing of temperature coincides with the position of the low-temperature steep part—jump/drop—of this peak in NBT [21]. Based on the detected anomalies of elastic properties, Hiruma et al. [30] have also assumed the existence of a rhombohedral–tetragonal phase transition in the region of this jump/drop. It follows from their report, but they did not pay attention to this correlation.

Diaz et al. have jointly analyzed temperature dependence of dielectric permittivity, Young's modulus, and thermally stimulated depolarization currents in NBT ceramics [31]. The authors have managed to describe temperature dependences of the studied properties, including dielectric permittivity at one particular frequency (1 MHz) in the temperature region starting slightly below the T_c and extending to the high-temperature range as the result of accumulation of free charges at grain boundaries due to variations in conductivity between grains and grain boundaries. Certainly, increased conductivity at elevated temperatures plays an essential role in electrical properties [32]. However, the offered model is based purely on an RC equivalent circuit. This means that, in the framework of this model, dielectric properties are of Debye type, which does not correspond to the experimentally observed nature of the dielectric permittivity peak. It is also not quite clear how to adopt this model for single crystals, where dielectric properties are similar [33,34].

A common feature distinguishing NBT-based compositions from other ferroelectrics is the existence of a so-called shoulder in dielectric permittivity temperature dependence below the diffused peak at T_c. Contrary to the peak, which is frequency-independent, this shoulder has a well-expressed frequency dependence, resembling the behavior of relaxor ferroelectrics. In spite of the clearly observable frequency dependence, this shoulder is sometimes associated with depolarization. The latter stems from the fact that the location of the shoulder coincides with the temperature range where the poled state is destroyed upon heating. Due to this, the dielectric permittivity shoulder was used to identify the depolarization temperature T_d in NBT-based compositions instead of determining it from heating experiments of poled samples [31,35–37]. However, such a coincidence, if it is not substantiated by other evidence of a phase transition (the term "depolarization" in the unpoled state is not applicable), appears to be accidental. First of all, the jump in dielectric permittivity dependence observed in poled NBT-based ferroelectrics, in fact, is related to the destruction of a ferroelectric domain state. Therefore, there is no reason to address the shoulder appearing upon heating in the unpoled NBT-based samples in the same manner as this jump if there is no evidence of ferroelectric domains. Such is the case of NBT [3]. In addition, there are NBT-based compositions in which destruction of the poled state, depolarization, occurs at temperatures clearly below the temperature range of the shoulder [38,39].

The fact that dispersion of dielectric permittivity in the temperature range of the shoulder resembles relaxor ferroelectrics, as well as the above mentioned hypothesis that antiferroelectric state is present in the temperature range between the shoulder and the peak, has inspired the idea of antiferroelectric nanoregions embedded in nonpolar matrix in $(1-x)$NBT-x$BaTiO_3$ solid solutions in the concentration range of the morphotropic phase boundary [40,41]. These nanoregions were assumed to be responsible for the observed frequency-dependent dielectric permittivity shoulder. Meanwhile, after detecting the simultaneous presence of superlattice reflections attributed to phases of R3c and P4bm symmetry by TEM in 0.95NBT-0.06$BaTiO_3$, Jo et al. have assumed the existence of two kinds of polar nanoregions (PNRs) with different crystalline symmetry—rhombohedral and tetragonal [42]. The shoulder mainly contains contributions from relaxation of the rhombohedral PNRs, whereas the peak of dielectric permittivity at T_c is related to contribution from a specific combination of two processes: transformation of the rhombohedral PNRs to the tetragonal ones and their relaxation. Taking into account later comprehensive studies of the crystalline structure of NBT [8–10,13,14], which have confirmed the presence of a polar R3c phase among several coexisting phases, at least a part of this concept—related to the PNRs of rhombohedral symmetry—appears to be well substantiated.

Recently, Fan et al. have focused their studies on the dielectric permittivity shoulder after different treatments (rapid cooling, cooling at applied electric field, and mechanical stress) [43]. They have found that the applied treatment has an impact on the shoulder, leaving the part of dielectric permittivity dependence below it practically without changes. The authors have explained the observed features using a so-called "breathing" model, where movement of PNR domain walls pinned by defects is considered to be the reason of the dielectric dispersion, instead of changes in the polarization direction of PNRs [44].

Despite the fact that the frequency dependence of dielectric properties of NBT-based compositions is well documented, only a few studies were focused on deeper analysis of the dielectric dispersion. Besides the above-mentioned work of Jo et al. [42], we would like to pay attention to the research by Petzelt et al. [45]. In this work, the authors have extracted several relaxation mechanisms, extending below and above T_c, from measurements of temperature dependence of the high-frequency dielectric dispersion in NBT. Remarkably, the mechanism located in the range of the longest relaxation times had quite a smooth relaxation time temperature dependence with values ranging from 10–11 at 900 K down to 10^{-8} at 100 K, derived using the Cole–Cole formula on the dielectric spectra. This dependence does not reveal any signs of freezing. However, dielectric strength has a well-expressed peak at T_c with a shape resembling the dielectric permittivity peak. Similar results were obtained when studying the frequency dispersion in 0.7NBT-0.3$Sr_{0.2}Bi_{0.7}TiO_3$ solid solution, which also possesses the temperature dependence of dielectric permittivity characteristic of NBT [46]. Again, the application of the Cole–Cole formula to the dielectric dispersion did not reveal any anomalies in temperature dependence of the extracted relaxation time. At the same time, temperature dependence of dielectric permittivity around T_c, including even the minimum observed between the peak at T_c and the shoulder, was concluded to be determined solely by static dielectric permittivity. It should be noted that dielectric dispersion in 0.7NBT-0.3$Sr_{0.2}Bi_{0.7}TiO_3$ at the considered temperatures is apparently limited to the frequency range from the GHz region and to the highest measurement frequencies [34,46].

It is important to note that the steep part of $\varepsilon'(T)$ peak has revealed the thermal hysteresis. Recently, based on comparison of NBT with 0.95NBT-0.05$CaTiO_3$, where the steep change in $\varepsilon'(T)$ even resembles a jump and obviously corresponds to the phase transition between orthorhombic and tetragonal phases, we have made an assumption that such an interpretation can be extended also to NBT [10]. Namely, the steep change in $\varepsilon'(T)$, possessing thermal hysteresis, corresponds to the first-order phase transition between the low-temperature state of unclear symmetry or rather coexisting phases and the high-temperature tetragonal phase. This conclusion corresponds well to the results reported by Ge et al. [21]. In comparison with 0.95NBT-0.05$CaTiO_3$, the phase transition in NBT is more diffused. We have compared anomalies of several physical properties at this phase transition for both compositions [47]. One of our significant observations was that temperature dependence of Young's modulus, which is acknowledged as a sensitive tool for detecting phase transitions, reveals a clear anomaly close to the phase transition to the tetragonal phase with thermal hysteresis, corresponding to the hysteresis of dielectric permittivity. In the present study, dielectric dispersion is compared for the unpoled and poled NBT and 0.95NBT-0.05$CaTiO_3$ ceramics. A new interpretation is suggested for temperature dependence of dielectric permittivity in NBT-based solid solutions in relation with the non-ferroelectric phase transition.

2. Materials and Methods

$Na_{0.5}Bi_{0.5}TiO_3$ (NBT) and 0.95$Na_{0.5}Bi_{0.5}TiO_3$–0.05$CaTiO_3$ (0.95NBT–0.05CT) ceramics were synthesized by the conventional solid-state reaction method with a two-stage calcination, as described in detail in our previous work [10]. Chemical-grade Na_2CO_3, Bi_2O_3, $CaCO_3$, and TiO_2 powders (purity > 99.5%) were used as starting materials. The mixed and milled powders were calcined first at 850 °C and then at 1000 °C, followed by sintering at 1160 °C for NBT and at 1180 °C for 0.95NBT-0.05CT for 2 h. To minimize Bi evaporation, calcination and sintering were carried out in covered platinum crucibles with the samples embedded in powders of the same composition.

The sintered ceramics were cut and polished to a thickness of approximately 0.3 mm, and the plates were electroded by applying a silver paste and firing it at 400 °C for further examination. In the present work, dielectric permittivity measurements were performed for the unpoled and poled NBT and 0.95NBT-0.05CT ceramics. Poling of both ceramics was performed by applying an electric field of 70 kV/cm at room temperature.

The broadband dielectric spectroscopy measurements were performed using a custom-made ultrabroadband spectrometer. In this spectrometer, the sample is placed in a coaxial line constructed from platinum-coated ceramic tubes. This design allows heating of the coaxial line to high temperatures (up to 800 °C or higher). Heating is achieved by applying a DC current to a wire wrapped around the outer shell of the coaxial line. This part of the sample compartment is well isolated from the environment. A K-type thermocouple, positioned inside the inner tube of the coaxial line, measures the temperature at the point where the sample surface contacts the inner tube. Measurements in this study were taken in 2 K increments, with the sample temperature stabilized at each step.

Coverage of the 10 Hz–100 MHz frequency range was achieved by combining two experimental methods. An electrical relay was used to switch between the two setups, enabling simultaneous measurement of the entire spectrum.

The low-frequency method (10 Hz–10 MHz) employed a TiePie Handyscope HS3 oscilloscope (TiePie Engineering, Sneek, The Netherlands) with a built-in signal generator. It generates sinusoidal waves and measures the current flowing through the sample. A detailed schematic of this method is provided in the literature [48].

The higher frequencies (10–100 MHz) were covered by using vector network analyzer Agilent E5071C (Keysight Technologies, Santa Clara, CA, USA). In this setup, the complex reflection and transmission coefficient are measured. The full description of the experimental technique can be found in the literature [49].

X-ray diffraction (XRD) patterns and scanning electron microscopy (SEM) micrographs were also obtained to confirm that the structural and microstructural properties of the samples are consistent with our previous studies, indicating a single-phase composition and a well-developed microstructure. Detailed analyses of the structure, microstructure, and other physical properties can be found in our earlier publications [10,47].

3. Results

In general, temperature–frequency dependences of the real part of dielectric permittivity ε′(T, f) for both NBT and 0.95NBT-0.05CT in the unpoled state resemble the behavior characteristic of the whole family of NBT-based compositions, as depicted in Figure 1. They possess a frequency-dependent shoulder at lower temperatures and a frequency-independent anomaly at higher temperatures. In the case of NBT, the frequency-independent anomaly is represented by a diffused peak with a maximum at a temperature T_c and a steep change in its low-temperature part (Figure 1a), which corresponds well to the literature. Further, we will denote this steep change as a diffused jump/drop upon increasing and decreasing of temperature, respectively. Whereas, in the case of 0.95NBT-0.05CT, it seems that solely the steep change—the jump/drop—is observed, without the maximum (Figure 1b). However, in our previous study of (1−x)NBT-xCT solid solutions, where the measurements have covered a wider temperature range, we have observed a diffused hump in ε′(T) dependence around 450 °C, which earlier we have interpreted as surplus of this diffused part with a maximum at T_c [10] (see the additional graph in Figure 1b). (It should be noted that it was not possible to achieve high temperatures, reaching 450 °C, in the current study, due to technical limitations of the wide-frequency range equipment. Therefore, our previous work is mentioned in this context.)

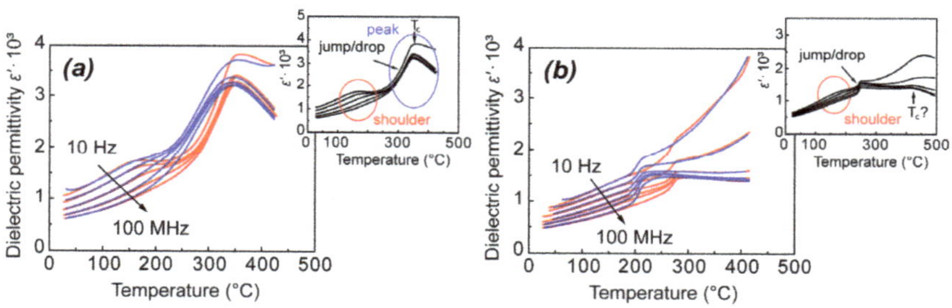

Figure 1. Temperature dependences of the real part of dielectric permittivity ε′(T) for the unpoled NBT (**a**) and 0.95NBT-0.05CT (**b**) samples at several frequencies upon heating (red lines) and followed cooling (blue lines), grasping the most characteristic features of NBT-based solid solutions. The additional graphs represent a schematic view of these dependences denoting the characteristic anomalies. The additional graph for 0.95NBT-0.05CT is presented from our previous study [10], covering a wider measurement temperature range.

For both NBT and 0.95NBT-0.05CT, the characteristic frequency-independent jump on heating and drop on cooling mark the region of pronounced temperature hysteresis, which is slightly wider than the jump and drop themselves. Particularly, in 0.95NBT-0.05CT, this hysteresis appears between 270 °C and 203 °C, while in NBT, it is observed just in the temperature range between 320 °C and 230 °C. Further, we focus mostly on this hysteresis region, giving a significant characterization to the studied samples. While the diffused part of the peak might have just a subordinate meaning, as it will be discussed below.

No thermal hysteresis was observed for the ε′(T) shoulder in the case of the unpoled samples. However, unambiguous differences in its behavior were observed upon heating of the poled samples (Figure 2). In this case, a well-expressed frequency-independent sharp increase in ε′(T) was observed at the depolarization temperature T_d—141 °C for NBT and 103 °C for 0.95NBT-0.05CT. Taking into account that T_d is located in the low-temperature side of the shoulder, poling did not destroy the part of the shoulder located above T_d.

On the contrary, the shoulder after poling became much more expressed for both studied compositions, especially for 0.95NBT-0.05CT.

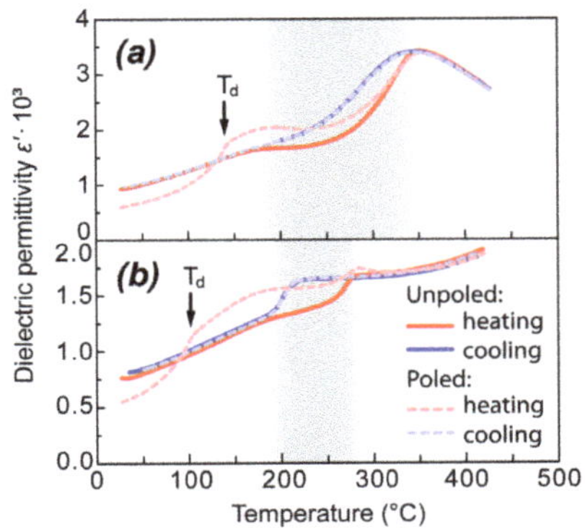

Figure 2. Temperature dependences of the real part of dielectric permittivity ε′(T) for the unpoled and poled NBT (**a**) and 0.95NBT-0.05CT (**b**) samples upon heating and cooling at a particular frequency of 860 Hz. The temperature hysteresis range is shaded grey. Rather low frequency is chosen in order to clearly demonstrate the ε′(T) shoulder which is more expressed in direction of low frequencies.

Mutual comparison of temperature dependences of dielectric permittivity ε′(T) on heating and on cooling of the unpoled and poled samples allows us to extract the following features of them for both NBT and 0.95NBT-0.05CT (in Figure 2, this comparison is illustrated at one chosen frequency—860 Hz):

1. In the case of the unpoled samples:
   ε′(T) has temperature hysteresis only in the temperature range marked by the jump and drop on heating and on cooling, respectively; outside this range, ε′(T) does not depend on the direction of temperature change and the heating and cooling curves coincide;
2. When the unpoled and poled samples are compared:
   a. The heating curves coincide for the poled and unpoled samples only above the thermal hysteresis region, whereas in the temperature range below it, they differ significantly: (1) From room temperature up to the depolarization temperature T_d, ε′(T) values in the poled sample are lower in comparison with the unpoled sample; (2) after the sharp increase at T_d up to the thermal hysteresis region, ε′(T) values in the poled sample are higher than in the unpoled sample;
   b. The cooling curves for the poled and unpoled samples coincide completely. This confirms that heating above the temperature region of thermal hysteresis erases influence of poling on dielectric permittivity.

The real part of dielectric permittivity as a function of frequency logarithm ε′(log(f)) also has similar behavior for both NBT and 0.95NBT-0.05CT and possesses features characteristic for relaxor ferroelectrics. This can be well seen from the examples of ε′(log(f)) for the unpoled NBT, depicted in Figure 3a at several temperatures, which represent three temperature ranges with different characteristic frequency dependences:

1. Range I (below the frequency-dependent shoulder)—with linear ε′(log(f)) in the whole measurement frequency range;

2. Range II (in the temperature region of the frequency-dependent shoulder)—with linear $\varepsilon'(\log(f))$ at high frequencies and diffused non-linear transition to the practically frequency-independent $\varepsilon'(\log(f))$ at low frequencies;
3. Range III (above the frequency-dependent shoulder)—practically frequency-independent $\varepsilon'(\log(f))$ in the whole frequency range, supplemented by a contribution from the high-temperature conductivity, giving significant rise in ε' at low frequencies.

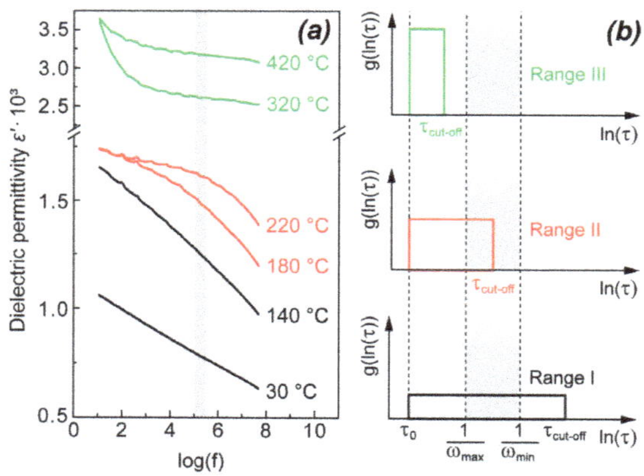

Figure 3. (**a**) Dielectric permittivity as a function of frequency logarithm $\varepsilon'(\log(f))$ at several temperatures, representing three characteristic temperature ranges for unpoled NBT as examples: Range I (30 and 140 °C), where $\varepsilon'(\log(f))$ is linear in the whole frequency range; Range II (180 and 220 °C), where a diffused transition between linear and practically frequency-independent $\varepsilon'(\log(f))$ occurs; Range III (320 and 420 °C), where $\varepsilon'(\log(f))$ is practically frequency-independent in the whole frequency range, but it has a significant contribution from the high-temperature conductivity at low frequencies. (**b**) Distribution functions of relaxation times $g(\ln\tau)$ in three temperature ranges directly corresponding to those mentioned in context of the $\varepsilon'(\log(f))$ measurements: High-temperature range (Range III) where the whole distribution function is outside the measurement frequency range; middle-temperature range (Range II) where the inverse maximal relaxation time $1/\tau_{\text{cut-off}}$ is within the measured frequency range; and low-temperature range (Range I) where $1/\tau_{\text{cut-off}}$ is below the measured frequency range. $1/\omega_{\max}$ and $1/\omega_{\min}$ mark the measurement frequency range on the $\ln\tau$ scale.

Temperature dependences of the absolute values of slope of $\varepsilon'(\log(f))$ (further denoted simply as "slope of $\varepsilon'(\log(f))$" for brevity) are determined for a particular frequency range of 150–370 kHz upon heating and cooling of the unpoled and poled NBT and 0.95NBT-0.05CT and are shown in Figure 4. This frequency "window" is shaded grey in Figure 3a. The choice of the frequency range here was performed in such a way to avoid the electromechanical resonance range, observed for the poled samples in MHz region, as well as to reduce the influence of conductivity, which gives a significant contribution to dielectric permittivity in the high-temperature range at low frequencies.

Upon heating of the unpoled samples, the $\varepsilon'(\log(f))$ slope increases, reaching a maximum at a temperature Ta. It should be noted that Ta depends on the frequency range where the slope is determined. In the chosen frequency "window", Ta for NBT is 160 °C (Figure 4a), while, for 0.95NBT-0.05CT, it is 130 °C (Figure 4b). At lower temperatures (Range I, according to the notation above), the increasing of the $\varepsilon'(\log(f))$ slope reflects the fact that the linear dependence of $\varepsilon'(\log(f))$ becomes steeper. In the case of NBT, it can be well seen for the temperatures 30 °C and 140 °C, shown as examples, in Figure 3. As it follows from Figure 3a, the practically frequency-independent part of ε′(log(f)) appears from the low-frequency side upon increasing temperature. We start to observe the decreasing of the ε′(log(f)) slope in Figure 4, marking Ta, as soon as the diffused transition to the region of the frequency-independent part of ε′(log(f)) approaches the considered frequency "window", starting to give contribution to the calculated slope. It can be seen also qualitatively for 180 °C and 220 °C, shown as examples, in Figure 3a. In the case of NBT, this continues until the practically frequency-independent part of ε′(log(f)) completely determines ε′ in the whole considered frequency "window". At this point, the determined slope value becomes equal to 0 and does not change upon further increasing temperature (Figure 4a). The transition between the linear part and the frequency-independent part of ε′(log(f)) is very diffused, as it can be seen qualitatively in Figure 3a for the temperatures 180 °C and 220 °C and as it is reflected in Figure 4a by the wide temperature range between Ta and the temperature where ε′(log(f)) becomes equal to 0. In the case of NBT, this transition extends in approximately a 100 °C range.

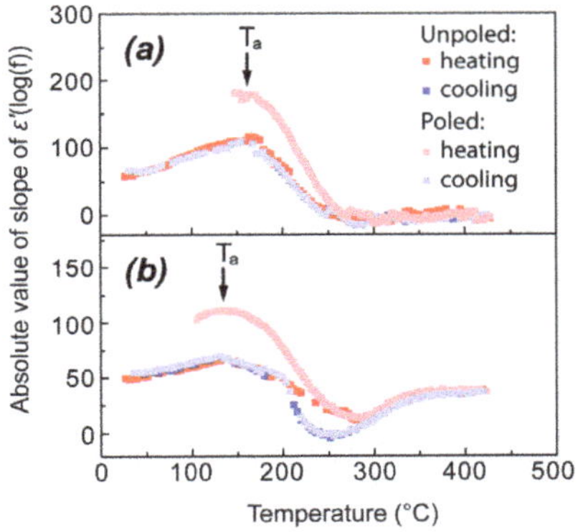

Figure 4. Temperature dependences of the absolute value of slope of ε′(log(f)), calculated for the frequency range 150–370 kHz, for the unpoled and poled NBT (**a**) and 0.95NBT-0.05CT (**b**) samples upon heating and cooling. The particular frequency range is chosen in such a way to avoid the contribution from the low-frequency dispersion, appearing at high temperatures, as much as possible, on the one hand, and the electromechanical resonance range, observed for the poled samples, on the other hand.

It should be also noted that, at high temperatures, dielectric permittivity experiences increase in the low-frequency region, which can be seen, for example, for the curve at 420 °C in Figure 3a. This low-frequency dispersion is attributed to the characteristic growth of conductivity at high temperatures, which starts to give contributions to ε′(log(f)). The range of this dispersion is outside the frequency "window" where we have quantitatively determined the slope. Therefore, its traces are not visible in Figure 4a.

In the case of unpoled 0.95NBT-0.05CT, the temperature range with almost frequency-independent ε′(log(f)), and thus ε′(log(f)) slope equal to 0, could not be distinguished. On the contrary, upon further increasing of temperature, above the range of continuous decreasing of ε′(log(f)) slope, the value of the slope increases again (Figure 4b). This behavior is obviously related to the significantly more expressed growth of conductivity at high temperatures compared to NBT, giving rise to the low-frequency dispersion and its contribution to the total ε′(log(f)) dependence, including the considered frequency "window".

For the unpoled samples, the temperature dependences of the ε′(log(f)) slope in the cooling runs are very similar to the heating runs in the whole temperature range outside the thermal hysteresis region, where clear differences are observed in the case of 0.95NBT-0.05CT (Figure 4b). In the case of NBT, the thermal hysteresis of $\varepsilon'(T)$ is not so expressed. This could be the reason why the heating and cooling curves of the $\varepsilon'(\log(f))$ slope seem to coincide practically in all considered temperature ranges (Figure 4a).

Taking into account that, when poled, both NBT and 0.95NBT-0.05CT are in ferroelectric state, it is reasonable to consider the $\varepsilon'(\log(f))$ slope of the poled samples only above T_d—in the context of relaxor behavior. However, also in a certain temperature region above T_d, the slope of $\varepsilon'(\log(f))$, determined in the 150–370 kHz frequency "window", significantly differs from the slope of the unpoled samples (Figure 4a,b). The observed differences could have been expected, as dielectric permittivity in the part of the shoulder remaining above T_d is significantly higher in comparison with the unpoled state.

Cooling curves of the $\varepsilon'(\log(f))$ slope both for the poled NBT and 0.95NBT-0.05CT completely coincide with the cooling curves of the unpoled samples (Figure 4a,b). This, once again, confirms that heating above Tm completely erases the signs of poled state.

4. Discussion

The obtained results allow us to discuss behavior of dielectric properties which could be common for a large number of NBT-based compositions. They refer to both dependence on temperature and frequency. Concerning the temperature dependence of the real part of dielectric permittivity $\varepsilon'(T)$ at any particular frequency, for now, we will address two characteristic properties observed for NBT and 0.95NBT-0.05CT:

Thermal hysteresis of $\varepsilon'(T)$ for the unpoled samples of both compositions is limited to the temperature range between the dielectric permittivity jump on heating and drop on cooling and does not influence $\varepsilon'(T)$ outside it. This allows us to consider dielectric permittivity behavior inside and outside this range separately.

As it is indicated by comparison of $\varepsilon'(T)$ for the poled and unpoled samples, especially in the case of 0.95NBT-0.05CT, the $\varepsilon'(T)$ shoulder and depolarization are two separate features on the temperature scale. In the poled state, dielectric permittivity in the region of the shoulder at temperatures above T_d is significantly higher.

Three $\varepsilon'(\ln f)$ dispersion features can be distinguished within the measured temperature–frequency range:

1. Quasilinear $\varepsilon'(\ln f)$ dispersion at low temperatures, observed in the whole frequency range, which transforms into almost frequency-independent ε' above the $\varepsilon'(T)$ shoulder;
2. Dielectric dispersion which appears at high temperatures and is especially expressed in the low-frequency range. This dispersion is obviously related to the increased dc conductivity at high temperatures.
3. In the poled samples, diffused resonance-type dispersion, observed at temperatures below T_d in the range of several MHz, is obviously created by the electromechanical resonance.

In our further discussion, we will focus on the quasilinear $\varepsilon'(\ln f)$ dependence (the first $\varepsilon'(\ln f)$ dispersion feature), which is apparently related to relaxor behavior in the studied compositions. Similar dispersion was observed in NBT single crystals [34] and 0.7NBT-0.3$Sr_{0.7}Bi_{0.2}TiO_3$ ceramics, as well [46]. Like in other relaxor ferroelectrics, such dielectric dispersion can be explained by flipping of dipole moments of PNRs due to thermal agitation.

Introduction of distribution of relaxation times g(lnτ) allows us to express the dielectric dispersion as follows [50]:

$$\varepsilon'(\omega) = \varepsilon_\infty + \Delta\varepsilon \int_{-\infty}^{\infty} \frac{g(\ln\tau)d(\ln\tau)}{1+\omega^2\tau^2} \quad (1)$$

where dielectric strength $\Delta\varepsilon = \varepsilon_s - \varepsilon_\infty$ is the difference between ε' in the low-frequency side (ε_s) and the high-frequency side (ε_∞) of the dielectric dispersion, and ω is angular frequency, equal to $2\pi f$.

Macutkevic et al. have attempted to construct the shape of the arbitrary distribution function g(lnτ) presented in Equation (1) for particular relaxor ferroelectrics using the experimentally obtained data of dielectric dispersion and applying the Tikhonov regularization method [51]. However, the main features of dielectric dispersion in relaxor ferroelectrics can be reflected even assuming a much more simple shape of distribution function—as follows:

$$\begin{gathered} g(\ln(\tau)) = D = \frac{1}{\ln(\tau_{cut-off}/\tau_0)},\ \text{if}\ \tau_0 < \tau < \tau_{cut\text{-}off}, \\ g(\ln(\tau)) = 0,\ \text{if}\ \tau < \tau_0\ \text{or}\ \tau > \tau_{cut\text{-}off} \end{gathered} \quad (2)$$

If these assumptions are made and if, additionally, $1/(1+\omega^2\tau^2)$ is approximated by a step function with value 0, if $\omega\tau > 1$, and with value 1, if $\omega\tau < 1$ (this approximation influences $\varepsilon'(\ln\omega)$ only close to $1/\tau_{cut\text{-}off}$ and $1/\tau_0$, making this dependence smoother), a linear dependence of $\varepsilon'(\ln\omega)$ is obtained within the frequency range $1/\tau_{cut\text{-}off} < \omega < 1/\tau_0$:

$$\begin{gathered} \varepsilon'(\omega) = \varepsilon_\infty + \Delta\varepsilon,\ \text{if}\ \omega < 1/\tau_{cut\text{-}off}, \\ \varepsilon'(\omega) = \varepsilon_\infty + \frac{\Delta\varepsilon}{\ln\tau_{cut-off} - \ln\tau_0}(-\ln\omega - \ln\tau_0),\ \text{if}\ 1/\tau_{cut\text{-}off} < \omega < 1/\tau_0 \\ \varepsilon'(\omega) = \varepsilon_\infty,\ \text{if}\ \omega > 1/\tau. \end{gathered} \quad (3)$$

Remarkably, the linear dependence of $\varepsilon'(\ln\omega)$, together with the frequency-independent imaginary part of dielectric permittivity ε'', observed in relaxor ferroelectrics at temperatures significantly below T_m, naturally follows from this concept if $\tau_{cut\text{-}off} >> 1/\omega_{min}$. Meanwhile, unsurpassable difficulties are met when trying to describe these experimental observations by the frequently used relaxation functions, such as the Cole–Cole and the Havriliak–Negami relations [51].

The simplified dependence in Equation (3) corresponds to the main characteristics of the experimentally measured $\varepsilon'(\ln\omega)$, represented by the examples at several temperatures in Figure 3a. In the whole studied frequency and temperature range, $\omega_{max} < 1/\tau_0$. In the temperature region where $\omega_{max} > 1/\tau_{cut\text{-}off}$ and $\omega_{min} < 1/\tau_{cut\text{-}off}$, $\varepsilon'(\ln\omega)$ consists of two parts with a diffused transfer between them: (1) the frequency-independent part at frequencies $\omega < 1/\tau_{cut\text{-}off}$ and (2) the linearly decreasing part, following Equation (2), at frequencies $\omega > 1/\tau_{cut\text{-}off}$ (the curve at 220 °C and to some extent also the curve at 180 °C, as examples, in Figure 3a). In the temperature region where $\omega_{max} < 1/\tau_{cut\text{-}off}$, dielectric permittivity is frequency-independent in the whole measured frequency range if the contribution from conductivity is ignored (the curves at 320 and 420 °C, as examples, in Figure 3a). Such an approach is supported by the behavior of dielectric permittivity above the temperature range of the shoulder in NBT single crystals, where the frequency dependence of dielectric permittivity vanishes below MHz range [21,34], whereas in the temperature range in which $\omega_{min} >> 1/\tau_{cut\text{-}off}$, the linear dependence of $\varepsilon'(\ln\omega)$ is observed in the entire frequency range (the curves at 30 and 140 °C, as examples, in Figure 3a).

It is widely assumed that, in relaxor ferroelectrics, temperature dependence of the maximal relaxation time $\tau_{cut\text{-}off}(T)$ follows the Vogel–Fulcher law [52] up to complete freezing at a certain temperature T_f, while τ_0 is only weakly dependent on temperature. Just increasing $\tau_{cut\text{-}off}$ in the direction of low temperatures determines the shift in the $\varepsilon'(T)$ maximum towards the low temperatures upon reducing the measurement frequency. In the temperature range above $T_m(\omega_{min})$, dielectric permittivity at lower frequencies is outside the dispersion range and is equal to ε_s according to Equation (3). In relaxor ferroelectrics, ε_s increases at low temperatures. Nevertheless, the substantial widening of the distribution function $g(\ln\tau)$ leads to the reduction in the $\varepsilon'(\ln\omega)$ slope [53], which depends on $\tau_{cut\text{-}off}$ and $\Delta\varepsilon$ according to Equation (3). In the temperature range below $T_m(\omega_{min})$, where the frequency-independent part of dielectric permittivity at low frequencies is no longer observed in the measured frequency range, $\varepsilon_s(T)$ cannot be directly determined. On the other hand, increasing ε_s upon cooling also in the temperature region between T_f and $T_m(\omega_{min})$ seems to be quite natural. Moreover, as per some estimates [53], due to the extremely rapid widening of $g(\ln\tau)$ upon approaching T_f according to the Vogel–Fulcher law, $\varepsilon_s(T)$ must exhibit significant growth towards lower temperatures to make it possible to fit the experimentally observed $\varepsilon'(T, \omega)$ below $T_m(\omega)$ under Equation (3). Otherwise, the slope of $\varepsilon'(T)$ below T_m at low frequencies is much steeper compared to the experimentally observed slope [53]. However, the dielectric permittivity maximum at T_c in NBT-based compositions in most cases does not possess a clear shift with measurement frequency, characteristic of the conventional relaxor ferroelectrics. Therefore, the reduction in $\varepsilon'(T)$ below T_c in the low-temperature direction cannot be explained based on the above considerations. This difference is highlighted by the findings in two reports of dielectric dispersion in NBT-based compositions: NBT single crystals [45] and $0.7NBT\text{-}0.3Sr_{0.7}Bi_{0.2}TiO_3$ ceramics [46], mentioned in Section 1. In both of these studies, it was concluded that dielectric strength $\Delta\varepsilon$ is responsible for the maximum at T_c in the temperature dependence of dielectric permittivity.

In the present research, due to the highly pronounced contribution of conductivity to dielectric permittivity at low frequencies (see the curves at 320 and 420 °C in Figure 3a), direct determination of εs, in the way it was succeeded in $(1-x)NBT\text{-}xSr_{0.7}Bi_{0.2}TiO_3$, is not possible. However, we assume that presence of the maximum in $\varepsilon_s(T)$ is a common feature for all NBT-based compositions possessing the considered temperature dependence of dielectric permittivity: exhibiting the maximum at T_c, which does not shift in dependence on measuring frequency, and the frequency-dependent shoulder below this maximum. In the present study, the absence of dependence on frequency for dielectric permittivity in NBT in a range of hundreds of kHz (Figure 4a), extending from the high-temperature range to some temperature region below T_c, can also be attributed to the remaining part of the frequency-independent feature. Therefore, it can be considered as a signature of reduction in ε_s upon decreasing of temperature below T_c.

Taking into account that the observed behavior of ε_s below T_c cannot be related to the relaxor behavior, we have to consider other reasons explaining it. In 0.95NBT-0.05CT, the temperature of the jump/drop of $\varepsilon'(T, \omega)$ below T_c and its thermal hysteresis coincide with the temperature and thermal hysteresis of the first-order structural phase transition between two nonpolar phases—orthorhombic and tetragonal [10]. Therefore, in this case, treating this jump/drop as a result of a phase transition between phases with different $\Delta\varepsilon$ values seems to be appropriate. As it was pointed out in our earlier publication [10], temperature dependence of $\varepsilon'(T, \omega)$ in NBT is similar to that of 0.95NBT-0.05CT—it also has pronounced thermal hysteresis in a certain temperature range above the shoulder, containing the jump/drop (Figure 1a). In the case of NBT, the link to a structural phase transition is not easy to establish because the phase transition is much more diffused [4,19,20]. However, Ge et al. [21] have observed that the temperature of the phase transition to the tetragonal phase upon heating corresponds to the jump in $\varepsilon'(T, \omega)$. At the same time, the thermal hysteresis between the jump and the drop of $\varepsilon'(T, \omega)$ coincides with that of elastic compliance [47], which indirectly confirms the relation of the jump/drop of $\varepsilon'(T, \omega)$ to the phase transition.

The discussed phase transition changes only ε_s (or $\Delta\varepsilon$) above the frequency-dependent shoulder, leaving the dielectric relaxation untouched. Taking this into account, we can imagine that the peak in $\varepsilon'(T, \omega)$ could be eliminated by reduction in ε_s or $\Delta\varepsilon$, leading to the pure relaxor behavior, not influenced by this phase transition. Such compensation can be achieved at an appropriate value of a coefficient k introduced to reflect the ratio between the reduced value of ε_s and the real one $k = \frac{\varepsilon_{s,\ reduced}}{\varepsilon_{s,real}}$. At the same time, k is also the ratio between the corrected and the measured ε' at any frequency within the relaxation range. This easily follows from Equation (3).

To determine the coefficient k as a function of temperature from our measurements, we have considered temperature dependence of dielectric permittivity at one particular chosen frequency—360 Hz. The low-temperature segment of the $\varepsilon'(T)$, including the shoulder, was left unchanged (the blue segment in Figure 5a,c)—multiplication coefficient k here is equal to 1 (Figure 5b). The high-temperature segment of the $\varepsilon'(T)$ curve starting slightly above T_c (the red segment in Figure 5a) was multiplied by a constant coefficient k, chosen in such a way that this segment is shifted downwards to visually form a smooth diffused peak together with the low-temperature segment (the red segment together with the blue segment in Figure 5c). Further, the missing part of the constructed $\varepsilon'(T)$ curve between the blue and the red segments was interpolated (green dashed segment in Figure 5c). The coefficient k in this range was determined as the ratio between the values on the interpolated curve and the experimental values at 360 Hz. Thus, determined temperature dependence of the coefficient k at 360 Hz is shown in Figure 5b. Further, the determined values of k were used to calculate the transformed dielectric permittivity dependences on temperature (Figure 5c) from the experimentally measured curves (Figure 5a) at all other frequencies by multiplying the experimental data at each particular frequency by the values of k determined at 360 Hz (Figure 5b).

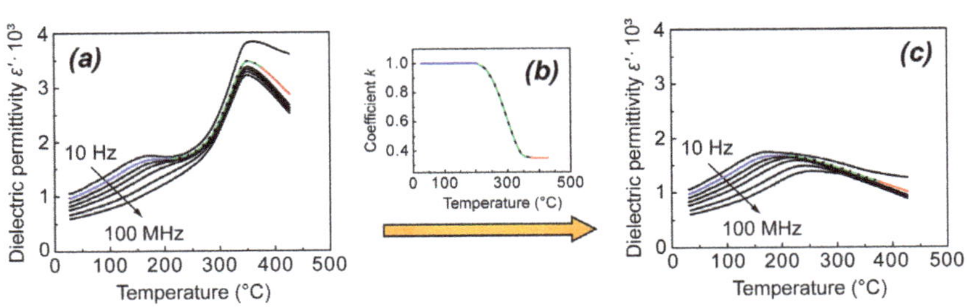

Figure 5. $\varepsilon'(T)$ at different frequencies for the unpoled NBT on heating (**a**), correction to static dielectric permittivity ε_s due to nonferroelectric phase transition, represented by multiplication coefficient k (**b**) and constructed $\varepsilon'(T)$ without influence of phase transition (**c**).

Figure 5c represents the results obtained by this approach for the heating curves of the unpoled NBT. Indeed, in such a way, the experimental curves are transformed into the smooth $\varepsilon'(T, \omega)$ dependences, characteristic of relaxor ferroelectrics, in the whole frequency range. We assume that such dependences describe the dielectric dispersion in NBT in the absence of the phase transition. However, it should be noted that the temperature dependences of dielectric permittivity obtained in this way are suitable for illustrative purposes only—the choice of k is, to some extent, arbitrary, as the only criterion to derive it is obtaining a smooth behavior or $\varepsilon'(T)$ between the temperature range of the frequency-dependent shoulder and the temperature range above the thermal hysteresis region.

Similar results were obtained also after applying the same approach to the heating curves of the poled NBT sample, with the difference that the maximal values of the transformed ε′(T) are significantly higher (see Figure 6). Another difference is the break in ε′(T) below the maximum, which is related to the depolarization. In the framework of the proposed concept, the depolarization temperature can be considered the temperature of the irreversible phase transition from the ferroelectric state to the relaxor state upon heating of a previously poled traditional relaxor ferroelectric, possessing solely frequency-dependent maximum of ε′(T) [50,54,55]. As it is known for relaxor ferroelectrics, the depolarization temperature T_d can be located far below the frequency-dependent ε′(T) maximum, which is known not to be related to any phase transition. In addition to the arguments mentioned in Section 1, the discussed similarities between NBT-based compositions and traditional relaxor ferroelectrics support the hypothesis that the frequency-dependent ε′(T) shoulder in NBT-based compositions does not correspond to the depolarization temperature T_d.

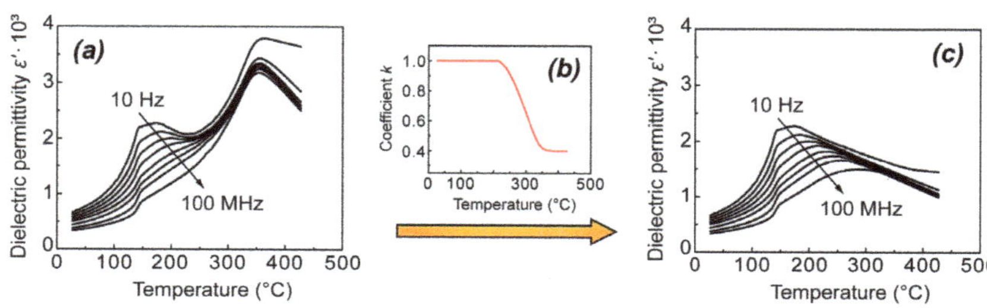

Figure 6. ε′(T) at different frequencies for the poled NBT on heating (**a**), correction to static dielectric permittivity ε_s due to non-ferroelectric phase transition, represented by multiplication coefficient k (**b**) and constructed ε′(T) without influence of phase transition (**c**).

At first glance, similar treatment of dielectric permittivity in 0.95NBT-0.05CT could be even simpler, taking into account that the jump/drop is less diffused and dielectric permittivity only weakly depends on temperature at high frequencies above this jump/drop. However, a much larger contribution of conductivity creates strong frequency dependence of dielectric permittivity in high temperature region. For this reason, the simple approach, used for elimination of jump contribution in NBT, is not applicable in this case. However, from our point of view it does not mean that the presented interpretation of temperature dependence of dielectric permittivity fails in this case.

Considering common properties of relaxor ferroelectrics and the studied NBT and 0.95NBT-0.05CT compositions, we have to stress also a difference between them: in NBT and 0.95NBT-0.05CT, poling leads to appearance of significantly more pronounced relaxor behavior above T_d compared to the unpoled samples, while, in traditional relaxor ferroelectrics, this does not happen. This difference could be related to the fact that, in NBT and 0.95NBT-0.05CT, the concentration of PNRs remaining after destruction of the poled state above T_d is higher than the concentration of PNRs in the unpoled state in the same temperature range. In our previous study, comparison of the X-ray diffraction pattern obtained for the poled and the unpoled 0.95NBT-0.05CT sample in the temperature region above T_d revealed clear differences. However, we were not able to relate these differences to the increased concentration of the rhombohedral phase in the poled samples above T_d [10].

In the framework of the proposed interpretation, the characteristic dielectric permittivity maximum of NBT at T_c could have no independent meaning but is rather a side effect of the diffused jump of Δε at the phase transition between two nonferroelectric phases.

This maximum was interpreted in the same way in studies of NBT-$PbTiO_3$ [56], where it is located clearly above the structural phase transition. We also came to a similar conclusion in studies of NBT-$BaTiO_3$ solid solutions in the tetragonal side from the morphotropic phase boundary [57]. Since the frequency-dependent $\varepsilon'(T, \omega)$ shoulder reflects the relaxor-characteristic behavior, it is not indicative of a phase transition but just reflects temperature dependence of the relaxation time. The values of $\varepsilon'(T, \omega)$ in its vicinity are too high to be related to pure nonferroelectric phases. However, we have to take into account the relaxor nature of the dielectric properties. First of all, it concerns the phase in the temperature region below the jump/drop of $\varepsilon'(T)$. Presence of PNRs in this phase is usually assumed, and the high values of dielectric permittivity could be related to the contribution of these PNRs, like in traditional relaxor ferroelectrics. Apparently, preserving of PNRs should be assumed also after the transfer into the high-temperature phase. At first glance, considering flipping of dipole moments of PNRs, it could be quite natural to relate the jump/drop of $\varepsilon'(T)$ with the change in crystalline structure inside PNRs, just as it was assumed for NBT-based compositions by Jo et al. in their study [42]. Thus, the jump/drop of the unit cell dipole moment values is a consequence of this change. However, it is not clear why the dipole moments of PNRs should have higher values in the high-temperature phase. Taking into account that the P4bm phase in NBT is only weakly polar [4], rather the opposite could be assumed—PNRs in the low-temperature phase possessing the R3c symmetry should have larger dipole moments than the ones in the high-temperature P4bm phase. In addition, in our previous studies of second harmonic generation (SHG), monotonous reducing of SHG signal intensity was observed upon increasing temperature both below and above the phase transition temperature either in NBT or in 0.95NBT-0.05CT, which contradicts with such a change in the dipole moments [47].

Based on our observation regarding SHG, we have suggested that the reason for the jump/drop of dielectric permittivity in the low-temperature side of T_c could be the change in interaction between PNRs and the surrounding matrix due to the symmetry change in the matrix [47]. This idea could more correspond to the "breathing" model of PNRs [58] than to the flipping model. It also allows assuming that symmetry of PNRs does not change at the phase transition, avoiding the above-mentioned contradiction. The change in $\Delta\varepsilon$ in this case can be explained by differences in surface tension of PNR boundaries in both matrix phases. These differences result in variations in bending of PNR boundaries by the electric field and, as a consequence, in changes in PNR volume and total dipole moment. It should be pointed out that the characteristic logarithmic frequency dependence of dielectric permittivity (Figure 3a), which can be described by Equation (3), also corresponds to the "breathing" model [44]. In this model, the distinction between the cases of linear $\varepsilon'(\ln\omega)$ dependence and the frequency-independent dielectric permittivity (quasistatic regime) depends on the relation between the size of PNRs and the pinning free distance. If PNRs are of R3c symmetry, their existence above the nonferroelectric phase transition does not contradict with the assumed coexistence of P4bm and R3c in this temperature region [4,19].

It is likely that the above-described concept is applicable also for other NBT-based compositions, where $\varepsilon'(T, f)$ behavior characteristic of NBT is observed. Moreover, it could be even extended beyond such behavior, assuming that the sequence of the non-ferroelectric phase transition and the temperature region where the dielectric dispersion appears on temperature axis could differ depending on the composition and, presumably, even depending on variations in the producing route of the material. This could explain the differences in $\varepsilon'(T, f)$ between NBT, on the one hand, and behavior in $(1-x)$NBT-x$PbZrO_3$ solid solutions, on the other hand [22]. Unlike the case discussed in this study, the maximum dielectric permittivity in these compositions shifts with frequency, closely aligning with traditional relaxor ferroelectrics. Apparently, the upper temperature limit, at which the distribution of the relaxation times is significant, in the case of (1−x)NBT-x$PbZrO_3$ and Bi-over stoichiometric NBT, extends above the temperature of the structural phase transition, to a large extent hiding the features of it. Similar superimposion of the structural phase transition to the maximum in the temperature dependence of dielectric permittivity characteristic for relaxor ferroelectrics can exist also in (1−x − y)NBT-x$Sr_{0.7}Bi_{0.2}TiO_3$-y$PbTiO_3$ solid solutions [58]. The specific thermal hysteresis of this maximum can be related to thermal hysteresis of the structural phase transition, which extends in the temperature region of the dielectric dispersion.

In the present study, elimination of the jump/drop contribution through calculations leads to the transformed $\varepsilon'(T, f)$ dependences with pure relaxor-characteristic behavior, whereas in the case of (1−x)NBT-x$Sr_{0.7}Bi_{0.2}TiO_3$, it can be assumed that such elimination occurs in reality upon increasing of $Sr_{0.7}Bi_{0.2}TiO_3$ content—the maximum at T_c continuously reduces upon increasing of x until it finally disappears, leaving just the pure relaxor-characteristic behavior [27,46]. Apparently, rise of $Sr_{0.7}Bi_{0.2}TiO_3$ concentration leads to continuous weakening of features of the structural phase transition and its influence on PNRs until they disappear. The opposite scenario is observed in (1−x)NBT-x$PbTiO_3$ solid solutions where an already small amount of $PbTiO_3$ completely suppresses the shoulder. At the same time, the well-expressed jump in $\varepsilon'(T)$ dependence upon heating corresponds to a convincingly detected structural phase transition between the rhombohedral and the high-temperature phase with presumably tetragonal phase [56]. While, as it was noted above, further increasing $\varepsilon'(T)$ up to the maximum is interpreted as a "diffused region", which also belongs to a ferroelectric–tetragonal phase transition but is already inside the structure of the high-temperature tetragonal phase. Behavior of $\varepsilon'(T)$, observed in both cases, confirms mutual independence of the jump and the shoulder, observed in NBT-based compositions, as well as correspondence of the jump in $\varepsilon'(T)$ to the structural phase transition.

Thus, the presented concept, implementing mutually independent non-ferroelectric phase transition, on the one hand, and dielectric dispersion characteristic for relaxor ferroelectrics, on the other hand, allows explaining the specific behavior of dielectric properties of NBT-based compositions. The sequence of these two mechanisms on the temperature axis can vary, resulting in significant differences in the temperature dependences of dielectric permittivity, which are experimentally observed. The proposed concept seems to be appropriate also with respect to phase diagrams of these compositions available in the literature.

5. Conclusions

New interpretation of temperature dependence of dielectric permittivity in NBT-based compositions is proposed, based on the investigation of dielectric properties in two compositions: NBT and 0.95NBT-0.05$CaTiO_3$. The temperature dependence of dielectric permittivity in both compositions is described by two contributions: one from the structural phase transition between two non-ferroelectric phases, and one from pure relaxor behavior, which is performed in this way for the first time. The steep change (jump/drop) in the temperature dependence of dielectric permittivity above the frequency-dependent shoulder, which possesses well-expressed thermal hysteresis, is explained as the result of the first-order phase transition between two non-ferroelectric phases, while PNRs are preserved both below and above the transition. The latter explains the high values of dielectric permittivity in these phases. The jump/drop, characterizing dielectric properties by relaxation mechanism, is reflected solely in change in dielectric strength. The so-called "breathing" model is applied. The difference in dielectric strength in both non-ferroelectric phases, in the framework of this concept, is explained by the change in the interaction between PNRs and matrix in these phases.

If the contribution of the phase transition between two non-ferroelectric phases is excluded from the dielectric strength, behavior of dielectric permittivity in the studied compositions corresponds well to the one characteristic of classical relaxor ferroelectrics. First of all, the frequency-dependent shoulder in the temperature dependence of dielectric permittivity transforms into a frequency-dependent maximum, corresponding to the general concept of relaxor ferroelectrics. Secondly, steep change in $\varepsilon'(T)$ at heating of the poled samples below the shoulder corresponds to the phase transition from the ferroelectric to relaxor state.

Author Contributions: Conceptualization, E.B. and M.D.; methodology, E.B., M.D., Š.S. and A.K. (Algimantas Kežionis); validation, M.D., Š.S. and E.B.; formal analysis, E.B.; investigation, M.D., Š.S. and A.K. (Algimantas Kežionis); resources, J.B. and A.K. (Andrei Kholkin); data curation, M.D.; writing—original draft preparation, E.B. and M.D.; writing—review and editing, E.B., M.D. and A.K. (Andrei Kholkin); visualization, M.D.; supervision, E.B. and J.B.; project administration, E.B.; funding acquisition, E.B., J.B. and A.K. (Andrei Kholkin). All authors have read and agreed to the published version of the manuscript.

Funding: This research has been funded by the European Union HORIZON-WIDERA-2023-TALENTS-01 program within the framework of the project "Low-dimensional ferroelectrics for advanced electronic and biomedical devices" (FeLow-D) (Grant agreement number 101186499). It was also co-funded by the European Union through the project No. 10-052-P-0001 "To increase the level of internationalization of science by implementing the activities of the Eureka Network Program". Institute of Solid State Physics, University of Latvia, was supported by the European Union's Horizon 2020 Framework Programme H2020-WIDESPREAD-01-2016-2017-TeamingPhase2 (#739508, CAMART^2).

Data Availability Statement: The original contributions presented in this study are included in the article. Further inquiries can be directed to the corresponding author.

Conflicts of Interest: The authors declare no conflicts of interest. The funders had no role in the design of the study; in the collection, analyses, or interpretation of data; in the writing of the manuscript; or in the decision to publish the results.

References

1. Rödel, J.; Webber, K.G.; Dittmer, R.; Jo, W.; Kimura, M.; Damjanovic, D. Transferring lead-free piezoelectric ceramics into application. *J. Eur. Ceram. Soc.* **2015**, *35*, 1659–1681. [CrossRef]
2. Rao, B.N.; Ranjan, R. Electric-field-driven monoclinic-to-rhombohedral transformation in $Na_{1/2}Bi_{1/2}TiO_3$. *Phys. Rev. B* **2012**, *86*, 134103. [CrossRef]
3. Yao, J.; Ge, W.; Yang, Y.; Luo, L.; Li, J.; Viehland, D.; Bhattacharyya, S.; Zhang, Q.; Luo, H. Observation of partially incoherent {110} boundaries between polar nanodomains in $Na_{1/2}Bi_{1/2}TiO_3$ single crystals. *J. Appl. Phys.* **2010**, *108*, 064114. [CrossRef]
4. Jones, G.O.; Thomas, P.A. Investigation of the structure and phase transitions in the novel A-site substituted distorted perovskite compound $Na_{0.5}Bi_{0.5}TiO_3$. *Acta Cryst. B* **2002**, *58*, 168–178. [CrossRef] [PubMed]
5. Picht, G.; Töpfer, J.; Henning, E. Structural properties of $(Bi_{0.5}Na_{0.5})_{1-x}Ba_xTiO_3$ lead-free piezoelectric ceramics. *J. Eur. Ceram. Soc.* **2010**, *30*, 3445–3453. [CrossRef]
6. Craciun, F.; Galassi, C.; Birjega, R. Electric-field-induced and spontaneous relaxor-ferroelectric phase transition in $(Na_{1/2}Bi_{1/2})_{1-x}Ba_xTiO_3$. *J. Appl. Phys.* **2012**, *112*, 124106. [CrossRef]
7. Gorfman, S.; Thomas, P.A. Evidence for a non-rhombohedral average structure in the lead-free piezoelectric material $Na_{0.5}Bi_{0.5}TiO_3$. *J. Appl. Cryst.* **2010**, *43*, 1409–1414. [CrossRef]
8. Aksel, E.; Forrester, J.S.; Kowalski, B.; Jones, J.L.; Thomas, P.A. Phase transition sequence in sodium bismuth titanate observed using high-resolution x-ray diffraction. *Appl. Phys. Lett.* **2011**, *99*, 222901. [CrossRef]
9. Rao, B.N.; Fitch, A.N.; Ranjan, R. Ferroelectric-ferroelectric phase coexistence in $Na_{1/2}Bi_{1/2}TiO_3$. *Phys. Rev. B* **2013**, *87*, 060102. [CrossRef]

10. Birks, E.; Dunce, M.; Ignatans, R.; Kuzmin, A.; Plaude, A.; Antonova, M.; Kundzins, K.; Sternberg, A. Structure and dielectric properties of $Na_{0.5}Bi_{0.5}TiO_3$-$CaTiO_3$ solid solutions. *J. Appl. Phys.* **2016**, *119*, 074102. [CrossRef]
11. Liu, L.; An, Z.; Wang, R.; Zhou, B.; Zhang, Z.; Ming, B.; Zhu, L.; Zhu, M.; Sui, M. Atomic-scale polar vortices in $Na_{0.5}Bi_{0.5}TiO_3$ grains. *Ceram. Int.* **2022**, *48*, 11830–11835. [CrossRef]
12. Trolliard, G.; Dorcet, V. Reinvestigation of Phase Transitions in $Na_{0.5}Bi_{0.5}TiO_3$ by TEM. Part II: Second Order Orthorhombic to Tetragonal Phase Transition. *Chem. Mater.* **2008**, *20*, 5074–5082. [CrossRef]
13. Dorcet, V.; Trolliard, G. A transmission electron microscopy study of the A-site disordered perovskite $Na_{0.5}Bi_{0.5}TiO_3$. *Acta Mater.* **2008**, *56*, 1753–1761. [CrossRef]
14. Roukos, R.; Romanos, J.; Dargham, S.A.; Chaumont, D. Evidence of stable tetragonal (P4bm) phase in rhombohedral (R3c) complex perovskite $(Na_{1/2}Bi_{1/2})_{1-x}Ca_xTiO_3$ lead-free piezoelectric materials. *Mater. Sci. Eng. B* **2023**, *288*, 116196. [CrossRef]
15. Usher, T.-M.; Forrester, J.S.; Cruz, C.R.d.; Jones, J.L. Crystal structure of $0.96(Na_{0.5}Bi_{0.5}TiO_3)$-$0.04(BaTiO_3)$ from combined refinement of x-ray and neutron diffraction patterns. *Appl. Phys. Lett.* **2012**, *101*, 152906. [CrossRef]
16. Dorcet, V.; Trolliard, G.; Boullay, P. The structural origin of the antiferroelectric properties and relaxor behaviour of $Na_{0.5}Bi_{0.5}TiO_3$ *J. Magn. Magn. Mater.* **2009**, *321*, 1758–1761. [CrossRef]
17. Levin, I.; Reaney, I.M. Nano- and Mesoscale Structure of $Na_{1/2}Bi_{1/2}TiO_3$: A TEM Perspective. *Adv. Funct. Mater.* **2012**, *22*, 3445–3452. [CrossRef]
18. Beanland, R.; Thomas, P.A. Symmetry and defects in rhombohedral single-crystalline $Na_{1/2}Bi_{1/2}TiO_3$. *Phys. Rev. B* **2014**, *89*, 174102. [CrossRef]
19. Suchanicz, J.; Kwapulinski, J. X-ray diffraction study of the phase transitions in $Na_{1/2}Bi_{1/2}TiO_3$. *Ferroelectrics* **1995**, *165*, 249–253. [CrossRef]
20. Tu, C.-S.; Huang, S.-H.; Ku, C.-S.; Lee, H.-Y.; Chien, R.R.; Schmidt, V.H.; Luo, H. Phase coexistence and Mn-doping effect in lead-free ferroelectric $Na_{1/2}Bi_{1/2}TiO_3$ crystals. *Appl. Phys. Lett.* **2010**, *96*, 062903. [CrossRef]
21. Ge, W.; Devreugd, C.P.; Phelan, D.; Zhang, Q.; Ahart, M.; Li, J.; Luo, H.; Boatner, L.A.; Viehland, D.; Gehring, P.M. Lead-free and lead-based ABO_3 perovskite relaxors with mixed-valence A-site and B-site disorder: Comparative neutron scattering structural study of $Na_{1/2}Bi_{1/2}TiO_3$ and $Pb(Mg_{1/3}Nb_{2/3})O_3$. *Phys. Rev. B* **2013**, *88*, 174115. [CrossRef]
22. Dorcet, V.; Marchet, P.; Mercurio, J.P. Dielectric Properties of $Na_{0.5}Bi_{0.5}TiO_3$-$PbZrO_3$ Ceramics. *Ferroelectrics* **2006**, *339*, 29–35. [CrossRef]
23. Hiruma, Y.; Nagata, H.; Takenaka, T. Thermal depoling process and piezoelectric properties of bismuth sodium titanate ceramics. *J. Appl. Phys.* **2009**, *105*, 084112. [CrossRef]
24. Wang, S.-F.; Tu, C.-S.; Chang, T.-L.; Chen, P.-Y.; Chen, C.-S.; Schmidt, V.H.; Anthoniappen, J. Structural stability and depolarization of manganese-doped $(Bi_{0.5}Na_{0.5})_{1-x}Ba_xTiO_3$ relaxor ferroelectrics. *J. Appl. Phys.* **2014**, *116*, 154101. [CrossRef]
25. Chen, P.-Y.; Chen, C.-S.; Tu, C.-S.; Chang, T.-L. Large E-field induced strain and polar evolution in lead-free Zr-doped 92.5%$(Bi_{0.5}Na_{0.5})TiO_3$-7.5%$BaTiO_3$ ceramics. *J. Eur. Ceram. Soc.* **2014**, *34*, 4223–4233. [CrossRef]
26. Zuo, R.; Ye, C.; Fang, X.; Li, J. Tantalum doped $0.94Bi_{0.5}Na_{0.5}TiO_3$-$0.06BaTiO_3$ piezoelectric ceramics. *J. Eur. Ceram. Soc.* **2008**, *28*, 871–877. [CrossRef]
27. Ang, C.; Yu, Z. Dielectric and ferroelectric properties in $(Sr,Ni,Na)TiO_3$ solid solutions. *J. Appl. Phys.* **2010**, *107*, 114106. [CrossRef]
28. Ge, W.; Li, J.; Viehland, D. Influence of Mn Doping on the Structure and Properties of $Na_{0.5}Bi_{0.5}TiO_3$ Single Crystals. *J. Am. Ceram. Soc.* **2010**, *93*, 1372–1377. [CrossRef]
29. Sakata, K.; Masuda, Y. Ferroelectric and antiferroelectric properties of $(Na_{0.5}Bi_{0.5})TiO_3$-$SrTiO_3$ solid solution ceramics. *Ferroelectrics* **1974**, *7*, 347. [CrossRef]
30. Hiruma, Y.; Nagata, H.; Takenaka, T. Phase Transition Temperatures and Piezoelectric Properties of $(Bi_{1/2}Na_{1/2})TiO_3$-$(Bi_{1/2}K_{1/2})TiO_3$-$BaTiO_3$ Lead-Free Piezoelectric Ceramics. *Jpn. J. Appl. Phys.* **2006**, *45*, 7409–7412. [CrossRef]
31. Diaz, J.C.C.; M'Peko, J.-C.; Venet, M.; da Silva, P.S., Jr. Unveiling the high-temperature dielectric response of $(Na_{0.5}Bi_{0.5})TiO_3$. *Sci. Rep.* **2020**, *10*, 19491. [CrossRef]
32. Yang, F.; Li, M.; Li, L.; Wu, P.; Pradal-Velazquez, E.; Sinclair, D.C. Defect chemistry and electrical properties of sodium bismuth titanate perovskite. *J. Mater. Chem. A* **2018**, *6*, 5243–5254. [CrossRef]
33. Yao, J.; Ge, W.; Luo, L.; Li, J.; Viehland, D.; Luo, H. Hierarchical domains in $Na_{1/2}Bi_{1/2}TiO_3$ single crystals: Ferroelectric phase transformations within the geometrical restrictions of a ferroelectric inheritance. *Appl. Phys. Lett.* **2010**, *96*, 222905. [CrossRef]
34. Svirskas, S. Lattice Dynamics and Electromechanical Response in Disordered Perovskites. Ph.D. Thesis, Vilnius University Center for Physical Sciences and Technology, Vilnius, Lithuania, 2018. Available online: https://talpykla.elaba.lt/elaba-fedora/objects/elaba:30868113/datastreams/MAIN/content (accessed on 11 June 2025).
35. Zhang, S.-T.; Kounga, A.B.; Aulbach, E.; Jo, W.; Graznow, T.; Ehrenberg, H.; Rödel, J. Lead-free piezoelectrics with giant strain in the system $Bi_{0.5}Na_{0.5}TiO_3$-$BaTiO_3$-$K_{0.5}Na_{0.5}NbO_3$. II. Temperature dependent properties. *J. Appl. Phys.* **2008**, *103*, 034108. [CrossRef]

36. Chen, M.; Xu, Q.; Kim, B.H.; Ahn, B.K.; Ko, J.H.; Kang, W.J.; Nam, O.J. Structure and electrical properties of $(Na_{0.50}Bi_{0.50})_{1-x}Ba_xTiO_3$ piezoelectric ceramics. *J. Eur. Ceram. Soc.* **2008**, *28*, 843–849. [CrossRef]
37. Ge, W.; Cao, H.; Li, J.; Viehland, D.; Zhang, Q.; Luo, H. Influence of dc-bias on phase stability in Mn-doped $Na_{0.5}Bi_{0.5}TiO_3-5.6$ at.%$BaTiO_3$ single crystals. *Appl. Phys. Lett.* **2009**, *95*, 162903. [CrossRef]
38. Wu, L.; Xiao, D.-Q.; Lin, D.-M.; Zhu, J.; Yu, P.; Li, X. Temperature dependence of electric properties of $[Bi_{0.5}(Na_{1-x}Ag_x)_{0.5}]_{1-y}$BayTiO$_3$ ceramics. *Jpn. J. Appl. Phys.* **2007**, *46*, 7382–7387. [CrossRef]
39. Sapper, E.; Schaab, S.; Jo, W.; Graznow, T.; Rödel, J. Influence of electric fields on the depolarization temperature of Mn-doped $(1-x)Bi_{1/2}Na_{1/2}TiO_3$-x$BaTiO_3$. *J. Appl. Phys.* **2012**, *111*, 014105. [CrossRef]
40. Ma, C.; Tan, X. In situ transmission electron microscopy study on the phase transitions in lead-free $(1-x)(Bi_{1/2}Na_{1/2})TiO_3-xBaTiO_3$ ceramics. *J. Am. Ceram. Soc.* **2011**, *94*, 4040–4044. [CrossRef]
41. Guo, Y.; Gu, M.; Luo, H.; Liu, Y.; Withers, R.L. Composition-induced antiferroelectric phase and giant strain in lead-free $(Na_y,Bi_z)Ti_{1-x}O_3(1-x)$-x$BaTiO_3$ ceramics. *Phys. Rev. B* **2011**, *83*, 054118. [CrossRef]
42. Jo, W.; Schaab, S.; Sapper, E.; Schmitt, L.A.; Kleebe, H.J.; Bell, A.J.; Rödel, J. On the phase identity and its thermal evolution of lead-free $(Bi_{1/2}Na_{1/2})TiO_3$-6 mol%$BaTiO_3$. *J. Appl. Phys.* **2011**, *110*, 074106. [CrossRef]
43. Fan, Z.; Randall, C.A. Engineering the nature of polarization dynamics in lead-free relaxors based on $(Bi_{1/2}Na_{1/2})TiO_3$. *Appl. Phys. Lett.* **2021**, *119*, 112904. [CrossRef]
44. Glazunov, A.E.; Tagantsev, A.K. A "Breathing" Model for the Polarization Response of Relaxor Ferroelectrics. *Ferroelectrics* **1999**, *221*, 57–66. [CrossRef]
45. Petzelt, J.; Nuzhnyy, D.; Bovtun, V.; Pasciak, M.; Kamba, S.; Dittmer, R.; Svirskas, Š.; Banys, J.; Rödel, J. Peculiar Bi-ion dynamics in $Na_{1/2}Bi_{1/2}TiO_3$ from terahertz and microwave dielectric spectroscopy. *Phase Transit.* **2014**, *87*, 953–965. [CrossRef]
46. Dunce, M.; Olšauskaite, A.; Birks, E.; Svirskas, Š.; Kežionis, A.; Bikshe, L.; Sternberg, A.; Banys, J. Revision of the freezing concept in relaxor ferroelectrics: The case of $Na_{0.5}Bi_{0.5}TiO_3$-$Sr_{0.7}Bi_{0.2}TiO_3$ solid solutions. *Ferroelectrics* **2020**, *569*, 266–279. [CrossRef]
47. Dunce, M.; Birks, E.; Bikse, L.; Ignatans, R.; Fuith, A.; Kabelka, H.; Nitiss, E.; Kundzins, M.; Sternberg, A. Novel approach in analyzing phase transitions in $Na_{0.5}Bi_{0.5}TiO_3$-Comparison with 0.95$Na_{0.5}Bi_{0.5}TiO_3$-0.05$CaTiO_3$. *J. Appl. Phys.* **2022**, *131*, 224101. [CrossRef]
48. Kežionis, A.; Butvilas, P.; Šalkus, T.; Kazlauskas, S.; Petrulionis, D.; Žukauskas, T.; Kazakevičius, E.; Orliukas, A.F. Four-electrode impedance spectrometer for investigation of solid ion conductors. *Rev. Sci. Instrum.* **2013**, *84*, 013902. [CrossRef]
49. Kezionis, A.; Kazlauskas, S.; Petrulionis, D.; Orliukas, A.F. Broadband method for the determination of small sample's electrical and dielectric properties at high temperatures. *IEEE Trans. Microw. Theory Technol.* **2014**, *62*, 2456–2461. [CrossRef]
50. Kamba, S.; Nuzhnyy, D.; Veljko, S.; Bovtun, V.; Petzelt, J.; Wang, Y.L.; Setter, N.; Levoska, J.; Tyunina, M.; Macutkevic, J.; et al. Dielectric relaxation and polar phonon softening in relaxor ferroelectric $PbMg_{1/3}Ta_{2/3}O_3$. *J. Appl. Phys.* **2007**, *102*, 074106. [CrossRef]
51. Macutkevic, J.; Kamba, S.; Banys, J.; Brilingas, A.; Pashkin, A.; Petzelt, J.; Bormanis, K.; Sternberg, A. Infrared and broadband dielectric spectroscopy of PZN-PMN-PSN relaxor ferroelectrics: Origin of two-component relaxation. *Phys. Rev. B* **2006**, *74*, 104106. [CrossRef]
52. Courtens, E.; Vogt, H. Observation of freezing in a structural glass over an exceptionally broad range of frequencies. *Z. Phys. B Condens. Matter* **1986**, *62*, 143–151. [CrossRef]
53. Dunce, M.; Birks, E.; Antonova, M.; Kundzins, M. Description of relaxor state in $Na_{1/2}Bi_{1/2}TiO_3$-$SrTiO_3$-$PbTiO_3$ system of solid solutions. *Integr. Ferroelectr.* **2011**, *123*, 40–46. [CrossRef]
54. Cheng, Z.-Y.; Katiyar, R.S.; Yao, X.; Bhalla, A.S. Temperature dependence of dielectric constant of relaxor ferroelectrics. *Phys. Rev. B* **1998**, *57*, 8166–8177. [CrossRef]
55. Cordero, F.; Corti, M.; Craciun, F.; Galassi, C.; Piazza, D.; Tabak, F. Polar and nonpolar atomic motions in the relaxor ferroelectric $Pb_{1-3x/2}La_xZr_{0.2}Ti_{0.8}O_3$ from dielectric, anelastic and NMR relaxation. *Phys. Rev. B* **2005**, *71*, 094112. [CrossRef]
56. Hong, K.S.; Park, S.-E. Phase relations in the system of $(Na_{1/2}Bi_{1/2})TiO_3$-$PbTiO_3$. II. Dielectric property. *J. Appl. Phys.* **1996**, *79*, 388–392. [CrossRef]
57. Plaude, A.; Ignatans, R.; Birks, E.; Dunce, M.; Antonova, M.; Sternberg, A. Structure and dielectric properties at phase transition of $Na_{1/2}Bi_{1/2}TiO_3$-$BaTiO_3$ solid solutions. *Ferroelectrics* **2016**, *500*, 47–53. [CrossRef]
58. Dunce, M.; Birks, E.; Perantie, J.; Hagberg, J.; Antonova, M.; Ignatans, R.; Sternberg, A. Phase transitions in $Na_{0.5}Bi_{0.5}TiO_3$-$(Sr_{0.7}Bi_{0.2})TiO_3$-$PbTiO_3$ solid solutions. *Ferroelectrics* **2016**, *498*, 94–101. [CrossRef]

MDPI AG
Grosspeteranlage 5
4052 Basel
Switzerland
Tel.: +41 61 683 77 34

Materials Editorial Office
E-mail: materials@mdpi.com
www.mdpi.com/journal/materials

www.ingramcontent.com/pod-product-compliance
Lightning Source LLC
Chambersburg PA
CBHW061508220926
51850CB00004B/206
* 9 7 8 3 7 2 5 8 6 6 3 2 8 *